Air Pollution: Engineering, Science, and Policy

©2015 by College Publishing. All rights reserved. No portion of this publication may be reproduced or transmitted in any form or by any means electronic or mechanical, including photocopying, recording, or any information storage and retrieval system, without permission in writing from the publisher.

College Publishing books are printed in the United States of America on acid-free paper.

International Standard Serial Number: 978-1-932780-07-9 (Print)
International Standard Serial Number: 978-1-932780-08-6 (E-Book)
Library of Congress Control Number: 2015938135

College Publishing
12309 Lynwood Drive
Glen Allen, VA 23059
Phone (804) 364-8410 outside the U.S. or (800) 827-0723 within the U.S.
Fax (804) 364-8408
Email: collegepub@mindspring.com
Internet: www.collegepublishing.us

Air Pollution: Engineering, Science, and Policy

DR. STEVEN P.K. STERNBERG, PE

College Publishing
Glen Allen, Virginia

ALSO BY COLLEGE PUBLISHING

Fate and Transport of Contaminants in the Environment (ISBN: 1-932780-04-1)

Journal of Environmental Solutions for Oil, Gas, and Mining
(http://www.journalofenvironmentalsolutionsforoilgasandmining.com)

Journal of Green Building (http://www.journalofgreenbuilding.com)

Writing Style and Standards in Undergraduate Reports, 2e (ISBN: 1-932780-04-1)

DEDICATION

This book is dedicated to Jean and Leo – thanks for always believing in me; to Margaret – thanks for always being there for me; and Joseph, Regina, and Eleanor – for keeping me motivated to make the world a better place.

ACKNOWLEDGMENTS

I would like to thank all of my undergraduate students for letting me use the initial drafts of this text in class and for all their feedback on how to make it better. I would especially like to thank Emily (Bell) Anderson, Sarah Anderson, Aaron Boothe, Justin Boucher, Kayla Brown, Ryan Clark, Luke Dahlin, Doug Eli, Justin Finke, Caitlin Leach, Marcus Thompson, and Josie Wise for the detailed critiques, creation of figures and tables, the well thought out example problems, and the detailed case studies. I would also like to thank all my other students who provided proof reading and worked-through the examples and homework problems.

I would also like to thank the manuscript reviewers for the time and thought they put into helping me create a better text – Ray Carter, University of Kansas; John Cimbala, Pennsylvania State University; Jeffrey Collett, Colorado State University; Michael Roberts, John Hopkins University; Stephen R. Turns, Pennsylvania State University; and John Vandenberg, Duke University.

Finally, I want to thank my editor/publisher, Stephen Mosberg, for creating the opportunity for me to write this book. I appreciate the patience, guidance, and encouragement. The book would not have happened without your help.

PREFACE

Rationale

Air Pollution: Engineering, Science, and Policy. This textbook explores the sources and sinks, health effects, regulatory methodologies, history, and control technologies for air pollutants. Chapters are organized by the physical and chemical nature of the pollutants, which determines the technologies useful for their control. The book includes chapters on each of the US Clean Air Act criteria air pollutants (particulate matter, sulfur oxides, nitrogen oxides, ozone, carbon monoxide, and lead). Additional topics include mercury, reactive carbon (methane, volatile organic compounds), indoor air quality, hazardous air pollutants, mobile sources, stratospheric ozone depletion, and carbon dioxide induced climate change. Each section also explores the impacts of control technologies on society, industry, and the local, regional, and global atmosphere.

The text is aimed at upper level undergraduates in the sciences and engineering who have completed courses in general chemistry, general physics, and calculus. It provides the background (or review) information needed to help students that have not had (or have forgotten) courses in: problem solving calculations, mass and energy balances, particle technology, fluid mechanics, meteorology, dispersion modeling, and green engineering. These chapters provide the necessary background so any student in a science or engineering discipline can understand and use this book.

The book discusses the *science* of what is known about each pollutant – the natural and anthropogenic sources; what happens to it once it is released into the environment; and the ultimate fate of the release. *Policy* issues include: how humans have organized systems to minimize harm, and the effects of these systems. *Engineering* issues include: control methodologies and technologies, such as controlling inputs, changing processes, and emission removal equipment. Many chapters include case studies from particular industries and situations that may describe the relevant air pollutant issues; provide short and relevant science summaries; describe the current practices for pollutant removal; and explain why some alternative technology is not chosen.

Highlights

- Include global issues and potential solutions for global problem pollutants (carbon dioxide, mercury, stratospheric ozone depletion). These problems are the most relevant, significant, and contemporary issues in air pollution.
- The successes and remaining problems surrounding the criteria pollutants (SOx, NOx, PM, VOC's, Pb, and tropospheric ozone). Readers new to the field of air pollution will want to learn about these past victories, but will also need to know what the current issues are that they may be expected to address.

- Explore issues on the scale of each problem. Scales range from the personal scale, to buildings (indoor air quality) to cities (particulate matter) to regions (acid rain and smog) to global (CFCs, methane, and carbon dioxide). A scale based hierarchy or categorization helps readers move beyond the traditional command and control of point source solutions and see that other solutions can be viable.
- Encourage readers to think about air quality issues in terms of their life as a citizen and consumer, not just as scientist or engineer. When these problems are explored at the personal scale, it provides additional motivation to learn about the subject.
- Explore engineering and non-engineering solutions to air quality problems. Sometimes the best solution is to not make the mess in the first place rather than looking for ways to clean up the mess. This aspect of the text is blended in as green chemistry, green engineering, and pollution prevention. It has a place in each section of the text.
- Includes international perspectives. Other countries have similar goals and regulations to the USA, yet they solve the problems in other ways. This information is included in many chapters and in the discussion of specific problem areas and solutions.
- The text contains more information than could be completed in a typical 15 week course. This gives the instructor some choices as to what to cover. The basic core chapters are
    - Chapter 1. Introduction
    - Chapter 3. Laws and Regulations
    - Chapter 5. Meteorology
    - Chapter 6. Dispersion
    - Chapter 8. Particulate matter
    - Chapter 9. Sulfur
    - Chapter 10. Nitrogen
    - Chapter 12. Ozone
    - Chapter 14. Carbon Dioxide

Other chapters either provide review information or expand the types of pollutants and control technologies.

I have taught courses in air pollution for over 20 years. Students are mostly undergraduate engineers and environmental scientists; however, students from every science and engineering discipline have taken the course and found it useful. Many had not ever considered the topic as an area for their career until taking this course. They have then gone on to successful careers in air pollution control with industrial facilities, power companies, consulting firms, and state government.

CONTENTS

1 Air Pollution: Introduction 1
- 1.1 Introduction 1
- 1.2 Air Pollution Problems 2
- 1.3 Control of Air Pollutants 14
- 1.4 Public Information and Tools 17
- 1.5 Questions 19
- 1.6 Bibliography 20

2 Basic Problem Solving 23
- 2.1 Dimensional Analysis 23
- 2.2 Basic Chemistry Review 26
- 2.3 Reaction Stoichiometry 28
- 2.3.1 Atomic and Molecular Weight 28
- 2.3.2 Stoichiometry of Chemical Reactions 30
- 2.4 Concentrations, Flow Rates, and Efficiencies 31
- 2.4.1 Concentrations 31
- 2.4.2 Flow Rates 36
- 2.4.3 Removal Efficiency 38
- 2.5 Material Balances 39
- 2.5.1 Flow Systems 39
- 2.5.2 Unsteady Flow Systems 42
- 2.5.3 Flow Reactor Systems 44
- 2.5.4 Batch Reactor Systems 52
- 2.5.5 General Systems 54
- 2.6 Problems 59
- 2.7 Bibliography 61

3 Environmental Laws and Regulations 63
- 3.1 System of US Law 64
- 3.1.1 Law Hierarchy 64
- 3.1.2 Process for Creating a Law 65

	3.1.3	Process for Using a Law 65
	3.1.4	Process for Creating a Regulation 67
	3.2	The Clean Air Act 69
	3.2.1	Air Quality Standards 71
	3.2.2	Emission Limits 75
	3.2.3	Delegation to States and Tribes 81
	3.2.4	New Source Review Permits (Construction Permits) . . . 82
	3.2.5	Citizen Participation 89
	3.2.6	Mobil Sources and Clean Fuels 90
	3.2.7	Air Toxics and Hazardous Air Pollutants (Title III) 90
	3.2.8	Operating Permits (Title V) 92
	3.2.9	Compliance and Enforcement 94
	3.2.10	Interstate Air Pollution 95
	3.3	ISO14000 96
	3.4	Questions 99
	3.5	Problems 101
	3.6	Group Project Ideas 103
	3.7	Bibliography 104

4 Fluid Mechanics Review 107

	4.1	Terms and Units 107
	4.2	Fluid Statics 114
	4.3	Fluid Dynamics 116
	4.3.1	Friction During Internal Flow 123
	4.3.2	Friction During External Flow 124
	4.4	Bibliography 126

5 Air Pollution Meteorology 127

	5.1	Introduction 127
	5.2	Composition 127
	5.3	Structure 131
	5.3.1	Layers 131
	5.3.2	Pressure Profile 132
	5.4	Solar Radiation 134
	5.5	Global Wind Circulation 135
	5.6	Local Wind Circulation 138
	5.6.1	Wind Roses and Wind Speeds 140

5.6.2	Vertical Profiles	143
5.7	Stability	144
5.7.1	Lapse Rate	144
5.7.2	Stability Analysis	146
5.7.3	Inversions	149
5.7.4	Mixing Depth	151
5.7.5	Stack Plumes	152
5.8	Questions	153
5.9	Problems	154
5.10	Group Project Ideas	156
5.11	Bibliography	157

6 Air Pollution Dispersion 159

6.1	Introduction	159
6.2	Modeling	159
6.3	Gaussian Model	160
6.4	Common Applications of the Gaussian Model	166
6.4.1	Case 1. Ground Level Source	166
6.4.2	Case 2. Elevated Source	167
6.4.3	Case 3. Elevated Source with Reflection	168
6.4.4	Case 4. Ground Level Concentrations from Elevated Source with Reflection	170
6.4.5	Case 5. Line Source	175
6.4.6	Case 6. Puff source	176
6.4.7	Case 7. Multiple sources	179
6.5	Effect of Averaging Time	181
6.6	Effective Stack Height and Plume Rise	182
6.6.1	Holland Formula	183
6.6.2	Concawe Formula	183
6.6.3	Briggs Model	184
6.6.4	Stack Tip Downwash	188
6.7	Other Air Dispersion Models	189
6.8	Long Range Transport (>50 km)	189
6.9	Questions	190
6.10	Problems	191
6.11	Group Project Ideas	193
6.12	Bibliography	193

7 Properties of Particulate Matter 195
- 7.1 Physical Properties of Particulate Matter 195
- 7.1.1 Size 195
- 7.1.2 Shape 195
- 7.1.3 Density 195
- 7.1.4 Distributions 198
- 7.1.5 Graphical Description of Distributions 206
- 7.2 Properties of Atmospheric Aerosols 210
- 7.3 Particle Motion 212
- 7.3.1 Particle Motion Due to Gravity 214
- 7.3.2 Particle Motion due to Centrifugal Forces 216
- 7.4 Interactions with Radiation 218
- 7.5 Questions 221
- 7.6 Problems 221
- 7.7 Bibliography 223

8 Particulate Matter 225
- 8.1 Sources 226
- 8.2 Particle Formation 230
- 8.2.1 Primary Formation 231
- 8.2.2 Secondary Formation 232
- 8.3 Health and Welfare Effects 233
- 8.3.1 Effects on Public Health 233
- 8.3.2 Effects on Public Welfare 236
- 8.3.2.1 Visibility, Haze and Air Color 236
- 8.3.2.2 Environmental Damage 237
- 8.4 Regulation and History 238
- 8.4.1 Regulation 238
- 8.4.2 Historical Trends in US - Particulate Matter 240
- 8.4.3 Regulatory Trends 241
- 8.5 Control Technologies 241
- 8.5.1 Collection Efficiency 242
- 8.5.2 Settling Chamber 246
- 8.5.2.1 Flow Regime 247
- 8.5.2.2 Laminar Flow 247
- 8.5.2.3 Turbulent Flow 249
- 8.5.3 Cyclones 250

8.5.4	Filters	256
8.5.5	Electrostatic Precipitators	263
8.5.6	Wet Scrubbing	268
8.5.7	Atmospheric Removal Mechanisms	270
8.5.7.1	Atmospheric Removal by Dry Deposition	270
8.5.7.2	Atmospheric Removal by Wet Deposition	271
8.6	Choosing a Particle Control System	272
8.7	Questions	277
8.8	Problems	278
8.9	Group Project Ideas	281
8.10	Bibliography	281

9 Sulfur Emissions . . . 285

9.1	General Information	285
9.2	Sources	286
9.3	Health and Welfare Effects	291
9.3.1	Effects on Public Health	291
9.3.2	Effect on Public Welfare	292
9.3.2.1	Acid Precipitation	293
9.3.2.2	Geographic Distribution of Acid Rain	297
9.4	History and Regulation	300
9.4.1	Regulatory Methodology - US Approach	302
9.4.2	Trends in US SO_2 Concentrations	307
9.4.3	Regulatory Methodology - European Environment Agency	311
9.5	Control Technologies	316
9.5.1	Fuel Replacement	316
9.5.2	Fuel Desulfurization	318
9.5.3	Flue Gas Desulfurization	322
9.5.3.1	FGD Cost	327
9.6	Questions	332
9.7	Problems	333
9.8	Group Project Ideas	336
9.9	Bibliography	336

10 Nitrogen Emissions . . . 339

10.1	General Information	339

10.2	Sources and Sinks	341
10.2.1	Nitrogen (N_2)	341
10.2.2	Nitrous Oxide (N_2O)	342
10.2.3	Ammonia (NH_3)	343
10.2.4	Nitrogen Oxides (NO_x)	344
10.2.4.1	Formation of NO_x from Combustion	345
10.2.4.2	Formation of NO_x from Nitric Acid Manufacture and Uses	347
10.2.4.3	Formation of N_2O and NO_x from Inorganic Fertilizer (Synthetic Fertilizer)	348
10.2.4.4	Formation of N_2O and NO_x from Wastewater Treatment	349
10.3	Health and Welfare Effects of NO_x	349
10.3.1	Effects on Public Health	349
10.3.2	Effects on Public Welfare	351
10.4	History and Regulation	352
10.4.1	NH_3	352
10.4.2	N_2O	352
10.4.3	Nitrates	352
10.4.4	NO_x	352
10.4.5	Regulatory Methodology – US Approach	353
10.4.6	Regulatory Methodology – Japan	360
10.5	Control Technologies	362
10.5.1	NO_x Control Technologies for Agriculture	363
10.5.2	NO_x Control Technologies for Combustion Processes	363
10.5.2.1	Fuel Switching	368
10.5.2.2	Combustion Control Techniques	371
10.5.2.2.1	Low Excess Air (LEA) Firing	371
10.5.2.2.2	Staged Combustion (SC)	372
10.5.2.2.3	Air Staging	373
10.5.2.2.4	Fuel Staging	373
10.5.2.2.5	Cyclonic Low NO_x Burner	373
10.5.2.2.6	Flue Gas Recirculation (FGR)	374
10.5.2.2.7	Fuel Induced Recirculation	374
10.5.2.2.8	Steam/Water Injection	375
10.5.2.3	Flue Gas Treatment Methods	376
10.5.2.3.1	Selective Non-Catalytic Reduction (SNCR)	376
10.5.2.3.1.1	Ammonia based SNCR	376

10.5.2.3.1.2	Urea based SNCR	0
10.5.2.3.2	Selective Catalytic Reduction (SCR)	378
10.5.2.3.3	Fuel Reburning	379
10.5.3	Choosing a NO_x Reduction Technology	380
10.6	Questions	383
10.7	Problems	384
10.8	Group Project Ideas	386
10.9	Bibliography	386

11 Reactive Carbon Compounds 391

11.1	Carbon Monoxide (CO)	392
11.1.1	General Information	392
11.1.2	Sources, Sinks, and Ultimate Fate	392
11.1.3	Health and Welfare Effects	393
11.1.3.1	Human Health	393
11.1.3.2	Environmental Welfare	395
11.1.4	History and Regulation	395
11.1.4.1	Control of Motor Vehicle Emissions	396
11.1.4.2	CO Air Quality Trend	398
11.1.5	Control Strategies	399
11.1.5.1	Good Combustion Practices	399
11.1.5.2	Catalytic Oxidation	399
11.2	Methane (CH_4)	399
11.2.1	General Information	399
11.2.2	Sources, Sinks, and Ultimate Fate	400
11.2.3	Health and Welfare Effects	406
11.2.3.1	Human Health	406
11.2.3.2	Environmental Welfare	406
11.2.4	History and Regulation	408
11.2.5	Control Strategies	408
11.2.5.1	Wastewater Treatment	409
11.2.5.2	Livestock	409
11.2.5.3	Landfill Methane	410
11.2.5.4	Natural Gas Production	410
11.2.5.5	Coal Production	411
11.3	Volatile Organic Compounds (VOC)	412
11.3.1	General Information	412
11.3.2	Sources, Sinks, and Ultimate Fate	412

11.3.3	Health and Welfare Effects	418
11.3.3.1	Human Health	418
11.3.3.2	Environmental Welfare	418
11.3.4	History and Regulations:	418
11.3.5	Control Strategies	419
11.3.5.1	Adsorption	421
11.3.5.2	Absorption	430
11.3.5.3	Thermal Destruction	438
11.3.5.3.1	Flaring	438
11.3.5.3.2	Incineration	440
11.3.5.4	Biological Control	449
11.3.5.5	Condensers	452
11.4	Questions	455
11.5	Problems	456
11.6	Group Project Ideas	459
11.7	Bibliography	459

12 Ozone ... 463

12.1	Tropospheric Ozone	465
12.1.1	General Information	465
12.1.2	Sources and Sinks	465
12.1.2.1	Daily Variations	470
12.1.2.2	Seasonal Variations	470
12.1.2.3	Temperature Dependency	470
12.1.2.4	Location Dependency	471
12.1.3	Health and Welfare	471
12.1.3.1	Effect on Human Health	471
12.1.3.2	Effect on Human Welfare	472
12.1.4	History and Regulation	474
12.1.5	Control	481
12.2	Stratospheric Ozone	485
12.2.1	Sources and Sinks	485
12.2.2	The Ozone Hole	488
12.2.3	Health and Welfare	490
12.2.4	History and Regulation	491
12.2.5	Control	496
12.2.6	Alternatives to Chlorofluorocarbons	501

12.3	Questions	502
12.4	Problems	503
12.5	Group Project Ideas	504
12.6	Bibliography	505

13 Hazardous Air Pollutants ... 509

13.1	Hazardous Air Pollutants - General Information	509
13.1.1	Emission Sources	509
13.1.2	Health and Welfare	512
13.1.3	History and Regulations	515
13.1.4	Control Strategies	518
13.2	Benzene	519
13.2.1	General Information	519
13.2.2	Sources, Sinks, and Ultimate Fate	519
13.2.3	Health and Welfare	520
13.2.3.1	Human Health	520
13.2.3.2	Environment Welfare	521
13.2.4	History and Regulation	522
13.2.5	Control Strategies	523
13.3	Lead	525
13.3.1	General Information	525
13.3.2	Sources, Sinks, and Ultimate Fate	526
13.3.3	Health and Welfare	528
13.3.3.1	Human Health	528
15.3.3.2	Environmental Welfare	530
13.3.4	History and Regulation	530
13.3.5	Control Strategies	533
13.3.5.1	Source Reduction	533
13.3.5.2	Emission Reduction	534
13.4	Mercury	535
13.4.1	General Information	535
13.4.2	Sources, Sinks, and Ultimate Fate	536
13.4.3	Health and Welfare Effects	540
13.4.3.1	Human Health	542
13.4.3.2	Environmental Welfare	544
13.4.4	History and Regulation	544
13.4.4.1	US Federal	545

13.4.4.2	US States	550
13.4.4.3	International	551
13.4.5	Control Technologies	553
13.4.5.1	Product Substitution	555
13.4.5.2	Process Modification, Work-Practice Standards, and Material Separation	556
13.4.5.3	Flue Gas Treatment Technologies	556
13.4.5.4	Powered Activation Carbon	557
13.4.5.5	Multi-Pollutant Control	558
13.4.5.6	Alternative Strategies	559
13.5	Questions	560
13.6	Problems	561
13.7	Group Project Ideas	563
13.8	Bibliography	564

14 Carbon Dioxide and Climate Change 567

14.1	General Information and Basic CO_2 Chemistry	567
14.2	Sources and Sinks	569
14.2.1	Sources	570
14.2.2	Sinks	573
14.3	Health Effects	576
14.3.1	CO_2 and Climate Change Science	576
14.3.1.1	Black Body Model	577
14.3.1.2	One-Layer Model	580
14.3.1.3	Greenhouse Gases	581
14.4	Regulatory Considerations	592
14.4.1	US Regulatory Considerations	592
14.4.1.1	Actions for New Power Plants	595
14.4.1.2	Actions for Existing Power Plants	597
14.4.2	International Regulatory Considerations	598
14.5	Control Technologies	599
14.5.1	Scale of the Required Yearly Carbon Emission Changes	599
14.5.2	Scale of the Required Energy Production Changes	601
14.5.3	Control Strategies	606
14.5.3.1	Mitigation	607
14.5.3.2	Adaptation	620
14.5.3.3	Cost	621
14.6	Questions	624

14.7	Problems	625
14.8	Group Project Ideas	628
14.9	Bibliography	628

15 Indoor Air Quality 631

15.1	History and General Information	631
15.1.1	Occupational Indoor Air Quality	633
15.2	Sources and Health Effects	635
15.2.1	Carbon Monoxide	635
15.2.2	Carbon Dioxide	636
15.2.3	Ozone	636
15.2.4	Radon	637
15.2.5	Asbestos	639
15.2.6	Volatile Organic Compounds	640
15.2.7	Metals	641
15.2.7.1	Mercury	641
15.2.7.2	Lead	642
15.2.8	Biological Air Pollutants – Molds, Allergens, Bacteria	643
15.2.9	Tobacco Smoke	644
15.3	Regulation	645
15.3.1	National Laws	645
15.3.2	State Laws	646
15.4	Control Technologies	648
15.4.1	Ventilation and Equipment	649
15.4.1.1	Air Exchange Rate	649
15.4.1.2	Hoods	652
15.4.1.3	Ductwork	654
15.4.1.4	Fans	661
15.4.1.5	Venting Stacks	663
15.4.2	Filtration and Adsorption	664
15.4.3	Maintenance and Housekeeping	664
15.4.4	Green Engineering and Indoor Air Quality	664
15.5	Questions	666
15.6	Problems	667
15.7	Group Project Ideas	669
15.8	Bibliography	669

16 Mobile Sources 673

16.1	Introduction	673
16.2	Engines	676
16.2.1	Internal Combustion Reciprocating Engines	676
16.2.1.1	Two-Stroke Reciprocating Engine	676
16.2.1.2	Four Stroke Reciprocating Engines	678
16.2.1.3	Turbine Engines	681
16.3	Fuel Characteristics	684
16.3.1	Air to Fuel Ratio	686
16.4	US Regulation of Mobile Sources	687
16.4.1	Highway Vehicles	688
16.4.1.1	Emission Factors	688
16.4.1.2	Emission Standards	690
16.4.2	Heavy Duty Vehicles	691
16.4.3	Non-Road Land-Based Diesel	692
16.4.4	Non-Road Spark Ignition Engines	693
16.4.5	Marine Diesel Engines	695
16.4.5.1	Background:	695
16.4.5.2	Regulations:	695
16.4.6	Locomotives	699
16.4.6.1	Background:	699
16.4.6.2	Regulations:	700
16.4.7	Aircraft	701
16.4.7.1	Background:	701
16.4.7.2	Regulations:	703
16.5	Fuel Regulations	704
16.6	Engine Emissions and Control	705
16.6.1	Spark Ignition and Diesel Engine Design	705
16.6.1.1	Air to Fuel Ratio	705
16.6.1.2	Combustion Ignition Timing	706
16.6.1.3	Compression	707
16.6.2	SI and Diesel Emission Controlling Equipment	708
16.6.2.1	Catalytic Converter	708
16.6.2.2	Diesel Engine Emissions: Catalysts and Filters	711
16.6.2.3	Positive Crankcase Ventilation (PCV)	712
16.6.2.4	Evaporative Emission Control (EVAP)	713
16.6.2.5	Exhaust Gas Recirculation (EGR)	714

16.6.2.6	Onboard Diagnostic System (OBDII)	716
16.6.2.7	Inspection and Maintenance (I/M) Programs	716
16.6.3	Refueling Losses	717
16.6.4	Transportation	721
16.7	Questions	723
16.8	Problems	725
16.9	Group Project Ideas	727
16.10	Bibliography	727
16.11	Appendices	730

17 Green Engineering 737

17.1	Risk Assessment	739
17.2	Environmental Impact Assessment	745
17.3	Life-Cycle Analysis	746
17.4	Conservation of Life	751
17.5	Process Integration (System Analysis)	755
17.6	Evaluation of Alternatives Analysis	757
17.7	Questions	761
17.8	Problems	762
17.9	Group Project Ideas	762
17.10	Bibliography	763

Appendices		765
I.	Conversion Factors	765
II.	New Source Performance Standards (NSPS)	769
III.	Emission Factors	781
IV.	Dispersion Tables	784
V.	List of US-EPA Hazardous Air Pollutants	787
VI.	Data Sets	795
	a. Viscosity of Gases	
	b. Henry's law Constants	
	c. Heat Capacity and Heat of Reaction Data	
	d. Vapor Pressure (Antoine Equation)	
	e. Vapor Pressure of Water	
	f. Auto-Ignition Temperatures	
Index		805

LIST OF FIGURES

Figure 1-1. US Military Observing a Nuclear Test. 4
Figure 1-2. Skyline Showing Particulate Matter and Smog from Air Emissions. Background Shows the Ford River Rouge Plant in Dearborn, MI, USA, July 1973. 7
Figure 1-3. Uncontrolled Emission from Ore Carrier Contributes to Smog over Lake Superior, June 1973. 8
Figure 1-4. Smog in Beijing, China. May, 2014 9
Figure 1-5. Smog-Damaged and Clean-Air Plants Shown Together. 10
Figure 1-6. False Color Views of Total Ozone Over the Antarctic Pole. 11
Figure 1-7. Sign posted in The Everglades National Park 11
Figure 1-8. Steps in Addressing an Air Pollution Problem. 15

Figure 2-1. Schematic of Simple Flow System 40
Figure 2-2. System Sketch for Mixing Three Streams in Example 2-17 41
Figure 2-3. Sketch of Mixer System for Example 2-18. 42
Figure 2-4. Unsteady Flow: Filling a Vessel 43
Figure 2-5. Sketch of Continuously Stirred Tank Reactor (CSTR). 49
Figure 2-6. Sketch of Flow System with Chemical Reaction 51
Figure 2-7. Sketch of Batch Reactor 52
Figure 2-8. Big Bend Power Station, Apollo Beach, FL, USA - An Example of a Coal Fired Power Plant 55
Figure 2-9. Sketch for Case Study 2-1 showing Inputs and Outputs from a Coal Fired Power Plant 56

Figure 3-1. US Court System Hierarchy. 65
Figure 3-2. Air Quality Monitoring Station, Oregon Department of Environmental Quality.. 74
Figure 3-3. Application of PSD Increment. 86
Figure 3-4. Map of US, July 2014, showing regions in non-attainment (air pollution exceeds levels defined in NAAQS) or in maintenance (exceeded NAAQS in the past, but may not currently be in exceedance). 87
Figure 3-5. Comparison of US Trends in Growth with Air Pollution. 94

List of Figures

Figure 4-1. Force Components on a Control Volume 112
Figure 4-2. Sketch for Example 4-7 118
Figure 4-3. Sketch for Example 4-8. 120
Figure 4-4. Boundary Layer Thickness (δ) and Velocity Profile (u_x) Across a Flat Plate . . . 125

Figure 5-1. Structure of the Atmosphere. 132
Figure 5-2. Example of Coriolis Effect Around a Low Pressure Cell. 136
Figure 5-3. Three Cell Model of Air Circulation Around a Smooth Homogeneous Sphere. . . 137
Figure 5-4. Land-Sea Wind Develops from Uneven Heating and Cooling of Different Surfaces. 139
Figure 5-5. Mountain -Valley Wind Develops from Quicker Heating of Mountain Side During Day and Quicker Cooling at Night. 139
Figure 5-6. Wind Rose Showing Wind Data (speed and direction) for February in Duluth, MN, Averaged Over 1961 - 1990. 142
Figure 5-7. Comparison of Adiabatic and Environmental Lapse Rates Showing Stable, Neutral, and Unstable Comparisons. 146
Figure 5-8. Frontal Inversion Schematic and Vertical Temperature Profile. 149
Figure 5-9. Radiation Inversion Schematic and Vertical Temperature Profile. 150
Figure 5-10. Subsidence Inversion Schematic and Vertical Temperature Profile. 150
Figure 5-11. Maximum Mixing Depth (MMD) for Different Atmospheric Conditions. . . 151

Figure 6-1. Stack-Based Emission Source and Gaussian Model Approximation of Plume . . . 162
Figure 6-2. Horizontal Mixing Length for Rural and Urban Areas as a Function of Pasquill Atmospheric Stability Class. 163
Figure 6-3. Vertical Mixing Length for Rural and Urban Areas as a Function of Pasquill Atmospheric Stability Class. 164
Figure 6-4. Change in Downwind Concentration due to the Ground as a Reflection Source. . 165
Figure 6-5. Downwind, Centerline Concentration Profile for Example 6-3. 170
Figure 6-6. SO_x Concentration Contour Plot for Example 6-5. 173
Figure 6-7. Value and Distance to Maximum Ground Level Concentration from an Elevated Source with Reflection (rural areas). 174
Figure 6-8. Concentration Profile at Three Downwind Positions from an Instantaneous Puff Release. 179
Figure 6-9. Sketch for Example 6-10. 180
Figure 6-10. Example of normal plume and a plume experiencing downwash. 182

Figure 7-1. Range of Particle Sizes for Several Example Materials and Processes 196
Figure 7-2a/b. Particle Number Distribution and 7-2.b Natural log Particle Number Distribution. . 199

List of Figures

Figure 7-3a/b. Particle Mass Distribution and 7-3.b Natural Log Particle Mass Distribution . . . 202
Figure 7-4. Comparison of Results from Example 7-1 and Example 7-2 207
Figure 7-5. Log-Probability Plot of Particle Size from Table 1 209
Figure 7-6. Log-Probability Paper 210
Figure 7-7. Force Balance and Resultant Particle Velocity during Free-fall 214
Figure 7-8. Terminal Velocity (up) of Particles Settling in Air at 1 atm and 20 °C for Various Particle Densities (ρp) 215
Figure 7-9. Scattering and Absorption of light by a large particle 219
Figure 7-10. Basic packing arrangements of uniform spheres 221

Figure 8-1. Particulate Matter Size Fractions 227
Figure 8-2. Eyjafjallajökull volcanic eruption showing plume, April 17th, 2010, Holsvöllur Iceland 228
Figure 8-3. Estimated Extent of Ash Cloud from Eyjafjallajökull Volcano, April 15, 2010 . . 228
Figure 8-4. Prototypical Size and Number Distribution of Tropospheric Particles with Selected Sources and Pathways of Particle Formation 231
Figure 8-5. Particle deposition as a function of particle diameter in various regions of the lung 234
Figure 8-6. Visibility Impairment from Air Pollution, Badlands National park, South Dakota, USA 237
Figure 8-7. Hong Kong Skyline Obscured by PM and Haze 238
Figure 8-8. US Particulate Matter Historical Trend 240
Figure 8-9. Examples of Particle Collection by Interaction with a Target 242
Figure 8-10. Collection Efficiency Ranges by Size 243
Figure 8-11. Grade Efficiency for Various Types of Particle Control Equipment 244
Figure 8-12. Settling Chamber Schematic 247
Figure 8-13. Basic Dimensions for Particle Control Equipment 248
Figure 8-14. Cyclone Schematic 251
Figure 8-15. Comparison of Cyclone Collection Efficiency Models 255
Figure 8-16. Filter Bag Equipment Schematic 256
Figure 8-17. Cyclical Nature of Pressure Drop of a Filter Bag Collection Chamber 260
Figure 8-18. Line fit of Equation 34 to Experimental Data 261
Figure 8-19. Electrostatic Precipitator (ESP) Schematic 264
Figure 8-20. Counter-Current Wet Scrubber Schematic 269
Figure 8-21. Flow Chart for Choosing a Particle Control System 273
Figure 8-22. Sketch of one corner of a filter bag compartment 277

Figure 9-1. The Atmospheric Sulfur Cycle 286
Figure 9-2. Restoration of a Market Cross in the UK Showing Architectural Damage from Air Pollution 292

List of Figures

Figure 9-3.	The pH Scale.	295
Figure 9-4.	pH Isopleths in the Continental US Maps.	298-299
Figure 9-5.	Average Winning Bid for a One Ton SO_2 Emission Credit.	304
Figure 9-6.	US National Air Quality Trend for SO_2 Concentration (ppmv) – Annual Average.	307
Figure 9-7.	US National Air Quality Trend for SO_2 Concentration (ppmv) – 1-Hour Average.	308
Figure 9-8.	US National Air Quality Trend for Total SO_2 Emissions.	309
Figure 9-9.	Annual Average Sulfate Concentration (SO_4^{-2}) Isopleths in the Continental US.	310-311
Figure 9-10.	4th Highest 24-Hour Mean SO_2 Concentration Observed at Urban Stations, EEA Member Countries, 1997-2010 [$\mu g / m^3$].	313
Figure 9-11.	Map of Europe Showing the EU-28 Member States.	315
Figure 9-12.	Process Flow Diagram for the Claus Process.	320
Figure 9-13.	Wet Spray Chamber Schematic.	325
Figure 10-1.	The Atmospheric Nitrogen Cycle.	343
Figure 10-2.	Fuel NO_x Combustion Mechanism.	347
Figure 10-3.	Health and Welfare Effects of Nitrogen Emissions.	350
Figure 10-4.	Timeline of US-EPA Regulatory Programs to Control NO_x in the Ambient Air	353
Figure 10-5.	US National Air Quality Trend for Annual Average NO_2 Concentration (ppm).	357
Figure 10-6.	US National Air Quality Trend for One-Hour Average NO_2 Concentration (ppb).	358
Figure 10-7.	Annual Average Inorganic Nitrogen Deposition (NO_3 and NH_3) Isopleths in Continental US.	359
Figure 10-8.	US National Air Quality Trend for Permitted, Stationary Sources of NO_2.	360
Figure 10-9.	Annual Average Ambient Air NO_2 concentration [ppm], Japan.	362
Figure 10-10.	Schematic Plan View of Simplified Firebox with Water Tube Boiler.	364
Figure 10-11.	Schematic Showing the Heat Transfer Sections of a Water Tube Boiler.	365
Figure 10-12.	Major NO_x Reduction Techniques.	368
Figure 10-13.	Effect of Excess Oxygen in Emissions on NO_x.	371
Figure 10-14.	Feed Arrangements in Staged Combustion.	372
Figure 10-15.	Schematic Showing Flue Gas Recirculation (FGR) and Fuel Induced Recirculation (FIR) Systems.	375
Figure 10-16.	Selective Non-Catalytic Reduction (SNCR).	376
Figure 10-17.	Selective Catalytic Reduction of NOx with Ammonia.	379
Figure 11-1.	Figure 11 1. US Average Fuel Efficiencies and Standards.	397
Figure 11-2.	US National Air Quality Trend for CO Concentration (ppm).	398

List of Figures

Figure 11-3. Historical Trend in Atmospheric Methane Concentration. 407
Figure 11-4. Example of Breakthrough Curve and Adsorption Wave. 422
Figure 11-5. Methane and Ethane on Activated Carbon Adsorbent (Isotherm at T=301.4 K) . 423
Figure 11-6. Schematic of Adsorption System with VOC Recovery. 426
Figure 11-7. Example 11-3 Model fits of Ethane Isotherm Data at Temperature of 301.4 K. 429
Figure 11-8. Schematic System Design for Absorption System. 431
Figure 11-9. Absorption Design Diagram - CO_2 in Water at 10°C and 25 atm.. 434
Figure 11-10. Absorption Design Diagram for Example 11-4 Ethene Absorption in Water... 437
Figure 11-11. Elevated Flare in Operation. 439
Figure 11-12. Interior of Combustion Chamber. 440
Figure 11-13. Schematic of A) an Incinerator and B) an Incinerator with Recuperative Heat Exchanger. 441
Figure 11-14. Schematic of a Basic Biofilter. 449

Figure 12-1. Ozone Distribution in the Atmosphere. 463
Figure 12-2. Global Total Ozone Distribution for March 18, 2011. 464
Figure 12-3. Sources and Sinks of Ozone (O_3) in the Troposphere. 465
Figure 12-4. US National Air Quality Trend for Ambient O_3 Concentration (ppm). 478
Figure 12-5. Calculated Ozone Isopleths Resulting from Constant Emission Rates of NO_x and VOCs.. 483
Figure 12-6. South Pole Stratospheric Ozone Hole: Size and Minimum Measured Concentration, 1979 to 2009. 488
Figure 12-7. United Nations Environment Program Projection of the Effect of the Montreal Protocol Amendments on Stratospheric Chlorine Concentration and the Resulting Excess Number of Skin Cancers. 494
Figure 12-8. Projection of Global Atmospheric Concentration of CFC-11 over Next 50 Years, Solution to Example 12-2.. 501

Figure 13-1. US National Air Quality Trend for Benzene Concentration [$\mu g/m^3$] 523
Figure 13-2. US Reported Annual Lead Emissions [x10^3 ton/year]. 527
Figure 13-3. US National Air Quality Trend for Pb Concentration, as Total Suspended Particles [$\mu g/m^3$]. 533
Figure 13-4. Historical Atmospheric Concentration of Mercury from Ice-Cores. 539
Figure 13-5. Schematic of Global Mercury Cycle. 541

Figure 14-1. Historical Values of CO_2 Concentration in Atmosphere from the Siple Station Ice Core in Antarctica (1744 – 1953).. 569
Figure 14-2. Concentration of CO_2 in the Atmosphere. 570

List of Figures

Figure 14-3.	Annual Anthropogenic Emissions of Carbon to Atmosphere (1800 – 2008)	572
Figure 14-4.	Global Carbon Cycle, 2005 Data.	574
Figure 14-5.	Global Energy Flux Models.	578
Figure 14-6.	Model IR Spectrum Emitted from Earth Compared with Blackbody Radiation from Various Temperature Surfaces.	584
Figure 14-7.	Simplified Model of a Path a Photon (wavenumber 667.3 cm^{-1}) may Take between Emission from the Earth's Surface until it Escapes into Space.	585
Figure 14-8.	Effect of CO_2 Concentration on the Required Surface Temperature of Earth to Achieve Isun = Iearth = 260 W/m^2.	587
Figure 14-9.	Various Pathways for Reducing Fossil Fuel Carbon Emissions to Minimize Harm to Human Health and the Environment.	600
Figure 14-10.	Geometry of a cone-shaped pile.	600
Figure 14-11.	US Primary Energy Consumption by Fuel Source 1800 – 2010.	602
Figure 14-12.	Simplified Schematic of a Steam Turbine Electricity Generating Unit.	609
Figure 14-13.	Simplified Schematic of Combustion Turbine Electricity Generating Unit.	610
Figure 14-14.	CO_2 Phase Diagram.	617
Figure 15-1.	US-EPA Map of Radon Zones.	638
Figure 15-2.	Three Basic Types of Hoods.	653
Figure 15-3.	Basic Sizing Relationships for Airflow in a Duct.	654
Figure 15-4.	System Sketch for Example 15-5.	656
Figure 15-5.	System Sketch for Example 15-6.	659
Figure 15-6.	Example 15-7, System Operating Point.	663
Figure 16-1.	Mobile Source Pollutants Reported to US-EPA, 2012.	675
Figure 16-2.	Schematic of Two-Stroke Engine.	677
Figure 16-3.	Diagram Showing the Stages of a Four-Stroke Engine in Operation.	680
Figure 16-4.	Timing of Major Events of a 4-Stroke, Spark-Ignition Engine.	681
Figure 16-5.	Schematic of aTurbofan Engine.	682
Figure 16-6.	Annual World Shipping by Marine Vessels.	696
Figure 16-7.	The Effects of Air-Fuel Ratio on Hydrocarbon, Carbon Monoxide, and Nitric Oxide Exhaust Emissions.	706
Figure 16-8.	Pictures of a) Catalysts for Motorcycles and Small Engines, and b) Three-way catalyst for Automobiles.	708
Figure 16-9.	Schematic of Catalytic Emission Control System.	709
Figure 16-10.	Diagram of a Particle Filter.	711
Figure 16-11.	Schematic of Positive Crankcase Valve (PCV) System.	713
Figure 16-12.	Schematic of Evaporative Emission Control (EVAP) System.	714

Figure 16-13. Schematic of Exhaust Gas Recycle (EGR) System. 715
Figure 16-14. Effect of EGR on NO_x Emissions in SI Engine. 715
Figure 16-15. Stage I and II Vapor Recovery Systems. 717
Figure 16-16. Stage II Coaxial Dispenser a) Nozzle Schematic b) Photo 719

Figure 17-1. Risk Assessment Steps 739
Figure 17-2. General Dose-Response Characteristic Curves. 741
Figure 17-3. Flow Diagrams before and after Process Integration. 756

LIST OF TABLES

Chapter 1
Table 1-1. Summary of Above Ground Nuclear Device Testing....................3
Table 1-2. The Air Quality Index...18

Chapter 2
Table 2-1. A Problem Solving Strategy...23
Table 2-2. Different Types of Units to Describe Concentration 31
Table 2-3. Sets of Standards Conditions.......................................34
Table 2-4. Material Balance System Types......................................39
Table 2-5. Quantity of Atomic Species in One mol of Coal for Case Study 2-1 56
Table 2-6. Exhaust Gas Composition and Molar Composition for Case Study 2-1 58
Table 2-7. Mass concentrations of the Exhaust in Case Study 2-1 58

Chapter 3
Table 3-1. Websites with Information on US Environmental Laws and Regulations.....69
Table 3-2. Organization of 1990 Clean Air Act and Amendments..................71
Table 3-3. National Ambient Air Quality Standards (as of October, 2011)...........71
Table 3-4. Example 3-1 CO data (2007 – 2010):.................................73
Table 3-5. Example 3-1 SO_2 data (1998 – 2010):..............................73
Table 3-6. NSPS for Electric Utility Steam Generating Units [40CFR60 subparts D et seq.]:..76
Table 3-7. PSD Increments as of July 2013 (adapted from 40 CFR Part 52.21).........85
Table 3-8. Sources Requiring a Title V Operating Permit........................93
Table 3-9. Mass of Each Constituent in the Coal for Case Study 3-3................98

Chapter 4
Table 4-1. Classification of Flow by Reynolds Number..........................122
Table 4-2. Surface Roughness of Common Materials............................124

Chapter 5
Table 5-1. Composition of Dry Air, Global Average.............................128
Table 5-2. Beaufort Wind Scale..141
Table 5-3. Stability Exponent, p, for Vertical Wind Profiles (Equation 5.12).........143

Table 5-4.	Descriptions of the Pasquill Stability Classes by Environmental Lapse Rate.	147
Table 5-5.	Meteorological Descriptions of Pasquill Stability Classes.	147

Chapter 6

Table 6-1.	Example 6-3 Data and Calculations.	169
Table 6-2.	Example 6-5 Data and Calculations.	172
Table 6-3.	Mixing Length Formulas for Instantaneous Releases.	177
Table 6-4.	Example 6-9 Calculations for Instantaneous Release of Butane.	178

Chapter 7

Table 7-1.	Particle Shapes.	196
Table 7-2.	Specific Gravity of Common Particulate Matter in the Atmosphere.	197
Table 7-3.	Example Distribution of Particles by Number and Bin Size.	198
Table 7-4.	Example 7-1 Data and Worksheet.	204
Table 7-5.	Example 7-2 Data and Worksheet.	206
Table 7-6.	Example 7-3 Data and Worksheet.	208
Table 7-7.	Extinction and Visual Range in Air Containing 18 g/m^3 of Water in Different Forms.	220

Chapter 8

Table 8-1.	Particle Size Terminology.	225
Table 8-2.	Global Particle Sources (10^6 ton/year)	229
Table 8-3.	US Particle Emissions (10^3 ton/year)	230
Table 8-4.	Particle Formation Mechanisms	230
Table 8-5.	Ambient Air Quality Standards for PM in Several Countries	239
Table 8-6.	Forces used in Various Particulate Matter Control Equipment	243
Table 8-7.	Collection Efficiency Calculation for a Gravity Settler.	245
Table 8-8.	Collection Efficiency Calculation for a Cyclone	246
Table 8-9.	Data and Worksheet for Example 8-5	255
Table 8-10.	Filter Fabric Characteristics.	257
Table 8-11.	Data Set for Example 8-6.	261
Table 8-12.	Size Distribution for Cyclone Furnaces Burning Bituminous Coal.	275

Chapter 9

Table 9-1.	Sources and Annual Flux of Sulfur in the Atmosphere (TgS / year)	287
Table 9-2.	Human Caused Sources of SO_2 in USA (2005).	287
Table 9-3.	SO_2 Emission Leaders, Total and Per Capita, 2012 data.	288
Table 9-4.	Health Effects of SO_2 Exposure.	291
Table 9-5.	pH Tolerance Levels of Some Aquatic Organisms.	297

Table 9-6.	Ambient Air Quality Standards for SO$_2$ in Several Countries	301
Table 9-7.	SO$_2$ Emissions and Allowances (106 ton) from Acid Rain Program Sources	306
Table 9-8.	Energy and Sulfur Content and Cost of Several Fossil Fuels	317
Table 9-9.	Flue Gas Desulfurization (FGD) Methods for Coal Combustion	323
Table 9-10.	Cost Information for SO$_2$ Scrubbing Systems	328
Table 9-11.	Characteristics of Fuels for Paper Mill in Case Study 9-4	329
Table 9-12.	Emission Limits for Paper Mill in Case Study 9-4	329
Table 9-13.	Fuel Costs and Sulfur Emissions for Paper Mill in Case Study 9-4	329
Table 9-14.	Sulfur Removal Requirements and Total Costs for Paper Mill in Case Study 9-4	330

Chapter 10

Table 10-1.	Global Nitrogen Budget in Tg N/yr	342
Table 10-2.	U.S. Nitrous Oxide (N$_2$O) Emissions by Source (Tg CO$_2$ Equivalents), 2008	344
Table 10-3.	U.S. Anthropogenic Nitrogen Oxide Emissions by Source, 2008	345
Table 10-4.	National Ambient Air Quality Standards for NO$_2$ in Several Countries (2010)	353
Table 10-5.	Performance Summary for NO$_x$ Control Techniques on Industrial Boilers	380

Chapter 11

Table 11-1.	US Carbon Monoxide Emission Sources [tons/year], 2011	393
Table 11-2.	Milestones in Motor Vehicle Emissions Controls	396
Table 11-3.	Global Estimates for Methane in the Atmosphere [Tg CH$_4$/yr]	401
Table 11-4.	US Methane (2010) Emissions Reported by Source [Tg CH$_4$/yr]	404
Table 11-5.	Daily Waste and Methane Production by Dairy, Beef, and Swine per 1,000 Pounds of Animal Weight	410
Table 11-6.	Annual Global Average of Natural VOC Emissions	413
Table 11-7.	List of Major Biogenic VOC Species	413
Table 11-8.	Anthropogenic Sources of VOC Emissions	414
Table 11-9	US VOC Emissions Reported by Source, 2011	416
Table 11-10.	Calculated Lifetimes for Several NMVOC in the Atmosphere	417
Table 11-11.	VOC Limits in US Consumer Products	420
Table 11-12.	Typical Industrial Emissions	421
Table 11-13.	Unit Operations for Control of VOCs	421
Table 11-14.	Example 3 Best Fit Line Parameters	428
Table 11-15.	Example 3 Isotherm Model Parameters	428
Table 11-16	Stoichiometric Relationships between Molar Flows In and Out of Combustion Chamber	443
Table 11-17.	Flow Rates for Example 11-5	446

Table 11-18.	Mean Heat Capacities for Example 11-5.	447
Table 11-19.	Flow Rates with Additional Reaction for Example 11-5.	447
Table 11.20.	Additional Heat Capacities Needed in Example 11-5.	448
Table 11.21.	Refrigerated Condenser Operating Fluids.	453

Chapter 12

Table 12.1.	Photochemical Lifetime (days) of Ozone in the Troposphere.	469
Table 12.2.	Ozone Ambient Air Quality Standards in Several Countries.	474
Table 12.3.	History of US-EPA Ground-level Ozone Standards.	475
Table 12.4.	US Motorcycle Emission Standards.	477
Table 12.5.	Photons Characterized by their Wavelength (λ).	486
Table 12.6.	US-EPA UV Index.	496
Table 12.7.	Ozone Depleting Substances with their associated atmospheric lifetime (years), Ozone Depletion Potential (ODP), and Global Warming Potential (GWP).	498

Chapter 13

Table 13-1.	US Background Concentrations of Several Hazardous Air Pollutants (HAPs).	511
Table 13-2.	Quantities of HAPs released in the US.	511
Table 13-3.	2005 NATA Health Effects Drivers and Contributors Risk Characterization.	512
Table 13-4	2005 National Air Toxics Assessment Results.	513
Table 13-5.	Causes of Death and Associated Risk in US, 2009.	514
Table 13-6.	US-EPA Calculated Cancer Risk from Inhalation of Benzene.	521
Table 13-7.	Ambient Air Quality Standards for Pb in Several Countries.	532
Table 13-8.	Replacement for Lead.	534
Table 13-9.	Global Mercury Sources and Sinks.	537
Table 13-10.	US Anthropogenic Sources of Mercury (1999).	538
Table 13-11.	Global Anthropogenic Mercury Emissions by World Region, 2012.	538
Table 13-12.	Hospital/Medical/Infectious Waste Incinerators.	546
Table 13-13.	Large Municipal Waste Combustors.	546
Table 13-14.	Small Municipal Waste Combustion.	546
Table 13-15.	Other Solid Waste Incineration Units.	546
Table 13-16.	Sewage Sludge Incineration Units.	547
Table 13-17.	Portland Cement Plants.	547
Table 13-18.	Mercury Cell Chlor-Alkali Plants MACT.	547
Table 13-19.	Gold Mine Ore Processing MACT.	547
Table 13-20.	Boilers and Process Heaters MACT.	548
Table 13-21.	Coal and Oil Electric Generating Units (EGU) MACT.	548
Table 13-22.	Mercury Emission Factors for Several Industrial Processes.	554
Table 13-23.	Two Models for Estimating the Adsorption of Mercury Chloride on Adsorbent x-7000.	557

Table 13-24. Average Mercury Capture Efficiency by Existing Post-Combustion Control Equipment for Pulverized Coal Fired Boilers, by Coal Type..............559

Chapter 14
Table 14-1. Cumulative Worldwide Anthropogenic Emissions of Carbon to the Atmosphere (1750 – 2008)................................571
Table 14-2. US Anthropogenic Emissions of Carbon to Atmosphere by Source (1990 – 2009) [Megaton C/ year]..................................573
Table 14-3. Albedo of Several Materials......................579
Table 14-4. The Major Greenhouse Gases, 2011..................581
Table 14-5. IR Wavenumber (cm-1) of Various Excited States for Three Greenhouse Gases. 583
Table 14-6. Intensity of Earthlight, Iearth [W/m^2], for Surface Various Temperatures and Atmospheric CO_2 Concentrations......................586
Table 14-7. The US-EPA Six Major Greenhouse Gases..............592
Table 14-8. Storage Considerations for a Few Forms of Carbon......601
Table 14-9. Energy Density and Carbon Content of Several Fuels......603
Table 14-10. US Energy Consumption by Economic Sector, 2010......603
Table 14-11. Large Hydroelectric Electricity Generating Units......605
Table 14-12. Energy Yields of Various Energy Crops..............606
Table 14-13. Power Plant Efficiencies.......................610
Table 14-14 Cost Ranges for Components of a CCS System Applied to a Given Source, 2011...................................619

Chapter 15
Table 15-1. Typical Indoor Air Quality Materials of Concern......635
Table 15-2. Typical Methods to Control Indoor Air Quality......648
Table 15-3. Pressure Change as a Function of Flow for Example 15-7......662

Chapter 16
Table 16-1. Numbers of Mobile Sources (2010)..................674
Table 16-2. US-EPA Mobile Source Categories..................676
Table 16-3. Fractions of Crude Oil (with range of carbon chain sizes) [and other names]...685
Table 16-4. US-EPA Vehicle Categories for Highway Light Duty Vehicles......688
Table 16-5. LDV Emissions 1990 - 2020, US-EPA MOVES Estimate......689
Table 16-6. Emissions Factors for LDVs (Gas Powered)..............689
Table 16-7. US-EPA Tier 1 and 2 Emission Standards for LDV at 100,000 miles [g/mile]..690
Table 16-8. California Diesel Engine Emission Factors...........692
Table 16-9. Tier 4 Emission Standards—Engines Above 560 kW [g/kWh]......693

Table 16-10.	Small Spark Ignition (SI) Engine Classifications.	694
Table 16-11.	Number of Ocean Going Ships Subject to International Air Pollution Regulations, 2008.	695
Table 16-12.	Marine Engine Categories Based on Engine Cylinder Displacement Volume.	696
Table 16-13.	MARPOL Annex VI NOx Emission Limits.	698
Table 16-14.	MARPOL Annex VI Sulfur Limits in Fuel [wt%].	698
Table 16-15.	Locomotive Uses.	699
Table 16-16.	Estimated Emission Factor Baseline In-Use Emission Rates [g/bhp-hr].	700
Table 16-17.	Aircraft Fuel Usage (kg) and Emission Factors (kg/LTO or kg/ton fuel used).	701
Table 16-18.	Emissions for LTO and Cruise in kg for Example 16-5.	702
Table 16-19.	US Regulatory Tiers for Aircraft.	703
Table 16-20.	The Saturation Factor in Equation 16.13.	718

Chapter 17

Table 17-1.	The Twelve Principles of Green Engineering.	737
Table 17-2.	Safety Factor Multipliers used in Extrapolating Dose-Response from Experimental Data.	742
Table 17-3.	Properties of Paper and Plastic Grocery Sacks.	749
Table 17-4.	The Conservation of Life Principles.	752
Table 17-5.	Case Study 17-4 Project Stakeholders.	758
Table 17-6.	Case Study 17-4 Project Objectives.	759
Table 17-7.	Case Study 17-4 Stakeholder Weights for Objectives and Final Values.	759
Table 17-8.	Case Study 17-4 Alternative Scores.	761
Table 17-9.	Case Study 17-4 Alternative Rankings.	761

LIST OF ABBREVIATIONS

Chapter 1
AQI	Air Quality Index
CFC	chlorofluorocarbons
CO_2	carbon dioxide
CO	carbon monoxide
EPA	Environmental Protection Agency
HAP	hazardous air pollutant
NO_x	nitrogen oxide compounds
OSHA	Occupational Health and Safety Administration
PM	particulate matter
SO_x	sulfur oxide compounds
TEL	tetraethyl lead
VOC	volatile organic compounds

Chapter 3
AQA	Air Quality Analysis
BACM	Best Available Control Measures
BACT	Best Available Control Technology
CAA	Clean Air Act
CAAA	Clean Air Act Amendments
CAFE	Corporate Average Fuel Economy
CFR	Code of Federal Regulations
EPA	Environmental Protection Agency
FIP	Federal Implementation Plan
GPO	Government Printing Office
HAP	Hazardous Air Pollutants
ISO	International Standards Organization
LAER	Lowest Achievable Emission Rate
MACT	Maximum Achievable Control Technology
NAAQS	National Ambient Air Quality Standards
NHTSA	National Highway Traffic Safety Administration
NSPS	New Source Performance Standards
NSR	New Source Review
OTC	Ozone Transport Commission

PSD	Prevention of Significant Deterioration
RACM	Reasonably Available Control Measures
RACT	Reasonable Available Control Technology
RCRA	Resource Conservation and Recovery Act
SIP	State Implementation Plan
TIP	Tribal Implementation Plans
USC	United States Code

Chapter 4

cP	centi-Poise, a unit for viscosity
F	friction
M	mass flow [mass/time]
m	mass
mw	molecular weight
n	number of moles
P	pressure
Q	volumetric flow [volume/time]
R	ideal gas constant (several values are given in appendix 1)
Re	Reynolds number
T	temperature
V	volume
W	work

Chapter 5

MMD	maximum mixing depth
RH	relative humidity
VP	vapor pressure
Γ	dry adiabatic lapse rate
Λ	actual or environmental lapse rate

Chapter 6

ADE	advection dispersion equation
C	concentration
CO	carbon monoxide
H	effective stack height
HAP	hazardous air pollutant
ISC3	industrial source complex (an EPA dispersion model)

List of Abbreviations

NO_x	nitrogen oxides
PM	particulate matter
Q	emission rate [mass/time],
SO_x	sulfur oxides
u	wind speed
VOC	volatile organic compound
σ_y and σ_z	mixing lengths

Chapter 7

D_{eff}	effective diameter
f	probability function
K_c	Cunningham correction factor
PM	particulate matter

Chapter 8

ESP	electrostatic precipitators
PM	particulate matter, any size
$PM_{0.1}$	particulate matter with a diameter of less than 0.1 μm
$PM_{2.5}$	particulate matter with a diameter of less than 2.5 μm
PM_{10}	particulate matter with diameters of less than 10 μm
TSP	total suspended particles

Chapter 9

ARP	Acid Rain Program
CAIR	Clean Air Interstate Rule
CCN	cloud condensation nuclei
EU	European Union
FGD	flue gas desulfurization
NAAQS	National Ambient Air Quality Standards
NAPAP	National Acid Precipitation Assessment Program
NSPS	New Source performance Standards
NO_x	nitrogen oxides
PM	particulate matter
SO_x	sulfur oxides
UK	United Kingdom

LIST OF ABBREVIATIONS

Chapter 10

ARP	Acid Rain Program
BOOS	burners-out-of-service
CAIR	Clean Air Interstate Rule
FGR	flue gas recirculation
FIR	fuel induced recirculation
HNO_3	nitric acid
HONO	nitrous acid
LEA	low-excess air
LNB	low NO_x burners
NBP	NO_x Budget Trading Program
NH_3	ammonia
NH_4^+	ammonium
N_2	molecular nitrogen
NO	nitrogen oxide
NO_2	nitrogen dioxide
NO_x	nitrogen oxide compounds (NO and NO_2)
NO_3^-	nitrate
NO_2^-	nitrite
N_2O	nitrous oxide
N_2O_5	dinitrogen pentoxide
OFA	over-fire-air
OTC	Ozone Transport Commission
RACT	Reasonably Available Control Technology
SC	staged combustion
SCA	staged combustion air
SCR	selective catalytic reduction
SIP	state implementation plan
SNCR	selective non-catalytic reduction

Chapter 11

CAFE	Corporate Average Fuel Efficiency
CH_4	methane
CMM	coal mine methane
CO	carbon monoxide
LFG	landfill gas
NHTSA	National Highway and Traffic Safety Administration

NMTOC	non-methane total organic carbon
NMVOC	non-methane volatile organic compounds
VOC	volatile organic compounds
VRU	vapor recovery units

Chapter 12

CAIR	clean air interstate rule
CFC	chlorofluorocarbons
CH4	methane
CMAQ	community multi-scale air quality
CMAS	community modeling and analysis system
CO	carbon monoxide
DU	Dobson units
EKMA	empirical kinetic modeling approach
GWP	global warming potential
HFC	hydrofluorocarbons
HCFC	hydrochlorofluorocarbon
NMVOC	non-methane volatile organic compounds
ODP	ozone depletion potential
ODS	ozone-destroying substance
PAN	peroxyacetyl nitrate
PSC	polar stratospheric clouds
VOC	volatile organic compounds

Chapter 13

BACT	best available control technology
CAA	clean air act
EGU	electricity generating unit
HAP	hazardous air pollutants
MACT	maximum achievable control technology
MATS	mercury and air toxics standards
MSAT	mobile source air toxics
NATA	national air toxic assessment
NESHAP	national emissions standards for hazardous air pollutants
PAC	powdered activated carbon
VOC	volatile organic compounds

Chapter 14

CAFE	corporate averaged fleet efficiency
CCS	carbon capture and sequestration
CFC	chlorofluorocarbons
CFL	compact fluorescent lights
CO_2	carbon dioxide
CT	combustion turbine
EGU	electricity generating unit
GCC	global climate change
GHG	greenhouse gases
GtC	gigaton carbon
IGCC	integrated gasification combined cycle
IPCC	Intergovernmental Panel on Climate Change
LCO_2	liquid carbon dioxide
LED	light emitting diodes
NGO	non-government organizations
PC	pulverized coal
PSD	prevention of significant deterioration
SCO_2	supercritical fluid carbon dioxide

Chapter 15

ATSDR	Agency for Toxic Substances and Disease Registry
FSP	fan static pressure
HVAC	heating, ventilating, and air conditioning
IAQ	indoor air quality
NIOSH	National Institute for Occupational Safety and Health
OSHA	Occupational Safety and Health Administration
Re	Reynolds number
TSCA	Toxic Substances Control Act

Chapter 16

A/F	air to fuel
ALVW	adjusted loaded vehicle weight
ATDC	after top dead center
BDC	bottom dead center
BTDC	before top dead center
CARB	California Air Resources Board
CO	carbon monoxide
CO_2	carbon dioxide

List of Abbreviations

CR	compression ratio
DOC	diesel oxidation catalyst
DPF	diesel particle filter
ECA	emission control areas
EGR	exhaust gas recirculation
EVAP	evaporative emission control
FAA	Federal Aviation Administration
GVWR	gross vehicle weight rating
HAP	hazardous air pollutant
HC	hydrocarbon
HDV	heavy duty vehicles
HFO	heavy fuel oil
HLDT	heavy light-duty truck
ICAO	International Civil Aviation Organization
ILEV	inherently low emission vehicles
I/M	inspection and maintenance
IMO	International Maritime Organization's
LLDT	light light-duty truck
LDT	light-duty truck
LDV	light duty vehicles
LEV	low emission vehicles
LPG	liquefied petroleum gas
LTO	landing/take-off
MDPV	medium-duty passenger vehicle
MIL	malfunction indicator lamp
MOVES	Motor Vehicle Emission Simulator
NMOG	non-methane organic gas
NO_x	nitrogen oxide compounds
OBDII	onboard diagnostic system
PCV	positive crankcase ventilation
PM	particulate matter
RFG	reformulated gasoline
RPM	revolutions per minute
SCR	selective catalytic reduction
SI	spark ignition
SO_x	sulfur oxide compounds
SUV	sport utility vehicles
TDC	top dead center

UEGO	universal exhaust gas sensor
ULEV	ultra-low emission vehicles
VOC	volatile organic compounds
ZEV	zero emission vehicles

Chapter 17

CDC	Center for Disease Control and Prevention
COL	conservation of life
EIA	environmental impact assessment
GE	green engineering
HAZOP	hazard and operability analysis
LCA	life cycle analysis
MSDS	material safety data sheets
NEPA	National Environmental Policy Act
NIOSH	National Institute for Occupational Safety and Health
OSHA	Occupational Safety and Health Administration
PEL	permissible exposure limits
PI	process integration
REL	recommended exposure limits

CHAPTER **1**

Air Pollution: Introduction

1.1 Introduction

All living things interact with their surroundings. They consume resources and emit wastes. These wastes become pollutants when they accumulate in concentrations beyond the normal, self-cleaning or dispersive capabilities of the environment. Air pollution occurs when such wastes accumulate in the atmosphere.

There are many definitions of air pollution; each is somewhat different, but most definitions include the idea of the presence of substances in outdoor air that cause harm to humans, animals, plants and/or property, and exist in concentrations above that found in clean air. See table 5-1 for a list of the constituents of clean air, and which constituents are currently changing.

The substances can be dusts, fumes, mists, liquids, smokes, vapors, gases, odorous substances, or any combination thereof while excluding uncombined water vapor. Typical substances that are considered pollutants include:

- Particle matter (PM),
- Sulfur Oxides (SO_x),
- Volatile Organic Compounds (VOC),
- Nitrogen Oxides (NO_x),
- Carbon Monoxide (CO),
- Fluorine, Chlorine, Bromine, and their compounds,
- Radioactive substances,
- Hazardous substances (carcinogens, mutagens, teratogens, ...),
- Photochemical substances (chemically react in sunlight to form ozone and smog), and
- Metals (mercury, lead, nickel, zinc, ...).

Harm includes discomfort or damage to the body, or prevention of the enjoyment of one's property. The harm from air pollution has been a concern of people and governments for hundreds of years. The first known air pollution law occurred in 1306 when King Edward I of England banned the burning of sea-coal in London's craftsman furnaces due to the foul smelling fumes.

There are four basic categories or types of air:
- Outdoor air, also called ambient air, is monitored nationwide and regulated at the national level by the Environmental Protection Agency (EPA). It is the most studied and most regulated type of air. The majority of this textbook focuses on the study and control of ambient air quality.

Air Pollution: Engineering, Science, and Policy

- Indoor air within public buildings or vehicles includes privately owned buildings where the public has access, such as restaurants, banks, sport stadiums, schools, universities, government buildings, and vehicles such as buses, trains, and airplanes. State and local governments typically oversee indoor air with very limited federal government oversight. See Chapter 15 for more information.
- Workplace or occupational air includes any air, indoor or outdoor, where people work. It is the right and responsibility of the worksite owner/operator to maintain air quality that does not cause harm to any employee. The Occupational Health and Safety Administration (OSHA) oversee the monitoring and regulation for workplace air quality. See Chapter 15 for more information.
- Indoor air within private building or vehicles where the public does not have general access, such as residential homes and automobiles. There are few if any national or local regulations concerning private indoor air. However, many state and local building codes either require or encourage builders to include provisions for creating and maintaining indoor air quality. Leadership in Energy and Environmental Design, LEED, is a green building certification program that recognizes best practices during design and construction (US-GBC, 2014). Indoor environmental quality is an area with such standards. The standards promote better indoor air quality and access to daylight.

A particular air space may include multiple categories. For example, a college classroom is public air for the students and workplace air for the faculty and staff that work there. A restaurant can be private air for the owner, public air for the customers, and workplace air for the employees. In such cases, the most stringent requirements for air quality must be met.

Some additional concepts in air pollution definitions, though not as common, include: the pollution must be caused by human activities, the pollutant must cause harm, and that the harm (not the pollutant) must be controllable. All of these concepts concern identifying and solving air pollution problems. An example of an air pollution source that would be excluded from these definitions is the emissions from a volcano, which cause genuine harm and generate airborne pollutants, but laws against volcanic eruptions do not work, nor is it possible to control these emissions.

1.2 Air Pollution Problems

There are many air pollution problems. The most serious affect people and their possessions everywhere, other problems are more localized such as in urban areas or near particular sources. A few problems have been solved and are no longer a concern; most problems that have been identified are being improved. A few problems are just starting to be addressed and may still be getting worse. This section will introduce several of the most important air pollution problems.

Above ground testing of nuclear weapons.

The United States, The Union of Soviet Socialist Republics, United Kingdom, France, The People's Republic of China, and South Africa detonated more than 500 nuclear devices in the atmosphere between 1945 and 1980, see Table 1-1. These tests released vast quantities of radioactive material into the atmosphere and lifted huge quantities of radioactively contaminated earth high into the atmosphere (more than 15 km above the Earth's surface during some tests). By the early 1960's these materials could be found everywhere on Earth (Simon, et al., 2006).

Table 1-1. Summary of Above Ground Nuclear Device Testing.

Nation	Number of Above Ground Detonations	Years	Total Yield megatons TNT
United States of America	216	1945-1962	153.8
Union of Soviet Socialist Republics	214	1949-1962	281.6
United Kingdom	21	1952-1958	10.8
France	46	1960-1974	11.4
People's Republic of China	23	1964-1980	21.5
South Africa	1	1979	0.003

The released materials usually start as a vertical moving column or cloud. Once it reaches its stabilization height (where the initial velocity and heat induced buoyancy are balanced by gravity) it forms a mushroom-shaped cloud, see Figure 1-1. Next, vertical and horizontal winds begin to disperse it, and the materials move downwind. The material spreads over large areas because the winds speed and direction vary over the height of the cloud. The largest particles settle first, in the local area near the test (within 500 km). Medium sized particles fallout regionally (within 3,000 km). The smallest particles, gases, and vapors travel more than 3,000 km and become global fallout. Highly local and concentrated fallout may occur at great distances due to rainfall.

People, animals, and plants are exposed to the fallout once it deposits on the earth's surface. Exposure may be external or internal. Gamma radiation emitted from the various radioactive particles causes external irradiation. Internal irradiation results from inhaling or ingesting the fallout materials, which release radiation inside the organism when the radioactive atoms decay. A study by the US National Cancer Institute found that any person living in the United States since 1951 has been exposed to radioactive fallout from these above ground tests, and all of a person's organs and tissues have received some exposure (NCI, 1997). The most common contaminants of concern are cesium-137, iodine-131, and strontium-90. Cesium-137 has a half-life of 30 years. It reacts with water to form a soluble salt. When ingested it is uniformly distributed throughout the body. It releases gamma radiation when it decays, which damages DNA and can cause a variety of cancers. Stron-

FIGURE 1-1. US Military Observing a Nuclear Test. This photo shows the November 1, 1951, "Dog" detonation, conducted as part of the Buster/Jangle test series between October and November of 1951. The test was an airdrop with a yield of 21 kilotons of TNT. The troops were located approximately 10 km from the blast. (NDEP, 2011).

tium-90 has a half-life of 28 years. Results from a study of baby teeth from children in St. Louis, MO, USA, showed increasing levels of strontium-90 throughout the 1950's. Children born in 1963 had levels 50 times greater than children born before nuclear testing began (Gould, 2000), (Mangano, 2003). Iodine-131 has a half-life of eight days. When released it can accumulate on plants and become ingested by the animals feeding on the plants. A typical human exposure comes from ingesting milk from animals that have consumed plants contaminated by fallout. It tends to concentrate in the thyroid gland and can cause cancer and other diseases in the thyroid. If it is inhaled or ingested by a nursing mother, it can be transferred to the infant via the mother's breast milk.

In 1963 the Treaty Banning Nuclear Weapons Tests in the Atmosphere, in Outer Space, and Under Water was signed in Moscow by the United States, the Soviet Union, and the United Kingdom as well as over 100 non-nuclear nations. Notably, France and China have not signed the treaty, but France has not conducted an above ground test since 1974 and China has not since 1980. The treaty bans all tests of nuclear weapons except those conducted underground.

The levels of radioactive contaminants in the environment have decreased since the above-ground test ban, with noticeable increases after the nuclear accidents at Chernobyl, Ukraine, (1986) and Fukushima, Japan (2011). The long-term reduction is directly attributable to the test ban, and this can be considered an air pollution problem that has been solved.

Lead pollution from fuel additives.

Lead, in the chemical form of tetraethyl lead (TEL), was used as a fuel additive for gasoline powered engines beginning in the 1920s. It boosted fuel octane rating, increased engine power and performance, and decreased pre-ignition problems (often called engine knocking or pinging). TEL began to be phased out in the US between the 1970s and 1995. As of 2011 only a few countries still allow its use (Afghanistan, Algeria, Burma, Georgia, Iraq, The Democratic People's Republic of Korea, and Yemen). It is a known neurotoxin, and poisoned the man who discovered this application, as well as many of the workers in its manufacturing facilities. Combustion destroys the TEL compound, but the lead is still released in either the gas phase or particle phase (as atomic lead or lead oxide). After release, it remains harmful to living things. See Section 13.3 for greater detail.

The levels of lead in the atmosphere have decreased significantly in the US and worldwide as more countries banned the use of lead fuel additives. The remaining amounts of lead originate from re-emission of previously deposited lead and the few remaining users of lead additives, chiefly in aviation and racing fuels. The reduction is directly attributable to the ban, and this air pollution problem can be considered mostly solved.

Acid Rain

Acid rain, or more appropriately named acid precipitation, originates from the byproducts of combustion of fossil fuels. The combustion process generates heat and/or electricity, which has great demand worldwide, and polluting byproducts. The most important byproducts that generate acid rain are sulfur oxides (SO_x) and nitrogen oxides (NO_x). Sulfur is often a contaminant in the coal and oil used in combustion. Nitrogen may be a fuel contaminant or may be from the air used to obtain the oxygen needed to combust the fuels. Uncontrolled emissions of these compounds cause the acidification of water and soil downwind from the sources. If many sources are located close together, huge regions may be contaminated. As much as one-third of the United States saw significant acidification by the late 1980s. The levels of acidification did not pose a direct threat to human health – only the most sensitive skin would notice the difference between acidified rain and normal rain, and such skin may develop a rash that would usually go away after a short time. However, the acidification causes harm to plants and animals. The acid leaches nutrients from soil and solubilizes metals such as aluminum in soils that cause toxic effects to nearby organisms. This reduced ability to collect nutrients and resources combined with increased exposure to greater levels of toxins leads to a decrease in ecosystem diversity.

The US and other developed nations started to control the emissions of these byproducts from fossil fuels in the 1990s. Strategies include replacing high-sulfur fuels with low-sulfur fuels, removing sulfur from the fuels, or capturing the sulfur oxides from the emissions. By 2010, the US had reduced sulfur emissions by over 50% even while increasing the amount of electricity produced. Nitrogen emissions are more difficult to control, though they have also been reduced by 50% by 2010. Control methods include changing the way fuel and air are added to the furnace, changing the flame zone shape, and/or reducing the amount emitted by treating the emissions. The amount of land that is still being acidified is greatly reduced. This problem is not yet fully solved, but impacts from it have been greatly reduced. A continuation of current strategies and controls on additional sources could see reductions from 1980 emission levels of 75 – 90% by 2020. See Chapters 9 and 10 for greater details about this problem.

Particulate Matter

Particulate Matter (PM) refers to any liquid or solid particle in the air. Almost every activity people do such as mining, burning, transportation, grinding, sanding, and cleaning generates PM. There are also many natural sources such as plant pollen, sea salt, volcanoes, wind erosion, and forest or grass fires. PM can limit visibility and can cause respiratory problems to people that are exposed to them.

Primary PM form during the activity and are emitted directly to the atmosphere. Secondary PM form after the release of the emissions – they form due to many mechanisms including coagulation, condensation, precipitation, chemical reactions, and agglomeration. Figure 1-2 and Figure 1-3 show typical releases of PM and other common air pollutants from normal human activity before implementation of the Clean Air Act regulations.

PM are removed from the atmosphere by either wet or dry deposition. Wet deposition involves the particles being swept from the air by water (e.g. rain, snow, fog) and then being deposited onto the ground. Dry deposition occurs when the particles deposit on the ground due to gravity or diffusion. Large particles, greater than 100 μm diameter (1 meter = 10^6 μm), settle very close to the source and only cause localized problems. Medium particles, 2.5 - 100 μm diameter, take longer to settle and are easily returned to the atmosphere. They may create a regional problem. Smaller

FIGURE 1-2. Skyline Showing Particulate Matter and Smog from Air Emissions. Background Shows the Ford River Rouge Plant in Dearborn, MI, USA, July 1973.
(Clark, 1973)

particles, 0.1 – 2.5 μm diameter, can remain in the atmosphere long enough to circle the globe (20 days) and thus create a global problem. The smallest sizes, less than 0.1 μm diameter, are removed at the local to regional scale by dry deposition, as they can diffuse quickly to the Earth's surface. It is entirely likely that the particles will change size while in the atmosphere. Generally, particles tend to increase in size by joining together or with other constituents in the air, particularly water. Larger particles are generally more quickly removed.

Historically, coal-fueled power plants were the largest human caused sources of PM. Coal is a mixture that includes minerals that form ash during the combustion process. A coal may contain 1 to 10% ash. A large power plant may use 100 railway cars worth of coal in a day. The emission of the ash caused black plumes from each site and deposited ash downwind for miles. Emission controls are capable of capturing more than 99% of the particles, and this source has greatly reduced its impact on air pollution. However, not all sources can be controlled, and many remain unregulated. Much progress has been made on this problem, but it remains one of the main causes for unhealthy air in the US. See Chapters 7 and 8 for more detail about this issue.

FIGURE 1-3. Uncontrolled Emission from Ore Carrier Contributes to Smog over Lake Superior, June 1973. (Emmerich, 1973)

Smog and Ozone

Smog is a word made by combining the words smoke and fog. There are two types of smog. A *London type* results from sulfur dioxide emissions that combine with high humidity or foggy weather to form liquid particles that remain in the atmosphere for extended times. This smog consists of particles such as fly ash and sulfur compounds. It is also called classic smog, and it is reducing. A *Los Angeles type* smog forms during sunny weather due to photochemical reactions between other pollutants - particulate matter, nitrogen oxides, and volatile organic compounds. It creates a yellowish-brownish haze in the air. This type of smog is oxidizing. The reactions happen within the lowest part of the atmosphere called the troposphere (ground level to an elevation of 10 km). These substances react in the presence of sunlight to form ozone (O_3) and often creates peroxy acetyl nitrate (PAN). Beijing, China, is one of the most polluted cities in the world today. It has air quality that frequently causes severe health problems for its people, see Figure 1-4. The smog in Beijing is mostly caused by particulate matter, sulfates, and nitrates that combine with moisture during stagnant weather conditions that last several days to several weeks.

FIGURE 1-4. Smog in Beijing, China. May, 2014. The visibility in this photo only allows buildings one mile from the photographer to be seen. Photo courtesy of Mitchell Kruse, 2014.

Ozone is more difficult to control than most other air pollutants. It is not discharged directly into the atmosphere, so it is considered a secondary pollutant. Ozone is a known respiratory irritant, harms animals, damages plants, see Figure 1-5, and causes the degradation of plastics and textiles. Control relies on reducing the emissions of the precursor emissions, primarily volatile organic compounds and nitrogen oxides. Both of these pollutants have large natural sources as well as anthropogenic sources. It is also a seasonal problem, occurring during sunny and warm conditions. One significant problem in controlling it is that the chemical precursors can be transported hundreds of miles from the source before the conditions for the formation of ozone lead to their degradation. The time delay and travel distance means a particular source helping to cause an ozone event in a location one day may not contribute at all several days later due to a shift in wind patterns. It also means that the emissions may not cause a problem in the local region where they are emitted, but instead cause the problem hundreds of miles down-wind. In the US, the intermingling of pollutants between different States creates many legal issues, as

**FIGURE 1-5
Smog-Damaged and
Clean-Air Plants
Shown Together.**
(Daniels, 1972)

one State does not have authority to control emissions in another state. Ozone is one of the biggest current air pollution problems and much work, legal and technical, is focused on its control. See Chapters 10, 11, and 12 for more information.

Ozone Hole

The ozone in smog is harmful to human health. However, the same ozone in the region of the atmosphere called the stratosphere, which is about 10 km to 60 km above the Earth's surface, provides a very beneficial service. Stratospheric ozone absorbs ultraviolet (UV) radiation from the Sun. UV radiation can damage living organisms and is capable of breaking DNA molecules. The ozone is destroyed by reactions involving chlorofluorocarbons (CFCs) and sunlight. CFCs are nontoxic and have excellent properties that make them useful for refrigeration and cleaning, including being stable, non-toxic, easy to manufacture, and relatively inexpensive. It is interesting to note that the discoverer of these compounds was the same person who discovered the use of tetraethyl lead as a fuel additive.

The CFCs do not react in the troposphere, and are thus able to disperse and accumulate throughout the lower atmosphere until they cross into the stratosphere. In the stratosphere, they encounter UV radiation that destroys the molecule and releases chlorine. The chlorine catalyzes a reaction which destroys ozone molecules. One CFC molecule can destroy thousands of ozone molecules. This destruction creates what is called the ozone hole, which is just a region of the atmosphere with a reduced concentration of ozone (sometimes as much as 50%, see Figure 1-6). This reduction allows more UV radiation to pass through the stratosphere and impact organisms in the troposphere. The increase in UV radiation leads to skin cancers and eye disease in the people exposed to them.

Concern for this problem led the nations of the world to ban the use and manufacture of CFCs. A process started by the United Nations treaty called the Montreal Protocol (1987). This treaty was only the first step and by itself, it would not solve the problem. However, there have been many revisions to this document as more information became available and with the creation of new technologies. The rate of destruction of ozone has leveled out, and it is expected to begin recovering within the next few decades. The time lag is caused by the very long lifetime CFC molecules have in the atmosphere, 15 – 150 years. The ozone layer will not fully recover until all of these compounds have been destroyed, and then only if no new ozone-destroying compounds take their place. See section 12.2. for more information.

FIGURE 1-6. False Color Views of Total Ozone Over the Antarctic Pole. Views compare changes between 1979 to 2008 maximum extent of ozone hole. Darker shading has lower total ozone. The lowest level in 1979 was 280 DU and in 2008 it was 100 DU. The extent in 2008 is more than 10 times larger.

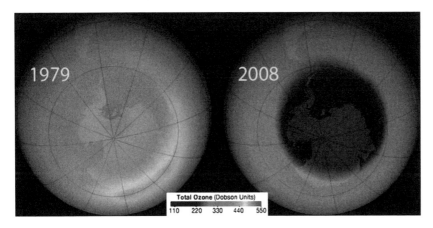

Mercury

Mercury is a known toxic material and is considered to be a hazardous air pollutant (HAP). Emissions occur as a direct result from mercury mining and as a byproduct of burning coal and mining gold. Once emitted it inter-converts between its elemental, +2 ion, and particle forms. The ionic form is water soluble and can be in either organic or inorganic forms. The water-soluble forms can be chemically converted to insoluble forms, and it then can be re-emitted into the atmosphere. It is only quite slowly removed from the environment by

FIGURE 1-7. Sign posted in The Everglades National Park. (Dragonfly, 2010)

burial in sediments. There is no known safe dose for humans, and even minuscule doses (micrograms) can cause harm. Toxic effects on plant and animal life are less well understood, but again very low doses lead to harm for these organisms.

An additional concern for mercury is its ability to bio-accumulate. Bio-accumulation means the mercury stored in plants, animals and fish, is passed on to the consumers of these organisms and stored within them at much higher amounts than in the material eaten. This process can occur over many levels of consumer, and top predators (bears, birds of prey, and humans) can receive very high doses of mercury from certain foods. Many lakes in the US have mercury advisories that inform fish consumers to limit their intake of fish to a few fish a week or month. This mercury can be passed from mother to fetus or to an infant through breast milk.

Mercury has only very recently (since the 2000s) been the subject of regulation and control. All major sources of mercury have been subject to control since 2012 in the US. The primary control methods are reduction of use or capture and removal from emission sources. It is not well understood how long it takes until contaminated soils and surface waters recover from mercury pollution or when it will be safe to eat fish from such areas as often as one likes. See section 13.4 for more information.

Visibility and Haze

Haze is one of the easiest to notice forms of air pollution. It reduces visibility and can add color to the air – most often whites, browns and yellows, but may include greenish-reds. Haze results from sunlight encountering tiny pollution particles in the air. Some light is absorbed by particles. Other light is scattered away before it reaches an observer. The absorption and scattering of the light reduces the clarity and color of distant landscapes. More pollutants mean more absorption and scattering of light, which reduce the clarity and color of what we see. Haze is similar to smog and shares may of the same causes - PM, ozone, sulfates and nitrates. Haze can occur anywhere the air is contaminated by pollutants.

Some haze-causing particles are directly emitted to the air. Others form when gases emitted to the air create particles as they react with other pollutants. Formation and effects depend on the types of pollutants and the weather. Sulfur emissions combined with humid conditions form small droplets. Nitrogen oxides react with organic compounds during warm, sunny weather to form particulates and other pollutants. The wind can transport either the haze or the precursor pollutants by hundreds of miles from where they originated. In the eastern United States, average visual range has decreased from 90 miles to 15-25 miles. In the western United States, average visual range has decreased from 140 miles to 35-90 miles.

The pollutants that form haze are linked to serious health problems and environmental damage. Exposure to small particles contributes to increased respiratory illness, decreased lung function, and premature death. Particles of nitrates and sulfates contribute to acid rain formation that makes lakes, rivers, and streams unsuitable for many fish, and erodes buildings, historical monuments, and paint on cars. Control of haze requires control of the emissions of the pollutants that create the haze.

Climate Change

The release of carbon from fossil fuel sources is causing an increase in environmental levels of carbon dioxide (CO_2) and methane (CH_4). CO_2 has increased in the atmosphere by over 40% from preindustrial times, and methane has increased by over 100%. An additional climate changing chemical compound, N_2O, is also increasing – in large part due to combustion and agricultural use of fertilizer. CO_2 is also increasing in the oceans and soil, but not at a rate sufficient to prevent a change in the atmosphere. The main problem is that there is no natural removal mechanism to counterbalance this increase, meaning that the extra carbon remains in the environment for thousands of years. This extra carbon partitions between the atmosphere, ocean, and soil with each accepting part of the increased emissions, and it causes changes in each. The only known long-term removal mechanism is geological burial (sequestration) in deep ocean sediments

CO_2 and CH_4 are greenhouse gases, which means they absorb infrared radiation and later re-emit it, but in a random direction. The random re-emission causes some of the radiation to return to the earth, which leads to an increase in atmospheric and ocean temperatures. The phenomenon was first discussed in the 1890s. Atmospheric models suggest that a doubling of CO_2 concentration will cause the average temperature at the earth's surface to increase by 4 to 8°F (2 to 5°C), with larger increases at the poles and less in the tropics. Like the increase in CO_2, the increase in temperature also partitions between the atmosphere, soil, and ocean. A change of this magnitude is extremely rare in geological history, especially over the given timescale of a few hundred years. More commonly geological changes occur over hundreds of thousands of years. The full impact is unknowable, although it is very clear that the impact will be large and worldwide.

Human society greatly values the energy produced from using fossil fuels. This valuation leads many people to not accept the problem as real, or to assume it can be dealt with in the future. No nation has addressed this problem at the level required to make an impact. The only long-term solution is to stop using fossil fuels and to somehow capture the already released carbon and remove it from the environment. These tasks are non-trivial, especially given the timescale (30 – 80 years) required to prevent a larger than 2°C increase in global temperature. It is believed that the global impacts of a 2°C will change the global climate and ecosystems in a serious way, but that humans will be able to adapt to the changes. It is unknown what will happen if this threshold is exceeded.

Solution to this problem require new energy sources, new ways to govern the use of resources, the creation of laws and regulations on a multinational scale, and the development of methods for capturing carbon, transporting it, and storing it. Combined these ideas will ultimately restructure the entire world economy. While work has started in each area, much more needs to be done. See Chapter 14 for more information.

1.3 Control of Air Pollutants

Solutions to air pollution problems are complex and require that many different groups in society work together. The control of air pollution is based on four assumptions (AAAS, 1965):

1. Air is in the public domain. No one owns the air, nor can one control the flow of the atmosphere, even if allowed to control the airspace over a property. Release of an air pollutant will inevitably trespass onto another's property and become a public problem
2. Air pollution is an inevitable concomitant of modern life. It is not possible to maintain today's lifestyle without using resources and creating waste.
3. Scientific knowledge can be applied to the shaping of public policy. The best tool for understanding the material world is the scientific method. It provides reliable, reproducible, and consistent information.
4. Methods to reduce air pollution must not increase pollution in other sectors of man's environment. Simple tradeoff between air pollution and water pollution do not solve a problem, it just shifts the burden from one group to another. Such methods may at times be used, but they do not help the environment to sustain life.

Figure 1-8. Shows a simplified flow chart for controlling pollution.

The first step is discovering that a problem exists. It is often scientists that make this discovery. Sometimes the problem is well known but poorly understood (i.e. smog), other times it is discovered while looking for something else (stratospheric ozone depletion). Scientists then attempt to build a model of the phenomenon to determine the causes and effects. The results of these models are used to show how important the problem is.

Once the importance of the problem is understood, the members of society must decide that the problem needs to be addressed. Most governments do not work on solving an issue unless there is consensus among many groups of citizens. There will always be some groups opposed to doing anything and other groups that want to do more. The more groups in favor of doing something (or not doing it) the more likely that will happen.

After it is decided to address the issue, the government creates a law. Laws are somewhat vague and do not specify every action nor when those actions must occur. Overall, the law sets the basic framework, guidelines, and goals.

The details of how to make a law work are detailed in a set of regulations. A law may be five to twenty pages long, and the pages of associated regulation may number in the thousands. Regulations have public comment periods, and the issuing government agency then addresses all comments and may make modifications before issuing the official regulation. One of the guiding principles of the US-EPA in creating environmental regulations is to achieve environmental justice, such that the benefits and burdens of implantation of pollution control should be shared by everyone, and not just a particular group. The EPA's definition is the following:

FIGURE 1-8. Steps in Addressing an Air Pollution Problem.

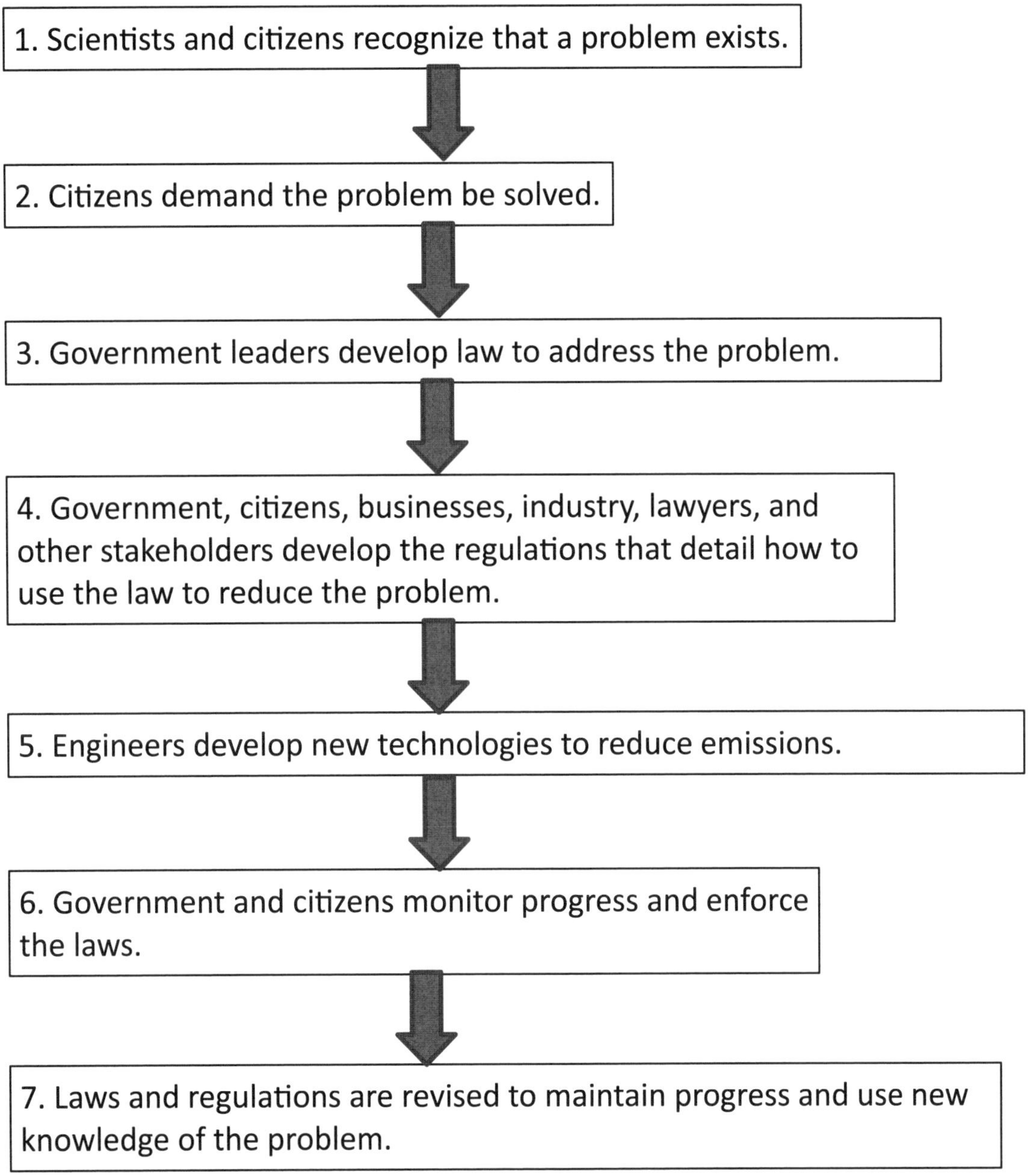

1. Scientists and citizens recognize that a problem exists.

2. Citizens demand the problem be solved.

3. Government leaders develop law to address the problem.

4. Government, citizens, businesses, industry, lawyers, and other stakeholders develop the regulations that detail how to use the law to reduce the problem.

5. Engineers develop new technologies to reduce emissions.

6. Government and citizens monitor progress and enforce the laws.

7. Laws and regulations are revised to maintain progress and use new knowledge of the problem.

"Environmental Justice is the fair treatment and meaningful involvement of all people regardless of race, color, national origin, or income with respect to the development, implementation, and enforcement of environmental laws, regulations, and policies. EPA has this goal for all communities and persons across this Nation. It will be achieved when everyone enjoys the same degree of protection from environmental and health hazards and equal access to the decision-making process to have a healthy environment in which to live, learn, and work" (US-EPA, 2013a).

The equipment, systems, and policies needed to comply with a regulation do not exist before the law is passed. Regulations typically provide a number of years between implementation and enforcement. The time is needed to develop the infrastructure for compliance. Regulations are frequently adapted to this learning process and very rarely do they require activities that are found to be impossible or unrealistic.

Methods for reducing harm from air pollution include:

Substitution – change the nature of emissions by using different materials or methods that cause less harm.

Zoning – allowing emissions to occur in limited areas, usually away from population centers.

Market mechanisms – whereby an emitter is charged some amount to remedy the problems caused by the emissions. The amount should reflect the cost of the harm and can also be used to change behavior.

Banning – general prohibitions to prevent harmful emissions.

Each of these methods can be used to control air pollution, and usually one works better for certain pollutants than another one, but no single one is always the best choice.

The progress made in controlling emissions requires that every emitter be held to the same standard and that every emitter monitors all their emissions. Equal enforcement prevents the creation of unfair market advantages by a subset of favored emitters. It also means that the cost of complying with the law is passed onto those that consume the products and services. Without the consumers, the business would not be polluting.

Most regulations and some laws require timely review. Such review allows the methods to be adapted to actual circumstances and allows new information to drive the improvement of future regulation.

1.4 Public Information and Tools

There are many sources of information available to anyone who is interested. Data sets and maps for other air pollution related issues are available from the US-EPA at www.epa.gov/airdata/ . A similar site for the European Union is www.eea.europa.eu/data-and-maps. These sites provide information ranging from very basic to extremely detailed on the topics of air quality statistics, monitoring site data including location, daily measurement values and summary information, air pollution deposition maps, and pollutant concentration plots. Detailed current and historical information of the six most common air pollutants in the US (carbon monoxide, lead, nitrogen oxides, sulfur oxides, particulate matter, and ozone) can be found at www.epa.gov/air/emissions . This site provides basic information, air quality trends, laws and regulations, and lists of regulated sources. Other information sources in the US include:

National Aeronautics and Space Administration – www.nasa.gov

National Oceanic and Atmospheric Administration – www.noaa.gov

Department of Energy – www.energy.gov

Center for Disease Control and Prevention – www.cdc.gov

One of the primary tools the US-EPA uses to communicate information for citizens to use in their daily life is called the Air Quality Index (AQI). It is a tool used to communicate simple information on air quality to local communities (US-EPA, 2003). It provides an estimate of how polluted the air is, what the health concerns are, and who may be affected. It focuses on the health effects a person may experience within a few hours or days after breathing the outdoor air. The index is based on measurements of five major air pollutants - ozone, particulate matter, carbon monoxide, sulfur dioxide, and nitrogen dioxide. Each of the five AQI substances is also a criteria pollutant, as defined by the US-EPA, see section 3.2.1. The sixth criteria pollutant, lead, is not included as it no longer creates much health risk, except in a few locations.

The AQI is a value that ranges from 0 to 500. High values represent greater health concerns. The number is calibrated such that an index level of 100 corresponds to the national ambient air quality standard for each of the five pollutants. The AQI is a national index, which means that the values and colors are the same everywhere in the US. The AQI is divided into six categories:

Table 1-2. The Air Quality Index.

AQI Value	Level of Health Concern	Color Code
"When the AQI is in this range ...	...air quality conditions are	... as symbolized by this color:"
0 to 50	Good	Green
51 to 100	Moderate	Yellow
101 to 150	Unhealthy for Sensitive Groups	Orange
151 to 200	Unhealthy	Red
201 to 300	Very Unhealthy	Purple
301 to 500	Hazardous	Maroon

The six levels are:
> *Good* – The air quality is considered satisfactory, and air pollution poses little or no risk
>
> *Moderate* – The air quality is considered acceptable, however there may be some concern for a small number of people who are unusually sensitive to the pollutant.
>
> *Unhealthy for Sensitive Groups (USG)* – These levels are of concern for some people, and they may experience health effects. This group includes those with lung disease when the pollutant is ozone and heart disease when the pollutant is ozone or particulate matter. Children, the elderly, and those who exercise outdoors are more susceptible to any pollutant at this level. The general population is not likely to be affected when the AQI is in this range.
>
> *Unhealthy* – Everyone may begin to experience health effects, and sensitive groups may experience more serious health effects.
>
> *Very Unhealthy* – This level triggers a health alert, and everyone may experience more serious health effects.
>
> *Hazardous* – This level triggers emergency conditions, and the entire population is likely to be affected.

Air quality is measured from concentrations of each pollutant every day at a set of more than 1000 monitors across the country. The AQI value is calculated for each pollutant in an area, and the highest value of the five is the AQI value for that day. The model assumes that there are no synergies between pollutants. Some communities also provide a forecast for the next day's AQI to help residents plan their outdoor activities to protect their health.

All communities (or statistical metropolitan areas) with a population of 350,000 or more are required to report the AQI to the public daily. When AQI is above 100 the state or local agencies are also required to report which groups (such as children, the elderly, and people with asthma, heart disease or lung disease) may be sensitive to the specific pollutant. If more than one pollutant has an AQI value above 100, the agencies must report all the groups that are sensitive to those pollutants. Many smaller communities also report the AQI as a public health service. Most communities in the US have AQI values below 100 most of the time. Values above 100 happen just a few times each year. Larger cities have more severe air pollution problems, and their AQI may exceed 100 more often. Values above 200 are infrequent, and values above 300 are extremely rare. In addition to local communication in each community, the US-EPA maintains a central database that shows the AQI across the country, see www.airnow.gov (US-EPA, 2013b). A similar site for the European Union is located at www.airqualitynow.eu/. Many other nations also provide air quality index values using similar schemes.

1.5 Questions

* – Questions and problems may require additional information not included in the textbook.

1. Work with another student to generate a definition of air pollution. Try to identify several possible stakeholders that will use the law and make it fair for each.

2. Find a process flow diagram of a fossil fuel powered electricity generating facility. Note any pollution control equipment that is shown. Try to think of three other ways the facility could reduce emissions of air pollutants.*

3. Visit the US-EPA RAdNet Air Monitoring site (http://www.epa.gov/radnet/index.html). Find the location of the nearest monitoring location to where you live. Explore the data from the monitor. Does the site help you interpret the information? Write a paragraph or two discussing your findings.*

4. What substances are currently used to replace lead additives in gasoline? Compare their short and long term potential for harm to human health.*

5. Collect a sample of rain water and measure its pH (your instructor should provide pH test paper to help). Compare your results with your classmates and with the National Atmospheric Deposition Program data (http://nadp.sws.uiuc.edu/). Note the location and time of your sample.*

6. When was the last time the air where you live had an ozone pollution episode, defined as an Air Quality Index greater than 100 for ozone. US data is available at www.airnow.gov.*

7. Use your state government's environmental quality website and determine if any near by surface water has a mercury / fish consumption advisory.*

8. Do you think the smog in Beijing China is the London type or the LA type? Why?

9. Pick one of the four assumptions that form the basis for modern air pollution control. Write a paragraph explaining why the assumption is necessary. Consider how things could be different if the assumption is not used.

10. Find an environmental justice issue in the region you live. Prepare a five minute discussion to share with your class.

11. Make a list of three questions you have about this chapter or air pollution concerns you have.

1.6 Bibliography

AAAS. 1965. *Air Conservation*. Washington DC, USA : American Association for the Advancement of Science, 1965.

Clark, Joe, and US-EPA. 1973. *Dearborn Skyline. Ford River Rouge Plant in Background, 07/1973*. Environmental Protection Agency, College Park, MD, USA : National Archives, 1973. ARC Indentifies 549710 / Local Identifier 412-DA-7225.

Daniels, Gene, and US-EPA. 1972. *SMOG-DAMAGED AND "CLEAN" PLANTS SHOWN TOGETHER AT THE STATEWIDE AIR POLLUTION RESEARCH CENTER, 05/1972*. US Environmental Protection Agency, College Park, MD, USA : National Archives, 1972. ARC Identifier 542692 / Local Identifier 412-DA-199.

Dragonfly, 777. 2010. *Everglades Signage*. Photobucket, Florida.

Emmerich, Donald, and US-EPA. 1973. *ORE CARRIER HEADS INTO LAKE SUPERIOR FROM DULUTH HARBOR AHEAD IS A LAYER OF SMOG HANGING IN THE AIR OVER THE CITIES OF DULUTH MINNESOTA AND SUPERIOR, WISCONSIN, SMOKE FROM ORE BOAT STACKS CONTRIBUTE TO SMOG, 06/1973*. Environmental protection Agency, College Park, MD, USA : National Archives, 1973. ARC Identifier 551542 / Local Identifier 412-DA-9057.

Gould, J. M., E. J. Sternglass, J. D. Sherman, J. Brown, W. McDonnell, and J. J. Mangano. 2000. Strontium-90 in Deciduous Teeth as a Factor in Early Childhood Cancer. *International Journal of Health Services*. 2000, Vol. 30, 3.

Mangano, J. 2003. An Unexpected Rise in Strontium-90 in US Deciduous Teeth in the 1990s. *The Science of the Total Environment*. s.l. : Elsevier Press, 2003.

NCI. 1997. *Estimated Exposures and Thyroid Doses Received by the American People from Iodine-131 in Fallout Following Nevada Atmospheric Nuclear Bomb Tests*. Washington DC, USA : National Cancer Institute at the National Institutes of Health, 1997. http://www.cancer.gov/i131/fallout/contents.html.

NDEP. 2011. Photos: Atmospheric Nuclear Testing - Nevada Test Site. [Online] Nevada Division of Environmental Protection, October 19, 2011. [Cited: July 24, 2013.] http://ndep.nv.gov/boff/photo01.htm.

Simon, S, Bouville, A and Land, C. 2006. Fallout from Nuclear Weapons Tests and Cancer Risks. *American Scientist*. 1, 2006, Vol. 94, p. 48.

US-EPA. 2003. *Air Quality Index: A Guide to Air Quality and Your Health*. Washington DC, USA : US Environmental Protection Agency Office of Air and Radiation, 2003. EPA-454/K-03-002.

—. 2013a. Environmental Justice. [Online] Environmental Protection Agency, May 20, 2013a. [Cited: July 26, 2013.] http://www.epa.gov/environmentaljustice/.

—. 2013b. Today's AQI Forecast. *AirNow*. [Online] Environmental Protection Agency, July 26, 2013b. http://www.airnow.gov/.

US-GBC. 2014. LEED. [Online] United States Green Building Council, 2014. http://www.usgbc.org/leed.

US-NASA. 2013. September 25, 2006. *Ozone Hole Watch*. [Online] National Aeronautics and Space Administration, July 27, 2013. [Cited: July 27, 2013.] http://ozonewatch.gsfc.nasa.gov/index.html.

CHAPTER **2**

Basic Problem Solving

This section provides some background and examples to help the student practice their problem solving skills. Solving problems is the key to understanding how things work. Quantification of these relationships allows one to get a feel for how important different variables are and how one can alter the problem to better understand how the variables behave. There are many good books that focus on problem solving, and here are two the author found especially helpful – [(Polya, 1945), (Fogler, et al., 1995)].

Table 2-1. A Problem Solving Strategy.

> I. Understand the Problem
>   Read the problem, identify the unknown and known data, sketch a figure, write down the conditions, and read the problem again (and again).
> II. Devise a Plan
>   Find the connection between the unknown and the known. You may need to develop auxiliary relationships to create a path linking known to unknown.
> III. Carry out the plan.
>   Do it! Use your math skills (finally).
> IV. Look back

Check to make sure your answer is reasonable. Unreasonable answers would include nonsensical values like negative absolute pressure or temperature, water that flows uphill without adding energy, equipment that stretches to the moon, or is smaller than a molecule!

2.1 Dimensional Analysis

The primary problem solving skill needed for this textbook is dimensional analysis. The technique is also called the unit factor, unit conversion, or the factor label method. It is typically introduced to students in high school level chemistry and physics and is extensively used in their college sequels. It uses the fact that any number may be multiplied by one. The trick is to find different values of 'one' with comparable units. Create the factors from the ratio of any two terms that describe the same or equivalent "amounts" of something.

■ **Example 2-1.**
Find the number of seconds in 20 years.

Solution:
Create a chain of unit sets that cancel, leaving only the desired unit. In this example each set of units that cancel are crossed out with a unique cross-out to help the reader see exactly what is being cancelled. In future problems, the cross-outs are omitted and are implied.

$$X \sec = 20 \, year * \frac{365 \, day}{year} * \frac{24 \, hour}{day} * \frac{60 \, min}{hour} * \frac{60 \sec}{min} = 630,720,000 \sec$$

The method is even more useful when multiple sets of unit factors are applied.

■ **Example 2-2.**
Find the speed of a car in meter/ sec if its speedometer reports it is moving at 50 mile/ hour.

Solution:

$$X \frac{m}{s} = 50 \frac{mile}{hr} * \frac{hr}{3,600 s} * \frac{5,280 \, ft}{mile} * \frac{m}{3.28083 \, ft} = 22.4 \frac{m}{s}$$

Appendix 2 provides a set of useful conversions. Many more conversion factors are available from web search engines or scientific and engineering handbooks.

An interesting problem in atmospheric science is calculating the average time a chemical spends in the atmosphere. There are many ways to perform this calculation, one very simplistic method is to determine the ratio between the amount entering and leaving the atmosphere (i.e. the flux) and the total amount in the atmosphere.

2.1
$$\tau_{total} = \frac{M_{global}}{F_{total}}$$

Where: M_{global} is the total global atmospheric burden [mass], and
F_{total} is the total addition (or leaving) rate [mass/time] from the atmosphere

Note that F_{total} may have multiple causes, each of which may have a different rate. The losses may happen due to atmospheric reactions, uptake by soil, and loss to the oceans.

Basic Problem Solving

■ Example 2-3.
Determine the average lifetime of carbon in the atmosphere. Assume the amount of carbon in the atmosphere is 762 giga-ton (Gt), and emissions from the land and ocean are 218.2 Gt/yr. Assume that the emissions rate is equal to the removal rate.

Solution:
Divide the total amount by the emission rate

$$\frac{762 Gt}{218.2 Gt/yr} = 3.5\,yrs$$

Fermi Problems are an interesting sub-class of dimensional analysis techniques. The goal of these problems is to develop an order of magnitude type answer to the problem. The technique may be done to estimate a solution very quickly (also known as the back of an envelope solution) and to check if an answer makes sense (step IV from table 2-1). The technique is very useful in teaching dimensional analysis, helping develop a sense for approximating answers with little or no data, and to logically connect different ideas.

■ Example 2-4.
The classic Fermi problem is "How many piano tuners are there in Chicago?"

Solution:
The solution to this problem involves approximating several factors, which when multiplied yield the answer.

$$X_1 People_{Chicago} * \frac{Household}{X_2 People} * \frac{X_3 Piano}{Household} * \frac{X_4 tunings}{Piano} * \frac{X_5 tuning\ time}{tunings} * \frac{Worker}{X_6 time}$$

The answer is only as good as the estimates. For example, we might make the following estimates:
1. Population of Chicago: 5,000,000 people.
2. Household density: 2 people per household.
3. Piano density: 5% households have pianos.
4. Piano tuning frequency: Once per year.
5. Tuning time: About 2 hours per piano, including travel to and from shop to piano.
6. Worker time: 8 hour days, 5 day weeks, 50 weeks per year.

Now complete the multiplication

$$5 \times 10^6 People_{Chicago} * \frac{Household}{2 People} * \frac{0.05 Piano}{Household} * \frac{1 tunings/yr}{Piano} * \frac{2 hour - tuning\ time}{tunings}$$

$$* \frac{Worker}{(8)(5)(50) hr/yr} = 125\ Workers$$

The problems are easier if the dimensional quantities (not including the number value) are determined first, and then determine estimates for the quantities.

2.2 Basic Chemistry Review

Many air pollution regulations and controls recognize that chemical reactions must be used to determine emissions and emission control levels. Knowledge of the chemical reaction allows the quantification of the materials needed for the reaction to occur and to determine the amount of products formed. There are many introductory chemistry books such as [(Kotz, et al., 2006)] if a more detailed review is desired. A generalized form of a chemical reaction, which will be used in developing the general mass and energy balance equations, is:

2.2 $$aA + bB \Leftrightarrow cC + dD$$

Where: A and B are reacting chemical species
C and D are product chemical species, and
a, b, c, and d are the stoichiometric coefficients for each species.

Determination of a, b, c, and d requires knowledge of each of the chemical species involved in the reaction (A, B, C, and D). Choose coefficients such that there are equal numbers of each atom type between the reactants and products.

■ Example 2-5.
Find the stoichiometric coefficients for the neutralization reaction between sulfuric acid and sodium hydroxide.

Solution:
Determine the chemical formulae for all reactants and products, where the products are the result of neutralization. Then, balance the number of each atom type between the reactants and products.

Reactants
Sulfuric Acid = $H_2(SO_4)$
Sodium Hydroxide = NaOH

Products
Water = H_2O
Sodium Sulfate = $Na_2(SO_4)$

The reaction is:

2.3 $$aH_2SO_4 + bNaOH \Leftrightarrow cH_2O + dNa_2SO_4$$

There are 4 types of atoms, H, S, O, and Na; so 4 atom balances can be written:

The H atom balance can be obtained by noticing there are 2 H in sulfuric acid, 1 H in sodium hydroxide, 2 H in water and none in sodium sulfate:

2.4 $$H: \quad 2a + b = 2c$$

Similarly for the other atoms:

2.5
$$S: \quad a = d$$
$$O: \quad 4a + b = c + 4d$$
$$Na: \quad b = 2d$$

Solving the four equations yields a=1, b=2, c=2, and d=1, and equation 2.3 becomes:

2.6 $$1H_2SO_4 + 2NaOH \Leftrightarrow 2H_2O + 1Na_2SO_4$$

Sometimes there is not a unique solution (too few equations for the number of unknowns) in which a solution can be found by assuming a basis. A basis is just an assumed value for any one of the variables. Only one basis may be chosen in any problem. For example, just choose any one of the coefficient values and set it to one. This allows the remaining coefficients to be solved. These values are all based on having the chosen coefficient be equal to one. The resulting solution may lead to some of the unknowns to be non-integers. If so, it is typical to multiply by a common factor, so all coefficients are integers.

■ *Example 2-6.*
Write out the balanced stoichiometric coefficients for the combustion of methanol.

Solution:
Methanol is written as CH_3OH. Combustion implies the chemical reaction is the complete oxidation to form carbon dioxide and water.

2.7
$$aCH_3OH + bO_2 \Leftrightarrow cCO_2 + dH_2O$$

There are 3 types of atoms, C, H, and O; so 3 atom balances can be written. However there are four unknowns a, b, c, and d. This requires that one variable be guessed, we will choose a =1:

2.8
$$C: \quad a = c$$
$$O: \quad a + 2b = 2c + d$$
$$H: \quad 4a = 2d$$

Solving the three equations and using a =1 yields a=1, b=3/2, c=1, and d=2, and equation 2.7 becomes:

2.9
$$CH_3OH + \frac{3}{2}O_2 \Leftrightarrow CO_2 + 2H_2O$$

To obtain integer coefficients, multiply through by 2.

2.10
$$2CH_3OH + 3O_2 \Leftrightarrow 2CO_2 + 4H_2O$$

2.3 Reaction Stoichiometry

2-3.1 Atomic and Molecular Weight

Molecular weight represents the mass associated with a given number of atoms or molecules of a single species. The atomic weights (in atomic weight units) for each atomic species are listed in the periodic table (see inside cover).

The weight of a single atom or molecule is quite small and is of limited value in determining the amount of reactants needed to obtain a reasonable amount of product. It is more interesting to consider large numbers of atoms or molecules. The standard number for this determination is called the **mole**. There are many forms of the mole, each with slightly different units. The most common mole is the gram-mole. It may be abbreviated as g-mol or even just mol. Another common unit is the pound-mole or #-mol.

The gram-mole is the number of atoms in an atomic weights number of grams of that particular species. This number is known as **Avogadro's number, 6.022 x 10^{23}**. For example, the molecular weight of sulfuric acid is 98, so 98 grams of sulfuric acid contain 6.022 x 10^{23} molecules. This relationship allows calculation of the mass associated with any chemical reaction.

■ *Example 2-7.*
Find the Molecular Weight of the Following Compounds a) nitrogen, b) oxygen, c) sulfuric acid, d) limestone, and e) calcium phosphate.

Solution:

The molecular weight of a compound is the sum of the atomic weights of each atom making up the compound. To find the molecular weight you need to have the correct chemical formulae. We will use the periodic table to obtain the atomic weights.

a) Nitrogen (N_2) 2 nitrogen atoms at 14.0067 g/mol   = 28.0134 g/mol

b) Oxygen (O_2) 2 oxygen atoms at 15.9994 g/mol   = 31.9988 g/mol

c) Sulfuric Acid (H_2SO_4)
    2 hydrogen atoms at 1.00794   = 2.01588
    1 sulfur atom at 32.065   = 32.065
    4 oxygon atoms at 15.9994   = 63.9976
    TOTAL   = 98.0785 g/mol

d) Limestone ($CaCO_3$)
    1 calcium at 40.078   = 40.078
    1 carbon at 12.0107   = 12.0107
    3 oxygen at 15.9994   = 47.9982
    TOTAL   = 100.087 g/mol

e) Calcium Phosphate [$Ca_3(PO_4)_2$]
    3 calcium at 40.078   = 120.234
    2 phosphorus at 30.9738   = 61.9536
    8 oxygen at 15.9994   = 127.9952
    TOTAL   = 310.183 g/mol

We can also calculate the molecular weight of a mixture, as long as the amount of each component of the mixture is known. The mixture molecular weight is just the mole-fraction weighted average of the individual components.

■ Example 2-8.

Find the molecular weight of a mixture. The mixture is air, where we assume air-is composed of 79 mole% nitrogen and 21 mole % oxygen.

Solution:

The molecular weight of a mixture is found in a similar way as for the compound. However, the answer will depend on the actual quantity of each component of the mixture. A compound always has the exact same number of constituent atoms. In the case of a mixture, the molecular weights of each component are multiplied by the mole fraction and then summed.

Component	Mole%	Mole Fraction	Molecular Weight
Nitrogen – N_2	79%	0.79	28.013
Oxygen – O_2	21%	0.21	31.999

$$28.013*0.79 + 31.999*0.21 = 28.85 \frac{g}{mol}$$

The molecular weight of dry air is 28.85 g/mol.

2.3.2 Stoichiometry of Chemical Reactions

The stoichiometric coefficients represent the ratio of molecules for each species involved in the reaction. The number could represent individual molecules or large groups of molecules. The chemical reaction equation describes a very precise recipe for quantifying the amount of each reactant and product. The description works for any amount as long as the balanced ratio is used.

■ **Example 2-9.**

How much sodium hydroxide is required to react with 10 g of sulfuric acid? How much sodium sulfate forms?

Solution:

Obtain the balanced reaction coefficients from Example 2-5. Determine the molecular weight of each species in the reaction. Multiply the number of moles (the stoichiometric coefficient) by the molecular weight to obtain the stoichiometric mass in the reaction. Finally, obtain the actual mass by setting up proportional masses.

	$H_2(SO_4)$	NaOH	H_2O	$Na_2(SO_4)$
Moles in reaction	1 mol	2 mol	2 mol	1 mol
Mol weight	98 g/mol	40 g/mol	18 g/mol	142 g/mol
Stoich mass	98g	80g	36g	142g
Actual mass	10g	$10*\frac{80}{98} = 8.16g$	$10*\frac{36}{98} = 3.67g$	$10*\frac{142}{98} = 14.49g$

Detailed calculation of the actual mass for NaOH:

$$X \, g \cdot NaOH = 10g \cdot H_2SO_4 \frac{2mol * 40 \frac{g \cdot NaOH}{mol}}{1mol * 98 \frac{g \cdot H_2SO_4}{mol}} = 8.16g \cdot NaOH$$

Calculations are similar for the other species. Check the answer by summing the mass of reactants and products to make sure they are equal:

Reactants : $10g + 8.16g = 18.16g$
Products: $3.67g + 14.49g = 18.16g$

2.4 Concentrations, Flow Rates, and Efficiencies

2.4.1 Concentrations

Most problems do not involve pure substances; rather the chemical species are mixed together. The species present in the larger amount is called the *solvent*, all other species in the mixture are called *solutes*. Together they create a *solution*. Gases are completely soluble and form a single phase when mixed. Liquids and solids may have incomplete solubility and may form two or more phases when mixed. When a solvent contains a maximum amount of solute, the solution is saturated. Any additional solute added will precipitate into a new phase. An example of this is water vapor in air. If water is added to air such that the partial pressure it exerts exceeds its vapor pressure, then some of the water must precipitate out as rain, fog, or snow in the atmosphere.

There are many ways to describe the amounts of materials in a solution. The units can involve mass, moles, and volumes. Table 2-2 lists several of the more common units.

Table 2.2 Different Types of Units to Describe Concentration.

Units	Symbol	Name
$Mass_A/mass_{total}$	X	Mass fraction
Mol_A/mol_{total}	x	Mole fraction
$Volume_A/volume_{total}$	y	Volume fraction
$Mass_A/volume_{total}$	C	Mass Concentration
$Mol_A/volume_{total}$	c	Mole Concentration
$Mol_A/mass_{solvent}$	Π	Molality

Conversion between mass and moles uses the molecular weight:

2.11 $\qquad m = n*MW$

Where: m = mass,
n = moles, and
MW = molecular weight.

Example 2-10.

Determine the number of moles contained in 3 g of sulfuric acid.

Solution:

The molecular weight of sulfuric acid is 98 g/mol, see Example 2-7, and apply equation 2.11

$$n = \frac{m}{MW} = \frac{3g}{98 g/mol} = 0.0306 \, mol$$

Conversion to volume depends on the phase of the materials, and requires an appropriate equation of state (EOS). Liquids and solids do not have a single EOS to use, and typically use a measured quantity, density, for conversions. Gases at high temperatures and low pressures, such as the atmosphere, use the ideal gas law EOS:

2.12 $$PV = nRT$$

Where: P = pressure [M/Lt^2],
V = volume [L^3],
n = number of moles [mol],
R = ideal gas constant (see appendix 1) [L^{4}M/t^2 mol K], and
T = absolute temperature [K]

This relationship allows calculation of the density of an ideal gas:

2.13
$$\rho_{molar} = \frac{n}{V} = \frac{n \cdot P}{n \cdot R \cdot T} = \frac{P}{R \cdot T}$$

$$\rho_{mass} = \frac{m}{V} = \frac{m \cdot P}{n \cdot R \cdot T} = \frac{mw \cdot P}{R \cdot T}$$

Example 2-11.

Calculate the molar and mass density of an ideal gas (nitrogen) at standard conditions.

Solution:

Use equation 2.13 to determine the molar and mass density. Standard conditions are listed in Table 2-3: T=298K, P=101,325 Pa = 1.0 atm. Obtain a value for R from appendix 1, R= $0.082057 \frac{liter \cdot atm}{mol \cdot K}$

$$\rho_{molar} = \frac{P}{R \cdot T} = \frac{1.0 atm}{0.082057 \frac{liter \cdot atm}{mol \cdot K} \cdot 298K} = 0.0409 \frac{mol}{liter}$$

$$\rho_{mass} = \frac{mw \cdot P}{R \cdot T} = \frac{28 \frac{g}{mol} \cdot 1.0 atm}{0.082057 \frac{liter \cdot atm}{mol \cdot K} \cdot 298K} = 1.145 \frac{g}{liter}$$

Note that the molar density does not depend on what the actual gas is – all ideal gases have the same molar density at a given temperature and pressure. The mass density requires identification of the particular gas (in this case nitrogen).

The inverse of the molar density is called the molar volume. The molar volume in example 2-11 is:

2.14
$$\hat{V} = \frac{1}{\rho} = \frac{1}{0.0409 \frac{mol}{liter}} = 24.45 \frac{liter}{mol}$$

The two most common air pollution related concentrations for a gas are ppmv (parts per million by volume) and $\mu g/m^3$ (mass of one species/ volume of all species). The concentration of an ideal gas in ppmv can be calculated using either volumes or moles:

2.15
$$ppm_v = \frac{V^A}{V} * 10^6 = \frac{n^A}{n} * 10^6$$

Where: V^A = volume associated with species A
V = total volume including A and all other species
n^A = moles of species A
n = total moles

A similar unit is *parts per billion by volume* (ppbv). Its measurement is similar, except the ratio is multiplied by 10^9 instead of 10^6. These measures of concentration are independent of the gas conditions of temperature and pressure (as long as the ideal gas assumptions are still valid).

Measures of mass concentration relate the mass of a species contained in the gas to the volume of gas. The volume of a gas depends on the temperature and pressure. Typically, the concentration is measured at the actual conditions, and the units given are mass per actual cubic meter ($\mu g/acm$). A correction to standard conditions allows different measures to be comparable, and the units are mass per standard cubic meter ($\mu g/scm$), see Table 2-3 (Wikipedia, 2014). **Standard conditions**

for this textbook will be assumed to be pressure of 1.0 atmosphere and temperature of 298 K.
The correction uses the ideal gas law:

2.16 $$X \text{ scm} = acm * \frac{298K}{\text{Actual Temperature (K)}} * \frac{\text{Actual Pressure (atm)}}{1 atm}$$

Table 2-3. Sets of Standards Conditions.

Establishing Entity	Temperature	Pressure	Notes
	°C	kPa	
US-EPA	20	101.325	for NSPS
US-EPA	25	101.325	for NAAQS
EEA	15	101.325	0% RH
IUPAC	0	100.000	
NIST	0	101.325	
SPE	15	100.000	
ISO 5011	20	101.3	50% RH
	°F	psia	
SPE, US-OSHA	60	14.696	
OPEC, US-EIA	60	14.73	
US-Army	59	14.503	78% RH
	°F	in Hg	
US-FAA	59	29.92	

Where, US-EPA = United States Environmental Protection Agency
NSPS = New Source Performance Standards
NAAQS = National Ambient Air Quality Standards
EEA = European Environment Agency
RH = Relative Humidity
IUPAC = International Union of Pure and Applied Chemistry
NIST = National Institute of Standards and Technology
SPE = Society of Petroleum Engineers
ISO = International Organization for Standardization
US-OSHA = United States Occupational Safety and Health Administration
OPEC = Organization of the Petroleum Exporting Countries
US-EIA = Energy Information Administration
US-FAA = Federal Aviation Administration

Basic Problem Solving

The conversion between the two sets of concentration units for an ideal gas is:

2.17
$$X \frac{\mu g}{acm} = \frac{ppmv * MW * P}{R * T}$$

Where: R = 8.2057x10⁻⁵ (m³ atm / mol K) or other appropriate set of units
P is given in atm
T is given in K
MW is given in g/mol
ppmv = concentration value in ppm/ 10⁶

This can be simplified at standard conditions (P = 1 atm, and T = 298 K) as:

2.18
$$X \frac{\mu g}{scm} = \frac{ppmv * MW}{0.02445}$$

■ Example 2-12.

The concentration of carbon monoxide is 30 ppmv. Convert this value to a) $\mu g/scm$ and b) $\mu g/acm$ at T=50°F and 0.985 bar.

Solution:
Recall that concentrations in ppm do not need to specify a temperature or pressure. Part a uses equation 2.18, part b uses equation 2.17. For part b) the temperature and pressure need to be converted to K and atm. The molecular weight of carbon monoxide is 28 g/mol.

a) $X \dfrac{\mu g}{scm} = \dfrac{30\, ppm_v * 28 \dfrac{g}{mol}}{0.02445 \dfrac{ppm_v \cdot scm \cdot g}{mol \cdot \mu g}} = 34,340 \dfrac{\mu g}{scm}$

b) Conversions: T = 50°F = 10°C = 283 K
P = 0.985 bar = 0.972 atm

$X \dfrac{\mu g}{acm} = \dfrac{\dfrac{30}{10^6} ppm_v * 28 \dfrac{g}{mol} * 0.972 atm}{8.2057x10^{-5} \dfrac{m^3 \cdot atm}{mol \cdot K} * 283K} * \dfrac{10^6 \mu g}{g} = 33,931 \dfrac{\mu g}{acm}$

■ Example 2-13.

Convert a concentration of nitrogen at 0.064 lb-mol/acf (320 K, 0.985 atm) to [$\mu g/scm$] using the standard conditions for the EEA, IUPAC, US-OSHA, and US-FAA.

Solution:

Use dimensional analysis to change the given nitrogen concentration to [μg/acm], then apply Equation 2.16 to complete the conversion.

$$0.0015 \frac{lb-mol}{acf} * \frac{35.315 \, ft^3}{m^3} * \frac{454 \, mol}{lb-mol} * \frac{28g}{mol} * \frac{10^6 \, \mu g}{g} = 6.734 \times 10^8 \frac{\mu g}{acm}$$

Next, convert the denominator units to the various standard conditions.

$$6.734 \times 10^8 \frac{\mu g}{acm} * \frac{1}{\frac{298K}{\text{Actual Temperature (K)}} * \frac{\text{Actual Pressure (atm)}}{1 atm}}$$

a) EEA STP = (15 °C, 101.325 kPa)

$$6.734 \times 10^8 \frac{\mu g}{acm} * \frac{1}{\frac{298K}{(320)\,(K)} * \frac{0.985 atm * 101.325 kPa / atm}{101.325 kPa}} = 7.341 \times 10^8 \frac{\mu g}{scm_{EEA}}$$

b) IUPAC STP = (0 °C, 100.000 kPa)

$$6.734 \times 10^8 \frac{\mu g}{acm} * \frac{1}{\frac{298K}{(320)\,(K)} * \frac{0.985 atm * 101.325 kPa / atm}{100.000 kPa}} = 7.269 \times 10^8 \frac{\mu g}{scm_{IUPAC}}$$

c) US-OSHA STP = (60 °F, 14.696 psia)

$$6.734 \times 10^8 \frac{\mu g}{acm} * \frac{1}{\left[(60-32)*\frac{5}{9}+273K\right] \Big/ (320)\,(K)} * \frac{0.985 atm * 14.696 psia / atm}{14.696 psia} = 7.578 \times 10^8 \frac{\mu g}{scm_{US-OSHA}}$$

d) US-FAA = (59 °F, 29.92 in Hg)

$$6.734 \times 10^8 \frac{\mu g}{acm} * \frac{1}{\left[(59-32)*\frac{5}{9}+273K\right] \Big/ (320)\,(K)} * \frac{0.985 atm * 29.92 psia / atm}{29.92 psia} = 7.593 \times 10^8 \frac{\mu g}{scm_{US-FAA}}$$

2.4.2 Flow Rates

Many problems involve the rate at which a substance enters or leaves a process over a given amount of time. The standard unit of time is the second, but many air pollution problems are more

Basic Problem Solving

concerned with hourly, daily, or annual quantities. Much process data from industry uses minutes. There are three flow rates to consider - mass flow, molar flow, and volumetric flow, as given by:

2.19
$$\text{Mass Flow} = M = \frac{m}{t} = \frac{mass}{time}$$
$$\text{Molar Flow} = N = \frac{n}{t} = \frac{moles}{time}$$
$$\text{Volumetric Flow} = Q = \frac{V}{t} = \frac{Volume}{time}$$

■ Example 2-14.

100 Liter/min of a gas is emitted from a process at T=50°C and 1.05 atm. Calculate the mass flow rate of carbon monoxide if the gas stream contains 30 ppmv CO.

Solution:

This problem can be solved using the ideal gas law to convert from volume to moles, and then applying dimensional analysis to convert from moles to mass from the compounds molecular weight.

$$M \frac{g_{CO}}{min} = 100 \frac{L}{min} * \frac{1.05 atm}{0.08208 \frac{L \cdot atm}{mol \cdot K}} * \frac{30 mol_{CO}}{10^6 mol_{gas}} * \frac{28 g_{CO}}{mol_{CO}} = 1.07 \frac{g_{CO}}{min}$$

Knowledge of the concentration of a species can be used to convert mass or molar flows into volumetric flows.

2.20
$$M = CQ$$
$$N = cQ$$

The units for these equations are defined in Table 2-2 and equation 2.19.

■ Example 2-15.

Use the result from Example 2-14 to find the mass concentration of CO. Then determine the molar flow rate.

Solution:
Use equation 2.14.

$$C = \frac{M}{Q} = \frac{1.07 \frac{g_{CO}}{min}}{100 \frac{L}{min}} = 0.0107 \frac{g_{CO}}{L}$$

$$N = cQ = \frac{CQ}{MW} = \frac{0.0107 \frac{g_{CO}}{L} * 100 \frac{L}{min}}{28 \frac{g_{CO}}{mol_{CO}}} = 0.0382 \frac{mol_{CO}}{min}$$

2.4.3 Removal Efficiency

Air pollution control equipment is used to separate contaminants from a gas stream before it is emitted to the atmosphere. An air discharge permit quantifies the amount of separation required. These permits are issued by local, state, or national governments and typically specify performance in one of two ways – a fixed percentage of contaminant must be removed (removal efficiency) or that the emission stream must meet a certain threshold value. The second method quantifies the allowed amount in the output, typically as a percent, ppmv, μg/scm, or other specified units. Removal efficiency may also be called *separation efficiency*, η, and is quantified as:

2.21
$$\eta = \frac{M_{in}^C - M_{out}^C}{M_{in}^C}$$

Where: M^C = mass flow rate of the contaminant in or out of the control equipment.

Note that the units do not need to be mass flow rates, but can be whatever is most logical for the problem. For example, many efficiencies use the units of the emission standard. The only constraint is that the units of each term must match.

■ Example 2-16.

A cyclone separator is used to remove particulate matter (PM) from a gas stream. A 20 m³/min inlet stream (400 K, 1 atm) contains 1550 mg/acm of PM. Find the required removal efficiency if the cleaned outlet stream must contain no more than 24 mg/scm.

Solution:

First, correct the inlet concentration to scm, so it is comparable to the outlet condition. This correction allows the gas flow rates to be equal for the inlet and outlet (Qin=Qout when at the same conditions). Then apply equation 2.21 to solve for the efficiency.

$$X \frac{mg}{scm} = 1550 \frac{mg}{acm * \frac{298K}{400K} * \frac{1atm}{1atm}} = 2080.5 \frac{mg}{scm}$$

$$\eta = \frac{M_{in}^C - M_{out}^C}{M_{in}^C} = \frac{C_{in}^{PM} Q_{in} - C_{out}^{PM} Q_{out}}{C_{in}^{PM} Q_{in}} = \frac{C_{in}^{PM} - C_{out}^{PM}}{C_{in}^{PM}} = \frac{2{,}080.5 - 24}{2{,}080.5} = 0.988$$

Therefore, the cyclone must remove 98.8% of the particles.

2.5 Material Balances

The Law of Conservation of Mass, also known as continuity, describes the amounts and types of materials moving through a system. The following equation describes the distribution of mass as it moves through a system, and relies on the fact that mass is a conserved quantity. Energy and momentum are also conserved quantities, with similar conservation laws. A good material and energy balance textbook is (Felder, et al., 2000). The general form for this equation is;

2.22   Input – Output + Generation = Accumulation

Where: Input = mass moving into system from surroundings,
Output = mass moving from system to surroundings,
Generation = mass of a particular chemical species created or consumed in a system by a chemical reaction (may be positive or negative),
Accumulation = time dependent collection of mass within the system (may be positive or negative).

This expression is valid for the total amount of material or mixture, for an individual chemical species, or the subparts of a mixture. It is a powerful tool that allows calculation of control-equipment inflows and outflows. These mass flows are used to determine equipment efficiencies and equipment sizes. There are many specialized forms of this equation and it is helpful to learn how to recognize and use them to quantify control systems. Table 2-4 lists several possible forms, showing which terms to expect when describing a particular system type.

Table 2-4. Material Balance System Types.

System type	Input	Output	Generation	Accumulation
Flow	X	X		
Unsteady Flow	X	X		X
Flow Reactor	X	X	X	
Batch Reactor			X	X
General System	X	X	X	X

2.5.1 Flow Systems

These systems include only inlets and outlets. No reaction or accumulation occurs within the system. Examples include flow in a pipe or duct, flow through a tee fitting, multiple streams joining in mixer, or separation equipment based on physical forces.

FIGURE 2-1. Schematic of Simple Flow System.

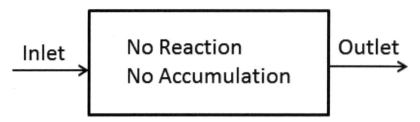

Applying this case to equation 2.22 yields:

2.23 Input − Output = 0

The input term is from the sum of the mass of each species flowing in, M_{in}. Similarly, the output is the sum of each species flowing out, M_{out}. The input and output streams can include one or more different streams. The units for M are [mass/time]. There are many ways to represent these mass flow rates:

2.24 $$M = \sum_{i}^{N} M_i = M_1 + M_2 + M_3 + M_4 + ... + M_N$$

Each stream can be composed of several species:

2.25 $$M_1 = \sum_{j}^{A,B,C,D,...} M_1^j = M_1^A + M_1^B + M_1^C + M_1^D + ...$$

Where the mass of each species can be obtained from the mole fraction:

2.26 $$M_1^A = M_1 X_1^A$$

or concentration (Note that the Q is generally not conserved and may not be summed in any simple way, see Example 2-17):

2.27 $$M_1^A = C_1^A Q_1$$

The other unit systems shown in Table 2-2 may be used as long as the units for each term are [mass] or [mass/time].

■ Example 2-17.

Three air streams containing solid particles are blended together before being transported to a solids separator. Stream 1 contains 2.8% solids and flows at 2.3 kg/min. Stream 2 contains 0.5% solids and flows at 1.1 kg/min. Stream 3 contains 5.6% solids and flows at 1.9 kg/min. Find the total flow and solids content of the blended stream.

Solution:
Sketch

FIGURE 2-2. System Sketch for Mixing Three Streams in Example 2-17.

Apply the mass conservation law around the blender. The unknowns are M_{out} and X^{PM}_{out}. M_{out} can be calculated using equation 2.24:

$$M_{out} = M_{in1} + M_{in2} + M_{in3} = 2.3\frac{kg}{min} + 1.1\frac{kg}{min} + 1.9\frac{kg}{min} = 5.3\frac{kg}{min}$$

X^{PM}_{out} can be calculated by combining equations 2.26 and 2.24 into equation 2.23:

$$M_{in1}X^{PM}_{in1} + M_{in2}X^{PM}_{in2} + M_{in3}X^{PM}_{in3} = M_{out}X^{PM}_{out}$$

Solving for X^{PM}_{out}:

$$X^{PM}_{out} = \frac{2.3(0.028)\frac{kg}{min} + 1.1(0.005)\frac{kg}{min} + 1.9(0.056)\frac{kg}{min}}{5.3\frac{kg}{min}} = 0.033 = 3.3\%$$

■ Example 2-18.

A heated mixer is used to heat and blend two gaseous waste streams together. Stream 1 flows at 3 acm/min [acm = actual cubic meters], has a temperature of 25°C, and contains 1000 μg/acm acetaldehyde. Stream 2 flows at 5 acm/min, has a temperature of 100°C, and contains 1000 μg/acm acetaldehyde. The blended streams leave the heated mixer at 240°C. Find the concentration of acetaldehyde and the volumetric (acm) and mass flows of the blended stream. Assume all three streams have the same pressure (1 atm) and the same molecular weight (MW=29).

Solution:

Apply the mass balance around the mixer; streams 1 and 2 are inputs and stream 3 is the output. Next, convert the volumetric flows to scm. Their sum is the total flow. Next, convert this flow back to the outlet operating conditions (T_3). Finally, combine the mass flow and volumetric flow to obtain the outlet concentration. A sketch such as Figure 2-3 helps organize the problem information.

FIGURE 2-3. Sketch of Mixer System for Example 2-18.

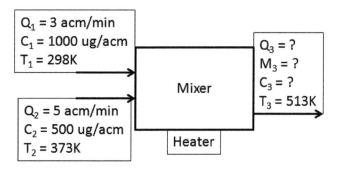

$$M^A_{out} = M^A_{in} = M^A_1 + M^A_2 = C^A_1 Q_1 + C^A_2 Q_2$$

$$M^A_{out} = 1000 \frac{\mu g}{acm} 3 \frac{acm}{min} + 500 \frac{\mu g}{acm} 5 \frac{acm}{min} = 5500 \frac{\mu g}{min}$$

Converting volumetric flows to standard conditions:

$$Q_1 = 3 \frac{acm * \frac{298K}{273K} * \frac{1atm}{1atm}}{min} = 3.27 \frac{scm}{min}$$

$$Q_2 = 5 \frac{acm * \frac{298K}{373K} * \frac{1atm}{1atm}}{min} = 3.99 \frac{scm}{min}$$

$$Q_3 = Q_1 + Q_2 = (3.27 + 3.99) \frac{scm}{min} = 7.26 \frac{scm}{min}$$

Next, convert Q_3 to actual conditions:

$$Q_3 = 7.26 \frac{scm}{min} * \frac{513K}{298K} = 12.50 \frac{acm}{min}$$

Finally, obtain the concentration from the values obtained in the above calculations:

$$C^A_3 = \frac{5500 \frac{\mu g}{min}}{12.50 \frac{acm}{min}} = 440 \frac{\mu g}{acm}$$

2.5.2 Unsteady Flow Systems

These problems allow accumulation in the system. No reaction occurs within the system, but mass may accumulate within part of the system. Examples include the filling of a vessel; draining of a vessel; and the running and cleaning of an adsorption column.

Basic Problem Solving

FIGURE 2-4. Unsteady Flow: Filling a Vessel.

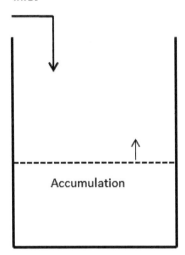

The accumulation term can be written as a derivative of the mass with respect to time:

2.28
$$Accumulation = \frac{dm}{dt}$$

Applying this case to equation 2.22:

2.29
$$\frac{dm}{dt} = M_{in} - M_{out}$$

■ Example 2-19.
Find the time to fill a compressed air tank from a pressure of 1 atm to 5 atm. The tank has a constant volume of 5 liters and remains at a constant temperature of 298 K. Air enters the tank at a rate dependent on the pressure within the tank $M_{fill} = c(10 - P)$, where $c = 0.37$ g/(atm min).

Solution.
The unknown is the fill time, which is part of the derivative in equation 2.29. Additional information includes: M_{fill} (the flow of mass in) and that the flow of mass out is zero. Summarizing:

2.30
$$\frac{dm}{dt} = M_{fill} = c(10 - P)$$

This equation has three variables: m, t, and P, and a known final state. One more relationship is needed before solving. The mass in the tank is related to the pressure through the ideal gas law:

$$m = \frac{MW \cdot V \cdot P}{R \cdot T}$$

Substitution into equation 2.30 and rearranging yields:

$$\frac{dP}{dt} = \frac{cRT(10-P)}{V \cdot MW}$$

Separate and integrate from P=1 atm at t=0 to P = 5 atm at t=t' (t' is the unknown time):

$$\int_{1atm}^{5atm} \frac{dP}{(10-P)} = \frac{cRT}{V \cdot MW} \int_0^{t'} dt$$

Solving:

$$-\ln(10-P)\Big|_{1atm}^{5atm} = \frac{\ln(9)}{\ln(5)} = \frac{0.37 \frac{g}{atm \cdot min} \cdot 0.08208 \frac{l \cdot atm}{mol \cdot K} \cdot 298K}{51.29 \frac{g}{mol}} (t'-0)$$

$$t' = 21.9 \, min$$

2.5.3 Flow Reactor Systems

Some pollution removal technologies require a chemical reaction to reduce/ eliminate the contaminant from the emissions. The stoichiometry of the reaction can be used to determine the quantity of reactants needed and the amount of products formed. The reaction kinetics can be used to determine the size and operating characteristics of the equipment needed.

Mass or Mole Balance Using Reaction Stoichiometry

These problems require the use of a balanced chemical reaction and knowledge of the amount of the reactants or products or some combination of both. The mass balance equation includes the flow in and out terms, plus the generation term. Generation happens due to the chemical reactions in the system that transforms one species into a different species. The reaction stoichiometry is easiest to perform in mole based units, and later converted into any other set of units, including mass-flow based units. The solution technique requires the use of a mole balance for each individual species. Mole balance equations are written for each species in the problem including reactants, products, and inerts.

Basic Problem Solving

2.31
$$moles_{out} = moles_{in} - moles_{reacted}$$

The number of moles reacted is related to the stoichiometric coefficients, symbolized as lower case letters, and the amount reacted of the basis species, symbolized as ξ - the Greek letter xi [pronounced as zie].

2.32
$$moles^i_{reacted} = \frac{i}{a}\xi^A$$

The unit for each term is moles of the species. The basis species should be chosen as the species where the amount reacted is known, and it is typically in terms of the limiting reactant. This information may be specified by saying the reaction goes to completion, or that a certain percentage reacts. An example set of equations for the general reaction $aA + bB \Leftrightarrow cC + dD$ with inert, denoted as "I", follow. An inert is any substance that does not participate in the reaction. A common example is nitrogen from air used for combustion. These equations assume that species A is the basis species.

2.33
$$n^A_{out} = n^A_{in} - \frac{a}{a}\zeta^A$$
$$n^B_{out} = n^B_{in} - \frac{b}{a}\zeta^A$$
$$n^C_{out} = n^C_{in} - \frac{c}{a}\zeta^A$$
$$n^D_{out} = n^D_{in} - \frac{d}{a}\zeta^A$$
$$n^I_{out} = n^I_{in}$$

Note that the stoichiometric coefficients for reactants are negative (-), products are positive (+), and inerts are zero. Reactants are considered negative because they are consumed during the reaction.

■ Example 2-20.
Determine the flow rate of oxygen (g/min and scm/min) needed to burn ethane completely at a rate of 1,000g/min. Assume the addition of 50% excess oxygen.

Solution:
Generate a balanced chemical reaction for ethane (C_2H_6), convert mass flow to mole flow, determine the moles of oxygen needed factoring in the excess, and finally, convert moles to mass and to volume.

Reactants	Products
C_2H_6	CO_2
O_2	H_2O
	O_2

The balanced chemical equation is:

2.34
$$2C_2H_6 + 7O_2 \rightarrow 4CO_2 + 6H_2O$$

Note that the excess oxygen does not show up as a product in the balanced chemical reaction. It is accounted for in the mass or mole balance equations (Example 2-21). The molar flow of ethane is:

$$N^{C_2H_6} = 1000\frac{g}{min} * \frac{mol}{30g} = 33.3\frac{mol \cdot C_2H_6}{min}$$

The actual molar flow of oxygen is the stoichiometric amount required times 1.5 [1 + xs%/100):

$$N^{O_2-actual} = 33.3\frac{mol \cdot C_2H_6}{min} * \frac{7mol \cdot O_2}{2mol \cdot C_2H_6} * 1.5 = 175\frac{mol \cdot O_2}{min}$$

Converting to mass flow:

$$M^{O_2} = N^{O_2} MW = 175\frac{mol \cdot O_2}{min} * \frac{32g \cdot O_2}{mol} = 5600\frac{g \cdot O_2}{min}$$

Converting to volumetric flow:

$$Q^{O_2} = \frac{N^{O_2} RT}{P} = 175\frac{mol \cdot O_2}{min} \frac{0.08208\frac{l \cdot atm}{mol \cdot K} 298K}{1 atm} * \frac{m^3}{1000 l} = 4280\frac{scm}{min}$$

■ **Example 2-21.**
Determine the concentrations (in mole fraction) of all species in the exhaust from Example 2-20. Air (79 mol% N_2, 21 mol% O_2) is the source for the oxygen.

Solution:
Use the information from Example 2-20 (molar flow rates) and equation 2.33. The basis species is ethane (assume complete consumption).

$$n_{out}^A = n_{in}^A - \frac{a}{a}\zeta^A = 33.3\frac{mol_{C_2H_6}}{min} - \frac{(-2\ mol_{C_2H_6})}{(-2\ mol_{C_2H_6})}33.3\frac{mol_{C_2H_6}}{min} = 0\frac{mol_{C_2H_6}}{min}$$

$$n_{out}^B = n_{in}^B - \frac{b}{a}\zeta^A = 175\frac{mol_{O_2}}{min} - \frac{(-7mol_{O_2})}{(-2\ mol_{C_2H_6})}33.3\frac{mol_{C_2H_6}}{min} = 58.45\frac{mol_{O_2}}{min}$$

$$n_{out}^C = n_{in}^C - \frac{c}{a}\zeta^A = 0\frac{mol_{CO_2}}{min} - \frac{(4mol_{CO_2})}{(-2\ mol_{C_2H_6})}33.3\frac{mol_{C_2H_6}}{min} = 66.6\frac{mol_{CO_2}}{min}$$

$$n_{out}^D = n_{in}^D - \frac{d}{a}\zeta^A = 0\frac{mol_{H_2O}}{min} - \frac{(6mol_{H_2O})}{(-2\ mol_{C_2H_6})}33.3\frac{mol_{C_2H_6}}{min} = 100\frac{mol_{H_2O}}{min}$$

$$n_{out}^I = n_{in}^I = \frac{79mol_{N_2}}{21mol_{O_2}}175\frac{mol_{O_2}}{min} = 658.3\frac{mol_{N_2}}{min}$$

These answers yield the following mole fractions in the exhaust:
- x_{C2H6} = 0.0
- x_{O2} = 0.066
- x_{CO2} = 0.076
- x_{H2O} = 0.113
- x_{N2} = 0.745

Equipment Design Using Mass Balance and Reaction Kinetics

Reaction kinetics describe the speed with which a reaction occurs. The speed is dependent on the reaction temperature. The kinetic information may be used to size a reactor. In general the longer a reaction takes, the larger the reactor must be. However, if reactions happen too quickly, they may become uncontrollable and cause an explosion. Kinetic information must be obtained by experimental observation.

These systems are similar to those in the previous section, see equation 2.22, except the generation term becomes more complex. Instead of a constant value for the generation (extent), it is now a function of the reaction volume, concentrations, temperature, and time. Generation happens due to chemical reactions in the system that transforms one species into a different species. The generation term may be quantified with an integral of the rate of the reaction throughout the system volume:

2.35
$$Generation = \int_{Volume} rate \cdot dV$$

The rate is determined from the kinetics of the chemical reaction; it is negative for reactants and positive for products. There are many equation forms to describe rate, one common form, a power law is as follows:

2.36
$$rate = r^A = k\left(c^A\right)^n$$

Where: k = rate constant [units depend on n, the reaction order]
c^A = concentration of species A [moles/volume]
p = reaction order, or power, with respect to species A.

The units for the reaction rate are moles/time/volume. Application of the species molecular weight converts the rate to mass/time/volume as required.

The rate constant is temperature dependent, usually described by the Arrhenius equation:

2.37
$$k(T) = A\exp\left[\frac{-E_A}{RT}\right]$$

Where: A = constant (sometimes called frequency factor), with the same units as k(T),
exp = exponential function,
E$_A$ = activation energy, with typical units of [J/mol]
R = ideal gas constant, with a value of 8.31 [J/(mol K)] and
T = absolute temperature [K].

The rate equation may be written in terms of any of the species in the reaction. They are related through the reaction stoichiometry. For the general chemical reaction given in equation 2.2 the relationship is:

2.38
$$\frac{r_A}{-a} = \frac{r_B}{-b} = \frac{r_C}{c} = \frac{r_D}{d}$$

When a reaction occurs within the system, it is often easier to work in terms of molar flow rates rather than mass. Converting each term to molar flow rates and substitution of equation 2.35 into 2.22 and applying it to species A gives the molar balance:

2.39
$$N^A_{in} - N^A_{out} + \int_{Volume} r^A dV = 0$$

There are two ways we can apply this equation to systems with reactions. The first is to assume the system is well mixed and that the rate of reaction is the same everywhere within the system. The second assumes that there is little or no mixing within the system and that the species move through it as a slug. Slug flow, or plug flow, is used to describe flow when the fluids do not mix or change position with respect to each other.

CASE 1. Well mixed or constantly stirred tank reactor (CSTR).

FIGURE 2-5. Sketch of Continuously Stirred Tank Reactor (CSTR).

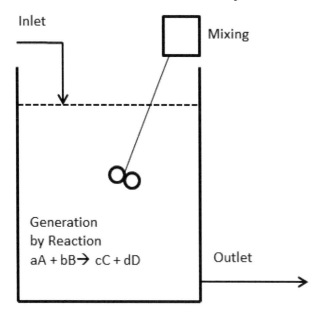

Assume the reaction rate is constant, and then it can be pulled out of the integral of equation 2.35 simplifying it to $r^A \int dV = r^A V$. Substitution into equation 2.39 yields:

2.40 $$N_{in}^A - N_{out}^A + r^A V = 0$$

Next, substitute N = QC to work with concentration:

2.41 $$Q_{in} C_{in}^A - Q_{out} C_{out}^A + r^A V = 0$$

We can assume that $Q_{in} = Q_{out} = Q$ because there is no accumulation in the system. Also, because the system is well mixed the concentration will be the same everywhere, including the outlet, so CA_{out} = CA which is the system variable:

2.42 $$Q\left(C_{in}^A - C^A\right) + r^A V = 0$$

Solving for V, obtain the design equation for a CSTR:

2.43 $$V = \frac{-Q\left(C_{in}^A - C^A\right)}{r^A} = \frac{Q\left(C^A - C_{in}^A\right)}{r^A}$$

This equation allows the calculation of the size of a CSTR needed to reduce the concentration of the reacting species from the inlet concentration to the desired concentration.

Space-time or *residence time* is a measure of the average time any molecule spends within the system. It has units of time and is defined as:

2.44
$$\tau = \frac{V}{Q}$$

Dividing equation 2.43 by Q, substituting equation 2.44 and equation 2.36 and solving for τ yields another form of the design equation:

2.45
$$\tau = \frac{C^A - C^A_{in}}{k(C^A)^n}$$

■ **Example 2-22.**
Nitric oxide (NO) can be produced by reacting nitrogen and oxygen at high temperature and pressure in the presence of a catalyst. A stoichiometric feed of 60 mol% nitrogen and 40 mol% oxygen at 3 atm and 500 K is fed to a CSTR reactor at a volumetric flow of 0.1 acm/min. How large must the reactor be to produce a product stream that is 50% NO? The rate law is $r^{N_2} = 40 \min^{-1}(c^{N_2})$.

Solution:
Use equation 2.43 to determine the volume. Solution requires the concentration of nitrogen in the feed and the outlet. The reaction is:

2.46
$$N_2 + O_2 \rightarrow 2NO$$

Inlet concentration of nitrogen can be determined from the ideal gas law:

$$C^{N_2}_{in} = \frac{X^{N_2} P}{RT} = \frac{0.6(3 atm)}{0.08208 \frac{l \cdot atm}{mol \cdot K} 500K} = 0.0438 \frac{mol N_2}{l}$$

Outlet concentration of NO can also be determined from the ideal gas law:

$$C^{NO}_{out} = \frac{X^{NO} P}{RT} = \frac{0.5(3 atm)}{0.08208 \frac{l \cdot atm}{mol \cdot K} 500K} = 0.0365 \frac{mol NO}{l}$$

The amount of nitrogen remaining can be determined by subtracting the amount of nitrogen used to form nitric oxide from the amount added at the inlet. Assume a basis of 1.0 liter for ease of calculation:

$$X \text{ mol } N_2 = 0.0365 \text{ mol } NO * \frac{1 \text{ mol } N_2}{2 \text{ mol } NO} = 0.0183 \text{ mol } N_2$$

The amount of nitrogen in the outlet per mole of feed is:

$$C^{N_2}_{out} = C^{N_2} = (0.0438 - 0.0183)\frac{mol_{N_2}}{l} = 0.0255\frac{mol_{N_2}}{l}$$

Finally, solve for the volume:

$$V = \frac{Q(C^{N_2} - C^{N_2}_{in})}{r^{N_2}} = \frac{Q(C^{N_2} - C^{N_2}_{in})}{-k(C^{N_2})^p} = \frac{100\frac{l}{min}(0.0255 - 0.0438)\frac{mol}{l}}{-40\,min^{-1}\left(0.0255\frac{mol}{l}\right)^1} = 1.79l$$

Note that r^{N_2} is negative because nitrogen is a reactant and it decreases during the reaction.

Case 2. The plug flow rector (PFR) system.

FIGURE 2-6. Sketch of Flow System with Chemical Reaction.

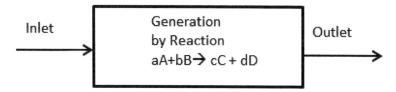

The amounts of products and reactants vary with position in this system. Therefore, the rate of reaction also varies with position. The mass balance equation is equation 2.39. The design equation, which defines the volume needed to conduct the reaction, can be obtained by taking the derivative with respect to the volume and assuming that the molar flow into the system is constant:

2.47
$$\frac{d}{dV}\left[N^A_{in} - N^A_{out} + \int_{Volume} r^A dV\right] = -\frac{dN^A}{dV} + r^A = 0$$

Separating:

2.48
$$dV = \frac{dN^A}{r^A}$$

and integrating from $N^A = N^A_{in}$ at $V = 0$ to $N^A = N^A_{out}$ at $V = V_{out}$ yields the design equation for a plug flow reaction system.

2.49
$$V_{out} = \int_{N^A_{in}}^{N^A_{out}} \frac{dN^A}{r^A}$$

Example 2-23.
Rework Example 2-22 to find the volume needed for the reaction in a PFR.

Solution:
Substitute N = cQ in equation 2.49 and solve using the previously calculated terms from Example 2-22:

$$V_{ut} = \int_{C_{in}^A}^{C_{out}^A} \frac{dQC^{N_2}}{-k(C^{N_2})} = -\frac{Q}{k}\left[\ln(C^{N_2})\right]_{0.0438}^{0.0255}$$

$$= -\frac{100 \frac{l}{\min}}{40 \min^{-1}}\left[\ln(0.0255) - \ln(0.0438)\right] = 1.35 l$$

Note that the PFR reactor is smaller than a CSTR when the power in the rate law is greater than zero assuming the rate law is given as a power law equation, as in equation 2.36.

2.5.4 Batch Reactor Systems

The independent variable for a batch system is time, whereas for the flow systems the independent variable is volume (size). This means that the design of a batch system focuses on how long to run the reaction. The required time must include the required maintenance between runs. A standard operation will include the time to fill the reactor, to run the reaction, to empty the reactor, and to clean the reactor. Once cleaned it can be run again. The time for the maintenance operations (fill, drain, and clean) is typically on the order of four to eight hours. The design however is based on only the time required to run the reaction. The design equation simplifies from equation 2.22 to include only the generation (reaction) and accumulation terms. There are no inputs or outputs.

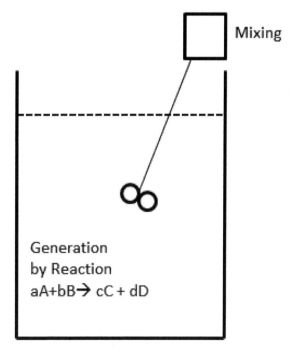

FIGURE 2-7. Sketch of Batch Reactor.

Mixing

Generation by Reaction
aA+bB→ cC + dD

Batch systems are run with excellent mixing (the concentration of every species is the same everywhere), so the generation term is similar to the CSTR case:

2.50 $$Generation = \int_{Volume} r^A dV = r^A \int_{Volume} dV = r^A V$$

The accumulation term, written in terms of moles, is:

2.51 $$Accumulation = \frac{dn}{dt}$$

Substitution into equation 2.22 yields:

2.52 $$r^A V = \frac{dn}{dt}$$

If the volume remains a constant throughout the reaction, it can be moved into the derivative to obtain the design equation for a batch reaction system:

2.53 $$r^A = \frac{1}{V}\frac{dn}{dt} = \frac{d\frac{n}{V}}{dt} = \frac{dc}{dt}$$

This equation allows determination of the time a batch system must run in order to generate (consume) a set amount of a particular chemical species.

■ Example 2-24

Find the time needed for the following reaction to reduce the reactant by 95% in a batch reactor. Assume the reactor volume remains constant at 2 liters, temperature is 320 K and the pressure is 1 atm.

2.54 $$2A \rightarrow B + C$$

$$r^A = k(C^A)^2, \text{ where } k = 15\frac{l}{mol \cdot min}$$

Solution:
Find the initial and final concentrations of A:

$$C^A_{initial} = \frac{X^A P}{RT} = \frac{0.1(1 atm)}{0.08208\frac{l \cdot atm}{mol \cdot K} 320K} = 0.003807\frac{mol A}{l}$$

$$C^A_{final} = (1-0.95)*0.00381 = 0.00019\frac{mol A}{l}$$

Next, apply the rate law data to equation 2.53: $\frac{dC^A}{dt} = -k(C^A)^2$

Separate and integrate: $$\int_0^{t'} dt = -\frac{1}{k}\int_{C_{initial}}^{C_{final}} \frac{dC^A}{(C^A)^2} = -\frac{1}{k}\left[\frac{1}{C_{start}} - \frac{1}{C_{final}}\right]$$

Solving: $$t' = -\frac{1}{15\frac{l}{mol\cdot min}}\left[\frac{1}{0.003807\frac{mol}{l}} - \frac{1}{0.00019\frac{mol}{l}}\right] = 333\,min = 5\,hour\,33\,min$$

Note that the volume of the batch reactor plays no role in determining the time required for the reaction.

2.5.5 General Systems

These systems use the most general form of equation 2.22. Rewritten in terms of molar flow rates:

2.55 $$N^A_{in} - N^A_{out} + \int_{Volume} r^A dV = \frac{dn}{dt}$$

The utility of this equation depends on what is known about the actual system:
  o If well mixed, apply equation 2.50 to the generation term.
  o If steady state, the accumulation term can be neglected (set to zero).
  o Are either the input or output terms zero?
  o Is the rate of accumulation known or determinable with other given information?
  o Are any of the terms small enough to neglect?
Until some additional information is determined, these systems cannot be fully specified.

● **Case Study 2-1.**
Coal Fired Power Plant built in 2012. Determine the concentrations (ppm and g/m³) of all products formed from the combustion of coal with air. Assume the coal is represented by the chemical formulae $C_{70}H_{100}O_2N_{0.2}S_{0.3}Ash_{(7wt\%)}$. Assume dry air enters at 50% excess and apply standard conditions to the exhaust (25°C, 1.0 atm). Further, assume all the nitrogen in the fuel forms gaseous NO_2 and all the sulfur forms gaseous SO_2. Finally, assume the ash consists of 65% bottom ash (MW = 150 g/mol and 35% fly ash (85 g/mol).

FIGURE 2-8. Big Bend Power Station, Apollo Beach, FL, USA - An Example of a Coal Fired Power Plant. (Wknight94, 2014)

Solution:
This is a complex problem so we will break it into steps:
1. Make a sketch.
2. Write out balanced chemical reactions for each atomic species.
3. Determine the amount of product formed from each reaction.
4. Determine the total amount of oxygen needed to fully combust the coal.
5. Determine the amount of actual air needed to obtain the required amount of oxygen.
6. Calculate the ppm and g/m³ concentrations of each constituent. Account for the fly ash in the exhaust gas and bottom ash in the solids waste from the power plant.

1. Make a sketch:

FIGURE 2-9. Sketch for Case Study 2-1 showing Inputs and Outputs from a Coal Fired Power Plant.

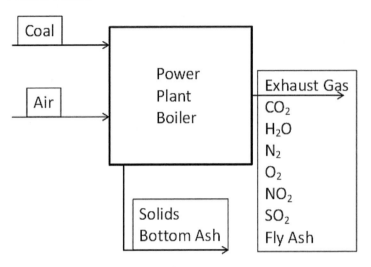

2. The atomic chemical reactions are:

    a) $C + O_2 \rightarrow CO_2$
    b) $4H + O_2 \rightarrow 2 H_2O$
    c) $N_{fuel} + O_2 \rightarrow NO_2$
    d) $S_{fuel} + O_2 \rightarrow SO_2$

3. Determine the amount of oxygen needed for each reaction. We must choose a basis so that we can determine quantities of each species. A simple basis to choose is one gram-mole of the fuel.
a) 1 gram-mole fuel contains 70 mol of C and uses 70 mol O_2 and forms 70 mol CO_2.

The other reactions are similar, and we can summarize the findings in a table:

Table 2-5. Quantity of Atomic Species in One mol of Coal for Case Study 2-1.

Species	Amount in 1 mol coal	Amount of Product	Required Oxygen
C	70 mol	70 mol CO_2	70 mol O_2
H	100 mol	50 mol H_2O	25 mol O_2
N_{fuel}	0.2 mol	0.2 mol NO_2	0.2 mol O_2
S_{fuel}	0.3 mol	0.3 mol SO_2	0.3 mol O_2

Note that for the hydrogen (H) reaction the stoichiometric coefficients are not one, so the proper ratios must be used.

4. Determine the required oxygen by summing the amounts required from the table above. Note that the coal already has some oxygen in it, so it can be deducted from the sum before accounting for the excess amount.

(70 mol O_2 + 25 mol O_2 + 0.2 mol O_2 + 0.3 mol O_2) – 1 mol O_2 = 94.5 mol O_2 required

Actual O_2 added is equal to the required O_2 needed * excess ratio (1 + %excess/100) =

94.5 * (1 + 50/100) = 141.75 mol O_2

5. Assuming dry air is 79 mol% N_2 and 21 mol% O_2:

Amount of air = 141.75 mol O_2 / 0.21 = 675.0 mol air
Amount of N_2 = 141.75 mol O_2 * (0.79/0.21) = 533.3 mol N_2

6. Determine the amount of fly ash. This is a bit tricky, since the formula for coal included the ash as a weight per-cent, and the ash is composed of two different fractions. Fly ash is lighter and is carried out with the exhaust gases. Bottom ash is heavier and is removed from the boiler as a solid waste.

Mass of 1 mole of coal contains:
- C: 70 mol * 12 g/mol = 840 g C
- H: 100 mol * 1 g/mol = 100 g H
- O: 2 mol * 16 g/mol = 32 g O
- N: 0.2 mol * 14 g/mol = 2.8 g N
- S: 0.3 mol * 32 g/mol = 9.6 g S

Total: 984.4 g coal without ash

The mass of the atomic constituents account for 93% of the total weight, so the total weight is 984.4/0.93 = 1058.5 g/mol coal

Total ash is 7% of this value or 74.1 g, of which 65% is bottom and 35% is top:
Ash_{bottom} = 48.2 g/mol coal
Ash_{Fly} = 25.9 g/mol coal

Neither takes up any appreciable volume and can be excluded from the ppm concentration calculations. The composition of the exhaust is:

Table 2-6. Exhaust Gas Composition and Molar Composition for Case Study 2-1.

Species	Amount in Outlet [mol]	Mole Fraction	ppm
CO_2	70	0.100	99,850
H_2O	50	0.071	71,300
O_2	47.25	0.067	67,400
N_2	533.25	0.761	760,700
NO_2	0.2	0.000285	285
SO_2	0.3	0.000428	428
Total	701.0		

Note: Obtain the amount of O_2 in the exhaust by comparing the amount in and the amount used: 141.75 mol O_2 in – 94.5 mol O_2 used = 47.25 mol O_2 out.

We can also determine the mass/volume based concentrations. Assume ideal, standard conditions for the exhaust gas. Determine the total moles and the volume of the exhaust gas.

Total moles = 701.0, from **Table 2-6**.

2.56
$$V = \frac{nRT}{P} = \frac{(701.0 mol)\left(8.2057 \times 10^{-5} \frac{m^3 \cdot atm}{mol \cdot K}\right)(298K)}{1 atm} = 17.14 m^3$$

Next, determine the mass of each constituent, then the concentration is obtained by dividing by the total volume (17.1 m³):

Table 2-7. Mass concentrations of the Exhaust in Case Study 2-1.

Species	Amount in Outlet [mol]	MW [g/mol]	Mass in Exhaust [g]	Concentration [g/m³]
CO_2	70	44	3080	180
H_2O	50	18	900	52
O_2	47.25	32	1512	88
N_2	533.3	28	14932	871
NO_2	0.2	46	9.2	0.54
SO_2	0.3	64	19.2	1.12
Ash			25.9	1.51

Later chapters (i.e. 3, 8, 9, and 10) will use this case study to explore more details of the use of coal in power plants and to determine the appropriate controls for each pollutant.

2.6 Problems

1. Rewrite Table 2-1 for your own personal style of problem solving, If you are aware of certain issues you have, make notes in the table to double check. For example, some students have trouble writing out the correct units when doing a calculation. You should add a step in parts II, III, and IV to check your units. Other students may have trouble with negative signs. Again, add a specific sub-step to each part of your problem solving strategy.

2. Use dimensional analysis to determine the number of seconds you have lived. If you don't know what time you were born, assume noon.

3. How old will you be when you have lived for 10^9 seconds?

4. How many power plants does your state need? [This is a Fermi problem.]

5. How many bicycles are sold each year in the USA? [This is a Fermi problem.]

6. Balance the following chemical reactions:
   a) $aH + bO_2 \rightarrow cH_2O$
   b) oxidation of hydrogen sulfide
   c) oxidation of methane
   d) combustion of ethanol
   e) production of gypsum from sulfur dioxide and lime slurry (eqn 9.6)

7. Determine the mass of all reactants and products for the following reactions:
   a) $100 \text{ g } H_2 + \_\_O_2 \rightarrow \_\_H_2O$
   b) oxidation of 100 g hydrogen sulfide
   c) oxidation of 100 g methane
   d) combustion of 100 g ethanol
   e) production of 100 g gypsum from sulfur dioxide and lime slurry (eqn 9.6)

8. The concentration of sulfur dioxide is 8,800 µg/scm. Convert this value to a) ppmv and b) µg/acm at T=10°C and 1.085 bar.

9. Convert a concentration of ozone at 5,370 µg/acm (280 K, 0.964 atm) to standard conditions using the US-EPA NAAQS, IUPAC, SPE, US-OSHA, and US-FAA standards.

10. 1.3 m³/min of a gas is emitted from a process at T=20°C and 1.165 atm. Calculate the mass flow rate of nitrogen dioxide if the gas stream contains 2 ppmv NO_2.

Chapter 2

11. A scrubber separator removes sulfur dioxide (SO_2) from a gas stream. A 0.5 m³/sec inlet stream (350 K, 1 atm) contains 1.2 ppmv SO_2. Find the required removal efficiency if the cleaned outlet stream must contain no more than 240 µg/scm.

12. A filter bag system is used to remove lead particulate matter (Pb) from a gas stream. A 7 m³/min inlet stream (320 K, 1 atm) contains 1.5 µg/acm of Pb. Find the required removal efficiency if the cleaned outlet stream must contain no more Pb than 100 ppt [parts per trillion = 1 in 10^{12}].

13. Two gas streams containing methane, nitrogen, and water are blended in a mixer before entering an activated carbon adsorption tower. Stream 1, flowing at 3 scm/min, contains 10% methane and 80% nitrogen. Stream two, flowing at 1.3 scm/min, contains 5% methane and 25% water. Determine the amount and composition of the stream leaving the mixer.

14. A diluent gas is added to the feed of a blower to reduce the concentration of solids in the blended stream. The diluent contains no solids. The feed stream flows at 100 acf/sec (100°F, 1 atm) and contains 0.5 wt% solids. The blended stream must never exceed 0.08% solids. What volumetric flow rate [acf/sec] of the diluent should be used? If you need more information – state what is needed and then assume a reasonable value. Clearly justify your assumptions.

15. The Los Angeles basin floor covers approximately 700 square miles (2×10^{10} ft²) and is almost completely surrounded by mountain ranges. If one assumes an inversion height in the basin of 2000 feet, the corresponding volume of air in the basin is 4×10^{13} ft³. Using this system volume we can model the accumulation and depletion of air pollutants. As a very rough first approximation, we shall treat the Los Angeles basin as a well-mixed container (similar to a CSTR) in which there are no spatial variations in pollutant concentrations. We will also simplify the system by considering only the pollutant carbon monoxide and assuming that its only source is from automobile exhaust. On the average we will further suppose that there are 400,000 cars operating in the basin at any one time. Each car gives off approximately 3,000 standard cubic feet of exhaust each hour containing 2 mol% carbon monoxide. Problem excerpted from (Fogler, 1992).

    Perform an unsteady-state mole balance on CO as it is depleted from the basin area by a Santa Ana wind. Santa Ana winds are high-velocity winds that originate in the Mojave Desert just to the east of Los Angeles. This clean desert air flows into the basin through a corridor assumed to be 20 miles wide and 2000 ft high (inversion height) replacing the polluted air, which flows out to sea or toward the south. The concentration of CO in the Santa Ana wind entering the basin is 0.08 ppm (2.04×10^{-10} lbmol/ft³).

a) How many pound moles of gas are in the system chosen for the Los Angeles basin if the temperature is 75°F and the pressure is 1 atm?

b) What is the rate, $F_{CO}A$, at which all autos emit carbon monoxide into the basin (lb mol CO/h)?

c) What is the actual volumetric flow rate (ft^3/h) of a 15 mph wind through the corridor 20 miles wide and 2000 feet high?

d) At what rate, $F_{CO}S$, does the Santa Ana wind bring carbon monoxide into the basin (lbmol CO/h)?

e) Assuming that the volumetric flow rates entering and leaving the basin are identical $v = vo$ show that the unsteady mole balance on CO within the basin, starting at equation 2.22 becomes

$$V \frac{dC^{CO}}{dt} = F_A^{CO} + F_S^{CO} - v_0 C^{CO}$$

Note that there is no generation of CO within the system as chosen.

f) Verify that the solution to the previous equation is

$$t = \frac{V}{v_0} \ln \left[\frac{\left(F_A^{CO} + F_S^{CO}\right) - v0C_0^{CO}}{\left(F_A^{CO} + F_S^{CO}\right) - v0C_f^{CO}} \right]$$

g) If the initial concentration of carbon monoxide in the Los Angeles basin before the Santa Ana wind starts to blow is 8 ppm (2.04 x 10^{-8} lb mol/ft^3), calculate the time required for the carbon monoxide to reach a level of 2 ppm.

16. Determine the actual in flow rate of oxygen (g/min and scm/min) needed to completely burn 700 g/min of ethanol. Assume the addition of 35% excess oxygen.

17. Determine the concentrations (in mole fraction) of all species in the exhaust from the complete combustion of propane (C_3H_8). Assume 50% excess air (79 mol% N_2, 21 mol% O_2) is the source for the oxygen.

18. Redo case study 2-1 for an oil with chemical formula $C_{16}H_{25}S_{0.1}$. Assume 50% excess air is used and that there are no PM or NO_x formed in the power plant. Assume standard conditions for all gas streams.

2.7 Bibliography

Felder, Richard M and Rousseau, Ronald W. 2000. *Elementary Principles of Chemical Processes*. New York, NY, USA : John Wiley and Sons, Inc., 2000. ISBN 0-471-53478-1.

Fogler, H Scott and LeBlanc, Steven E. 1995. *Strategies for Creative Problem Solving*. Upper Saddle River, NJ, USA : Prentice Hall, Inc, 1995. ISBN 0-13-179318-7.

Fogler, H.S. 1992. *Elements of Chemical Reaction Engineering, 2nd Edition*. Englewood Cliffs, NJ, USA : Prentice Hall P T R, 1992. ISBN 0-13-263534-8.

Kotz, John C, Treichel, Paul M and Townsend, John. 2006. *Chemistry and Chemical Reactivity, 7th Edition*. Belmont, CA, USA : Brooks/ Cole Cengage Learning, 2006. ISBN 978-0-495-38703-9.

Polya, G. 1945. *How to Solve It*. Princeton, NJ, USA : Princeton University Press, 1945. ISBN 0-691-08097-6.

Wikipedia. 2014. Standard conditions for temperature and pressure. *Wikipedia*. [Online] June 15, 2014. [Cited: July 1, 2014.] http://en.wikipedia.org/wiki/Standard_conditions_for_temperature_and_pressure.

Wknight94. 2014. *Big Bend Power Station*. Licensed under CC BY-SA 3.0 via Wikimedia Commons.

CHAPTER **3**

Environmental Laws and Regulations

This chapter summarizes the fundamentals of US environmental law as it applies to the ambient (outdoor) air. It also provides examples of how to use and apply this information towards air pollution questions. All engineers and scientists working with environmental issues need to be familiar with the concepts and language of laws and regulations since such work focuses on activities required by laws. Later chapters explore additional areas of the law and include some law/regulation perspectives from other countries and non-government organizations.

Laws are the primary tool a country uses to address problems. They are an extension of the will of the groups allowed to have a say in the creation of laws. A general trend across countries is that as more stakeholder groups are included in governance, there are higher standards for preventing harm from pollution. Laws are used to address environmental problems, but not necessarily eliminate them, because there is little economic benefit for an organization to remove all pollutants from a process. While there are benefits from removing pollution, such as better worker health, happier neighbors, and a sense of stewardship, most organizations do not value these benefits at the same level as the costs for removing pollution. Occasionally, voluntary regulations are tried, but since they do not create a level playing field, they rarely work for long. When everyone has to take on the cost of pollution control, then the economics favor the most efficient process. When only a few or no-one takes on the burden, then economics favor the do-nothings and the old technologies.

Laws tend to drive environmental technologies. Without a law to require the separation of waste from an emission stream, the separation technologies remain theoretical and undeveloped. This reality is why those opposed to environmental regulation claim the required cleanup will be too expensive, and that their estimates are usually very high when compared to the actual cost. It is difficult to estimate the cost of the needed equipment before developing that technology, and it is natural to estimate or extrapolate the cost based on currently known technology, even if it is inappropriate for the new situation.

The creation of an environmental law may take considerable time. Much information must be gathered or researched before the nature of the problem is understood. Once identified, more information is needed on causes, effects, and solutions of the problem. Sometimes an initial environmental law is passed to determine such information because, at an early stage, the problem is that there is not enough information.

The implementation of a law requires the creation of regulations to address the how, when, where, and cost issues associated with the law. The regulations also have to be approved after creation. It is not uncommon for the approval of regulations to take several years after creation of the law. This time allows many groups within a community to examine the law and its proposed implementation, and to provide comment that may improve how the law is carried out. Environmental laws typically require their regulations be re-examined every several years to account for new information.

3.1 System of US Law

The federal government of the United States (US) is composed of a triad of powers:
- Executive branch – controlled by an elected President,
- Legislative branch – made up of elected Representatives (Congress) who are split into an upper house (Senate) and a lower house (House of Representatives), and
- Judicial branch – headed by Presidential appointees to the Supreme Court.

Each branch has its own powers in a system of checks and balances between the three groups. For example, the Legislative branch is given the authority to create laws, the Judicial branch has the authority to interpret the laws, and the Executive branch has the authority to enforce the laws.

3.1.1 Law Hierarchy

The United States uses a hierarchy of laws. More important laws supersede less important ones when there is a conflict between the laws. The hierarchy of laws in the US is the following:

I. *US Constitution*. The Constitution is the highest law and courts overrule any other law that contradicts or violates it.
II. *Treaties*. Treaties must be ratified by the US Senate and are secondary only to the Constitution. Many environmental problems are now known to be global in effect, and are addressed through international treaties such as the Montreal Protocol (see section 12.2.5).
III. *Federal Statutes*. Federal statutes (laws) have been used to codify most of the major environmental programs of the last 70 years. However, federal authority is limited by the Constitution, which specifies, or enumerates, the powers given the federal government. All other powers (not enumerated) are given to the States. Thus, any federal law that infringes on the states' rights can be ruled unconstitutional and, therefore, unenforceable. Environmental laws are typically authorized under Congress's right to regulate interstate commerce.
State Statutes. States typically have similar systems to the federal system for promulgating, publishing, and codifying their laws and regulations. However, each state is different, and it is imperative to understand state specific procedures for applying the law within the state's jurisdiction. Many states have developed their environmental laws in parallel with federal laws. Many of the federal laws require the states to have environmental protections at least as stringent as the federal law, while other laws forbid states from setting standards higher than federal laws. In some areas, there are few or no federal environmental laws -

such as indoor air pollution and non-hazardous solid waste disposal – and states set their own standards and provide enforcement. Indeed, for many environmental problems, the States initiated the development and enforcement of laws.

IV. *Local (city/ county) Statutes*. Local environmental laws and regulations vary greatly both within a state and between states. Each state delegates environmental authority differently. Some local government agencies play critical roles in achieving clean air while in other areas within the same state the local government plays only a minimal role.

3.1.2 Process for Creating a Law

Federal environmental laws are created by the Legislative branch. A member of Congress must propose a bill before it can become law. These bills, or proposed laws, can be reviewed on the Library of Congress' website (THOMAS, 2011). The next step is for both houses of Congress to approve the bill. Once approved the bill then goes to the President who has the option to either approve it or veto it. If approved, the new law is called an act, and the text of the act is known as a **public statute.** Once an act is passed, the House of Representatives standardizes the text of the law and publishes it in the United States Code, which is the official record of all federal laws.

3.1.3 Process for Using a Law

The Judicial system of courts is used to render a legal decision when there is a question about the law. Such questions may be of the form: Is the law valid? Does it apply to the given situation? Would a different law be more applicable? Has someone actually violated the law? and What did lawmakers intend? Like laws, there is a hierarchy of courts with the US. The Supreme Court is the final, or highest, court of appeal. Figure 3-1 shows a simple schematic of this hierarchy, which, as the system of laws, also has a split between State and Federal authority (US-Courts, 2011).

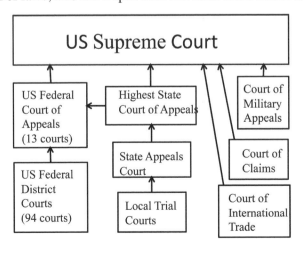

FIGURE 3-1. US Court System Hierarchy.

The state and local courts are independent of the federal courts. These courts decide 95% of the nation's legal cases. When either of the parties involved in a lawsuit believe a court did not rule correctly, they may appeal the decision to a higher court. The higher court must decide if it is willing to hear the appeal, or that they agree with the lower court. Cases that originate in state courts can be appealed to a federal court only if it involves a federal issue, and usually only after all avenues of appeal in the state courts have been tried.

The Federal District Courts decides most federal cases (questions about federal law). There are ninety-four of these courts across the country, with at least one in every state (larger states have up to four). District courts are the only courts in the federal system that uses juries and most cases at this level present before a single judge. Appealed cases from district courts go to a federal court of appeals. Courts of appeals do not use juries or witnesses and do not allow presentation of new evidence; appellate courts base their decisions on a review of lower-court records. The US Supreme Court is the final court of appeal. It can also: declare laws invalid if they violate the Constitution, overrule state and local laws using the supremacy of federal laws or treaties, and provide final authority on the meaning of the Constitution.

The courts use three basic ideas in rendering a legal decision:
1. *Statute law* considers the actual language and the intention of the law. Decisions may require the use of the notes and other writings, as well as verbal discussions, of the lawmakers.
2. *Case law* involves determining which laws apply and how they pertain to the case in question. The previous use and application of the law by other courts forms the basis of case law and these rulings establish legal precedents that lower courts are bound to follow.
3. *Common law* is the collection of well-established legal principles established by the courts over the years. It applies at the federal, state, or local court level depending on the specific case. It reflects the judicial concepts of fairness and equity applied in the absence of statutes and regulations. Most environmental concerns have been entirely codified and rarely do the courts need to resort to common law.

Civil law describes cases involving private rights and disputes. Criminal law describes cases where at least one party in the dispute is the government (also called the state). Environmental cases can be either type. Criminal cases involve direct violations of the law, including citizens suing the government to obey or enforce an environmental law. Civil cases usually involve tort, which assume a wrong or injury, other than breach of contract, for which a court can provide a remedy in the form of an action for damages. There are several types of tort actions that are important for pollution issues – nuisance, trespass, and negligence. Nuisance is the harm or injury to the rights of others caused by the unreasonable, unwarranted, or unlawful use by a person of their own property. Examples include actions for smoke or noise. Trespass is the unlawful interference with another person, or with their property or rights. While similar to nuisance, it involves intrusion on another's property. Negligence is acting in a careless or unreasonable manner. Courts may invoke the concept of a 'reasonable prudent person.' An example is that a person places a fence around his or her property if there is any potential someone could be injured by accessing that property.

3.1.4 Process for Creating a Regulation

Laws often do not include all the details needed to put them into practice. Typically, a 40 page law can require 1,000 or more pages of regulations to clearly and precisely explain how an individual, business, state or local government, or others might follow the law. Congress authorizes certain government agencies, such as the Environmental Protection Agency (EPA), to create regulations. Regulations set specific rules about what is legal, what isn't, how to achieve the law, and when to achieve it. The regulation may also detail how to enforce the law and enumerate the fines or penalties if the requirements of the law are not met. For example, a regulation issued by EPA to implement the Clean Air Act might list what levels of carbon monoxide are safe. It would tell industries how much they can legally emit into the air, and what the penalty is if they emit too much. Once the regulation is in effect, EPA then works to help people comply with the law and to enforce it (US-EPA, 2011a).

The first step for the EPA to create an environmental regulation is to decide if a regulation is needed. The agency researches it and, if necessary, proposes a regulation. Many laws require the governing agency to review the law every few years and to propose additional regulations as necessary. The proposal is listed in the Federal Register (US-GPO, 2011a) so that members of the public can consider it and send their comments to the agency. Next, the agency considers all the comments and revises the regulation accordingly in order to issue a final rule. At each stage in the process, the agency publishes a notice in the Federal Register. These notices include the original proposal, requests for public comment, notices about meetings where the proposal will be discussed (open to the public), and the text of the final regulation.

Once completed, the regulation is printed in the Federal Register as a final rule, and it is "codified" by being published in the **Code of Federal Regulations** (CFR). The CFR is the official record of all regulations created by the federal government. It consists of 50 volumes, called titles, each of which focuses on a particular area. Almost all environmental regulations appear in Title 40. The CFR is not the same as United States Code (where laws are recorded), which can be a source of confusion.

Once a regulation is in effect, the EPA works to help citizens and corporations comply with the law and to enforce it. This activity includes education, creation of forms and documentation, and site visits.

● **Case Study 3-1. The United States Environmental Protection Agency**
Congress passed the Clean Air Act as well as many other environmental laws in 1970. President Richard Nixon (R) proposed to create a new federal agency to write and enforce the necessary regulations based on the environmental laws passed by Congress. This agency is called the **Environmental Protection Agency** *(EPA or US-EPA) and it began operation on December 3, 1970. The EPA is led by an Administrator, who is appointed by the President and approved by Congress. The EPA works with states, tribes, local governments, industry, and environmental groups to establish a variety of programs to reduce pollution levels across America. The EPA headquarters is located in Washington, DC. It also has 10 regional offices and operates 27 laboratories around the country.*

The EPA's mission is to protect human health and the environment. To achieve this mission, the EPA conducts environmental assessment, research, and education. This work is designed to achieve the agency's mission by ensuring that:
- *All Americans are protected from significant risks to human health and the environment where they live, learn, and work;*
- *National efforts to reduce environmental risk are based on the best available scientific information;*
- *Federal laws protecting human health and the environment are enforced fairly and effectively;*
- *Environmental protection is an integral consideration in U.S. policies concerning natural resources, human health, economic growth, energy, transportation, agriculture, industry, and international trade, and these factors are similarly considered in establishing environmental policy;*
- *All parts of society—communities, individuals, businesses, and state, local and tribal governments—have access to accurate information sufficient to effectively participate in managing human health and environmental risks;*
- *Environmental protection contributes to making our communities and ecosystems diverse, sustainable, and economically productive;*
- *The United States plays a leadership role in working with other nations to protect the global environment; and*
- *The agency encourages a wide variety of voluntary pollution prevention programs and energy conservation efforts.*

3.1.4.1 Law numbering

When a bill becomes a law it is assigned a Public Law number, which has the format of the prefix PL, then the number of the congressional session in which passed it, followed by a specific number for the law. For example, the Resource Conservation and Recovery Act (RCRA) was passed in 1976 by the 94th session of Congress and is designated as PL-94-580. All federal law is also codified in the United States Code (USC). The USC is regularly updated with the latest amendments to the statutes and annotations are added describing case law interpretations of the statute. RCRA, as amended, is referenced as 42 USC 6901 et seq. (read as Title 42 of the United States Code, Section 6901 and following sections). Laws may also have a common name, such as the Clean Air Act (aka PL-91-604 or 42 USC §7401 et seq.).

3.1.4.2 Researching Laws and Regulations

Federal statutes and regulations are almost all available on web sites, see Table 3-1. Each federal agency maintains a website with a library of downloadable copies of all their publications and documents. The information includes current and proposed regulations, guidelines on how to follow the law and worksheets to help various users comply with permits and data reporting. State statutes and regulations are also available online, though some knowledge of how the state organizes its environmental agencies is usually required.

Table 3-1. Websites with Information on US Environmental Laws and Regulations.

Code Federal Regulations	www.gpo.gov/fdsys/
General law information	www.lexispublishing.com
Environmental Protection Agency	www.epa.gov
Proposed and approved regulations	www.regulations.gov

3.2 The Clean Air Act

The law most relevant to outdoor air pollution is the **Clean Air Act** (CAA) 42 U.S.C. §7401 et seq. The CAA is the federal law that regulates air emissions from stationary and mobile sources in order to protect public health and public welfare. The goal of this statute is to maintain outdoor air quality through the regulation of emissions of air pollutants.

The CAA was signed into law by President Richard Nixon (R) in 1970. Additional **Clean Air Act Amendments** (CAAA) followed in 1977 and 1990, primarily to set new goals (dates) for achieving air quality outcomes. The 1990 CAAA are, as of 2014, the most current set. This section explores the basic concepts of how the CAA works in order to develop a feel for the regulatory process. Note that the CAA is not the only environmental statute, and there may be several other applicable federal and/ or state environmental laws pertaining to a project. Finally, a word of caution – laws and regulations can change at any time. The values and methods listed in this text are correct as of the listed dates, but that is no guarantee that everything in the book will still be valid at a future time. Therefore, references are provided for how to find the most current information.

The CAAA contains a number of sections (called titles) addressing outdoor air concerns and national focus priorities. Several are listed in Table 3-2 with reference to additional information in this text. The EPA has many regulatory tools to achieve the CAA goals. These tools include the authority to:

- Set outdoor air quality standards. These standards are concentration based values chosen to ensure basic health and environmental protection from air pollution for all Americans.
- Set emission limits. Limits have been set for certain sources of air pollution such as automobiles, chemical plants, utilities, and steel mills. Emission limits are typically given in terms of emission concentrations, total mass, or mass per unit of production.
- Delegate authority to states and tribes. This authority includes the local oversight, data collection, enforcement and compliance of air quality laws through State Implementation Plans.
- Require construction permits from proposed air pollution sources. These permits allow the agency to determine what, if any, impact a new project will have on the ambient air.
- Allow and encourage citizen participation. The US-EPA actively encourages citizen input into all EPA activities, including citizen lawsuits against polluters or the EPA for failing to protect the environment.
- Set standards for mobile sources. Standards may include emission controls, fuel quality, and efficiency.
- Control air toxins. Standards identify, regulate, and set emission controls for hazardous air pollutants.
- Require operating permits. All regulated sources must provide data on how their facility is achieving emission regulations. These permits are required to be completed after one year of operation and updated every five years thereafter.
- Help regulated sources comply with the law and enforce the law. The purpose of the CAA is to achieve clean air.
- Address interstate and international concerns. Work with regional government groups to address problems from regional and global pollution problems.

Table 3-2. Organization of 1990 Clean Air Act and Amendments.

Title I:	National Ambient Air Quality Standards (see section 3.2.1)
Title II:	Mobile Sources (see chapter 16)
Title III:	Air Toxics (see chapter 13)
Title IV:	Acid Deposition Control (see chapter 9)
Title V:	Permits (see section 3.2.8)
Title VI:	Stratospheric Ozone and Global Climate Protection (see chapter 12)
Title VII:	Enforcement (see section 3.2.9)

3.2.1 Air Quality Standards

The CAA (40 CFR part 50) requires the EPA to set **National Ambient Air Quality Standards** (NAAQS) for pollutants considered harmful to public health and the environment. The standards are set and revised every five years based on the current best scientific data. Congress has stated that these standards are to be health based and are not to be based on economic concerns. The US Supreme Court has repeatedly stated that this is the clear intent of the law. NAAQS have been set for six pollutants as shown in Table 3-3, (US-EPA, 2011b). It lists both primary standards to protect human health – including the health of sensitive populations such as asthmatics, children, and the elderly, and secondary standards to protect welfare and the environment – including protection against decreased visibility, damage to animals, crops, vegetation, and buildings. These six pollutants are called 'criteria' pollutants and are (or were) the most common pollutants in US air.

Table 3-3. National Ambient Air Quality Standards (as of October, 2011).

Pollutant	Primary Standards Level	Averaging Time	Secondary Standards
Carbon Monoxide	9 ppmv (10 mg/m^3)	8-hour	
	35 ppmv (40 mg/m^3)	1-hour	
Lead	0.15 µg/m^3	rolling 3-month	Same as primary
Nitrogen Dioxide	53 ppbv	Annual	Same as primary
	100 ppbv	1-hour	
Particulate Matter (PM$_{10}$)	150 µg/m^3	24-hour	Same as primary
Particulate Matter (PM$_{2.5}$)	12.0 µg/m^3	Annual	15.0 µg/m^3
	35 µg/m^3	24-hour	Same as primary
Ozone	0.075 ppmv (2008 std) [A]	8-hour	Same as primary
	0.08 ppmv (1997 std) [B]	8-hour	Same as primary
	0.12 ppmv (revoked 1997)	1-hour	Same as primary
Sulfur Dioxide	0.075 ppmv	1-hour	0.5 ppmv, 3-hour

A. The EPA is in the process of reconsidering these standards (set in March 2008). EPA is expected to propose a new standard of 0.060 or 0.065 ppmv in 2015 or 2016.
B. The 1997 standard—and the implementation rules for that standard—remain in place for implementation purposes as EPA undertakes rulemaking to address the transition from the 1997 to the 2008 ozone standard.
PM10 refers to particulate matter with an effective diameter of 10 μm or less.
PM2.5 refers to particulate matter with an effective diameter of 2.5 μm or less.

A violation, called *nonattainment*, occurs when the air in an area measures above these standards, with some leeway for statistical considerations of an ordinary level of unforeseen troubles. Areas where the NAAQS have not exceeded these levels are listed as *attainment* for that pollutant. Areas where there is insufficient data to determine attainment or nonattainment are called *unclassifiable*. It is entirely possible for an area to be in attainment for one pollutant, nonattainment for another, and unclassifiable for a different pollutant. Unclassifiable areas are assumed to have air quality at the attainment level, for example see Figure 3-5 in section 3.2.4.2.

Each pollutant has different averaging requirements in how the decision of attainment or nonattainment is made:
- The CO standards may not be exceeded more than once per year; that is nonattainment occurs on the second occurrence in a year.
- The NO_2 standard requires the 3-year average of the 98th percentile of the daily maximum 1-hour averages at each monitor within an area to be less than 100 ppbv.
- The PM_{10} standards may not be exceeded more than once per year on average over 3 years.
- The $PM_{2.5}$ standard requires the 3-year average of weighted annual means from an area's monitors to not exceed the listed level, nor allow the 3-year average of the 98th percentile of 24-hour concentrations at *each* monitor to exceed the listed 24-hour level.
- The ozone standard requires the 3-year average of the fourth-highest daily maximum 8-hour average at each monitor in an area to not exceed the listed 8-hour average.
- The SO_2 3-hour standard (secondary) may not be exceeded more than once per year, nor may the 3-year average of the 99th percentile of the daily maximum 1-hour level at each monitor within an area exceed the listed 1-hour level primary).

■ Example 3-1.
Use the following data sets to determine if the area is in attainment for carbon monoxide and sulfur dioxide.

Solution:
Examine the two data tables and apply the averaging requirements to determine if there are violations of the standards.

Table 3-4. Example 3-1 CO data (2007 – 2010):
Five highest recordings of 1-hour averages for each year, units are ppmv.

2007	2008	2009	2010
28	34	34	36
25	31	34	26
32	38	30	29
26	27	32	36
27	26	31	23

In 2007, no readings were above the 35 ppmv 1-hour standard. In 2008, one reading was above the standard, so the region is still in attainment. In 2009, no readings were above the standard. In 2010 two readings were above the standard, so the region would be nonattainment in 2010.

Table 3-5. Example 3 1 SO2 data (1998 – 2010):
99th percentile of the daily maximum 1-hour level for each year at a monitor, units are ppbv.

1998	1999	2000	2001	2002	2003	2004	2005	2006	2007	2008	2009	2010
68	71	76	72	76	78	80	75	72	69	63	61	62

The three-year running average must be calculated to determine attainment. This average is calculated as:

3.1
$$[SO_2]_{year_i}^{3-year} = \frac{[SO_2]_{year_{i-2}} + [SO_2]_{year_{i-1}} + [SO_2]_{year_i}}{3}$$

This calculation for each year yields:

1998	1999	2000	2001	2002	2003	2004	2005	2006	2007	2008	2009	2010
-	-	71.7	73.0	74.7	75.3	78.0	77.7	75.7	72.0	68.0	64.3	62.0
ND	ND	Att	Att	Att	Non	Non	Non	Non	Att	Att	Att	Att

Where ND means non determinable, Att means attainment, and Non means nonattainment. Note that in 2002, the actual value was above the National Ambient Air Quality Standards (NAAQS) standard, but the area was still considered in attainment because the three-year average was less than the NAAQS value. Similarly, in 2006 the value was below NAAQS but it was still in nonattainment. This averaging method helps for a one-time excursion, but can drag out consequences if the excursions repeatedly happen.

The EPA subdivides the country by counties (or parishes) and/ or standard metropolitan statistical areas in order to assess the air quality in each, often at multiple locations. These areas are sometimes referred to as *air quality control regions*. Monitoring locations are chosen to be as representative as possible. Each criteria pollutant is measured in each region, though not all air monitoring equipment evaluates each pollutant, nor is each pollutant sampled at the same frequency. Locations, test procedures, and sampling procedures and times depend on local conditions and history. The federal regulations provide some details whereas state implementation plans provide other details.

The siting and design of air quality monitoring equipment is primarily defined by 40 CFR Part 58 Appendices D & E. The objectives of the siting criteria are to provide air pollution data to the public, to support forecasts and advisories, ensure compliance with the National Ambient Air

FIGURE 3-2. Air Quality Monitoring Station, Oregon Department of Environmental Quality. (OR-DEQ, 2011)

Quality Standards (NAAQS) and to support research studies. The goal of monitoring is to locate sites that will be representative of an area (e.g. urban, rural, residential, industrial, wet, dry). Recent studies have suggested that nearness to roadways is important, as many people spend considerable time around roads. Thus new monitors have been added to the national network to increase the number of measurements near and adjacent to roads. Figure 3-2 shows an example of an air quality monitoring station.

Where: 1. Meteorological sensors for wind speed and direction (10 m height),
        2. Temperature sensor,
        3. Temperature sensor,
        4. Air intake for nephelometer for measuring visibility,
        5. Sample ports for measuring PM_{10}, $PM_{2.5}$, and other atmosphere constituents.

The 'nonattainment' classification for ozone is further refined into several levels: marginal, moderate, serious, severe, or extreme. The level of classification triggers varying requirements for future compliance. Marginal areas are required to conduct an inventory of their emissions and must institute a permit program. The more serious classifications are required to implement various control measures. The worse the air quality, the more controls an area will need to implement.

Similar nonattainment classifications exist for carbon monoxide and particulate matter, with levels listed as moderate or serious. Areas that exceed the carbon monoxide standard are required to implement programs introducing oxygenated fuels and/or enhanced emission inspection programs, among other measures. Areas exceeding the particulate matter standard will, depending upon their classification, have to implement either **reasonably available control measures** (RACM) or **best available control measures** (BACM), among other requirements.

Once an area has been listed in nonattainment, it must perform certain actions before it can be relisted as attainment (or maintenance). The change is not automatic even if the air quality returns to levels below NAAQS.

3.2.2 Emission Limits

The CAA requires EPA to set emission standards for new, modified, and reconstructed facilities through the **New Source Performance Standards** (NSPS) program. NSPS are technology-based standards that apply to specific categories of stationary sources, as listed in 40 CFR part 60 (US-GPO, 201 1b). The setting of these emission standards is coupled with achieving the goals of NAAQS.

NSPS requires facilities to attain emission levels that "reflect the degree of emissions limitation achievable through the application of the best system of emissions reduction that the US-EPA Administrator determines has been adequately demonstrated." The EPA uses engineering evidence to determine what is achievable in terms of reducing emissions for each regulated source. The EPA can factor in the cost as well as energy, health, and environmental impact of a technology in developing the standards, unlike NAAQS.

In setting this type of performance standard, EPA has some discretion to distinguish among classes, types, and sizes within source categories. However, the limit EPA sets must take the form of a standard and may not prescribe a particular technology itself. Under the law, EPA is required to review the technological options available for emission reductions and, if appropriate, establish a new standard every eight years. In practice, most standards remain in place for longer than eight years.

NSPS are only applicable to facilities constructed or modified after the date listed in the regulation. A facility built before the listed date, and that has not undergone modifications since, is not subject to newer NSPS. There is considerable uncertainty as to the meaning of modifications[1]:

How big must the modification or reconstruction be?
Does it include routine maintenance/replacement of worn out units with upgraded ones?
Does it include changing fuel sources?
Do multiple small modifications act the same as one large modification? or
Are costs based on the entire facility or some sub-part?

The majority of CAA lawsuits between the EPA and industrial facilities are caused by this uncertainty because it is nearly impossible to clearly define for each facility. The best advice to avoid a lawsuit is to ask the state or federal agency about a modification before starting the work.

All industries subject to NSPS must meet certain general requirements, such as monitoring and record keeping. In addition, certain specific requirements apply to each industry. Each NSPS defines the facilities subject to it and prescribes emission limits for specified pollutants, compliance requirements, monitoring requirements, and test methods and procedures. Table 3-6 lists the emission limits for a large (>73MW) electric utility steam generator. Appendix 2 lists the NSPS for additional types of facilities.

Table 3-6. NSPS for Electric Utility Steam Generating Units [40CFR60 subparts D et seq.]:

Started operation after 08/17/1971 and have a heat input >73 MW
PM: 0.10 lb/MMBTU
SO_2: 0.80 lb/MMBTU – liquid fossil fuel, or
1.20 lb/MMBTU – solid fossil fuel

---

1. Specific statutory and regulatory provisions define what constitutes a modification or reconstruction of a facility. 40 CFR 60.14 provides that an existing facility is modified, and therefore subject to an NSPS, if it undergoes "any physical change in the method of operation . . . which increases the amount of any air pollutant emitted by such source or which results in the emission of any air pollutant not previously emitted." 40 CFR 60.15, in turn, provides that a facility is reconstructed if components are replaced at an existing facility to such an extent that the capital cost of the new equipment/components exceed 50 percent of what is believed to be the cost of a completely new facility.

NO$_x$:
    0.20 lb/MMBTU – gas fossil fuel, or
    0.30 lb/MMBTU – liquid fossil fuel, or
    0.70 lb/MMBTU – solid fossil fuel, except
    0.60 lb/MMBTU – lignite, except
    0.80 lb/MMBTU – lignite from ND, SD, MT with cyclone-fired unit.

Started operation after 09/18/1978 and have a heat input >73 MW

PM:
    0.03 lb/MMBTU and
    99% reduction for solid fuel or
    70% reduction for liquid fuel

SO$_2$:
    Solid fuel
    1.2 lb/MMBtu and 90% reduction; or
    0.6 lb/MMBtu and 70% reduction
    Liquid or gaseous fuel
    0.8 lb/MMBtu and 90% reduction; or
    0.2 lb/MMBtu and 0% reduction

NO$_x$:
    Gaseous fuel

• Coal derived:	0.5 lb/MMBtu
• All others:	0.2 lb/MMBtu

and 25% reduction of potential combustion concentration

    Liquid fuel

• Coal Derived	0.5 lb/MMBtu
• Shale Oil	0.5 lb/MMBtu
• All Others	0.3 lb/MMBtu

and 30% reduction of potential combustion concentration

    Solid fuel

• Coal Derived	0.5 lb/MMBtu
• Any fuel > 25% coal refuse	Exempt (No limit)
• Any fuel > 25% lignite[1]	0.6 lb/MMBtu
• Subbituminous coal	0.5 lb/MMBtu
• Bituminous coal	0.6 lb/MMBtu
• Anthracite coal	0.6 lb/MMBtu
• All Others	0.6 lb/MMBtu

and 65% reduction of potential combustion concentration

---

1. Except lignite from ND, SD, MT used in slag tap furnace, which has a limit of 0.8 lb/MMBtu

Started between 07/09/1997 – 02/28/2005, and have a heat input >73 MW
- NO_x: *Construction*
  1.6 lb/MW-hr gross energy output
  Reconstruction
  0.15 lb/MMBtu heat input

Started after 02/28/2005 and have a heat input >73 MW
- PM: 0.14 lb/MW-hr gross energy output; or
  0.015 lb/MMBtu heat input; or
  0.03 lb/MMBtu heat input, and
  99.9% reduction for a construction or reconstruction, or
  99.8% reduction for a modification
- SO_2: *Construction*
  1.4 lb/MW-hr gross energy output or 95% reduction
  Reconstruction
  1.4 lb/MW-hr gross energy output; or
  0.15 lb/MMBtu heat input; or 95% reduction
  Modification
  1.4 lb/MW-hr gross energy output; or
  0.15 lb/MMBtu heat input; or 90% reduction
- NO_x: *Construction*
  1.0 lb/MW-hr gross energy output
  Reconstruction
  1.0 lb/MW-hr gross energy output; or
  0.11 lb/MMBtu heat input
  Modification
  1.4 lb/MW-hr gross energy output; or
  0.15 lb/MMBtu heat input
- Hg: (Starting on or after 01/30/2004)

Bituminous fuel:	0.020 lb/GWh output
Subbituminous fuel and > 25 inch rain/year	0.066 lb/GWh
Subbituminous fuel and ≤ 25 inch rain/year	0.097 lb/GWh
Lignite fuel:	0.175 lb/GWh
Coal Refuse fuel:	0.016 lb/GWh
IGCC unit:	0.020 lb/GWh

The limits described in NSPS use a basis dependent on the concept of potential emissions. Potential emissions are the amount the facility could potentially produce by operating 24-hours, seven days/week, even if the facility runs only a fraction of this time (8-hour days and 5 day/week).

Pollution control efficiencies or removal efficiencies (η) quantify the reduction needed to comply with emission limits. The following equation shows the calculation:

$$3.2 \qquad \eta = \frac{Emission_{uncontrolled} - Emission_{limit}}{Emission_{uncontrolled}} * 100\%$$

Emission calculations vary depending on what information is known about the input to the facility and the basis units for the limit. Example 3-2 shows the calculation of emissions from a coal-fired power plant. Note that gross energy output means the gross amount of useful work, which in this case is the actual output (plant size 500 MW) divided by the efficiency percentage. This basis encourages high efficiencies.

■ Example 3-2.

Determine the (PM, SO_2, and NO_x) pollution control efficiencies needed for a new 500 MW electric utility steam generating unit (33 % efficient) which uses a sub-bituminous coal with 6 wt% fly ash, 2.1 wt% sulfur and heating value of 10,000 BTU/ lb. Assume uncontrolled NO_x emissions at 7.2 lb/ton of coal.

Solution:

First, determine the NSPS for this facility. Then, determine the amount of uncontrolled emissions for each pollutant on the appropriate basis and finally, calculate the required removal efficiencies. Calculate the uncontrolled emission of PM from assuming that all fly ash is emitted as PM. Calculate the uncontrolled emission of sulfur by assuming that all the sulfur in the coal is emitted as SO_2. Calculate the uncontrolled emission of nitrogen from the given emission factor.

This is a new facility, so the controlled emission limits from Table 3-6 are: PM = I) 0.14 lb/MWh gross energy output; or II) 0.015 lb/MMBtu heat input; or III) 0.03 lb/MMBtu heat input and 99.9% reduction. Since we have a choice, calculations for each basis are required (lb/hr per MW gross energy output and lb/MMBTU heat input). Similarly, for SO_2 the limit from Table 3-6 is 1.4 lb/MWh gross energy output or 95% reduction. The NO_x limit is 1.0 lb/MWh gross energy output.

PM potential emissions for case I using 33% efficiency between energy in and energy out:

$$E^I_{PM} = 3.413 \times 10^6 \frac{BTU}{MW-hr} * \frac{1_{in}}{0.33_{out}} * \frac{1 \text{ lb coal}}{10,000 BTU_{in}} * \frac{0.06 \text{ lb fly ash}}{1 \text{ lb coal}} = 62.05 \frac{\text{lb fly ash}}{MW_{out}-hr}$$

and for case II and III efficiency is not used, the calculation is based on energy input:

$$E^{II}_{PM} = \frac{1 \text{ lb coal}}{10,000 BTU} * \frac{10^6 BTU}{MMBTU} * \frac{0.06 \text{ lb fly ash}}{1 \text{ lb coal}} = \frac{6 \text{ lb fly ash}}{MMBTU}$$

So, the required control efficiencies for each case are:

$$\eta^I_{PM} = \frac{(62.05 - 0.14)}{62.05} * 100\% = 99.77\%$$

$$\eta^{II}_{PM} = \frac{(6.000 - 0.015)}{6.000} * 100\% = 99.75\%$$

$$\eta^{III}_{PM} = \frac{(6.00 - 0.03)}{6.00} * 100\% = 99.50\%$$

For case III, the required control level does not meet the additional requirement of 99.9% removal. Option III is better only if the amount of PM generated > 30 lb/MMBTU.

Case II is the easiest option to meet, though there is not much difference between case I and II. Actual design would use a greater removal so that a system upset would not cause the facility to exceed their emission limits.

SO_2 potential emissions:

$$E_{SO_2} = 3.413 \times 10^6 \frac{BTU}{MW-hr} * \frac{1_{in}}{0.33_{out}} * \frac{1 \text{ lb coal}}{10,000 BTU_{in}} * \frac{0.021 \text{ lb S}}{1 \text{ lb coal}} * \frac{64 \text{ lb SO}_2}{32 \text{ lb S}} = 43.44 \frac{\text{lb SO}_2}{MW_{out}-hr}$$

The removal efficiency for the limit value is

$$\eta_{SO_2} = \frac{(43.44 - 1.4)}{43.44} * 100\% = 96.8\%$$

This is greater than the other requirement of 95%, so the actual permissible emission limit would be

$$\text{Actual Emission Limit}_{SO_2} = 43.44 \frac{\text{lb SO}_2}{MW-hr} * (1 - 0.95) = 2.17 \frac{\text{lb SO}_2}{MW-hr}$$

NO_x potential emissions:

$$E_{NO_x} = 3.413 \times 10^6 \frac{BTU}{MW-hr} * \frac{1_{in}}{0.33_{out}} * \frac{1 \text{ lb coal}}{10,000 BTU_{in}} * \frac{7.2 \text{ lb NO}_x}{1 \text{ ton coal}} * \frac{1 \text{ ton coal}}{2000 \text{ lb coal}} = 3.72 \frac{\text{lb NO}_x}{MW_{out}-hr}$$

The removal efficiency for the limit value is

$$\eta_{NO_x} = \frac{(3.72-1.0)}{3.72} *100\% = 73.1\%$$

There are no choices on this limit, so this is the required NO$_x$ removal efficiency.

Result summary for required removal efficiencies:
  PM = 99.75%
  SO2 = 95%
  NOx = 73.1%

3.2.3 Delegation to States and Tribes

The CAA requires each state to develop a **State Implementation Plan** (SIP) (US-EPA, 2010a). Tribal areas may develop similar documents called **Tribal Implementation Plans** (TIP) (US-EPA, 2007). Congress recognized, in the 1990 CAAA, that Indian tribes have the authority to implement air pollution control programs. EPA's Tribal Authority Rule gives tribes the ability to develop air quality management programs, write rules to reduce air pollution and implement and enforce their rules in Indian Country. While state and local agencies are responsible for all Clean Air Act requirements, tribes may develop and implement only those parts of the Clean Air Act that are appropriate for their lands.

A SIP is a collection of the regulations, programs, and policies that a state or tribal area uses to clean up polluted areas. It must include a general plan to attain and maintain NAAQS and a specific plan to attain NAAQS for each area designated nonattainment. SIPs and TIPs must be developed with public input, be formally adopted by the state or tribe, and be submitted to the EPA by the Tribal or State Governor's designee. After reviewing submitted SIPs, EPA proposes to approve or disapprove all or part of each plan. The public then has an opportunity to comment on EPA's proposed action. EPA considers public input before taking final action on the state's plan. Once EPA approves all or part of the SIP or TIP, those control measures are enforceable in federal court. If a state fails to submit an approvable plan or if EPA disapproves a plan, EPA is required to develop a **Federal Implementation Plan** (FIP). For ease in notation, all further references of state, federal, and tribal plans will be listed as the SIP.

These plans must be at least as stringent as the federal laws and regulations, and the federal government is required to meet the approved state and local laws and regulations contained in the SIP. The SIP can require federally owned land, buildings, and activities to comply with state requirements. The SIP has two main purposes;

1. Demonstrate that the state has the basic air quality management program components in place to implement a new or revised NAAQS.
2. Identify the emissions control requirements the state will rely upon to attain and/or maintain the primary and secondary NAAQS.

When NAAQS change, the states must submit revised SIPs to demonstrate attainment and maintenance of the new or revised standards. EPA has two years from promulgation of the new standards to identify or 'designate' areas as attainment or nonattainment based on the most recent air monitoring data. States have three years to submit revised SIPs and some additional time if they have nonattainment areas. The SIP allows states and tribes to have local control of CAA activities, since they are better able to develop appropriate solutions for pollution problems that require special understanding of local industries, geography, housing, and travel patterns, as well as other factors.

3.2.4 New Source Review Permits (Construction Permits)

New Source Review (NSR) requires all new stationary sources and all existing stationary sources that are adding new equipment or modifying existing equipment that could potentially increase emissions of air pollution to obtain an NSR air permit before they start construction, reconstruction, or major modifications. These are also called construction or pre-construction permits. There are three types of NSR permit requirements, and sources may need to meet more than one of them:
1. **Prevention of Significant Deterioration** (PSD) permit is required for sources in an attainment or unclassifiable area,
2. Nonattainment NSR permit is required for sources in a nonattainment area, and
3. Minor source permits.

Either the state air pollution control agencies or the EPA issue NSR permits. These permits are legal documents that the source must follow. The permit specifies what construction is allowed, what emission limits must be met, and how the source must be operated. They may also contain conditions to make sure that the source is built to match parameters in the application that the permitting agency relied on in their analysis, such as stack height and diameter. Permits also contain monitoring, recordkeeping, and reporting requirements.

A New Source Review permit issued by EPA takes between six months and one year. The time frame for NSR permits issued by state and local air pollution control agencies varies, as specified in local regulations. For example, in California, state law requires agencies must issue NSR permits within 180 days. New Source Review permits can expire if not used within a certain amount of time. The expiration time varies according to local regulations. If the EPA issues a permit, sources must commence construction within 18 months of permit issuance. If construction does not commence within 18 months, the permit expires. Once construction begins, and is completed in a reasonable time, the permit lasts indefinitely and does not have to be renewed. If a state or local control agency issues a permit, the source must commence construction within the time frame established by the permitting authority's regulations, typically 12 to 24 months. As with EPA-issued permits, once construction begins, the permit lasts indefinitely and does not have to be renewed. Note that in some state and local regulations, NSR permits expire when an Operating Permit is issued.

One important consideration in the development and approval of an NSR permit is that the proposed facility cannot sample their emissions in order to determine the amount of a pollutant the facility will produce because it does not yet exist. The EPA has prepared a list of standard emission factors known as AP-42 (see appendix 3) for general use (US-EPA, 2011c). The permit must include a design that is stringent enough and clearly described to demonstrate that the proposed control equipment will meet the required standards.

3.2.4.1 Prevention of Significant Deterioration (PSD)

The CAA of 1970 was unclear as to what was permissible in terms of air quality changes in attainment and unclassifiable areas. Some interpretations suggested that anything that did not cause exceedance of NAAQS was acceptable and other interpretations said that no negative changes were allowed. Based on judicial and administrative actions from the 1970 CAA, the 1977 CAAA developed the Prevention of Significant Deterioration (PSD) regulations to address how states were to handle this issue.

PSD applies to new major stationary sources[2], major modifications[3], or reconstruction of existing sources for pollutants in NAAQS attainment or unclassifiable areas. It allows sources (new or existing) to increase emissions in these areas, but also to preserve good air quality. PSD allows a region to:
- protect public health and welfare;
- preserve, protect and enhance air quality in specially protected areas;
- insure economic growth occurs in a manner consistent with the preservation of existing clean air resources; and
- allow decisions to include careful evaluation of possible consequences to all stakeholders.

---

2. Major Stationary Sources: In PSD areas the cutoff level may be either 100 or 250 tons, depending upon the source. In a nonattainment area, any stationary pollutant source with potential to emit more than 100 tons per year is considered a major stationary source.
3. Modification -a project is a major modification for a regulated NSR pollutant if it causes two types of emissions increases—a significant emissions increase or a significant net emissions increase.

The EPA and states established an area classification scheme to cover all regions within their jurisdiction for PSD permit and control requirements. PSD defines three classes of air. **Class I** areas include international parks, national wilderness areas, memorial parks exceeding 5000 acres, and national parks exceeding 6000 acres. Although industrial projects rarely locate within them, their Class I status can affect projects in neighboring areas where meteorological conditions might result in the transport of emissions into them. **Class II** areas allow limited amounts of new emissions to encourage moderate industrial growth. **Class III** areas allow a greater amount of new emissions for major industrialization. Initially, all areas were categorized as Class II and states were authorized to reclassify specified areas as Class I or Class III. Reclassification to class III requires a public referendum. While many Class I areas have been designated, there are no class III areas as of 2014. Some class I areas have additional protection called a mandatory class I area, which means they may not be re-designated to a less protective classification.

PSD permits require 1) installation of Best Available Control Technology 2) an air quality analysis 3) an additional impacts analysis, and 4) public involvement.

Best Available Control Technology (BACT) is an emission limitation based on the maximum degree of achievable control. It considers the energy, environmental, and economic impact on the source. BACT can be add-on control equipment or modification of the production processes or methods. The changes include fuel cleaning or treatment and innovative fuel combustion techniques. BACT may be a design, equipment, work practice, or operational standard if imposition of an emission standard is infeasible. The EPA maintains an informational database on BACT in air permits [RACT/BACT/LAER Clearinghouse (US-EPA, 2011d)].

Air Quality Analysis (AQA) is used to demonstrate that new emissions will not cause or contribute to a violation of any NAAQS or PSD increment. The analysis involves an assessment of current air quality from existing or additional ambient monitoring data and predictions from dispersion modeling (see chapter 6) to assess the impact of the proposed source's emissions on air quality.

The **additional impacts analysis** assesses the impacts of air, ground, and water pollution on soils, vegetation, and visibility caused by any increase in emissions of any regulated pollutant from the source or modification under review, and from associated growth. Associated growth is industrial, commercial, and residential growth that will occur in the area due to the source, including increased auto, truck, and railroad traffic.

Public involvement with an NSR permit occurs after the permitting agency has received a complete permit application, and a draft permit has been created. The controlling agency will then publish a notice to solicit public comments (typically for 30 days, although longer comment periods may be used) and the deadline to request a public hearing on the draft permit. The notice can be published in the local newspaper or a state publication as well as the agency's website. Copies of the draft permit are typically made available at the local public library. The agency may then decide to revise the draft permit based on the comments received, and they may publish a notice and seek additional comments on the revised permit. The agency will issue a permit after the review process is complete. However, the permit may still be appealed. Each state has an established appeal process.

PSD allows new sources to increase emissions in an area, but it restricts the amount of increase to an increment above that area's baseline level. The EPA or state agency responsible for air quality determines the baseline levels. The agency uses the air quality measured on a given date as published in the Code of Federal Regulations. The values are different for each air quality area and each pollutant. The new source may increase the concentration of a pollutant by the amount listed in Table 3-7 or up to the NAAQS level, whichever is smaller.

Table 3-7. PSD Increments as of July 2013 (adapted from 40 CFR Part 52.21).

Pollutant	Class I area	Maximum Allowable Increase µg/m3	
		Class II area	Class III area
$PM_{2.5}$			
Annual arithmetic mean	1	4	8
24-hr maximum	2	9	18
PM_{10}			
Annual arithmetic mean	4	17	34
24-hr maximum	8	30	60
Sulfur dioxide			
Annual arithmetic mean	2	20	40
24-hr maximum	5	91	182
3-hr maximum	25	512	700
Nitrogen dioxide			
Annual arithmetic mean	2.5	25	50

For any period other than an annual period, the applicable maximum allowable increase may be exceeded during one such period per year at any one location.

As of January 1, 2011, emissions of condensable matter (primarily SO_x and NO_x) must be included in the determination of PM, PM_{10}, and $PM_{2.5}$ emissions. Emission limitations issued earlier do not need to include condensable particulate matter, this is known as a grandfather clause. A typical equivalency is 40 µg/m³ SO_2 or NO_2 = 10 µg/m³ $PM_{2.5}$

FIGURE 3-3. Application of PSD Increment. The baseline plus increment for area 1 source remains below NAAQS so the entire increment may be used. The baseline plus increment for area 2 source would exceed NAAQS, so only part of the increment is available [adapted from (Wark, et al., 1998)]

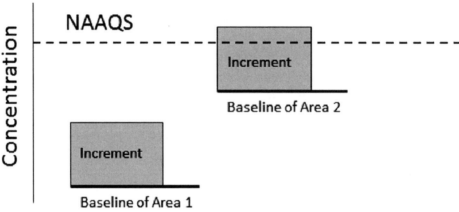

■ Example 3-3

A new source would like to obtain a permit to operate in a class II area. It has the potential to emit $PM_{2.5}$, PM_{10}, SO_2, and NO_x. The baseline for each is $PM_{2.5} = 12$ µg/m³, baseline $PM_{10} = 45$ µg/m³, baseline $SO_2 = 35$ µg/m³, and baseline $NO_x = 90$ µg/m³. What is the maximum annual average impact the facility may have on the existing air quality for each constituent?

Solution:
Consult Table 3-7 for the appropriate increments. Check that the increment plus baseline is less than the NAAQS values in Table 3-3, as shown in Figure 3-3. Note the need to convert to the proper units so that every term in the table below is in units of [µg/m³].

Pollutant	Area Baseline	PSD	Baseline + Increment	NAAQS	Maximum Increment Allowed
$PM_{2.5}$	12	4	16	15	3
PM_{10}	45	17	62	-	17
SO_2	35	20	55	78	20
NOx	90	25	115	100	10

Note that there is no annual average PM_{10} NAAQS, so the maximum allowed increment is always the full increment.

3.2.4.2 Nonattainment NSR

Nonattainment NSR applies to new major sources or major modifications at existing sources for pollutants where the area is in nonattainment with NAAQS. Nonattainment NSR requirements are customized for the nonattainment area. Nonattainment NSR programs have to require (1) the installation of the lowest achievable emission rate, (2) emission offsets, and (3) opportunity for public involvement. Figure 3-4 shows a map of the US counties coded for the number of pollutants in nonattainment (US-EPA, 2014). The US-EPA constantly updates the maps, as areas grow and change.

FIGURE 3-4. Map of US, July 2014, showing regions in non-attainment (air pollution exceeds levels defined in NAAQS) or in maintenance (exceeded NAAQS in the past, but may not currently be in exceedance).

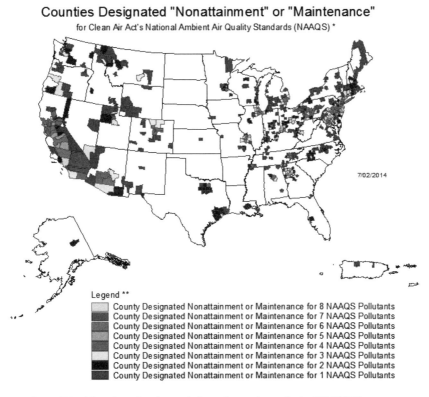

The Lowest Achievable Emission Rate (LAER) is the most stringent emission limitation derived from either of the following:
- the most stringent emission limitation contained in the implementation plan of any State for such class or category of source; or
- the most stringent emission limitation achieved in practice by such class or category of source.

LAER may result from a combination of emissions-limiting measures such as (1) a change in the raw material processed, (2) a process modification, and (3) add-on controls. The EPA maintains an informational database on LAER in air permits [RACT/BACT/LAER Clearinghouse (US-EPA, 201 1d)].

Reasonably Available Control Technology (RACT) is required on existing sources in areas that are not meeting national ambient air quality standards (i.e., non-attainment areas). BACT is required on major new or modified sources in clean areas (i.e., attainment areas).

Emission Offsets are emission reductions obtained from other existing sources located in the vicinity of the proposed source which must (1) offset the emissions increase from the new source or modification, and (2) provide a net air quality benefit. The obvious purpose for requiring offsetting emissions is to allow the area to move towards attainment of the NAAQS while still allowing some industrial growth.

Where an area is in ozone nonattainment, the volatile organic compounds and nitrogen oxides offsetting ratio must be greater than 1-to-1, depending on the severity classification. These offsets are a means of controlling (abating) existing pollution levels by requiring a reduction within the area. Offsets may include emissions trading within the same nonattainment area. A new or modified facility acquires offset credits by reducing emissions from other existing facilities in the area. It has been estimated that offsets increase the cost of a project by 1 – 6% (US-EPA, 2002)

■ Example 3-4.

A cement kiln would like to expand operations in an area that is in non-attainment for particulate matter. The kiln would emit 500 ton PM_{10}/yr, which modeling shows would cause a maximum ambient air increase of 8 $\mu g/m^3$. Suggest ways this facility could operate within the non-attainment area.

Solution:

Facility must use a PM removing technology that is certified to have the lowest achievable emission rate (LAER). The EPA database mentioned above lists such technologies. Also, the facility must find a way to offset their emissions such that the project creates a net air benefit. Reductions are achieved by lowering their own emissions from other sources such that they reduce the ambient air PM levels by at least 8 $\mu g/m^3$. Possible methods would be to alter the operations of their other kilns by installing or improving the PM collection rates, changing their vehicle fleet to less polluting sources - using electric powered equipment rather than diesel-powered, or changing the maintenance at their facility to reduce road dust. The change does not have to be from an identical source, nor from only a single source.

3.2.4.3 Minor NSR

Minor NSR is for pollutants from stationary sources that do not require PSD or nonattainment NSR permits. A minor source is one whose emissions are less than the major source threshold of a pollutant:
- less than 100 tpy (tons per year) any one pollutant or 250 tpy (tons per year) any combination of pollutants in attainment areas, or
- less than 100 tpy any combination of pollutants in non-attainment areas.

The purpose of a minor NSR is to make the emission source known to the air quality agency so as to prevent the construction of sources that would interfere with attainment or maintenance of NAAQS or violate the control strategy in nonattainment areas. Also, minor NSR permits often contain permit conditions to limit the sources emissions to avoid PSD or nonattainment NSR. States are able to customize the requirements of the minor NSR program as long as their program meets minimum requirements. The permit agency's minor NSR program is part of the State Implementation Plan (SIP).

3.2.5 Citizen Participation

The CAA allows and actively encourages citizen participation. Throughout the Act, different provisions give the public opportunities to take part in determining how the law is carried out:
- Review reports and permits issued by EPA or state air quality agencies,
- Access air monitoring information,
- Access information on how much pollution is released by source, area, or category,
- Comment on proposed rules,
- Comment on proposed permits,
- Participate in and comment on development of SIP,
- Have EPA take action against a polluter, and
- Take legal action to enforce the CAA requirements against a source, state, or the EPA.

EPA maintains websites that allow the public to access information on air emissions and monitoring data at *www.epa.gov/airtrends* and *www.airnow.gov*. An excellent citizen participation how to manual is provided by the EPA. (NY-PIRG, 2000)

3.2.6 Mobil Sources and Clean Fuels

The air-mobile sources program (US-EPA, 2009) includes emissions standards for all motor vehicles and mobile engines, and the fuels that are used in vehicles and engines. The mobile source standards apply to vehicles and engines of all sizes, ranging from engines that power large marine vessels and locomotives, to engines used in hand-held lawn and garden equipment. The fuels standards apply to all gasoline and diesel fuel used nationwide, including fuel that is produced at domestic refineries and fuel that is imported.

Compliance with the vehicle and engine emissions standards is the responsibility of the vehicle or engine manufacturer as detailed in the regulation 40 CFR 85 et seq. Vehicles and engines used in the United States must be manufactured under the terms of an emissions certificate of conformity issued by EPA. Imported vehicles and engines must be EPA-certified, with certain very limited exceptions. The removal or disabling of vehicle or engine emission controls by any person is a violation of federal laws.

Vehicle fuel economy, or mileage, is regulated through the **Corporate Average Fuel Economy** (CAFE) standards as enforced by the **National Highway Traffic Safety Administration** (NHTSA). CAFE is a sales-weighted average fuel economy of a manufacturer's fleet of passenger cars or light trucks. The standard is in terms of miles per gallon (mpg). The EPA sets the testing protocol. The current standard (2012) for passenger vehicles is 27.5 mpg. The new standard of 54.5 mpg is scheduled for model year 2025, with levels increasing beginning in 2017.

Compliance with the motor vehicle fuels standards is primarily the responsibility of the refiners and fuel importers, 40 CFR 79-80. In addition, parties in the fuel distribution system are responsible for ensuring that motor vehicle fuel is not contaminated and is used in the proper locations and times. For example, more stringent gasoline standards apply during the summer high-ozone season requiring reformulated gasoline for use in certain ozone nonattainment areas. Examples of fuel standards include the control of sulfur content, control of oxygen content (more in winter, less in summer), metal content (lead, nickel), and volatility (higher in winter, lower in summer). Refiners and importers are required to test all their gasoline and submit reports to the EPA. In addition, refiners and importers are required to use independent laboratories to conduct quality assurance testing of reformulated gasoline when produced, quality surveys of reformulated gasoline when sold at retail outlets, and to submit reports to the EPA of these tests and surveys.

3.2.7 Air Toxics and Hazardous Air Pollutants (Title III)

The EPA regulates other substances in addition to those listed in NAAQS. Prior to 1990, the CAA established a risk-based program under which only a few standards were developed. The 1990 CAAA established a new and complex program to regulate emissions of 188 **hazardous air pollutants** (HAPs) from particular industrial sources, see Chapter 13 for more details. The Act requires the EPA to regulate emissions of these HAPs by developing and promulgating technology-based standards based on the best-performing similar facilities in operation.

The resulting **National Emission Standards for Hazardous Air Pollutants** (NESHAPs) established by the U.S. Environmental Protection Agency (EPA) are commonly called **Maximum Achievable Control Technology** (MACT) standards. MACT standards are designed to reduce HAP emissions by a maximum achievable amount, taking into consideration the cost of reductions and other factors. EPA is required to review these standards every eight years to determine whether any residual risk exists for that source category and, if necessary, revise the standards to address such risk.

When developing a MACT standard for a particular source category, the EPA looks at the current level of emissions achieved by the best-performing similar sources through clean processes, control devices, work practices, or other methods. These emission levels set a baseline, often referred to as the "MACT floor" for the new standard. At a minimum, an MACT standard must achieve, throughout the industry, the level of emission control that is at least equivalent to the MACT floor. The EPA can establish a more stringent standard when it makes economic, environmental, and public health sense to do so. Wherever feasible, the EPA writes the final MACT standard as an emission limit - a percent reduction in emissions or a concentration limit that regulated sources must achieve. Emission limits provide flexibility for industries to determine the most effective ways to comply with the standards.

The MACT floor differs for existing sources and new sources:
- For existing sources, the MACT floor must equal the average current emissions limitations achieved by the best-performing 12 percent of sources in the source category if there are 30 or more existing sources. If there are fewer than 30 existing sources, the MACT floor must equal the average current emissions limitation achieved by the best-performing five sources in the category.
- For new sources, the MACT floor must equal the current level of emission controls achieved by the best-controlled similar source.

Sources subject to MACT standards are classified as either major sources or area sources. Major sources are sources that emit 10 tons per year of any of the listed HAPs, or 25 tons per year of a mixture of HAPs. The sources may include conventional emissions during discharge through emission stacks or vents, or accidental releases from equipment leaks and when materials are transferred from one location to another. Area sources consist of smaller-size facilities that emit less than 10 tons per year of a single HAP, or less than 25 tons per year of a combination of HAPs. Though emissions from individual area sources are often relatively small, collectively their emissions can be of concern, particularly where large numbers of sources are located in heavily populated areas.

● **Case Study 3-2. Urban Hazardous Air Pollutants (HAP) Regulations**
The CAAA required that, by November 15, 2000, the EPA promulgate emission standards to assure that area sources accounting for 90 percent of the aggregate area source emissions of the 30 most significant urban HAPs are subject to regulation. Similarly, the CAAA required the EPA to, by November 15, 2000, assure that sources accounting for not less than 90 percent of the aggregate emissions of alkylated lead compounds, polycyclic organic matter (POM), mercury, hexachlorobenzene, polychlorinated biphenyls (PCB), 2,3,7,8-tetrachlorodibenzofurans (TCDF) and 2,3,7,8-tetrachlorodibenzo-p-dioxin (TCDD) are subject to emission standards. As a result of lawsuits filed by Sierra Club alleging that EPA has failed to complete these actions by the statutory deadline, EPA was placed under a court order to complete these obligations. Under the order, which was most recently amended on January 20, 2011, EPA was to complete these obligations by February 21, 2011. On March 21, 2011 EPA published a notice that it had completed the emission standards required by the CAAA.

3.2.8 Operating Permits (Title V)

There are two types of air pollution permits: construction permits and operating permits. Section 3.2.4 discusses construction permits. Operating permits are submitted one year after operations begin using actual plant operation data. They are legally binding documents required by Title V of the CAAA for certain sources of air pollution. They establish limits on the amounts and types of air pollution emissions from a facility, operating requirements for pollution control devices or pollution prevention activities, monitoring and sampling methodology, record keeping and reporting requirements. Some permit conditions are general to all types of emissions, and some are specific to the source. State permitting authorities issue most Title V operating permits (40 CFR part 70 permits), however the EPA may also issue permits in special cases (40 CFR part 71). The Title V operating permit requirements are primarily procedural and are not intended by Congress to create new substantive requirements. There have been approximately 40,000 Title V permits in the US, and every year there are about 100 new sources required to obtain initial permits and about 3,000 sources required to obtain renewal permits.

Table 3-8 lists the sources required to have operating permits. The amounts listed are potential amounts, which means that the source would emit this much if it operated full time at full design capacity. The threshold values in the list below may be reduced in nonattainment areas.

Table 3-8. Sources Requiring a Title V Operating Permit.
- Major stationary sources:
    - Any source that emits 100 ton/yr of any air pollutant
    - Any source that emits 10 ton/yr or more of any one HAP
    - Any source that emits 25 ton/yr or more of any combination of HAPs
- Affected Sources under the acid rain provisions, regardless of size,
- Any source subject to NSPS or PSD regulations (see section 3.2.4)
- Solid waste incineration units, including municipal waste, hospital/medical/infectious waste, commercial and industrial solid waste, and sewage sludge.
- Any source in an EPA designated source category (none as of January, 2013)
- Non-major sources subject to NESHAP standards
    - Hazardous waste combustors,
    - Portland cement manufacturers,
    - Mercury cell chloralkali plants,
    - Secondary lead smelters,
    - Carbon black production,
    - Chromium compounds chemical manufacturing,
    - Primary copper smelting,
    - Secondary copper smelting,
    - Zinc, cadmium, & beryllium (nonferrous metals) area sources,
    - Glass manufacturing,
    - Electric arc furnace (EAF) steelmaking facilities,
    - Gold mine ore processing and production.

The amount of time needed to obtain a permit varies for many reasons – source complexity, the permitting authority, how controversial the project is, and if the permit is appealed. Also, the EPA may delay issuance of the permit on its own initiative or as the result of a citizen request. Therefore, the time needed depends on agency specific time frames and circumstances that are often beyond the applicant's control.

Title V permits must be renewed every five (5) years. The permit also needs to be revised if the facility undergoes major changes – construction, reconstruction, or major modification. These types of changes would also require a new construction permit under NSR. The permit must be revised if the law changes or additional requirements under the Clean Air Act become applicable. If three or more years remain prior to the expiration of the Title V permit, the permit must be reopened and revised. If two years or less remain, the changes can be made when the permit is renewed, at the end of its five-year term.

Operating permits are especially useful for businesses covered by more than one part of the Clean Air Act and additional state or local requirements since information about the source's air pollution is in one place. The permit program simplifies and clarifies businesses' obligations for cleaning up air pollution and can reduce paperwork. For instance, an electric power plant may be covered by the acid rain, toxic air pollutant, and smog (ground-level ozone) sections of the Clean Air Act. The detailed information required by those separate sections is consolidated into one place in an operating permit. Thousands of operating permits that have been issued across the United States are available to the public on the EPA or state websites (US-EPA, 2010b).

3.2.9 Compliance and Enforcement

Compliance and Enforcement is an integral part of environmental protection. Compliance with the nation's environmental laws is the ultimate objective, but enforcement is a vital part of encouraging governments, businesses, and other companies who are regulated to meet their environmental obligations. Enforcement and compliance actions are organized around environmental problems and broad patterns of non-compliance rather than provisions of single statutes. Figure 3-5 shows various trends in US growth and compares them with the change in emissions of the criteria pollutants.

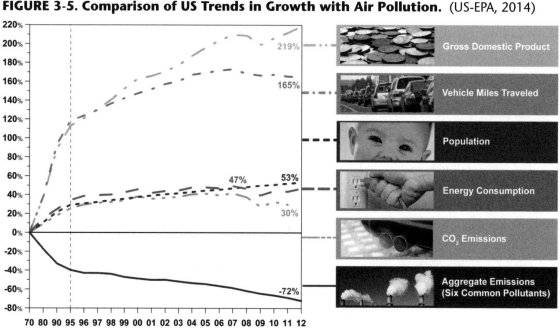

FIGURE 3-5. Comparison of US Trends in Growth with Air Pollution. (US-EPA, 2014)

Compliance means that a facility or source meets its requirements with environmental laws and regulations. The EPA has several tools to help, including compliance assistance, monitoring, and incentives. Assistance includes activities and tools that increase understanding of environmental obligations such as, one-to-one counseling, online resource centers, fact sheets, clearinghouse information, and case-studies highlighting cost effective ways to comply. Monitoring involves reviewing information submitted by governments, industry, and businesses, consolidating this information and presenting it to the public, as well as providing on-site inspections by qualified personnel to verify data collection and submission. Incentives are a set of policies and programs that eliminate, reduce, or waive penalties under certain conditions for business, industry, and government facilities that voluntarily discover, promptly disclose and correct noncompliance, and prevent future environmental violations. The incentives help prevent honest mistakes and accidents from causing larger problems.

Enforcement deters those who might otherwise profit from violating the law. There are three types of enforcement: civil, cleanup, and criminal. Civil enforcement promotes the cleanup of the most widespread types of violations that pose the most substantive health and environmental risks. It is used to return violators to compliance, deter misconduct in others, and to create a level playing field for responsible companies that obey the law. Cleanup enforcement includes the remediation and cleanup of abandoned waste sites, private facilities, and state and federal facilities. Intentional or deliberate violation requires criminal enforcement.

A source that violates one or more enforceable permit condition(s) is subject to enforcement action including, but not limited to, penalties and corrective action. The local permitting authority, EPA, or in many cases, citizens can initiate enforcement actions.

The CAAA gives the EPA enforcement powers. Prior to the 1990 CAAA, it was difficult for the EPA to penalize a company for violating the Clean Air Act because it had to go to court for even minor violations. The 1990 Amendments strengthened the power to enforce the Act by increasing the range of civil and criminal sanctions available. In general, when a violation has occurred, the agency can do the following: issue an order requiring the violator to comply, issue an administrative penalty order (use EPA administrative authority to force payment of a penalty), or bring a civil judicial action (sue the violator in court).

Public access to air quality monitoring data is available from the EPA at their air trends website, *www.epa.gov/airtrends*. This site allows access to EPA, state, local, or tribal monitoring reports. The data is archived so changes over time can be studied.

3.2.10 Interstate Air Pollution

States and tribes seeking to clean up air pollution are sometimes unable to meet the EPA's national standards because of pollution blowing in from other areas. The CAAA has a number of programs designed to reduce long-range transport of pollution from one area to another. The Act has provisions designed to ensure that emissions from one state are not contributing to public health problems in downwind states. Each State Implementation Plan (SIP) is required to contain provisions preventing emissions from the facilities or sources within its borders from contributing significantly to air pollution problems "downwind" – specifically in those areas that fail to meet

NAAQS. If a state or tribe has not developed the necessary plan to address this down-wind pollution, the EPA can require the state to do so. States and tribes can also petition the EPA to require the upwind areas to reduce air pollution.

The EPA has created several interstate projects for addressing regional concerns. Two examples include the **Ozone Transport Commission** (OTC) and haze in the national parks and wilderness areas. The OTC includes state and tribal governments from Maine to Virginia, the government of the District of Columbia, and the EPA working to reduce ground-level ozone along the east coast, (see chapter 12 for program details). Haze (reduction in visibility caused by air pollution) affects many parts of the country, and it can affect locations hundreds of miles from its source. The CAA requires the EPA to work with the National Park Service, U.S. Fish and Wildlife Service, the U.S. Forest Service, the states and others to develop and implement air quality protection plans. These plans are used by the EPA and states to reduce the regional haze that affects visibility in 156 affected national parks and wilderness areas.

3.3 ISO14000

The **International Standards Organization** (ISO) has established a set of managerial procedures voluntarily created by companies for the continuous minimization of pollution emissions. These standards, called ISO 14000 and 14001, are designed to assist companies achieve a standard level of environmental performance. Any company that wishes to participate in the global economy needs ISO14000 registration. Many governments insist on these standards before they approve international trade agreements. These standards include all aspects of the company's impact on the environment – air, water, solids, and hazardous materials.

The standard is not a set of environmental laws and regulations, and it does not enumerate air quality standards or emission performance requirements. Instead, it provides a framework to assist in developing an environmental management system. This system assists companies in continuously improving their environmental performance, while also complying with any applicable national legislation. The organization is responsible for setting their own targets and performance measures, with the standard serving to assist them in meeting objectives and goals and the subsequent monitoring and measurement. Two organizations that have different measures and standards of environmental performance can both comply with ISO 14000 requirements.

The basic idea of these standards is managing a process of continual improvement. One common management system is the Plan-Do-Check-Act method. Planning occurs before implementing. The organization identifies all current operations and future operations that interact with the environment. Then, measurable goals and targets can be created. Next, the plan is implemented. The resources, procedures, and processes are put into place, and measures (concentrations, materials used, materials wasted) are made part of the process. After some time has passed, the results from the measurements are checked to determine if the goals have been achieved. The check includes internal and external audits of the results and the process. Next, a review is performed to assess the process, and improvements are recommended either because of changes in circumstances or to implement better ways of operation. This process leads to, hopefully, a system that always improves. The most important factor in making this system work is to have all levels of management commit to the process.

The ISO 14000 standards do not conflict with US-EPA laws and regulations; rather, they provide a framework wherein the laws and regulations become the goals and targets of the organization. The EPA considers ISO 14000 registration as evidence of 'due diligence' for compliance.

● Case Study 3-3.
Reconsider the coal fired power plant from Case Study 2-1. Assume the coal has a heat content of 11,750 BTU/lb and the efficiency of the process is 34.5%. Determine the removal efficiency required for each pollutant listed in NSPS for this type of facility.

Solution:
Look at the NSPS listed in Table 3-6. Determine what pollutants the facility is likely to have and note the units. Then convert the units from Case Study 2-1 into the NSPS set of units. Determine efficiencies as per equation 2.20.

From Table 3-6 we see that facilities started after 2005 (the power plant in this problem was built in 2012) have emission controls on PM, SO_2, and NO_2.

The NSPS values are:

PM:	I	0.14 lb/MW-hr gross energy output; or
	II	0.015 lb/MMBtu heat input; or
	III	0.03 lb/MMBtu heat input, and
		99.9% reduction for a construction
SO_2:		*Construction*
		1.4 lb/MW-hr gross energy output or 95% reduction
NO_x:		*Construction*
		1.0 lb/MW-hr gross energy output

Calculations of the potential emissions for each pollutant follow:

PM potential emissions *for case I using 34.5% efficiency between energy in and energy out:*

$$E_{PM}^I = 3.413 \times 10^6 \frac{BTU}{MW-hr} * \frac{1_{in}}{0.345_{out}} * \frac{1 \text{ lb coal}}{11{,}750 BTU_{in}} * \frac{0.07 \text{ lb ash}}{1 \text{ lb coal}} * \frac{0.35 \text{ lb flyash}}{1 \text{ lb ash}} = 20.63 \frac{\text{lb fly ash}}{MW_{out}-hr}$$

and for case II and III efficiency is not used, the calculation is based on energy input:

$$E_{PM}^{II} = \frac{1 \text{ lb coal}}{11,750 BTU} * \frac{10^6 BTU}{MMBTU} * \frac{0.07 \text{ lb ash}}{1 \text{ lb coal}} * \frac{0.35 \text{ lb flyash}}{1 \text{ lb ash}} = \frac{2.09 \text{ lb fly ash}}{MMBTU}$$

So, the required control efficiencies for each case are:

$$\eta_{PM}^{I} = \frac{(20.63 - 0.14)}{20.63} * 100\% = 99.32\%$$

$$\eta_{PM}^{II} = \frac{(2.09 - 0.015)}{2.09} * 100\% = 99.28\%$$

$$\eta_{PM}^{III} = \frac{(2.09 - 0.03)}{2.09} * 100\% = 98.6\%$$

For case III, the required control level does not meet the additional requirement of 99.9% removal. Option II is the best choice for this example.

Before we can calculate the SO_2 and NO_2 removals, we need to know the weight fraction of S and N in the coal. To obtain these values, we need to determine the mass of each constituent in the coal:

Table 3-9. Mass of Each Constituent in the Coal for Case Study 3-3.

Basis: 1 mole coal			
Species	moles	MW	mass
C	70	12	840
H	100	1	100
O	2	16	32
N	0.2	14	2.8
S	0.3	32	9.6
BA		150	48.2
FA		85	25.9
Total			1058.5

SO_2 potential emissions:

Weight fraction of sulfur in the coal is 9.6/1058.5.

$$E_{SO_2} = 3.413 \times 10^6 \frac{BTU}{MW-hr} * \frac{1_{in}}{0.345_{out}} * \frac{1 \text{ lb coal}}{11,750 BTU_{in}} * \frac{9.6 \text{ lb S}}{1058.5 \text{ lb coal}} * \frac{64 \text{ lb } SO_2}{32 \text{ lb S}} = 15.27 \frac{\text{lb } SO_2}{MW_{out} - hr}$$

The removal efficiency for the limit value is

$$\eta_{SO_2} = \frac{(15.27-1.4)}{15.27}*100\% = 90.8\%$$

This is less than the other requirement of 95%, so choose the 90.8% removal.

NO$_x$ potential emissions:

Weight fraction of nitrogen in the coal is 2.8 / 1058.5

$$E_{NO_x} = 3.413x10^6 \frac{BTU}{MW-hr} * \frac{1_{in}}{0.345_{out}} * \frac{1 \text{ lb coal}}{11,750 BTU_{in}} * \frac{2.8 \text{ lb } NO_x}{1058.5 \text{ lb coal}} = 2.23 \frac{\text{lb } NO_x}{MW_{out}-hr}$$

The removal efficiency for the limit value is

There are no choices on this limit, so the required NOx removal efficiency is 55.2%.

$$\eta_{NO_x} = \frac{(2.23-1.0)}{2.23}*100\% = 55.2\%$$

Result summary for required removal efficiencies:
PM = 99.28%
SO2 = 90.8%
NOx = 55.2%

3.4 Questions

* - Questions and problems may require additional information not included in the textbook.

1. Visit the EPA website to determine if any changes have been made to the Table 3-3 NAAQS (40 CFR part 50).*

2. What are the emission standards (NSPS) for a new oil-fueled power generator with the output between 29 – 73 MW?

3. Describe in your own words how to determine a rolling 3-year average of the 98th percentile values of the daily maximum 1-hour averages based on samples taken every 10 minutes.

4. Determine if the air quality control area you live in is in attainment for all criteria pollutants. If not, which pollutants are in non-attainment? *

5. Find the locations of the ambient air quality monitoring stations near where you live, work, or go to school? *

6. What are the emission limits for A) a Portland cement kiln? B) a primary zinc smelter? C) a rubber tire manufacturing plant? D) a sulfuric acid plant? and E) a bulk gasoline terminal?

7. Which state government agency maintains your State's SIP? *

8. Why do you think Congress forbids the EPA from using cost considerations for developing NAAQS? *

9. Why do you think Congress allows the EPA to use cost considerations for developing NSPS emission limits? *

10. What is the difference between the NAAQS values and NSPS values?

11. Do NAAQS differ between class I, II, and III regions?

12. What is the difference between a SIP and FIP?

13. Describe the differences between BACT, LAER, MACT, and RACT.

14. Make a list of three ways to obtain an emission offset. *

15. Do all sources of PM emissions need a Title V permit to discharge into the air?

16. What is the current Corporate Average Fleet Efficiency (CAFE) mileage standard? What will it be in five years? *

17. Make a list of the Title V permit holders in the area where you live, work, or go to school? *

18. Make a list of three interstate or international air pollution problems besides the two listed in the text. *

3.5 Problems

1. Convert all the NAAQS values into units of ppmv and $\mu g/m^3$.

2. The following table lists the five highest ozone readings (ppmv) from an air quality monitor for six years. Determine if there is a violation for air quality standards in any year

Year 1	Year 2	Year 3	Year 4	Year 5	Year 6
0.115	0.120	0.100	0.110	0.070	0.080
0.110	0.115	0.085	0.090	0.070	0.075
0.085	0.100	0.085	0.085	0.065	0.065
0.065	0.070	0.085	0.075	0.065	0.060
0.065	0.065	0.080	0.065	0.060	0.060

3. The emissions from a stack were measured by collecting a sample on a filter. The sampler was operated for 15 minutes, had an inlet area of 12.5 cm², an inlet velocity of 20 m/s, and collected 0.23 grams of PM. The stack gas had a temperature of 340 K, pressure of 1.018 atm, and volumetric flow of 245 scm/sec. Determine the PM concentration [$\mu g/dscm$], and the total rate of emissions [kg/day].

4. Find the controlled emission rate in units of mass/time or mass/volume from the following sources, assuming they are in compliance with NSPS.
   a) Ammonium sulfate manufacturer producing 20,000 tons/yr
   b) Bulk gasoline terminal servicing 25 trucks of 5,000 gallon capacity per day, and operating 5 day/ week.
   c) Grain elevator (20' diameter x 40' height) with fan capacity of 3 air exchange/hour. (an air exchange replaces the air in an empty tank volume).
   d) PM from a new Portland Cement plant kiln producing 10 ton clinker per day.
   e) Nitric acid plant producing 50 ton/day of 90% HNO_3.
   f) Sewage treatment plant incinerator generating 30,000 ton dry sludge per year.

5. Determine the required removal efficiencies for PM and SO_2 from the primary production of lead. The emission factors are: PM = 1 kg/Mg ore input and SO_2 = 27.5 kg/Mg ore input. Assume the emission gas production is 50 kg-mol/Mg ore.

6. Primary copper is produced by reacting the Cu_2S in a concentrated ore with oxygen from air to form copper and SO_2. The concentrated ore also contains inert silica materials. A

particular facility reacts a concentrated ore that is 20 wt% S with 100% of the stoichiometric required amount of oxygen during production. Determine the required removal efficiencies for PM and SO_2 for this facility. Assume an emission factor of 8 kg PM/ Mg concentrated ore, and that the gas emissions are only from the sulfur reaction products and air feed.

7. A 425 MW power plant (28% efficiency) uses coal with a sulfur content of 2.0 wt% and a heating value of 9,500 BTU/lb. Determine the removal efficiency for SO_2 needed by the facility if a) it was constructed in 1974, b) it was constructed in 1998, and c) it is new. Assume that all the sulfur in the coal is released as SO_2.

8. A facility produces lime (CaO) from limestone ($CaCO_3$) using heat. Assume the procedure generates particulate emissions at a rate of 3 g per kg of lime produced. Will the facility need to control the emissions of PM? If so, what efficiency is required?

9. Determine the emission control efficiencies needed in a new 700MW coal fired electric utility steam generating plant (35% efficiency). The uncontrolled emission factors for this bituminous coal are: 1.2 lb SO_2/ MMBTU, 7.3 wt% ash (of which 60% is fly ash and 40% is bottom ash), and 8.8 lb NO_x / ton coal. It and has a heating value of 12,500 BTU/lb.

10. Recalculate the emission control efficiencies using the data in the problem above, but assume the facility was constructed in 2001 and has not been reconstructed or modified.

11. Determine the emission control efficiencies needed in a new 700MW No.2 fuel oil (2 wt% sulfur) fired electric utility steam generating plant (35% efficiency). This fuel oil has a density of 7.0 lb/gal. The uncontrolled emission factors for 1000 gallons of fuel oil are: PM=2 lb, SO_2=157S lb (where S is the wt% sulfur) NO_x=24 lb. It has a heating value of 140,000 BTU/gal.

12. Recalculate the emission control efficiencies using the data in the problem above, but assume the facility was reconstructed in 2001 and has not been modified since.

13. Determine the emission control efficiencies needed in a new 700MW natural gas fired electric utility steam generating plant (35% efficiency). The uncontrolled emission factors for a million cubic feet of gas are: PM=7.6 lb, SO_2=0.6 lb, and NO_x=190 lb. It has a heating value of 950 BTU/dscf.

14. Recalculate the emission control efficiencies using the data in the problem above, but assume the facility was constructed in 1976 and has not been reconstructed or modified since.

15. A city (pop. 175,000) is considering construction of a municipal solid waste combustor using a fluidized bed system to burn the city's trash. The proposed system includes some pollution control measures and is guaranteed to reduce the emissions of PM to 0.07 g/dscm, SO_2 to 35 ppmv, and NO_x to 110 ppmv. The exit gas will contain 5% oxygen. Will these emissions conform to the required NSPS values? Assume that the generation of waste is 5 lb/person/day.

16. An industrial facility is proposing to build a new steel plant using a basic oxidation furnace. The plant location is in a class II area currently in attainment and is more than 200 miles from the nearest class I area. The potential uncontrolled emissions of PM_{10} will be at least 250 ton/yr. Dispersion modeling shows that maximum ground level concentration will increase local levels by 30 $\mu g/m^3$ (24 hour average) assuming 99.5% PM_{10} control efficiency. The PM_{10} baseline air quality in this region is 125 $\mu g/m^3$. Would this source be allowed to operate as proposed? If not, what control efficiency would be required to allow it to operate?

17. Would the source in the above problem benefit from having the area of the proposed site reclassified as class III?

18. A publicly owned utility is proposing to build a new 350 MW natural gas fired steam generating unit. The plant location is in a class II area currently in attainment and is at least 100 miles from the nearest class I area. The potential uncontrolled emissions of all criteria pollutants will be at least 250 ton/yr. Dispersion modeling shows the plant will increase ground level concentrations of SO_2 by 18 $\mu g/m^3$, PM_{10} by 25 $\mu g/m^3$, and NO_x by 6 $\mu g/m^3$. The area base line values are $SO_2 = 65$ $\mu g/m^3$, $PM_{10} = 100$ $\mu g/m^3$, and $NO_x = 50$ $\mu g/m^3$. Would this source be allowed to operate as proposed? If not, how much additional reduction of each pollutant would be required to allow it to operate?

19. Redo the problem above for the case that the proposed site is located near a class I area.

3.6 Group Project Ideas

For each project the students should work in small groups and present their finding in either a short report (5-8 pages) or a 15 minute presentation:

1. The US-EPA is required to re-examine and revise the regulations for the criteria pollutants every few years. Choose one and summarize the latest work the EPA has done. Details can be found on the EPA website list of NAAQS.

2. Explore the method the US-EPA uses to generate MACT standards. Choose a local industry that is required to achieve this standard and explain how they can accomplish this.

3. The US, in 1987, signed onto the Montreal Protocol, an international treaty designed to reduce harm from a global pollution problem. How has the US-EPA regulated the requirements of these international treaty obligations? That is, specify the regulatory actions of the EPA.

3.7 Bibliography

NY-PIRG. 2000. Operating Permits. *The Proof is in the Permit*. [Online] 2000. [Cited: June 23, 2011.] http://www.epa.gov/oar/oaqps/permits/partic/proof.html. Part 2, Chapter 3.

OR-DEQ. 2011. Laboratory and Environmental Assessment. *Air Quality Monitoring*. [Online] 2011. [Cited: June 21, 2011.] http://www.deq.state.or.us/lab/aqm/airmonitoring.htm.

THOMAS. 2011. Legislation in Current Congress. *The Library of Congress THOMAS*. [Online] 2011. [Cited: June 15, 2011.] http://thomas.loc.gov/home/thomas.php.

US-Courts. 2011. Educational Resources. *United States Courts*. [Online] Administrative Office of the U.S. Courts, 2011. [Cited: July 15, 2011.] http://www.uscourts.gov/EducationalResources.aspx.

US-EPA. 2011b. Air and Radiation. *National Ambient Air Quality Standards (NAAQS)*. [Online] July 7, 2011b. [Cited: July 15, 2011.] http://epa.gov/air/criteria.html.

—. 2009. Air Mobile Sources Program. *Compliance Monitoring*. [Online] US-EPA, June 16, 2009. [Cited: June 23, 2011.] http://www.epa.gov/oecaerth/monitoring/programs/caa/mobile.html.

—. 2014. Air Quality Trends. *US Environmantal protection Agency*. [Online] April 21, 2014. [Cited: July 31, 2014.] http://www.epa.gov/air/airtrends/images/y70_12_lineStyles.png.

—. 2011a. Basic Information. *Compliance and Enforcement*. [Online] July 12, 2011a. [Cited: July 15, 2011.] http://www.epa.gov/compliance/basics/index.html.

—. 2011c. Emissions Factors & AP 42, Compilation of Air Pollutant Emission Factors. *Technology Transfer Network*. [Online] Feb 8, 2011c. [Cited: July 15, 2011.] http://www.epa.gov/ttnchie1/ap42/.

—. 2014. mapnpoll.pdf. *US-EPA Air Quality Green Book*. [Online] July 2, 2014. [Cited: October 23, 2014.] http://www.epa.gov/airquality/greenbook/mapnpoll.html.

—. 2002. *New Source Review: Report to the President*. Washington DC, USA : US Government Printing Office, 2002.

—. 2011d. RBLC. *RACT/BACT/LAER Clearinghouse (RBLC)*. [Online] June 23, 2011d. [Cited: June 23, 2011.] http://cfpub1.epa.gov/RBLC/.

—. 2010a. Six Common Air Pollutants. *State Implementation Plan Overview*. [Online] April 27, 2010a. [Cited: June 23, 2011.] http://www.epa.gov/air/urbanair/sipstatus/overview.html.

—. 2007. Tribal Air. *Developing a Tribal Implementation Plan (TIP)*. [Online] March 15, 2007. [Cited: June 23, 2011.] http://www.epa.gov/oar/tribal/tip2002/index.html.

—. 2010b. Where You Live. *Operating Permits*. [Online] November 23, 2010b. [Cited: July 16, 2011.] http://www.epa.gov/airquality/permits/whereyoulive.html.

US-GPO. 2011b. GPO Access. *Electronic Code of Fereral Regulations e-CFR.* [Online] National Archives and Records Administration, July 2011b. [Cited: July 15, 2011.] http://ecfr.gpoaccess.gov/cgi/t/text/text-idx?c=ecfr&tpl=/ecfrbrowse/Title40/40cfr60_main_02.tpl. 40 CFR part 60.

—. 2011a. U.S. Government Printing Office. *Federal Register.* [Online] 2011a. [Cited: June 15, 2011.] http://www.gpo.gov/fdsys/browse/collection.action?collectionCode=FR.

Wark, Kenneth, Warner, Cecil F and Davis, Wayne T. 1998. *Air Pollution: Its Origin and Control, 3rd Edition.* Menlo Park, CA, USA : Addison Wesley Longman, Inc., 1998.

CHAPTER 4

Fluid Mechanics Review

This chapter provides a general background in describing how fluids (liquids and gases) move. The study of this information is called fluid mechanics, and such study is used extensively by engineers, geologists, biologists, and physicists. In this textbook, it is used in helping to describe mixing and motion in the atmosphere, and to quantify how pollution control equipment works. This review section covers a few of the basic concepts of fluid mechanics and provides background material for understanding air pollution control devices. Because these ideas are central to understanding how things work, the student should expect to refer to this chapter multiple times while reading the remainder of the text. Note that there are many excellent fluid mechanics textbooks that offer a more complete and complex analysis. [(Fox, et al., 2006), (Wilkes, 2006), (Thomson, 2000), (Smits, 2000), (Shaugnessy, et al., 2005)].

These concepts are especially applicable to the material in Chapters 5, 6, and 15, which deal with movement and transportation of fluids.

4.1 Terms and units

A fluid is a substance that deforms continuously when subjected to an applied tangential or shear force. Air and water are common examples of fluids. Solids also deform under an applied force, but the solid reaches an equilibrium position in which the elastic forces in the solid counterbalance the applied force and no additional deformation or motion occurs due to the constant force.

Before describing the interaction of the forces and motion, we begin by describing several of the commonly used terms. Note that the symbol [=] means 'has the units of.' This section uses the MLTt system to define units in terms of four fundamental units (Mass, Length, Temperature and Time) before imposing a unit system (such as SI or US). Appendix 1 provides several unit conversions.

Density (ρ) describes the mass of a given volume of a material and does not depend on the amount of material that is present. It is described as;

4.1 $$\rho = \frac{\text{mass}}{\text{volume}} [=] \frac{M}{L^3}$$

Density of gases can be calculated from a thermodynamic equation of state. For the problems in this book, we assume that gases behave in an ideal manner, and use the ideal gas equation of state for any appropriate calculations. Non-ideal equations of state do exist, and the interested reader should consult any textbook in thermodynamics or physical chemistry for more information. The ideal gas equation is given as (also see equation 2.8)

CHAPTER 4

4.2
$$PV = nRT$$

Where: P = the absolute pressure,
V = the volume of the gas,
n = the number of moles of the gas (may be a mixture),
R = the gas constant (several values are given in appendix 1), and
T = the absolute temperature.

This equation can be rearranged to calculate the molar density of a gas

4.3
$$\rho_{molar} = \frac{n}{V} = \frac{P}{RT}$$

When combined with the molecular weight of the gas the mass density may also be determined (n=m/mw),

4.4
$$\rho_{mass} = \frac{n \cdot mw}{V} = \frac{P \cdot mw}{RT}$$

Ideal gases are assumed to mix perfectly so the molar density of pure gas and a mixture of many gases is the same under similar conditions of pressure and temperature, or number of moles and volume.

The density of liquids is not easy to calculate from an equation of state. Rather, values are tabulated, usually as a function of temperature. The density of a liquid does not change over moderate changes in pressure. [See reference texts like (Lide, 2009) or (Perry, et al., 1997)]

Another term used to describe the density of a substance is the **specific gravity**. It is a ratio that compares the density of a gas to the density of air at standard condition (typically 1 atm and 0°C). Note that there is no agreed upon international standard conditions for air, so caution is necessary when interpreting published values (see Table 2-3.) The specific gravity of a liquid or solid is usually referenced to that of water at standard conditions (1 atm, 4 °C which yields a density of 1.000 gram/cm^3).

■ **Example 4-1.**
Find the density of air at a pressure of 1 atm and temperature of 20°C.

Solution:
Apply the ideal gas equation. Convert answer from moles into mass using the molecular weight of air as 28.85 g/g-mole.

$$\rho = \frac{n\, mw}{V} = \frac{P\, mw}{RT} = \frac{(1\,\text{atm})\left(\dfrac{28.85\,\text{g}}{\text{gmol}}\right)}{\left(0.08206\,\dfrac{\text{liter atm}}{\text{gmol K}}\right)(273+20)\,\text{K}} = 1.199\,\dfrac{\text{g}}{\text{liter}}$$

Viscosity (μ) is a measure of the resistance of a substance to flow under an applied shear stress. A material with a larger value of viscosity requires a greater shear stress to move with the same speed as a material with lower viscosity. Newton's Law of Viscosity describes the relationship for a large class of materials, including air and water. It relates the shear stress and the change in velocity with respect to position with a proportionality constant. The equation is written as;

4.5
$$\tau = \mu \frac{du}{dx}$$

Where: τ = the shear stress, given as the force per area(F/A),
du/dx = the velocity gradient.

The units for viscosity are mass per unit length per unit time,

4.6
$$\mu [=] \frac{M}{Lt}$$

Values for gases are reported in the SI units called poise (1 P = 1 g/cm s). Appendix 6 lists values for gases as a function of temperature, as shown in equation (4.7). Equation (4.5) is not valid for all materials, and such materials are classified as non-Newtonian fluids (Bird, et al., 1960). The atmosphere and most emissions are considered ideal gases, and ideal gases are assumed to be Newtonian fluids.

4.7
$$\mu = \mu_0 \left(\frac{T}{T_0}\right)^n$$

Where: T = absolute temperature [K],
n = constant, listed in Appendix 6 [no units], and
μ_0 = viscosity at reference temperature T_0 = 273 K [cP, centi-Poise] from Appendix 6.

■ *Example 4-2.*
Determine the viscosity of air at 1 atmosphere pressure and 70°F.

Solution:
Use the relationship and data (μ_0 and n) from Appendix 6. Note that 70 °F = 295 K

$$\mu_{air} = \mu_0 \left(\frac{T}{T_0}\right)^n = 0.0171 cP \left(\frac{295 K}{273 K}\right)^{0.768} = 0.01815 cP$$

$$= 1.815 \times 10^{-5} \frac{kg}{m\, sec}$$

Kinematic viscosity ($v=\mu/\rho$) is the ratio of viscosity and density. It is used because of the frequency with which these two properties appear as a ratio in fluids problems. It will have units of $[L^2/t]$, and in the metric system is given the name of Stokes (1 St = 10^4 m^2/sec)

Velocity (u) describes how far something moves from a given location in a unit of time. It is assumed to be defined at every point in time and space for the entire system. This is usually referred to as the continuum hypothesis, which, while we know does not hold at molecular scales, yet works fine for human-sized systems. The term velocity is usually assumed to describe the bulk or average motion. It is important to realize that small sub-sections in a system that is moving may each move at different velocities. It is possible that none of the sub-units of a system are moving at the average velocity. Consider water motion around a rock in a stream – as the water moves around the rock some flows left or right of the rock, and some may go over the rock. The average is downstream, but parts of the water are moving in almost any direction. Velocity can be described mathematically as,

4.8 $$\mathbf{u} = \frac{\Delta x}{\Delta t} = \frac{dx}{dt} [=] \frac{\text{length}}{\text{time}} = \frac{L}{t}$$

The first expression describes an average velocity, the second an instantaneous velocity, and the final describes the units for velocity. Velocities are vectors because they have direction and magnitude. **Bold** symbols are used to identify the term as a vector, a shorthand way of writing the motion in each of the three directions. The motions in each of the three directions are independent and listed as (u_x, u_y, u_z). When not listed in bold, the velocity direction is unimportant for the problem and it is as a scalar quantity.

Volumetric Flow (Q) describes the volume of fluid moving in a given unit of time.

4.9 $$Q = \frac{\text{Volume}}{\text{time}} = uA [=] \frac{L^3}{t}$$

Mass Flow (M) describes the mass of fluid moving in a given unit of time.

4.10 $$M = \frac{\text{Mass}}{\text{time}} [=] \frac{M}{t}$$

These two forms can be interconverted using the fluid mass density.

4.11 $$Q = \frac{M}{\rho_{mass}}$$

Acceleration (a) describes the time rate of change of velocity. Similar to velocity, it is a vector. We can describe an average value and an instantaneous one as;

4.12 $$\mathbf{a} = \frac{\Delta u}{\Delta t} = \frac{du}{dt} = \frac{d^2 x}{dt^2} [=] \frac{L}{t^2}$$

One of the most important acceleration terms is gravity. It has a value of 9.8 m/s² in the down direction.

Momentum (M) describes the product of an object's mass and velocity. Momentum is a conserved quantity, meaning that the total momentum of any closed system (one not affected by external forces) cannot change. Momentum is either linear momentum or angular momentum. Linear relates to motion between two locations (translation), and angular refers to rotational (or spinning) motions. Linear momentum is a vector quantity, and is calculated as;

4.13 $$\mathbf{M} = m\mathbf{u}[=]\frac{ML}{t}$$

If an object is moving in any reference frame, then it has momentum in that frame. It is important to note that momentum is frame dependent. That is, the same object may have a certain momentum in one frame of reference, but a different amount in another frame. For example, when you are driving a car at a constant speed, you do not have momentum with respect to the car, but you have momentum with respect to the road over which you are moving.

Force (F) is the rate of transfer of momentum. The mathematical formulation is

4.14 $$\mathbf{F} = m\mathbf{a} = m\frac{d\mathbf{u}}{dt}[=]\frac{ML}{t^2}$$

The total force on an object may be determined by summing all forces on the object. It is important to note that this is a vector quantity, and so direction is important. The SI unit for force is the Newton, where 1 N = 1 kg m / sec².

Forces may act on just the exterior of an object (surface forces), or the force can act on every part of the volume of the object (body forces).

Surface forces – There are two types of surface forces – normal and shear. When a force is applied in a perpendicular direction to the surface, it is a normal force, see F_z in Figure 4-1. If it pushes on the surface it is a compressive normal force and is considered negative (-). If the normal force pulls on the surface, it is a tensile normal force and is considered positive (+). If the force is parallel to the surface, it is a shear force, see F_x and F_y in Figure 4-1. Shear forces are also sometimes referred to as tangential forces. Forces are vectors, just like velocities and accelerations and so may act in up to three directions upon a surface. Of the three components, two are shear forces and one is normal.

Body forces – when a force acts throughout the volume and not just on the surfaces, it is called a body force. The most common body force is from gravity. Others include electric, magnetic, and centrifugal forces.

FIGURE 4-1. Force Components on a Control Volume.

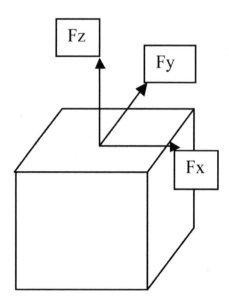

Pressure (P) is a normal force applied over an area. It is one of the primary descriptive properties of a fluid, along with temperature. It is described as:

4.15 $$P = \frac{F}{A} [=] \frac{M}{Lt^2}$$

The SI unit is the Pascal (Pa). 1 Pa = 1 N/m² = 1 kg/(m s²). Pressure is often reported in atmospheres (atm), where 1 atm = 101,325 Pa.

The pressure of an ideal gas varies linearly with temperature, volume, and amount (or moles) of a substance, see equation (4.2) - the ideal gas law. The pressure of a liquid also depends on these variables, as well as other variables such as surface tension, but the relationship is not as straightforward (Smith, et al., 1987).

Gauge pressure is the pressure relative to the local atmospheric pressure. It can be calculated with the following relationship:

4.16 $$P_{guage} = P_{absolute} - P_{atmosphere}$$

Where: $P_{absolute}$ = the pressure measured in the system referenced to a vacuum, and
$P_{atmosphere}$ = the local atmospheric or ambient pressure.

Note that P_{guage} can be a negative number, but $P_{absolute}$ can never be negative. When $P_{absolute}$ is zero, the system is a perfect vacuum.

■ Example 4-3:
Find $P_{absolute}$ in a tire if P_{guage} is 32.0 psig and the atmospheric pressure is 14.6 psia.

Solution:
Apply equation (4.16), where P_{guage} = 32.0 and $P_{atmosphere}$ = 14.6.

$P_{absolute} = P_{guage} + P_{atmosphere} = 32.0 + 14.6 = 46.6 psia$

Vapor pressure or equilibrium vapor pressure is the pressure of a vapor in thermodynamic equilibrium with its liquid or solid phase in a closed volume. All liquids and solids have a tendency to evaporate into a gaseous form, and all gases have a tendency to condense back to their liquid or solid form. The equilibrium vapor pressure is an indication of a liquid's evaporation rate. It describes the tendency of molecules to escape from the liquid (or solid). A substance with a high vapor pressure at room temperatures is called volatile.

Energy (e) is a scalar physical quantity (has magnitude but not direction) and is a conserved quantity, meaning that it is constant when considering a system and its surroundings. There are many forms of energy, and they are all considered equivalent – energy in one form can disappear, but the same amount of energy appears in another form. Some of the forms of energy include kinetic, potential, chemical, thermal, gravitational, elastic, and electromagnetic. The forms of energy are often named after their related force. The SI unit is the Joule (J). 1 J = 1 N-m.

Work (W) or mechanical energy is the amount of energy transferred by a force acting over a distance. It is a scalar quantity. When force and the displacement are parallel and in the same direction, the mechanical work is considered positive. When force and the displacement are parallel but in opposite directions, the mechanical work is considered negative. When force and the displacement act perpendicular to each other, no work results from the force. Mechanical work can be calculated for a constant or variable force as follows;

4.17 $$W = \mathbf{Fd} = \int_{x0}^{x1} F(x)dx \,[=]\, \frac{ML^2}{t^2}$$

The SI unit is the Joule (J) or N-m, same as energy.

Power ($\mathbb{P}$) is the rate of work, or the rate of energy conversion between forms. The average and instantaneous power are described as

4.18 $$\mathbb{P} = \frac{\Delta Work}{\Delta t} = \frac{dW}{dt} \,[=]\, \frac{ML^2}{t^3}$$

The SI unit is the Watt (W). 1W = 1 J/s. Another unit for power is horsepower, where 1 hp = 746 W.

4.2 Fluid Statics

Fluid Statics describes the characteristics of the fluid when it is not moving. It is mostly concerned with determining pressures, forces, and densities throughout the system. The pressure at a given location results from the weight of the fluid above acting like a normal force and pressure is a function of elevation (height). This relationship is described with the hydrostatic equation:

4.19
$$\frac{dP}{dh} = -\rho g$$

This equation shows that the pressure decreases with increasing elevation. Density can be assumed constant for small elevation changes, or it can be allowed to vary as the following examples illustrate. The equation can be used to determine the atmospheric pressure atop a mountain, the water pressure at the bottom of a storage tank, or even to explain the high pressures in the deep ocean.

■ Example 4-4.
Determine the pressure as a function of height for a fluid in a vertical pipe, H meters tall.

Solution:
Assuming H is small, the change in elevation causes only a very small change in density, so assume it is constant. Equation (4.19), upon rearranging and integration, yields the desired answer.

4.20
$$\int_{P_0}^{P} dP = -\rho g \int_{h=0}^{h=H} dh$$

Solving:

4.21
$$P - P_0 = -\rho g H \quad \text{or} \quad P = P_0 - \rho g H$$

■ Example 4-5.
Determine the barometric pressure at the top of Mt Everest (8848 meters). Assume that the temperature is constant and that the atmospheric pressure at sea level is one atmosphere.

Solution:
We may no longer assume that density is constant because the elevation change is large. We must relate density to either of the variables (P, h) in the derivative. One way is to use the relation which exists between pressure and density from the ideal gas equation of state, equation (4.2), This is substituted into equation (4.19) and solved by separation and integration. Alternatively, if we had direct

measurements of density as a function of height, we could solve equation (4.19) numerically between each data point, a method explored in Chapter 5 where we relax the assumption of constant temperature. The first alternative is presented here. Recall the formula for the density of an ideal gas (4.2):

$$\rho = \frac{n \; mw}{V} = \frac{P \; mw}{RT}$$

Substituting into (4.19):

$$\frac{dP}{dh} = -\frac{P \; mw \; g}{RT}$$

Separate and integrate:

$$\int_{P_0}^{P} \frac{dP}{P} = -\frac{mw \; g}{RT} \int_{h=0}^{h=H} dh$$

Solving yields:

4.22 $$\ln\left(\frac{P}{P_0}\right) = -\frac{mw \; g \; H}{RT}$$

Assuming P_0 is 1 atm at h=0, the pressure at 8848 m is

$$P = P_0 \left[\exp\left(-\frac{mw \; g \; H}{RT}\right)\right] =$$

$$(1 \; atm) \exp\left(-\frac{\left(28.84 \frac{g}{gmol}\right)\left(9.8 \frac{m}{s^2}\right)(8{,}848 \; m)}{(8.314 \frac{m^3 Pa}{mol \; K})(290 \; K)} \frac{1 \; Pa}{1{,}000 \frac{g}{ms^2}}\right) = 0.354 \; atm$$

There are only a few known measurements of barometric pressure atop Mt Everest – the values range from 251 – 253 torr (0.330 atm) (West, 1999). The discrepancy between the calculated value and the measurement is caused by assuming temperature is constant rather than allowing it to vary.

If we had erroneously assumed that density was constant, the pressure could be calculated using equation (4.21), which yields a pressure of -0.026 atm, which is obviously impossible since absolute pressure must be positive.

4.3 Fluid Dynamics

Dynamics deals with the characteristics of the fluid when it is moving. It relates several forms of energy that a moving fluid may have (pressure, velocity, elevation, work and friction). The macroscopic equation, based on either an energy balance or a momentum balance, consists of five terms, which we will call the **mechanical energy balance**. It is a macroscopic equation because it is used to describe the bulk or macroscopic properties of the fluid. It would not be used to explore the behavior of fluids at the molecular level.

4.23
$$\frac{\Delta P}{\rho} + \frac{\Delta u^2}{2} + g\Delta h + W + F = 0$$

Where: ΔP = the pressure change between two points in the system,
ρ = the fluid density,
Δu = the change in velocity (term describes the kinetic energy),
g = gravity.
Δh = the change in height (term describes the potential energy),
W = the work term, and
F = the frictional loss in the system.

The derivation of this equation assumes: flow is steady (it does not vary with time), flow is incompressible (the density is constant), there is no temperature change in the fluid, and that the system is taken along a streamline. A streamline is a flow path that a control volume (or particle of fluid) can move along. Students interested in the derivation of this equation or its limitations are encouraged to explore a reference on fluid mechanics, for example, (Fox, et al., 2006) or (Wilkes, 2006).

Note when using this equation to ensure that each term has identical units and to use the proper signs. Each term has units of energy per unit mass. It is helpful to sketch a picture of the system that clearly labels points 1 and 2. It may also be helpful to chop the system into several smaller pieces and use the output from one section as the input to the next.

The *first term*, $\frac{\Delta P}{\rho}$, in equation (4.23) represents the system head or internal energy of the fluid (assuming that the temperature does not change). In a non-moving system only terms one and three would be non-zero, and would yield the hydrostatic equation, see equation (4.19), which shows that pressure decreases with increasing elevation. In general, pressure decreases along the flow path due to friction, assuming no work or heat crosses the system boundary, there is no elevation change, and the cross-sectional area of the flow remains constant. Pressure increases across a fan or pump (adding work), and decreases across a turbine.

The *second term*, $\frac{\Delta u^2}{2}$, in equation (4.23) represents the kinetic energy. Velocity can be calculated from the mass continuity equation, or mass balance, which is described as;

4.24 $\quad\quad\quad\quad \text{Mass}_{in} = \text{Mass}_{out}$

Mass flow rate can also be calculated from the density of the fluid, the fluid velocity, and the area available for flow;

4.25 $\quad\quad\quad\quad \rho A_1 u_1 = \rho A_2 u_2$

Velocity can be determined from the volumetric flow rate and the cross-sectional area of the system by rearranging equation (4.9). Note that the volumetric flow rate is constant in a steady system.

4.26 $\quad\quad\quad\quad u_1 = \frac{Q}{A_1}$

Where: Q = the volumetric flow rate (units: L³/t), and
A = the cross-sectional area that is perpendicular to the flow direction (units: L²).

■ Example 4-6.

Air flow enters a 10 cm square duct at a rate of 1 m³/min. The duct then expands to 18 cm square. Find the velocity of the air at the inlet and after the expansion.

Solution:
Apply equation (4.26) at the inlet and equation (4.25) after the expansion.

Inlet velocity: $\quad u_{in} = \frac{Q}{A_{in}} = \frac{1 \frac{m^3}{min}}{(0.1m \times 0.1m)} = 100 \frac{m}{min} = 1.67 \frac{m}{sec}$

Outlet velocity: Rearrange equation (4.25) to solve for u_{out}, assuming constant density:

$$u_{out} = \frac{A_{in} u_{in}}{A_{out}} = \frac{(0.1m \times 0.1m) 1.67 \frac{m}{sec}}{0.18m \times 0.18m} = 0.51 \frac{m}{sec}$$

The *third term*, gΔh, in equation (4.23) represents the potential energy of the fluid. It is the energy needed to move the fluid against gravitational force. It is positive for an increase in elevation.

Example 4-7.

Consider steady flow down through a cone. The fluid ($\rho=1.2$ kg/m³) enters the cone at a rate of 1 m³/min and a pressure of 120 kPa. The cone has a circular cross section whose diameter decreases linearly from 10 cm to 5 cm over a length of 500 cm. Determine the pressure and velocity of the fluid at 1) the entrance, and at 2) the exit. Assume steady, incompressible flow with no work or friction.

Solution:

Sketch the system, see Figure 4-2. Determine the velocity at each point using equation (4.25) and (4.26). Apply equation (4.23) at each position to determine the pressure from the known elevations and velocities.

FIGURE 4-2. Sketch for Example 4 7.

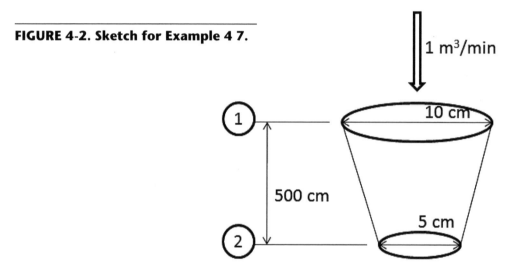

At point 1: h = 500 cm, P_1 = 120,000 Pa, velocity can be determined:

$$u_1 = \frac{Q}{A_1} = \frac{1 \frac{m^3}{min}}{(0.1 \text{ m})^2} = 127.3 \frac{m}{min} = 2.12 \frac{m}{s}$$

At point 2: h = 0 cm, P_2 is unknown, velocity can be determined:

$$u_2 = \frac{A_1 u_1}{A_2} = \frac{\frac{\pi}{4}(0.1)^2 2.12 \frac{m}{s}}{\frac{\pi}{4}(0.05)^2} = 8.48 \frac{m}{s}$$

Next, find P_2 by rewriting equation (4.23) setting W and F to zero and expanding the delta terms:

$$\frac{P_2 - P_1}{\rho} + \frac{u_2^2 - u_1^2}{2} + g(h_2 - h_1) = 0$$

Solve for P_2

$$\frac{P_2 - 120{,}000 \frac{kg}{ms^2}}{1.2 \frac{kg}{m^3}} + \frac{(8.48^2 - 2.12^2) \frac{m^2}{s^2}}{2} + 9.81 \frac{m}{s^2}(0m - 0.5m) = 0$$

$$P_2 = 1.2 \frac{kg}{m^3}\left[100{,}000 \frac{m^2}{s^2} - 33.71 \frac{m^2}{s^2} + 4.91 \frac{m^2}{s^2}\right] = 119{,}965\, Pa$$

We note that the pressure change due to velocity and elevation changes are quite small in this example.

■ Example 4-8.

Consider the system shown below. At point 1, water is held in a tank at a constant level. It drains through a pipe and is discharged to the air at point 3. Find the pressure at point 2 expressed in terms of the elevation, as shown. Assume steady, incompressible flow without friction.

Solution:
Use equation (4.23) with work and friction set to zero. Rewrite the equation to show that the energy at each of the three points is equal, though each term may be different:

4.27
$$\frac{P_1}{\rho} + \frac{u_1^2}{2} + gh_1 = \frac{P_2}{\rho} + \frac{u_2^2}{2} + gh_2 = \frac{P_3}{\rho} + \frac{u_3^2}{2} + gh_3 = \text{a constant}$$

The pressure is atmospheric at point 1 in Figure 4-3, which we quantify as 0 gauge pressure. The velocity is also zero since the tank is at a constant level. Only the elevation term is non-zero, and we can let $gh_1 = H$.

At point 3, the pressure is also atmospheric, since it is open to the air, and give it a value of 0 gauge pressure. The elevation is zero, relative to point 1. Only the velocity term is non-zero. Since the total energy must be constant, the term $u_3^2/2$ must be equal to H.

Finally, at point 2, the elevation term is H/2. The velocity term must be H, the same as at point 3 as required by mass continuity, equation (4.25), when the drain pipe has a constant area. Since the sum of the three terms must be H, the pressure term must be equal to (–H/2). The negative sign

FIGURE 4-3. Sketch for Example 4-8.

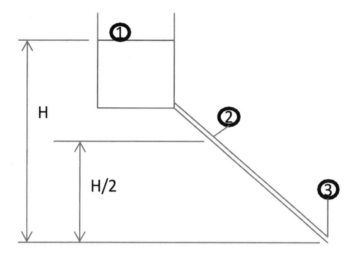

implies negative gauge pressure, which is acceptable as long as the absolute pressure is still positive. An interesting observation for this flowing system is that if a small hole developed in the pipe at point 2, the low pressure would force air to enter the pipe, rather than having water squirt out.

The fourth term, W, in equation (4.23) represents the energy transferred into/ out of the system due to mechanical work. It originates with mechanical devices such as fans, pumps, and turbines. The work term is only included when such a device is in part of the system, and that it transfers energy between the system and surroundings. It is positive (+) when the work is done by the fluid to the surroundings (such as in a turbine), and is negative (-) when the surroundings does work to the fluid (such as a pump or fan). There is always some inefficiency in transferring energy from the device to the fluid or from the fluid to the device.

■ Example 4-9.
Determine the amount of work done to 1.0 m³/sec of air (ϱ=1.2 kg/m³) from a 60% efficient fan with a 1 hp motor.

Solution:
Recall that the units for each term in equation (4.23) are in energy per unit mass.

4.28
$$W = \frac{Motor\ Power * efficiency}{density * Flow} = \frac{w\dot{\eta}}{\rho Q}$$

$$W = \frac{1hp * \dfrac{746 Watt}{hp} \dfrac{\frac{kgm^2}{s^3}}{Watt}(0.6\,efficiency)}{1.2\dfrac{kg}{m^3} 1.0 \dfrac{m^3}{s}} = -373\frac{m^2}{s^2}$$

Since the work is done by a fan (it is added to the system from the surroundings), the value is negative (-).

The *fifth term*, F, in equation (4.23) is the energy wasted by the system due to friction. It is always positive (+) in the system, and manifests as a temperature increase in the fluid. This energy is typically lost to the surroundings at a rate equal to its generation (isothermal assumption). Friction losses occur throughout the system, and are not avoidable (though clever design can minimize this wasted energy). Typical friction losses occur when the fluid;
- scrapes along the flow boundary (pipe or duct surfaces, or around the surface of an object in the flow field),
- changes direction,
- speeds up or slows down due to a size or shape change in duct or pipe works, or
- mixes with other fluids such as when multiple fluid streams join.

Calculation of the amount of frictional loss in a fluid requires more knowledge of the fluid and the system, specifically the nature of the flow, and the type of flow system. There are two main types of flow nature – laminar or turbulent. There are three main types of flow system - internal flow, boundary layer flow, or external flow.

Flow nature may be laminar or turbulent. The type depends on whether viscous or inertial forces dominate in the flow system. **Laminar flow** occurs when viscous forces dominate. The flow is slow, controlled, and the point velocity correlates strongly with the average velocity and distance from boundaries. **Turbulent flow** occurs when inertial forces dominate. The flow is fast, chaotic, and the local or point velocity correlates poorly with the average velocity or the presence of a flow boundary. The Reynolds number quantifies this characteristic. It is a dimensionless number describing the ratio of the inertial effects to viscous effects in a fluid.

4.29
$$\text{Re} = \frac{\rho_f u_f D}{\mu_f}$$

Where ρ_f = the fluid density (M/L³),
u_f = the fluid velocity (L/t),
D = a characteristic length of the system (L), and
μ_f = the fluid viscosity (M/Lt).

Typical values for ϱ for a gas can be calculated from the ideal gas law or looked up in a table of fluid properties. Appendix 6 provides values for viscosity. The choice for the characteristic length is somewhat arbitrary. The inside diameter of a pipe or duct may be used, or the effective diameter of an object the fluid is moving around. Difficulties arise when the boundary shapes are irregular or change.

Internal flow occurs when a fluid is moving inside an object, such as a pipe, duct, or stack. The average direction of flow is determined by the size, shape, and orientation of the equipment. **External flow** occurs when the fluid flows around an object, such as the wind moving past an object (drag) or over a solid surface (boundary layer). Note that both internal and external flow may occur in some situations, and that one does not necessarily determine the nature of the other. For example, consider when a particle falls through a gas stream that is moving through a pipe - the gas flow is internal with respect to the pipe, but is external with respect to the particle. **Boundary layer flow** occurs when a fluid is moving close to a solid or liquid surface. It assumes that the flow is zero at the surface (or at least with respect to the surface, which could also be moving). The velocity transitions from zero to the bulk flow velocity within this boundary layer. These flows may be either laminar or turbulent as previously described.

The nature of the flow for each system can be determined by examining the value of the Reynolds number, see Table 4-1. It should be noted that these values are approximate and were determined from observation of systems with well developed flow fields. The behavior may differ near transitions (entrance and exit points, as well as changes in direction). It is likely that the flow near a transitions will be turbulent, even in the low Reynolds number case.

Table 4-1. Classification of Flow by Reynolds Numbe

	Internal	Boundary Layer	External
Laminar	Re < 2,000	Re < 105	Re < 1
Turbulent	Re > 4,000	Re > 105	Re > 1,000

There may be a range of values in which the flow transitions between laminar and turbulent. This range is between 2,000 and 4,000 for internal flow and 1 and 1,000 for external flow. The flow flip-flops between laminar and turbulent in these Reynolds Number regions, and operating conditions are unpredictable which can make control difficult.

4.3.1 Friction During Internal Flow

This energy loss occurs at the boundary between the pipe or duct wall and the fluid flowing inside of it. It assumes that the fluid at the boundary has zero velocity and that this exerts a drag on the interior fluid. This drag force is what causes the frictional loss. The calculation of the friction term during internal flow is given by:

4.30 $$F = f_M \frac{L}{D} \frac{u^2}{2} = 8 f_M \frac{Q^2 L}{\pi^2 D^5}$$

Where: f_M = the Moody friction factor ($f_M = 4 f_F$ Fanning friction factor),
L = the length of the pipe or duct system,
D = the effective diameter,
u = the average velocity in the system, and
Q = the volumetric flow rate.

There are two similar sets of friction factors, Moody - f_M and Fanning - f_F. They differ by a factor of 4, so it is important to know which one is needed when looking up values on a chart or table. The choice is made based on the available chart, which should list which friction factor is presented. The friction factor depends on the Reynolds number and, for turbulent flow, the surface roughness of the pipe. The value for surface roughness (ε) depends on the surface material, see Table 4-2. The following equations estimate the Moody friction factor:

4.31 $$f_M = \frac{64}{Re} \quad \text{for laminar flow}(Re < 2,000)$$

4.32 $$f_M = \frac{0.25}{\left[\log_{10}\left(0.27\frac{\varepsilon}{D} + \frac{5.74}{Re^{0.9}}\right)\right]^2} \quad \text{for turbulent flow}(4,000 < Re < 10^8)$$

Graphical forms of this relationship may be found in any text of fluid mechanics.

Table 4-2. Surface Roughness of Common Materials. (Perry, et al., 1997)

Material	Roughness (mm)	Material	Roughness (mm)
Copper, Brass, Aluminum	0.0015	Corroded Steel (Rusted)	0.15 – 4.00
PVC and Plastic	0.004	Cast Iron	0.25
Epoxy, Vinyl Ester	0.005	Black pipe (Stove Iron)	0.5
Stainless Steel	0.045	Smooth Concrete	0.3
Commercial Steel	0.07	Ordinary Concrete	1
Galvanized Steel	0.15	Wood	5

The calculations and formulae assume that the ductwork is circular in cross section. If it is not circular, an effective diameter should be used. The effective diameter can be calculated as;

4.33 $$D_{eff} = 4\frac{Area}{Perimeter}$$

Where area and perimeter refer to the cross section of the duct through which the fluid is flowing. For rectangular shapes that are not nearly square, an additional multiplier may be needed (Crawford, 1976).

4.3.2 Friction During External Flow

This energy loss occurs as a fluid passes around an object or over a surface. The loss occurs to a shear force exerted by the surface onto the fluid – again it assumes that the fluid at the boundary has zero velocity. The two cases are called **drag** – flow around an object, and **boundary layer** – flow over the surface.

4.3.2.1 Drag

Friction during flow around an object is called drag. The drag force can be determined as;

4.34 $$F_D = \frac{1}{2}C_D A_p \rho V_\infty^2$$

Where: C_D is the drag coefficient
A_p is the projected area of the object in the direction of flow (same as a shadow),
ρ is the object density, and
V_∞ is the free stream velocity (velocity that has not been impacted by the object).

The free stream velocity is often taken to be the upstream velocity. The Reynolds number calculation also uses this value. The *drag coefficient for a sphere* can be approximated as follows:

4.35 $$C_D = \frac{24}{Re} \text{ for laminar flow } (Re < 1)$$

4.36 $$C_D = 18 \, Re^{-0.6} \text{ for transition flow} (1 < Re < 10^3)$$

4.37 $$C_D = 0.44 \text{ for turbulent flow} (10^3 < Re < 10^5)$$

Additional values (including charts and equations) for larger values of Re, or for different shapes, can be found in a number of sources [(Blevins, 1984), (Vogel, 1981), (Young, et al., 2001)].

These concepts are especially applicable to the material in Chapter 7 and Chapter 8, which deal with particulate matter in the atmosphere. Chapter 7 has an example problem for drag flow in terms of terminal velocity of a particle falling in a gas.

4.3.2.2 Boundary Layer

Friction between a surface and the flowing fluid manifests itself as a boundary layer, which is a region of high shear where viscous and inertial terms are both important. Within this layer, the fluid velocity varies from zero at the surface and increases with distance from the surface until it matches the free stream velocity. The distance over which this change occurs is the boundary layer. The thickness of the layer (δ) increases in the direction of flow. The boundary layer starts out as laminar and eventually becomes turbulent as the Reynolds number increases to 100,000. The Reynolds number is calculated as in equation (4.29) except that the characteristic length becomes the distance from the front of the object to the location of interest.

FIGURE 4-4. Boundary Layer Thickness (δ) and Velocity Profile (u_x) Across a Flat Plate.

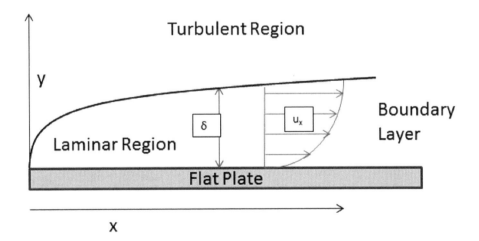

The laminar region of the boundary layer is thinner and has a more gradual change in the velocity profile. The turbulent region has a thicker boundary layer, but the velocity profile changes more quickly near the surface and then slowly approaches the free stream velocity.

Boundary layer concepts are useful during the discussion of wind speeds in the atmosphere, see Chapter 5 – Air Pollution Meteorology and Chapter 6 – Dispersion.

4.4 Bibliography

Bird, R. Byron, Stewart, Warren E. and Lightfoot, Edwin N. 1960. *Transpot Phenomena*. New York : John Wiley and Sons, 1960.

Blevins, R. D. 1984. *Applied Fluid Dynamics Handbook*. New York : Van Nostrand Reinhold, 1984.

Crawford, M. 1976. *Air Pollution Control Theory*. New York : McGraw Hill, 1976.

Fox, Robert W and McDonald, Alan T. 2006. *Introduction to Fluid Mechanics, 6th ed*. New York : John Wiley and Sons, 2006.

Lide, David R. 2009. *CRC Handbook of Chemistry and Physics, 90th Ed*. Boca Raton, Fl, USA : CRC Press, 2009.

Perry, Robert H. and Green, Don W. 1997. *Perry's Chemical Engineers' Handbook, 7th*. New York : McGraw Hill, 1997.

Shaugnessy, Jr, Edward J., Katz, Ira M. and Schaffer, James P. 2005. *Introduction to Fluid Mechanics*. New York : Oxford University Press, 2005.

Smith, J M and Van Ness, H C. 1987. *Introduction to Chemical Engineering Thermodynamics, 4th Ed*. New York, NY, USA : McGraw Hill, Inc, 1987. ISBN 0-07-058703-5.

Smits, Alexander J. 2000. *A Physical Introduction to Fluid Mechanics*. New York : John Wiley and Sons, 2000.

Thomson, William J. 2000. *Introduction to Transport Phenomena*. Upper Saddle River, NJ, USA : Prentice Hall, 2000.

Vogel, J. 1981. *Life in Moving Fluids*. Boston : Willard Grant Press, 1981.

West, John B. 1999. Barometric pressures on Mt. Everest: new data and physiological significance. *Journal of Apllied Physiology*. 1999, Vol. 86, 3, pp. 1062-1066.

Wilkes, James O. 2006. *Fluid Mechanics for Chemical Engineers, 2nd ed*. Upper Saddle River, NJ, USA : Prentice Hall, 2006.

Young, Donald F., Munson, Bruce R. and Okiishi, Theodore H. 2001. *A Brief Introduction to Fluid Mechanics, 2nd*. NEw York : John Wiley and Sons, 2001.

CHAPTER 5

Air Pollution Meteorology

5.1 Introduction

Meteorology is the field of science devoted to studying the Earth's atmosphere. Air pollution meteorology is the study of the fate and transport of pollutants in the atmosphere. The understanding of meteorological processes is useful in determining where emissions go (transport), how they dilute (dispersion), how they transform (reactions), how they are removed (fate), as well as where they are removed. Such information is essential to managing the quality of ambient air.

5.2 Composition

Air is a mixture of gases and particles. Table 5-1 lists the contents and composition of dry air. The table does not include water because the amount varies considerably from place to place as well as time of day and with temperature. The table excludes particles because they also vary so much. Clean air may contain less than 10 μg/m^3 particulate matter whereas during an intense sandstorm the air may contain more than 100 g/m^3. Typical particles include water droplets, dust, ash, soil, sea salt, organic carbon, elemental carbon (soot), ammonium nitrate, ammonium sulfate, nitrate salts, and sulfuric acid. See Chapter 8 for more discussion of particles.

The composition of dry air remains approximately constant until a height of about 85 km. Below this height (homosphere) there is enough turbulence to mix the constituents. However, some constituents may undergo rapid chemical reactions and thus create regions of localized non-homogeneity. Above 85 km, (heterosphere) the density of air is so small that the distance the molecules travel before colliding (mean free path) is large compared to the size of motions causing atmospheric mixing. The large mean free paths allow the gases to stratify by molecular weight, with the heavier ones (oxygen and nitrogen) located closer to earth and the lighter ones (hydrogen) farther from the earth. The total dry mass of the atmosphere is estimated to be 5.13 x 10^{18} kg. The nearest 10 km of atmosphere contains about 90% of the atmospheric mass.

Table 5-1. Composition of Dry Air, Global Average. Adapted from (Brasseur, et al., 1999)

Species	Symbol	MW g/mol	Volume Fraction ppm$_v$ Dry	Weight Fraction ppm$_w$ Dry
Nitrogen	N$_2$	28.0134	780840	755176
Oxygen	O$_2$	31.9989	209400	231330
Argon	Ar	39.9481	9340	12881
Carbon Dioxide*	<u>CO$_2$</u>	44.0096	<u>391</u>	<u>594</u>
Neon	Ne	20.1798	18.18	12.7
Helium	He	4.0026	5.240	0.72
Methane*	<u>CH$_4$</u>	<u>16.0426</u>	<u>1.800</u>	<u>1.00</u>
Krypton	Kr	83.7982	1.140	3.30
Nitrous Oxide*	<u>N$_2$O</u>	44.0129	<u>0.300</u>	<u>0.46</u>
Hydrogen	H$_2$	2.0159	0.550	0.04
Xenon	Xe	131.2936	0.090	0.41
Ozone*	O$_3$	47.9983	0.050	0.08
Nitrogen Dioxide*	NO$_2$	46.0056	0.020	0.03
Sulfur Dioxide*	SO$_2$	64.0644	0.005	0.01
Carbon Monoxide*	CO	28.0102	0.100	0.10

* denotes substances whose concentration changes rapidly,
Underlining denotes substances currently accumulating in the atmosphere.

The amount of water in the atmosphere varies from near zero to 4%. However, unlike other constituents, water in the atmosphere is measured as relative humidity rather than volume or weight fraction. The relative humidity is calculated by comparing the actual amount of water in the air to the maximum amount in the air at a given temperature:

5.1
$$RH = \frac{P_{water}^{Actual}}{P_{water}^{Maximum}(T)} = \frac{P_{water}^{Actual}}{VP_{water}(T)}$$

Where: P_{water}^{Actual} = partial pressure of water at atmospheric conditions

$VP_{water}(T)$ = vapor pressure of water at atmospheric conditions.

It is relative because the maximum amount of water in the air is highly dependent on the temperature. As a rule of thumb, the average relative humidity of the air is 50%, although it ranges from 0% to 100%. The dew point occurs when the partial pressure is equal to the vapor pressure (i.e. the

the RH is 100%). The dew point of air is the temperature that when actual air is lowered to it, the water vapor in it starts to condense. Appendix V1e lists values of the vapor pressure of water as a function of temperature.

■ Example 5-1.
Determination of water content of air at 30°C and 45% RH.

Solution:
Use equation 5.1 and data from Appendix 1e to find the partial pressure of water at the given temperature. Next divide by the atmospheric pressure to find the mole fraction (which is equivalent to the volume fraction). The vapor pressure is in units of millimeters mercury (mmHg), so the entire problem also uses this unit. Note that 1 atm = 760 mmHg.

$$P_{water}^{Actual} = RH * VP_{water}(30°C) = 0.45 * 31.824 mmHg = 14.32 mmHg$$

$$\text{mol fraction} = \frac{P_{water}^{Actual}}{P_{atm}} = \frac{14.32 mmHg}{760 mmHg} = 0.0188 \text{ or } 1.88 \text{ mol\%}$$

■ Example 5-2.
What is the Dew Point for the air in the example 1?

Solution:
Compare the actual partial pressure of water with the vapor pressure listed in Appendix 1e. Find the temperature associated with the vapor pressure equal to actual partial pressure.

An interpolated temperature of 17°C has a vapor pressure of 14.4 mmHg, which is very close to the actual value of 14.3 mmHg. If the air temperature in the above sample of air were lowered to 17°C, the water would start to condense and form a fog or mist.

■ Example 5-3.
What is the mw_{air} at 20°C and relative humidity of a) 0%, b) 50%, and c) 100%.

Solution:
Use data from Table 5-1 and for b) and c) add in the water content. Ignore components with fraction below 100 ppm. (Note the answers will have only four significant figures.)

5.2
$$mw_{air} = \sum_{i=N_2,O_2,Ar,CO,H_2O} mw_i * y_i$$

Where $\sum y_i = 1$, if this constraint is not true (as in the case when we include water, since the volume fractions in Table 5-1 are for dry air) it can be renormalized by dividing each y_i by the sum of dry parts plus the water.

a) $mw_{DryAir} = 0.7808*28.0 + 0.2094*32.00 + 0.00934*39.95 + 0.000391*44.01 = 28.96 g/mol$

b) Use the method shown in Example 5-1 to find the partial pressure of the water:

$$\text{mol fraction} = \frac{P_{water}^{Actual}}{P_{atm}} = \frac{RH*VP_{water}(20°C)}{760 mmHg} = \frac{0.5*17.535}{760} = 0.0115$$

$$\sum_{i=N_2,O_2,Ar,CO,H_2O} y_i = 0.7808 + .2094 + .00934 + .000391 + 0.0115 = 1.0114$$

$$mw_{50\% RH} = \frac{0.7808}{1.0114}*28.01 + \frac{0.2094}{1.0114}*32.00 + \frac{0.00934}{1.0114}*39.95 +$$

$$\frac{0.000391}{1.0114}*44.01 + \frac{0.0115}{1.0114}*18.02 = 28.84 g/mol$$

Similarly, for part c:

$$\text{mol fraction} = \frac{P_{water}^{Actual}}{P_{atm}} = \frac{RH*VP_{water}(20°C)}{760 mmHg} = \frac{1.0*17.535}{760} = 0.0231$$

$$\sum_{i=N_2,O_2,Ar,CO,H_2O} y_i = 0.7808 + .2094 + .00934 + .000391 + 0.0231 = 1.0230$$

$$mw_{100\% RH} = \frac{0.7808}{1.0230}*28.01 + \frac{0.2094}{1.0230}*32.00 + \frac{0.00934}{1.0230}*39.95 +$$

$$\frac{0.000391}{1.0230}*44.01 + \frac{0.0115}{1.0230}*18.02 = 28.72 g/mol$$

The difference amounts to a decrease of less than 1%.

5.3 Structure

5.3.1 Layers

The atmosphere is composed of several layers, each associated with different temperature gradients, see Figure 5-1. The troposphere is the layer closest to the surface, with a depth of approximately 7 km at the poles, 12 km at midlatitudes, and 18 km in the tropics. The difference in depth is mainly caused by the differences in surface temperature, but also on the earths rotational speed, which is fastest at the equator. The troposphere contains about 90% of the atmospheric mass. It is very well mixed due to convection currents associated with rapid vertical exchanges of mass and energy. It has a decreasing temperature as elevation increases, from about 20°C at the surface to -60°C at the top of the layer. The lowest portion of the troposphere (near the earth's surface) is called the planetary boundary layer. The planetary boundary layer has a depth of about 1 km, which varies strongly with temperature, time of day, local meteorology, and surface features. The surface generates friction with the atmosphere, and different surface conditions cause different amounts of friction. Flat areas, like oceans, seas, and plains offer little friction, whereas areas like mountains and forests create large amounts of friction. Mixing of mass and energy between the planetary boundary layer and the free troposphere depends on the local atmospheric stability, as explained in section 5.7. The top of the troposphere is a thin layer called the tropopause where the temperature does not vary with height. This lack of thermal gradient repels vertical motion by convection, reducing the transfer of mass across this boundary.

The stratosphere is the next highest layer, with a top (stratopause) around 50 km. The temperature does not change much in the lower 20 km of this layer, remaining around -50°C, and then it increases steadily to about 0°C at the stratopause. The positive increase in temperature with increasing elevation creates an inversion and reduces vertical mixing, as will be discussed in section 5.7. The stratosphere layer contains about 90% of the atmosphere's ozone, which absorbs ultra-violet radiation from the sun, and is called the ozone layer (see Chapter 12). The absorption of UV light adds energy to the stratosphere and is the main cause for the increasing temperature in this layer.

The mesosphere is above the stratosphere. The temperature again decreases in this layer, which can reach values of -100°C at the top of the layer (mesopause). The negative temperature gradient allows instability to cause swift vertical motions. Above this layer is the thermosphere, where the temperature gradient becomes positive again, due to interactions between the gas molecules and solar radiation.

FIGURE 5-1. Structure of the Atmosphere. Adapted from (Brasseur, et al., 1999).

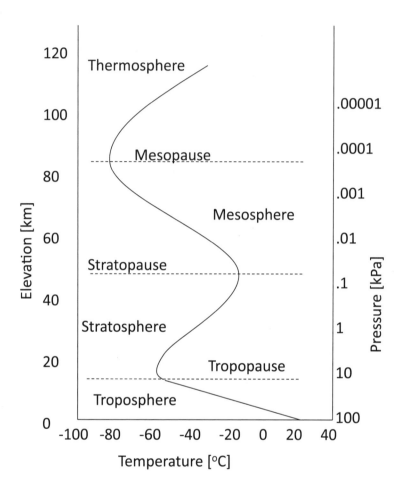

5.3.2 Pressure Profile

Pressure in the atmosphere decreases with elevation as described by the hydrostatic equation:

5.3
$$\frac{dP}{dh} = -\rho g$$

Where: P = atmospheric pressure [Pa],
h = elevation, h=0 at the surface and increases above the surface [m],
ρ = density of air [kg/m³], and
g = gravitational constant [9.81 m/s²].

The density of the atmosphere is not constant across large elevations. An expression can be obtained using the ideal gas equation of state, which relates the mass density to pressure and temperature:

5.4
$$\rho_{air} = \frac{n\, mw_{air}}{V} = \frac{P\, mw_{air}}{RT}$$

Where: n = moles of air [mol],
mw$_{air}$ = molecular weight [28.96 g/mol],
V = volume [m^3],
R = gas constant [8.314 $\frac{m^3 Pa}{mol\, K}$], and
T = absolute temperature [K].

Combining equations 5.3 and 5.4 and then rearranging yields;

5.5
$$dh = -\left[\frac{R}{mw_{air}\, g}\right]\frac{T}{P} dP$$

This equation contains two parameters that vary with elevation in the atmosphere, P and T. Pressure is the dependent variable. The temperature changes, so is included as a variable. The solution of equation 5.5 requires knowledge of temperature as either a function of pressure or elevation. Direct measurements of T and P from a weather balloon can generate this function, see homework problem 17. Otherwise, inspection of Figure 5-1 shows an idealized relationship for temperature in the troposphere – a linear decline:

5.6
$$\frac{dT}{dh} = -\gamma$$

Where: γ = a constant called the lapse rate [°C/ m].

The lapse rate shows temperature decreases with increasing elevation. The term dT/dh is approximately -1°C/100 m for dry air in the troposphere. The value changes to -0.6°C/100 m for saturated air (100% humidity). This quantity plays an important role in determining atmospheric stability (section 5.7, equation 5.20). If we assume a temperature T_0 at the surface (h=0) and constrain our solution to the troposphere, we can obtain a relationship between temperature and elevation:

5.7
$$T(h) = T_0 + (-\gamma) * h$$

Substitution of equation 5.7 into 5.4 and simplifying yields:

5.8
$$\rho = \frac{P \cdot mw}{R(T_0 - \gamma h)}$$

The only variables in the relationship are P, the dependent variable of equation 5.3, and h, the independent variable. All other terms are known. Substitution of 5.8 into 5.3 yields:

5.9
$$\frac{dP}{dh} = -\frac{P \cdot mw \cdot g}{R(T_0 - \gamma h)}$$

Applying a boundary condition $P = P_0$ at $h=0$, separating and integrating yields:

5.10
$$\int_{P_0}^{P} \frac{dP}{P} = -\frac{mw \cdot g}{R} \int_{0}^{h} \frac{dh}{T_0 - \gamma h}$$

Which yields:

5.11
$$P = P_0 \left[\frac{T_0 - \gamma h}{T_0} \right]^{\frac{mw \cdot g}{R \cdot \gamma}}$$

■ Example 5-4.

Determine the pressure at the top of Mt Everest (8848 meters). Assume that at sea level (h=0) the temperature is 32°C (305 K), and atmospheric pressure is 1 atm.

Solution:
Apply equation 5.11, using the conditions specified in the problem:

$$P = 1atm \left[\frac{305K - (0.6\frac{K}{100m})8848m}{305K} \right]^{\left[\frac{28.9\frac{g}{mol} \cdot 9.81\frac{m}{s^2}}{8.314\frac{m^3 Pa}{mol\,K} \cdot 0.6\frac{K}{100m} \cdot 1{,}000\frac{g}{ms^2}} \cdot \frac{1\,Pa}{} \right]} = 0.337\,atm$$

Compare this solution with example 4 in chapter 4 where temperature was assumed constant. There are only a few known measurements of barometric pressure atop Mt Everest – the values range from 251 – 253 torr (0.330 – 0.333 atm) when the temperature was (-21°C) (West, 1999).

5.4 Solar Radiation

The primary source of energy for atmospheric processes (motion and reactions) is from the sun. The energy transfers from the sun to the earth as electromagnetic radiation. The peak energy of this radiation occurs in the visible region of the light spectrum though there are significant contributions in the infrared and ultraviolet regions. The amount of energy received by the earth is called insolation (short for incoming solar radiation). Over the course of a year the solar radiation arriving at the top of the earth's atmosphere averages 1,350 - 1370 W/m² (watts per square meter). The amount absorbed by the earth's surface and atmosphere varies due to changes in the solar constant, the transparency of the atmosphere (which varies by wavelength of the radiation), daily sunlight duration (insolation is zero at night), and the angle of the sun relative to the earth's surface.

The solar constant represents the amount of energy available from the sun. It varies up to a few percent mostly due to the elliptical nature of earth's orbit – currently being closest to the sun in January and furthest in July. This solar energy can be absorbed or reflected by the atmosphere before it reaches the earth's surface.

The transparency is a measure of the amount of sunlight that makes it to the surface. Different substances can absorb this energy, for example approximately 98% of the ultraviolet radiation from the sun is absorbed by oxygen (O, O_2, and O_3) in the stratosphere. Transparency is greatly altered by clouds as well as the depth of atmosphere the light must travel through to reach the surface, which varies with the time of day due to the angle between the sun and the earth's surface.

Some materials reflect a portion of the energy back into space, excluding it from the earth's energy budget. This reflectivity is referred to as albedo (α). Different atmospheric constituents and regions of the surface will have different albedo values ($\alpha_{soil} = 0.1$, $\alpha_{snow} = 0.9$, $\alpha_{forest} = 0.2$, $\alpha_{city} = 0.1$, $\alpha_{cloud} = 0.5$; Note that these values are approximate and may vary considerably). Overall, the earth has an albedo of 0.3, which means 70% of the incoming radiation is absorbed by the earth's surface and atmosphere, and 30% is reflected back into space. About 15% of the total incoming energy is absorbed by the atmosphere, 5% by clouds, and 50% by the earth's surface. Local values may vary considerably – for example, nearly all the incoming energy in polar regions is reflected. Various constituents in the atmosphere may change the albedo of the air – notably many fine particles (e.g., sulfates) are very reflective and increase the albedo while some (e.g., soot) absorb solar radiation and warm the atmosphere.

The effect of this absorbed energy on surface temperature depends on the surface material. Land has a smaller specific heat*[1] than water, so when it absorbs the same amount of energy as water, it will have a higher temperature. An additional consideration for water in lakes and oceans is that the absorbed energy can set up a thermal current (convection) which can carry the energy to deeper regions of the water, whereas for solids the heat only moves by conduction. These factors contribute to the common observation of large differences between water and land temperatures.

5.5 Global Wind Circulation

There are four main factors determining wind speed and direction – pressure, Coriolis force, centrifugal force, and friction.

---

1. Specific heat represents the amount of energy needed to increase the temperature of a mass of material by a certain amount. It has units J/mol·K or cal/mol·K. This energy is stored in the material as vibrational energy of the constituent atoms and molecules

Uneven heating of the atmosphere and earth's surface causes pressure gradients. The uneven heating leads to the creation of high and low pressure areas (recall the ideal gas equation of state). The temperature differences between the poles and equator or over continents and oceans cause global scale pressure gradients to form. Differences between land and lakes or from changes in surface elevation lead to regional and local scale motions that may be very different from the global patterns. When pressure acts on a surface it generates a force, (F_p). The surface can be any boundary of any size or shaped region of the atmosphere and may include the boundary between the land or water and the atmosphere.

The Coriolis force (F_{Cor}) arises from the earth's rotation. Without it, the motion of air would tend in the direction from the high to the low pressure areas. However, the earth's rotation causes the surface to move fast at the equator and slow to zero at the poles. As air moves north or south this change in surface speed deflects the air's direction. In the Northern Hemisphere, the air deflects to the right of an observer facing the direction of the air motions (such that the air hits their back). In the Southern Hemisphere, the air is deflected to the left. The magnitude of this deflection is a function of the wind speed and the latitude (which correlates with the earth's rotational speed). The Coriolis force is always perpendicular to the wind direction. It is an inertial force that helps correct for motions due to having a rotating frame of reference (the earth's surface). Figure 5-2 shows the impact of this force as air moves towards a low pressure region. If the pressure is very low, this phenomenon can generate strong winds and storms including hurricanes and cyclones.

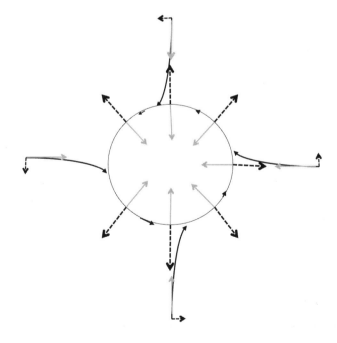

FIGURE 5-2. Example of Coriolis Effect Around a Low Pressure Cell. The curved arrows show flow streamlines. Pressure gradient forces are represented by the inward pointing (gray) arrows. The Coriolis force is always perpendicular to the velocity, as shown by the outward pointing (dashed) arrows. Illustration by Roland Geider, GNU Free Documentation License, 2007.

The centrifugal force (F_{Cen}) occurs when the wind's path curves. It is proportional to the wind speed and inversely proportional to the curvature of the flow path. Its direction is away from the center of curvature. The centrifugal force is also an inertial force.

A final force, friction (F_f) is only common within the planetary boundary layer (near the surface). Friction always acts in a direction opposite to the direction of the wind. There is very little friction above this layer (1 km) until the tropopause. Friction increases as wind speed increases and with the roughness of the surface. Its effect decreases as elevation increases.

These forces balance (sum to zero) when there is no acceleration of the air, as commonly happens in the upper troposphere, but rarely within the planetary boundary layer. The determination of actual directions requires a non-steady state force balance between the various forces acting on the air.

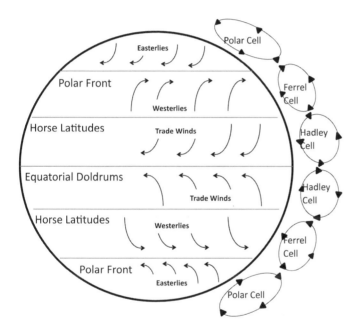

FIGURE 5-3. Three Cell Model of Air Circulation Around a Smooth Homogeneous Sphere. (Adapted from (Lazaridis, 2011).

Figure 5-3 shows the generalized Global scale horizontal and vertical motions can be generalized with a three-cell model. In this model the earth is a smooth sphere, and the Sun remains over the equator at all times. The greatest heating of the earth and atmosphere occurs at the equator. The heating causes the air to rise and move toward the poles. This air at the equator is warm, has small horizontal temperature gradients, and has slow surface winds. The region is known as the doldrums because of this slow wind speed. As the warm air rises it cools and the humidity condenses to generate Cumulus clouds. These clouds produce strong storms due to the release of the water vapor's latent heat. The air continues to rise until it reaches the tropopause, which acts as an obstacle, and diverts it toward the poles.

The air then moves north or south until it reaches about 30° latitude, where it descends and separates to return towards the equator or continue toward the pole. This air is very dry since it released its moisture when rising near the equator. The atmosphere generally has clear skies, low surface wind speeds, and warm temperatures. These locations are named the horse latitudes. Most of the world's deserts occur around the 30° latitudes. The wind flowing from the equator to the 30° latitude in the upper troposphere and back to the lower troposphere forms a closed circulation region known as a *Hadley cell*.

Some of the air continues moving in the lower troposphere toward the poles. Near 60° latitude it encounters cold air from the poles, known as the polar front, forcing it to rise. As this air moves across the surface between 30° and 60°, it picks up moisture from the ocean or land surface. When it ascends near 60° latitude, this moisture condenses and forms cloudy, stormy weather. This circulation region (30° – 60°) is called a *Ferrel cell*. The region between 60° – 90° is called the polar cell.

The surface winds that form in each region are diverted from their equator to pole direction by the Coriolis force. The surface winds formed in the Hadley cells are called the trade winds, because they helped ships traveling from Europe to the Americas during the Age of Sail. The direction of the *trade winds* is easterly because of the effect of the Coriolis force on the lower troposphere winds of the Hadley cell, which move in the direction towards the equator. The winds slow near the equator. The surface winds in the Ferrel cell are called westerlies, which were useful in sailing from the Americas back to Europe. Again, the westerly direction is caused by the Coriolis force acting on the lower troposphere winds in the Ferrel cell, which move in the direction from 30° latitude to 60°. The surface winds in the polar cell are *easterlies*.

Due to the uneven distribution of land and sea, the changing elevations of the Earth's surface, the Earth's axis of rotation, and many other parameters, this three cell model is only an approximation of actual wind patterns. These global trends can be greatly enhanced or changed by local patterns.

5.6 Local Wind Circulation

There are three common local scale wind patterns. These patterns must be considered when siting an emission source. These patterns result from localized uneven heating of the earth's surface between a lake and its shore, between a mountain and nearby valleys, or around an urban area.

Land-Sea winds originate from the difference in specific heat of land and water. The land heats faster than sea water when the sun rises and cools faster when the sun sets. The air rises over the warmer surface and settles over the cooler surface, thus setting up a local wind pattern, see Figure 5-4.

FIGURE 5-4. Land-Sea Wind Develops from Uneven Heating and Cooling of Different Surfaces. Adapted from (Lazaridis, 2011).

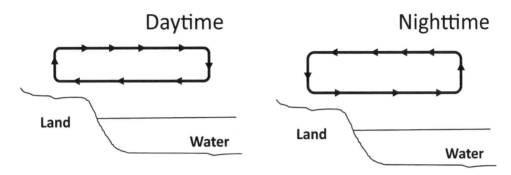

Mountain-Valley winds originate with the uneven heating and cooling between mountain slopes and nearby valleys or lowlands. After sunrise the sun-facing sides of a mountain absorb more solar energy and heats faster than the low-lying areas. The uneven surface heating causes the air to rise and then to settle over the cooler lowlands. In the evening the mountain side cools faster and the process is reversed, see Figure 5-5.

FIGURE 5-5. Mountain -Valley Wind Develops from Quicker Heating of Mountain Side During Day and Quicker Cooling at Night. Adapted from (Lazaridis, 2011).

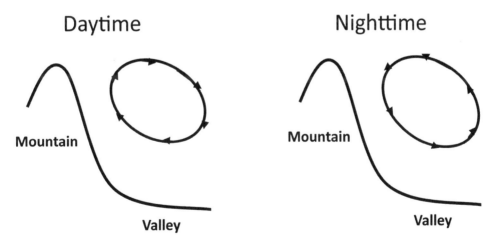

Urban areas have different surface roughness and thermal characteristics than their surrounding regions. Building materials such as concrete, brick, and asphalt absorb and hold heat more efficiently than soil and vegetation. After sunset, the urban area radiates more heat, which creates a warm area over the city called the heat island effect. It can cause the larger scale air masses to move around or over the city, which can lead to areas of stagnant air within the urban area. In opposition to this effect is the greater turbulence during air flow around buildings and the channeling along street corridors. Both trends make it extremely difficult to predict wind and weather in urban regions.

5.6.1 Wind Roses and Wind Speeds

Wind speed is measured 10 m above ground level. Measurements should be made in an area where other surface features (buildings, trees, hills) are unlikely to impact the measurement. The Beaufort Wind Scale, shown in Table 5-2, describes various wind speeds and their effects on land and over water.

Historical wind speed and direction trends are often considered in sighting emission sources such that they do not cause undue burden on the local population. These historical records are easily accessed in a wind rose. A wind rose (see Figure 5-6) provides a simple, but information-laden, view of wind speed and direction at a particular location. The wind rose shows the frequency of winds blowing FROM particular directions in a circle graph format. The length of each "spoke" around the circle is related to the frequency of time that the wind blows from a particular direction. The concentric circles quantify the frequency, starting from zero at the center to increasing values away from the center. Many wind roses contain additional information, in that each spoke is broken down into discrete frequency categories that show the percentage of time those winds blow from a particular direction and within a certain speed range. Other variations may include day and night differences, or they may include data over a season rather than a month. The wind rose shown here uses the 16 cardinal directions - N, NNE, NE, ENE, E, ESE, SE, SSE, S, SSW, SW, WSW, W, WNW, NW, NNW. Other wind rose data sets may use circular angles, where 0 is N, 90 is E, 180 is S and 270 is W. An excellent database for monthly US wind roses is hosted by the US Department of Agriculture – Natural Resources Conservation Service at [*http://www.wcc.nrcs.usda.gov/climate/windrose.html*].

Figure 5-6 shows the February wind rose for Duluth, MN, based on 30 years (1961 –1990) of hourly wind data (all hours of the day). The wind rose shows that the Duluth winds rarely blow from the northeast. In fact, winds from the NNE, NE and ENE account for less than 10% of all the hourly winds. This value is calculated by taking the sum of the frequencies of each of these directions (2.5 + 1.5 + 3.5 = 7.5%). The plot shows that the most common direction was from the NW at 14% of the time. The wind rose also categorizes the wind by speed in each direction. Each color represents a different speed, and the length of each color band in each direction represents the time the wind moves within the speed range.

Table 5-2. Beaufort Wind Scale. Adapted from (US-NOAA, 2011)

Beaufort Number	Wind Speed m/s	Wind Speed mile/hr	Description	Land Effects	Water Effects
0	<0.5	<1	Calm	Still, calm air, smoke rises vertically.	Water is mirror-like.
1	0.5 - 1.4	1 - 3	Light Air	Rising smoke drifts, wind vane is inactive.	Small ripples appear on water surface.
2	1.5 - 3.0	4 - 7	Light Breeze	Leaves rustle, can feel the wind on your face, wind vanes begin to move.	Small wavelets develop; crests are glassy.
3	3.1 - 5.3	8 - 12	Gentle Breeze	Leaves and small twigs move; lightweight flags extend.	Large wavelets, crests start to break, some whitecaps.
4	5.4 - 7.8	13 - 18	Moderate Breeze	Small branches move, raises dust, leaves and paper.	Small waves develop, becoming longer, whitecaps.
5	7.9 - 10.6	19 - 24	Fresh Breeze	Small trees sway.	White-crested wavelets (whitecaps) form, some spray.
6	10.7 - 13.6	25 - 31	Strong Breeze	Large tree branches move; telephone wires begin to "whistle," umbrellas are difficult to keep under control	Larger waves form, whitecaps prevalent, spray.
7	13.7 - 17.0	32 - 38	Moderate or Near Gale	Large trees sway, becoming difficult to walk.	Larger waves develop; white foam begins to form.
8	17.1 - 20.6	39 - 46	Gale or Fresh Gale	Twigs and small branches break from trees, walking is difficult.	Moderately large waves with blown foam.
9	20.7 - 24.4	47 - 54	Strong Gale	Slight damage occurs to buildings; shingles are blown off of roofs.	High waves (six meters), rolling seas, dense foam, Blowing spray reduces visibility.
10	24.5 - 28.3	55 - 63	Whole Gale or Storm	Trees are broken or uprooted; building damage is considerable.	Large waves (6-9 meters), overhanging crests, sea becomes white with foam, heavy rolling, reduced visibility.
11	28.4 - 32.5	64 - 72	Violent Storm	Extensive, widespread damage.	Large waves (9-14 meters), white foam, visibility further reduced.
12	> 32.5	73+	Hurricane	Extreme destruction, devastation.	Large waves over 14 meters, air filled with foam, sea white with foam and driving-spray, little visibility.

FIGURE 5-6. Wind Rose Showing Wind Data (speed and direction) for February in Duluth, MN, Averaged Over 1961 - 1990. (USDA, 2011).

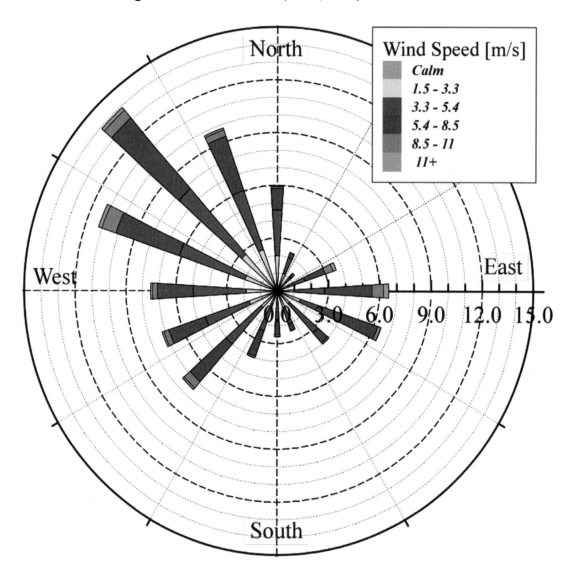

5.6.2 Vertical Profiles

Horizontal wind velocity generally increases with elevation within the boundary layer. The change can be modeled with a power law equation:

5.12
$$\frac{u_2}{u_1} = \left(\frac{h_2}{h_1}\right)^p$$

Where: u_1, u_2 = wind velocity at the lower and higher elevation,
h_1, h_2 = the lower and higher elevation, and
p = stability exponent.

Table 5-3 lists the stability exponent (p) values. The value depends on the surface characteristics and atmospheric stability, as described in the next section.

Table 5-3. Stability Exponent, p, for Vertical Wind Profiles (Equation 5.12).

Stability Class	Rural Exponent	Urban Exponent
A	0.07	0.15
B	0.07	0.15
C	0.10	0.20
D	0.15	0.25
E	0.35	0.30
F	0.55	0.30

■ Example 5-5.

Calculate the wind speed at the top of a 55 m high emission stack. The stack is in an urban area; the atmosphere is stability class D; and the wind speed at 10 m is 8 miles per hour.

Solution:
Apply the given information to equation 5.12. u1 = 10 mph, h_1 = 10 m, h_2 = 50 m, and p = 0.25.

$$u_2 = u_1 \left(\frac{h_2}{h_1}\right)^p = 10 mph \left(\frac{55m}{10m}\right)^{0.25} = 15.3 mph$$

As expected from Figure 4.2, the wind velocity increases with increasing elevation.

5.7 Stability

Atmospheric stability is related to the ease of vertical motion of the air. Stable air resists small vertical motions, so a parcel of air that is displaced vertically returns to its initial position. Unstable air assists small vertical motions, so a parcel of air that is displaced vertically continues moving away from its initial position. Unstable air is considered a better case for air pollution since it increases the mixing and dilution of an emission. Stable air may trap emissions near the source, leading to unhealthy air concentrations.

Stability is a function of the vertical temperature profile and water content of the air. Cold and dry air is denser than warm or humid air. When a warm parcel is emitted into colder or dryer air, it tends to ascend. The warm air parcel in this case expands and cools as it rises until eventually its temperature equals the air around it, and it loses its driving force. The vertical motion of air is usually assumed to be *adiabatic* – this means the air parcel does not exchange energy with its surroundings. Rather the cooling happens from expansion due to the reduction in atmospheric pressure as it rises.

5.7.1 Lapse Rate

The thermodynamic process in which pressure, volume, and temperature all change for a constant number of moles within a parcel is called a *polytropic* process. The basic thermodynamic laws, combined with the ideal gas equation of state, create a relationship between vertical motion and temperature.

The differential form of the first law of thermodynamics (energy is conserved) for an ideal gas, non-flow system is:

5.13 $$dU = dQ + dW$$

Where: U = internal energy of the parcel,
Q = energy flow (heat) across the system boundary, not associated with mass flow, and
W = work energy across the system boundary.

The internal energy is a function of temperature:

5.14 $$dU = C_v dT$$

Where: C_v = heat capacity under constant volume conditions, and
T = absolute temperature.

Similarly, work can be expressed for a mechanically reversible process:

5.15
$$dW = Pd\hat{V}$$

Where: P = pressure, and
$\hat{V}$ = volume per mol.

Substitution of 5.14 and 5.15 into 5.13 yields:

5.16
$$dQ = C_v dT + Pd\hat{V}$$

The partial derivative of the ideal gas law (PV=nRT) for 1 mol of gas when P, V, and T are allowed to vary gives:

5.17
$$Pd\hat{V} + \hat{V}dP = RdT$$

Substitution into 5.16:

5.18
$$dQ = (C_v + R)dT - \hat{V}dP = C_p dT - \hat{V}dP$$

Recalling that, for an ideal gas, $C_v+R = C_p$, the heat capacity under constant pressure conditions.

The heat flow across the system boundary is zero for an adiabatic process (dQ = 0), a good assumption between air parcels with small temperature differences. Combine this assumption with the hydrostatic equation (dP= -ϱgdh) and equation 5.18, yields:

5.19
$$\frac{dT}{dh} = -\frac{\hat{V}g\rho}{C_p}$$

Recognize that the volume per mole is equal to the inverse of the molar density ($\hat{V}$=1/ϱ), yielding:

5.20
$$\frac{dT}{dh} = -\frac{g}{C_p} = -\gamma$$

Where γ is the *adiabatic lapse rate*. For dry air, this lapse rate is given the symbol Γ. Its value can be calculated using the values for gravity = 9.81 m/sec^2, and the heat capacity (assuming 1 mole of dry air at 298 K as listed in appendix 6c:

5.21
$$C_p = R(3.355 + 5.75*T - 1600/T^2) = 8.314\frac{J}{K}*3.5083 = 1006.5\frac{m^2}{K \cdot \sec^2}$$

The calculation gives a value for the dry adiabatic lapse rate of Γ = 1°C/100m. Note that this value means the temperature decreases 1°C for every increase in elevation of 100 m. The lapse rate changes for humid air to (0.6°C/100 m). The International Standard Atmosphere (ISA) has a lapse rate of (0.65°C/100 m). This value was used in section 5.3.2 to determine the effect of temperature on atmospheric pressure due to elevation changes. These values represent the change in temperature for an adiabatic vertical motion of a gas parcel that is not accelerating. Most calculations of stability use the dry adiabatic lapse rate.

5.7.2 Stability Analysis

A comparison between Γ and the actual change in temperature with elevation, Λ, also called the environmental lapse rate, is used to describe atmospheric stability.

$\Gamma < \Lambda$, unstable
$\Gamma = \Lambda$, neutral
$\Gamma > \Lambda$, stable

where $\Lambda = \left(-\dfrac{\Delta T_{actual}}{\Delta h}\right)$ and $\Gamma = \left(-\dfrac{\Delta T_{adiabatic}}{\Delta h}\right) = 1.0\dfrac{^\circ C}{100m}$

FIGURE 5-7. Comparison of Adiabatic and Environmental Lapse Rates Showing Stable, Neutral, and Unstable Comparisons.

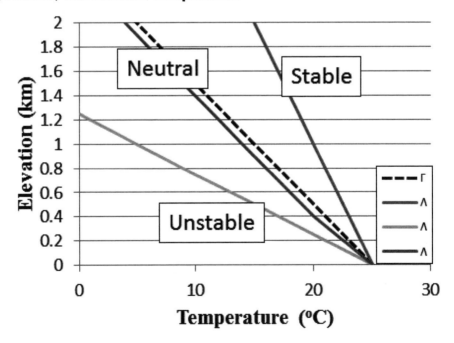

Note that this diagram places the independent variable (T) on the x-axis and the dependent variable (h) on the y-axis, an inversion of the traditional graphical representation. The slopes on this plot are $\Delta h/\Delta T$.

Figure 5-7 shows a comparison between the dry air adiabatic lapse rate, Γ, and several values of the environmental lapse rate, Λ. This figure can be understood by considering a small parcel of air that is nudged upward. As it moves upward, it changes temperature along the adiabatic lapse rate line (dashed line). Under stable conditions, this parcel cools faster than the surrounding air, making it denser and heavier. It stops its ascent and moves back to its original position. Under unstable conditions, this parcel cools slower than the surrounding air, making it even less dense and lighter than the surrounding air. It continues its ascent. A similar behavior exists if the original parcel is nudged downward instead, and it continues to descend. Under neutral conditions, the movement of the parcel is neither helped nor hindered.

Another method of characterizing atmospheric conditions, the Pasquill Stability Classes (Pasquill, 1961) is commonly used. Table 5-4 and Table 5-5 classify stability based on either the values of the environmental lapse rate or actual meteorological conditions of insolation and wind speed at 10 m above the surface. These classifications will be used in Chapter 6 for calculation of the impact stability has on the dispersion of a contaminant released into the atmosphere.

Table 5-4. Descriptions of the Pasquill Stability Classes by Environmental Lapse Rate.

Stability Class	Description	Λ, Environmental lapse rate °C/100m
A	Extremely unstable	Greater than 1.9
B	Moderately unstable	1.9 to 1.7
C	Slightly unstable	1.7 to 1.5
D	Neutral (Γ=1)	1.5 to 0.5
E	Slightly stable	0.5 to -1.5
F	Moderately stable	-1.5 to -4.0
G	Extremely stable	Less than -4.0

Table 5-5. Meteorological Descriptions of Pasquill Stability Classes.

	Daytime Isolation[1]			Night-time Conditions[2,3]	
Wind Speed [m/s][4]	Strong	Moderate	Slight	Clear - ½ low cloud	≥ ½ cloudiness
< 2	A	A – B	B	E	F
2 – 3	A - B	B	C	E	F
3 – 5	B	B – C	C	D	E
5 – 6	C	C – D	D	D	D
> 6	C	D	D	D	D

NOTES:

$$\Lambda = \left(-\frac{\Delta T_{actual}}{\Delta h}\right)$$

1. Strong insolation corresponds to sunny midday in midsummer in England; slight insolation to similar conditions in midwinter.
2. Night refers to the period from 1 hour before sunset to 1 hour after sunrise.
3. The neutral category D should also be used, regardless of wind speed, for overcast conditions during day or night and for any sky conditions during the hour preceding or following nightfall, as defined in Note 2.
4. Wind speed measured at 10 m above ground level.

■ Example 5-6.

A weather balloon measures the temperature as 12°C and pressure as 0.9625 atm at some elevation. If the ground level temperature was 16°C and pressure was 1.0021 atm, determine a) the elevation of the measurement and b) the environmental lapse rate assuming that the lapse rate is constant.

Solution:

Equation 5.11 can not be used without knowing h or γ. Instead, use equation 5.5 and solve it numerically with the trapezoid rule [see any introductory calculus text]. The first line gives the general form for use with many data points, the second line applies the formula to two specific points.:

5.22
$$\int_{h_0}^{h_1} dh = -\left[\frac{R}{mw_{air}g}\right]\int_{P_0}^{P_1} \frac{T}{P} dP = -\frac{R_{air}}{g}\left(\left[\frac{T_{i+1}}{P_{i+1}} + \frac{T_i}{P_i}\right]\left[\frac{P_{i+1} - P_i}{2}\right]\right) = h_{i+1} - h_i$$

$$= -\frac{R_{air}}{g}\left(\left[\frac{T_1}{P_1} + \frac{T_0}{P_0}\right]\left[\frac{P_1 - P_0}{2}\right]\right) = h_1 - h_0$$

Where h_1 = the unknown T_0 = 273 + 16°C = 289 K
h_0 = 0 m T_1 = 273+12°C = 285 K
P_0 = 1.0021 atm g = 9.81 m/s²
P_1 = 0.9625 atm R_{air} = R/mw$_{air}$ = 287.5 m² / [K s²]

Solving:

$$h_1 = -\frac{287.5 \frac{m^2}{Ks^2}}{9.81 \frac{m}{s}}\left(\left[\frac{285K}{0.9625 atm} + \frac{289K}{1.0021 atm}\right]\left[\frac{0.9625 atm - 1.0021 atm}{2}\right]\right) = 340m$$

Calculate the environmental lapse rate:

$$\Lambda = -\frac{\Delta T}{\Delta h} = -\frac{T_1 - T_0}{h_1 - h_0} = -\frac{285 - 289}{340} = \frac{1.18K}{100m}$$

This value is very close to the dry adiabatic lapse rate so the atmosphere stability is class D, neutral.

5.7.3 Inversions

The most important weather phenomenon when considering air pollution meteorology is an inversion layer. These occur when temperature increases with elevation. They may happen at the surface or in a layer above the surface. There are three types of inversions; frontal, radiation, and subsidence.

Frontal inversions (Figure 5-8) happen when a cold front displaces warm air, as occurs during storms. The inversion rarely lasts more than 10 – 15 minutes. It is not important for air pollution considerations because it is short lived and usually occurs with rain.

FIGURE 5-8. Frontal Inversion Schematic and Vertical Temperature Profile.

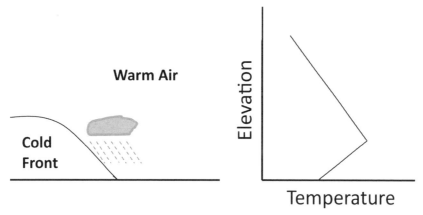

Radiation inversions (Figure 5-9) typically happen after sun-set on evenings with clear skies and little or no wind. They occur when warm surface air is cooled by a colder ground surface. The ground may be cooler for many reasons – the warm air has not had a chance to heat it such as in winter and spring, it may have evaporating moisture such as after a rainstorm or from vegetation. The inversion may last for several hours and will trap ground level emissions from automobiles, buses, trucks, and trains at the surface. The inversion layer may extend from the surface to 100 m. This type of inversion may make air unhealthy near roadways, airports, harbors, and train stations during the inversion times (sunset until after midnight). These conditions would make it highly advisable to avoid physical exercise near such areas.

FIGURE 5-9. Radiation Inversion Schematic and Vertical Temperature Profile. This phenomenon is caused when the ground is colder than the air above it. It typically occurs between sunset and midnight.

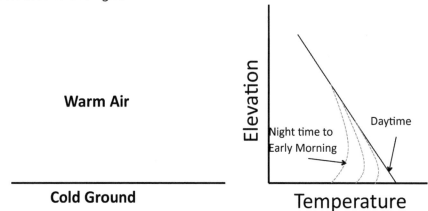

Subsidence inversions (Figure 5-10) happen in areas dominated by a stagnant high pressure cell. The inversion forms when subsiding air undergoes adiabatic heating while the air nearer the surface remains cool. If the air sinks low enough, the air at higher altitudes becomes warmer than at lower altitudes, producing a temperature inversion. These inversions may last for many days depending on when the high pressure cell moves or breaks up. A typical subsidence inversion layer is two to three thousand feet thick, and the base or lower limit of the layer may vary anywhere from 4 to 6 thousand feet above ground level to near 16 to 18 thousand feet above ground level. However, most subsidence inversions hover near the 6 to 10 thousand foot mark. Many of the most severe air pollution episodes occur during subsidence inversions due to their strength (layer thickness) and the amount of time they persist.

FIGURE 5-10. Subsidence Inversion Schematic and Vertical Temperature Profile.

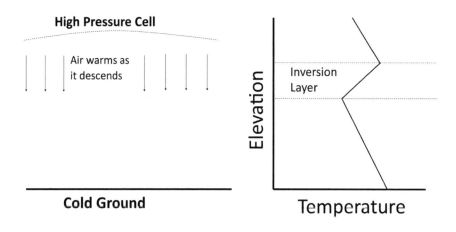

5.7.4 Mixing Depth

The amount of air available for mixing and diluting emissions varies with time of day, season, and by topographic features. The larger the volume, the smaller impact emissions have on health and welfare. The Maximum Mixing Depth (MMD) represents the height of the atmosphere available for mixing. It is the maximum elevation a parcel of air rises due to convection. The air below the MMD is called the mixing layer.

The MMD can be determined by comparing the dry adiabatic lapse rate of a heated air parcel at the surface to the actual temperature profile. The heating may be due to solar radiation, infrared radiation from the earth's surface, or from the process of generating an emission, such as the exhaust of a car or discharge from a power plant. In general, the two rate lines have different initial temperatures, T_0^{Env} and T_0^{Adb}, and different slopes (i.e. Λ and Γ) where the actual temperature profile has an initial temperature of the ground surface, and the adiabatic rate has the actual parcel temperature at its elevation of release. The intersection of these two rate lines is the MMD. Figure 5-11 shows several cases a) stable air, b) radiation inversion, and c) subsidence inversion layer.

Values of MMD are lowest at night and increase during the day by a factor of about 2. It is also a minimum in winter and a maximum in summer, differing by a factor of 1.5 to 3.0. During a radiation inversion, the MMD can be zero, depending on the initial temperature of the release. The MMD can be very high for unstable air, even reaching the tropopause. MMD values of 1500 m and less correlate with severe air pollution episodes.

FIGURE 5-11. Maximum Mixing Depth (MMD) for Different Atmospheric Conditions.

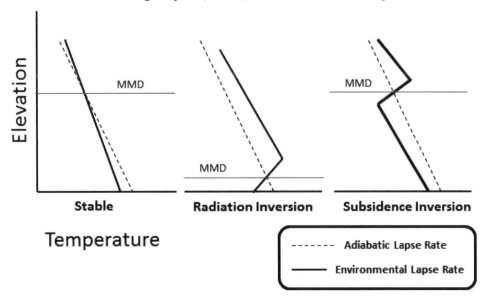

5.7.5 Stack Plumes

The atmospheric stability and resulting mixing depth have a large effect on the concentration of air pollutants. Wind speed, direction, and the amount of change in direction also play essential roles in determining concentrations. The amount of mixing and dilution caused by these processes is called dispersion, as discussed in Chapter 6. This chapter concludes with a discussion of the effects the different stability conditions have on a point-source emission from a stack. These emissions form a cloud or plume and take on characteristic shapes depending on the atmospheric stability.

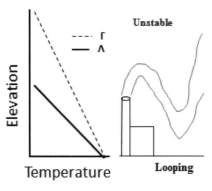

Looping plumes occur in highly unstable conditions and are due to the turbulence from the ease with which the air may overturn. The unstable condition is usually good for mixing and diluting emissions; however, it can cause brief episodes of high ground level concentration if the plume is diverted to the surface. Looping plumes typically occur in clear, daytime conditions with weak surface winds.

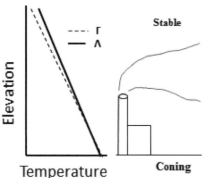

Coning plumes occur in neutral or slightly stable conditions. It is likely to occur on cloudy days, at dawn and dusk, and after the breakup of a radiation inversion. Wind speed may be moderate to strong. The shape of the cone expands approximately within a 20o arc centered on the point source for a wind of constant direction.

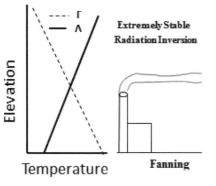

Fanning plumes occur in stable conditions. The lack of vertical motion can hold the plume at an elevation for long times / distances downwind from the source. The plume can still spread horizontally and can take on a fan shape. Fanning plumes often occur during radiation inversions. Ground level concentrations are unlikely to be impacted by a fanning plume.

Air Pollution Meterology

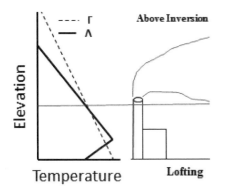

Lofting plumes occur when the plume is released above the top of an inversion layer into an unstable region. The plume continues to rise at its top, but the bottom layer is prevented from descending by the inversion. These plumes are unlikely to affect ground level concentrations. Lofting can happen when there is a ground level inversion.

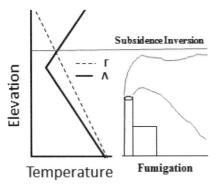

Fumigation plumes occur below an inversion layer. The top of the plume is prevented from rising by the inversion, but the bottom is free to mix in the unstable air below the inversion. Such conditions can lead to significant ground level concentrations. Tall stacks are used to prevent this from creating unhealthy ground level air. Fumigation plumes typically occur during the summer when skies are clear and winds are slow. The conditions typically last less than an hour.

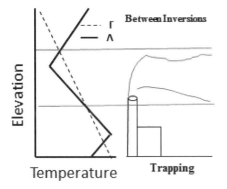

Trapping plumes occur to emissions released between two inversion layers. The top of the plume can not rise past the bottom of the upper inversion, and the bottom can not descend below the top of the lower inversion. The result is similar to the fanning case, but the plume thickness can be greater.

5.8 Questions

* - Questions and problems may require additional information not included in the textbook.

1. Describe in your own words why does the wind move?

2. Is vertical mixing possible in stable air?

3. Why does the temperature of the stratosphere increase with elevation, whereas in the troposphere it decreases?*

4. Describe what may happen to an air pollutant that is emitted into the air at the dew point.

5. Look at a map of the Atlantic Ocean. Use Figure 5-3 to determine the best latitudes to cross the ocean in an east direction and in a west direction.*

6. How might Figure 5-4 and Figure 5-5 inform a decision on locating a factory with air polluting emissions?

7. Find a wind rose for your hometown. How does today's weather forecast compare to the long term average?

8. When this book was published, the CO_2 concentration in the air had just surpassed 400 ppmv. What is the concentration today? What is the percentage change in this value? A different textbook reports that the CO_2 concentration is 360 ppmv. What year was that book published?

9. Make a list of three questions you have about this chapter or air pollution concerns you have.

5.9 Problems

1. What is the volume fraction of water in air at 30°C, 1 atm, and RH = 75%? What is the dew point?

2. Calculate the mass of air in 100 m³ at P = 750 mmHg, T = 15°C, and RH = 45%.

3. What would be the thickness of the atmosphere if it had a constant pressure of 760 mmHg and a constant temperature of 24°C? Note that the total dry mass of the atmosphere is estimated to be 5.13 x 10^{18} kg, and the radius of the Earth is 6.4 x 10^6 m.

4. Find the atmospheric pressure at 2000 m given that the surface conditions are 20°C and 1 atm, assuming a) dry air and b) humid air.

5. The surface temperature is 22°C, and surface pressure is 765 mmHg. If the pressure at a height of 1500 m is 640 mmHg, find the actual (or environmental) lapse rate, assuming dry air.

6. Sketch a wind rose from the following dataset.

Wind Direction	Beaufort Number					
	1	2	3	4	5	6+
N	0.8	4.0	4.7	3.2	0.5	0.2
NE	0.8	3.9	3.9	1.3	0.0	0.0
E	0.8	3.1	3.1	0.8	0.0	0.0
SE	0.8	3.7	6.3	6.6	0.7	0.1
S	0.8	5.0	7.1	6.4	1.3	0.3
SW	0.8	2.7	2.9	1.2	0.1	0.0
W	0.8	2.9	2.6	2.0	0.6	0.0
NW	0.8	4.0	4.7	3.3	1.0	0.2

7. Find the wind velocity at the top of a 100 m emission stack in an urban area, if the velocity at 10 m is 3.2 m/s, and the weather is overcast.

8. Given the wind speed is 1.5 m/s at a height of 10m, calculate the wind speed at 200 m, 400m, and 800 m for each stability class for rural conditions.

9. Wind speed was measured to be 2.3 m/s at a 10m height. Determine the difference in wind speed at 500 m between an urban and rural location for a B Pasquill stability class.

10. Determine if the atmosphere is stable, neutral, or unstable for the following conditions: 1) by comparing with the dry adiabatic lapse rate and 2) using the Pasquill Stability Classes:

Case	h_0 [m]	T_0 [°C]	h [m]	T [°C]
A	0	22	1000	26
B	0	32	1500	26
C	0	10	2000	35
D	500	25	2000	10
E	1000	35	2500	8
F	1500	32	2500	12

11. Estimate the temperature at 1000 m given a surface temperature of 22°C for a clear sunny day with a gentle breeze.

12. Find the maximum mixing depth (MMD) of air at 35°C released into the atmosphere and that rises at the dry adiabatic lapse rate, given that the actual surface temperature is 25°C, and the temperature decreases at a rate of 0.5°C/100m.

13. The atmosphere at a certain location has an actual air temperature of 23°C at the surface, 11°C at 1000 m, and 14°C at 1500m. Find the maximum mixing depth for a dry air parcel with a surface temperature of 25°C.

14. Determine the difference in MMD for two different air emissions released into the air that is 20°C at the surface, 9°C at 1000 m, 13°C at 1200 m, and 2°C at 3000 m. Assume that the temperature of air parcel one is 22°C, and air parcel two is 26°C.

15. A weather balloon is released and later reports a temperature of 8°C and pressure of 0.900 atm. The surface temperature was 16°C and pressure 0.995 atm. Determine the atmosphere stability class, assuming the temperature profile is constant.

16. Find the temperature at 800m when the ground level temperature is 10oC. Assume the environmental lapse rate is constant, and the stability class is B.

17. A weather balloon released at ground level (T=15°C and P=0.996 atm) reports the data in the table below. Estimate the elevation associated with each reading. Plot the temperature and pressure as a function of elevation.

Reading	P	T
1	0.961	12.5
2	0.927	10
3	0.905	14
4	0.863	15
5	0.833	14
6	0.8	12.3
7	0.76	9.6

18. Using the profile obtained in problem 17, estimate the MMD and the temperature profile effect on a stack releasing emissions at 100 m with a temperature of 19°C.

5.10 Group Project Ideas

For each project, the students should work in small groups and present their finding in either a short report (5-8 pages) or a 15 minute presentation:

1. Make a comparison between the local weather conditions and the weather predictions for your area. What is the difference between the actual and projected temperature, humidity, and rainfall? Does this value differ for a two day projection? a five-day projection? What would you consider to be an acceptable difference? You will need several weeks' worth of data to complete this project.

5.11 Bibliography

Brasseur, G P, Orlando, J J and Tyndall, G S. 1999. *Atmospheric Chemistry and Global Change*. New York, NY, USA : Oxford University Press, 1999. ISBN 0-19-510521-4.

Lazaridis, Mihalis. 2011. *First Principles of Meteorology and Air Pollution*. New York, NY, USA : Springer, 2011. ISBN 978-94-007-0161-8.

Pasquill, F. 1961. The Estimation of the Dispersion of Windborne Material. *Meteorological magazine*. 1961, Vol. 90, pp. 33-49.

USDA. 2011. Wind Rose Data. *Natural Resources Conservation Service*. [Online] 2011. [Cited: May 10, 2011.] http://www.wcc.nrcs.usda.gov/climate/windrose.html.

US-NOAA. 2011. Beaufort Wind Scale. *Storm Prediction Center*. [Online] National Oceanic and Atmospheric Administration., July 26, 2011. [Cited: July 26, 2011.] http://www.spc.noaa.gov/faq/tornado/beaufort.html.

West, John B. 1999. Barometric pressures on Mt. Everest: new data and physiological significance. *Journal of Apllied Physiology*. 1999, Vol. 86, 3, pp. 1062-1066.

CHAPTER 6

Air Pollution Dispersion

6.1 Introduction

Air pollution dispersion models are used to predict the concentrations of emissions as a function of distance from the discharge source. There are two main types of variables – meteorological and source conditions. Meteorological variables include wind speed and direction, atmospheric stability, and temperature. Source conditions include location, elevation, emission rate, velocity, temperature, and pollutant (PM, NO_x, SO_x, VOC, HAP, ...). Models incorporate some or all of these variables to estimate the amount of mixing a plume undergoes as it is moves from the source. Mixing has many causes: diffusion, wind speed and direction changes (tortuosity), friction within the boundary layer, and turbulence. Advanced models can include effects from local terrain and buildings, and changes due to chemical reactions, surface deposition, and washout by rain or snow.

Dispersion models are used by regulating agencies to determine if a new or modified emission will have an impact on ambient air quality. Not all emission sources are required to run dispersion models. State or national guidelines define which sources need to model their emissions. Identification of sources that are required to model will depend on the type of emission, the amount of emissions of a particular substance or group of substances, or the current ambient air conditions. Also, not all sources that are required to perform models need to install pollution control equipment. The decision depends on the results of the modeling.

The US EPA and state regulatory agencies require emission modeling for major sources. Major sources (such as fossil fuel-fired power plants, kraft paper mills, smelters, steel mills and oil refineries) are those that have the potential to emit 100 ton/ yr or more of any one pollutant regulated by the Clean Air Act. Other facilities that can potentially emit 250 ton/yr. or more of any combination of pollutants are also classified as major sources. Potential emissions assume that the source operates at maximum capacity for a full year. These values may be modified if the source is in a region that has pollutants that exceed the National Ambient Air Quality Standards (NAAQS), or if the source is near a protected area (recreational area, national park, or wilderness area). Different limits may also be imposed under Prevention of Significant Deterioration rules (see chapter 3.2.4 New Source Review for more details).

6.2 Modeling

There are two types of models: screening level models and refined models. Initially, a screening level model is used. These models use relatively simple estimation techniques and assumptions that tend to overestimate pollutant concentrations downwind. If the screening model suggests there may be an air quality impact, then refined models are used to reassess the emission. Results from such models are used to justify the requirement of emission reduction, site relocation, or another appropriate action.

There are four types of screening level models: Gaussian, numerical, statistical, and physical - **Gaussian** models assume a single point source plume that follows a Gaussian or normal probability distribution equation. This type of model system is straight-forward in terms of calculations (steady state, non-reactive, single component) and is the most widely used. **Numerical** models are used for area sources and may include chemical reactions. They also allow multiple constituents to be included and modeled simultaneously. They relax many of the assumptions used in the Gaussian model, but require much more detailed information. **Statistical** models are used to allow modeling when the system information is poorly known. Correlations and approximation methods replace the detailed knowledge required in other models. **Physical** models involve the construction of scaled models to allow physical observations of fluid flow within the model system. This method can model very complex terrain or building effects, but it is very time consuming.

The choice of screening level model depends on the type of pollutants being emitted, the complexity of the source, the type of topography surrounding the facility, and the type and amount of known information. Many pollutants react within ambient air to transform into different forms - particles can coagulate (fluid to solid), agglomerate (two or more dissimilar solids join to form a larger solid), conglomerate (two or more similar solids join to form a larger solid), or dissolve into water; NO_x and VOCs photo-react to form ozone; SO_x and NO_x can react with water to form acid gases and particles. Source complexity, such as multiple point sources, line sources, or area sources complicate calculations and require greater knowledge of source type and geometry. Topography issues, such as the presence of a river or mountain valley, changes in land use (rural to urban transformations), hills, and large buildings can bias the local weather patterns by constraining local atmospheric movement. Elevated terrain heights downwind may experience higher pollutant concentrations since they are closer to the plume centerline. Proposed sources in complex environments may be difficult to precisely model, and the model may need to improvise for missing information by using correlations, regional approximations, and history match methods.

A detailed discussion of each model is beyond the scope of this text. The most commonly used model system, the Gaussian dispersion model, serves to introduce the concepts. Additional information about other models may be found in sources such as (US-EPA, 2011), (EEA, 2007), and (Moussiopoulos, et al., 1996).

6.3 Gaussian Model

The Gaussian model equation is based on a solution to the Advection Dispersion Equation (ADE):

6.1
$$\frac{\partial C}{\partial t} = -u_i \frac{\partial C}{\partial x_i} + \frac{\partial}{\partial x_i}\left[D_{ij}\frac{\partial C}{\partial x_j}\right]$$

Where: C = concentration at any point in the system,
t = time,
x_i = position vector (x, y, z),
u_i = velocity vector (v_x, v_y, v_z), and
D_{ij} = dispersion tensor, a 3 x 3 tensor describing mixing in each direction caused by forces acting in each direction.

The ADE is used to describe the transport of a contaminant due to both advection (transport due to the bulk motion of the fluid) and dispersion (transport due to mixing motions such as diffusion, flow path tortuosity, and turbulence). The solution includes the following assumptions:
1. Steady state, $\partial C / \partial t = 0$,
2. Advection occurs only in the x-direction ($u_i = u_x$) and is constant,
3. Advection has a much larger effect than dispersion, and
4. Contaminant does not react or otherwise change form.

Assumption 2 allows all the non-diagonal dispersion terms to be set to zero. Assumption 3 allows the D_{xx} term to be ignored, which implies dispersion is zero in the x-direction. Applying these assumptions simplifies the ADE to:

6.2
$$-u_x \frac{\partial C}{\partial x} + D_{yy} \frac{\partial^2 C}{\partial y^2} + D_{zz} \frac{\partial^2 C}{\partial z^2} = 0$$

Applying the additional conditions:
C = C_0 at (x=0, y=0, z=0) = 0 [contaminant is discharged at a constant value from the point (0, 0, 0)]
C = 0 as x, y, z $\to \infty$ [contaminant has background concentration of zero]

The general form of the solution is:

6.3
$$C(x,y,z) = \frac{Q}{u} \frac{F_y}{\sigma_y \sqrt{2\pi}} \frac{F_{z1} + F_{z2} + F_{z3}}{\sigma_z \sqrt{2\pi}}$$

Where: C = downwind, steady state concentration [μg/m3],
x = distance downwind from source [m],
y = horizontal distance from plume centerline [m],
z = vertical distance from plume centerline [m],
Q = emission rate [μg/s],
u = wind speed at effective stack height, always in x direction [m/s],
σ_y = mixing length in horizontal direction [m],
σ_z = mixing length in vertical direction [m],
F_y = horizontal mixing (see equation 6.5), and
F_z = vertical mixing (see equation 6.7).

Figure 6-1 provides a sketch of a Gaussian model from a stack source. This model assumes that the wind speed is constant in direction and magnitude, that wind speed is not a function of elevation, the plume has a Gaussian or normal distribution, the emission rate (Q) is continuous and constant, that there are no other emission sources, and there are no additional sources of turbulence (nearby buildings, for example). While the assumptions are limiting, these models still provide an order of magnitude estimate and allow identification of when and where a potential problem may occur. The Gaussian models are quite conservative in that they overestimate down-wind ground level concentrations by factors of 2-3 or even 10 fold. They are excellent first models to determine if the source could be a problem because it is quick and easy to use and provides an overly conservative estimate.

FIGURE 6-1. Stack-Based Emission Source and Gaussian Model Approximation of Plume.

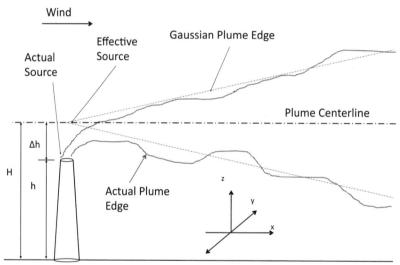

The mixing lengths, σ_y and σ_z, are functions of downwind distance and atmospheric stability (Pasquill stability class, section 5.7.2). The horizontal mixing length, σ_y, can be read from Figure 6-2 (easiest) or calculated for rural areas using equation 6.4 and Appendix IV Table 1, or for urban areas use the formulas in Appendix IV Table 2. These equations originate from experimental observation of measurements over flat, rural terrain (Turner, 1994).

6.4 $$\sigma_y = 465.116 * x * \tan\left[0.0174533(a - b * \ln[x])\right]$$

Where: x = distance downwind in km,
a and b = rural area parameters, values are given in appendix IV Table 1, and
tan = tangent function, with argument in radians.

Horizontal mixing is symmetric about the plume centerline:

6.5
$$F_y = \exp\left(-\frac{y^2}{2\sigma_y^2}\right)$$

The horizontal mixing value increases with downwind distance, increases with increasing atmospheric instability, and is generally larger in urban environments.

The vertical mixing length, σ_z, can be read from Figure 6-3 or calculated for rural areas using equation 6.6 and Appendix IV Table 3, or for urban areas use the formulas in Appendix IV Table 4. Note that Figure 6-2 and Figure 6-3 were generated using these equations, yielding good agreement between the two methods. Also note that the equations have greater precision, but both methods have the same accuracy (±25%).

6.6
$$\sigma_z = cx^d$$

Where: x = distance downwind in km, and
c and d – rural area parameters, values given in Appendix IV Table 3.

FIGURE 6-2. Horizontal Mixing Length for Rural and Urban Areas as a Function of Pasquill Atmospheric Stability Class. The data for this figure is from the ISC3 dispersion model (US-EPA, 1995).

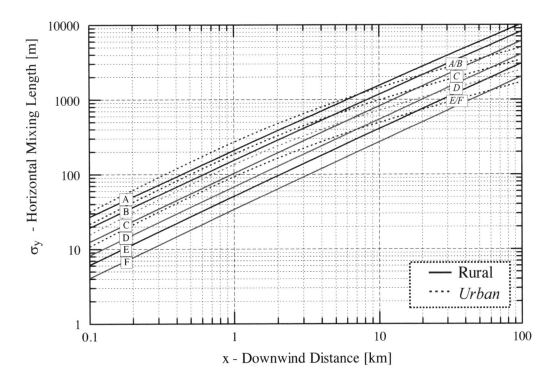

FIGURE 6-3. Vertical Mixing Length for Rural and Urban Areas as a Function of Pasquill Atmospheric Stability Class. The data for this figure is from the ISC3 dispersion model (US-EPA, 1995).

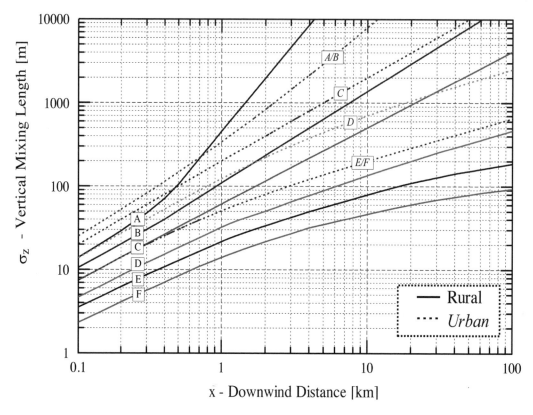

Vertical mixing is more complex than the horizontal counterpart. The first consideration is that many sources are released above ground level, typically by means of a stack of height H (see Figure 6-1). The second consideration is that the ground forms a barrier to the movement of the contaminant. Some contaminants, such as particulate matter deposit onto the ground. Other contaminants, such as sulfur oxides (SO_x), nitrogen oxides (NO_x), and volatile organic compounds (VOCs), bounce or reflect from the ground surface, causing increased concentrations downwind beyond the location where the plume hits ground level, see Figure 6-4. It is also possible for additional reflections to occur from an inversion layer above the plume. Equation 6.7 describes these considerations, which represent the three vertical dispersion terms in equation 6.3. F_{z1} accounts for the material directly released from a stack. F_{z2} accounts for additional mass available downwind from the material added by ground reflection (if the pollutant does not reflect, $F_{z2} = 0$). F_{z3} accounts for reflections from an inversion (if there is no inversion, $F_{z3} = 0$). It is usually adequate to end the sum in F_{z3} after n = 2. Note that this model assumes that the substance either is completely depos ited or completely reflects. Actual materials do some of each. Particles can be re-entrained into the atmosphere by surface winds. Gaseous compounds may adsorb onto moisture on the ground or plants. More information allows for better modeling, but adds complexity that is rarely useful for a screening level model.

FIGURE 6-4. Change in Downwind Concentration due to the Ground as a Reflection Source. Ambient air concentrations downwind of x=R are enhanced as the material reflects from the ground back into the air above ground level.

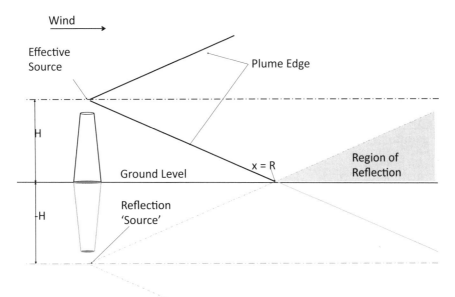

6.7
$$F_{z1} = \exp\left(-\frac{(z-H)^2}{2\sigma_z^2}\right)$$

$$F_{z2} = \exp\left(-\frac{(z+H)^2}{2\sigma_z^2}\right)$$

$$F_{z3} = \sum_{m=1}^{n} \exp\left(-\frac{(z-H-2mL)^2}{2\sigma_z^2}\right) + \exp\left(-\frac{(z+H+2mL)^2}{2\sigma_z^2}\right)$$
$$+ \exp\left(-\frac{(z-H+2mL)^2}{2\sigma_z^2}\right) + \exp\left(-\frac{(z+H-2mL)^2}{2\sigma_z^2}\right)$$

Where: H = effective stack height [m], and
L = elevation of inversion above ground level, not shown in Figure 6-4 [m].

Note: F_{z3} is only applied when the plume is expected to interact with an inversion layer. The plume will 'bounce' or reflect off the layer similar to how it reflects from the ground.

The vertical position of the plume centerline is higher than the physical height of the stack due to two characteristics of the emissions: it is launched with some vertical velocity as it exits the stack (momentum), and it is warmer than the ambient air, which gives it buoyancy. A simple description of this *effective stack height* is:

6.8
$$H = h + \Delta h$$

Where: h = physical elevation of the top of the stack [m], and
Δh = additional plume rise [m], see section 6.6.

Determination of inversion elevation (L) is described in section 5.7.4 – Mean Mixing Depth. Also, L must be greater than H. When L is less than H, the plume can not disperse to ground level.

6.4 Common Applications of the Gaussian Model

This section explores several common applications of the Gaussian model. Each case describes possible scenario(s) and the form of the applicable Gaussian model solution. All calculations assume one hour averaging time as suggested by the US EPA models.

6.4.1 Case 1. Ground Level Source

This scenario occurs from emissions at ground level, such as open pit burning, evaporation from an open tank or pond, or slow release from a spill or leak. Downwind ground level concentrations are given by combining equations 6.3, 6.5, and 6.7 while setting H =0:

6.9
$$C(x,y,z) = \frac{Q}{2\pi u \sigma_y \sigma_z} \exp\left(-\frac{y^2}{2\sigma_y^2}\right) \exp\left(-\frac{z^2}{2\sigma_z^2}\right)$$

This result should be multiplied by a factor of 2 if the emission material reflects from the ground surface rather than being deposited (in which case $F_{z2} = F_{z1}$ rather than zero for the no reflection case.

Example 6-1.

Calculate the PM_{10} concentration 500 m downwind from the open burning of 100 pounds per hour of wood using a ground level wood stove. The weather is overcast with a wind speed of 3 m/s in a rural area.

Solution:
Assume that PM_{10} do not reflect so use equation 6.9 as is, with x = 500m, y = 0, and z = 0. Determine σ_y and σ_z from Figure 6-2 and Figure 6-3 at x = 0.5 km. Use atmospheric stability class D because the weather is overcast. Obtain Q from the emission factors in Appendix 3 which shows PM emissions are 30.6 pounds per ton of wood burned.

$$C(500,0,0) = \frac{694.6 \times 10^6 \frac{\mu g}{hr}}{2\pi \left(3\frac{m}{s}\right)(35m)(18m)} \frac{hr}{3600s} = 16.3 \frac{\mu g}{m^3}$$

It should be noted that the assumption of zero reflection is not completely true. Smaller particles can be returned to the atmosphere by a slight breeze. We use this assumption to model certain behaviors, realizing that the final answers are accurate within +/- 25% - 50%.

6.4.2 Case 2. Elevated Source

This scenario is for non-reflecting pollutants, such as PM or vapors that are readily adsorbed or absorbed by the ground, such as SO_x over open water. Downwind concentrations are given by combining equations 6.3, 6.5, and 6.7 and ignoring F_{z2} and F_{z3}:

6.10 $$C(x,y,z) = \frac{Q}{2\pi u \sigma_y \sigma_z} \exp\left(-\frac{y^2}{2\sigma_y^2}\right) \exp\left(-\frac{(z-H)^2}{2\sigma_z^2}\right)$$

Large particles (diameter > 10 µm) settle quickly due to gravity and thus are transported to the ground faster than this equation predicts. A modification of H can help account for this difference:

6.11 $$C(x,y,z) = \frac{Q}{2\pi u \sigma_y \sigma_z} \exp\left(-\frac{y^2}{2\sigma_y^2}\right) \exp\left(-\frac{\left(z-\left(H-\frac{v_t x}{u}\right)\right)^2}{2\sigma_z^2}\right)$$

Where: v_t = terminal velocity of the particle [m/sec], see figure 7.8 in section 7.3.1 Particle Motion Due to Gravity.

Example 6-2.

Repeat the calculation from Example 6-1, except with the source elevated by a chimney 10 m in effective height.

Solution:

Use equation 6.10 with H = 10m and the remaining data from Example 6-1.

$$C(500,0,0) = \frac{694.6 \times 10^6 \frac{\mu g}{hr}}{2\pi \left(3 \frac{m}{s}\right)(35m)(18m)} \exp\left[-\frac{(0-10m)^2}{2(18m)^2}\right] \frac{hr}{3600s} = 13.9 \frac{\mu g}{m^3}$$

The chimney reduces the concentration at 500m by about 15% although it does not change the total amount of PM emitted by this source.

6.4.3 Case 3. Elevated Source with Reflection

This scenario is for reflecting pollutants, such as SO_x, NO_x, CO, and VOCs that do not interact with the ground. Downwind concentrations are given by combining equations 6.3, 6.5, and 6.7 and ignoring F_{z3}:

$$6.12 \quad C(x,y,z) = \frac{Q}{2\pi u \sigma_y \sigma_z} \exp\left(-\frac{y^2}{2\sigma_y^2}\right) \left[\exp\left(-\frac{(z-H)^2}{2\sigma_z^2}\right) + \exp\left(-\frac{(z+H)^2}{2\sigma_z^2}\right)\right]$$

Example 6-3.

55 g/s of NO_x is emitted in an urban area from a stack of 35m (effective height). The weather is sunny with a wind speed of 4 m/s at the stack height. Graph the downwind centerline concentration from 0.5 to 10 km.

Solution:

NO_x is a reflecting compound so use equation 6.12 to determine the concentrations. The solution must allow x to vary from 0.5 to 10 km, and set y = 0 and z = H which follows the downwind centerline of the plume. Q and u are given. The atmospheric stability is found using Figure 6-2 and Figure 6-3 to determine σ_y and σ_z. To find atmospheric stability we need to find the wind speed at

10 m and estimate the daytime insolation condition. Use Table 5-5, we initially guess the stability to be class B, which allows choosing the stability exponent p = 0.15 (Class B, urban).

$$u_{10m} = u_{35m}\left(\frac{10}{35}\right)^{0.15} = (4.0\frac{m}{s})(0.285)^{0.15} = 3.3\frac{m}{s}$$

Returning to Table 5-5 we see our initial guess is confirmed (wind speed 3.3 and strong daytime insolation). Atmospheric stability is class B.

Now we use Figure 6-2 and Figure 6-3 to estimate σ_y and σ_z at several downwind distances, and to calculate the concentration at each x position. A spreadsheet can help with the repetitive calculations and to create a plot of the results:

Table 6-1. Example 6-3 Data and Calculations.

x	y	z	σ_y	σ_z	Conc
km	m	m	m	m	$\mu g/m^3$
0.5	0	35	83	51	720.3
0.6	0	35	97	62	551.6
0.7	0	35	112	74	433.5
0.8	0	35	126	86	347.8
0.9	0	35	140	97	284.1
1	0	35	154	109	235.8
2	0	35	286	234	64.1
3	0	35	409	365	29.1
4	0	35	527	500	16.5
5	0	35	641	639	10.7
6	0	35	752	780	7.4
7	0	35	861	924	5.5
8	0	35	967	1070	4.2
9	0	35	1071	1218	3.4
10	0	35	1174	1367	2.7

FIGURE 6-5. Downwind, Centerline Concentration Profile for Example 6-3.

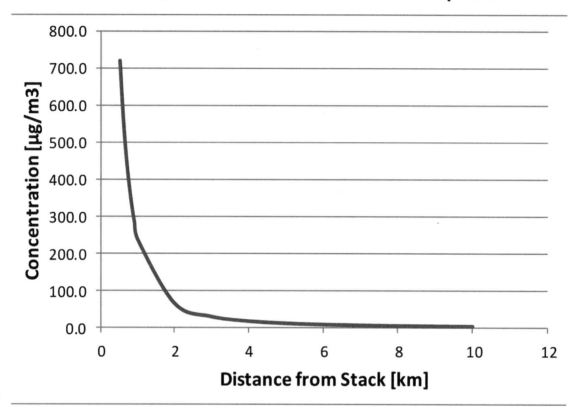

6.4.4 Case 4. Ground Level Concentrations from Elevated Source with Reflection

This subset of solutions quantifies the impact an emission may have on the local air. It focuses on concentrations at ground level, where people, animals, and plants are. This case is also used to determine the impact a source has on ambient air quality. Downwind ground level concentrations are given by:

6.13
$$C(x,y,z) = \frac{Q}{2\pi u \sigma_y \sigma_z} \exp\left(-\frac{y^2}{2\sigma_y^2}\right)\left[2\exp\left(-\frac{(H)^2}{2\sigma_z^2}\right)\right]$$

Note that the terms F_{z1} and F_{z2} are combined since $(z-H)^2$ and $(z+H)^2$ are equal when $z = 0$. *Do not use this equation for any other elevation.*

Example 6-4.

SO_x are emitted at a rate of 125 g/s from a stack with an effective height of 70 m. The atmospheric stability is class C and the wind speed at 10 m is 5 m/s in a rural area. Determine the downwind ground level concentration at x = 1 km and y = 100m.

Solution:

This example is solved using equation 6.12 with x=1 km, y=100, and z =0 or equation 6.13 with x=1 km, y=100. The wind speed must be corrected to the effective stack height using equation 5.12 and p=0.1 from Table 5.3. Then determine σ_y and σ_z from Figure 6-2 and Figure 6-3 for rural areas at x=1 km.

$$u_{70m} = u_{10m}\left(\frac{70}{10}\right)^{0.1} = (5.0\frac{m}{s})(1.215) = 6.1\frac{m}{s}$$

$$C(1000,100,0) = \frac{125 \times 10^6 \frac{\mu g}{s}}{2\pi\left(6.1\frac{m}{s}\right)(100m)(60m)} \exp\left(-\frac{100^2}{2(100m)^2}\right)$$

$$\left[\exp\left(-\frac{(0-(70m))^2}{2(60m)^2}\right) + \exp\left(-\frac{(0+(70m))^2}{2(60m)^2}\right)\right] = 334\frac{\mu g}{m^3}$$

Example 6-5.

Use the data from Example 6-4 to create an isopleth (concentration contour) map of downwind SO_x concentrations within boundaries of (0<x<3000m, 0<y<400m, with z=0).

Solution:

This problem is best solved using a spreadsheet program with graphing capability. The plot is generated by calculating the concentration at several downwind locations. Determine σ_y and σ_z from Figure 6-2 and Figure 6-3 or equations 6.4 and 6.6 for rural areas at each x-distance. Concentrations are then calculated using equation 6.12. Note that this equation is symmetric about the centerline (y = 0), so the solution at y = 100 m is the same as y = -100 m. Therefore, only the positive y-axis is shown.

Table 6-2. Example 6-5 Data and Calculations.

Q =	125	g/s
u =	5	m/s at 10 m
u at H =	6.1	m/s at 70 m
H =	70	m
Atm Class =	C	

x [m]	y [m]	σ_y [m]	σ_z [m]	C [ug/m3]
0	0	1	1	0.0
250	0	29	17	3.3
500	0	55	32	358
750	0	79	47	577
1000	0	103	61	537
3000	0	279	167	128
250	100	29	17	0.0
500	100	55	32	68
750	100	79	47	261
1000	100	103	61	336
3000	100	279	167	120
250	200	29	17	0.0
500	200	55	32	0.5
750	200	79	47	24
1000	200	103	61	82
3000	200	279	167	99
250	400	29	17	0.0
500	400	55	32	0.0
750	400	79	47	0.0
1000	400	103	61	0.3
3000	400	279	167	46

This data can be used to generate a surface or contour plot. Note that σy and σz are not defined at x=0, so all concentrations are set to zero at this location.

FIGURE 6-6. SO$_x$ Concentration Contour Plot for Example 6-5. Distances are in meters. The concentration (in µg/m³) contours are not smooth in this plot due to the limited data and the simplified interpolation scheme used in generating the contours.

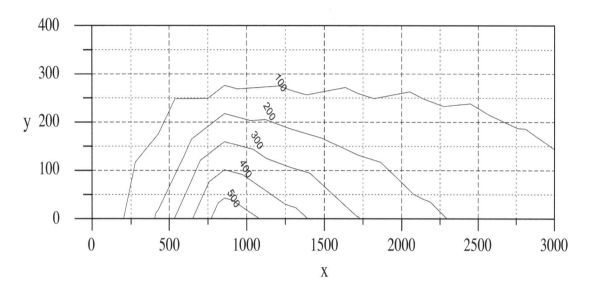

One of the more important calculations for this equation is finding the maximum ground level concentration (C_{max}) and its position (x_{max}). Figure 6-7 provides a graphical representation of solutions for these two variables as functions of effective stack height and atmospheric stability class. The figure is used by determining the effective stack height (H) and atmosphere stability class (A, B, C, D, E, or F) using the Pasquill classification scheme (section 5.7.2 – Stability Analysis). Locate these two points in the net superimposed on the log-log grid, and then the two coordinates (x_{max} and $C_{max}u_x/Q$) are obtained. C_{max} is found by multiplying the y-coordinate by Q/u_x.

Letters within the grid correspond to Pasquill atmospheric stability class. Numbers within the grid correspond to the effective stack height in meters. Use ½ of the calculated Cu/Q value if the contaminant does not reflect. Adapted from (Turner, 1994).

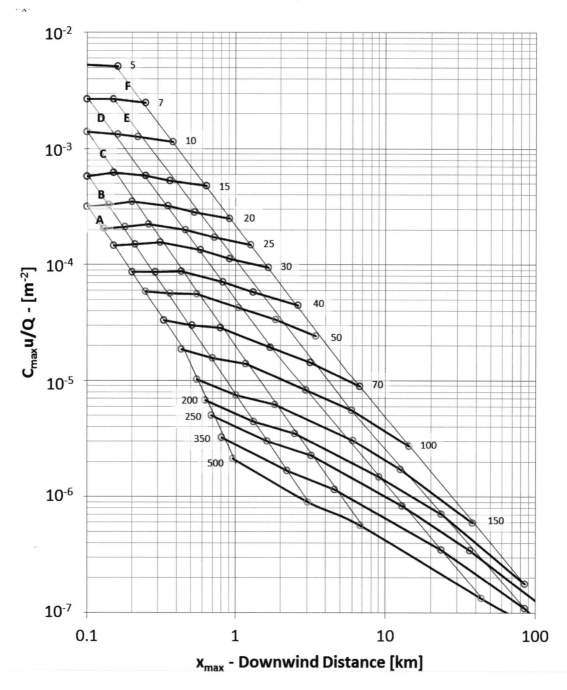

FIGURE 6-7. Value and Distance to Maximum Ground Level Concentration from an Elevated Source with Reflection (rural areas).

Example 6-6.
Find the position and value of the maximum downwind concentration for the data in Example 6-5.

Solution:
Use Figure 6-7 with the points H = 70 m and atmospheric stability class C. The x-coordinate yields xmax = 0.8 km or 800 m. The y-coordinate yields $C_{max}\, u_x/Q = 3 \times 10^{-5}$ m^{-2}.

$$C_{max} = \left(\frac{C_{max} u_x}{Q}\right)\frac{Q}{u_x} = 3 \times 10^{-5} \frac{1}{m^2}\left(\frac{125 \times 10^6 \frac{\mu g}{s}}{6.1 \frac{m}{s}}\right) = 615 \frac{\mu g}{m^3}$$

Compare to Figure 6-6 where the distance to the maximum concentration is around 800 m, and has a value around 580 µg/m³, which is well within the implied accuracy of 25%.

6.4.5 Case 5. Line Source

A line source emission models pollutants from sources such as a long road with heavy traffic, or a series of sources along a river or harbor. Downwind concentrations are given by combining equations 6.3, 6.5, and 6.7. Assume the wind (u_x) is perpendicular to the road, the road runs in the y-direction. The F_y term is set to 1.0 because the line source distributes the emissions equally along its length, so there is no net dispersion in this direction. The units on the source, Q, become mass per unit length per unit time.

6.14
$$C(x,z) = \frac{2Q}{(2\pi)^{1/2} u_x \sigma_z} \exp\left(-\frac{z^2}{2\sigma_z^2}\right)$$

The z term can be modified to (z-H) for elevated sources.

Example 6-7.
Determine the concentration of CO 300 m downwind of a busy highway. The highway carries 3200 cars/hour at an average speed of 45 miles/hour. A 4 m/s wind perpendicular to the road. The atmospheric stability is class B. Note that Light Duty Gasoline Vehicles (passenger cars) have an average emission factor of 3.8 g CO/mile.

Solution:

Calculate Q from the emission factor and road use data. Note that for a line source the units of Q are [mass/(length·time)]. Determine σ_z from Figure 6-3. Finally, apply the given information to equation 6.14 for ground level (z=0):

$$Q = 3.8 \frac{gCO}{Car \cdot mile} \frac{mile}{1600m} 3200 \frac{Car}{hour} \frac{hour}{3600 \sec} = 0.0021 \frac{gCO}{m \cdot \sec}$$

$$C(300,0) = \frac{2\left(0.0021 \frac{gCO}{m \cdot \sec}\right)}{(2\pi)^{1/2} \left(4 \frac{m}{\sec}\right) 30m} = 14 \frac{\mu g}{m^3}$$

This model assumes an infinitely long line source (road). Corrections can be applied at finite lengths to account for the edge effects (turns or other changes in direction). Interested readers may want to investigate the US-EPA website for air dispersion models at 'http://www.epa.gov/scram001/dispersion_alt.htm'.

6.4.6 Case 6. Puff source

A puff source is an emission where the material is released over a very short time, such as the spill of a volatile liquid, filling a tank, or during an explosion. Model these conditions as instantaneous releases from the location (0,0,0). The model equation is:

6.15 $$C(x,y,z) = \frac{2Q}{(2\pi)^{3/2} \sigma_x \sigma_y \sigma_z} \exp\left(-\frac{(x-u_x t)^2}{2\sigma_x^2}\right) \exp\left(-\frac{y^2}{2\sigma_y^2}\right) \exp\left(-\frac{z^2}{2\sigma_z^2}\right)$$

Note that this equation includes a σ_x term, representing the mixing length in the x-direction. It has a value the same as σ_y. Q in this equation is the total amount released and has units of mass. The maximum concentration always occurs along the downwind centerline at a position of x = ut. The concentration observed downwind will suddenly rise, and then quickly decrease as the puff passes a location. The last term becomes $[F_{z1} + F_{z2}]/2$ from equation 6.7 for the case of an elevated puff source with reflection.

The short times associated with an instantaneous release require modified mixing length calculations to reflect the differences from a continuous release. Table 6-3 lists the formulas to use for this situation (Slade, 1968).

Table 6-3. Mixing Length Formulas for Instantaneous Releases.

Mixing Length [m]	Atmospheric Stability	Formula
σ_x, σ_y	Unstable	$0.14(x)^{0.92}$
	Neutral	$0.06(x)^{0.92}$
	Very Stable	$0.02(x)^{0.89}$
σ_z	Unstable	$0.53(x)^{0.73}$
	Neutral	$0.15(x)^{0.70}$
	Very Stable	$0.05(x)^{0.61}$

■ **Example 6-8.**

A spill during the filling of a tank releases 24 kg of butane. Determine the maximum downwind concentration at 500 m and the time at which this occurs under very stable conditions with 3 m/s wind speed. Assume the entire spill vaporizes instantaneously.

Solution:

The maximum concentration will occur at $x = ut$, $y=0$, and $z=0$. Calculate the unknowns at these coordinates using equation 6.15 and the mixing lengths from Table 6 3.

$$\sigma_x = \sigma_y = 0.02(500)^{0.89} = 5m$$

$$\sigma_z = 0.05(500)^{0.61} = 2.2m$$

$$t = \frac{x}{u_x} = \frac{500m}{3\frac{m}{s}} = 167 \sec = 2.8 \min$$

$$C(500,0,0) = \frac{2(24,000g)}{(2\pi)^{3/2}(5m)(5m)(2.2m)} = 55.4\frac{g}{m^3}$$

The explosive limits of butane in air are LEL = 1.8% (lower explosive limit) and UEL = 8.4% (upper explosive limit). Air has a density of 1,200 g/m³ at 1 atm and 20 °C. The fraction of butane at 500 m would constitute 55.4 / 1,200 *100% = 4.6%. This value is between the lower and upper explosive limits. The emission would certainly explode if any ignition source is in the area.

■ **Example 6-9.**

Plot the concentration-time profile at 300m, 450 m, 600 m in the downwind direction for the data in Example 6-8.

Solution:

These calculations are similar to that shown in the previous example. A spreadsheet is useful to handle the repetitive calculations and to create the plot.

Table 6-4. Example 6-9 Calculations for Instantaneous Release of Butane.

Q	24000	g
u_x	3	m/s

x	σ_x	σ_y	σ_z	t	$C(g/m^3)$
300	3.2	3.2	1.6	92.5	0.0
300	3.2	3.2	1.6	97.5	11.8
300	3.2	3.2	1.6	99.0	118.1
300	3.2	3.2	1.6	100.0	183.1
300	3.2	3.2	1.6	101.0	118.1
300	3.2	3.2	1.6	102.5	11.8
300	3.2	3.2	1.6	107.5	0.0
450	4.6	4.6	2.1	138.8	0.0
450	4.6	4.6	2.1	146.3	3.5
450	4.6	4.6	2.1	148.5	43.0
450	4.6	4.6	2.1	150.0	69.5
450	4.6	4.6	2.1	151.5	43.0
450	4.6	4.6	2.1	153.8	3.5
450	4.6	4.6	2.1	161.3	0.0
600	5.9	5.9	2.5	185.0	0.0
600	5.9	5.9	2.5	195.0	1.4
600	5.9	5.9	2.5	198.0	21.0
600	5.9	5.9	2.5	200.0	34.9
600	5.9	5.9	2.5	202.0	21.0
600	5.9	5.9	2.5	205.0	1.4
600	5.9	5.9	2.5	215.0	0.0

FIGURE 6-8. Concentration Profile at Three Downwind Positions from an Instantaneous Puff Release.

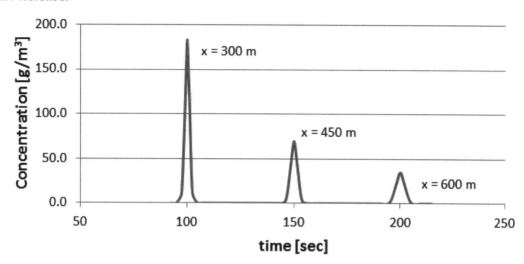

The three peaks represent the concentration profiles at different distances and times.

6.4.7 Case 7. Multiple sources

Multiple source problems can become very complex, especially if the various plumes interact. A simplified approach is to assume superposition between the plumes such that the downwind concentration at any given location is the simple sum of all sources. The exact locations of the sources must be known. These solutions are also dependent on the wind direction, which could cause increases or decreases at a location depending on the source geometry.

■ Example 6-10.

Two sources of NO_x emissions are located close together. Source one emits 220 g/s at an effective stack height of 48 m. Source two emits 55 g/s at an effective height of 38 m. Source two is located 250 m south and 400 m east of source one. Determine the ground level NO_x concentration at a receptor site located 1 km downwind and 150 m south of source one. The wind moves from the west at a speed of 2.5 m/s, measured at 10m. The atmospheric stability class is E.

Solution:
The concentration at the receptor site is obtained by summing the contributions at this location from each of the two sources. Equation 6.13 – Ground Level Concentrations from Elevated Source with

Reflection is used to determine the concentration from each source. A sketch of the locations of the two sources and the receptor site is useful in determining the coordinates for each calculation.

FIGURE 6-9. Sketch for Example 6-10.

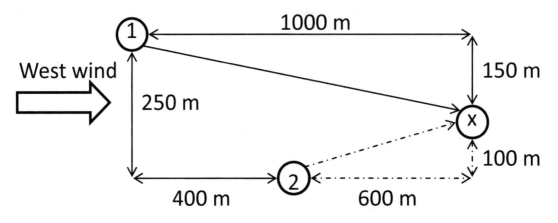

The contribution from source one is $C_1(+1000,-150,0)$. The contribution from source two is $C_2(+600,+100,0)$. The calculation of these concentrations follows:

Source 1
$\sigma_y = 50m$
$\sigma_z = 21m$
$u_{48} = 2.5(48/10)^{0.35} = 4.3 m/s$

$$C_1(1000,-200,0) = \frac{220 \times 10^6 \frac{\mu g}{s}}{2\pi \left(4.3 \frac{m}{s}\right)(50m)(13m)} \exp\left(-\frac{(-150)^2}{2(50)^2}\right)$$

$$\left[\exp\left(-\frac{(0-48)^2}{2(21)^2}\right) + \exp\left(-\frac{(0+48)^2}{2(21)^2}\right)\right] = 16.3 \frac{\mu g}{m^3}$$

Source 2
$\sigma_y = 32m$
$\sigma_z = 13m$
$u_{38} = 2.5(38/10)^{0.35} = 4.0 m/s$

$$C_2(600,100,0) = \frac{55 \times 10^6 \frac{\mu g}{s}}{2\pi \left(4.0 \frac{m}{s}\right)(32m)(13m)} \exp\left(-\frac{(100)^2}{2(32)^2}\right)$$

$$\left[\exp\left(-\frac{(0-38)^2}{2(13)^2}\right) + \exp\left(-\frac{(0+38)^2}{2(13)^2}\right)\right] = 2.4 \frac{\mu g}{m^3}$$

The increase at the receptor site is the sum of these two sources:
$C_{receptor} = C_1 + C_2 = 16.3\ \mu g/m^3 + 2.4\ \mu g/m^3 = 18.7\ \mu g/m^3$

The analysis can be used during project planning stages to determine if a new source can be expected to have an impact on ambient air quality.

6.5 Effect of Averaging Time

The estimates of horizontal and vertical mixing lengths are calculated from concentration measurements averaged over 10-minute intervals (Turner, 1994). These values for mixing length over-estimate the concentrations at longer time intervals, and under-estimate at shorter time intervals. The determination of concentrations at different averaging times requires the value to be rescaled. One rescaling equation is given as (Nonhebel, 1960):

6.16
$$C_2 = C_1 \left[\frac{t_1}{t_2}\right]^{0.17}$$

Where: C_x = the concentration at a given averaging time, and
t_x = averaging time in minutes

The models used and approved by the US-EPA assume the results from the above mixing length correlations are for 1-hour time-averaged concentrations not the 10-minute averages from the mixing length correlations. Thus, the US-EPA models predict conservative estimates of dispersion and associated concentrations.

■ *Example 6 -11.*
Determine the one-hour averaged PM^{10} concentration that would not exceed the US-EPA NAAQS.

Solution.

The NAAQS for PM10 is 150 µg/m³ averaged over 24 hours. Use equation 6.16 to determine the one-hour average:

$$C_1 = C_{24}\left[\frac{24}{1}\right]^{0.17} = \left(150\frac{\mu g}{m^3}\right)(1.716) = 257\frac{\mu g}{m^3}$$

6.6 Effective Stack Height and Plume Rise

The most effective stacks are those which provide sufficient lift to the emissions to prevent them from interacting with ground level until they are dispersed and diluted to the point that they have little impact on ground level concentrations.

The main function of the stack is to launch the emissions far above ground level. It does not decrease the amount of emissions, but does allow the emissions to dilute within the atmosphere. The two mechanisms used to launch the plume are buoyancy and momentum. Heated emissions are less dense than the surrounding air, causing it to be buoyant. The heat originates within the process which generated the emissions (i.e. fuel combustion). The need for buoyancy limits the amount of heat recovery available from the emission stream. The speed of the gas exiting the stack provides it with vertical momentum. The exit velocity is determined from the volumetric flow of emissions and the diameter of the stack top. The diameter is chosen to give an exit velocity of at least 1.5 times the maximum expected wind speed at the stack height. Slower exit velocities may lead to stack tip downwash (see section 6.6.4), which causes the emissions to move downward behind the stack leading to elevated ground level pollution concentrations near the source, see Figure 6-10.

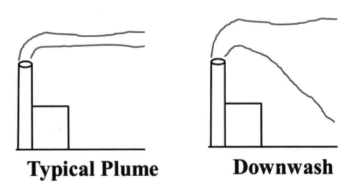

FIGURE 6-10. Example of normal plume and a plume experiencing downwash.

The effective stack height, H, is the sum of the physical stack height (h) and the plume rise (Δh):

6.17
$$H = h + \Delta h$$

Figure 6-1 shows an example of these terms. The stack height should be 2.5 times higher than the tallest nearby buildings. Airflow becomes turbulent as it moves over and around buildings, leading to enhanced mixing up to twice the height of the building. If the plume were to enter one of these turbulent zones it could mix with it, causing elevated pollution concentrations within the turbulent and wake zones.

There are many empirical formulae and theoretical models that can be used to determine the plume rise. These calculations use specific cases of meteorological and stack conditions. Differences in calculated values of plume rise may be as large as 10-fold (Briggs, 1975).

6.6.1 Holland Formula

One model that was once recommended by the US-EPA is the Holland formula (Holland, 1953):

6.18
$$\Delta h = \frac{v_s D_s}{u} \left[1.5 + 2.68 \times 10^{-3} \left(\frac{T_s - T_a}{T_a} \right) P_a D_s \right]$$

Where: v_s = vertical stack gas velocity [m/s],
D_s = stack inside diameter at exit [m],
u = wind speed at stack height [m/s],
T_s = stack gas temperature [K],
T_a = ambient atmospheric temperature [K],
P_a = ambient atmospheric pressure [mbar],

This formula was developed for neutral atmospheric stability conditions. Holland suggested a correcting multiplier of 1.2 for Class A, 1.1 for B, 0.9 for E, and 0.8 for F.

6.6.2 Concawe Formula

Another model is the modified Concawe formula (Thomas, et al., 1970) which assumes that the plume rise is driven by buoyancy effects. It is calculated as:

6.19
$$\Delta h = \frac{4.71 \left[m c_p (T_s - T_a) \right]^{0.444}}{u^{0.694}}$$

Where: m = flow rate of exit gas [kmol/s]
c_p = specific heat of exit gas [kJ/kmol/K]

The mass flow rate in equation 6.19 can be determined from the area of the stack exit, the exit velocity, and the ideal gas law for converting volume to mass:

6.20
$$m = \left(\frac{\pi}{4}D_s^2\right)v_s\left(\frac{P}{RT}\right)$$

6.6.3 Briggs Model

The current US-EPA method is the Briggs model, as used in the Industrial Source Complex (ICS3) dispersion models (US-EPA, 1995). This model determines plume rise by accounting for buoyancy and momentum driven plumes in stable and unstable or neutral conditions. In each situation, the plume rise is not instantaneous, rather it develops over time. This time is converted to distance using the wind speed. For each case a calculation will be presented of the horizontal distance the plume travels before it reaches its effective height and of the plume rise.

6.6.3.1 Unstable or Neutral Conditions

Plume rise predictions depend on whether the plume is dominated by buoyancy or momentum. A determination of which term is dominant is needed when the stack gas temperature, Ts, is greater than or equal to the ambient air temperature, Ta. The buoyancy flux parameter [m⁴/s³] is:

6.21
$$F_b = gv_sD_s^2\left[\frac{T_s - T_a}{4T_s}\right]$$

Where: g = gravity [9.81 m/s²]

The momentum flux parameter [m⁴/s²] is:

6.22
$$F_m = v_s^2 D_s^2\left[\frac{T_a}{4T_s}\right]$$

The determination is made by calculating a crossover temperature (ΔT_c) and comparing it to the temperature difference between the stack gas and the ambient air. The calculation depends on the value of the buoyancy flux:

6.23
$$\Delta T_c = 0.297 \cdot T_s \frac{v_s^{1/3}}{D_s^{2/3}} \quad F_b < 55\frac{m^4}{s^3}; \text{ and unstable or neutral conditions}$$

$$\Delta T_c = 0.00575 \cdot T_s \frac{v_s^{2/3}}{D_s^{1/3}} \quad F_b \geq 55\frac{m^4}{s^3}; \text{ and unstable or neutral conditions}$$

The comparison between the actual temperature difference and the crossover temperature determines if the plume rise is buoyancy or momentum driven:

6.24
$$(T_s - T_a) = \Delta T \geq \Delta T_c \quad \text{Buoyancy dominated plume rise}$$
$$(T_s - T_a) = \Delta T < \Delta T_c \quad \text{Momentum dominated plume rise}$$

6.6.3.1.1 Unstable or Neutral Conditions - Buoyancy Dominant

The calculations for the plume rise and horizontal distance to final rise are:

6.25

or

$$x_f = 49 F_b^{5/8} \quad F_b < 55$$
$$\Delta h = 21.425 \frac{F_b^{3/4}}{u}$$

6.26

$$x_f = 119 F_b^{2/5} \quad F_b \geq 55$$
$$\Delta h = 38.71 \frac{F_b^{3/5}}{u}$$

6.6.3.1.2 Unstable or Neutral - Momentum Dominant

Use the following calculations for the momentum dominated case as described above, or for cases where the stack temperature, Ts, is less than the ambient temperature, Ta:

6.27

$$x_f = 4 D_s \frac{(v_s + 3u)^2}{v_s u} \quad \text{for } F_b \leq 0$$

$$x_f = 49 F_b^{5/8} \quad \text{for } 0 < F_b \leq 55 \frac{m^4}{s^3}$$

$$x_f = 119 F_b^{2/5} \quad \text{for } F_b > 55 \frac{m^4}{s^3}$$

6.28
$$\Delta h = 3 D_s \frac{v_s}{u} \quad \text{All } F_b$$

(Briggs, 1969) suggests this equation fits best when $v_s/u > 4$.

6.6.3.2 Stable Conditions

For stable conditions (Class E or F) the relationships use a stability parameter, S [s^{-2}], given as

6.29
$$S = \frac{g}{T_a}\left(\frac{\Delta T}{\Delta h}\right)$$

Where: $\left(\dfrac{\Delta T}{\Delta h}\right)$ = Potential Temperature Gradient [K/m],
g = gravity (9.81 m/s2), and
Ta = ambient temperature.

If data are not available, the average potential temperature gradient is used; 0.02 K/m for class E, and 0.035 K/m for class F. The crossover temperature is:

6.30
$$\Delta T_c = 0.019582 \cdot T_s v_s S^{1/2}$$

The determination for buoyancy or momentum dominated plume rise is the same as given in equation 6.24.

6.6.3.2.1 Stable Conditions- Buoyancy Dominant

The plume characteristics are:

6.31
$$x_f = 2.0715 \dfrac{u}{S^{1/2}}$$

$$\Delta h = 2.6 \left(\dfrac{F_b}{uS}\right)^{1/3} \quad \text{for } u \geq 1.5 \text{ m/s}$$

$$\Delta h = 5 F_b^{1/4} S^{-3/8} \quad \text{for } u < 1.5 \text{ m/s}$$

6.6.3.2.2 Stable Conditions- Momentum Dominant

The plume characteristics are:

6.32
$$x_f = 0.5 \left(\dfrac{u}{S^{1/2}}\right)$$

$$\Delta h = 1.5 \left(\dfrac{F_m}{uS^{1/2}}\right)^{1/3}$$

Compare this value with the unstable-neutral momentum dominant plume rise, equation 6.28, and use the smaller of the two, because the plume rise in stable air should never exceed that in unstable or neutral conditions.

There is no single best correlation, and the predictions may vary by more than 50%. The choice depends on the specific local conditions and the assumptions used in generating the formula and model equations. As such the accuracy is limited.

Example 6-12.

Calculate the plume rise and the distance to maximum plume rise, using the Holland, modified Concawe, and Briggs models to make the estimates. The mass flow rate is 2.5 kmol/s with an average heat capacity of 35 kJ/kmol/K. The stack exit temperature is 450K; the ambient pressure is 980 mb; and the atmospheric ambient temperature is 27°C (300K). The gas exit velocity is 35 m/s. The wind speed is 5 m/s at the stack height in an unstable atmosphere (class B). The stack inside diameter at the exit is 3m.

Solution:
Apply the known data to equations 6.18, 6.19, and perform the calculations for each model.

Holland Formula:

$$\Delta h = \frac{\left(35\frac{m}{s}\right)(3m)}{\left(5\frac{m}{s}\right)}\left[1.5 + 2.68 \times 10^{-3}\left(\frac{450-300}{300}\right)(980\, mbar)(3m)\right] = 114.2m$$

Modified Concawe Formula:

$$\Delta h = \frac{4.71\left[\left(2.5\frac{kmol}{s}\right)\left(35\frac{kJ}{kmol \cdot K}\right)(450-300)K\right]^{0.444}}{\left(5\frac{m}{s}\right)^{0.694}} = 103.8m$$

Briggs Model:

$$F_b = \left(9.81\frac{m}{s^2}\right)\left(35\frac{m}{s}\right)(3m)^2\left[\frac{450-300}{4(450)}\right] = 257.5\frac{m^4}{s^3}$$

$$F_m = \left(35\left[\frac{m}{s}\right]\right)^2 (3m)^2 \left[\frac{300}{4(450)}\right] = 1837.5\frac{m^4}{s^2}$$

Use equation 6.23 with Fb>55

$$\Delta T_c = 0.297 \cdot (450K)\frac{(35)^{1/3}}{(3)^{2/3}} = 210.2K$$

Comparing the critical temperature with the stack-air difference we obtain:

$(450 - 300) = 150 < 210$

Since the critical temperature is greater than the stack-ambient air temperature difference, the plume is momentum dominated. Finally, calculate the distance and plume rise from equations 6.27 and 6.28 for unstable or neutral, momentum dominated conditions:

$$x_f = 119(257.5)^{2/5} = 1,100 \text{ m}$$

$$\Delta h = 3(3m) \frac{\left(35 \frac{m}{s}\right)}{\left(5 \frac{m}{s}\right)} = 63m$$

The large differences in values are normal. In this case, the Briggs model is probably better as the other two models always assume buoyancy dominated plume rise, yet the critical temperature showed the system is momentum dominated.

6.6.4 Stack Tip Downwash

Downwash can occur when the exit velocity is less than 1.5 times the wind speed. As the wind moves past a structure, it can create a turbulent wake zone on the downwind side. The wake causes recirculating eddies and regions of very high vertical mixing. The turbulent zone can extend twice the building height and extend five to ten times the height downwind. (Briggs, 1974) suggested that downwash be modeled as a reduction to the physical height of the stack. One model equation is:

6.33
$$H = h + 2D_s \left[\frac{v_s}{u} - 1.5 \right] \quad v_s < 1.5u$$

Where: h = physical stack height [m],
D_s = inside diameter of stack at exit [m],
v_s = vertical velocity of gas at stack exit [m/s], and
u = wind speed at stack height h [m/s].

This equation yields an effective height, H, which is less than h.

In addition to designing high exit velocity, the stack itself can be designed to reduce the size and magnitude of the wake caused by the stack. Two design devices to accomplish this are a large flat disk located circumferentially at the stack exit with a diameter of 3 times the stack exit diameter, and helical wind turning vanes located around the stack (Cooper, et al., 2011).

6.7 Other Air Dispersion Models

The US EPA maintains an air dispersion modeling database (US-EPA, 2010). It is a model clearinghouse that contains screening and refined models for use in developing air permits. The site provides guidance in the use of these models and alternative models. Most of the models include computer code (typically written in the FORTRAN language) for running or compiling the model. User guides accompany each to help understand the input and output for each model. Specialized models include photochemical effects, multi-pollutant applications, receptor analysis, meteorological concerns, and permit guidance.

6.8 Long Range Transport (>50 km)

The Gaussian model is adequate for estimating pollutant concentrations up to approximately 50 km from the source. Beyond this limit, large-scale weather systems, variations in sunlight, and the presence of precipitation have larger effects on the plume. Most pollutants have atmospheric lifetimes that exceed the time needed to move 50km. A different type of model is used at these greater distances to account for the greater variety of conditions a plume encounters. These methods for studying atmospheric transport over long distances use Lagrangian models of fluid flow (Lin, et al., 2011). The models work by simulating atmospheric flow of air and following the motions of parcels, or particles, embedded in the atmospheric flow. Lagrangian models track the movement of the parcels using a moving frame of reference. The movement of the parcel tends toward the average flow of the fluid, but diverts from this average flow field due to turbulence, washout, chemical reactions, or other changes in the parcels environment. The atmospheric flow field is determined from meteorological observations and fluid flow models. The models are used to track intercontinental transport of pollutant plumes or radioactive releases. Running the models backward in time, using historical meteorological data, can help determine the source regions of observed contaminants.

One long-range transport model developed by the US-NOAA and Australia's Bureau of Meteorology is called HYbrid Single-Particle Lagrangian Integrated Trajectory or HYSPLIT (US-NOAA, 2011). It is a modeling system that computes air parcel trajectories that include complex dispersion, deposition, advanced advection algorithms, meteorological stabilities, and chemical transformations. The model is based on either puff or particle dispersion. In the puff model, puffs expand until they exceed the size of the meteorological cell (vertical or horizontal). Then they split into several new puffs, each with its share of the pollutant mass. In particle mode, a fixed number of particles are advected through the model domain by the mean wind velocity field and spread by a turbulent mixing component. Another model, called FLEXTRA/ FLEXPART, (Stohl, 2001) is a kinematic trajectory model used to calculate several types of trajectories such as 3D, isentropic, isobaric, and boundary layer trajectories. It is developed by the Institute of Meteorology and Geophysics in Vienna by Andreas Stohl, Gerhard Wotawa, and Petra Seibert.

6.9 Questions

* - Questions and problems may require additional information not included in the textbook.

1. Why are screening level models based on estimations that are very conservative. That is why is it acceptable that they always predict higher concentrations than are observed?

2. The x-direction mixing was ignored in the derivation of equation 6.2. Explain how and comment on the effect this assumptions would have on the actual concentrations.

3. Describe the effect on ground level concentrations if wind speed (ux) was allowed to vary in the z-direction around an elevated source (see equation 5.12).

4. Suggest an addition to the Gaussian model that would account for the reduction in concentration of a pollutant that reacts in the atmosphere.

5. Sketch a concentration contour of a long plume that is suddenly exposed to a wind direction shift of 90°. How would this differ from a line source?

6. Sketch the elevation view of a plume from an elevated source that undergoes reflection from the ground and an inversion layer located at an elevation of L. Identify the sources of the terms given in equation 6.7.

7. Why are Gaussian screening models only useful up to 50 km? Would it be possible to improve the long-range predictions by adding an additional term to the Gaussian model?

8. What would be the effect on ground level concentrations if plume rise were ignored?

9. How could the Gaussian model be altered to model a contaminant that only partially reflects?

10. Visit the US-EPA technology transfer network website. Find the names of 3 alternative screening level models (not ISC3) for predicting dispersion.* [http://www.epa.gov/scram001/dispersion_screening.htm]

11. Make a list of three questions you have about this chapter or air pollution concerns you have.

6.10 Problems

1. Consider the following cases. Would they need to model their air emissions. Assume that each case is located in a class II region that is currently in attainment, with no nearby class I regions.
   a) A new coal fired boiler, burning 0.5 tons/hr of coal with an 8% ash content. Assume the only emission is PM.
   b) A power generator burns 5000 gal/ hr of No. 2 fuel oil in an existing facility, and undergoes an expansion that doubles its use of fuel oil. Assume the only emission is PM.
   c) A new coal fired boiler which burns 20 ton/hr of bituminous coal with 3% sulfur, and with equipment that will remove 80% of SO_x. Assume SO_x is the only emission.
   d) A medical waste incinerator burns 1 ton/hr and removes 97% of the PM generated. Assume the only emission is PM.

2. Determine the y-direction mixing length, σ_y, at 2 km for each atmospheric stability class in rural and urban areas.

3. Determine the z-direction mixing length, σ_z, at 2 km for each atmospheric stability class in rural and urban areas.

4. Determine the downwind concentration [$\mu g/m^3$] of PM 800 m from a rural ground level source if the emission rate is 30.0 g/s. The midday weather is sunny with a gentle breeze.

5. Determine the emission rate of SO_x [kg/hr] from an urban ground level source if the 1,000 m downwind concentration is 5.0 µg/m³. The evening weather is overcast with a light breeze.

6. Determine the 1-km downwind ground level concentration of PM [µg/m³] from an urban power plant that burns 3 ton/hr of bituminous coal containing 5.5% ash, and having 99% PM removal. The source emits from a stack with effective height 100m. The wind at the stack height is 6 m/s and the atmospheric stability is class C.

7. Determine the effective stack height needed to reduce a rural emission of 1.5 g/s of PM to an ambient ground level concentration of 5 µg/m3 at a rural location 700 m downwind. The wind is 3.2 m/s at 10 m elevation, in class D conditions.

8. Calculate the 1 km downwind rural concentration at (a) stack height and (b) ground level of 20 µm particulates ($\varrho=1.5 g/cm^3$) emitted at a rate of 2.2 g/s from a stack with an effective height of 50 m. The wind speed at stack height is 5 m/s in class D conditions. Hint: These are large particulates and they may settle more quickly than a gas phase emission.

Chapter 6

9. Consider a Hawaiian rural sugar mill that consumes 100 ton/hr of sugar cane. The cane contains about 15 wt% sucrose solution. The remaining plant matter is a waste called bagasse, which can be burned to generate steam. This mill collects 90% of the PM generated from the process. The remaining PM and other waste gasses (2.5 kmol/s) are discharged at 340 K, have a heat capacity of 30 kJ/ kmol/ K, and are emitted from a 25 m stack. The local weather is 305 K on a day with slightly unstable air and a wind speed of 4 m/s measured at a height of 10 m.
   a) Calculate the PM emission rate [g/s],
   b) Determine the effective stack height using the modified Concawe formula,
   c) Plot the ground level downwind concentrations for $0 < x < 5000$ m,
   d) Find the maximum downwind ground level concentration and its position,
   e) Draw a ground level isopleth map with -500 m $< y < 500$ m and x as in part c.

10. Determine the effective height of a stack needed to keep the maximum ground level ambient air concentration below 10 µg/m³ from a source that emits 12 g/s of NO_x in a rural area with wind speed 4 m/s measured at the stack height and moderately stable air.

11. The critical wind speed is the wind speed that yields the largest value of C_{max}. Note that the downwind concentration is inversely proportional to wind speed, which will decrease the concentration as wind speed increases. The effective stack height is also inversely proportional to wind speed, but this will tend to increase the concentration as wind speed increases. Since the two terms are in opposition, the concentration will go through a maximum with increasing wind speed. Determine this critical speed and concentration for a 85 g/s emission of SO_x from a stack of 40 m physical height, under atmospheric stability class D in an rural area. Use the modified Concawe formula with the term $[mcp(T_s-T_a)] = 14,000$ kJ/s. Hint: Figure 6-7 will greatly reduce the calculations.

12. Determine the effect of atmospheric stability class on the ground level concentration of CO at a location 500 m downwind from a busy urban highway for a 3 m/s cross wind perpendicular to the road. Assume the traffic emits 0.003 g/m/sec of PM.

13. An explosion releases 5 kg of dust instantaneously into the air at ground level. The atmospheric conditions are unstable with a 4.5 m/s wind. Estimate the maximum concentration 1 km down wind. How long after the release will this maximum occur?

14. Prepare an isopleth map $[C(x,y,0)]$ of the release in the problem 13 above at $t = 5$ min, $t = 10$ min, and $t = 15$ min.

15. Consider the NO_x emissions from two rural sources. The second source is located 500 m south and 500 m east of the first source. Source 1 emits 20.5 g/s and has an effective height of 100 m.

Source 2 emits 35 g/s and has an effective height of 150 m. Sketch the locations of the stacks and the receptor site. Determine the concentration at a receptor located 4 km east and 200 m south of source one. Atmospheric conditions are neutral with a westerly wind of 5 m/s measured at 10 m.

16. Using the emission sources and data from problem 15, estimate the ground level concentration at a receptor location 4 km downwind of source one, assuming a NW wind at 5 m/s.

17. Using the emission sources and data from problem 15, estimate the ground level concentration at a receptor location 4 km downwind of source one, assuming a SW wind at 3 m/s.

18. Assume the result in Example 6-6 is a 10 minute averaged concentration. Would this concentration exceed the one hour SO_x standard in the US-EPA NAAQS?

19. Consider a 450K emission source that exits a 2 m diameter, 50 m high stack. The emissions leave at 20 m/s into a rural, neutral atmosphere with wind (at 10 m) of 5 m/s, 300 K, and 980 mbar. The heat capacity of the emission is 35 kJ/(kmol K). Determine the plume rise using a) Holland formula, b) modified Concawe formula, and c) Briggs method. Hint: R = 0.083145 (m^3 bar) / (kmol K).

20. Calculate the effective stack height for the data in problem 19, except let u at H be equal to v_s.

6.11 Group Project Ideas

For each project, the students should work in small groups and present their findings in either a short report (5-8 pages), a poster, or a 15 minute oral presentation.

1. Combine the information from the controlled emissions from a particular local emission source of your choice (such as your university power plant), a standard wind rose, a map of your location, and your own calculations of dispersion in each direction. Superimpose your ground level concentrations onto the map for each wind rose direction. How would you expect air quality to be impacted from the site? Be sure to weight your calculations by the wind rose percentages at each velocity and each direction. Emissions data is available from the source's website or EPA.gov/state environmental quality websites.

6.12 Bibliography

Briggs, G A. 1974. *Diffusion Estimation for Small Emissions*. Oak Ridge, TN, USA: US Atomic Energy Commission, 1974. USAEC Report ATDL-106.

Briggs, G.A. 1975. *Plume Rise Predictions*. s.l. : American Meteorological Society, 1975.

—. 1969. *Plume Rise, AEC Critical Review Series*. Washington, DC, USA : Atomic Energy Commission, 1969. TID-25075.

Cooper, C David and Alley, F C. 2011. *Air Pollution Control: A Design Approach*. Long Grove, IL, USA : Waveland Press, INC., 2011. ISBN 978-1-57766-678-3.

EEA, European Environment Agency. 2007. EIONET - European Topic Centre on Air Pollution and Climate Change Mitigation. *MDS - Model Documentation System*. [Online] June 4, 2007. [Cited: June 6, 2011.] http://acm.eionet.europa.eu/databases/MDS/index_html.

Holland, J.Z. 1953. *A Meteorological Survey of the Oak Ridge Area*. Washington, DC, USA : Atomic Energy Commission, 1953. Report ORO-99.

Lin, J C, Brunner, D and Gerbig, C. 2011. Studying Atmospheric Transport Through Lagrangian Models. *EOS Transactions, American Geophysical Union*. May 24, 2011, Vol. 92, 21.

Moussiopoulos, Nicolas, et al. 1996. *AMBIENT AIR QUALITY, POLLUTANT DISPERSION AND TRANSPORT MODELS*. Copenhagen, Denmark : European Environment Agency, 1996.

Nonhebel, G. 1960. *Journal of the Institue of Fuel*. 1960, Vol. 33, 4.

Slade, D.H., editor. 1968. *Meteorology and Atomic Energy*. Washington DC, USA : US Department of Energy, 1968. TID-24190.

Stohl, Andreas. 2001. The FLEXTRA homepage. [Online] November 2, 2001. [Cited: June 7, 2011.] http://zardoz.nilu.no/~andreas/flextra.html.

Thomas, F W, Carpenter, S G and Colbaugh, W C. 1970. Plume Rise Estimates for Electric Generating Stations. *Journal of the Air Pollution Control Association*. 1970, Vol. 20, 3.

Turner, D B. 1994. *Workbook of Atmospheric Dispersion Estimates, 2nd Ed*. Boca raton, FL, USA : CRC Press, Inc., 1994.

US-EPA. 2010. Dispersion Modeling. *Technology Transfer Network*. [Online] May 13, 2010. [Cited: May 28, 2011.] http://www.epa.gov/ttn/scram/dispersionindex.htm.

—. 2011. TTN - Support Center for Regulatory Atmospheric Modeling. *Screening Models*. [Online] May 2011. http://www.epa.gov/scram001/dispersion_screening.htm.

—. 1995. *User's Guide for the Industrial Source Complex (ISC3) Dispersion Models: Volume II - Description of Model Algorithms*. Research Triangle Park, NC, USA : Office of Air Quality Planning and Standards, 1995. EPA-454/B-95-003b.

US-NOAA. 2011. HYSPLIT - Hybrid Single Particle Lagrangian Integrated Trajectory Model. ARL - *Air Resources Laboratory*. [Online] National Oceanic and Atmospheric Administration, Feb 11, 2011. [Cited: June 7, 2011.] http://www.arl.noaa.gov/HYSPLIT_info.php.

CHAPTER 7

Properties of Particulate Matter

7.1 Physical Properties of Particulate Matter

Particles can have a very wide range of physical properties, including the variables of size, shape, density, phase (liquid or solid), distribution, and composition. The *particulate matter* (PM) in a gas may be composed of materials similar in physical and chemical properties (homogeneous), or they may vary in some or all variables (heterogeneous). *Aerosols* are mixtures of liquids or solids in a gas. Typical concentrations may range from 1 – 1000 grams of particles per cubic meter of gas. The particles are too small to settle out of the gas by gravity and are free to move within the gas. The aerosol mixture is the form of PM that causes air pollution problems. In this chapter, we review the basic concepts for describing, measuring, and quantifying the properties of particles and aerosols. Chapter 8 explores particulate matter as an air pollutant, and how to remove it.

7.1.1 Size

The size of a particle may range from visibly large to microscopically small. Large particles may include sand, grit, snow, and rain. Small particles may include smoke, bacteria, clay, and mists. Figure 7-1 (see page 196) lists several particle size classifications or fractions. Also included are common substances in each size range, typical formation mechanisms, and removal technologies.

7.1.2 Shape

Particles may come in any shape – from nearly spherical to complex irregular shapes that are difficult to describe with standard geometric tools. Table 7-1 (on page 196) lists several regular shapes. All shapes can be described with an effective diameter, D_{eff}. The effective diameter is the diameter of a spherical particle that behaves the same as the actual particle. D_{eff} depends on the measurement device more than on the actual particle shape. For example, the effective diameter for particles classified with a sieve is the size of the sieve opening through which it may pass. If measured with a sedimentation technique, such as sand grains settling in water, then the effective diameter is the size of a sphere with the same sedimentation velocity as the falling grain. The choice of measurement method should reflect how the information is used.

7.1.3 Density

Particles have densities similar to the material from which they form, see Table 7-2. Bulk density refers to the collection of particles and not the aerosol. Bulk density is lower than the pure material

FIGURE 7-1. Range of Particle Sizes for Several Example Materials and Processes.
Adapted from (Lapple, 1961)

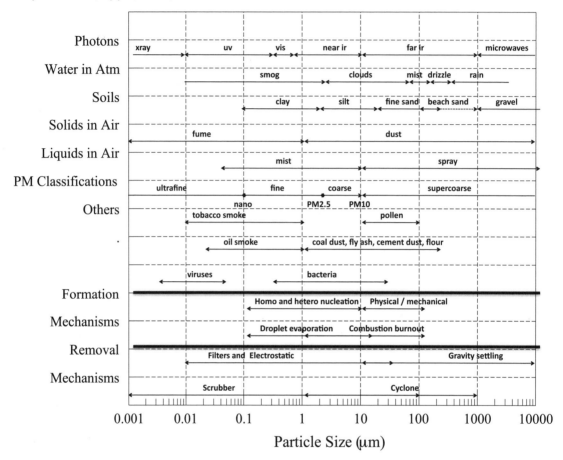

Table 7-1. Particle Shapes.

Regular shapes	Important size descriptions	Example
Sphere-solid	Diameter	Glass beads
Sphere-hollow	Outside diameter	Hollow glass beads
Cube	Length of side	Sugar, calcite
Cylinder/Fiber	Length and Radius	Asbestos
Cone	Height and Radius at base	Crushed minerals
Ellipsoid	Major and Minor axis	Abraded solids, atomized drops
Flake	Effective Diameter	Talc
Plate	Diameter and Thickness	Clay

because of the voids between the particles. Another name for the void space is porosity. A typical porosity for a bulk collection of spheres of identical size is 33%. When a distribution of sphere sizes is present, the porosity tends to go down. Shape can also play a large role in porosity, making prediction or typical values very difficult to determine.

Table 7-2. Specific Gravity of Common Particulate Matter in the Atmosphere.
Adapted from (Perry, et al., 1997). Bulk values apply to piles of the material and include void spaces.

Material		SG*	Material		SG*
Earth			Wastes		
	Clay (dry)	1	Fly Ash		0.6 - 0.8
	Clay (wet)	1.8	Tire wear, particles from		0.4 – 0.6
	Sand	1.4 - 1.7	Asbestos Break wear		2.0 – 2.2
	Soil	1.1 - 1.8	Non-asbestos Break wear		1.8 – 2.0
	Volcanic ash, particles	1.8 – 3.0	Slag		1.9 - 2.3
	Bulk Dry, un-compacted	0.5 – 1.3	Wet Sludge		1.2 - 2.0
	Bulk Wet, compacted	1.0 – 2.0	Soot		0.6 – 0.9
Rock and Mineral			Cinders, Coal, Ashes & Clinker, bulk		0.6
	Asbestos	2.1 - 2.8	Cinders, Blast Furnace, bulk		0.9
	Bauxite	2.5 - 2.6	Clinker Dust, bulk		1.4
	Chalk	1.8 - 2.8	Coal – Pulverized, bulk		0.6
	Lime	0.8 - 1.0	Coke dust, bulk		0.2
	Limestone	2.0 - 2.7	Fuels		
	Marble	2.3 - 2.7	Coal	Anthracite	1.4 - 1.8
	Sandstone	1.9 - 2.5		Bituminous	1.2 - 1.5
	Ammonium Nitrate	0.78		Lignite	1.1 - 1.4
	Ammonium Sulfate	1.1		Peat	0.6 - 0.9
Construction				Charcoal	0.3 - 0.6
	Cement (dust)	1.5 - 2.0	Petroleum	Crude	0.8 - 0.9
	Cement (set)	3.1 - 3.2		Gasoline	0.7 - 0.8
	Concrete	2.2 - 2.4	Tar		1.2

* SG, specific gravity, is the ratio of material density to density of water, which is 1,000 kg/m3.

The density of particles in air is measured in either number of particles per unit volume, or mass of particles per unit volume. There is tremendous range in these values. For example, air in an industrial clean room for the manufacture of electronics or pharmaceuticals may be as low as 1 – 10 particles per cm^3, whereas urban air near a heavily traveled roadway may exceed 100,000 particles per cm^3. A typical background value over the ocean is 100 particles per cm^3.

7.1.4 Distributions

Particles will rarely all have the same characteristics; instead, their properties have a range of values. Typically, the range is more important than the average. Knowing the range of particle properties (size, density, shape) is important in air pollution control because the collection efficiency of different types of particle control equipment is heavily dependent on these properties.

The range of sizes is commonly referred to as the size distribution. Table 7-3 lists the results from a measurement of diameters of particles collected from a filter. The diameters are given as a range between two sizes (start and end) due to the method of measurement. It would be very tedious to measure and list every particle diameter individually, so the results are binned. The range between the start size and end size is called the bin size. We use this distribution for several of the worked example problems in this chapter. Each line of the table represents a bin, and the bin size can be the start size, the end size, or the average size. The choice of which size to use is relatively unimportant, but it is important to be consistent for all bins.

Table 7-3. Example Distribution of Particles by Number and Bin Size.

Start Size (μm)	End Size (μm)	Number of Particles
2.5	5	640
5	10	7789
10	20	34909
20	40	57555
40	60	48100
60	80	34909
80	105	22000
105	121	16002
121	160	7789
160	210	3500
210	320	700
320	560	40

Figure 7-2a and b are histograms of the particle distribution given in Table 7-3. Note that the largest sizes are easily removed from the air due to gravity settling. Figure a is a plot of the information directly as a function of bin size (diameter). Figure b is the same distribution plotted as a function of the natural log of the bin size (diameter). Examination shows that figure b takes on a normal (or Gaussian) bell shape curve. This characteristic is common for particle size distributions and for many other properties where multiple random factors contribute to creating the overall distribution. The distribution is sometimes referred to as having a log-normal distribution.

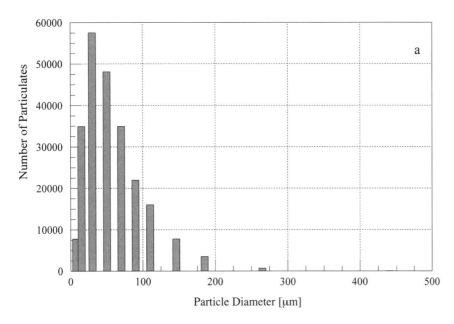

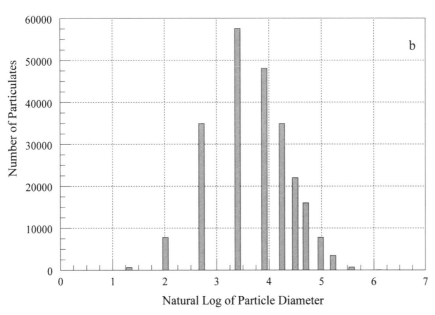

FIGURE 7-2.a Particle Number Distribution and **FIGURE 7-2.b Natural log Particle Number Distribution.** These figures show two different ways to represent the information in Table 7-3.

A complete description of a normal distribution requires only two numbers – the average and the standard deviation. The two-term description is mathematically convenient for characterizing a distribution, and greatly simplifies calculations involving distribution information. Equation 7.1 describes the probability, f_i, that a given particle has a certain size, assuming that the property has a normal or Gaussian distribution.

7.1 $$f_i = \frac{1}{\sigma\sqrt{2\pi}} \exp\left[-\frac{(x_i - x_{ave})^2}{2\sigma^2}\right]$$

Where: x_i represents the particle diameter size,
x_{ave} represents the average particle diameter size, and
σ represents the standard deviation of the particle diameters.

This function has its maximum value when x_i is equal to x_{ave}.

Determining the number fraction of particles (F_i) within a size range (x_i to x_{i+1}) requires integrating equation 7.1 over the size range, as shown below:

7.2 $$F_i = \frac{N_i}{N} = \int_{x_i}^{x_{i+1}} f_i \delta(x) = \int_{x_i}^{x_{i+1}} \frac{1}{\sigma\sqrt{2\pi}} \exp\left[-\frac{(x_i - x_{ave})^2}{2\sigma^2}\right] \delta(x)$$

Where: N_i represents the number of particles in a size range, and
N represents the total number of particles.

If the particle distribution is discretized into bins, it may be easier to integrate f_i numerically than to use equation 7.2. Equation 7.3 provides a simple numerical approximation to equation 7.2 using Simpson's 1/3 rule:

7.3 $$F_{i+1} = \int_{x_i}^{x_{i+2}} f_i \delta(x) = \frac{(x_{i+2} - x_0)}{2}\left[\frac{f(x_{i+2}) + 4f(x_{i+1}) + f(x_i)}{3}\right]$$

Be careful when using this equation to piecewise integrate an entire distribution. If equation 7.3 is applied to every bin ($x_0, x_1, x_2, \ldots$), rather than every other bin ($x_1, x_3, x_5, \ldots$), it yields twice the value because each range is counted twice. Applying to every point may be useful when only a coarse discretization of the data is available (less than 16 bins), but the solution must be reduced by a factor of 2 (Example 2 demonstrates this technique).

There are many ways to calculate the average diameter, so it is important to be clear about which average to use. The most commonly used averages for particles are the arithmetic and geometric. Other averages exist, such as quadratic, cubic, and harmonic. The standard deviation must use the same basis as the average diameter. It is important to know how the distribution is used before deciding which values to calculate. Typically, use arithmetic values when the property of interest is additive. Use geometric values when the property of interest is multiplicative. When a distribution is log-normal, it suggests that the relevant property is multiplicative.

The arithmetic average diameter, X^A_{ave}, is given as:

7.4 $$x^A_{ave} = \frac{1}{N}\sum_{i=1}^{N} x_i = \frac{(x_1 + x_2 + x_3 + \ldots + x_N)}{N}$$

The arithmetic standard deviation, σ^A, is given as:

7.5 $$\sigma^A = \sqrt{\frac{\sum_{i=1}^{N}(x_i - x^A_{ave})^2}{N-1}}$$

The geometric average diameter, x_{ave}, is given as:

7.6 $$x_{ave} = \sqrt[N]{\prod_{i=1}^{N} x_i} = (x_1 * x_2 * x_3 * \ldots * x_N)^{\frac{1}{N}}$$

Geometric standard deviation, σ, is given as:

7.7 $$\sigma = exp\left[\sqrt{\frac{\sum_{i=1}^{N} \ln(x_i) - \ln(x_{ave})^2}{N}}\right]$$

A distribution may be in terms of mass of particles rather than the number. When the particle density is constant for all sizes, the mass and number of particles of a given diameter are related by the volume and density of the particles, given as:

7.8 $$n_{x_i} = \frac{m_{x_i}}{\rho_p \frac{\pi}{6} x_i^3}$$

Where: n_{xi} represents the number of particles of a given diameter,
m_{xi} represents the mass of the particles of that diameter,
ρ_p represents the particle density, and
x_i is the particle diameter.

Note that n_{xi} is dimensionless, so all the units on the right hand side must cancel. Figure 7-3a and b show plots based on mass distribution for the same data presented in Figure 7-2a and b.

FIGURE 7-3.a Particle Mass Distribution and **FIGURE 7-3.b Natural Log Particle Mass Distribution.**

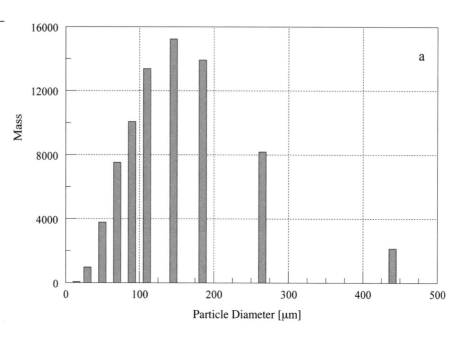

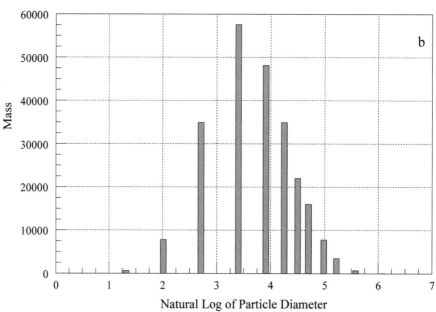

Example 7 1.

Determine the average and standard deviation from the particle count analysis given in Table 7-3.

Solution:

Columns 1 and 2 of Table 7-4 list the given data from Table 7-3. Column 3 calculates the number fraction (the ratio of particles to the total number). Begin by summing column 2 to obtain the total number of particles. Next, for each size range, divide the number in that size range by the total number. Note that in this example the general subscript, i, is replaced by a particular value, 1 or 2 or 3…, to note exactly which bin is being characterized.

$$F_1 = 640/233934 = 0.00274$$

Column 4 calculates the average diameter of the particles in the given size range. This example uses the average bin size. Similar calculation could have used the start size or the end size and found very similar results.

$$x_1 = (2.5+5)/2 = 3.75 \, \mu m$$

Column 5 calculates the natural log of each of the average diameter sizes

$$\ln(x_1) = \ln(3.75) = 1.322 \, \mu m$$

Column 6 is the product of the natural log of the average diameter (column 5) and the number fraction (column 3) for each size range. Sum all the values. The geometric average diameter (xg) is determined by taking the antilog of this sum. The general form of this equation follows. To use it in this example, sum the values in column 6.

$$x_g = \exp\left[\sum_{i=1}^{12} \ln(x_i) * F_i\right] = \exp[SumColumn6] = \exp[3.755] = 42.74 \, \mu m$$

Note that to use equation 7.6 for this calculation; you could multiply all 233,934 particle sizes and then take the 233,934th root, which is not practical.

Column 7 begins the procedure for determining the standard deviation. For each bin (row), multiply the number fraction of particles by the square of the difference between the average size and the geometric average:

$$F_1 * \left(\ln(x_1) - \ln(x_g)\right)^2 = 0.00274 * (1.322 - 3.755)^2 = 0.01620$$

To find the standard deviation, sum the values in column 7, calculate the square root of this sum and then take its antilog

$$\sigma_g = \exp\left[\sqrt{(SumColumn7)}\right] = \exp\left[\sqrt{0.5560}\right] = 2.11 \, \mu m$$

Table 7-4. Example 7-1 Data and Worksheet.

Bin Number	col. 1 x Particle Diameter Size Range	col. 2 Ni – Number of Particles	col. 3 Fi – Number Fraction of Particles	col. 4 xi – Average Particle Diameter	col. 5 ln(xi)	col. 6 ln x_i * F_i	col. 7 Square of Difference
1	2.5 - 5	640	0.0027	3.75	1.32176	0.00362	0.0162
2	5 - 10	7789	0.0333	7.50	2.01490	0.06709	0.10083
3	10 - 20	34909	0.1492	15.0	2.70805	0.40411	0.16359
4	20 - 40	57555	0.2460	30.0	3.40120	0.8368	0.03081
5	40 - 60	48100	0.2056	50.0	3.91202	0.80436	0.00507
6	60 - 80	34909	0.1492	70.0	4.24850	0.63398	0.03633
7	80 - 100	22000	0.0940	90.0	4.49981	0.42318	0.05216
8	100 -120	16002	0.0684	110	4.70048	0.32154	0.06114
9	120 - 160	7789	0.0333	140	4.94164	0.16454	0.04688
10	160 -210	3500	0.0150	185	5.22036	0.0781	0.03212
11	210 -320	700	0.00299	265	5.57973	0.0167	0.00996
12	320 +	40	0.00017	440	6.08677	0.00104	0.00093
		Sum N_{bin} = 233,934	sum = 1.000			sum = 3.755 x_g = 42.74	sum = 0.556 stddev = 2.11

■ **Example 7-2.**

Generate a number distribution from the distribution described in Example 7-1 as a function of the particle diameter. The average particle diameter is 42.74 μm and the standard deviation is 2.11 μm.

Solution:

In this problem, we use the result from the last example to reconstruct the original data. The difference between the sets is due to the assumed Gaussian model.

In column 1, we choose the average diameter for each size range. For ease in comparison, we use the same ranges as in example 1 plus an extra size at the small and large side for calculation convenience. Column 2 calculates the natural log of the size range average diameter (column 1). Column 3 calculates the Gaussian distribution for each size:

$$f_1 = \frac{1}{\ln(\sigma_g)\sqrt{2\pi}} \exp\left[-\frac{(\ln(x_1) - \ln(x_g))^2}{2(\ln\sigma_g)^2}\right] =$$

$$\frac{1}{\ln(2.11)\sqrt{2\pi}} \exp\left[-\frac{(\ln(3.75) - \ln(42.74))^2}{2\ln(2.11)^2}\right] = 2.61 \times 10^{-3}$$

Column 4 calculates the area under the curve generated when plotting df/dx as a function of the natural log of the particle diameter, using Simpson's 1/3 rule for numerical integration. Note that F_i is not calculated for the extra rows (0 and 13), but these rows are useful for the determination in row 1 and 12, respectively.

$$F_1 = \int_{dp0}^{dp2} fd(\ln(x)) = \frac{1}{2} \frac{\ln(x_2) - \ln(x_0)}{2} \left[\frac{f_0 + 4f_1 + f_2}{3}\right]$$

$$= \frac{1}{2} \frac{2.015 - (0.916)}{2} \left[\frac{3.81 \times 10^{-4} + 4 * 2.61 \times 10^{-3} + 3.51 \times 10^{-2}}{3}\right] = 0.0042$$

In column 5, we determine the number of particles in each size range by multiplying the size fraction (column 4) by the number of particles in total. Note that column 4 should sum to one, but due to the limited number of size ranges and the inaccuracy of the numerical integration scheme it sums to 1.006 (it can have a value above or below one). The values in column 5 need to be divided by this sum in order to have the particles sum to the required number.

$$N_1 = \frac{F_1 * N}{\sum_{i=1}^{12} F_i} = \frac{0.0042 * 233934}{1.006} = 978$$

Table 7-5. Example 7-2 Data and Worksheet.

Bin Number	col. 1	col. 2	col. 3	col. 4	col. 5
	x_i	$\ln(x_i)$	f_i	F_i	N_i
0	2.5	0.916	3.81E-04		
1	3.75	1.322	2.61E-03	0.0042	978
2	7.5	2.015	3.51E-02	0.0396	9210
3	15	2.708	2.00E-01	0.1515	35245
4	30	3.401	4.78E-01	0.2644	61491
5	50	3.912	5.23E-01	0.2119	49284
6	70	4.248	4.30E-01	0.131	30471
7	92.5	4.527	3.13E-01	0.0762	17732
8	113	4.727	2.29E-01	0.048	11158
9	140.5	4.945	1.50E-01	0.0372	8647
10	185	5.22	7.76E-02	0.0257	5988
11	265	5.58	2.68E-02	0.0136	3171
12	440	6.087	4.03E-03	0.0024	557
13	500	6.214608	0.002322		

Sum = 1.006

Figure 7-4 (See page 207) shows a comparison of Column 5 from example 2 and Column 2 from example 1. The difference represents the error from assuming that the particle distribution is Gaussian.

7.1.5 Graphical Description of Distributions

A graphical method, using a log-probability plot of size and number fraction or mass fraction can also be used to determine the average and standard deviation. Example 7-3 demonstrates this method. Figure 7-6 is a blank graph for use in homework problems.

FIGURE 7-4. Comparison of Results from Example 7-1 and Example 7-2.

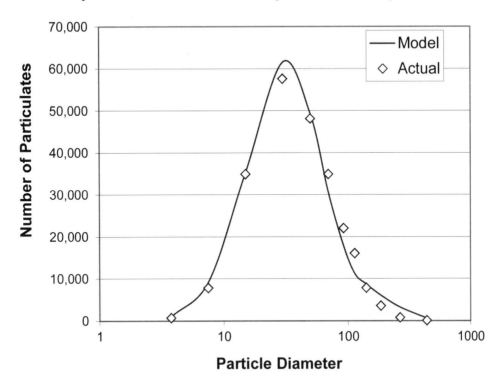

■ Example 7-3.

Determine the geometric average diameter and standard deviation from the data in Table 7-1 by plotting the variables on log-probability graph paper.

Solution:

Transform the number of particles in each bin in Table 7-6 into number fraction (Column 3) by dividing the number in each bin by the total (233933), and then make a running sum of the cumulative amount (Column 4). The cumulative fraction is the same as the probability.

Table 7-6. Example 3 Data and Worksheet.

col. 1	col. 2	col. 3	col. 4
Size μm	Number	Fraction	Cumulative Fraction
5	640	0.0027	0.0027
10	7789	0.0333	0.036
20	34909	0.1492	0.1853
40	57555	0.246	0.4313
60	48100	0.2056	0.6369
80	34909	0.1492	0.7861
105	22000	0.094	0.8802
121	16002	0.0684	0.9486
160	7789	0.0333	0.9819
210	3500	0.015	0.9968
320	700	0.003	0.9998
560	40	0.0002	1
sum =	233933		

Plot results (Figure 7-5) and fit with a line. Read the average diameter from the graph where the 0.5 probability of N_f intersects the line, recalling that the 50% probability value is the average. We see that the average (50% value) is about 42 μm. Determine the standard deviation by recalling that one standard deviation is the difference between the 50 and 84.1 (or 15.9) percentile. Determine the standard deviation value from:

7.9
$$\sigma_g = \frac{d_{84.1}}{d_{50}} = \frac{d_{50}}{d_{15.9}}$$

Reading $d_{84.1}$ from the graph yields 83 μm, so the geometric standard deviation is 2 μm. Note that it is difficult to read the plot to more precision than 1 μm.

FIGURE 7-5. Log-Probability Plot of Particle Size from Table 1.

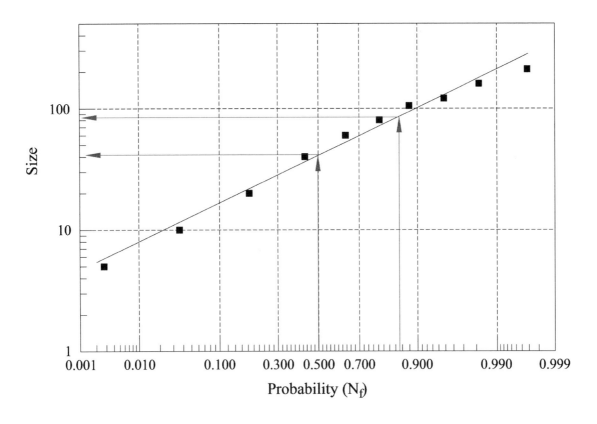

FIGURE 7-6. Log-Probability Paper.

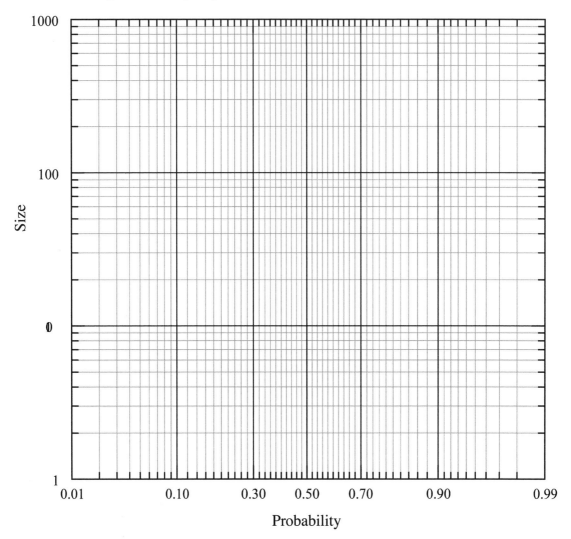

7.2 Properties of Atmospheric Aerosols

An *aerosol* is a mixture of particulate matter (PM) and air, although sometimes the word aerosol implies just the particles. Atmospheric PM has many sources, natural and anthropogenic. It includes materials like water, smoke, soot, soil, minerals, dust, pollen, spores, organic and inorganic acids, and sea-salt. Composition varies by location, time of day and time of year.

Particles play an important role in determining the behavior and the chemical and physical properties of the atmosphere. They
- Allow transport of non-volatile matter from one place to another,
- Alter the optical properties of the atmosphere,
- Act as nuclei for the formation of clouds and fog, and
- Alter the chemical reactions occurring in the atmosphere.

Transport of minerals, soil, and nutrients allows global distribution of some trace nutrients, including into the ocean. These nutrients may allow organisms to thrive in otherwise unfavorable conditions. The optical properties of the atmosphere include its ability to transmit or absorb radiation from the sun, or from the earth. These changes will alter the distribution of energy in the atmosphere, on land, and in the ocean. The changes can affect the Earth's climate. Clouds depend on the existence of PM for drop formation, and may create PM when drops evaporate. The distribution of clouds strongly influences the distribution of water on Earth. Finally, atmospheric particles may participate in atmospheric reactions or act as catalysts by mixing non-volatile species throughout the atmosphere.

The distributions of atmospheric PM include several groupings: the *Aitken Nuclei Range* (diameter < 0.1 μm), the *Accumulation Range* (0.1 to 2.0 μm), and the *Course Range* (> 2.0 μm). The groupings are based on number of particles and their diameter. The smallest grouping forms from the condensation of hot gases. These particles can grow in size due to coagulation, agglomeration, and conglomeration. The growth acts to remove particles from this size range, though not necessarily from the atmosphere. The coarse grouping particles include those created by mechanical abrasion as wind-blown dust, sea-salt, volcanic emissions, tire and brake wear, and biological matter. Atmospheric removal of the coarse material occurs by gravity settling (also called sedimentation). The particles in-between these sizes, the accumulation range, are too small to be removed by gravity, yet too large to undergo the random motions that lead to coagulation. These particles tend to accumulate in the atmosphere until removal by wet deposition (rain, fog, or snow).

Number density of particulates in the atmosphere range from a background level of 100 particles per cm^3, to tens of thousands in a desert sandstorm, to hundreds of thousands along an urban highway. In general, the number of particles is greatest in the accumulation range (0.1 to 2.0 μm), and the number decreases as diameter increases or decreases from this range.

Particles may interact with water in the atmosphere – a property called *hygroscopic*. Most soluble salts are hygroscopic because they readily absorb and retain water from the air. Such particles swell and expand when exposed to humid air and may even form droplets. The humidity at which deliquescence occurs (when the particle begins to absorb water) and the actual amount of moisture absorbed depends on the chemical composition. It is quite challenging to predict these properties since most atmospheric particles are heterogeneous mixtures of a variety of substances, including some that are non-hygroscopic.

7.3 Particle Motion

A review of particle motion in a gas can help explain the behavior of particles within the atmosphere and inside control equipment. The motion of particles depends on the forces acting upon them. When the forces are not in balance, the particle undergoes acceleration and it undergoes a velocity change. When the forces are in balance, there is no acceleration and no change in velocity. However, the velocity does not need to be zero.

Several common forces that act on a particle in the atmosphere or within control equipment include drag, gravity, buoyancy, and centrifugal forces.

Drag forces occur when a gas or fluid exerts a force on a particle when the particles velocity is not the same as the gas velocity. It can be very difficult to determine the exact drag force for complex shapes moving swiftly through a fluid. However, if we make a set of reasonable assumptions, we can describe the slow movement of a simple shape. The necessary assumptions follow, and when these assumptions are valid, the motion is called Stokes Flow.

1. Particle shape is spherical,
2. The particle moves slowly relative to the gas, and
3. The gas is of infinite extent and moves uniformly.

The drag force is a surface force, and its magnitude depends on the cross-sectional area of the sphere and on the relative difference in velocities between the gas and the particle. The equation for drag force is:

7.10 $$F_{drag} = \frac{\pi}{4} x_i^2 \left[\frac{1}{2} \rho_g u_p^2 C_D \right]$$

Where: x_i = particle diameter,
ρ_g = density of the gas,
u_p = relative velocity of the particle (compared to the gas velocity), and
C_D = drag coefficient.

The drag coefficient correlates with the Reynolds number when the Reynolds number is less than one (see Chapter 4 – Fluid Mechanics Review for additional detail). The experimentally observed correlation is:

7.11 $$C_D = \frac{24}{Re}$$

Where: Re = Reynolds number, given as:

7.12 $$Re = \frac{\rho_g u_g x_i}{\mu_g}$$

Where: ϱ_g = gas density,
 u_g = gas velocity,
 x_i = characteristic length (the particle diameter in this case), and
 μ_g = gas viscosity.

Note that the particle diameter is the effective diameter (D_{eff}). It is the diameter of a spherical particle that behaves the same as the actual particle, see section 7.1.2.

Substitution of equations (7.11) and (7.12) into equation (7.10) yields:

7.13 $$F_{drag} = 3\pi\mu_g u_p x_i$$

Gravity is another force that acts on a particle. It is a body force – meaning that gravity acts on the entire object, not just on the surface of the object. This force, $F_{gravity}$, can be calculated by taking the product of the particle mass and the gravity constant. It is described mathematically as:

7.14 $$F_{gravity} = \frac{\pi}{6} x_i^3 \rho_p g$$

Where: ϱ_p = particle density, and
 g = gravitational constant.

Buoyancy is a force that acts on all objects in a gas. It operates in a direction opposite to gravity and is due to the volume of gas displaced by the object. It is described mathematically as:

7.15 $$F_{bouyancy} = -\frac{\pi}{6} x_i^3 \rho_g g$$

Where: ϱ_g = gas density

Centrifugal forces occur either when an object moves in a circular direction around another object due to a flow constraint or to connection with another object. The force appears when a rotating frame of reference is used, rather than an inertial frame of reference. If the object were not constrained it would move in a linear manner, not circular. Centrifugal force can be described mathematically as:

7.16 $$F_{centrifugal} = \frac{\frac{\pi}{6} x_i^3 \rho_p u_t^2}{r}$$

Where: r = distance between the object and the center of rotation, and
 u_t = tangential velocity (typically the fluid is moving tangential).

7.3.1 Particle Motion Due to Gravity.

Figure 7-7 shows the forces acting on a falling particle. The forces are gravity, buoyancy, and drag. Choosing down as the positive direction (which allows the particle velocity to be positive) their sum is:

7.17 $$F_{gravity} - F_{bouyancy} - F_{drag} = M_p \frac{du_p}{dt}$$

A particle that begins with no velocity difference from the gas accelerates downwards quickly due to gravity initially. As the velocity increases, so will the drag force. After a short time, the particles increased velocity increases such that the drag and buoyancy forces counterbalance gravity, and the acceleration becomes zero. The balance occurs within about 0.033 seconds for a 30 μm grain of sand, a negligible time period for control equipment design considerations. Setting the acceleration term (the right hand side) to zero, we can solve for the particle velocity by combining equations (7.13), (7.14), and (7.15). This velocity is the *Stokes Law terminal* velocity, or the settling velocity, and it is described mathematically as:

7.18 $$u_p = \frac{x_i^2 (\rho_p - \rho_g) g}{18 \mu_g}$$ (only for particles larger than 10 μm and Re<1)

Where: x_i = particle size expressed as the diameter of a sphere having the same properties,
ϱ = density (p = particle, g = gas),
g = gravitational constant, and
μ = viscosity (g = gas).

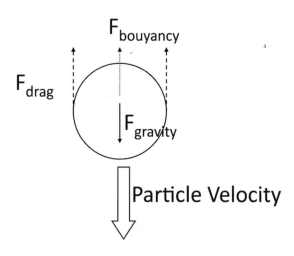

FIGURE 7-7. Force Balance and Resultant Particle Velocity during Free-fall.

For very small particles, the particle diameter approaches the mean free path of a gas molecule (how far it travels before hitting another molecule), which allows the particle to 'slip' past the gas, and avoid part of the drag force. The Cunningham correction factor (K_c), (Wark, et al., 1998), (Strauss, 1966) accounts for the difference for sizes of 10 μm and smaller. At standard temperature and pressure, the *Cunningham correction factor* is (when x_i is in μm):

7.19 $$K_c = 1 + \frac{0.171}{x_i} + \frac{0.0544}{x_i}\exp[-8.09 x_i]$$

At 10 μm it increases the terminal velocity by 2%, at 1 μm by 17% and at 0.1 μm by 190%. The settling velocity for small particles is:

7.20 $$u_p = \frac{x_i^2 K_c \left(\rho_p - \rho_g\right) g}{18 \mu_g} \quad \text{(for particles less than 10 μm and Re<1)}$$

FIGURE 7-8. Terminal Velocity (u_p) of Particles Settling in Air at 1 atm and 20°C for Various Particle Densities (ρ_p).

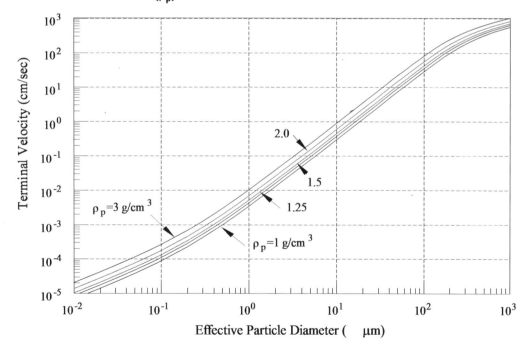

Equation (7.20) or Figure 7-8 can be used to obtain a settling velocity for a particle of a given diameter and density.

Example 7-4.

Determine the terminal velocity of a particle of ammonium sulfate. The particle has an effective diameter of 0.5 μm, the dry air temperature is 20°C and pressure of 1 atm.

Solution:
We can solve equation 7.20 or use figure 7-8. Using the equation – find values for each variable

$$u_p = \frac{x_i^2 Kc (\rho_p - \rho_g) g}{18 \mu_g}$$

$x_i = 0.5 \mu m = 0.5 \times 10^{-6}$ m
$\rho_p = 1,100$ kg/m³ from Table 7-2.

$$\rho_g = \frac{mw \cdot P}{R \cdot T} = \frac{28.96 \frac{kg}{kmol} \cdot 1.0 atm}{0.082057 \frac{m^3 \cdot atm}{kmol \cdot K} \cdot 293K} = 1.20 \frac{kg}{m^3}$$ from example 5-3.

$g = 9.81$ m/s2 from appendix 1.
$\mu_g = 1.83 \times 10^{-5}$ kg/m/s from appendix 6.

$$K_c = 1 + \frac{0.171}{x_i} + \frac{0.0544}{x_i} \exp[-8.09 x_i] = 1.304$$ from equation 7.19 and $x_i = 0.5 \mu m$.

Solving

$$u_p = \frac{x_i^2 Kc (\rho_p - \rho_g) g}{18 \mu_g} = \frac{(0.5 \times 10^{-6} m)^2 (1.304)(1,100 - 1.2) \frac{kg}{m^3} \left(9.81 \frac{m}{s^2}\right)}{18 \left(1.83 \times 10^{-5} \frac{kg}{m \cdot s}\right)} = 9.85 \times 10^{-6} \frac{m}{s}$$

From Figure 7-8 $u_p = 10^{-3}$ cm/s or 10×10^{-6} m/s.

7.3.2 Particle Motion due to Centrifugal Forces

The particle velocity is determined by equating the centrifugal and drag forces on the particle in a radial direction. This balance ignores gravity and particle-particle interactions. It also assumes that the particles are spherical and that Stokes flow assumptions are valid. The force balance becomes:

7.21 $$F_{centrifugal} - F_{drag} = 0$$

Substituting $F_{centrifugal}$ as defined in equation (7.16), F_{Drag} as defined in equation (7.10), assuming Stokes Law is valid, and solving for the particle velocity yields:

7.22 $$u_p = \frac{\rho_p x_i^2 u_g^2}{18 R \mu_g}$$

Where: ρ_p = density of the particle,
x_i = particle diameter,
u_g = the gas velocity ($u_g = Q/A$)
Where: Q = volumetric flow rate,
A = cross sectional area,
μ_g = gas viscosity, and
R = distance particle travels.

■ Example 7-5.

Determine the centrifugal speed of a 0.5 um particle of ammonium sulfate flowing in a 10 inch diameter cyclone (see Chapter 8 section 5.3) just as it touches the outer wall (distance traveled is 5 inch). The air speed is 2.5 m/s, temperature is 20°C, and pressure is 1 atm.

Solution: Apply equation 7.22. $u_p = \dfrac{\rho_p x_i^2 u_g^2}{18 R \mu_g}$

ρ_p = 1,100 kg/m³ from Table 7-2.
x_i = 0.5μm = 0.5 x10⁻⁶ m
u_g = 2.5 m/s as given.
R = 5 inch = 0.127 m, as given.
μ_g = 1.83 x 10⁻⁵ kg/m/s from appendix 6.

Solving

$$u_p = \frac{\rho_p x_i^2 u_g^2}{18 R \mu_g} = \frac{\left(1,100 \frac{kg}{m^3}\right)(0.5 \times 10^{-6} m)^2 \left(2.5 \frac{m}{s}\right)^2}{18(0.127 m)\left(1.83 \times 10^{-5} \frac{kg}{m \cdot s}\right)} = 41.1 \times 10^{-6} \frac{m}{s}$$

It would take the particle nearly one hour to travel the 5 inch distance at this speed.

7.4 Interactions with Radiation

Atmospheric particles can scatter and absorb electromagnetic radiation. This reduction in light between a source (e.g. sunlight entering the atmosphere) and the receptor (e.g. the Earth's surface) is called extinction. Overall extinction is quantified, in the absence of multiple scattering, by the Beer-Lambert Law,

7.23 $$\frac{I}{I_0} = \exp[-\beta_e L]$$

Where: I = light intensity observed at the receptor,
I_0 = light intensity incident to the sample (from the source),
βe = extinction coefficient, and
L = path length through the sample (between the source and receptor).

The extinction coefficient is a complicated function of the particle diameter, particle shape, the ability of the particle to interact with each wavelength of the source light, and whether the interaction is scattering or absorbing. It is beyond the scope of this text to perform the calculation for βe, and the interested reader should consult advanced physical chemistry texts [such as (Atkins, et al., 2009)] or atmospheric chemistry texts [such as chapter 4 of (Brasseur, et al., 1999)]. However, the basic concept of extinction can be further explained. Extinction occurs when light interacts with the atmosphere in a way that prevents the source light from reaching the receptor (e.g. your eye). The interaction can be with particles or with certain molecules.

Light interacts with particles in a variety of ways, see Figure 7-9.
- *Scattering* refers to light that is altered by a particle but retains the same wavelength. Scattered light may take on a new direction after the interaction. Scattering occurs by three mechanisms – *diffraction*, *refraction*, and *phase shifting*.
- *Absorption* refers to light that interacts with a particle, alters the internal molecular energy of that particle, and is later re-radiated in a random direction and perhaps at a different wavelength.

The direction that light scatters depends on the size of the particle it interacts with. The type and amount of scattering depends on the particle diameter. A size parameter can help describe the interaction:

FIGURE 7-9. Scattering and Absorption of light by a large particle.

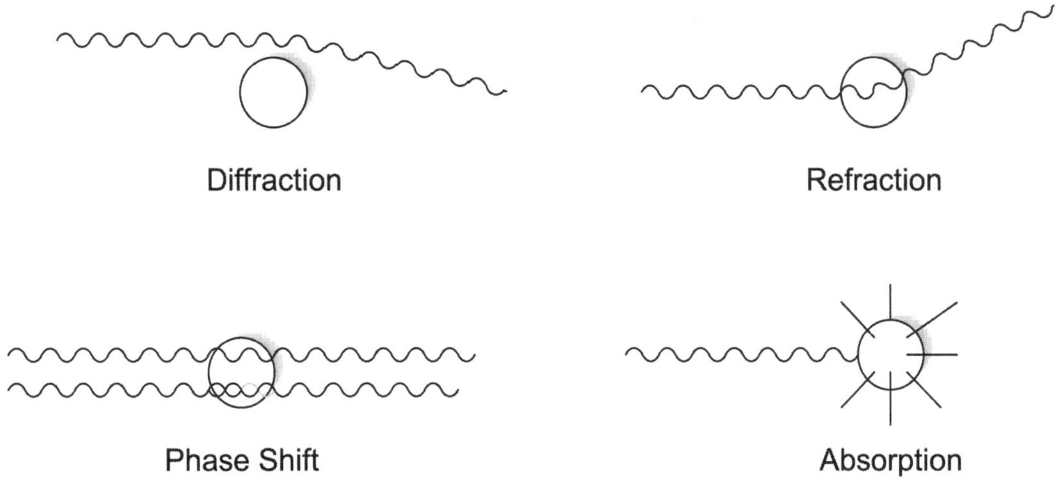

Diffraction

Refraction

Phase Shift

Absorption

7.24
$$\alpha = \frac{\pi x_i}{\lambda}$$

Where: α = size parameter,
x_i = particle diameter [μm], and
λ = radiation wavelength [μm].

At small values, $\alpha \ll 1$, when the particle is smaller than the wavelength of the light, the scattering will be in a random direction - equally as often forward as backward. This type of interaction is described as Rayleigh scattering. At large values, $\alpha \gg 1$, when the particle is much larger than the wavelength, the scattering is more often in the forward direction, relative to its direction before the interaction. The tendency becomes nearly 100% for the largest particles. These interactions are described by geometric optics. In between these limiting cases, when the wavelength and particle diameter are similar, the scattering phenomena is much more complex. Small particles are less efficient at scattering than large particles. Particles that are the same size as the wavelength of light are the most efficient scatters.

Table 7-7 provides an example of how particle size affects extinction and the associated visual range.

Table 7-7. Extinction and Visual Range in Air Containing 18 g/m³ of Water in Different Forms. Adapted from (Hinds, 1982).

Particle Size	Extinction	Visual Range
μm	$1 - I/I_0$	km
Vapor	1.8×10^{-7}	220
0.01	3.8×10^{-5}	1
0.1	0.29	1.1×10^{-4}
1	0.64	3.8×10^{-5}
10	0.052	7.4×10^{-4}
1 mm (rain)	5.3×10^{-4}	0.074

■ Example 7-6.

Determine the extinction coefficient for air containing 18 g/m³ of water in the form of 1 μm droplets (particles).

Solution:

Apply equation 7.23 and the data from Table 7-7.

$$\frac{I}{I_0} = \exp[-\beta_e L] \quad \text{Rearrange to} \quad \beta_e = \frac{-1}{L}\ln\left[\frac{I}{I_0}\right]$$

where $1 - I/I_0 = 0.64$, thus $I/I_0 = 0.36$
$L = 3.8 \times 10^{-5}$ km or 3.8×10^{-2} m.

$$\beta_e = \frac{-1}{L}\ln\left[\frac{I}{I_0}\right] = \frac{-1}{3.8 \times 10^{-2} m}\ln[0.36] = 26.9 m^{-1}$$

7.5 Questions

* Questions and problems may require additional information not included in the textbook.

1. Use a microscope to determine the shapes and relative sizes of several common particles, such as sand, salt, sugar, flour, and dust.*

2. Why is the grouping of particles between 0.1 to 2.0 um called the accumulation range?

3. Sketch a force balance (See Figure 7-7) for a) a particle moving horizontally, and b) a particle moving in an upward spiral.

4. Explain in your own words why the Cunningham correction factor is needed for small particles.

5. Consider Table 7-7. Describe the relationship between particle size and the ability to see distant objects. Why do you think this relationship occurs?

7.6 Problems

1. Calculate the porosity for a set of uniform size spheres packed regularly in a a) simple cubic, b) body center cubic arrangement, and c) a face centered cubic arrangement.

FIGURE 7-10. Basic packing arrangements of uniform spheres.

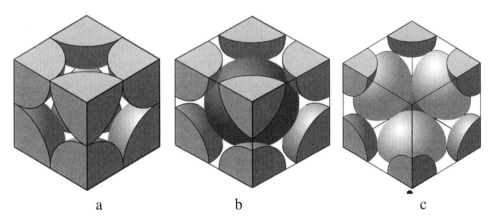

a) Simple cubic, b) body centered cubic, and c) face centered cubic (with corner and front face atoms removed)

2. Determine the average size and standard deviation of the following particle distribution.

Bin Range			Ni
Diameter um			
0.1	0.5		55
0.5	2		867
2	7		5442
7	25		13591
25	50		8983
50	85		4608
85	150		3113
150	200		1040
200	300		1288
300	500		588
500	800		279
800	1000		54

3. Generate a particle distribution for 100,000 particles that have an average size of 100 μm and a standard deviation of 10 μm.

4. Use Figure 7-6 to plot the data in problem 2. Then determine the average and standard deviation of the particle size distribution.

5. Derive equation 7.18 – Stokes Law terminal velocity.

6. Calculate the terminal velocity for a 100 μm diameter particle of sand (ρp=2.25 g/cm³) falling through air using equation (7.18). Compare your answer with the value obtained using Figure 7-8.

7. Calculate the visibility range for air with a PM concentration of 200 μg/m³. Assume the extinction coefficient for a particular emission source is given by the following equation, and I/I0 for visibility should be at least 0.02.

$$\beta_e = 0.045 km^{-1} + \left(0.0063 \frac{m^3}{\mu g \cdot km}\right) * C_{PM}$$

7.7 Bibliography

Atkins, Peter and de Paula, Julio. 2009. *Physical Chemistry, 9th edition*. New Yourk, NY, USA : Oxford University Press, 2009. ISBN-13: 978-1429218122.

Brasseur, Guy P, et al. 1999. *Atmospheric Chemistry and Global Change*. New York, NY, USA : Oxford University Press, 1999.

Hinds, W C. 1982. *Aerosol technology: Properties, Behavior, and Measurement of Airborn Particles*. New York, NY, USA : Wiley, 1982.

Lapple, C.E. 1961. 1961, Stanford Research Institute Journal, Vol. 5, p. 95.

Perry, Robert H and Green, Don W. 1997. Densities of Miscellaneous Materials. *Perry's Chemical Engineering Handbook, 7th Ed*. New York : McGraw Hill, 1997, pp. 2-119.

Strauss, W. 1966. *Industrial Gas Cleaning*. London : Pergamon Press, 1966.

Wark, K, Warner, C and Davis, W. 1998. *Air Pollution: Its Origin and Control*. 3rd. s.l. : Addison Wesley Longman, Inc., 1998.

CHAPTER 8

Particulate Matter

Particles and particulate matter (PM) refer to the solids and liquids found in a gas. The term *aerosol* refers to the mixture of particles and gas, although it is sometimes used to refer to just the particle fraction. All air contains some PM. Excessive PM changes normal atmospheric processes such as cloud formation and photochemical reactions. PM can also alter how the atmosphere transmits light, leading to reduction in visibility, changes in color, and alteration of the energy balance between incoming sunlight and outgoing infrared earthlight. Changes in the energy balance lead to changes in weather and can even alter climate. Types of particulate matter include acids, organic chemicals, combustion byproducts (soot), metals, salts, soil, and dust. The most important gases include atmospheric air (nitrogen, oxygen, and water), and sulfur, nitrogen, and organic compounds.

Table 8-1 lists the US-EPA terminology for air quality particle size, where the given sizes are in terms of a particle's aerodynamic diameter. *Aerodynamic diameter* refers to the particle diameter of a sphere that falls through air at the same speed as the actual particle. For the purposes of air pollution, particles with sizes larger than 100 μm are ignored as they quickly settle from the atmosphere by gravity and have only a localized spatial impact. Additional terminology includes TSP, PM_{10}, $PM_{2.5}$, and $PM_{0.1}$, see Figure 8-1.

Table 8-1. Particle Size Terminology.

Size Range in microns (μm)	Description
>10	Supercoarse
2.5 to 10	Coarse
0.1 to 2.5	Fine
<0.1	Ultrafine

TSP, total suspended particles, includes all the particles from all the sizes described in Table 8-1. It is measured by weighing a filter before and after it filters a known volume of air. The additional weight on the used filter is the TSP. Units are mass of particles per volume of filtered air. It is calculated as:

8.1
$$TSP = \frac{mass_{particles}}{V_{air}}$$

PM$_{10}$ includes all the particulate matter with diameters of less than 10 μm collected by a PM$_{10}$ collection device. However, it is easier to consider PM$_{10}$ as all the particles having an aerodynamic diameter of 10 μm or less. Course particles, such as crushed rock, sawdust, metal filings, and sanding debris typically form from material handling or abrasion. They are relatively easy to remove from a gas stream and will settle out of the air within a few hours.

PM$_{2.5}$ includes all the particulate matter with a diameter of less than 2.5 μm collected by a PM$_{2.5}$ collection device. However, it may also be useful to consider PM$_{2.5}$ as all the particles having an aerodynamic diameter of 2.5 μm or less. Fine particles settle slowly from the air and may take days for removal from the atmosphere. PM$_{2.5}$ typically has a very different chemical composition than the coarse and super-coarse fractions. These smaller sizes are composed primarily of sulfates, nitrates, organic compounds, and metals occurring from combustion, waste incineration, and metallurgical sources. Air emission testing and air pollution control methods for PM$_{2.5}$ particles are different from those for coarse and super-coarse particles. Particles in the range of 0.5 to 2 μm are especially challenging to collect.

PM$_{0.1}$ describes particles with an aerodynamic diameter of less than 0.1 μm or 100 nm. They are also called ultrafine particles or nano-materials. For comparison, the particles may be only 100 times larger than a gas molecule. Particles of this size may be stable in the atmosphere, but are likely to join with other particles through coagulation (small particles colliding with a large particle and adhering), agglomeration (adhere to similar particles to form a homogeneous particle) or conglomeration (adhere to different particles to form a heterogeneous particle such as a droplet of sulfuric acid pulling water vapor into the droplet). Their removal mechanisms will then coincide with those of the larger particles they have formed. PM$_{0.1}$ originate as part of the combustion process, condensation from combustion products, and various gas phase reactions involving acids, nitrogen compounds, organic compounds, and humidity which occur after these gases have entered the atmosphere.

Figure 8-1 shows a simplified device for collecting and measuring the particulate matter in a sample of air. It uses several filters, each of which is capable of capturing particles of the listed size or larger. After completing the filtering process, the number and mass of collected particles can be determined. The information is used to develop the particle distribution.

8.1 Sources

There are many sources of particles in the atmosphere; some are natural, and others are due to human activities. Natural sources include wind erosion, sea salt, fires, emissions from biomass, and vented gases from earth processes like volcanoes. The single largest natural particle source is an

FIGURE 8-1. Particulate Matter Size Fractions.

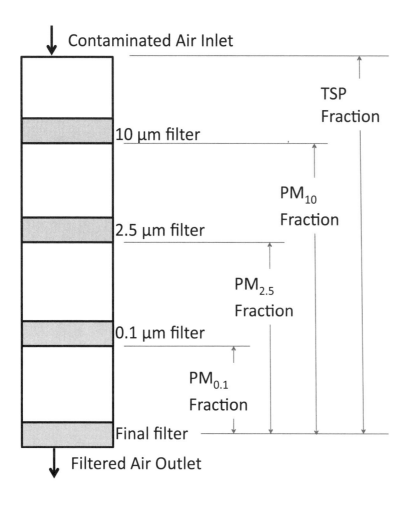

erupting volcano. One volcano can add as much particulate matter to the atmosphere as all other sources combined for a year. Figure 8-2 and Figure 8-3 show the plume from the Icelandic volcano, Eyjafjallajökull, at the source and its impact on Europe.

Anthropogenic (human-caused) sources include material handling, combustion products, resource extraction, and transportation activities. The largest source from human activities is road dust, which includes dirt from unpaved roads, dust from paved roads, and tire and brake wear (Vallero, 2008). Particulate matter source categorizations include primary and secondary emissions.

FIGURE 8-2. Eyjafjallajökull volcanic eruption showing plume, April 17th, 2010, Holsvöllur Iceland. (Boaworm, 2010)

FIGURE 8-3. Estimated Extent of Ash Cloud from Eyjafjallajökull Volcano, April 15, 2010. (Bourrichon, 2010)

Primary emissions are solids and liquids emitted directly into the atmosphere. They usually originate from combustion or mechanical process. The emissions may be either natural or anthropogenic. Primary particles are emitted near the Earth's surface and generally remain in the lower troposphere. Note that some volcanoes have ejected particulate matter directly into the stratosphere. Typically, primary particles are coarse to fine in size.

Secondary emissions originate as a gas that later transforms by chemical or physical processes into condensable particulate matter. The gas may be the result of a combustion process, exhaust gases from a stack, atmospheric gas-phase reactions between natural and/or anthropogenic materials, liquids that evaporate after entering the atmosphere, or caused by the absorption/ desorption between gases and already existing particles. Secondary particles can form in different parts of the atmosphere, e.g near a cloud, near the tropopause, or near the Earth's surface. Typically, secondary particles are fine or smaller.

Table 8-2 provides an annual estimate of the amount of particles added to the earth's atmosphere (Kreidenweis, et al., 1999). Removal of the larger sizes occurs within hours to days, and the smaller sizes may take days to weeks. Table 8-3 lists the particle emissions reported to the United States

Table 8-2. Global Particle Sources (10^6 ton/year).

Global Sources Manmade		Total Emissions $x_i < 25\ \mu m$	Fine Emissions $x_i < 1\ \mu m$
Primary	Industrial Activity	40 - 130	20 - 60
	Combustion byproducts	10 - 30	10 - 30
	Biomass Burning	50 - 190	50 - 190
	Wind erosion	850	140
Secondary	SO_x	120-180	120 - 180
	NO_x	20 - 50	5 - 10
	VOC	5 - 25	5 - 25
Natural			
Primary	Fires	5 - 150	2 - 75
	Sea Spray	1,000 - 10,000	20 - 100
	Volcanoes	5 - 10,000	1 - 100
	Wind Erosion	1,000 - 3,000	250 - 300
Secondary	SO_x	50 - 140	50 - 140
	NO_x	10 - 40	10 - 40
	VOC	40 - 200	40 - 200

Environmental Protection Agency (US-EPA) in 1990, 2000 and 2010 (US-EPA, 2015). The miscellaneous category includes road dust, open burning, and wildfires. The data are from the National Emission Inventory website, a database maintained by the US EPA (http://www.epa.gov/ttnchie1/trends/). There is poor agreement between the global and US sources due to the different methods of data collection and different definition of sources. Also, as Table 8-3 shows, the amounts change over time. The most variable PM source in the US is due to wildfires. In fire conducive years (hot and dry) wildfires may account for 20% of all PM emissions, whereas emissions are closer to 4% under average conditions.

Table 8-3. US Particle Emissions (10^3 ton/year).

Source	PM2.5			PM10		
Year	1990	2000	2010	1990	2000	2010
Transportation	623	468	469	715	552	579
Fossil Fuel Combustion	909	716	217	1,196	893	371
Industrial Processes	795	638	430	1,306	874	1,245
Miscellaneous	5,233	4,681	2,171	24,536	20,642	16,005
Total Amount [1000 tons]	7,560	6,502	3,286	27,753	22,961	18,199

8.2 Particle Formation

Table 8-4 lists several common air pollutant particle formation mechanisms along with typical sources. Each of these mechanisms creates particles with common characteristics, such as composition, particle size, and size distribution. It is possible to estimate the general size range of a source by knowing which particle formation mechanisms a given process uses, for examples see US-EPA AP-42 emission factor listings (www.epa.gov/ttnchie1/ap42/).

Table 8-4. Particle Formation Mechanisms.

Mechanism	Source
Physical attrition/ mechanical dispersion	Manufacturing
Combustion particle burnout	Power generation from combustion
Homogeneous and heterogeneous nucleation	Fuel combustion and metal smelting
Droplet evaporation	Wet-dry separations
Coagulation	Particle interaction in atmosphere

Figure 8-4 shows the general evolution of particle sizes from various sources. The largest particles originate from mechanical processes such as erosion, grinding, crushing, and sanding. The large particles are removed from the atmosphere by sedimentation and gravity settling. The smallest particles start as high temperature vapor which nucleate (condense) as they cool. The nuclei may grow when other hot gases condense on them, or when they collide with other particles - a process called coagulation. The particles grow until they are removed from the atmosphere by rain or snow.

FIGURE 8-4. Prototypical Size and Number Distribution of Tropospheric Particles with Selected Sources and Pathways of Particle Formation. Dashed line is 2.5 µm diameter. Adapted from: (Whitby, K T; Cantrell, B, 1976).

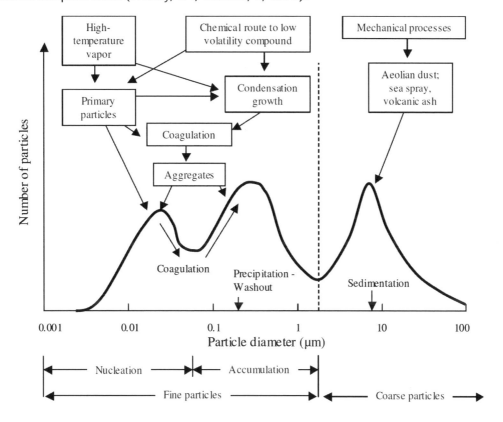

8.2.1 Primary Formation

Physical attrition occurs when two or more solid surfaces rub against each other, such as the brake pads and rotors or drums on an automobile or the grinding, crushing, and pulverizing operations performed in the minerals industry. Rubbing induces shear forces, which cause heating and allows particles of the solids to break away. These particles have similar compositions and densities to the original material. Particle sizes are typically 10 to 1,000 μm. Particles may be a by-product of the process and treated as waste, such as sawdust from wood operations, dusts from grinding of a metal part, or sanding a surface. Particles may also be created intentionally to increase a material's surface area in the hope of increasing rates of dissolution or chemical reaction such as the generation of limestone slurry and the crushing/ pulverizing of coal before burning.

Combustion particle burnout occurs after injection of fuel particles into the burner of a combustion process, such as a coal-fired power plant. This process vaporizes and oxidizes volatile organic compounds in the fuel while burning away the non-volatile materials. As combustion completes, the fuel particles become continually smaller and less dense, until all that remains is incombustible matter (ash). This residual material is about ten times smaller than the starting fuel particle, with a size range of 1 to 100 μm in size. Two types of ash are produced and collected during this process. Ash that leaves the combustion chamber with the product gas is *fly ash*. The ash that settles to the bottom of the burner is *bottom ash*. Different fuels have different ratios of fly ash to bottom ash. Changing the operation methods can lead to changes in this ratio.

Droplet evaporation occurs when water that contains solids (as particles or in solution) evaporate such as sea spray, clouds, or during an industrial process where a water spray is used to cool a gas stream. Sea spray and cloud droplets grow and shrink depending on the temperature and relative humidity. When droplets encounter warm dry air, the water evaporates. If enough water evaporates, dissolved minerals in the droplet can crystalize. The crystal can reabsorb water if it encounters humid air before removal from the atmosphere A water spray injects an atomized water stream into a hot gas stream. The small droplets of water cool the gas stream by evaporation and release the suspended or dissolved solids as small particles. The particle size may range from 0.1 to 20 μm

8.2.2 Secondary Formation

Homogeneous and heterogeneous nucleation involves the conversion of vapor phase materials to a particle form. The gas to particle conversion takes place by three processes: condensation, nucleation, and coagulation. The particle sizes from these mechanisms may range from 0.01 to 10 μm.

Condensation involves gaseous vapors condensing into, onto, or within existing small nuclei, usually called condensation nuclei. The vapor-containing gas stream must cool below its saturation temperature before condensation can occur. Each vapor phase element may have different nucleation temperatures and condensation patterns. An example is the emissions from a combustion process: some materials will nucleate in the relatively hot gas zones within the emission stack, while others remain in the vapor phase until the gas stream has left the stack and cooled to ambient temperature.

Nucleation occurs when gases combine with other droplets. The process is called homogeneous nucleation, or agglomeration, when the gas and droplet are the same substance. The process is called heterogeneous nucleation, or conglomeration, when the gas and droplet are of different substances. An example of the heterogeneous process transforming gases into particulate matter is the reaction of gaseous ammonia (NH_3) with nitric acid (HNO_3) to form ammonium nitrate particles (NH_4NO_3).

Coagulation occurs when gases or particles collide and adhere to each other. This mechanism is not a source of PM, rather it leads to changes in particle number, shape, size, and distribution. Coagulation occurs most efficiently between small particles (nanometers) and large particles (micrometers). Small particles collide with large ones due to Brownian motion (the irregular motion of small particles in a gas due to the bombardment of the particle by the molecules of the gas). The large particles barely change in size or mass, while the small particles are removed. This creates a pathway for reducing the number of particles in the atmosphere (and hence their shape, size, and distribution). It does not change the total mass of PM. Particles smaller than 0.02 μm coagulate and disappear from the atmosphere after just a few hours, particles smaller than 0.1 μm disappear within a few days. Coagulation has little effect on the size distribution of particles larger than 0.5 μm.

Another important consideration for estimating the size and distribution of PM is the relative humidity of the air. At high humidity a salt particle may spontaneously absorb water and grow in size (known as the deliquescence relative humidity). As humidity further increases, these particles can absorb more water and continue to grow in size. If humidity decreases, some water will evaporate and particle size will shrink.

8.3 Health and Welfare Effects

8.3.1 Effects on Public Health

Particle emissions create a serious air pollution problem worldwide because of the variety and severity of effects they have on human health and welfare. They are such a problem that most countries have laws concerning particle emissions and their control, and such laws (when enforced) are the easiest and quickest ways to improve health in polluted areas. Particles can also reduce crop yield, lower growth rates, and cause permanent cell damage in plants. Particulate matter pollution causes an estimated 22,000-52,000 indirect human deaths per year in the United States (Mokdad, 2004) and 200,000 deaths per year in Europe.

Numerous scientific studies have linked particle pollution exposure to a variety of problems, including:

- increased respiratory problems such as irritation of the airways, coughing, or difficulty breathing,
- decreased lung function,
- aggravated asthma,
- development of chronic bronchitis,
- irregular heartbeat,
- nonfatal heart attacks, and
- premature death in people with heart or lung disease.

Note that these effects are for particles in general. Additional problems arise when the particles include toxic substances, such as heavy metals, organic compounds, and certain minerals.

Toxicology Assessment - The effect of particles on a human mainly depends on where the particle interacts with the body. Particle size is the main determinant of where in the respiratory tract the particle lodges when inhaled. Figure 8-5 shows how the respiratory system interacts with the different size ranges of particles. Nearly all large particles (larger than 10 μm) are filtered by the nose and throat. Coarse particles (approximately 2.5 to 10 μm in diameter) can settle in the bronchi and lungs and cause health problems of a respiratory nature. Coarse particles increase the susceptibility to respiratory infections and can aggravate existing respiratory diseases, such as asthma and chronic bronchitis. They can also increase the number and severity of asthma attacks, cause or aggravate

FIGURE 8-5. Particle deposition as a function of particle diameter in various regions of the lung.

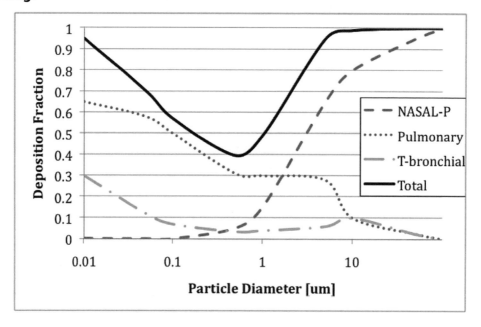

bronchitis and other lung diseases, and reduce the body's ability to fight infections. Exposure correlates to the premature death of people who already have heart and lung disease, especially the elderly (Peng, et al., 2008). Because of this, the 10 μm size has been agreed upon for monitoring of airborne particulate matter by most regulatory agencies, although it should be noted that this size does not represent a strict boundary between respirable and non-respirable particles.

The nasopharyngeal region (NASAL-P) consists of the nose and throat; the tracheobronchial (T-bronchial) region consists of the windpipe and large airways; and the pulmonary region consists of the small bronchi and the alveolar sacs. Adapted from (Dynamics, 1966).

Fine particles (approximately 0.1 to 2.5 μm in diameter) tend to penetrate into the gas-exchange regions of the lung, where they interact with lung tissue. It may even enter the bloodstream, causing an immune system response. It has been hypothesized to lead to increased plaque deposits in arteries, causing vascular inflammation and hardening of the arteries. These effects can lead to heart attacks and other cardiovascular problems. Research suggests that even short-term exposure to elevated concentrations of fine particles could significantly contribute to heart disease (Pope, et al., 2002)

Exposure Assessment - Although particulate matter can cause health problems for everyone, certain people are especially vulnerable to its adverse health effects. These "sensitive populations" include children, the elderly, exercising adults, and those suffering from lung ailments such as asthma and bronchitis. The elderly are very sensitive to particle exposure and are at increased risk of admission to hospitals or emergency rooms during times of high-particle concentrations in the air. Children are especially sensitive because they have small lungs, are usually more active than adults, and their breathing zone is closer to ground sources of PM. Children breathe 50 percent more air per pound of body weight than adults, thus exposing their bodies to higher concentrations of airborne particles. Because children's respiratory systems are still developing, they are more susceptible to environmental threats than healthy adults are. The average adult breathes 13,000 liters of air per day - about 12 to 20 breaths per minute. During exercise, this rate may increase to 30 - 35 breaths per minute and even go as high as 60 -70 breaths during intense exercise. The increase in breathing makes that person especially susceptible to air pollutants. People with existing heart or lung diseases are at increased risk of premature death and or admission to hospitals when exposed to even slightly elevated levels of particles.

Epidemiology - A positive correlation between a large number of deaths and other health problems associated with an increase in particle pollution was first demonstrated in the early 1970's (Lave, 1973) and has been reproduced many times since. Health effects have been associated with exposures to particulate matter over both short (acute exposure) and long periods (chronic exposure).

Epidemiological evidence for an association between air pollution and hospital admissions due to cardiovascular disease, uncovered in the 1990's, showed an association between particulate matter and heart disease (Dockery, et al., 1993), (Gold, et al., 2000). These studies explore the correlation between PM concentrations, hospital admission, and mortality data. Current epidemiological studies demonstrate associations between daily changes in PM concentrations and cardiovascular disease symptoms, hospital admissions, and mortality (acute exposure). Evidence also shows that annual average $PM_{2.5}$ exposures are associated with increased risks for mortality caused by heart disease (chronic exposure) (Pope, et al., 2004).

8.3.2 Effects on Public Welfare

Particles also affect public welfare; they reduce visibility, cause haze, alter air color, damage outdoor structures, and may cause environmental damage over large areas.

8.3.2.1 Visibility, Haze, and Air Color

The historical definition of visibility is "the greatest distance at which an observer can just see a black object viewed against the horizon sky." A more modern definition of visibility is the conditions that allow appreciation of the inherent beauty of landscape features - shape, color, texture, form, contrast, and brightness. It is recognized that visibility is a process that includes the viewer and their values, as well as the physical interaction of light with particles in the atmosphere. Important factors involved in seeing an object include:
- Illumination from sunlight as well as the sunlight reflected and scattered from the atmosphere, clouds, the ground and vegetation,
- The optical properties of the atmosphere - the scattering, absorption, refraction, and reflection as light passes through the atmosphere toward the observer,
- The viewer's response to the incoming light.

It is well known that particles and certain gases in the atmosphere alter a viewer's ability to observe the landscape. A clear atmosphere, that is one with few pollutants, may allow an observer to see landscape details a hundred miles away. A polluted landscape may reduce the distance to ten miles or even less. *Haze* is the general name for the mixture of particles and gases that alter visibility. Measuring, monitoring, modeling, and controlling these substances requires an understanding of how pollutants interact with light, how particles evolve within the atmosphere, and how these substances are dispersed and removed from the atmosphere.

Scattering and absorption are responsible for the colors of haze in the sky, see section 7.4 for a basic description of the interaction between light and the atmosphere. A clear sky appears blue because blue photons have wavelengths that are closest in size to nitrogen and oxygen molecules. Green and red photons are less likely to be scattered. The sky takes on a blue haze to an observer because more blue photons get scattered into their view-path. When a pollutant is in the air and it has a distribution of sizes, it can scatter all the wavelengths of sunlight, creating a white or gray haze. Some pollutant molecules absorb a particular wavelength of light, causing it to disappear to an observer. For example, nitrogen dioxide (NO$_2$) absorbs blue photons, so white light that passes through a plume containing NO$_2$ will create a reddish brown haze because the blue photons have been removed.

The most common PM$_{2.5}$ pollutants that cause scattering are sulfates, nitrates, organic molecules, and soil. The most common absorber is elemental carbon, also known as soot or carbon black. The relative importance of each to overall visibility depends on season and location. The highest PM$_{2.5}$ concentrations occurs in the summer, while winter has the lowest. Nitrates tend to be higher in winter and spring than in summer and fall. Trends in soil are variable, while elemental carbon shows little variation from season to season. Sulfates are by far the single largest contributor to PM$_{2.5}$ in the eastern United States, while in the Northwest organics contribute most to fine mass. Nitrates edge out organics and sulfates in southern California, while in the Southwest sulfates, organics, and soil all contribute about equally to PM$_{2.5}$.

In many parts of the US, the range of visibility has been reduced by 70% from natural conditions, see Figure 8-6 (US-EPA, 2012). In the east, the current range is 14-24 miles vs. a natural visibility of 90 miles. In the west, the current range is 33-90 miles vs. a natural visibility of 140 miles. Note that fine particles can remain suspended in the air and travel long distances. For example, one-third of the haze over the Grand Canyon can be attributed to air emissions occurring in Southern California (US-NPS, 2008). Figure 8-7 shows haze around the city of Hong Kong, where visibility is often reduced to just a few miles.

FIGURE 8-6. Visibility Impairment from Air Pollution, Badlands National park, South Dakota, USA.

A Clear Day   A Hazy Day

8.3.2.2 Environmental Damage

Particles can move over long distances by wind and then settle on ground or water. The effects of this deposition include acidification of lakes and streams, alteration of the nutrient balance in

FIGURE 8-7. Hong Kong Skyline Obscured by PM and Haze. Photo courtesy of Andrew Ziminski, Copyright 2009. (Ziminski, 2009)

coastal waters and large river basins, depletion of the nutrients in the soil, and damage to plants. Plant damage can be due to direct deposition on the plants, or to changing the chemistry of the soil. Particle pollution can also stain, soil, or encourage corrosion of stone and other materials, including culturally important objects such as statues and monuments.

8.4 Regulation and History

8.4.1 Regulation

Most countries have clean air regulations that set a standard or goal for the quantity of a pollutant allowed in ambient air, see Chapter 3 for details of US regulations. The level of particles is monitored at specific locations at set intervals (hourly, daily, weekly, monthly). Specialized sampling equipment measures the air quality within each air quality control region. Locations are chosen to represent typical ambient air available to the people in the area. These measurements are not the same as measuring the emissions from a smokestack, but rather they try to measure the impact all the emission sources in the area have on the ambient air. Table 8-5 lists the ambient air quality goals for particles in several countries. It is interesting to note the variation in regulation levels. The addition of $PM_{2.5}$ standards has been recent, and more countries have plans to add standards for them.

Many urban areas in the U.S. and Europe still frequently violate their particle standards, though urban air has become cleaner, on average, with respect to particles over the last 40 years. Much of the developing world, especially Asia, exceeds standards by such a wide margin that even brief visits to these places may be unhealthy (World Bank, 2007). It is estimated that living in an environment with constant high sources of particle pollution can reduce an average person's life expectancy by two years. Concentration decreases of 10 $\mu g/m^3$ of fine particulate matter for a 20 year span are associated with an estimated increase in life expectancy of 0.6 year (Pope, et al., 2009).

Table 8-5. Ambient Air Quality Standards for PM in Several Countries.

Pollutant	PM_{10}		$PM_{2.5}$	
Time Period Average Country \ Units	Annual $\mu g/m^3$	Daily (24-hour) $\mu g/m^3$	Annual $\mu g/m^3$	Daily (24-hour) $\mu g/m^3$
US		150	15	35
EU	40	50	25	
Japan		100		
India	60	100	40	60
China (Class I/II/III)*	40/100/150	50/150/250		
Brazil	50	150		
Mexico	50	120	15	65
Russian Federation		150		
South Africa	60	75		
WHO		50		

*Class I: Tourist, conservation area; Class II: Residential area; Class III: Industrial and Heavy Traffic area.

In the US, the Clean Air Act established two types of national air quality standards for particle pollution. Primary standards set limits to protect public health, including the health of "sensitive" populations such as asthmatics, children, active adults, and the elderly. Secondary standards set limits to protect public welfare, including protection against visibility impairment, damage to animals, crops, vegetation, and buildings.

8.4.2 Historical Trends in US - Particulate Matter

The US-EPA has developed ambient air quality trends for particulate matter (PM) based on measurements from a nationwide network of monitoring sites. Under the Clean Air Act (CAA), the US-EPA sets and reviews national air quality standards for PM every five years. The US-EPA, state, tribal and local agencies use this data to ensure that PM in the air is at levels that protect public health and the environment. Trends from 1990-2010 (US-EPA, 2015) are shown in Figure 8-8 for $PM_{2.5}$ and PM_{10}. National monitoring for $PM_{2.5}$ began in 1999. The data points show the average, or 50% value, where half of all air quality control regions reported values below that level. The data also shows the 10% and 90% levels for how many of the nation's air quality control regions were below that concentration. Nationally, average concentrations have decreased during this period. Almost no national data was collected before the passage of the CAA and the creation of the US-EPA so it is difficult to make comparisons of before and after. However, it seems very likely that the improvements shown in this data are due to this law and its enforcement.

FIGURE 8-8. US Particulate Matter Historical Trend.

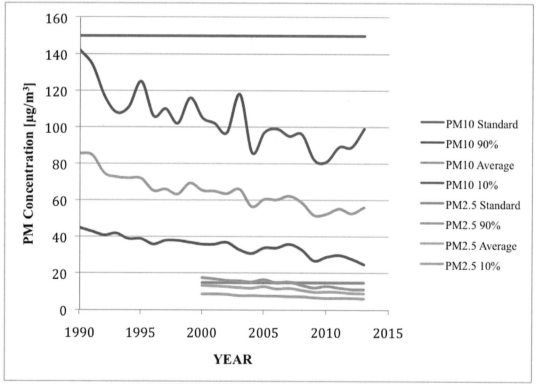

8.4.3 Regulatory Trends

Regulations concerning PM have evolved over time. Between 1971 and 1987, the US-EPA's standards regulated total suspended particles (TSP), including those larger than 10 micrometers (μm). By 1987, research had shown that the particles of greatest health concern were those equal to or less than 10 μm that can penetrate into sensitive regions of the respiratory tract. The US-EPA and state administrators took action to monitor and regulate particulate matter 10 μm (PM_{10}) and smaller, and to stop regulating TSP (particles larger than 10 μm). Ten years later the US-EPA revised the PM standards again, setting separate standards for $PM_{2.5}$ based on their link to serious health problems ranging from increased symptoms, hospital admissions and emergency room visits, to premature death in people with heart or lung disease. The 1997 standards also retained but slightly revised standards for PM_{10}, which were intended to regulate "inhalable coarse particles" that ranged from 2.5 to 10 micrometers in diameter. The US-EPA again revised the air quality standards for particle pollution in 2006. The 2006 standards tightened the 24-hour fine particle standard from 65 to 35 $\mu g/m^3$ and retained the annual fine particle standard at 15 $\mu g/m^3$. The EPA decided to retain the existing 24-hour PM_{10} standard of 150 $\mu g/m^3$. However, the EPA revoked the annual PM_{10} standard, because the available evidence did not suggest a link between long-term exposure to PM_{10} and health problems.

Current public health studies clearly show that large particles are less harmful than small ones. It is becoming increasingly clear that the legislative limits for particle emissions based on the total mass of a given size range are not a proper measure of the health hazard. One particle of diameter 10 μm has approximately the same mass as 1 million particles of 100-nm diameter, but it is much less hazardous since it may be filtered by the nose and throat before it enters the lungs. An idea proposed in some countries suggests new PM regulations use the number of particles, the particle surface area, the particle size or some combination of these variables rather than the particle mass.

8.5 Control Technologies

This section provides an overview of the main types of equipment available, how they work, and the basic information needed during the preliminary design. Additional details for completing a design include temperature, pressure, humidity, nature of the gas phase (e.g. corrosivity, reactivity), particle phase (mass loading, size distribution, solubility), and the emission limits. Typically, the equipment manufacturer performs the detailed mechanical, structural, electrical, and process control designs. The following equipment discussion should provide the reader with the basic theory and practice to make a preliminary process design of particle control equipment.

The key to removing a particle is to encourage it to hit a target and get captured, see Figure 8-9. Hitting a target occurs by changing the particle velocity relative to the gas stream velocity. The velocities are changed by applying a force to the particle or the gas. The force causes an acceleration of the particle relative to the gas, which is exploited to cause the particle to hit a target. A good target collects any particle that touches it. The gas must be free to move around or through the target and leave the collection device without the particles. Also, the particles must be removed from the target during a cleaning cycle, so they should not attach too firmly. Target materials include liquid drops, fibers, larger particles, plates, and walls.

FIGURE 8-9. Examples of Particle Collection by Interaction with a Target. Particle A has a direct hit with target and is captured (*Impaction*). Particle B has an indirect collection since only part of the particle hits the target, but it is still captured (*Interception*). Particle C misses the target. Particle D should miss the target when following its streamline, but diffusion or turbulence in the flow around the target cause it to change its streamline, and it is captured (*Diffusion or Turbulence*). A streamline describes the flow path of an ideal particle, shown as dotted lines in the figure.

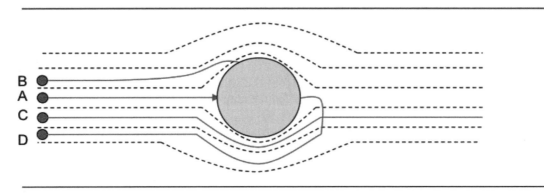

It is important to note that the size of the particles will be important in determining which of the collection mechanisms will be most effective. Large particles, which change course slowly, will likely be captured by impaction or interception. Smaller particles, considering either size or mass, can change their direction and speed more easily. They can more closely follow the gas streamlines around targets. However, these small particles are more likely to be affected by Brownian motion and thus become captured by diffusion or turbulent mechanisms. Electrostatic forces may occur between particles and targets which alters the probability of a collision and capture.

All particle control technologies use physical forces to separate particles from the gas stream. Table 8-6 lists the equipment described in this text and the forces used to separate the particles from the gas.

8.5.1 Collection Efficiency

Collection efficiency describes the fraction of incoming particulate matter removed by the control equipment. It is usually described as a function of the particle diameter, even though it also depends on several other variables (such as the temperature, pressure and viscosity of the gas; the size distribution, shape, density, composition, and surface properties of the particles). For all the particle

Table 8-6. Forces used in Various Particulate Matter Control Equipment.

Equipment type	Forces used
Settling Chamber	Gravity, buoyancy and drag
Cyclone	Centrifugal and drag
Filters	Inertial impact and diffusion
Electrostatic Precipitator	Electrostatics and drag
Wet Scrubber	Inertia and diffusion

control equipment discussed below, larger diameter particles are collected with greater efficiency. Above 100 µm, particles are collected with very high efficiency by inertial impaction, electrostatic attraction, and even settling due to gravity. Efficiency remains high in the range of 10 to 100 µm due to the ability of inertial and/or electrostatic forces to act on a particle mass. For particles less than 10 µm, the limits of inertial forces and electrostatic forces become apparent, and the efficiency drops as the drag force and shear force due to gas viscosity overwhelm inertial forces. Efficiency decreases to low levels between 0.3 and 3 µm depending on the equipment design factors such as gas velocities (inertial forces) and electrical field strengths (electrostatic attraction). Particles less than 1 µm in diameter are the most difficult size to collect. Brownian diffusion becomes important for particles below 0.3µm. At these sizes, diffusion allows the particle to fluctuate randomly in its position relative to the gas stream. These fluctuations allow a particle to collide with a target even in the absence of the other forces causing an increase in efficiency to occur. Figure 8-10 shows the general trends to expect for the typical range of particle sizes.

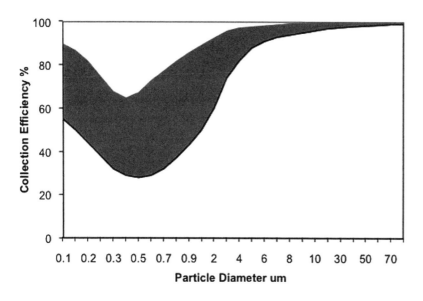

FIGURE 8-10. Collection Efficiency Ranges by Size.

The equation describing the collection efficiency is:

8.2
$$\eta_{overall} = 1 - \frac{M}{M_0} = \frac{(M_0 - M)}{M_0}$$

Where: M = mass of particles collected by the target, and
M_0 = mass originally in the gas.

There are three types of collection efficiency to consider - single particle, grade, and overall.

Single particle efficiency is the collection efficiency for a particle of a certain diameter. It represents the likelihood that a collector would remove a particle of that diameter.

Grade efficiency is the collection efficiency for a given particle size range, or bin size. It can be directly measured, or inferred from the calculation of a single-size particle representing the average size within the range. Direct measurement would compare the mass of particles within a size range entering and leaving a given collector. Figure 8-11 shows the grade efficiencies for several types

FIGURE 8-11. Grade Efficiency for Various Types of Particle Control Equipment. ESP is an Electrostatic Precipitator.

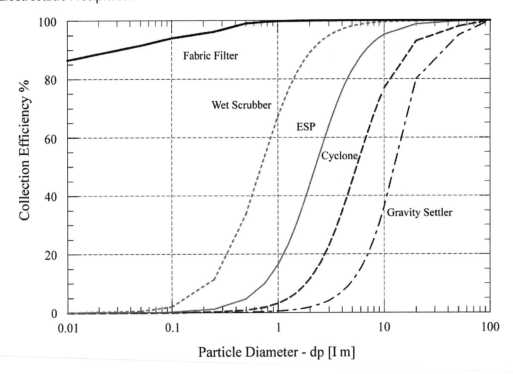

Particulate Matter

of collectors. They all show collection increasing with particle diameter. It also shows that fabric filters have the highest collection efficiency. However, fabric filters are the most expensive to build and operate.

Overall efficiency is the collection efficiency for all the particle sizes. It is determined by summing the grade (or bin) efficiencies weighted by the mass fraction of particles within the grade as given in equation (8.3). It represents the total mass captured by a collector. This value is the same as that given in equation (8.2).

8.3 $$\eta_{overall} = \Sigma \eta_i M_i$$

■ Example 8-1.
Determine the overall efficiency for each of the five collection devices shown in Figure 8-11 for the particle distribution given below.

Solution:
Prepare a calculation table (shown in Table 8-7 and Table 8-8) that lists the mass fraction as a function of the diameter, and add columns for the collector grade efficiency (Eff) for each size range. Then apply equation (8.3) to calculate the overall efficiency.

Table 8-7. Collection Efficiency Calculation for a Gravity Settler.

Particle Diameter x_i (μm)	Mass Fraction M_i	Gravity Settler Eff	Product Eff*M_i
0.1	0.0005	0	0
0.3	0.002	0	0
0.7	0.028	0.005	0.00014
1	0.167	0.01	0.00167
3	0.359	0.04	0.01436
7	0.323	0.2	0.0646
10	0.118	0.38	0.04484
25	0.0023	0.85	0.001955
50	0.0002	0.92	0.000184

Sum = 13%

The calculation for the cyclone is similar, except that the grade efficiencies change:

Table 8-8. Collection Efficiency Calculation for a Cyclone.

Particle Diameter x_i (μm)	Mass Fraction M_i	Cyclone Eff	Product Eff*M_i
0.1	0.0005	0	0
0.3	0.002	0.005	0.00001
0.7	0.028	0.01	0.00028
1	0.167	0.05	0.00835
3	0.359	0.2	0.0718
7	0.323	0.6	0.1938
10	0.118	0.78	0.09204
25	0.0023	0.92	0.002116
50	0.0002	0.98	0.000196

Sum = 37%

Similarly, the overall efficiency for the electrostatic precipitator is 64%, the wet scrubber is 90%, and the fabric filter is 99.95%. the detailed calculation is left to the reader for practice. The actual choice of equipment depends on the required removal efficiency and cost to build and operate the equipment.

8.5.2 Settling Chamber

Settlers use gravity to remove particles by accelerating them away from the main gas flow path, see Figure 8-12. The additional velocity component caused by gravity is called the settling velocity. Gravity settlers have poor removal efficiencies for small particles (less than 30 μm diameter). Efficiency improves if the settler is made larger, which is achieved by increasing the cross-sectional area or decreasing the total flow rate. The settler size becomes even larger when the flow is laminar (see chapter 4 for definition of laminar flow). However, they are inexpensive to install and maintain, use very little energy, and are forgiving of system upsets. They are sometimes used as a preliminary treatment to remove the largest particles.

FIGURE 8-12. Settling Chamber Schematic. Light lines show typical flow paths for particles. The largest and heaviest particles collect in the first chamber. Smaller particles collect in the second chamber. The smallest particles do not collect and remain in the gas stream.

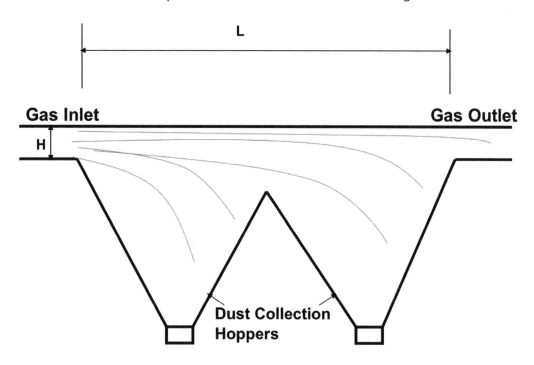

8.5.2.1 Flow Regime

The gas stream flow regime must be determined (laminar or turbulent flow) after calculating the settling velocity. The flow regime determines the form of the single-particle collection efficiency equation. Equation 4.29 or 7.12 describes the dimensionless Reynolds number, which is used to determine the flow type. Use caution when interpreting the value – gas flow within equipment is internal flow, but flow around a particle within the equipment is external flow. See Chapter 4 - Fluid Mechanics Review for more detail.

8.5.2.2 Laminar Flow

Particle removal efficiency during laminar gas flow depends on the amount of time needed for the particle to move to the collector, Δt_{move}, and the residence time of the gas in the control equipment, $\Delta t_{residence}$. Figure 8-13 depicts these quantities within a generalized particle control device. Relevant formulae are:

8.4 $$\Delta t_{move} = \frac{H}{u_p}$$

8.5 $$\Delta t_{residence} = \frac{L}{u_g}$$

Where: H = maximum distance the particle must move to hit the collector,
u_p = particle velocity towards collector and away from main gas flow,
L = length of the collector (in the direction of gas flow), and
u_g = velocity of gas flow.

FIGURE 8-13. Basic Dimensions for Particle Control Equipment. Where L is the length of the collector, H is the maximum distance a particle must travel to the collector, u_g is the gas stream velocity, and u_p is the velocity of the particle towards the collector.

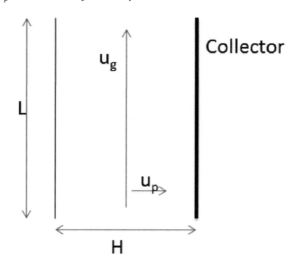

The single particle efficiency for laminar flow is the ratio of the residence time to the move-to-collector time:

8.6 $$\eta = \frac{u_p L}{u_g H}$$

If this number is greater than one, then the device collects all particles of that size. Even if this efficiency is a larger number, do not use values above 100% efficiency, instead just use 100%.

8.5.2.3 Turbulent Flow

Turbulent gas flow causes continuous mixing between the gas and particles, except at the fluid – wall interface. This interface contains a boundary layer. Gas flow is laminar within the boundary layer, and the velocity reduces to zero at the wall. Collection depends on getting a particle into the boundary layer. The capture efficiency correlates with the ratio of the boundary layer volume to the overall volume. Most collection equipment uses turbulent flow conditions, because the collection efficiency is higher for small particles, and thus allows the equipment to be much smaller. The efficiency equation for turbulent flow is:

8.7 $$\eta = 1 - \exp\left[\frac{-u_p L}{u_g H}\right]$$

■ Example 8-2.

Determine the length of a settling chamber needed to collect 99% of a) 100μm and b) 10μm particles of density 1.25 g/cm³. The chamber cross section is 15 cm high by 25 cm deep. The air flow rate is 150 m³/hr, and it is at 20°C and 1 atm.

Solution:

First, determine the gas flow regime from the Reynolds number. The calculation requires the effective diameter (D_{eff}) of the settling chamber and the density and viscosity of air at the stated temperature and pressure (which are from data tables in Appendix 6, or using the ideal gas law). We also need to determine the air velocity from the flow rate and settling chamber cross-sectional area.

$$D_{eff} = \frac{4A}{P} = \frac{4(15\,cm \times 25\,cm)}{(15+15+25+25)\,cm} * \frac{1\,m}{100\,cm} = 0.1875\,m$$

Air Density = ρ_g = 1.20 kg/m³
Air Viscosity = μ_g = 1.83 x 10⁻⁵ kg/m/s

$$\text{Air Velocity} = u_g = \frac{Q}{HD} = \frac{150 \frac{m^3}{hr} \frac{hr}{3600\,sec}}{(0.15\,m)(0.25\,m)} = 1.11 \frac{m}{sec}$$

The Reynolds Number, from equation 7.12 is

$$\text{Re} = \frac{\rho_g u_g D_{eff}}{\mu_g} = \frac{\left(1.2 \frac{kg}{m^3}\right)\left(1.11 \frac{m}{sec}\right)(0.1875\,m)}{1.83 \times 10^{-5} \frac{kg}{m\,sec}} = 13,650$$

This value of Reynolds number is larger than 4,000 so flow is turbulent (see chapter 4), so use equation (8.7) for efficiency. Length is solved for by rearrangement. Note that the value of 100% efficiency cannot be used since the natural log of zero is not defined; instead use a value of 99% in the calculation. Use equation 7.20 to determine the terminal velocity of the particles. For particles with a density of 1.25 g/cm³ and a diameter of a) 100 μm it is 35.6 cm/s and b) 10 μm it is 0.38 cm/s.

$$L_{100} = -\frac{u_g H}{u_p} \ln(1-\eta) = -\frac{(1.11\frac{m}{sec})(0.15m)}{0.36\frac{m}{sec}} \ln(1-0.99) = 2.1 m$$

$$L_{10} = -\frac{u_g H}{u_p} \ln(1-\eta) = -\frac{(1.11\frac{m}{sec})(0.15m)}{0.0038\frac{m}{sec}} \ln(1-0.99) = 202 m$$

8.5.3 Cyclones

Cyclones are used to remove particles larger than 10 μm, and some high-efficiency models remove particles down to 5 μm. They are common in mining and mineral processing industries. Cyclones use centrifugal forces to accelerate the particles away from the main gas flow path and onto the inside surface of the outer wall. This force is applied by causing the gas to travel in a helical path down the inside of the cylinder. Since the particle has a larger size and mass, its inertia does not change as fast as the smaller and lighter gas molecules, so it tends to move towards the outside wall of the cylinder. Once the particle hits the outside wall, it slides down the wall and collects in a storage hopper at the cyclone base. Figure 8-14 shows a basic schematic and dimensions of a cyclone.

Advantages of using cyclones include a low-pressure drop, handles high concentrations of solids, classification by size of the solids if run in series, low-capital cost, low-maintenance cost, and operates over a wide range of temperatures. Disadvantages include the low-collection efficiency for particles smaller than 5μm, and they cannot handle sticky or tacky materials. Figure 8-11 includes a general grade efficiency for a cyclone. Section 7.3.2 discusses the general motion of particles due to centrifugal forces. The operation of a cyclone is based on exploiting centrifugal forces to separate particles from the air stream.

The size efficiency of a cyclone can be determined by comparing the time required for a particle to move to the outside wall to the total amount of time spent in the cyclone. The time required to move to the outside wall depends on the particles location within the inlet stream. The farthest distance a particle must move is (R_o - R_i) as shown in Figure 8-14. This distance is also the width (W) of the inlet. The width (W) typically has a dimension of ½ R_o for a conventional cyclone. High

FIGURE 8-14. Cyclone Schematic.

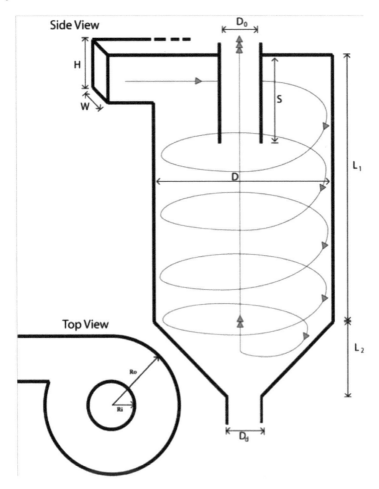

efficiency cyclones use a smaller value (0.3- 0.4 R_0), but this leads to an increased pressure drop. If a given diameter particle can move this distance or greater while it is in the cyclone, it is removed with 100% efficiency. If it moves a shorter distance, denoted as r, then only those size particles within the range ($R_0 - r$) collect, and the single-particle removal efficiency is:

8.8
$$\eta = \frac{R_0 - r}{W}$$

The distance ($R_0 - r$) is found by determining the particle velocity component normal to the gas velocity. The velocity of a particle that travels from the inside radius of the inlet to the outside radius wall during its residence time inside the cyclone is:

8.9 $$u_p = \frac{R_0 - R_i}{\Delta t_{residence}} = \frac{W}{\Delta t_{residence}}$$

The residence time describes the total amount of time an average particle spends in the cyclone. It depends on the path distance and the gas velocity:

8.10 $$\Delta t_{residence} = \frac{2\pi R_0 N_e}{u_g}$$

Where Ne is the number of effective turns in the helical path of the gas, approximated by:

8.11 $$N_e = \frac{1}{H}\left(L_1 + \frac{L_2}{2}\right)$$

Where: L_1 = length of the cylindrical portion of the cyclone, and
L_2 = length of the cyclone bottom cone, see Figure 8-14.

Combining the equations (8.8), (8.9), (8.10), (8.11) and solving for the particle diameter (x) yields the smallest particle size that moves ($R_0 - R_i$) during its residence time in the cyclone. This particle diameter is the smallest size collected with 100% efficiency:

8.12 $$x = \left(\frac{18W\mu_g R_i}{2\pi\rho_p u_g N_e R_0}\right)^{\frac{1}{2}}$$

The single particle efficiency can be determined by combining equations (8.8) and (8.12) to yield:

8.13 $$\eta = \left(\frac{\pi N_e \rho_p x_i^2 u_g}{9\mu W}\right)$$

Experimental evidence suggests that equations (8.12) and (8.13) are overly optimistic and that there is no sharp cutoff diameter, due to the turbulent flow profile of the gas stream.

An improved approach is to determine the size of a particle that collects at 50% efficiency, which can be done by using $\eta = 0.5$ and rearranging equation (8.13) to solve for x_{50}:

8.14 $$x_{50\%} = \left(\frac{9W\mu_g}{2\pi\rho_p u_g N_e}\right)^{\frac{1}{2}}$$

Other grade efficiencies can be determined from x_{50} using (Perry, et al., 1997):

$$\eta_i = \frac{\left(\dfrac{x_i}{x_{50\%}}\right)^2}{\left(1 + \left(\dfrac{x_i}{x_{50\%}}\right)^2\right)} \qquad 8.15$$

This equation has a more realistic tendency to approach 100% removal asymptotically.

The main operational cost variable for cyclones is the operation of a fan to overcome the pressure drop of the gas as it moves through the cyclone unit. Pressure drop increases proportionally with the square of the gas velocity (See Chapter 4 -Fluid Mechanics Review) and efficiency increase linearly with velocity. A typical pressure drop for a conventional design would be 3 to 6 inches of water. A high-efficiency cyclone would have a higher loss, around 5 to 10 inches of water, and low-efficiency cyclone would be lower, around 0.5 to 3 inches of water. Note a pressure of 1 inch of water is equivalent to 249 Pascal or 0.0361 psi or 0.00246 atm.

■ Example 8-3.

Consider a cyclone with a diameter of 0.5 m, a top section length of 1 m, and a bottom cone of length 1 m. The height of the inlet box is 0.25 m, and its width is 0.125 m. Determine the removal efficiency for a) 5 μm, b) 10 μm, c) 50 μm particle. Assume a particle density of 1250 kg/m³, air density of 1.2 kg/m³, air viscosity of 1.8×10^{-5} kg/m s, and an air stream volumetric flow rate of 150 m³/hr.

Solution:
Use equation (8.13) to find the efficiency for each particle diameter, after determining N_e and u_g:

$$N_e = \frac{1}{H}\left(L_1 + \frac{L_2}{2}\right) = \frac{1}{0.25m}\left(1m + \frac{1m}{2}\right) = 6$$

The air stream makes six revolutions while inside the cyclone. The gas velocity inside the cyclone is:

$$u_g = \frac{Q}{HW} = \frac{150\dfrac{m^3}{hr}}{(0.25m)(0.125m)} = 4800\frac{m}{hr} * \frac{1hr}{3600sec} = 1.333\frac{m}{sec}$$

Using these values in equation (8.13) yields:

$$\eta = \left(\frac{\pi N_e (\rho_p - \rho_g) x_i^2 u_g}{9\mu W}\right) = \frac{\pi 6 (1250-1.2)\frac{kg}{m^3}(5x10^{-6}m)^2 (1.333)\frac{m}{\sec}}{9\left(1.80x10^{-5}\frac{kg}{m\sec}\right)(0.125m)} = 0.039 = 3.9\%$$

Similarly for the b) 10 μm particle; η = 15.5% and for c) 50 μm, η = 139.5%. This last value is greater than 100% so use a value of 100% in further calculations.

■ Example 8-4.
Find the particle size removed with 50% efficiency for the cyclone described in Example 8-3.

Solution:
Use equation (8.14) to find the particle size collected at 50% efficiency.

$$x_{50\%} = \left(\frac{9W\mu_g}{2\pi(\rho_p - \rho_g)u_g N_e}\right)^{\frac{1}{2}} = \left(\frac{9(0.125m)\left(1.80x10^{-5}\frac{kg}{m\sec}\right)}{2\pi\left[(1250-1.2)\frac{kg}{m^3}\right]\left[1.333\frac{m}{\sec}\right](6)}\right)^{\frac{1}{2}} \frac{10^6 \mu m}{m} = 18\mu m$$

■ Example 8-5.
Create a figure of cyclone efficiency as a function of particle size using equations (8.13) and (8.15).

Solution:
Calculate the efficiency for several particle diameters (column 1) and make a plot for the two sets of efficiency (column 2 and column 3). Use the data from Example 8-3 and Example 8-4 to make the necessary calculations.

Table 8-9. Data and Worksheet for Example 8-5.

Col 1	Col 2	Col 3	Col 1	Col 2	Col 3
x_i μm	Eff (Eqn 8.25)	Eff (Eqn 8.27)	x_i μm	Eff (Eqn 8.25)	Eff (Eqn 8.27)
1	0.15	0.31	35	100	79.15
2.5	0.97	1.9	40	100	83.22
5	3.87	7.19	45	100	86.26
7.5	8.72	14.85	50	100	88.57
10	15.5	23.66	60	100	91.78
12.5	24.22	32.63	70	100	93.82
15	34.87	41.09	80	100	95.2
17.5	47.47	48.7	90	100	96.17
20	62	55.36	100	100	96.87
25	96.87	65.96	200	100	99.2
30	100	73.61			

FIGURE 8-15. Comparison of Cyclone Collection Efficiency Models.

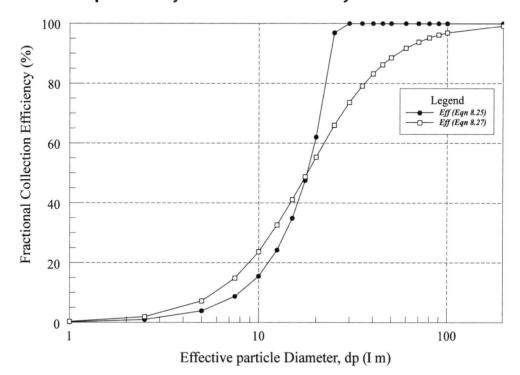

Results are similar at small sizes, but there are large differences above the size associated with 50% efficiency. The values associated with equation (8.15) better represent actual operation results.

8.5.4 Filters

Filters use inertia to decelerate a particle and remove it from a gas stream. Figure 8-16 shows a simplified schematic of a *filter bag house* – which consists of a filter-bag, a shaking mechanism, equipment for transporting the gas stream in and out, and a system for removing the particles. Filters consist of materials with large surface area and high porosity (many voids). They are made from fibers arranged either randomly or woven. The spaces between the fibers create pores, which allow the gas to move through it in a tortuous pathway. The particles in the gas cannot change direction as easily as the gas, so the particles' inertia causes it to collide with the filter fibers (or other particles once a cake of them builds up on the bag). The collision decelerates the particle, and the filter may capture it. Even if it bounces off the filter after the first collision, it loses much of its speed and can be captured after another collision. It is also possible that the filter openings can be smaller than the particles thus causing the particle to become lodged in the opening. The blocking of the small openings can be problematic as this reduces the porosity of the material and may lead to reduced permeability. Such blockage can cause very large pressure drops across the filter.

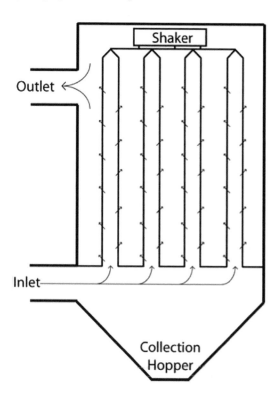

FIGURE 8-16. Filter Bag Equipment Schematic.

While the filter medium collects some of the particles, it is more useful when it acts as a support for the creation of a layer of the particles, which thicken and form a *filter cake*. The filter cake is very efficient at collecting even very fine particles, and improves the overall collection efficiency. Once a filter collects enough material it requires cleaning. The cleaning process begins by stopping the gas flow into the equipment. Next, the collected particles are dislodged from the fabric bag. Finally, the particles collect in a hopper below the bag. Three common methods for dislodging the particles from the fabric bags are shakers, reverse air, and pulse-jet. Shakers move the bag tops and loosen the filter cake from the bag. Reverse air blows air from the outlet side into the bags to loosen the cake. Pulse jets blow high-pressure air into the cake in order to loosen it.

There are several important design considerations for filter systems; fiber, weave (including random placement), size of the bag (diameter and length), arrangement and number of bags, pressure drop, and cleaning method. These choices occur after the characteristics of the aerosol (gas with entrained particles) are known. Table 8-10 lists some of the more common materials used as filters.

Table 8-10. Filter Fabric Characteristics.

Fiber Type	Important Characteristics
Cotton	Inexpensive, short life
Polypropylene	General utility for all conditions
Nylon Polyamide	Good for abrasives (best choice)
Acrylic	Good for solvents
Polyester	Good for abrasives, solvents
Nomex	Good for abrasives, solvents
Teflon	Good for high temperature (500°F), acids, alkalis
Fiberglass	Good for high temperature (500°F), acids, solvents

Advantages of filters include high overall collection efficiency even for very small particles; removal of dusts with a wide variety of properties with a single-bag house; design is straight forward, and expansion is easy.

Disadvantages include large floor areas; do not work on wet particles, hygroscopic materials, or with high-humidity gases, and they may create a fire or explosion hazard (depending on the type of dust). Winter operation with a warm, moist gas may cause some water to condense in the filter cake. This moisture can greatly reduce the void area available for flow, which can dramatically increase pressure drop and even blind the filter. The moisture can react with some materials to form a cement-like material.

The important design variables for fabric filters are pressure drop during operation, time or maximum pressure drop between cleaning cycles, and time between filter replacements. Pressure drop is affected by the gas flow rate, cake buildup, permeability of the filter, permeability of the filter cake, and arrangement of the filter equipment. The time between cleanings depends on pressure drop. The time may be forecast with some accuracy, and the cleanings can be scheduled rather than depending on pressure drop measurement. The time between filter replacements may also be periodic or depend on measured pressure drop after a cleaning cycle.

The single particle efficiency for a filter may be low (20%) when the fabric is new and clean. It typically increases to 95 – 99.9 % when loaded with cake. A filter that has been through several cycles of filtering and cleaning has a higher efficiency after cleaning (80 – 90%) than when it was new because it has some residual cake remaining. It also has a greater pressure drop. When the residual pressure drop becomes too great, the filter bags must be replaced.

The overall pressure drop is the sum of individual pressure drops due to the fabric, the cake, and the equipment:

8.16 $$\Delta P_{total} = \Delta P_{equipment} + \Delta P_{cake} + \Delta P_{fabric}$$

The equipment pressure drop is assumed constant and is often very small in comparison to the other pressure drops. It is ignored for the rest of this section.

The pressure drop [units N/m² or Pa] due to the fabric and the cake can be determined using Darcy's equation for fluid flow through a porous medium:

8.17 $$\Delta P_{fabric} = \frac{\mu_g T_{fabric} u_{sup}}{K_{fabric}}$$

8.18 $$\Delta P_{cake} = \frac{\mu_g T_{cake} u_{sup}}{K_{cake}}$$

Where: μ_g = gas viscosity [kg/m-s]
T_i = thickness of the material (fabric or cake) [m],
K_i = permeability of the materials [m²], and
u_{sup} = superficial gas velocity.

The superficial velocity (u_{sup}) is different from the gas velocity (u_g) in the inlet to the equipment or the stack:

8.19 $$u_{sup} = \frac{Q}{A_{bag}}$$

Where: Q = volumetric gas flow rate [m³/sec], and
A = cloth area available for flow [m²].

The superficial gas velocity is also called the filtration velocity or the Air-to-Cloth ratio. Typical values depend on the filter cleaning method, filter fabric, particle sizes, and the particle concentration. Values for the three main cleaning methods range from 0.01 – 0.013 m/sec for reverse air cleaning, 0.013 – 0.025 m/sec for shaker cleaning, to 0.03 – 0.1 m/sec for pulse jet cleaning (Siebert, 1977).

Values for permeability are dependent on many system specific characteristics, such as particle size and size distribution, particle packing factor, porosity, and flow rate and flow direction. Thus, it is difficult to determine from theory and is most often measured directly in a test unit.

The overall pressure drop varies over time because T_{cake} increases as the filter captures more of the particles. Assuming a constant flow rate and constant concentration of particles, the thickness increases linearly with time as:

8.20 $$T_{cake} = \frac{Mt}{A\rho_{cake}}$$

Where: M = mass loading of particles in the gas [kg/sec],
t = time [sec], and
ρ_{cake} = bulk density of the filter cake [kg/m³]

The bulk density of the cake depends on the particle density and the void volume of the cake, expressed as the cake porosity. The porosity, ε, is a value between 0 and 1 and represents the ratio of the void volume to the total cake volume:

8.21 $$\rho_{cake} = \rho_p(1-\varepsilon) = \rho_p\left(1 - \frac{V_{void}}{V_{cake}}\right)$$

Combining equations (8.16) - (8.21) yields an equation describing the overall pressure drop over time:

8.22 $$\Delta P_{total} = u_{sup}\mu_g\left(\frac{T_{fabric}}{K_{fabric}} + \frac{Mt}{A\rho_{cake}K_{cake}}\right)$$

This equation shows that pressure drop is directly related to the superficial gas velocity, gas viscosity, and the mass loading of particles in the gas. It also suggests that the pressure drop should increase linearly with time. However, this is not usually observed. Rather the pressure drop starts out low, increases quickly as cake builds up on the fabric, and then the slope decreases and becomes constant after the cake has built up, as shown in Figure 8-17. The sudden decreases in pressure drop occur due to cleaning at the end of a cycle.

FIGURE 8-17. Cyclical Nature of Pressure Drop of a Filter Bag Collection Chamber.

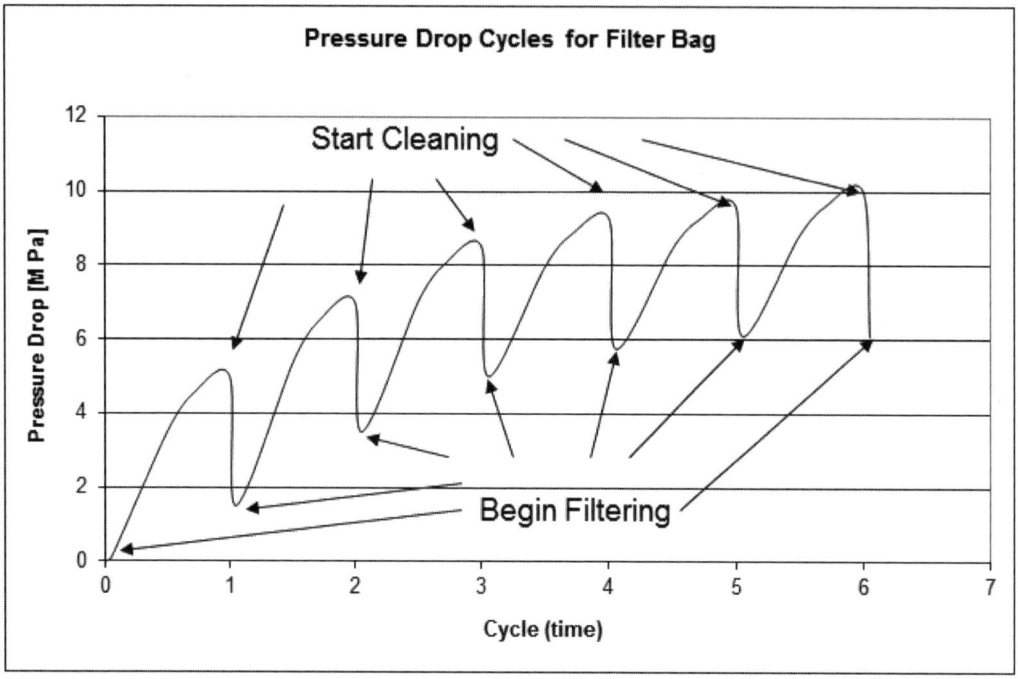

■ Example 8-6.

Determine the permeability of a fly ash collected in a bag filter from the following pressure drop data. The gas flow rate is 1000 m³/hr through a 25 m long bag with 20 cm diameter. The particles have a density of 2500 kg/m³, and the cake has a porosity of 0.10. The bag collects 330 g/ hr. The bag thickness is 5 mm.

Solution:

Plot the pressure data as a function of time and fit a line to the results, see Figure 8-18. Assume equation (8.22) is valid; we note that the intercept correlates with the pressure drop from the bag filter and that the slope correlates with the fly ash cake. Also, the initial point does not lie on the line that fits the rest of the data, this is common and occurs from the initial non-linear way the cake builds up on the fabric.

Table 8-11. Data Set for Example 8-6.

Time hours	Pressure Drop inch H_2O
0	0.5
1	4.9
2	5.3
3	5.7
4	6.4
5	6.7
6	7.2
7	7.5
8	7.9
9	8.5
10	8.9

FIGURE 8-18. Line fit of Equation 34 to Experimental Data.

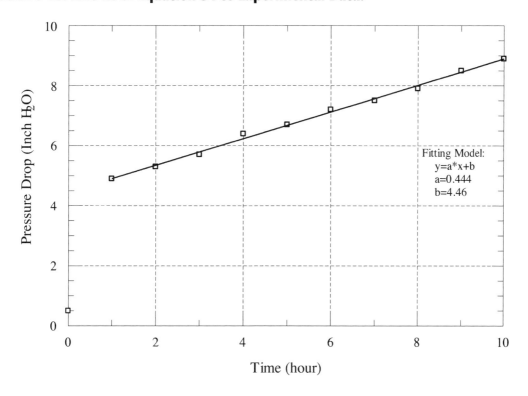

$$\text{Intercept} = \Delta P_{fabric} = u_{sup}\mu\left[\frac{T_{fabric}}{K_{fabric}}\right] = 4.46 \, in \, H_2O$$

$$\text{Slope} = \Delta P_{cake} = u_{sup}\mu\left[\frac{M}{A\rho_{cake}K_{cake}}\right] = 0.444\frac{in \, H_2O}{hour}$$

The superficial gas velocity is:

$$u_{sup} = \frac{Q}{Area_{bag}} = \frac{Q}{H\pi D_{bag}} = \frac{1000\frac{m^3}{hr}}{(25m)\pi(0.20m)}\frac{1\,hr}{3600\,sec} = 0.0177\frac{m}{sec}$$

The gas viscosity is $\mu = 1.83 \times 10^{-5}$ kg/m/s.
The bag area is A = 25 m (π) (0.20 m) = 15.71 m^2
The cake density is ρ_{cake} = (1 - Cake Porosity) * $\rho_{particle}$ = (1-0.1)2500 = 2250 kg/m^3
Solving the intercept equation for K_{fabric} and converting the pressure from inch H$_2$O to Pa:

$$K_{fabric} = u_{sup}\mu\left[\frac{T_{fabric}}{\text{Intercept}}\right] =$$

$$(0.0177\frac{m}{sec})\left(1.83\times10^{-5}\frac{kg}{m\,sec}\right)\left[\frac{0.005\,m}{4.46\,in\,H2O}*\frac{1\,in\,H_2O}{248.8\,Pa}*\frac{Pa}{1\frac{kg}{m\,sec^2}}\right] = 1.46\times10^{-12}\,m^2$$

Solving the slope equation for K_{cake}:

$$K_{cake} = u_{sup}\mu\left[\frac{M}{A\rho_{cake}\text{Slope}}\right] = \left(0.0177\frac{m}{sec}\right)\left(1.83\times10^{-5}\frac{kg}{m\,sec}\right)*$$

$$\left[\frac{0.33\frac{kg}{hr}}{(15.71m^2)\left(2250\frac{kg}{m^3}\right)(0.444\frac{in\,H2O}{hour})}*\frac{1\,inH_2O}{248.8\,Pa}*\frac{Pa}{1\frac{kg}{m\,sec^2}}\right] = 2.74\times10^{-14}\,m^2$$

8.5.5 Electrostatic Precipitators

Electrostatic Precipitators (ESP) use electrostatic forces to move a particle from the gas stream to a target collector. They are capable of removing particles at collection efficiencies from 90 to 99%. Particle size is the most important variable in determining collection efficiency, with larger sizes being collected at higher efficiency. Also important are the particle size distribution, the residence time of the particles within the ESP, the strength of the electric field, the particle's electrical resistivity, the gas stream temperature, and the chemical composition of the gas and the particle. Approximately 80% of all ESPs in the United States are used in coal fired electrical power plants (US-EPA, 2003). They are also used in the pulp and paper industry, cement manufacturing, mining, and mineral processing.

ESP operation occurs in several steps. An initial pretreatment removes large particles. The pretreatment equipment is either a cyclone or settler. After pretreatment, if any, the particles are ionized by flow through a corona (a region where gaseous ions flow between charged electrodes) where an ionized gas transfers electrical charges to the particles. After charging the particles, they are deflected by an applied electric field from the main flow path of the gas towards collection plates. The particles collide with the plates and are collected. The particles must be removed periodically from the plate and collected in a hopper. Removal is done by striking the plates with a hammer or by rinsing the plates with water.

The electrical resistivity of the particles is most strongly influenced by their temperature and chemical composition. If the particles are resistant to collecting an electrical charge in the corona or are acting to insulate the plate from the gas stream, a conditioner may be added upstream to the ESP to alter their resistivity. Typical conditioners include SO_3, H_2SO_4, ammonia, or water and are chosen based on experimental observations.

The forces used in this process act only on the particles and not on the gas, unlike all the other particle collection methods. Therefore, more of the energy used goes to separation, and less on moving the gas. Figure 8-19 shows a schematic of a plate and wire ESP. Tube and wire configurations are also possible.

The gas flow is usually horizontal. Gas flow rates of 100 to 500 m^3/s and velocities of 0.5 to 3 m/s are typical and result in turbulent flow of the gas within the ESP. The inlet concentration of particles is 2 to 100 g/m^3. Spacing between the plate and wire is 0.1 to 0.2 m. Plate size is typically 6 - 12 m in height and 1 to 5 m in length. Wires and plates are arranged in parallel, typically 2 to 8 sections, so that both sides of the plate may be used for collection (Wark, et al., 1998), (Cooper, et al., 2002). The specific collection area (SCA) ranges from 0.2 to 2 $m^2/(m^3/min)$. The process can be operated at temperatures up to 700°C, although thermal ionization may start to occur above 250°C.

The power supplied to the ESP needs to develop potentials of 10,000 to 100,000 volts. This voltage causes the air between the electrodes to ionize and form a corona – it may glow blue in color. The ions formed within this corona follow the electric field lines between the wires and plate electrodes. Each of the wires in the ESP creates a charging zone, through which the particles must pass. The particles collide with the ions, become charged and then they start to follow the electric field lines and move towards the plate collectors. Small particles can adsorb a small number of ions before they collect enough charge to repel other ions. Large particles can adsorb many more (up to thousands of times as many) before they become so charged. Therefore, the electrical forces are much stronger on larger particles.

Actual construction of an ESP chamber requires the plates and wires be well spaced. The wires are supported at both ends so that they cannot meet the plates, and that there is some clearance between the plates and the particle collection hopper. The clearance leaves regions where the gas can flow without passing through the charging zones. This bypass is called 'sneakage' and may account for 5 to 10% of the total flow. Anti-sneakage baffles can be placed in these regions to reduce the effect. The baffles act to channel the flow towards the collector plates rather than through these spaces.

FIGURE 8-19. Electrostatic Precipitator (ESP) Schematic. (Powerspan, 2003)

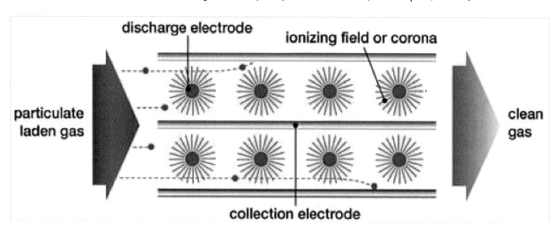

Advantages of ESPs include the high overall collection efficiency, good grade efficiency for fine particles, minimal pressure drop even with very large flow rates, particles can be wet, dry, or mixed, efficient operation over a wide range of temperatures and pressures, and low-operating costs.

Disadvantages include the unit size, high cost of construction, difficulty in adapting to changes in operating conditions, and that they do not work efficiently if particles have high electrical resistivity.

The particle size collection efficiency can be determined by setting up a force balance between the electrostatic force and the drag force on a particle. Equation 7.13 gives the drag force. The electrostatic force is:

Particulate Matter

8.23 $$F_{es} = qE_p$$

Where: E_p = collecting field strength, and
q = saturation field charge on a given particle, approximated by summing the diffusion charging and the field charging:

8.24 $$q = \left[\frac{3D}{D+2}\right]\pi\varepsilon_0 E_c x_i^2$$

Where: D is the dielectric constant for the particle,
ε_0 is the permativity constant (8.854×10^{-12} coulombs /volt-meter [C/Vm]),
E_c is the strength of the charging field strength (V/m).

Recall that a Coulomb has units of [J/V] or [kg m²/(s²V)].

Summing the electrostatic force and the drag force, setting equal to zero (no acceleration) and solving for the particle velocity yields:

8.25 $$u_p = \frac{E_p K_c}{3\mu_g}\left[\frac{3D}{D+2}\right]\varepsilon_0 E_c x_i$$

Where: K_c = Cunningham correction factor as defined in equation 7.19.

This velocity is the *drift velocity*. Typical values for fly ash may be 0.1 to 10 m/s.

The minimum particle size collected with 100% efficiency can be determined by finding the time that is needed to cause a particle to travel the distance between the wire and plate that is equal to the residence time a particle spends in the ESP:

8.26 $$\Delta t_{drift} = \frac{s}{u_p}$$

8.27 $$\Delta t_{residence} = \frac{L}{u_g}$$

Where: s = distance between the charging and collecting electrode (s=D/2)
Where: D = distance between the plates, and
L = length of the ESP in the direction of gas flow.

The residence time should include only the time that the gas spends in the actual collection chamber between the electrodes and not include the entry or exit regions.

The single particle efficiency can be calculated using the efficiency equation for turbulent flow:

8.28 $$\eta = 1 - \exp\left[\frac{-u_p L}{u_g H}\right]$$

We can multiply the argument by H/H to obtain the Deustch-Anderson form of the efficiency equation:

8.29 $$\eta = 1 - \exp\left[\frac{-A_p u_p}{Q}\right]$$

Where: A_p = area of the collection plate (2L * H), and
Q = gas flow rate (H*s*u_{gas}).

Equations (8.28) and (8.29) overestimate the collection efficiency. The equation assumes
- all the particles are smooth spheres of uniform composition,
- flow rates are constant,
- the field strengths are constant and uniform,
- there is no re-entrainment of particles from the plates due to gas flow or during cleaning (rapping), and
- particle sneakage past the collecting plates does not exist.

These equations are adequate for a preliminary design to find approximate size values. Further design would require the determination of an effective drift velocity from a pilot study or previous experience with similar particles and ESP equipment. In the following examples, we use the drift velocity as calculated by equation (8.25) and realize that the final design could differ by 25%.

■ Example 8-7.
Determine the particle velocity and the single-particle collection efficiency for a) 5 μm, b) 10 μm, and c) 50 μm in an ESP. The ESP is 10 m high, 15 m long, and is used to treat 4 m³/s of a gas stream. The dielectric constant for the particles is 6.5. The viscosity of the gas is 1.8x10-5 kg/(m s). The spacing between the collector plates is 15 cm, and the potential (collecting and charging field strengths) at the wire and plate are each 35,000 V/m.

Solution:
First, calculate the drift velocity towards the plate due to the electrical field (equation (8.25)). For the small particles, apply the Cunningham correction factor as described in equation 7.19.

$$u_{p,5\mu m} = \frac{E_p K_c}{3\mu}\left[\frac{3D}{D+2}\right]\varepsilon_0 E_c x_i =$$

$$\cfrac{35,000\frac{V}{m}(1.034)}{3\left(1.8x10^{-5}\frac{kg}{m\,sec}\right)}\left[\frac{3*6.5}{6.5+2}\right]\left(8.85x10^{-12}\frac{C}{V\,m}\right)\left(35,000\frac{V}{m}\right)\left(5x10^{-4}\,cm\right)=0.238\frac{cm}{sec}$$

Similarly for the other sizes: up(10 μm) =0.468 cm/sec and up(50 μm) =2.31 cm/sec.

Next, find the gas velocity:

$$u_g = \frac{Q}{sH} = \frac{2\frac{m^3}{sec}}{(0.15m)(10m)} = 1.33\frac{m}{sec} = 133\frac{cm}{sec}$$

Finally, determine the efficiency using equation (8.28):

$$\eta_{(x=5\mu m)} = 1-exp\left[\frac{-u_p L}{u_g s}\right] = 1-exp\left[\frac{-0.238\frac{cm}{sec}15m}{133\frac{cm}{sec}0.075m}\right] = 1-0.699 = 0.301$$

Or 30.1 % efficient

Similarly, for the other sizes:

b) $\eta_{(x=10\,\mu m)}$ = 50.5 %

c) $\eta_{(x=50\,\mu m)}$ = 96.9 %

■ **Example 8-8.**

Calculate the arrangement and number of plates (one plate has a height of 4 m and length of 2 m) required to collect 50% of 2.5 μm diameter particles in a waste stream. The particle's velocity toward the collection plates is 45 cm/min. The gas flow rate is 100 m³/s. The plates are arranged such that the total collecting length of the plates is 16 m.

Solution:
Determine the collection plate area needed by rearranging the Deustch-Anderson relationship, equation (8.29):

$$A_p = \frac{-Q}{u_p}\ln(1-\eta) = \frac{100\frac{m^3}{sec}}{45\frac{cm}{min}\frac{m}{100\,cm}\frac{min}{60\,sec}}\ln(1-0.5) = 9,240\,m^2$$

This area needs to be modified by geometric considerations – we can use both sides of the plate for the middle plates, but the end plates only have one useful side. Therefore, the actual plate area is somewhat larger. To determine the area we need to know the plate arrangement – number of rows and row length.

8.30 $$N_{plates} = N_L n_{row}$$

Where N_L is the number of plates per row and n_{row} is the number of plate rows. The useful area is:

8.31 $$A_p = 2HL(N_L)(n_{row} - 1)$$

Equation (8.31) can be rearranged to solve for the number of rows, since we already calculated the useful area, and we can determine the number of plates per row from the row length (16 m) and the length of a plate (2 m) which yields $N_L = 8$ plates per row. Solving for n_{row} gives:

$$n_{row} = \frac{A_p}{2HLN_L} + 1 = \frac{9240 m^2}{2(4*2\frac{m^2}{plate})(8\frac{plate}{row})} + 1 = 73.2 \, rows \quad \text{(round up to 74)}$$

Then $N_{plates} = N_L n_{row} = $ (8 plate/row)(74 row) = 592 plates. A typical plate spacing is 20 cm so the equipment would be 14.8 m wide, yielding a collection volume of (4-m x 16-m x 14.8-m) = 950 m³, without consideration for inlet and outlet structures, collection hoppers, or housing for electrical equipment.

8.5.6 Wet Scrubbing

Wet scrubbers use inertia and diffusion to collect particles, see Figure 8-20. They work by adding small drops of liquid (typically water) into the gas stream. The particles and the drops must have different velocities. The difference in velocity allows the two to come into contact by inertia and allow the transfer of the particles from the gas stream into the drops, see Figure 8-9. The separation efficiency increases as the difference in velocities increases. The other important variables include the size and density of the particles, the size of the droplet, and the viscosity of the gas stream. The contact between the drops and particles may be countercurrent or cross current.

It is reasonable to assume that there is an optimal size drop for a given size particle. A very small drop can slow down and be entrained in the gas stream, eliminating the velocity difference needed to contact particles. A very large drop has a small surface area to volume ratio, and much of the water is unused. However, determining the optimal size is quite challenging and is usually done using experimental observation rather than calculation.

FIGURE 8-20. Counter-Current Wet Scrubber Schematic.

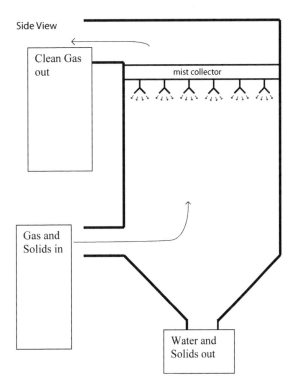

Several types of particle wet scrubbers exist throughout the chemical processing industry. The main differences between units are how the drops are added and the velocity difference between the drops and particles. They can be co-current, counter-current (see Figure 8-20), or cross-flow. The drops and particles can be introduced inside the scrubber, or into the feed to the scrubber.

Advantages of using wet scrubbers include removal of soluble gasses and particles, minimal risk in treating mixtures with explosive and flammable components, acids can be neutralized and removed, and the equipment can be easily adapted to changing inlet conditions.

Disadvantages include susceptibility to corrosion, generation of wet solid waste that probably cannot be reused or recycled, the equipment needs protection from winter temperatures, and high-maintenance costs due to plugging and fouling of spray nozzles.

It is possible to determine single-size particle collection efficiencies for a particular size drop from theoretical considerations. It is, however, difficult to translate that into a collection efficiency of the scrubber. This calculation would require knowledge of the size and size distribution of drops and particles, and quantification of the number of drops of each size a given particle approaches inside the scrubber. Such difficulties suggest a statistical approach may be useful. However, a better approach may be to use experimental analogies. These are usually available from equipment suppliers. In general, as the single-size efficiency increases, so will the collection efficiency of the scrubber. However, it is not a simple relationship and is beyond the scope of this book. Those interested in more information are encouraged to explore the following work *'Air Pollution Control Cost Manual - Sixth Edition. Section 6, Chapter 2 – Wet Scrubbers for Particulate Matter,'* (US-EPA, 2002a).

8.5.7 Atmospheric Removal Mechanisms

The main atmospheric removal mechanism for particles is deposition. There are two forms of deposition – dry and wet. Dry deposition occurs when particles interact with and are captured by the ground. Wet deposition occurs when the particles interact with water (e.g. rain, snow, fog) and are transported to the ground. Wet deposition is also called *washout*.

Particle size is important in understanding how deposition occurs. Small particles do not easily settle from the air because they are constantly mixed by the turbulent nature of the wind. These smaller particles remain in the atmosphere until they increase in size due to coagulation, agglomeration or conglomeration, see section 8.2. These mechanisms cause the particle size to increase, which increases the probability of particle removal.

The other important property to consider is the particles interaction with moisture in the atmosphere. Some materials are *hygroscopic* – they absorb water from humid air and form solution droplets. The absorption may occur with a chemical reaction, which changes the nature of the particle and water droplet – typically decreasing the pH. For example, sea-salt aerosols readily absorb moisture from the atmosphere. $MgCl_2$ begins to collect moisture when the relative humidity is 33%, NaCl begins at 75%, and $MgSO_4$ at 88%. Thus, at typical relative humidity, sea-salt aerosols may contain concentrated solutions of water-soluble salts. Other materials are *hydrophobic* and do not absorb water from the air, nor will they form solution droplets. Mixtures of both types of materials may sometimes form, making detailed modeling of atmospheric particles very complex.

8.5.7.1 Atmospheric Removal by Dry Deposition

Dry deposition occurs by sedimentation, interception, impaction, diffusion, and turbulence, see Figure 8-9. Sedimentation is the settling of particles by gravity. It removes the larger particles (> 10μm). Smaller particles generally remain aloft due to the natural turbulence of the wind. *Interception* occurs when small particles flow near an obstacle and are slowed enough to collide with it (e.g. a tree branch). *Impaction* is the direct collision of a particle with a larger object, and causes the removal of the smaller particle. If the particle bounces off the object, it may slow enough for capture by another object, or the particle may be re-entrained into the flowing air and not be captured. *Diffusion* is the process by which small particles move randomly due to collisions with gas molecules and are captured by colliding with larger objects. Turbulence in the air may cause random motions of particles, similar to diffusion but caused by eddies and vortices within the air, and also leads to capture by larger objects.

8.5.7.2 Atmospheric Removal by Wet Deposition

Wet deposition occurs when particles interact with liquid or solid water in the atmosphere. The water (atmospheric hydrometeors) collects the particles through the same mechanisms as dry deposition, interception, impaction, diffusion, and turbulence. The deposition of the particles occurs by gravity settling (sedimentation). Different types of wet deposition include below-cloud scavenging and in-cloud scavenging. Below-cloud happens as falling raindrops or snowflakes capture the particles. In-cloud happens when particles get into cloud droplets or cloud ice crystals as cloud condensation nuclei (CCN), or are captured by collision within a cloud. The particles deposit onto the ground after rain or snow forms in the cloud.

The overall deposition of particles, or other chemical constituents of air, can be described in terms of depositional flux. Flux describes the amount of something passing through a 2-dimensional surface in a given amount of time. It has units of [mass/m²/s]. The wet and dry deposition of particles can be modeled as a flux

8.32
$$F = F_d + F_w = C^A(v_d + v_w)z$$

Where: F_d = dry deposition flux [mass/m²/s]
F_w = wet deposition flux [mass/m²/s]
C^A = concentration of particles in the air [mass/m³]
v_d = dry depositional velocity [m/s]
v_w = wet depositional velocity [m/s]
z = reference height, typically 1 m.

The depositional velocities are usually determined by accounting for the resistance a particle will encounter within the reference height. It may include aerodynamic turbulence, molecular diffusion, and atmospheric interactions with ground cover. We can simplify the discussion by assuming the dry depositional velocity is proportional to the terminal velocity for a given particle size and relating the proportionality constant to the ground cover. The proportionality will be near one for water and gets smaller for more complex surfaces such as forests. The wet depositional velocity is more closely associated with the terminal velocity of the water / ice particles. Use a value of zero if there is no precipitation within the reference height.

Example 8-9.

Determine the depositional flux of 2 μm particles over a lake during a rainstorm. Assume the concentration of particles is 350 g/m³ and that rainfall consists of 10 μm particles.

Solution:
Apply equation (8.32).

Where: C^A = 350 g/m³
v_d = dry depositional velocity = 2.0x10⁻⁴ m/s, from figure 7-8
v_w = wet depositional velocity = 4x10⁻³ m/s, from figure 7-8
z = reference height, typically 1 m.

$$F = F_d + F_w = C^A(v_d + v_w)z = \left(350\frac{g}{m^3}\right)\left(2.0x10^{-4} + 4x10^{-3}\right)\frac{m}{s}(1m) = 1.47\frac{g}{m \cdot s}$$

8.6 Choosing a Particle Control System

The selection of a particle control system depends on many variables. Figure 8-21 shows a decision matrix that can help in choosing a system. When there is a high concentration of wet and/or sticky particulate matter, either a particle wet scrubber or wet electrostatic precipitator should be used. If wet or sticky materials are present with combustible materials or explosive gases or vapors, the particle wet scrubber is most appropriate. If the particulate matter is primarily dry, mechanical collectors, particle wet scrubbers, conventional electrostatic precipitators, and fabric filters can be used.

The next step in the selection process is to determine if the particulate matter and/or gases and vapors in the gas stream are combustible or explosive. If so, then mechanical collectors or particle wet scrubbers can be used because both of these categories of systems can be designed to minimize the risks of ignition. In some cases, a fabric filter can also be used if it includes the appropriate safety equipment. An electrostatic precipitator is not used due to the risk of ignition caused by electrical sparking in the precipitator fields. When selecting between mechanical collectors and wet scrubbers, mechanical collectors are usually an economical choice. They typically have a lower purchase cost and lower operating cost than wet scrubbers do.

If the dry particulate matter present in the gas stream is not combustible or explosive, the selection depends on the particle size range and the control efficiency requirements. A fabric filter is the most common choice if a significant portion of the gas stream is in the less than 0.5-micron size range and needs high-efficiency control. If a significant portion of the particulate matter is in the

0.5- to 5-micron size range and needs high-efficiency control, fabric filters, electrostatic precipitators, or particle wet scrubbers (certain types) could be used. If most of the particulate matter is larger than five micrometers, any of the four main types of particle control systems could be used.

FIGURE 8-21. Flow Chart for Choosing a Particle Control System.

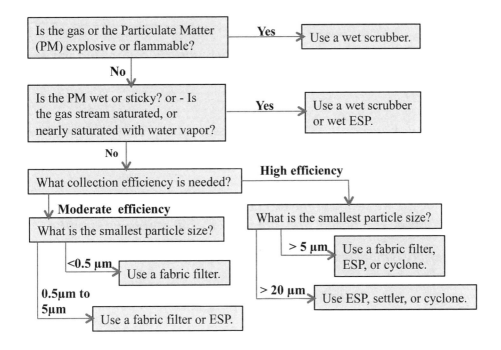

There are numerous exceptions to the general applicability of the information presented above due to site-specific process conditions and unique particulate matter control systems. Nevertheless, this chart provides a general indication of the uses and limitations of many commercially available particulate matter control systems. As always, a balance between cost and the required removal efficiency constrains the choice.

● Case Study 8-1. Particulate Matter Control in a Coal Fired Power Plant.

In this case study we will design a particle control system for the coal fired power plant introduced in Case Study 2-1, and further developed in Case Study 3-3. Additional information needed: The power plant is a 350 MW cyclone furnace burning bituminous coal (10,750 BTU/lb).

Case Study 2-1 shows the coal contained 25.9g fly ash/mol coal, and that one mol of coal had a mass of 1058.5 g. Case Study 3-3 shows the facility needs to remove 99.28% of PM to meet NSPS regulations. This number works well for the average, however if the coal has variations in ash content, or if the air has changes in temperature or humidity the facility could exceed their emission limit. The engineering team has determined that a slightly higher removal rate of 99.6% will protect the facility from minor deviations in the feeds.

The first step in our design is to determine which particle control systems would be practical using Figure 8-21. First, neither the PM nor the exhaust gas is flammable. Next, the PM is dry.

We do not know if the exhaust is saturated, or near saturation, so this must be determined. We will need to know the partial pressure and vapor pressure of the water in the exhaust.

From case Study 2-1 we saw that the exhaust contained 0.071 mole fraction of water (or 52 g/m³). This gives a partial pressure of .071 * 760 mmHg = 54 mmHg.

The vapor pressure is modeled with the Antoine Equation) as shown in appendix VI d:

$$\log_{10} P^{sat} = A - \frac{B}{T+C}$$

Where: P^{sat} [=] mmHg, and
              T [=] °C.
              A = 8.07121
              B = 1730.630
              C = 233.426

In this case we will determine the temperature when the vapor pressure is equal to the partial pressure (also known as the saturation temperature).

$$T = \frac{-B}{\left(\log_{10} P^{sat}\right) - A} - C = \frac{-1730.63}{\left(\log_{10} 54\right) - A} - 233.426 = 39.6°C \text{ or } 103°F.$$

The exhaust gas from the boiler will be much hotter than this, and we would not expect it to cool below this temperature until it leaves the system and enters the atmosphere. Typical power plant emissions leave the stack at approximately 150°C or 300°F, so we will assume that the gas is not wet or near saturation.

The next question requires that we describe the required collection efficiency. From Case Study 3-3 and the earlier discussion in this case study, we see that the required removal efficiency is 99.6%. This is considered a high efficiency since it is above 95%.

The final question requires an understanding of the particle size distribution. The distribution is difficult to know without more knowledge of the process. The US –EPA AP42 documents provide some information on the particle size distributions from many different sources, (US-EPA, 1995). The following table was generated using the information from chapter 1, volume 1 of the report.

Table 8-12. Size Distribution for Cyclone Furnaces Burning Bituminous Coal.

Bin		Uncontrolled
low [μm]	high [μm]	Mass %
15	50+	67
10	15	20
6	10	5
2.5	6	2.5
1.25	2.5	0.5
1	1.25	0
0.625	1	5
0	0.625	0

To achieve 99.6 % removal, all sizes must be removed, down to the 0.625 μm bin. The flow chart suggests using a fabric filter, ESP, or cyclone. We will choose the fabric filter since we know that cyclones and ESP have low removal efficiencies in the 1 – 2.5 μm range, which covers 8% of the total fly ash mass.

The basic design parameter for a fabric filter collection device is the required area of the filter bags that keep pressure drop to a manageable level. The area calculation requires the mass flow rate of coal used and the volumetric flow of the exhaust gas. The amount of coal needed depends on the power output of the power plant.

$$350 MW * \frac{3.413 \times 10^6 BTU}{MW \cdot hr} * \frac{1_{in}}{0.345_{out}} * \frac{1 \text{ lb coal}}{11,750 BTU_{in}} * \frac{454 g}{1 lb} * \frac{1 mol_{coal}}{1058.5 \text{ g coal}} = 126,400 \frac{mol}{hr}$$

Total mass of particulate matter to be collected:

$$350 MW * \frac{3.413 \times 10^6 BTU}{MW \cdot hr} * \frac{1_{in}}{0.345_{out}} * \frac{1 \text{ lb coal}}{11,750 BTU_{in}} * \frac{0.07 lb_{PM}}{1 lb} * \frac{ton}{2000 lb} = 10.3 \frac{ton_{PM}}{hr}$$

Total ash is composed of 65% bottom ash and 35% fly ash.
Ash$_{bottom}$ = 6.7 ton/hr
Ash$_{Fly}$ = 3.6 ton/hr

The gas flow rate can be determined by calculating the amount of coal used and using the conversion developed in Case Study 2-1 (1 g-mol of coal generates 17.15 scm of exhaust), and correcting the volumetric flow of exhaust gas at standard conditions into actual conditions (assume the operating temperature within the control equipment is 400K and 1 atm):

$$126,400 \frac{mol}{hr} * \frac{17.15 scm}{1 mol} * \frac{hr}{3,600 \sec} = 602 \frac{scm}{\sec} * \frac{450K}{298K} = 909 \frac{acm}{\sec}$$

To find the area of fabric needed, assume a superficial velocity (or air/ filter ratio) of 0.75 m/min, which keeps the pressure drop to a minimum. Also, assume the filter bags have a diameter of 25 cm and are 10 m long.

Total bag area needed is given by equation 8.18:

$$A_{bag} = \frac{Q}{u_{sup}} = \frac{909 \frac{m^3}{\sec} * \frac{60 \sec}{\min}}{0.75 \frac{m}{\min}} = 72,7200 m^2$$

One bag has an area of (bag circumference) * length = πDL = π (0.3m)(10m) = 7.85 m².

The number of bags needed is

$$N_{bags} = \frac{Area_{all-bags}}{Area_{1-bag}} = \frac{72,720 m^2}{7.85 m^2} = 9,300 bags$$

An estimate of the dimensions of the building used to house these bags can now be made. We can place the bags into several individual compartments. This allows one compartment to be taken offline so the bags can be emptied; such that at least one extra compartment is needed. As a quick estimate, assume the bags / compartments are placed into square arrangements, meaning a 100 x 100 series of bags (10,000 bags total). We split these into 16 compartments, each consisting of 25 x 25 bags so one can be offline for cleaning. Only 15 compartments will be operating at once or 15*(25x25) = 9,375 bags. This is sufficient and we will accept this value.

FIGURE 8-22. Sketch of one corner of a filter bag compartment. Circles represent the bags. 10,000 bags are arranged in a 100 x 100 square set of bags. Individual compartments contain 25 x 25 bags, and 16 compartments are needed.

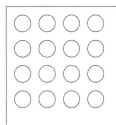

Next, we can determine the size of a single compartment. Each bag is 25 cm. Assume an extra 5 cm between bags to allow the filtered air to flow away from the bag. The dimensions of the compartment are 25 * (25 dia +5 extra cm) + 5 cm for the one remaining edge = 755 cm or 7.55 m. These are arranged into a 4 by 4 array of compartment to obtain the needed 16. This leads to a floor plan area of 30.2 m x 30.2 m. The height of a bag is 10 m. An extra 5 m will be needed for the auxiliary equipment needed to introduce the exhaust into the filter bags, to house the ash collection at the bottom, and to have some sort of equipment at the top to help remove the ash from the bags.

The design dimensions of our bag house are 15 m high by 30.2 m x 30.2 m (a 5 story building, that is 100 ft by 100 ft in floor plan). The actual size may be quite a bit larger to allow for access walkways, process control equipment, and other auxiliary needs.

8.7 Questions

* - Questions and problems may require additional information not included in the textbook.

1. How would changing PM regulations from mass to surface area affect the treatment of exhaust streams?

2. Visit the EPA AirData website (http://www.epa.gov/airdata/). Determine the main sources of particle emissions where you live. How much do they discharge?*

3. Visit a hardware store and examine the furnace air filters. What are the important characteristics as stated by the manufacturers as shown on the packaging? Is there a characteristic that cor relates with price besides area?*

4. What characteristics of an aerosol would increase the collection efficiency of a settling chamber? Which would decrease the efficiency?

5. What characteristics of an aerosol would increase the collection efficiency of a cyclone? Which would decrease the efficiency?

6. What characteristics of an aerosol would increase the collection efficiency of a mesh air filter? Which would decrease the efficiency?

7. What characteristics of an aerosol would increase the collection efficiency of an electrostatic precipitator? Which would decrease the efficiency?

8. How can the characteristics of an aerosol be altered to increase the removal rate of particles once it enters the atmosphere?

9. Determine an appropriate particle control technology for the following cases. Assume the given sizes represent the 10% and 90% sizes of the particle distributions.
    a. 5 - 20 μm water droplets in a natural gas pipeline.
    b. 0.5 – 50 μm emission from a cement plant kiln (dry, reactive with water).
    c. 10 – 100 μm emission from a grain storage silo (sticky and tacky).
    d. 20 – 80 μm emission from a rock mining crusher (wet process).
    e. 2 – 200 μm emissions from a coal fired power plant.

10. Make a list of three questions you have about this chapter or air pollution concerns you have.

8.8 Problems

1. A cement kiln with 500 tons per day feed releases 100 kg of dust into the air each day. Would this source be allowed under the New Source Performance Standards in the US?

2. A 1200 MW pulverized coal fired steam plant (32% efficient) uses coal with a heat content of 11,250 BTU/ lb. The coal has an ash content of 9%, and that 65% of this is fly ash. Determine the collection efficiency needed to meet US-EPA New Source Performance Standards for particulate matter in a coal fired power plant.

3. A power plant (32% efficient) burns No. 5 fuel oil (heat content is 125,000 BTU/gal). Will this plant need a particulate collector? If so, how efficient must it be? The particulate emission factor for No. 5 heating oil is 0.01 lbs/ gal burned.

4. A small (50 MMBTU/hr) industrial boiler (34% efficiency) uses natural gas to generate hot water. Will this facility be required to collect particulate emissions? The particulate emission factor for this system is 6.2 lb/10^6 ft^3 of gas burned.

5. Find the length of a settling chamber needed to collect 80% of 80 μm particulates (ρ_p = 1.5 g/cm^3). The chamber is 400 cm wide and 200 cm tall. The air flow rate is 25 m^3/sec, and it is at 20 °C and 1 atm.

6. Find the height of a settling chamber that will collect 95% of 65 μm particulates (ρ_p = 2.5 g/cm^3). The chamber is 120 cm wide and 10 m long. The air velocity is 110 m/min.

7. Find the width of a settling chamber needed to collect 50% of 45 μm particulates (ρ_p = 2.0 g/cm^3). The chamber is 210 cm high and 12 m long. The air flow rate is 5 m^3/s and is at standard conditions.

8. What collection efficiency will a 125 cm high x 150 cm wide x 8 m long settling chamber have for a 50 μm particulate (ρ_p = 3 g/cm^3) moving in turbulent air of 10°C and 1 atm at 1.5 m/s?

9. What collection efficiency will a 1 m high x 2 m wide x 15 m long settling chamber have for the particulate (ρ_p = 2 g/cm^3) distribution in problem 2 moving in turbulent 10°C and 1 atm air at 2.3 m/s?

10. Derive equation 8.8, the cyclone particulate velocity in radial direction, starting with a force balance, as shown in equation 7.22.

11. Find the radial velocity (up in equation 7.22) of a 50 μm particle (ρ_p = 2000 kg/m^3) moving in air through a 10 cm radius cyclone. The air enters at 20°C, one atm, and is moving at 2 m/s.

12. What is the maximum time needed for a 20 um particulate (ρ_p = 2200 kg/m^3) to travel 6 cm to the outer wall of a 12 cm radius cyclone? The air enters at 20°C, 1 atm, and is moving at 1.2 m/s.

13. Find the smallest diameter particulate (ρ_p = 1700 kg/m^3) to be collected with 50% efficiency in a 16 cm radius cyclone that has dimensions of L_1=2 m, L_2 = 1m, W = R_0/2, and H = 35 cm. The air enters at 20°C, 1 atm, and the gas velocity in the cyclone is 2 m/s.

14. Find the grade efficiencies for 10, 50, and 75 μm particulates in a cyclone if the dp_{50} is 35 μm.

15. What collection efficiency will a cyclone have for the particulate distribution in problem 7.2? The air inlet width is 10 cm, the cyclone radius is 22 cm, the particulates have a density of 2300 kg/m³, and the air speed in the cyclone is 1.5 m/s and it undergoes 6 revolutions within the cyclone.

16. What collection efficiency will the cyclone in problem 15 have for the following particulate distribution?

dpi	Ni
0 - 10	12000
10 - 20	8700
20 - 30	5400
30 - 40	3100
40 - 50	1500
50 - 70	600
70 -100	100

17. Find the area needed for an air filter that is 1 cm thick and has a permeability of 10^{-10} m². The air is at room conditions (T=293 K) and flows at a rate of 3 m³/sec. The initial pressure drop across the filter must be no more than 200 Pa.

18. A coal fired boiler emits 3 x 10⁶ acfm of exhaust that is filtered to remove particulates. The air/filter ratio is 3 ft/min (u_{sup}). The filter bags have a 12 inch inside diameter and are 24 feet long. Determine the number of bags needed for this operation.

19. Estimate the permeability of a filter and filter cake that is used to clean 225 m³/min of room temperature (293K) air. The filter is 16 m² and 0.75 cm thick. It collects dust at a constant rate of 4.1 kg/min. The particulates have a density of 1650 kg/m³ and a cake porosity of 0.18. Data shows that the pressure drop can be described with the following equation: ΔP (inch H_2O) = 1.3 inch H_2O + 2.8 t (inch H_2O/hr).

20. Estimate the expected pressure drop as a function of time from a filter and filter cake used to clean 10 m³/min of room temperature air through a 0.5 cm thick filter with an area of 3.7 m². The mass loading is 0.5 kg/min. The cake has a bulk density of 1880 kg/m³ as collected. Ignore initial and transient effects by assuming that the filter has a constant permeability of 4.5 x 10^{-11} m² and the cake has a constant permeability of 8.9 x 10^{-13} m².

21. Derive equation (8.25), ESP particle velocity starting from the force balance and using equation (8.23).

22. Find the drift velocity of a 25 μm particulate in an ESP whose collecting and charging field strengths are 45,000 V/m. The air stream is at 300 K, and the dielectric constant for the particulates is 8.5.

23. Find the smallest particulate size collected with 50% efficiency in an ESP with a collection plate area of 10 m^2, gas flow rate of 0.5 m^3/s, collecting and charging field strengths of 40,000 V/m, a dielectric constant of 7.2, which is used to clean air at room temperature.

24. Determine the grade efficiency of 20 um particulates flowing within air at 400 K. The air flows through an ESP with a collection plate area of 400 m^2, airflow rate of 1.25 m^3/s, collecting and charging field strengths of 25,000 V/m, and a dielectric constant of 4.6.

25. Determine the overall collection efficiency of the ESP in problem 24 for the particulate distribution in problem 7-2.

26. Determine the overall collection efficiency of the ESP in problem 24 for the particulate distribution given in problem 8-16.

27. Design the minimum volume arrangement of plates for an ESP to collect 50% of 8 μm particulates from a 350 K air stream flowing at 4 m^3/s. These particulates will move towards the plates at 5 cm/s. One plate has a dimension of 1 m by 2 m, and plates are spaced 20 cm apart.

8.9 Group Project Ideas

For each project, the students should work in small groups and present their finding in either a short report (5-8 pages) or a 15 minute presentation.

1. Choose a Source Category from the US-EPA AP-42 appendix B.2 Generalized Particle Size Distributions. Determine the expected performance of a) gravity settler, b) Cyclone, c) ESP, d) Wet Scrubber, and e) Fabric Filter using the efficiencies shown in Figure 8-11. Will any of these units achieve NSPS for a new facility? Choose one of the devices that you expect to work and perform basic designs as shown in this textbook. If none work, discuss how you could change the PM or the equipment in order to make it work.

8.10 Bibliography

Araujo, JA, et al. 2008. Ambient particulate pollutants in the ultrafine range promote early atherosclerosis and systemic oxidative stress. *Circ Research*. March 14, 2008, Vol. 102, 5, pp. 589-96.

Boaworm. 2010. *Eyjafjallajokull volcano plume 2010 04 18.JPG*. Wikimedia Commons, s.l. : 2010.

Bourrichon. 2010. *Estimated ash cloud from the Eyjafjallajökull eruption as of Apr 15, 2010 at 18:00*. Wikimedia Commons, s.l. : 2010.

Cooper, C David and Alley, F C. 2002. *Air Pollution Control: A Design Approach*. 3rd. Prospect Heights, Ill : Waveland Press, Inc., 2002.

Dockery, D W, et al. 1993. An Association between Air Pollution and Mortality in Six U.S. Cities. *New England Journal of Medicine*. 1993, 329, pp. 1753-1759.

Donaldson, Ken. 2004. *The toxicology of airborne nanoparticles*. Derbysgire, UK : First International Symposium on Occupational Health Implications of Nanomaterials, 2004.

Dynamics, Task Group on Lung. 1966. Deposition and Retention Models for Internal Dosimetry of the Human Respiratory Tract. *Health Physics*. 12, 1966, Vol. 2, pp. 173 - 207.

Gold, D R, et al. 2000. Ambient Pollution and Heart Rate Variability. *Circulation*. 101, 2000, pp. 1267-1273.

Kreidenweis, Sonia, et al. 1999. Aerosols and Clouds. [book auth.] Guy Brasseur, John Orlando and Geoffrey Tyndall. *Atmospheric Chemistry and Global Change*. New York : Oxford University Press, 1999.

Lave, Lester B. 1973. An Analysis of the Association between US Mortality and Air Pollution. *Journal American Statistical Association*. 1973, Vol. 68, p. 342.

Mokdad, Ali H. 2004. Actual Causes of Death in the United States, 2000. *Journal American Medical Association*. 2004, Vol. 291, 10, p. 1238.

Oberdorster, Gunter. 2004. *Inhaled Nano-sized Particles: Potential Effects and Mechanisms*. Derbyshire, UK : First International Symposium on Occupational Health Implications of Nanomaterials, 2004.

Peng, Roger D., et al. 2008. Coarse Particulate Matter Air Pollution and Hospital Admission for Cardiovascular and Respiratory Diseases Among Medicare Patients. *Journal American Medical Association*. May 14, 2008.

Perry, R H and Green, D W. 1997. *Perry's Chemical Engineers' Handbook*. 7th. New York : McGraw Hill, 1997.

Pope, C A, et al. 2004. Cardiovascular Mortality and Long-Term Exposure to Particulate Air Pollution : Epidemiological Evidence of General Pathophysiological Pathways of Disease. *Circulation*. 2004, 109, pp. 71-77.

Pope, C Arden and et, al. 2002. Cancer, cardiopulmonary mortality, and long-term exposure to fine particulate air pollution. *Journal American Medical Association*. 2002, Vol. 287, pp. 1132-1141.

Pope, C. Arden, Ezzati, Majid and Dockery, Douglan W. 2009. Fine Particulate Air Pollution and Life Expentancy in the United States. January 22, 2009, Vol. 360, 4, pp. 376-386.

Powerspan. 2003. *Top View of ESP Schematic Diagram*. Powerspan Corporation, s.l. : 2003.

Siebert, P.C. 1977. *Handbook on Fabric Filtration*. Chicago, Ill : ITT Research Institute, 1977.

Strauss, W. 1966. *Industrial Gas Cleaning*. London : Pergamon Press, 1966.

US-DOE. 2008. *Approach to Nanomaterial ES&H Revision 3a* . Washington DC : U.S. Department of Energy, 2008.

US-EPA. 2003. *Air Pollution Control Technology Fact Sheet*. Washington DC : U.S. Environmental Protection Agency, 2003. EPA-452/F-03-028.

—. 1995. *AP-42 Compilation of Air Pollution Emission Factors, Volume I: Stationary Point and Area Sources, 5th ed. Chapter 1: External Combustion Sources*. US Environmental Protection Agency. 1995. table1.1-8. Cumulative Size Distribution and Size Specific Emissionn Factors for Cyclone Furnaces Burning Bituminous Coal.

—. 2002a. *EPA Air Pollution Control Cost Manual - Sixth Edition. Section 6, Chapter 2 – Wet Scrubbers for Particulate Matter'*. Washington DC : U.S. Environmental Protection Agency, 2002a.

—. 2015. National Emissions Inventory (NEI) Air Pollutant Emissions Trends Data. *1970 - 2013 Average annual emissions, all criteria pollutants*. [Online] November 4, 2015. [Cited: November 5, 2015.] http://www.epa.gov/air/airtrends/pm.html.

—. 2012. *Visibility Impairment from Air Pollution*. US Environmental Protection Agency, s.l. : 2012.

US-NPS. 2008. *Annual Performance & Progress Report: Air Quality in National Parks*. Washington DC : US NAtional Park Service, 2008.

Vallero, D.A. 2008. *Fundamentals of Air Pollution*. 4th. San Francisco : Academic Press, 2008.

Wark, K, Warner, C and Davis, W. 1998. *Air Pollution: Its Origin and Control*. 3rd. s.l. : Addison Wesley Longman, Inc., 1998.

Whitby, K T; Cantrell, B. Particles, Fine. 1976. s.l. : Institute of Electrical and Electronic Engineers, 1976. Proceeding International Conference on Environmental Sensing and Assessment.

World Bank. 2007. *World Development Indicators*. Washington DC : International Bank for Reconstruction and Development, 2007.

Ziminski, Andrew. 2009. Minerva Conservation. [Online] 2009. [Cited: August 21, 2009.] http://www.minervaconservation.com/photos/devizes-marketcross.jpg.

CHAPTER **9**

Sulfur Emissions

Emissions of sulfur containing compounds can cause direct human health problems, and they create harm to the environment. Typical anthropogenic sulfur emissions are in the form of hydrogen sulfide (H_2S) or sulfur dioxide (SO_2), both are toxins. Once released into the environment these compounds will convert to sulfate and can grow into sulfate particles. The particles scatter and reflect light, which creates haze and can cause localized cooling of 0.01 – 1°C in areas downwind from a source. Sulfur compounds are removed from the atmosphere over weeks, which allow it to have a regional to global impact. Sulfur is removed from the atmosphere by wet or dry deposition. Both forms cause the degradation of outdoor objects made of stone, metal, and plastics. It also causes acidification of soil, surface water, and groundwater.

Emission reductions for sulfur have been quite successful. As of 2012, the US emitted less than 50% of the amount emitted in 1990, yet has seen the economy grow 63% in the same time frame. The success is due to a well thought out and mature set of regulations that began in the 1970's and have improved consistently. The regulations focus on strategies to encourage sulfur reduction in fuels, removal from emissions, and to find the lowest cost methods for doing so. A novel cap-and-trade program (see section 9.4.1) attempted to use market forces to encourage the development of cost efficient reduction strategies. Results of the experiment are good, although the current price of sulfur credits is very low due to faster than anticipated reductions in sulfur emissions.

The market approach requires the use of continuous emission monitoring of sulfur (as sulfur dioxide) to accurately account for each facilities sulfur emissions. These devices measure the overall emission flow rate, the amount of sulfur and nitrogen oxides, carbon dioxide, and opacity for regulated sources. The devices directly measure the actual amount of emissions in real-time. Such information has proven useful in understanding how operations impact emissions and in motivating reductions.

Sulfur removed from fuels or emissions can be collected and reused. The common sulfur products include elemental sulfur, sulfuric acid, and gypsum. Indeed, approximately one-half of the sulfur sold in the US originates from these sources.

9.1 General Information

Sulfur is an abundant element found in various forms throughout the environment. It is an essential component in all living organisms as a building block for proteins and participates in many biochemical reactions. Sulfur is used across the economy from manufacturing sulfuric acid to producing medicine, cosmetics, fertilizers and rubber products.

Sulfur exists in nature with oxidation states ranging from -2 to +6 and it readily participates in organic and inorganic reactions. The most important sulfur compounds for outdoor air quality are the *oxidized sulfur* compounds - sulfur dioxide (SO_2), sulfur trioxide (SO_3), and sulfuric acid (H_2SO_4). Other forms present in the atmosphere include dimethyl sulfide (($CH_3)_2S$), carbonyl sulfide (OCS), and hydrogen sulfide (H_2S) - which collectively are known as *reduced sulfur* (oxidation state is -2). Of these, only OCS is stable long enough (lifetime of approximately 1.5 years) to disperse into the stratosphere. Once there, it reacts with sunlight and free oxygen to form sulfate, which is transported back into the troposphere slowly. The other reduced forms oxidize in the atmosphere on a time scale of hours to days to form sulfur dioxide.

Atmospheric sulfate particles are the major source of cloud condensation nuclei (CCN). CCN allow water vapor in slightly saturated (tenths of a per-cent) air to condense into the droplets that form a cloud. Without CCN, air would need to be supersaturated by several 100% before water vapor could condense, which would dramatically alter the location and elevation at which clouds form. Anthropogenic sulfates also create CCN, and are believed to alter the distribution and formation of the earth's clouds.

9.2 Sources

FIGURE 9-1. The Atmospheric Sulfur Cycle. Depicts the atmospheric sulfur cycle. It shows the various sources, typical reactions, and removal mechanisms for sulfur.

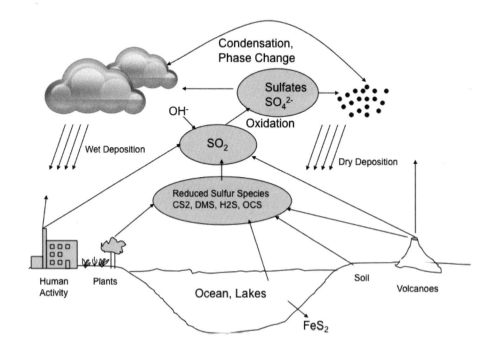

Sulfur enters the atmosphere from many sources – both natural and anthropogenic. Table 9-1 lists the worldwide sources of sulfur in the atmosphere and quantifies their annual flux, or annual movement into or out of the atmosphere. (Brasseur, et al., 1999). It is interesting to note the large variability of natural sources mostly due to one-time events related to volcanoes. Anthropogenic sources currently exceed natural sources into the atmosphere.

Table 9-1. Sources and Annual Flux of Sulfur in the Atmosphere (TgS/year).

Source	Species	Flux
Volcanoes [1]	SO_2, H_2S, OCS	7 - 10
Soil and Plants	H_2S, OCS, DMS, CS_2	0.5 - 1.5
Biomass Burning	SO_2, OCS, H_2S	2 - 3
Ocean [2]	H_2S, OCS, DMS, CS_2	10 - 40
Human Caused	SO_2, Sulfates	88 - 92
Sink	Species	Flux
Dry Deposition	SO_2, Sulfates	50 - 75
Wet Deposition	SO_2, Sulfates	50 - 75

Notes: 1 Tg = 10^{12} g
[1] Some volcanoes can produce much more, such as Tambora (Indonesia) which erupted in 1815 and emitted 50 TgS
[2] Does not include sulfate from sea salt. These sulfates typically are in the air a short time and quickly return to the ocean.

Table 9-2 lists the anthropogenic sources of sulfur dioxide emissions in the United States (US-EPA, 2012). It is very clear that the majority is due to the burning of fossil fuels (coal, oil, and natural gas) and industrial production.

Table 9-2. Human Caused Sources of SO_2 in USA (2005).

Source	SO_2 Emissions (Tons/year)
Electricity Generation	10,402,756
Other Fossil Fuel Combustion	2,172,886
Industrial Processes	1,116,099
Non-Road Equipment	362,447
On Road Vehicles	145,966
Fires	36,879
Waste Disposal	29,503
Residential Wood Combustion	5,263
Miscellaneous	688
Solvent Use	593

Sulfur from electricity production originates from the fossil fuel used to produce the electricity. Fossil fuels are also used to generate steam and heat. The sulfur from industrial processes is related to the refining of metal ores (typically sulfides) to make metal, the processing of petroleum (including the removal of sulfur from oil), and the production and use of sulfuric acid. Transportation related sulfur sources originate with power boilers (trains and ships) or large diesel engines that burn sulfur-containing fuels.

The amount of sulfur released in the US varies by state. The largest emitter in 2012, Pennsylvania, released more than the total from the 20 lowest emission states. Table 9-3 shows the amounts of SO_2 released from several of the states on a total mass basis and per capita basis (US-EPA, 2014). The emission leaders are the states that generate most of their electricity from coal.

Table 9-3. SO_2 Emission Leaders, Total and Per Capita, 2012 data.

State	Total SO_2 Emission (ton/yr)	State	Per Capita SO_2 Emission (Pounds/person/year)
PA	987,855	ND	428
OH	877,161	WY	400
IN	690,208	WV	378
TX	643,393	IN	213
GA	571,406	AL	186

The amount of sulfur released from combustion depends on the type of fuel (coal, oil, gas), rank (heat content), the sulfur composition, and the equipment used (size, load, presence of control technology, firing conditions, and level of maintenance). The rank of a fuel determines the mass needed to obtain a given amount of energy. A lower rank means more fuel is required to obtain the same amount of energy as a higher rank fuel. The sulfur composition depends on the type and source of fuel and level of pretreatment for sulfur removal. Sulfur composition and heat content do not correlate. The methods used to operate fuel burning equipment can also impact sulfur emissions, mostly through the use of pretreatments and sulfur control equipment.

There are three main types of *fossil fuels* – coal, oil, and natural gas. Table 9-8 lists several fuels, provides information about their sulfur content and emission factors, and notes some typical uses for the fuel. The sulfur in coal is either part of a separate physical phase, such as the mineral pyrite, or part of the coal's molecular structure. The sulfur in oil can be present as a dissolved gas of one of the reduced sulfur forms or as part of the oil molecular structure. The sulfur in natural gas is usually hydrogen sulfide. Each fuel can be cleaned with a physical separation to remove the sulfur that is not part of the molecular structure. The method of removal for chemically bound sulfur depends on the fuel. The removal of sulfur from coal occurs after combustion. The removal of sulfur in oil occurs before combustion. Natural gas usually does not have sulfur chemically bound in it. The reason for the difference in methods relates to the material's end use. Coal, used for electricity generation, is used at a few very large facilities where the exhaust is controllable. Oil products are used more for transportation which means use in many small mobile sources. Therefore, it is preferable to clean before sending it to the final user. The methods for controlling these sulfur sources are discussed later in this chapter.

Combustion of sulfur-containing fuel releases sulfur dioxide into the atmosphere with the other combustion byproducts. A simple example is the oxidation of methyl mercaptan (CH_3SH), the odorant added to natural gas to help detect leaks:

9.1 $$CH_3SH + 3\ O_2 \rightarrow CO_2 + 2\ H_2O + SO_2$$

This reaction only takes a fraction of a second at the high temperatures associated with combustion. The combustion products are emitted to the atmosphere where the sulfur dioxide can react with the hydroxyl radical, whose products react with oxygen (O_2, O_3, or H_2O_2) to form sulfur trioxide, a series of reactions that simplify to:

9.2 $$2SO_2 + O_2 \rightarrow 2SO_3$$

This reaction can take several hours if it occurs within the aqueous phase of clouds (80% of SO_2 oxidation) to many days when reacting with the hydroxyl radical in the gas phase (20% of SO_2 oxidation). SO_3 is hygroscopic, which means it readily reacts with water. This reaction forms sulfuric acid:

9.3 $$SO_3 + H_2O \rightarrow H_2SO_4$$

This reaction can take days to occur. H_2SO_4 has a very low vapor pressure and typically forms or joins particles quickly. The sulfuric acid slowly reacts over several days to form sulfate particles through homogeneous and heterogeneous nucleation (e.g. form ammonium bisulfate and ammonium sulfate). These particles leave the atmosphere by dry or wet deposition onto the earth's surface. Overall, the reactions may take 5 to 20 days between emission to the atmosphere and removal by deposition. A plume may travel thousands of miles during this time, depending on the dispersion properties of the atmosphere. It is this ability to travel that causes the emission of sulfur to be a regional and trans-national problem rather than a local problem.

The ultimate fate for sulfur released into the atmosphere is deposition as a sulfate onto land or water. Once deposited, it can be reduced and re-enter the atmosphere, or reduced into a metal-sulfide ore such as pyrite (FeS_2) that deposits into sediments and becomes part of the lithosphere (Earth's crust).

■ Example 9-1.

Determine the amount of sulfuric acid that is created by the emissions from 1,000 kg of coal that contains 3.07 wt% sulfur. Assume 95% of this sulfur is emitted as SO_2 in the exhaust gas and the other 5% of the sulfur remains in the ash and is not converted.

Solution:

First, determine the amount of sulfur emitted, then use the molecular weights of sulfur and sulfuric acid to calculate the mass of sulfuric acid.

$$1000\,kg\,coal * \frac{0.0307\,kg\,S}{kg\,coal} * 0.95\,emission\,rate = 29.165\,kg\,S$$

$$29.165\,kg\,S * \frac{98.07\,kg\,H_2SO_4}{32.06\,kg\,S} = 89.21\,kg\,H_2SO_4$$

The amount of SO_2 could also be determined, but it is not necessary for this problem.

■ Example 9-2.

Determine the amount of sulfur dioxide (SO_2) emitted from a 100 MW industrial boiler (constructed in 1975) using a No. 6 fuel oil with 3.19 wt% sulfur. Would this emission be covered under the NSPS for electric utility steam generating units? If so, what removal is required?

Solution:

The emission factor from appendix III for No. 6 fuel oil is 157*S lb/1000 gallon, where S is the weight percent sulfur (S=3.19 in this example) in the fuel. The heat content is 5.8 MMBTU/bbl (also from appendix 3)

The unit is larger than the minimum size applied to NSPS sources (73 MW) and needs to consider if its emissions exceed the limits in the regulation. The NSPS limit given in Table 3-6 lists the emission limit value of 0.8 lb SO_2/MMBTU for liquid fossil fuels. The amount of sulfur released from this fuel per million BTUs:

$$\frac{157*(3.19)\,lbS}{1000\,gal} * \frac{bbl}{5.8\,MMBTU} * \frac{42\,gal}{bbl} * \frac{64\,lb_{SO2}}{32\,lb_S} = 7.25\,\frac{lbS}{MMBTU}$$

Finally, calculate the removal needed to meet the required NSPS standards:

$$\frac{7.25 - 0.8}{7.25} * 100\% = 89\%$$

Because of the high level of sulfur and subsequent removal requirements, Fuel Oil no. 6 is rarely used to generate electricity or steam. It still finds use as a transportation fuel (esp. oceanic shipping.)

9.3 Health and Welfare Effects

Atmospheric sulfur compounds cause a wide variety of health and environmental impacts because of the way they react with other substances in the air. All forms, oxidized and reduced, can create environmental hazards. However, the oxidized forms, sulfur dioxide (SO_2) and sulfate particles (SO_4^{-2}), are the most ubiquitous sulfur pollutants and their effects on outdoor air have been the most studied [(US-EPA, 2009b), (US-DHHS, 2007), (Lippman, et al., 2006)]. Typically, only these oxidized forms are regulated.

9.3.1 Effects on Public Health

Sulfur dioxide (SO_2) is a moderate to strong irritant. Most inhaled SO_2 only penetrates as far as the nose and throat with minimal amounts reaching the lungs - unless the person is breathing heavily, breathing only through the mouth, or if the concentration of SO_2 is high. Typical levels of SO_2 in the ambient air are below 0.01 ppm in the US and below 0.015 ppm in Europe. Levels below 0.03 ppm do not appear to affect human health in long-term studies. Table 9-4 lists dose – response information for workers exposed to high levels of SO_2 [(US-CDC, 1984), (CCOHS, 1997)].

Table 9-4. Health Effects of SO_2 Exposure.

Concentration ppm	Exposure time	Effects
0.03	long term	Respiratory illness, aggravate heart disease
0.5	1 hour	Burning eyes, sore throat, chest tightness, cough, headache, nausea, shortness of breath, dizziness.
1	1 - 6 hours	Reversible decrease in lung function
1 - 2	one year (8 hr/day)	Decreased lung function
5	10 - 30 min	Constriction of bronchiole tubes (asthma)
5	long term	Permanent pulmonary impairment, chronic bronchitis
8	20 min	Reddening of throat, mild nose and throat irritation
500	minutes	Person cannot inhale a single deep breath

After release of SO_2 in the atmosphere, it reacts to form sulfuric acid which usually transforms into particles within the atmosphere. Sulfate can collect in the lungs when inhaled in the fine particle form (See Chapter 8). Such exposure is associated with increased respiratory symptoms and disease, difficulty in breathing, and for extremely high values - premature death. Particularly sensitive groups include children, the elderly, people with asthma, people who are active outdoors, and those with heart or lung disease.

9.3.2 Effect on Public Welfare

Sulfur emissions also have a serious effect on the public welfare and the environment. They create atmospheric particles that cause haze, reduce visibility, enhance corrosion and damage materials. Sulfur emissions are the major cause of acid rain globally (approximately 75%). Sulfate particles also cause haze in many parts of the world. Haze occurs when visibility is reduced due to the scattering and absorption of light by particles and gases. The scattering is enhanced by hygroscopic growth (See Sections 8.3.2.1 and 8.5.7).

Sulfur emissions can accelerate the decay of building materials and paints. Limestone is particularly susceptible to this decay. Many historic buildings and monuments are located in urban areas, where air has the highest acidity. In Europe, there is significant deterioration of buildings and monuments because the buildings are much older, and pollution levels have been ten times greater

FIGURE 9-2. Restoration of a Market Cross in the UK Showing Architectural Damage from Air Pollution. Picture shows the restoration of one of a group of architectural structures. Photo courtesy of Andrew Ziminski, Copyright 2009, (Ziminski, 2009).

than in the United States. Dry deposition of soot and acidic compounds can also dirty buildings and other structures, leading to increased maintenance costs. These effects significantly reduce the societal value of buildings, bridges, cultural objects (such as statues, monuments, and tombstones), and cars. Given enough time and neglect, it causes the loss of structural integrity of these objects.

9.3.2.1 Acid Precipitation

Acid precipitation, commonly referred to as acid rain, is a broad term referring to a mixture of wet and dry deposition of sulfuric and nitric acids from the atmosphere. Acid rain occurs when the oxides of sulfur and nitrogen (Chapter 10 discusses the creation and control of nitrogen oxides) react in the atmosphere with water, oxygen, and other chemicals to form acidic compounds and salts. This results in acidic gases (nitric) and acidic particles (sulfuric) that may form mild solutions when they interact with water. This acid may be deposited onto the ground in dry particle forms (the particles can be liquid and still be dry if they contain little or no water), or it may be deposited in a wet form as rain, fog, mist, or snow. Acid rain looks, feels, and tastes just like clean rain. The harm to people from acid rain is not usually direct. Walking in typical acid rain, or even swimming in an acidified lake, is no more dangerous than walking or swimming in clean water. The harm is indirect; the acid alters the properties of surface and ground water, and of the soil.

Dry deposition occurs everywhere at all times. Nitric acid gases and sulfate gases and particles can deposit onto any surface. Gases may contact a surface by diffusion or dispersion. Particles can become incorporated (coagulation mechanism) with dust or smoke and fall to the ground. The gases and particles can stick to the ground surfaces. Dry deposited gases and particles may interact with surface materials causing corrosion, increasing rates of deterioration and harming of plant surfaces. The compounds eventually wash from these surfaces by rainstorms, leading to sulfate salts and acidity in runoff, which either percolates into the nearby soil or runs into and affects nearby surface water.

Wet deposition occurs in areas when the weather is wet. These acidic materials are hygroscopic, so they tend to attract water from the air. Droplets can form within clouds, as the sulfates act as cloud condensation nuclei or are scavenged by water droplets through impaction, interception or diffusion (see figure 8-9). The droplets can also form below cloud level where sulfate can absorb water vapor to form new particles. The extent of gas uptake is only limited by the solubility of the gases. Once mixed with atmospheric water the sulfate compounds create acidic rain, fog, mist, and/or snow, which deposits onto the ground, buildings, cars, crops, and trees. The liquid forms can collect and immediately percolate into the soil or runoff into surface waters. Snow that collects until spring (e.g. at high elevations and high latitudes) causes large events of acid runoff when it melts. As these acidic waters flow over and through the ground, they affect the nearby soil, plants and animals. The strength of the effects depends on several factors, including how acidic the water is, the chemistry and buffering capacity of the soils involved, and the types of organisms that rely on the water.

The strength of an acid, or acidity, depends on the concentration of hydrogen ions in the water. Measurement is in terms of pH units, see Figure 9-3. A calculation of pH uses the following equation:

9.4 $pH = -log_{10} [H^+]$

[H+] is the concentration of hydrogen ions in molarity (gram-mole/liter). When the pH value of the solution is less than 7, the solution is considered acidic, when it is greater than 7, it is considered basic. Pure water has a pH of 7. Rainwater in clean air has a pH of 5.0 to 5.6. The acidity is due to the natural constituents in air – CO_2, sulfates, nitrates, and cloud condensation nuclei (CCN). The pH of rain, snow, and fog in air contaminated with sulfur may be as low as 2 (rarely), but in recent years in the US the lowest values are around 4.4 for rain, while some fogs and clouds can have lower pH. While acid rain has improved greatly within the US, there is increasing concern about deposition of excess reactive nitrogen (nitrate and ammonium) in many environments. Nitrogen deposition adversely affects pristine environments (e.g., the central Rockies) where the ecosystem evolved in a low nitrogen environment (see section 10.3.2).

It should be noted that the pH of rain is the result of a balance between acid and base inputs. Sulfuric and nitric acids are the most important acid inputs. Atmospheric bases, including gaseous ammonia and alkaline soil particles, are important sources of acid neutralization. The high pH values seen in the Great Plains, for example, reflect the large emissions of ammonia due to agricultural use of nitrogen fertilizers in the region.

■ **Example 9-3.**
Determine the volume of acid rain at pH = 4.5 that could be produced from the untreated combustion of 1 ton (1000 kg) of coal, as described in Example 9-1.

Solution:
From Example 9-1 we saw that the burning of one ton of coal would release 29.165 kg of sulfur (as S) which is equivalent to 89.21 kg H_2SO_4. Next, we need to determine the concentration of acid in water with a pH of 4.5. Using equation (9.4) and rearranging to determine concentration:

$$[H^+] = 10^{(-pH)} = 10^{(-4.5)} = 3.162 \times 10^{-5} \frac{mole\ H^+}{liter\ rain}$$

Moles of H+ produced from the coal:

$$1 ton\ coal * \frac{89.21 kg\ H_2SO_4}{ton\ coal} * \frac{1000g}{1kg} * \frac{1 mole\ H_2SO_4}{98.07 g\ H_2SO_4} * \frac{2 mole\ H^+}{mole\ H_2SO_4} = 1,819\ mole\ H^+$$

FIGURE 9-3. The pH Scale.

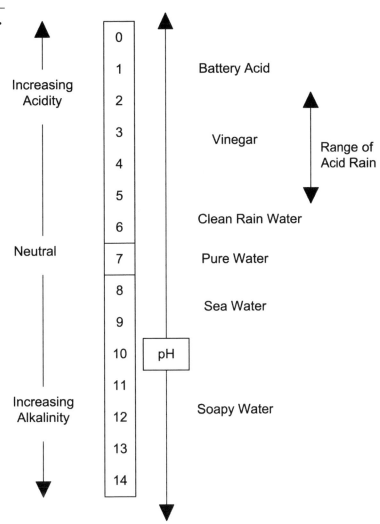

Liters of rain affected by the H+:

$$1{,}819\, mole\, H^+ * \frac{liter\ rain}{3.162 x 10^{-5}\, mole\, H^+} = 57.5 x 10^6\, liter\ rain$$

This volume of rain can cover almost 2 mile² with ½ inch of water. A large power plant (1000 MW) uses about 10,000 tons of coal each day.

When acidic water encounters a soil, it chemically reacts with it. The type of reactions and interactions depend on soil properties. Some soils contain materials that are slightly basic (pH > 7), such as limestone, that allow the soil to neutralize the acid rain's acidity. This ability to neutralize acids is referred to as buffering capacity. The soil *buffering capacity* determines the amount of damage an area undergoes from acid rain. The ability of soil to resist, or buffer, acidity depends on the presence of buffering materials, the thickness and composition of the soil, as well as the type of bedrock underneath. The effects of acid precipitation can be minimized by adding acid-neutralizing minerals to the soil or water to increase its buffering capacity.

Soils without buffering capacity, or where all the buffering material has been neutralized by acid precipitation, undergo significant and permanent changes when exposed to acidic deposition. The changes occur because acidic water dissolves the nutrients and minerals in the soil and carries them away before trees, and other plants can use them to grow. It also dissolves metals in the soil, such as aluminum that become toxic to trees and plants when in solution. The combination of loss of soil nutrients and increase of toxic metals weakens trees and plants, but is deadly only in extreme cases. The dissolved nutrients, minerals, and metals are transported in runoff or groundwater into surface waters (streams, rivers, and lakes). Surface water can become acidic when the water itself and its surrounding soil cannot buffer the incoming acid enough to neutralize it.

Direct damage from acid precipitation can seriously harm some plants and the animals that are dependent on them. Plants are damaged from the direct deposition of acid (wet or dry) onto their foliage. Damage occurs due to direct tissue damage or through the leaching of nutrients from the leaves and needles. The damage leads to reduced/slower growth of the plant and can also cause death. In areas that are heavily affected, entire forests can die. Forrest death correlates with areas of thin soils or high elevations. However, acid rain does not usually kill plants directly. Instead, it is more likely to weaken them. Death may occur when the effects of acid deposition combines with additional stresses, such as other air pollutants, insects, disease, drought, or very cold or warm weather. In most cases, the impacts on plants are due to the combined effects of acid rain and these other environmental stressors.

Many aquatic organisms have a hard time adapting to and reproducing in an acidic environment. Table 9-5 shows the pH levels below which several species cannot survive. This table only shows direct problems and does not account for the complex biological interactions between organisms in aquatic ecosystems. For example, frogs may tolerate relatively high levels of acidity (pH of 4), but if they eat insects like the mayfly (tolerates pH above 5.5); they may be affected because part of their food supply disappears due to the low pH. Because of the connections between the many fish, plants, and other organisms living in an aquatic ecosystem, changes in pH or aluminum levels that do not directly affect an organism may still alter its ability to survive in a region. The young of most species are more sensitive than adults. At pH 5, most fish eggs cannot hatch. At even lower pH levels, some adult fish die. Acidification due to acid precipitation has reduced the ability of some lakes to sustain fish life.

Table 9-5. pH Tolerance Levels of Some Aquatic Organisms.

Species	Lowest pH it can tolerate
Clams	6
Snails	6
Bass	5.5
Crayfish	5.5
Mayfly	5.5
Trout	5
Salamanders	5
Perch	4.5
Frogs	4

Food crops are rarely harmed by acid precipitation because farmers add fertilizers to the soil to replace nutrients that have washed away. They may also add crushed limestone to the soil to increase its buffering capacity. However, it has been observed that long-term exposures of SO_2 at levels above 0.05 ppm can significantly reduce crop yield of wheat and barley by 5 to 10% (Wilhour, et al., 1978).

An additional issue occurs when acid precipitation enters a lake or stream faster than the water body can neutralize it. Episodic acidification refers to brief periods during which pH levels decrease due to runoff from melting snow or heavy rain (Wigington, et al., 1992). Lakes and streams in many areas throughout the U.S. are sensitive to episodic acidification. Studies have shown that many lakes and streams in the Mid-Appalachians, the Mid-Atlantic Coastal Plain, and the Adirondack Mountains become temporarily acidic during storms and spring snowmelt (Rice, et al., 1999). For example, in the Adirondacks approximately three times as many lakes are sensitive to episodic acidification (those with low buffering capacity) as are chronically acidic (lakes with pH less than 5). Similarly, in the mid-Appalachians, approximately seven times as many streams are likely to become acidic during an episode compared to the number of chronically acidic streams. Such episodic acidification can cause "fish kills" because of the large number of fish that are negatively affected. A typical fish kill does not kill all the fish in an area, but may seriously affect them through injury, alteration of their food supply, and reduced ability to reproduce.

9.3.2.2 Geographic Distribution of Acid Rain

Figure 9-4 shows maps of the US detailing the distribution of pH in rainwater collected in 1994 and 2009. The very noticeable changes are due to the effects of the 1990 Clean Air Act Amendments (CAAA) that mandated requirements for the control of acid deposition, specifically the Acid Rain Program [40 CFR Parts 72 through 78]. The next section presents some details of the ARP. In these figures, the acidity in rain is measured by collecting samples and measuring their pH in a controlled laboratory setting. The rain samples are collected at sites all over the country (they appear as dots on the figures). Weather conditions are also monitored during sampling to help understand the distribution of acidity. The areas of the greatest acidity in the US have the lowest pH values and are located in the northeastern states. The large number of cities and the concentration of coal burning power and industrial plants generate this pattern. In addition, the prevailing wind direction brings storms and pollution to the Northeast from the Midwest.

FIGURE 9-4. pH Isopleths in the Continental US. Comparison shows the reduced level of acidity in precipitation from 1994 to 2009, (NADP, 2013).

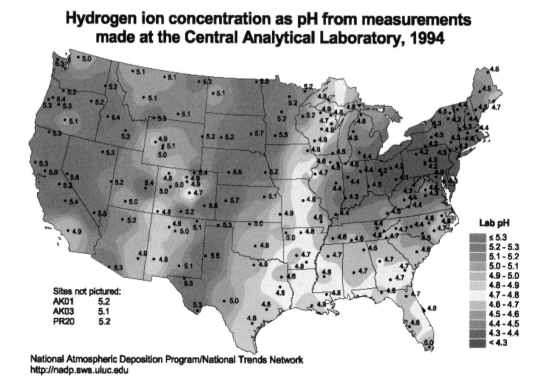

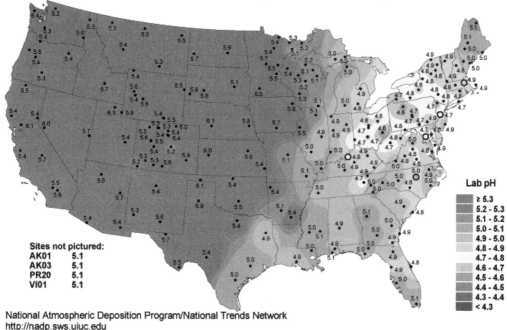

Hydrogen ion concentration as pH from measurements made at the Central Analytical Laboratory, 2009

Sites not pictured:
AK01 5.1
AK03 5.1
PR20 5.1
VI01 5.1

National Atmospheric Deposition Program/National Trends Network
http://nadp.sws.uiuc.edu

● **Case Study 9-1. Effects of Acid Rain on Automotive Coatings. (US-EPA, 2007)**
The US-EPA and the US automobile manufacturers have received numerous reports of damage to the paint and coatings on new automobiles. This damage appears to be due to environmental fallout, which refers to damage caused by air pollution (e.g., acid rain), decaying insects, bird droppings, pollen, and tree sap. The damage usually occurs on horizontal surfaces, appears as off-color irregular spots, and is especially noticeable on darker paint colors. It is easier to spot using fluorescent lights. It appears to occur after evaporation of a moisture droplet, and often has a drop-like shape. In addition, some evidence suggests damage occurs most frequently on freshly painted vehicles. Usually the damage is permanent; once it has occurred, the only solution is to repaint.

All forms of acid precipitation (wet and dry) may damage automotive coatings. The combination of dry deposition followed by dew or rain correlated with increased damage. However, it is difficult to quantify the damage due to acid rain when it is just one component of environmental fallout. The problem is confounded by the improper application of paint or deficient paint formulations. It appears that the best way of determining the cause of chemically induced damage is to conduct detailed, chemical analysis of the damaged area. The results of laboratory experiments and at least one field study have demonstrated that acid rain can scar automotive coatings. Chemical analyses of the damaged areas indicate elevated levels of sulfate, such as would be due to acid rain.

The problem does not affect all vehicles, even in geographic areas known to be subject to higher levels of acid rain. This suggests that technology exists to protect against this damage. Until the technology is identified and implemented, or until acid deposition is adequately reduced, the best ways to reduce the risk of damage are:
- *Frequently washing and drying.*
- *Covering the vehicle, or parking within a covered structure.*
- *Not parking under trees.*
- *Applying a protective coating.*

The auto and coatings industries are developing coatings that are more resistant to environmental fallout, including acid rain. Some manufacturers have developed an acid-resistant paint that adds an additional cost of $5- $15 to a new vehicle.

9.4 History and Regulation

Worldwide ambient air quality standards for SO_2 range from an annual average of between 15 and 80 µg/m³ (0.005 ppm to 0.03 ppm) to an hourly average of between 150 to 700 µg/m³ (0.056 ppm to 0.265 ppm). Table 9-6 shows the standards for several regulatory regions. Notice that each entry may contain values for more than one time averaging period, in which case all contribute in determining the air quality of the region. Concentrations below these values are assumed to have minimal impact on most people's health.

These values are determined by considering the health risk associated with different concentrations and exposures. The values are determined through animal testing, epidemiological analysis, and from incidents of accidental exposure to high levels. Differences in averaging times reflect the assumptions of health models on types of exposure. Higher concentrations are allowed over shorter times, with the understanding that the values fluctuate and do not represent the expected concentration at longer time intervals, see Example 9-4.

SO_2 pollution frequently occurs with particulate matter. Most air regulations before 1970 did not differentiate between PM, SO_x, NO_x, and ozone. The main measurement for air quality before

Table 9-6. Ambient Air Quality Standards for SO2 in Several Countries.

Time Period Average Country \ Units (unless noted)	Annual $\mu g/m^3$	Daily $\mu g/m^3$	1-hr $\mu g/m^3$	10 – min $\mu g/m^3$
US			0.075 ppm	
EU		125	350	
Japan		0.04 ppm	0.1 ppm	
India (Class I/II/III)*	15/60/80	30/80/120		
China (Class I/II/III)*	20/60/80	50/150/250	150/500/700	
Brazil	80	366		
Mexico	78	341		
South Africa	50	125		500
WHO		20		500

Notes: 1 ppm SO2 = 2.66 mg/m3
*Class I: Tourist, conservation area; Class II: Residential area; Class III: Industrial and Heavy Traffic area.

1970 was the color of a cloth filter after passing a set volume of outdoor air through it. The idea of dirty air and its harmful effects have been linked in the public's mind as synonymous, although it no longer is so strongly linked because air that looks clean may still contain a number of harmful substances. The idea that dirty air is linked to harmful effects is augmented by certain events of human suffering caused by SO_2:

- December 1930, Meuse River Valley, Belgium. A thermal inversion in the atmosphere (See chapter 5) trapped industrial fumes in a narrow valley, leading to air concentrations of 38 ppm. This event caused 63 people to die and injured thousands more.
- October 1948, Donora, PA, USA. A thermal inversion that lasted several days caused illness in over 40% of the 12,000 residents and killed 20 people. The measured atmospheric concentration of SO_2 was 2 ppm.
- December 1952, London, UK. A thermal inversion and extreme fog lasted for 5 days. SO_2 levels reached 1.3 ppm. There were approximately 3500- 4000 extra deaths over the normal levels for the time. Autopsies revealed irritation of the respiratory tract.

These events were frequently citied during the discussions that led to the creation of clean air laws. Many believe that the unfairness of these events were major drivers in the creation of environmental laws.

Acid deposition became an increasingly noticed and serious problem in the 1950's in the northeastern region of the United States and Scandinavia. The 1960 - 1980 acid rain debate in the US centered almost exclusively on the eastern portion of the United States and Canada. During the 1980's the discussion expanded to include the western parts of both countries, especially the Pacific coast. In 1991, The United States and Canada committed to addressing trans-boundary air pollution issues through a bilateral executive agreement (Canada, USA, 1991).

The 1972 United Nations Conference on the Human Environment (UNCHE) was one of the first global environmental meetings of the world's governments. The meeting was a response to Sweden's concerns that acid rain originating in Great Britain and Germany (East and West) was responsible for the acidification and death of Scandinavian lakes. At the time of the UNCHE, there was no scientific consensus or political acceptance of the idea that acid rain could fall 500 miles or more away from its pollution source. Nor was there much appreciation of the need for political action to address the trans-boundary and global nature of air pollution problems. Sweden used the UNCHE to bring international attention to the problem of trans-boundary acid rain and other increasingly pressing global environmental concerns.

The First International Symposium on Acid Precipitation and the Forest Ecosystem convened in 1975 in Columbus, Ohio, USA, to define the acid rain problem. Results of the meeting included:
- proposing a precipitation monitoring network in the United States,
- cooperating with the European and Scandinavian networks, and
- setting up protocols for collecting and testing precipitation.

In 1977, The US Executive Council of Environmental Quality was tasked to develop a national acid rain research program. The program became the basis for the National Acid Precipitation Assessment Program (NAPAP). The initiative eventually translated into legislative action in the US with the passage of the Energy Security Act (PL 96–264) in June 1980. Title VII of the Act (the Acid Precipitation Act) created the NAPAP and authorized federally financed support.

The first international treaty aimed at limiting air pollution was the United Nations Economic Commission for Europe (UNECE) Convention on Long-Range Trans-boundary Air Pollution, which went into effect in 1983. Thirty-eight of the fifty-four UNECE member states ratified it, which included not only European countries but also Canada and the United States. The treaty targeted sulfur emissions, and required the parties reduce emissions 30% from 1980 levels, the so-called 30% club.

9.4.1 Regulatory Methodology - US Approach

In 1990, the US Clean Air Act was amended for a second time, and included amendments designed specifically for reducing acid deposition. Title IV of the 1990 Clean Air Act Amendments (PL 101–549) set as its objective to achieve a ten-million-ton annual reduction in emissions from 1980 levels (17.3 million tons of SO_2 in 1980) by the year 2010 (actual emissions of 5.1 million tons of SO_2 in 2010). Traditionally, environmental regulation used the "command and control" approach in which the regulator specifies how to reduce pollution, by what amounts, and what technology to use. Title IV, however gave flexibility in choosing how to achieve these reductions. For example, emitters may reduce emissions by switching to low sulfur coal, installing pollution control devices called scrubbers, or shutting down older plants and increasing production in newer units. It also established the Acid Rain Program (ARP) to reduce the harmful effects of acid rain through reductions in emissions of SO_2 and NO_x. SO_2 reductions included a cap and trade program, which allows sources to buy or sell fixed amounts of SO_2 allowances on the open market while a limit, or cap, sets the total allowable amount of SO_2 emissions from all power plants. NO_x reductions used a regional emissions rate-based program.

The **cap and trade program** allows companies to buy, sell, trade, and bank sulfur dioxide pollution rights. Utility units are allocated allowances based on their historic fuel consumption and a specific emissions rate. Each allowance permits a unit to emit one ton of sulfur dioxide during or after a specific year. For each ton of sulfur dioxide discharged in a given year, one allowance is retired and can no longer be used. Companies that pollute less than the set standards may have allowances left over (banked allowances). They can then sell the difference to companies that pollute more than they are allowed, bringing them into compliance with overall standards. Companies that can clean up their pollution less expensively by changing fuel or persuading their customers to conserve energy would recover some of their costs by selling their pollution rights to other companies. Companies that emit more than their allowance must purchase additional allowances from other companies for the market established value.

Initially, (Phase I - 1995 to 2000) the program allocated allowances to electric power generating units at an emission rate of 2.5 pounds of SO_2/MMBtu (million British thermal units) of heat input, multiplied by the unit's baseline energy input, determined from the average fossil fuel consumed from 1985 through 1987. Alternative or additional allowance allocations for various units, including affected units in Illinois, Indiana, and Ohio, were allocated a pro rata share of 200,000 additional allowances each year from 1995 to 1999.

In Phase II (2000 to present), the program expanded to include virtually all electric power units over 25 MW in generating capacity, and tightened the allowance allocation. No facilities are grandfathered out of compliance in the acid rain program (unlike NSPS). Allowance allocation calculations accounted for various types of units, such as coal- and gas-fired units with low and high emissions rates or low-fuel consumption. EPA allocated allowances to each unit at an emission rate of 1.2 pounds of SO_2/MMBtu of heat input, multiplied by the unit's baseline. Beginning in 2010, the Act reduced the cap to 8.95 million (from 9.5 million) on the number of allowances issued to units each year. Units which began operating in 1996 or later are not allocated allowances. Instead, they have to purchase allowances from the market or the EPA auction to cover their SO_2 emissions.

Additional allowances are available upon application to three EPA reserves. In Phase I, units could apply for and receive additional allowances by installing qualifying Phase I technology (a technology that can be demonstrated to remove at least 90 percent of the unit's SO_2 emissions) or by reassigning their reduction requirements among other units employing such technology. A second reserve provides allowances as incentives for units achieving SO_2 emissions reductions through customer-oriented conservation measures or renewable energy generation. The third reserve contains allowances set aside for auction.

Anyone can participate in the annual allowance auction, which occurs at the end of March every year. The auction offers allowances at a price based on bids from potential buyers. Only the highest bidders can buy the allowances. The average price of one sulfur emission allowance in 2008 was $389.91 and in 2013 was $0.17. Figure 9-5 shows the historical trend of the average winning bid for the emission allowance (US-EPA, 2014). To supply the auctions with allowances, EPA sets aside an Auction Allowance Reserve of about 2.8 percent of the total annual allowances allocated to all units. During phase I, EPA auctioned 150,000 allowances each year. During phase II, EPA auctions 250,000 allowances each year. Of these allowances, half are sold in the spot auction, and winning bidders may use them immediately or bank for later. The other half must be banked for seven years before use, at which time they work exactly the same as any other allowance. Bidders must specify if they want spot or seven-year-out allowances. The market price has essentially collapsed since 2010 due to the over-allocation of allowances and uncertainty in the ARP replacement programs - Clean Air Interstate Rule (CAIR) that was vacated by the US Court of Appeals. The revised program (Cross-State Air Pollution Rule) was also vacated (2012), leaving CAIR in place, but creating great uncertainty in the market place about the regulations. The vacated regulations will not be discussed in detail, but their basic idea and flaw is that they have been trying to work with regions or groups of states together, as the SO_2 problem does not respect state borders. However, the legal basis for regulation is at the state level and is not regional.

FIGURE 9-5. Average Winning Bid for a One Ton SO_2 Emission Credit.

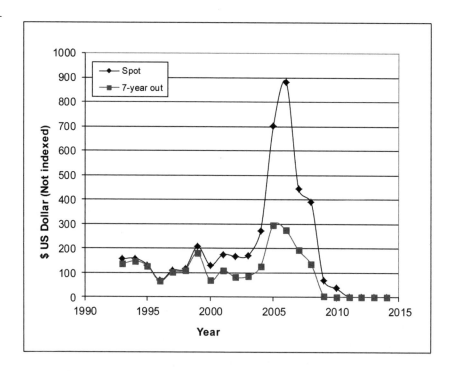

More than half of Phase I plants opted to switch to low-sulfur coal while 16% chose to install sulfur scrubbers to reduce sulfur emissions. The remaining producers relied on purchasing allowances, improving combustion processes and increasing efficiency to remain in compliance.

Some concerns with a cap and trade program include the *cap level*, over-allocation, and the creation of hot spots (localized regions of high concentration). The cap level determines the discharge level of SO_2 during the year. If the level is set lower than the actual amount produced, the producers will not have as many allowances as needed to operate and may have to pay fines up to $2,000 per ton of SO_2 emitted (ARP compliance is generally 100%). If the cap is set higher than the amount produced, there will be excess allowances in the market, which reduces their cost and provides little or no incentive for reducing emissions. These problems are overcome by consistently reviewing and adjusting the total number of allowances. The ARP regulations set the actual number of allowances. Allowances averaged 7.6 million per year during Phase I (1995-1999), 9.5 million during Phase II (2000-2009) and are currently set at 8.95 million until changed by Congress. The value now exceeds the annual emissions (*over-allocation*) and has created a tremendous glut of allowances – driving down the price of an allowance and reducing the market driven motivation to reduce emissions.

Because utilities generally reduce their sulfur emissions more than the minimum required, they have not used as many allowances as are allocated each year as shown in Table 9-7, (US-EPA, 2012). Unused allowances can be used to offset sulfur dioxide emissions in future years or placed into the auction. For example, in 2003, 9.5 million allowances were granted nationally for the year, 8.8 million carried over from prior years and 10.6 million were used. The 7.7 million unused allowances carried over into 2004. The total unused credits have been growing since 2005. The revision of the ARP allowances has not occurred in a timely manner as the replacement programs have been struck down by the US Court of Appeals. The CAIR program would have addressed the imbalance by requiring two credits per ton of SO_2. There is currently great uncertainty concerning the future regulation of SO_2 in the US.

There is also a possibility that a cap and trade system can create *hot-spots*. These are areas where the producers rely on purchasing emission credits rather than trying to reduce their emissions. The idea is that hot spots (areas with high levels of pollution) would occur in poor regions, and these areas would experience capital flight. The well capitalized companies that can afford pollution control would move to better areas. The capital poor companies get left behind, and the area attracts other companies that can not afford control equipment. Such poor areas would get more pollution, and wealthier areas would have cleaner air, newer facilities, and higher costs. Such inequities can be prevented by requiring all regions to meet a shared ambient air quality value and by providing low-cost capital to producers in regions with higher ambient levels. There is no evidence that the ARP has created SO_2 hotspots within the US. Instead, there has been significant improvement in ambient SO_2 concentration everywhere within the US. The lack of a problem with hot spots is probably due to the additional constraints from NAAQS and NSPS, which also regulate the emissions of SO_2.

Table 9-7. SO₂ Emissions and Allowances (10⁶ ton) from Acid Rain Program Sources.

Year	Phase I Emissions	Total Emissions	Allowances Allocated	Unused Year	Unused Total
1980	9.4	17.3[1]	-	-	-
1985	9.3	16.1	-	-	-
1990	8.7	15.7	-	-	-
1995	5.3	11.8	8.7	3.4	3.4
1996	5.4	12.5	8.3	2.9	6.3
1997	5.5	12.9	7.1	1.6	7.9
1998	5.3	13.1	7	1.7	9.6
1999	4.9	12.5	7	2.1	11.7
2000		11.2[2]	10	-1.2	10.5
2001		10.6	9.6	-1	9.5
2002		10.2	9.5	-0.7	8.8
2003		10.6	9.5	-1.1	7.7
2004		10.3	9.5	-0.8	6.9
2005		10.2	9.5	-0.7	6.2
2006		9.4	9.5	0.1	6.3
2007		8.9	9.5	0.6	6.9
2008		7.6	9.5	1.9	8.8
2009		5.7	8.95	3.25	12.1
2010		5.1	8.62[3]	3.52	15.6
2011		4.5	8.52[3]	4.02	19.6
2012		1.2	3.62[3]	2.42	22.0

Notes:

1 – Total emissions include all ARP phase I and phase II sources, and exclude any other source

2 – Phase II started in 2000.

3 – ARP allowance limit is 8.95 million tons. The 2009 Clean Air Interstate Rule (CAIR), which is currently in effect, set the values lower. However, CAIR has been vacated by the courts, and EPA has been ordered to address flaws in the law, while leaving CAIR in place temporarily. The first revision, the Cross-State Air Pollution Rule, was stayed in August 2012, and CAIR remains in place. The stay is currently (2014) being appealed by the US Government.

9.4.2 Trends in US SO2 Concentrations

The annual US ambient air concentrations of SO_2 decreased 79% between 1980 and 2010, and decreased 52% between 2000 and 2010 as shown in Figure 9-6. From 1970 until 2010 the standard was based on an annual average. In 2010, this was changed to 0.075 ppm hourly standard as shown in Figure 9-7, (US-EPA, 20014). The change resulted from a review of human health based observations which suggest that it is the peak value that is most strongly correlated to health problems rather than a long-term average. This change also reflects the fact that the long-term values have been well below the required standard for the entire period of regulation. The secondary standard for protecting the environment and human welfare remains unchanged as 0.5 ppmv three-hour average.

FIGURE 9-6. US National Air Quality Trend for SO_2 Concentration (ppmv) – Annual Average. Note that the NAAQS Standard was altered in 2010 from 0.03 ppm annual average level to 0.075ppm 1 hour average.

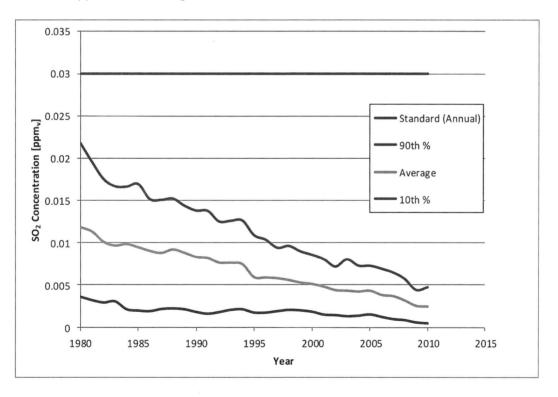

FIGURE 9-7. US National Air Quality Trend for SO₂ Concentration (ppmv) – 1-Hour Average. Dashed line denotes when NAAQS was changed to the one hour standard of 0.075 ppm.

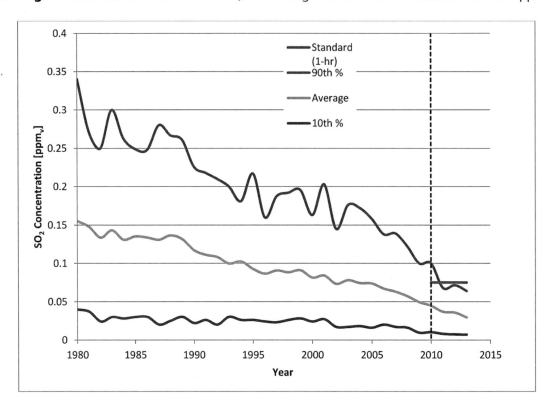

■ *Example 9-4.*

How does the 0.075 ppm 1-hour average compare with the 0.03 annual average? Use equation 6.16 to compare the sampling times.

Solution:

The values from different averaging times are not readily comparable. The correlation given in equation 6.16 is for dispersion plumes and accounts for variables such as wind direction change and intensity as well as changes in temperature over time. These variables may not necessarily be relevant to national air trends; however, it does provide some basis for comparing the two values.

$$C_2 = C_1 \left[\frac{t_1}{t_2}\right]^{0.17} = 0.075\,ppm \left[\frac{1\,hour}{1\,year} * \left(\frac{1\,year}{365\,day} * \frac{1\,day}{24\,hour}\right)\right]^{0.17} = 0.016\,ppm$$

A comparison of the two averages in this problem and of the data in Figure 9-6 and Figure 9-7 suggests this is too large of a discount between the two averages.

The total quantity of SO_2 emitted has also decreased as shown in Figure 9-8, (US-EPA, 2013). Both the total emissions as well as those covered by the Acid Rain Program (ARP) regulations have fallen. The total emissions decrease between 1980 and 2012 (most current year data is available) is 79%, and the ARP emissions decrease between 1980 and 2012 is 93%. This value exceeds the goal set by the EPA in 1980 of 50% reduction by 2010. Similarly, emissions from fuel combustion, industrial processes, and transportation sources decreased 41%, 40%, and 30%, respectively between 1990 and 2005. Note that there are no data for SO_2 emissions from ARP sources (power plants) prior to 1990. Emissions from 1980, 1985, and 1990 are estimates. Continuous emissions monitoring systems (CEMS) were required of all program sources in 1995, and they now provide data from every source.

FIGURE 9-8. US National Air Quality Trend for Total SO_2 Emissions.

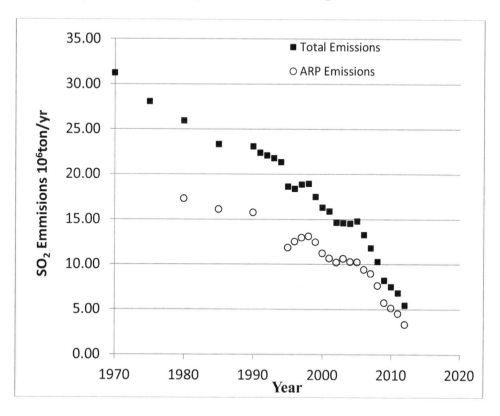

Another measure of the success of these regulations is the total sulfate deposition across the country, as shown in Figure 9-9. The final form of sulfur removed by wet or dry deposition from the atmosphere is either sulfate or sulfur dioxide, whereas sulfur dioxide is the most common initial form. These reductions are similar to those showing pH, Figure 9-4. These reductions have led to improving water quality in lakes and streams. Both the size of the affected region and magnitude of the highest concentrations have declined, with the largest decreases observed along the Ohio River Valley and in the Northeast.

The program has reduced SO_2 emissions faster and at far lower costs than anticipated, yielding wide-ranging health and environmental improvements. The 2003 Office of Management and Budget study (US-OMB, 2003) found that the ARP accounted for the largest quantified human health benefit, over $70 billion annually, of any US regulatory program implemented in the last 10 years, with annual benefits exceeding costs by more than 40:1. That means for every dollar spent on implementing the program, 40 dollars are returned in health and environmental benefits. A 2005 study estimated the program's benefits at $122 billion annually in 2010, while cost estimates are around $3 billion annually (in year 2000 dollars).

FIGURE 9-9. Annual Average Sulfate Concentration (SO_4^{-2}) Isopleths in the Continental US. Comparison shows the reduced level of sulfate in precipitation from 1994 to 2009. Dots show monitoring locations with an annual average in mg/L. Color gradations represent lines of constant concentration (isopleth). (NADP, 2013)

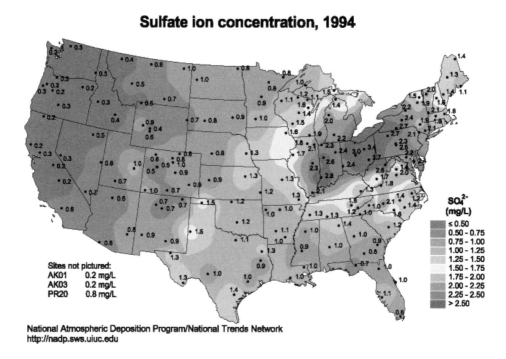

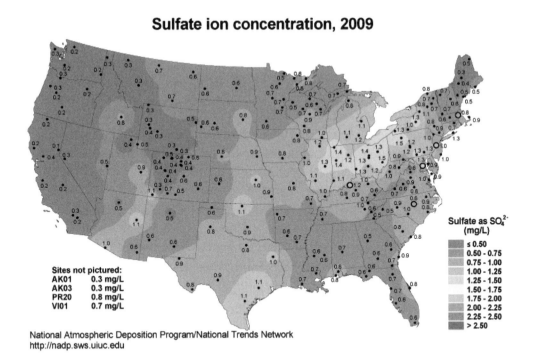

It should be noted that there has been a very recent and quick increase in coal use by China. These new coal plants have limited or no sulfur control equipment. These sulfur emissions have begun to contribute to an observed increased sulfate concentrations in the Western US.

9.4.3 Regulatory Methodology - European Environment Agency

It is interesting to compare and contrast the US environmental programs with that of other economies, such as the European Union (EU). Some member states (countries) of the EU have manufacturing and chemical industries similar to the US. Others are less developed, both economically and in terms of regulatory capabilities. The EU also has a very different history of interaction between sources of pollutants and their neighbors. Pollution laws have been centered more on common law as opposed to statute law in the US (see Chapter 3 for discussion). The EU developed pollution laws ten or more years after the US and Japan. However, the EU has reduced pollutants on a faster timeline, perhaps benefitting from the regulatory experience of other countries.

The European Environment Agency (EEA) is the EU's counterpart of the US-EPA. It began monitoring, measuring and collecting data of emissions of SO_2 in 1997 as part of the framework for a common strategy for improving ambient air quality based on the 1996 Air Quality Framework Directive. Since this time, there has been a 50% reduction in the fourth highest 24-hour mean SO_2 concentration observed at urban stations as shown in Figure 9-10 (European Environment Agency,

2012). The EEA chooses to monitor the fourth highest 24-hour mean value to prevent the occasional system upset from biasing the data set. Also during this time, the fraction of the urban population in EEA-32 member countries that are potentially exposed to ambient air concentrations of sulfur dioxide in excess of the EU limit value decreased to less than 1%.

In total 22 of the EEA-32 member countries included in the Urban Audit provided information on sulfur dioxide concentrations at 'urban background' and 'sub-urban background' stations as of 2008. The data is available in the European Union air quality database **AirBase** [http://air-climate.eionet.europa.eu/databases/airbase/]. Note that the majority of the information on sulfur dioxide concentrations results from stations in EU-15 Member States. The limit values tend to be more widely exceeded in the Central and Eastern European countries.

Several factors have contributed to the decrease in sulfur dioxide concentrations, mostly due to legislative work in the west and political changes leading to economic modernization in the central and eastern parts of Europe. The first (1985) and the second (1994) sulfur protocol under the *United Nations Economic Commission for Europe* (UNECE) *Long Range Trans-boundary Air Pollution* (LRTAP) Convention, together with the European Economic Community (defunct now, but transformed into the European Community when the European Union was created in 1993) set limit values in the 1989 Air Quality Directive (89/427/EEC). These limits resulted in emission reductions and correspondingly decreasing ambient concentrations. Political changes at the beginning of 1990's in central and eastern European countries led to economic restructuring, decline of heavy industry, and adoption of abatement measures on large point sources. These changes contributed to decreasing winter smog episodes in Europe. Additional policies and measures such as the *Large Combustion Plants Directive, the Integrated Pollution Prevention and Control Directive*, the standards regulating emissions from transport, the *National Emission Ceilings Directive*, and the reductions agreed under *LRTAP Convention* also bring significant health and environmental benefits and create a level playing field across the EU.

Across the EEA-32 region, reported SO_2 concentrations have declined by approximately 50% during 2001 – 2010, very similar to the US. Also, less than 1% of the population lives in areas exceeding the ambient air quality standard. This decrease occurred despite the increased rate of economic activity that occurred during this period.

Comparison between the US-EPA and the EU-EEA programs suggest that it takes approximately 10 to 20 years for regions adopting these types of regulations to delink sulfur emissions from energy production.

FIGURE 9-10. Fourth Highest 24-Hour Mean SO$_2$ Concentration Observed at Urban Stations, EEA Member Countries, 1997-2010 [μg/m³].

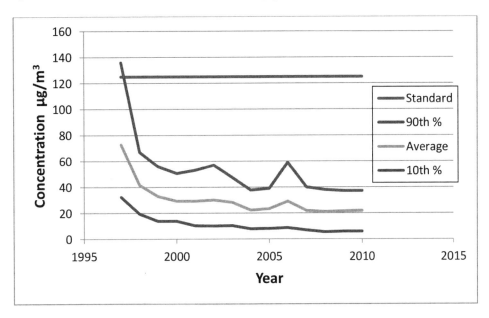

● **Case Study 9-2 What is the EU?**

The EU or European Union is an economic and political partnership between 28 (as of 2013) democratic countries or states, see Figure 9-11. The European Union is more than just a confederation of countries, but it is not a federal state. It is a new type of structure that does not fall into any traditional legal category. Its political system is historically unique and has been evolving over more than 60 years. It began in 1952 as an agreement between six western European countries to create centralized control over the production of coal and steel. It has grown and expanded to include most European countries, and it currently represents over 500 million people. It aims to provide peace, prosperity and freedom to its citizens in a fairer, safer world.

The governing institutions of the EU operate solely within the framework given to it through member state treaties and according to the principle of subsidiarity *which*

dictates that action by the EU should only occur where an objective cannot be sufficiently achieved by the member states alone.

The **European Commission** *acts as the EU's* **executive branch** *and is responsible for initiating legislation and the day-to-day running of the EU. It is intended to act solely in the interest of the EU as a whole, as opposed to the European Council, which consists of leaders of member states who reflect national interests.*

The **European Council** *provides the political leadership for the EU. Its makeup consists of one member from each member state. The European Council uses its leadership role to sort out disputes between member states and institutions, and to resolve political crises and disagreements over controversial issues and policies.*

The EU **judicial branch** *consists of two parts – The Council of Ministers and the European Parliament. The* **Council of Ministers** *is composed of a government minister from each member state and meets in different compositions depending on the policy area being addressed. It is not the same group as the European Council. In addition to its legislative functions, the Council of Ministers also exercises executive functions in relations to the common foreign and security policy. The* **European Parliament** *forms the other part of the EU's legislature. The 750 Members of the European Parliament (MEPs) are directly elected by EU citizens every five years. Although MEPs are elected on a national basis, they sit according to political groups rather than their nationality. Each country has a set number of seats, and in some cases are divided into sub-national constituencies. The Parliament and the Council of Ministers pass legislation jointly in nearly all areas under the ordinary legislative procedure, including the EU budget. The Commission is accountable to Parliament, requiring its approval to take office, having to report back to it and subject to motions of censure from it. The President of the European Parliament carries out the role of speaker in parliament and represents it externally. The president and vice presidents are elected by MEPs every two and a half years.*

The EU judicial branch, formally called the Court of Justice of the European Union, consists of three courts: the Court of Justice, the General Court, and the European Union Civil Service Tribunal. Together they interpret and apply the treaties and the law of the EU.

The EU air pollution laws are described by Directive 2008/50/EC of the European Parliament and the Council of 21 May 2008 on Ambient Air Quality and Cleaner Air for Europe (EU-EEA, 2008). This Directive lays down measures aimed at the following:

FIGURE 9-11. Map of Europe Showing the EU-28 Member States.

1. *Defining and establishing objectives for ambient air quality designed to avoid, prevent or reduce harmful effects on human health and the environment as a whole;*
2. *Assessing the ambient air quality in Member States based on common methods and criteria;*
3. *Obtaining information on ambient air quality in order to help combat air pollution and nuisance and to monitor long-term trends and improvements resulting from national and community measures;*
4. *Ensuring that such information on ambient air quality is made available to the public;*
5. *Maintaining air quality where it is good and improving it in other cases;*
6. *Promoting increased cooperation between the Member States in reducing air pollution.*

9.5 Control Technologies

Table 9-1 and Table 9-2 list the various natural and anthropocentric sources of sulfur emissions into the atmosphere. The largest fraction (85%) of controllable sources is due to the combustion of fossil fuels for electricity, heat, and transportation. Hence, the majority of control strategies focus on this category. Typically, SO_2 emissions are directly proportional to the amount of sulfur in the fuel. Growth in SO_2 emissions was proportional to increases in population and GDP until the 1990's in the US and Japan, and the 2000's in the EU. The correlation ended 10 to 20 years after adopting significant emission control regulations.

There are three classifications of control technologies for sulfur emissions from fuel:
1) fuel replacement - which involves using fuels with lower sulfur content,
2) fuel desulfurization - which involves treating a fuel before combustion to reduce the sulfur content, and
3) flue gas desulfurization – which involves removing sulfur from the gas emissions created from burning the fuel.

The first two are used when it is difficult or impractical to control the combustion emissions – such as in the generation of household heat or in automobiles. The third strategy is used for large fixed sources, such as electrical power generation.

9.5.1 Fuel Replacement

Fuel replacement occurs when emission levels are critical, such as when an area is in non-attainment (See section 3.2.1) or when it is economical considering cost, space, and equipment constraints. Switching to a fuel that has less sulfur (per energy content, not necessarily per unit mass) reduces the emissions of SO_2. Table 9-8 lists several fuels, their energy and sulfur content, and their cost in industry-specific units.

Note that most facilities that use a specific fossil fuel cannot easily switch to another fuel. Some level of retrofitting, replacement, and experimentation must occur before one fuel can replace another on a given unit (boiler, furnace, or heater)

■ *Example 9-5.*

Estimate the reduction in sulfur emissions by switching from a Northern Appalachia coal to a San Isabel natural gas for a 500 MW electric/ steam power plant that is 35% efficient in converting the chemical energy of the fuel into electricity. Estimate the difference in fuel cost associated with this change based on the values given in Table 9-8.

Table 9-8. Energy and Sulfur Content and Cost of Several Fossil Fuels.

	Source	Energy Content	Sulfur Content	Cost
Coal		BTU/lb	lbSO$_2$/MMBTU	$/short ton
	Central Appalachia	12,500	1.2	$ 55.92
	Northern Appalachia	13,000	3	$ 66.55
	Illinois Basin	11,800	5	$ 45.70
	Powder River Basin	8,800	0.8	$ 12.10
	Uinta Basin	11,700	0.8	$ 38.20
Oil		MMBTU/bbl	wt % S	$/bbl
	WTI, FOB Cushing OK	5.83	0.24	$ 56.91
	Brent, FOB	5.78	0.37	$ 58.87
	Dubai Fateh **		2.00	$ 76.73
	Isthmus, Mexico **		1.45	$ 71.43
	Loyd Blend, Canada		2.20	$ 57.27
	Tia Juana Light, Venezuela**		1.20	$ 71.99
	Kuwait, Kuwait **		2.50	$ 77.21
Natural Gas		BTU/scf	Mole % S	$/Mscf
	North Dakota Natural Gas		1.74	
	San Isabel, Colorado		0.3	
	Northwest Kansas		0.005	
	Great Falls MT		3.93	
	Average US	980		$ 3.21

Cost Date: December, 2014 unless otherwise noted.
** Cost Date: November, 2014.
MMBTU = 10^6 BTU
Short ton = 2,000 US pounds
Bbl = barrel = 42 US gallons
Scf = standard conditions, cubic foot
Mscf = 10^3 scf

Solution:

First, determine how much energy is needed per day to provide 500 MW generating capacity:

$$\frac{500\,MW}{0.35\,efficiency} * \frac{3,413,000\,BTU}{MW\,hr} * \frac{24\,hr}{day} = 117 \times 10^9 \frac{BTU}{day}$$

Next, calculate the amount of coal and natural gas needed to give this amount of energy.

Coal needed:
$$117 \times 10^9 \frac{BTU}{day} * \frac{lb\,coal}{13,000\,BTU} * \frac{ton\,coal}{2000\,lb\,coal} = 4,500 \frac{ton\,coal}{day}$$

Natural gas needed: $117 \times 10^9 \dfrac{BTU}{day} * \dfrac{scf}{930 BTU} = 125.8 \times 10^6 \dfrac{scf}{day}$

Now determine the amount of sulfur in the coal and gas.

Sulfur in the coal (3 wt %):

$$4,500 \dfrac{ton\,coal}{day} * \dfrac{0.03\,ton\,S}{ton\,coal} = 135 \dfrac{ton\,S}{day}$$

Sulfur in the natural gas (0.3 mole %):

$$117 \times 10^6 \dfrac{scf}{day} * \dfrac{1\,lbmol\,NG}{359\,scf} * \dfrac{0.003\,lbmol\,S}{lbmol\,NG} * \dfrac{32\,lb\,S}{lbmol\,S} * \dfrac{ton\,S}{2000\,lbS} = 15.6 \dfrac{ton\,S}{day}$$

Next, find the difference in sulfur emission and convert to SO_2

$$\left[135 \dfrac{ton\,S}{day} - 15.6 \dfrac{ton\,S}{day} \right] * \dfrac{64\,ton\,SO_2}{32\,ton\,S} = 238.8 \dfrac{ton\,SO_2}{day}$$

Finally, calculate the total daily cost of each fuel:

Coal cost: $4500 \dfrac{ton\,coal}{day} * \dfrac{55\$}{ton\,coal} = 247,500 \dfrac{\$}{day}$

Natural gas cost: $125.8 \dfrac{MMscf}{day} * \dfrac{3.64\$}{Mscf} * \dfrac{1000\,Mscf}{MMscf} = 460,000 \dfrac{\$}{day}$

The switch from coal to natural gas would add about 212,500 $/day for the fuel cost. From Figure 9-5 we see that the cost of a 1-ton SO_2 allowance is, on average, about $0.2/ton so the cost of allowances for the difference in sulfur emissions would reduce the cost for sulfur emissions by only $25 /day due to the fuel switch.

9.5.2 Fuel Desulfurization

Fuel desulfurization reduces the sulfur content of the fuel before combustion. The method used depends on the fuel source – natural gas, crude oil, or coal.

Natural Gas is mostly methane (CH_4) but when extracted from the ground it also contains other hydrocarbons (ethane, propane, butane, and pentanes), water vapor (H_2O), hydrogen sulfide (H_2S), carbon dioxide (CO_2), helium (He), nitrogen (N_2), and other trace gasses. If the natural gas contains less than 5.7 mg per normal m3 (0.25 grains per 100 scf) of H_2S, it is referred to as sweet gas. If it has more than this amount, it is called sour gas. A high concentration of H_2S gives the gas a rotten egg smell, and it can be harmful or even lethal to breath. Sour gas is also corrosive. Sweetening is the process for removing H_2S from natural gas. It is accomplished in a scrubbing tower by pumping the sour gas into the bottom of a closed tower that has an amine solution (containing water and monoethanolamine or diethanolamine) sprayed through the top. The sweetened gas is removed from the top of the tower, and the liquid solution that now contains the hydrogen sulfide is collected from the bottom. The amine solution absorbs the H_2S from the natural gas but does not absorb the natural gas. It is capable of removing 95% to 99.9% of the H_2S in the gas. This sulfur can be recovered using the Claus process (see Case Study 9-3) to form elemental sulfur (a bright yellow powder), or convert to sulfuric acid. Both of these products can be sold to reduce the overall cost of the sweetening process.

● Case Study 9-3. The Claus Process

The Claus Process is a multi-step reaction that recovers sulfur (S) from the gaseous hydrogen sulfide (H_2S) found in raw natural gas, from the by-product gases derived from refining crude oil, and from other industrial processes. Carl Friedrich Claus, a chemist working in England, invented the process and was issued a British patent in 1883. A German company (I.G.Farbenindustrie A.G.) later significantly modified the process. The Claus process is the industry standard for desulfurization (Goar, 1986).

The process is typically used on gas streams with H_2S concentrations over 25%, although lower concentrations can be converted in more specialized equipment. The overall chemical reaction is:

$$2 H_2S + O2 \rightarrow S_2 + 2 H_2O$$

The Claus technology consists of two process steps (see Figure 9-12), thermal and catalytic (Luinstra, 1999). The thermal step reacts the hydrogen sulfide containing gas in a substoichiometric combustion with oxygen at temperatures above 850°C. The thermal step converts approximately two-thirds of the hydrogen sulfide into elemental sulfur. Next, the catalytic step uses activated aluminum (III) or titanium (IV) oxide catalysts to react the remaining hydrogen sulfide with SO_2 formed during the thermal step to produce elemental sulfur. The sulfur can also be in the polymorphic molecular forms S_6, S_7, S_8 or S_9. The process yields over 97% of the sulfur from the input stream. In addition, because the reactions are exothermic, steam can be generated from the excess heat.

FIGURE 9-12. Process Flow Diagram for the Claus Process.

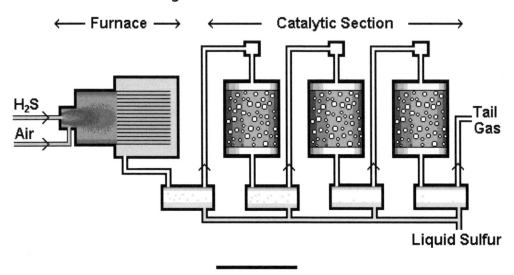

Example 9-6.

Calculate the annual amount of elemental sulfur produced at a natural gas sweetener facility that processes 5.0×10^6 scf gas/day. The gas has an incoming sulfur content of 1.74 mol% and 99% of this is removed in the sweetener process.

Solution:

Determine the daily mass of sulfur entering the facility in the gas from the given concentration and volume flow rate. Sulfur in the natural gas:

$$5.0 \times 10^6 \frac{scf}{day} * \frac{1\,lbmol\,NG}{359\,scf} * \frac{0.0174\,lbmol\,S}{lbmol\,NG} * \frac{32\,lb}{lbmol\,S} * \frac{ton}{2000\,lb} = 3.88 \frac{ton\,S}{day}$$

Next, factor in the unit efficiency and convert units to year basis. Yearly Sulfur Production:

$$3.88 \frac{ton\,S}{day} * \frac{0.99\,ton\,S\,Produced}{ton\,S\,in\,feed} * \frac{365\,day}{year} = 1{,}402 \frac{ton\,S}{year}$$

Crude oil also contains sulfur. It may be chemically bound to the various oil molecules in addition to having H$_2$S dissolved in it. Crude oil is sour when it contains >2.5 wt% of sulfur and sweet when it contains <0.5 wt% of sulfur. *Desulfurization*, sometimes also referred to as sweetening, is the process of removing sulfur from oil. However, it is not quite the same process as the one used to remove sulfur from natural gas. Crude oil desulfurization uses a technique called hydrotreating. The sulfur removal involves a chemical reaction between the sulfur-containing oil molecule and hydrogen at high temperature and pressure. The sulfur bond is broken, and two hydrogen atoms replace one sulfur atom. The removed sulfur reacts with excess hydrogen to form hydrogen sulfide (H$_2$S). The reaction occurs at lower temperatures and pressures when in the presence of a catalyst. The catalytic reaction typically takes place in a fixed bed reactor called a hydrotreater. After this reaction, the H$_2$S is collected in much the same way as from natural gas – an amine-based scrubbing tower. In addition, just like the natural gas case, the sulfur can be recovered using the Claus process to form elemental sulfur or sulfuric acid and sold. It is estimated that 46 of the 64 million tons of sulfur produced and sold in the US in 2005 originated from oil and gas refineries (USGS, 2009).

■ Example 9-7.

Calculate the amount of hydrogen needed to reduce the sulfur content of 1000 bbl oil (SG=0.8) from 2.83 wt% to 0.25 wt% S. Note that this process transforms the sulfur into gaseous hydrogen sulfide.

Solution:

First, calculate the moles of S removed:

$$1000\,bbl * \frac{42\,gal}{bbl} * \frac{ft^3}{7.48\,gal} * \frac{(0.8*62.4)\,lb\,oil}{ft^3} * \frac{(0.0283-0.0025)\,lb\,S}{lb\,oil} * \frac{lbmol\,S}{32\,lb\,S} = 226\,lbmol\,S$$

Next, determine the mass of hydrogen needed for the reaction to remove a sulfur atom from oil. Note that this problem assumes that two bonds are broken for each sulfur atom removed (i.e. that it is S^{+2} and not any other oxidation state) and each bond has a single hydrogen atom added. Also, two more hydrogen atoms are needed to convert the removed sulfur to H$_2$S, so four atoms of hydrogen or two H$_2$ molecules react for each sulfur atom removed, assuming 100% efficiency. Actual practice would require the addition of excess hydrogen.

$$226\,lbmol\,S * \frac{2\,lbmol\,H_2}{1\,lbmol\,S} * \frac{2\,lb\,H_2}{1\,lbmol\,H_2} = 904\,lb\,H_2$$

Removal of sulfur from coal is much more complex than for gas or oil fossil fuels. Some of the sulfur exists as a separate physical phase, such as inclusions of pyrite, hydrogen sulfide adsorbed onto the coal molecules, and some as part of the chemical structure of the coal molecules. Physical removal methods for sulfur minerals require crushing and grinding the coals to a fine size, and then separating by exploiting the differences in chemical properties and densities of the particles. The grinding is not really an extra step since fine-size coal particles are necessary for introducing the material into a coal burner. Chemical removal of organic sulfur is generally not done due to the complexities of the coal molecules, although it has been done successfully at pilot plant scales in the US (Smith, et al., 1995), (US-DOE, 2011). Sulfur removal is one of the necessary steps to implement clean coal technologies, but it is so far only at the research stage of implementation. Several novel biological treatments have also been explored. These technologies have not progressed to the point of being economically viable for sulfur reduction from coal.

9.5.3 Flue Gas Desulfurization

Flue gas desulfurization (FGD) uses a physical and/or chemical process to remove low concentrations (less than 2%) of sulfur oxides from the exhaust stream. FGD may be combined with other devices used to control PM and NO_x (see Chapters 8 and 10). There are two major classification types for FGD: 1) Throw-away or Regenerative and 2) Wet or Dry. The first refers to the ultimate fate of the treatment chemicals–disposed of into a landfill or chemically treated and recycled into the FGD. The second classification refers to the main phase in which the chemical reaction occurs – liquid water or the flue gas.

Methods for removing sulfur dioxide from boiler and furnace exhaust gases have been studied for over 150 years. Early concepts useful for flue gas desulfurization appear to have germinated in England circa 1850. The first major FGD installation on a power plant was put into operation in the United Kingdom at the Battersea Station, owned by the London Power Company, in 1931. Additional units in the UK include the Swansea Power Station in 1935 and the Fulham Power Station in 1938. The UK abandoned these FGD installations during World War II. Large-scale FGD units did not reappear in commercial operation at utilities until the 1970s, where most of the activity occurred in the United States and Japan (Biondo, et al., 1977).

In 2000, there were 678 FGD units operating on a total power plant capacity of about 229 GW (gigawatts) in 27 countries. About 45% of the units are in the US, 24% in Germany, 11% in Japan, and 20% in various other countries. The most commonly used methods worldwide are the wet throw-away process using lime or limestone (70%), dry throw away using lime (10%), and wet throw away using other alkaline sources (fly ash, sea water). Regenerative processes account for less than 10% of FGD capacity. The FGD methods used in the US include approximately 85% wet scrubbers, 12% spray dry systems and 3% dry injection systems.

Table 9-9. Flue Gas Desulfurization (FGD) Methods for Coal Combustion.

Process	Chemical Additive	Byproduct	Examples
Throw away			
Wet	Lime	$CaSO_3$ / $CaSO_4$	
	Limestone	$CaSO_3$ / $CaSO_4$	
	Dual Alkali	$CaSO_3$ / $CaSO_4$	
	Lime / Fly Ash	$CaSO_3$ / $CaSO_4$	Forced Oxidation
	Limestone / Fly Ash	$CaSO_3$ / $CaSO_4$	Mitsubishi
	Sodium Carbonate	Na_2SO_4	
	Sea Water	Na_2SO_4 / $CaSO_4$	
Dry	Lime	$CaSO_3$ / $CaSO_4$	
	Trona		
	Nahcolite		
Regenerative			
Wet	Lime	$CaSO_4$	
	Limestone	$CaSO_4$	
	Magnesium Oxide	Sulfuric Acid	
	Sodium Carbonate	Sulfuric Acid	Wellman Lord
Dry	Activated Carbon		

Dual Alkali uses a Na_2CO_3 solution and regeneration with lime or limestone

Approximately 95% of the sulfur contained in the coal oxidizes during combustion and is emitted as SO_2 and SO_3 in the flue gas. The remaining 5% of the sulfur reacts with alkalis already present in the fuel and exit the burning chamber as a solid sulfate with the ash. For a typical coal-fired power station, FGD removes 90% or more of the SO_2 in the flue gases. The Mitsubishi Flue Gas Treatment system has reported removal efficiencies greater that 99% (MHI, 2010).

Most FGD systems employ two stages: one for fly ash removal and the other for SO_2 removal. In *wet scrubbing systems* the flue gas first passes through a fly ash removal device, either an electrostatic precipitator or wet scrubber, and then into the SO_2 absorber. In *dry injection or spray drying operations*, the SO_2 is first reacted with the sorbent and then the flue gas passes through a particle control device. It is possible to combine both steps in a single unit.

SO_2 is an acid gas and thus the typical sorbent slurries used to remove the SO_2 from the flue gases are alkaline. Wet scrubbing reacts the SO_2 with a limestone ($CaCO_3$) slurry and produces calcium sulfite ($CaSO_3$). The simplified reaction is:

9.5 $\quad CaCO_3 \text{ (solid)} + SO_2 \text{ (gas)} \rightarrow CaSO_3 \text{ (solid)} + CO_2 \text{ (gas)}$

When wet scrubbing with lime ($Ca(OH)_2$) slurry, the reaction also produces calcium sulfite ($CaSO_3$):

9.6 $\quad Ca(OH)_2 \text{ (solid)} + SO_2 \text{ (gas)} \rightarrow CaSO_3 \text{ (solid)} + H_2O \text{ (liquid)}$

When wet scrubbing with a magnesium hydroxide ($Mg(OH)_2$) slurry, the reaction produces magnesium sulfite ($MgSO_3$):

9.7 $\quad Mg(OH)_2 \text{ (solid)} + SO_2 \text{ (gas)} \rightarrow MgSO_3 \text{ (solid)} + H_2O \text{ (liquid)}$

Some designs partially offset the cost of the FGD installation by further oxidizing the calcium sulfite ($CaSO_3$) to produce a marketable gypsum ($CaSO_4 \cdot 2H_2O$). This technique is known as forced oxidation:

9.8 $\quad CaSO_3 \text{ (solid)} + 2H_2O \text{ (liquid)} + \frac{1}{2}O_2 \text{ (gas)} \rightarrow CaSO_4 \cdot 2H_2O \text{ (solid)}$

A natural source of alkaline that absorbs SO_2 is seawater. A scrubber contacts the SO_2 with seawater and excess oxygen to form sulfate ions SO_4^- and free H^+. The surplus of H^+ reacts with the naturally occurring carbonates in the seawater, pushing the carbonate equilibrium to release CO_2 gas:

9.9 $\quad SO_2 \text{ (gas)} + H_2O + \frac{1}{2}O_2 \text{ (gas)} \rightarrow SO_4^{2-} \text{ (solid)} + 2H^+$

9.10 $\quad HCO_3^- + H^+ \rightarrow H_2O + CO_2 \text{ (gas)}$

Wet FGD systems use a number of scrubber designs to promote maximum gas-liquid surface area and residence time. Systems include spray towers, venturis, plate towers, and mobile packed beds. The current design trend is to use simple scrubbers such as spray towers (see Figure 9-13), instead of more complicated ones because of scale buildup, plugging, or erosion, which reduce FGD dependability and absorber efficiency. The configuration of the tower may be vertical or horizontal, and flue gas can flow co-currently, counter-currently, or cross-currently with respect to the liquid. The chief drawback of spray towers is that they require a higher liquid-to-gas ratio requirement for equivalent SO_2 removal than other absorber designs.

As explained above, alkaline sorbents are used for scrubbing flue gases to remove SO_2. Depending on the application, the two most frequently used are lime (CaO) and sodium hydroxide (NaOH, also known as caustic soda). Lime is obtained from limestone ($CaCO_3$) by heating it in a kiln, which causes a reaction that yields lime and CO_2. Large coal or oil fired boilers usually use lime as it is less expensive than caustic soda. The main problem with lime is it has a limited solubility in water

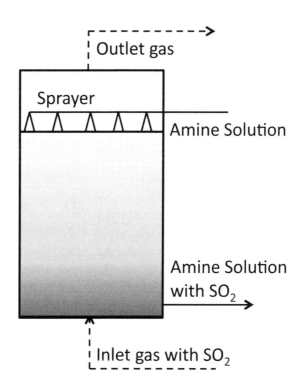

FIGURE 9-13. Wet Spray Chamber Schematic.

and forms a slurry (a mixture of solids and liquid), created by slaking[1] it in water. This mixture (slaked lime [$Ca(OH)_2$] and water) circulates through the scrubber. The solid limestone particles will stick to the equipment and pipe surfaces, requiring increased maintenance to clear the surfaces.

One other alternative to removing sulfur from the flue gases after burning is to remove the sulfur from the fuel during combustion. *Fluidized bed combustion* adds dry lime to the fuel during combustion. The lime reacts with the SO_2 to form solid sulfates that become part of the fly or bottom ash. It increases the contact time between the sulfur and lime, which increases lime utility, but reduces energy efficiency.

■ *Example 9-8.*
Determine the amount of limestone needed to achieve NSPS for an 800 MW coal burning power plant constructed in 1986. The plant uses a sub-bituminous coal with heat content of 10,150 BTU/lb, an ash content of 4.6 wt%, and sulfur content of 3.85 wt%. Assume the plant is 33% efficient in converting the energy of the coal into power. Further, assume 85% utilization of the limestone.

---

1. Slaking is the process of adding water to a substance. In this case, the water does not completely dissolve the limestone – it forms a mixture of limestone particles and limestone saturated water. The process is exothermic, and generates considerable heat.

CHAPTER 9

Solution:
First – determine the amount of sulfur removal required, which depends on the sulfur content of the fuel in lb S/MMBTU:

$$\frac{1 \text{ lb coal}}{10,150 \text{ BTU}} \frac{0.0385 \text{ lb S}}{\text{lb coal}} \frac{64 \text{ lb SO}_2}{32 \text{ lb S}} \frac{10^6 \text{ BTU}}{\text{MMBTU}} = 7.6 \frac{\text{lb SO}_2}{\text{MMBTU}}$$

Look up this value in the NSPS (or table 3.6) to find that the sulfur removal efficiency required is 90%.

Next, determine the mass of sulfur removal per ton of coal:

$$\frac{0.0385 \text{ lb S}}{\text{lb coal}} (0.9 \text{ removal reqd}) \frac{2000 \text{ lb coal}}{\text{ton coal}} = 69.3 \frac{\text{lb S}}{\text{ton coal}}$$

The amount of limestone needed using the molecular weights:

$$69.3 \frac{\text{lb S}}{\text{ton coal}} \frac{100 \text{ lb CaCO}_3}{32 \text{ lb S}} = 216.6 \frac{\text{lb CaCO}_3}{\text{ton coal}}$$

Now find the amount of coal used per day in tons of coal:

$$\frac{800 \text{ MW}}{0.33 \text{ efficiency}} \frac{1000 \text{ kW}}{\text{MW}} \frac{3413 \text{ BTU}}{\text{kW hr}} \frac{24 \text{ hr}}{\text{day}} \frac{1 \text{ lb coal}}{10,150 \text{ BTU}} \frac{\text{ton}}{2000 \text{ lb}} = 9782 \frac{\text{ton coal}}{\text{day}}$$

Finally, determine the total amount of limestone needed per day, note that utilization refers to the fact that some limestone fails to react:

$$216.6 \frac{\text{lb CaCO}_3}{\text{ton coal}} 9782 \frac{\text{ton coal}}{\text{day}} \frac{1}{0.85 \text{ utilization}} \frac{\text{ton}}{2000 \text{ lb}} = 1,246 \frac{\text{ton CaCO}_3}{\text{day}}$$

of which,

$$216.6 \frac{\text{lb CaCO}_3}{\text{ton coal}} 9782 \frac{\text{ton coal}}{\text{day}} \frac{\text{ton}}{2000 \text{ lb}} = 1059 \frac{\text{ton CaCO}_3}{\text{day}}$$

is used in the reaction.

■ **Example 9-9.**
Determine the mass of sludge produced from the solid waste (gypsum, unused limestone, and ash) in example 8, assuming the sludge is 65% water.

Solution:
Determine the amount of each product.

Ash:
$$9782 \frac{\text{ton coal}}{\text{day}} \frac{0.046 \text{ ton ash}}{\text{ton coal}} = 450 \frac{\text{ton ash}}{\text{day}}$$

Unused $CaCO_3$:
$$1246 \frac{\text{ton } CaCO_3}{\text{day}} - 1059 \frac{\text{ton } CaCO_3}{\text{day}} = 187 \frac{\text{ton } CaCO_3}{\text{day}}$$

Gypsum produced:
$$1059 \frac{\text{ton } CaCO_3}{\text{day}} \frac{172 \text{ ton } CaSO_4 \cdot 2H_2O}{100 \text{ ton } CaCO_3} = 1{,}821 \frac{\text{ton } CaSO_4 \cdot 2H_2O}{\text{day}}$$

Total mass from the sum of these three waste products (450+187+1821=2458 tons of solid waste) and water:
$$(450+187+1821) \frac{\text{ton waste}}{\text{day}} \frac{1 \text{ ton sludge}}{0.35 \text{ ton waste}} = 7023 \frac{\text{ton sludge}}{\text{day}}$$

A typical rail car for transportation of coal and limestone holds 50 tons. Based on these answers, this power plant requires a train with 175 coal cars and 25 limestone cars each day and fills 140 cars with waste sludge each day.

Another important design consideration associated with FGD systems is *corrosion*. The flue gas exiting the absorber is saturated with water and still contains some SO_2. (No system is 100% efficient.) Therefore, the gases are highly corrosive to any downstream equipment - e.g., fans, ducts, and stacks. Two methods that minimize corrosion are: (1) reheating the gases above their dew point to prevent the condensation of water and sulfuric acid and (2) choosing construction materials and design conditions that allow equipment to withstand the corrosive conditions. The selection of a reheating method or the decision not to reheat (thereby requiring the use of special construction materials) is a complex topic connected with FGD design. Both alternatives are expensive, and the choice depends on a site-by-site basis. In spray drying and dry injection systems, the flue gas temperature must be about 10-20°C above adiabatic saturation to avoid wet solids deposition and plugging of downstream equipment.

9.5.3.1 FGD Cost

Table 9-10 lists the capital, operating and maintenance (O and M), and annualized costs of SO_2 removal in 2001 US dollars (US-EPA, 2003). The annual cost combines the O and M cost with those associated with paying off the loan for the capital cost required for building the system.

Table 9-10. Cost Information for SO$_2$ Scrubbing Systems.

Scrubber Type	Unit Size	Capital Cost	O & M Cost	Annual Cost	Cost per ton SO$_2$ Removed
	MW	$/kW	$/kW	$/kW	$/ton
Wet	>400	100 to 250	2 to 8	20 to 50	200 to 500
	<400	250 to 1500	8 to 20	50 to 200	500 to 5,000
Dry	>200	40 to 150	4 to 10	20 to 50	150 to 300
	<200	150 to 1,500	10 to 300	50 to 500	500 to 4,000

● **Case Study 9-4. Sulfur Emission Control through Fuel Choice at a Paper Mill.**

A paper mill located on the Mississippi River produces 850 tons per day of lightweight coated and super-calendared papers for companies across the world. The mill has an energy requirement (heat, steam, and electricity) of 100 MW, and multiple fuels can be used to achieve this. The primary fuel choices include biomass (scrap wood and bark), coal, and tire-derived fuel (TDF). Table 9-11 describes each fuel in terms of historic use, water and sulfur content, and heating value.

Table 9-12 provides the potential emissions, actual emissions (2008), and Clean Air Act permit emission limits for this mill. The potential emissions were determined by assuming the mill operates at 100% rated capacity in terms of the boiler and that there is no sulfur emission reduction equipment. Actual emissions correspond to mill records in 2008, when a 25/75 coal (bituminous/sub-bituminous) blend was used, and it does not include flue gas desulfurization in the mill's wet scrubber (venturi type).

During 2008, plant engineers and managers explored fuel options as relates to their sulfur emissions. The facility is limited in the amount of TDF they can use due to market considerations. They also currently burn all their scrap biomass. Together these two fuel sources supply half of the facility need. Coal supplies the remaining energy. The mill's permit allows them to burn both bituminous [B] and sub-bituminous [SB] coal. Table 9-11 shows the differences in energy and sulfur content for each fuel and several blends. They wanted to determine if a different blend of coals would reduce costs. This study simplifies the cost to only the fuel and the cost to remove sulfur, assumed to be $350/ton sulfur. Additional assumptions include the complete conversion of sulfur to SO$_2$ and that all SO$_2$ produced is sent to the scrubber.

Table 9-11. Characteristics of Fuels for Paper Mill in Case Study 9-4.

Fuel Type	2008 Blend %	Moisture content (%)	Heating Value BTU/lb	Sulfur Content Wet Weight %
Tired Derived Fuel (TDF)	8%	1.10%	13800	1.5
Biomass	32%	46.00%	4449	0.0324
Coal	60%	-	-	-
100/0 B/SB Coal Mix	-	3.35%	14400	4.8
75/25 B/SB Coal Mix		8.30%	13200	3.7
50/50 B/SB Coal Mix	-	13.30%	11900	2.6
25/75 B/SB Coal Mix	-	18.20%	10600	1.5
0/100 B/SB Coal Mix	-	23.20%	9300	0.4

Table 9-12. Emission Limits for Paper Mill in Case Study 9-4.

Emission limit	1.2	lb SO_2/million BTU heat input or 90% of potential.
Potential Emissions	3,750	tons SO_2 per year
Actual Emissions	1,355	tons SO_2 per year (2008)

The cost of each fuel and the expected amount of sulfur produced from each fuel are described in Table 9-13. This table shows a difference of about $600,000/yr in fuel cost based on coal blend. It also shows the high production of sulfur dioxide from the bituminous coal. The current FGD can achieve a maximum of 90% sulfur removal.

Table 9-13. Fuel Costs and Sulfur Emissions for Paper Mill in Case Study 9-4.

Fuel Type	2008 Blend	Cost $/ton	Amount ton/yr	Cost $/yr	SO_2 Production ton/yr
TDF	8%	8	8,700	70,000	131
Biomass	32%	0	107,600	0	35
Coal	60%				
100/0 B/SB Coal Mix		14.45	62,300	901,000	2,990
75/25 B/SB Coal Mix		14.96	68,000	1,018,000	2,516
50/50 B/SB Coal Mix		15.48	75,400	1,167,000	1,960
25/75 B/SB Coal Mix		15.99	84,700	1,355,000	1,271
0/100 B/SB Coal Mix		16.5	96,500	1,593,000	386

Next, the amount of sulfur removal was determined, by comparing the actual emission to the requirement of 90% removal of the potential emissions. The mill is allowed to have emissions up to 375 tons/year. The total removal efficiencies (SO_2 removed/ total SO_2) - range from 88% for the 100/0 B/SB blend to 32% for the 0/100 B/SB blend, both of which are achievable with the current equipment. The last column in Table 9-14 provides the cost of the fuels (TDF + coal blend) and the cost to remove sulfur to the permitted level. This analysis shows that the somewhat more expensive fuel is less costly for mill operation.

Table 9-14. Sulfur Removal Requirements and Total Costs for Paper Mill in Case Study 9-4.

Fuel Type B/SB	Total SO_2 ton/yr	SO_2 removal ton/yr	SO_2 Cost $/yr	Total Cost $/yr
100/0 Coal Mix	3,156	2,781	974,000	1,875,000
75/25 Coal Mix	2,681	2,306	808,000	1,826,000
50/50 Coal Mix	2,126	1,751	613,000	1,780,000
25/75 Coal Mix	1,436	1,061	372,000	1,727,000
0/100 Coal Mix	551	176	62,000	1,655,000

This analysis did not account for other pollution control costs such as particles, nor did it include market dynamics in the price of the fuels. This type of calculation is repeated each time fuel is purchased. This mill contracts for fuel once per year and locks in the price at that time.

● **Case Study 9-5. Sulfur Emission Control in a Coal Fired Power Plant.**
In this case study we will design a sulfur control system for the coal fired power plant introduced in Case Study 2-1, and further developed in Case Study 3-3 and 8-1. Summary of given information: Fuel is a bituminous coal (10,750 BTU/lb) of elemental analysis $C_{70}H_{100}O_2N_{0.2}S_{0.3}Ash_{(7wt\%)}$. The fuel is burned with 50% excess air in a cyclone furnace of a 350 MW power plant with a 34.5% efficiency.

In case study 2-1 we assumed all the sulfur in the coal oxidized and exited with the exhaust gas. As this is the worst case, we will keep the assumption, although observation in real systems shows 5–10% of the sulfur will exit in the bottom ash solids. In Case Study 3-3 we determined the required removal efficiency to meet NSPS is a 90.8% reduction.

Next we need to choose a method for control of sulfur. The choices are fuel replacement, fuel desulfurization, or flue gas desulfurization. Fuel replacement and fuel desulfurization are not easy choices. Switching fuels generally requires replacement of the boiler system. Fuel desulfurization of coal is a difficult process generally requiring high temperatures, high pressures, and strong solvents since the sulfur atoms are contained within the coal molecular structure and not just physically mixed in the coal. This leaves flue gas desulfurization as the best choice. Desulfurization requires two additional choices: a wet or dry process, and a throw-away or regenerative process.

Wet or Dry? Both require the transport of chemical additives into the flue gas. The dry process requires moving powdered solids. The wet process requires moving the materials in a water based solution. It is also possible to inject the chemicals in a wet form, but have them dry out within the hot flue gas (spray dry system).

Throw-away or regenerative? The decision is based on the availability of chemical additives. Calcium in the form of lime or limestone is generally less expensive and more available than other choices using sodium or magnesium compounds. The answer depends on local circumstances.

In Example 9-8, we explored a wet limestone system. In this case study we will use a dry lime throw-away system. Lime (CaO) added directly to the combustion chamber will react with sulfur dioxide to form calcium sulfite.

CaO (solid) + SO_2 (gas) → $CaSO_3$ (solid)

This dry solid will be removed within the particle control system, rather than collecting it as a wet sludge from a spray chamber. This simplifies the solid waste collection, as long as collection and reuse or sale of calcium sulfite is not important. Note that the particle control system will need to be redesigned to handle the increased load. Finally, the lime will have a long time to react with the sulfur dioxide, so we can expect a utility of 90%.

Determine the amount of sulfur in the exhaust and the amount of removal needed:

$$350 MW * \frac{3.413 \times 10^6 BTU}{MW \cdot hr} * \frac{1_{in}}{0.345_{out}} * \frac{1\,lb\,coal}{11{,}750\,BTU_{in}} * \frac{454 g}{1 lb} * \frac{1 mol_{coal}}{1058.5\,g\,coal} * \frac{0.3 mol_S}{1 mol_{coal}} = 37{,}920 \frac{mol_S}{hr}$$

The amount of sulfur to be removed is 90.8 wt% to meet NSPS requirements, but we will plan to achieve 91%. Also, we need to account for the 90% utility of the lime (10% will pass through the process without reacting).

$$37,920\frac{mol_s}{hr}*\frac{64g_{SO2}}{1mol_s}*\frac{0.91g_{SO2}removed}{g_{SO2}produced}*\frac{ton}{10^6 g}=2.2\frac{ton_{SO2}}{hr}$$

$$2.2\frac{ton_{SO2}}{hr}*\frac{56ton_{CaO}}{64ton_{SO2}}*\frac{1mol_{CaO}aded}{0.9mol_{CaO}used}=2.15\frac{ton_{CaO}}{hr}$$

Next, determine the amount of calcium sulfite created:

$$2.15\frac{ton_{CaO}}{hr}*(0.9utility)*\frac{120ton_{CaSO3}}{56ton_{CaO}}=4.14\frac{ton_{CaSO3}}{hr}$$

The remaining, unused lime:

$$2.15\frac{ton_{CaO}}{hr}*\left[(1-0.9)utilized\right]=0.215\frac{ton_{CaO}}{hr}$$

So the additional solids load to the particulate control system is (4.14 + .215) = 4.4 ton/hr.

New amount of PM is [3.6 ton Fly Ash + 4.4 ton lime and calcium sulfite] = 8.0 ton PM.

The consumption of lime and production of calcium sulfite does not alter the gas emission production, so the size of the fabric filter area does not need to be increased. However, the rate at which the bags must be cleaned will increase. The addition of the lime and calcium sulfite will alter the particle size distribution (PSD) of the PM, and this new PSD should be determined. If the new distribution is much wider, an additional PM control, such as a set of cyclones, could be used to remove the larger particles and thereby narrow the distribution.

9.6 Questions

* - Questions and problems may require additional information not included in the textbook.

1. What is the difference between sweet and sour crude oil?

2. Visit the government environmental website (US or EU) and determine if any regions have ambient air that is currently not in compliance for SO_2.*

3. Find the current level of SO_2 in the air where you live.*

4. How is SO_2 measured in the ambient air? Is there more than one way to determine the concentration?*

5. List three industrial processes other than fossil fuel combustion, that release sulfur to the atmosphere.

6. What is the sulfur content of wood? Does it vary by tree species?*

7. Determine the largest source of SO_2 emissions near where you live.*

8. How do sulfur emissions lead to reductions in visibility?

9. Explain how an acid-tolerant aquatic species could still be adversely affected by acid rain.

10. Examine the ambient air quality standards for SO_2 between three different countries. How are they similar and how do they differ? Are the differences important?*

11. Some countries have different standards for different regions (India and China, for example). Why do they differentiate? Is this a better or worse practice, in your opinion, than having a single standard?

12. Why was there a spike in the auction price of an SO_2 emission allowance from 2004-2006?*

13. What is hydrotreating?

14. Why will reheating FGD exit gases above their dew point reduce corrosion in downstream equipment?

15. Make a list of three questions you have about this chapter or air pollution concerns you have.

9.7 Problems

1. Calculate the concentration of SO_2 in $\mu g/m^3$ for air with 0.016 ppm SO_2.

2. Calculate how much O_2 and H_2O are needed for the conversion of sulfur to sulfuric acid in the coal from Example 9-1.

3. Calculate the required SO_2 removal efficiency needed for a power plant burning a bituminous coal (2.7% S, 12,025 BTU/lb, 33 % efficient) to meet 2010 NSPS standards.

CHAPTER 9

4. Determine the amount of gypsum created from the capture and conversion of sulfur during copper production from copper sulfide (CuS) ore. The starting material is 1,000 tons of rock, which has a copper sulfide composition of 0.5 wt% of the ore. Do not consider the sulfur in the fuel needed to accomplish the conversion.

5. Determine the amount (kg) of SO_2 released from the combustion of 1,000 gallons of waste oil (135,000 BTU/gal) with 1.5 wt% S. The oil has a specific gravity of 0.86. How much would this need to be reduced to meet NSPS for a 125 MW electric utility steam generation source (34% efficiency)?

6. Determine the amount of SO_2 released per year from the combustion of lignite coal in a 500 MW electric utility steam generation unit. Assume the source emits SO_2 at 90% of the NSPS required level.

7. Find the pH of a solution composed of 3 moles of H_2SO_4 in 10^6 moles of water.

8. Find the volume (liters) of 18 M H_2SO_4 needed to yield a pH of 4.3 in 10,000 liters of pure water.

9. A new 600 MW power plant (0.35 efficiency) uses an Illinois Basin coal (see Table 9-8). Assume the FGD system removes enough sulfur so that the plant meets NSPS. Determine the volume of rainwater of pH = 5 resulting from one day of sulfur emissions. Assume all the sulfur that is emitted forms sulfuric acid, and that 65% of this acid is deposited as acid rain.

10. Calculate the cost and the change in SO_2 potential emissions (in %) due to switching from a Central Appalachia coal to a Uinta Basin coal, as described in Table 9-8.

11. Is there a cost advantage for a 500 MW electricity generating unit in switching from a Northern Appalachia coal to WTI oil? Use the costs described in Table 9-8. Assume the cost of sulfur removal is $250/ton and that the source must meet the NSPS emission limits for a new construction. Assume the facility is 31% efficient with either fuel. Oil density is 7.lb/gal.

12. Is there a cost advantage for a 500 MW electricity generating unit in switching from an Illinois Basin coal to a North Dakota gas? Use the costs described in Table 9-8. Assume the cost of sulfur removal from this coal is $350/ton and the cost to remove it from natural gas

is $20/ton. Also, source must meet the NSPS emission limits. Assume the facility is 34% efficient with either fuel.

13. An industrial boiler (33% efficient) constructed in 1995 uses Tia Juana Light oil to generate steam and is considering switching to a Northwest Kansas natural gas. What will be the difference in the required sulfur removal efficiency due to this change?

14. Estimate the yearly cost for FGD for a new 800 MW power plant (0.35 efficiency) that uses an 11,500 BTU/lb bituminous coal from Central Appalachia with 1.95 wt% S. Assume the cost of sulfur removal is $220/ton and that the source must meet the NSPS emission limits. How does this cost compare with the cost of the coal (see Table 9-8)?

15. Determine the amount (mol) of hydrogen needed to hydro-treat a 1000 bbl of crude oil with 1.8 wt % S to a sulfur content of 0.25 wt%. The oil has a specific gravity of 0.78.

16. How much sulfur can be recovered from sweetening 10^6 scf of natural gas (2.15 mole % S) by 98%?

17. Estimate the daily amount of limestone needed for FGD for a 700 MW power plant (0.33 efficiency) that uses a 12,500 BTU/lb bituminous coal with 1.05 wt% S. Assume an 85% utilization of limestone and that the plant meets NSPS.

18. Estimate the daily amount of lime needed for FGD for a 550 MW power plant (0.31 efficiency) that uses a 9,780 BTU/lb sub-bituminous coal with 2.35 wt% S. Assume a 90% utilization of lime and that the plant meets NSPS.

19. Estimate the daily amount of magnesium hydroxide needed for FGD for a 350 MW power plant (0.305 efficiency) that uses a 14,380 BTU/lb bituminous coal with 0.85 wt% S. Assume a 95% utilization of magnesium hydroxide and that the plant meets NSPS.

20. A 1000 MW power plant (0.335 efficiency) uses a Central Appalachian coal (6.5 wt% ash content). Estimate the daily amount of limestone (ton/day) needed for FGD. Assume an 85% utilization of the limestone and that the plant meets NSPS. Next, determine the daily amount of sludge (ton/day) produced (from ash, unused limestone, generated gypsum, and water). The sludge is 63 wt% water. Use Table 9-8 for additional information.

9.8 Group Project Ideas

For each project, the students should work in small groups and present their finding in either a short report (5-8 pages) or 15 minute presentation.

1. Are there any plans to modify the current SO_2 ambient air quality standards? Write a short paper summarizing the reasons for the revision. Note that the US EPA is required to examine the SO_2 NAAQS every five years.

2. The 2003 Office of Management and Budget study found that the Acid Rain Program accounted for the largest quantified human health benefits of any US regulatory program implemented in the previous 10 years, with annual benefits exceeding costs by more than 40:1. Determine how this value was obtained.

9.9 Bibliography

Biondo, S J and Marten, J C. 1977. A History of Flue Gas Desulfurization Systems Since 1850. *Journal of the Air Pollution Control Association*. 1977, Vol. 27, 10, pp. 948-961.

Brasseur, Guy, et al. 1999. Chapter 5. Trace Gas Exchanges and Biogeochemical Cycles. [book auth.] Guy P Brasseur, John J Orlando and Geoffrey S Tyndall. *Atmospheric Chemistry and Global Change*. New York, NY : Oxford University Press, 1999, pp. 195-201.

Canada, USA. 1991. AGREEMENT BETWEEN THE GOVERNMENT OF CANADA AND THE GOVERNMENT OF THE UNITED STATES OF AMERICA ON AIR QUALITY. [Online] August 26, 2002, March 31, 1991. [Cited: January 20, 2010.] http://www.ec.gc.ca/cleanair-airpur/caol/air/can_usa_e.html.

CCOHS. 1997. 2-Health Effects of Sulfur Dioxide. *Canadian Centre for Occupational Health and Safety*. [Online] December 29, 1997. [Cited: January 21, 2010.] http://www.ccohs.ca/oshanswers/chemicals/chem_profiles/sulfurdi/health_sul.html.

EU-EEA. 2008. Directive 2008/50/EC of the European Parliament and of the Council of 21 May 2008 on ambient air quality and cleaner air for Europe. *EUR-Lex Access to European Union Law*. [Online] 2008. [Cited: January 26, 2010.] http://eur-lex.europa.eu/LexUriServ/LexUriServ.do?uri=CELEX:32008L0050:EN:NOT.

European Environment Agency. 2012. AirBase - The European Air quality dataBase. *Data and Maps*. [Online] November 29, 2012. [Cited: June 13, 2013.] http://www.eea.europa.eu/data-and-maps.

Goar, B G. 1986. *Sulfur recovery technology*. Conference: American Institute of Chemical Engineers spring national meeting, New Orleans, LA, USA, 6 Apr 1986 : American Institute of Chemical Engineers,New York, NY, 1986.

Lippman, Morton and Ito, Kazuhiko. 2006. Chapter13. Sulfur Dioxide. [book auth.] World Health Organization - Europe. *Air Quality Guidelines - Global Update 2005*. s.l. : Druckpartner Moser, 2006, pp. 395-415.

Luinstra, Ed. 1999. Converting Hydrogen Sulfide. [Online] 1999. http://www.nelliott.demon.co.uk/company/claus.html.

MHI. 2010. Mitsubishi Heavy Industries, LTD. Flue Gas Treatment - Delivery Records. *Flue Gas Desulfurization Plant*. [Online] 2010. [Cited: February 11, 2010.] http://www.mhi.co.jp/en/products/pdf/delivery_record.pdf.

NADP. 2013. Annual Isopleth Maps. *National Atmospheric Deposition Program* (NRSP-3). [Online] 2013. [Cited: June 7, 2013.] http://nadp.sws.uiuc.edu/data/annualiso.aspx.

Rice, Karen C, Raffensperger, Jeff P and Webb, Rick. 1999. Hydrological and Geochemical Controls on Episodic Acidification of Streams in Shenandoah National Park, Virginia. *Virginia Water Research Symposium*. 1999.

Smith, G V, et al. 1995. *Desulfurization of Illinois Coals with Hydroperoxides of Vegetable Oils and Alkali*. s.l. : US-Department of Energy, 1995. DOE/PC/92521--T275.

US-CDC. 1984. Epidemiologic Notes and Reports Sulfur Dioxide Exposure in Portland Cement Plants. [ed.] Center for Disease Control and Prevention. *Morbidity and Mortality Weekly Report (MMWR)*. April 13, 1984, Vol. 33, 14, pp. 196-196.

US-DHHS. 2007. *Toxicological Profile for Sulfur Dioxide*. Atlanta, GA : US-GPO, 2007. Department of Health and Human Services.

US-DOE. 2011. Clean Coal and Natural Gas Power Systems. *Clean Coal Technology*. [Online] US Department of Energy, May 19, 2011. [Cited: July 29, 2011.] http://www.fossil.energy.gov/programs/powersystems/cleancoal/.

US-EPA. 2012. Acid Rain Program. *Clean Air Markets*. [Online] 2012. [Cited: June 12, 2013.] http://www.epa.gov/airmarkets/progsregs/arp/index.html.

—. 2003. *Air Pollution Control Technology Fact Sheet*. 2003. EPA-CICA Fact Sheet. http://www.epa.gov/ttn/catc/dir1/ffdg.pdf. EPA-452/F-03-034.

—. 2014. Annual Auction. *Clean Air Markets*. [Online] April 8, 2014. [Cited: November 4, 2014.] http://www.epa.gov/airmarkets/trading/auction.html.

—. 2007. Effects of Acid Rain - Automotive Coatings. *Acid Rain*. [Online] June 8, 2007. [Cited: January 26, 2010.] http://www.epa.gov/acidrain/effects/auto.html.

—. 2014. Generating Reports and Maps. *Air Data*. [Online] October 5, 2014. [Cited: November 8, 2014.] www.epa.gov/air/data/reports.html.

—. 2009b. Health Effects of Pollution. *Sulfur Dioxide (SO2)*. [Online] December 2, 2009b. [Cited: January 21, 2010.] http://www.epa.gov/Region7/programs/artd/air/quality/health.htm#so2.

—. 2013. *National Air Quality: Status and Trends*. Air Quality Assessment Division, Office of Air Quality Planning and Standards. s.l. : US-EPA, 2013. http://ampd.epa.gov/ampd/.

—. 2014. Sulfur Dioxide. *Air Trends*. [Online] October 8, 20014. [Cited: November 4, 2014.] http://www.epa.gov/air/airtrends/sulfur.html.

USGS. 2009. [Online] October 21, 2009. [Cited: February 2, 2010.] URL: http://minerals.usgs.gov/minerals/pubs/commodity/index.html.

US-OMB. 2003. *Informing Regulatory Decisions: 2003 Report to Congress on the Costs and Benefits of Federal Regulations and Unfunded Mandates on State, Local, and Tribal Entities.* Office of Information and Regulatory Affairs, Office of Management and Budget. s.l. : US-GPO, 2003. http://yosemite.epa.gov/SAB/sabcvpess.nsf/0/5143268e911789ba85256db900562c4b/$FILE/2003_costben_final_rpt.pdf.

Wigington, PJ, et al. 1992. Comparison of Episodic Acidification in Canada, Europe and the United States. *Environmental Pollution.* 1992, Vol. 78, 1-3, pp. 29-35.

Wilhour, Raymond G, et al. 1978. *Effects of Sulfur Dioxide on Cereal Grain Yields.* Corvallis Environmental Research laboratory, US Environmental Protection Agency. Corvallis, Oregon : US-EPA, 1978.

Ziminski, Andrew. 2009. Minerva Conservation. [Online] 2009. [Cited: August 21, 2009.] http://www.minervaconservation.com/photos/devizes-marketcross.jpg.

CHAPTER **10**

Nitrogen Emissions

The main forms of nitrogen emissions are oxides and ammonia compounds. The oxides are directly harmful to human health. NO_2 is a criteria pollutants (see Chapter 3). The oxides can chemically react to form other harmful compounds in the atmosphere, and they can participate in the formation of ozone, also a criteria pollutant (see Chapter 12). Significant progress has been made in most industrialized countries in reducing the nitrogen oxide forms, little has been achieved in reducing ammonia emissions.

The main anthropogenic sources are agriculture and combustion processes – irrespective of fuel. Natural sources include lightning, natural fires, and microbial activity. The main sink for these emissions is the chemical reaction (denitrification) to form diatomic nitrogen molecules (N_2). This form is quite stable; however, there are natural and anthropogenic processes that re-convert it into the other forms. These processes can occur in the atmosphere, in water, and in soils.

Emission reduction strategies include improved control over combustion processes and improved agricultural practices. The methods require sufficient control of the chemistry process to discourage the side reactions that generate the harmful forms – reducing temperature, controlling the amount of oxygen, and encouraging side reactions that destroy the harmful forms.

10.1 General Information

Nitrogen is the most abundant substance in the atmosphere. The main chemical form, N_2, is a very stable diatomic molecule. Other forms include:

- NO and NO_2: Nitrogen oxide and nitrogen dioxide, which are interrelated through chemical equilibria in the atmosphere. They are usually grouped together as nitrogen oxides, NO_x, and are referred to as active nitrogen or odd nitrogen.
- N_2O: Nitrous oxide is the dominant oxide of nitrogen in the stratosphere. It is considered to be a major greenhouse gas, after water and carbon dioxide.
- NO_3^-: The nitrate ion occurs in nitrate particles. It forms from the oxidation of NO_2.
- $NO_3\cdot$: The nitrate radical is very reactive due to its unpaired electron.
- N_2O_5: Dinitrogen pentoxide forms from the reaction of NO_3 and NO_2. The hydrolyzation reaction forms nitric acid and occurs on the surface of atmospheric aerosols.
- HNO_3: Nitric acid forms from the oxidation product of NO_x with OH radical or conversion from N_2O_5. It is important in the formation of nitrate particles. It can be deposited in wet or dry form from the atmosphere to the surface. Its deposition is the main removal mechanism of NO_x.

- HONO: Nitrous acid readily photolyzes in the atmosphere and can be an important source for the formation of OH radicals in areas with high concentrations of NO_x.
- NH_3: Ammonia readily dissolves into water aerosols, where it plays a role in acid-base equilibria.
- Organic nitrogen: Includes many important species of the form $R-NH_2$, where R is a complex organic group. Organic nitrogen may convert to ammonium and nitrates.

Other nitrogen species also exist and the forms all inter-react such that untangling their chemistry becomes a very complex process. The complexity explains why there is a very strong and active research area studying atmospheric nitrogen [(Asaf, et al., 2010), (Bishop, et al., 2010), (Russel, et al., 2010), (Skeen, et al., 2010), (Park, et al., 2009), (US-EPA, 2008), (Gibson, et al., 2006), (Grassian, 2002), and (US-EPA, 1993)]. The various forms of nitrogen chemically react with other atmospheric constituents through fixation, nitrification, denitrification, assimilation, and mineralization. The driving forces for these reactions include fires, biological growth and decay, industrial processes, the internal combustion engine, lightning, mineralization, and volcanic activity.

Nitrogen fixation is any process in which atmospheric nitrogen (N_2) reacts to form any other form of nitrogen. It may occur due to natural processes, such as the biological conversion of N_2 into NH_3, NH_4+ and NO_3- by microorganisms in the soil. It may also occur by anthropogenic activities, such as nitric acid (HNO_3) production. Biological fixation was the ultimate source of nitrogen for living organisms until industrial production of fertilizers began. Currently about ½ of the fixated nitrogen has an anthropogenic source.

Nitrification is the process in which fixed nitrogen oxidizes into nitrites (NO_2-) and nitrates (NO_3-). A simplified description of the reactions is:

10.1 $$2\,NH_4^+ + 3O_2 \rightarrow 2\,NO_2^- + 2H_2O + 4\,H^+$$

10.2 $$2\,NO_2^- + O_2 \rightarrow 2\,NO_3^-$$

Note that nitrification generates H+ ions, and, therefore, acidifies the soil or water where it occurs.

Denitrification is a process where nitrates (NO_3-) are reduced to any of the gaseous nitrogen species (N_2, N_2O, NO). This process can occur chemically or biologically, and it happens worldwide. Without this process, all nitrogen would reside in the ocean or sediments as nitrates, rather than as N_2 in the earth's atmosphere. The following chain of reactions summarizes the process:

10.3
$$NO_3^- \rightarrow NO_2^- \rightarrow NO \rightarrow N_2O \rightarrow N_2$$

The oxidation state of nitrogen reduces from 5 to 3 to 2 to 1 to 0 in these reactions. Denitrification occurs when nitrates are present in environments with limited oxygen and available reducing agents (usually organic carbon). If oxygen is very limited or absent, nitrate reduces all the way to nitrogen. However, when the supply of oxygen is greater than the supply of reducing agent, incomplete reduction occurs, and the intermediary compounds are released into the atmosphere in larger quantities.

Assimilation is the conversion of inorganic nitrogen (such as nitrate) into an organic form of nitrogen such as an amino acid. Enzymes reduce nitrate first to nitrite (by nitrate reductase), then to ammonia. Ammonia is incorporated into amino acids.

Mineralization is a process during which decomposers like earthworms, termites, slugs, snails, bacteria or fungi convert the organic nitrogen of dead plants into inorganic forms. The first step is the formation of ammonia and its salts ($NH_4^+X^-$). Assimilation and mineralization are opposed processes.

10.2 Sources and Sinks

Categories of nitrogen compounds in the atmosphere include molecular nitrogen, ammonia, nitrous oxide, and nitrogen oxides.

Figure 10-1 highlights the main sources and sinks of nitrogen in the atmosphere. Sources of each can be natural and/or human-caused. The only major sink (removal mechanism) is wet or dry deposition of nitrates onto the earth's surface. Table 10-1 lists approximate values for the sources and sinks of these nitrogen compounds (IPCC, 2007). Both the figure and table ignore molecular nitrogen, which is quite stable and rarely plays a chemical role in the troposphere.

10.2.1 Nitrogen (N_2)

Molecular nitrogen is the most abundant gas in the atmosphere, with a molar concentration of 78.084% by volume at sea level on a 15°C, 1 atm, dry air basis. It is chemically un-reactive at normal conditions but can undergo an exothermic reaction with oxygen at sufficiently high temperatures (2000 K). Indeed, there was some concern during the development of the atomic bomb that an atmospheric explosion could generate sufficient temperatures to ignite nitrogen in the atmosphere. However, within a year after the tests and use of the bomb, it was shown that such a catastrophe was extremely improbable (Konopinski, et al., 1946).

Table 10-1. Global Nitrogen Budget in Tg N/yr.

Amounts given in Tg N / yr 1 Tg = 10^{12} g	Ammonia NH_3	Nitrous Oxide N_2O	Nitrogen Oxides NO_x
Sources			
Natural			
Soil Emissions	6	6	20
Oceans	8	3	1
Tropospheric reactions			3
Anthropogenic			
Biomass burning	6	0.5	12
Industrial	0.2	1.3	0.5
Fossil fuel use	0.1	-	20
Agriculture	15	3.5	-
Animal husbandry	24	0.4	-
Sinks			
Stratospheric photochemistry		12.3	
Tropospheric reactions	3		
Wet deposition	21		27
Dry deposition	16		16

10.2.2 Nitrous Oxide (N_2O)

The global atmospheric nitrous oxide concentration has increased from a pre-industrial value of about 270 ppb to 320 ppb in 2010. The growth rate has been approximately constant since 1980 at a rate of 0.3%/yr [(IPCC, 2007), (WHO, 2003)]. It has a residence time in the atmosphere of 130 to 150 years. N_2O can absorb and re-radiate thermal radiation, which makes it a greenhouse gas (see Chapter 14). It contributes about 5% of the total atmospheric greenhouse effect from non-water gases. N_2O is 296 times stronger than CO_2 as a greenhouse gas on an equal mass basis assuming a 100-year time frame [(IPCC, 2001)]

The main source worldwide (about 50%) is due to denitrification in soils. Approximately one-third of all nitrous oxide emissions are anthropogenic and are primarily due to agriculture. The sink for nitrous oxide is transformation to NO_x from photochemical reactions with oxygen in the stratosphere, where it plays a major role in ozone depletion. N_2O is one of the largest contributors

FIGURE 10-1. The Atmospheric Nitrogen Cycle. This drawing shows the major (>1 TgN/yr) sources and sinks of NO_x (N_2O, NO, NO_2, and NO_3^-) and NH_3 (as NH_3 and NH_4^+) in the atmosphere. Note internal combustion includes many different sources such as automobiles, trucks, trains, ships, and generators.

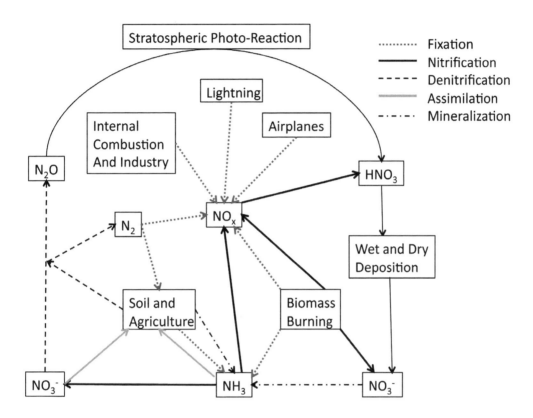

to stratospheric ozone depletion. Table 10-2 shows the level of N_2O emissions from US sources for the year 2008 given as CO_2 equivalents (US-EPA, 2010). A CO_2 equivalent is a measurement that relates molecules by their ability to absorb infrared radiation compared to CO_2. The CO_2 equivalent standard simplifies comparisons between greenhouse gases.

10.2.3 Ammonia (NH_3)

Ammonia concentrations in the atmosphere range from 0.1 to 10 ppb over land and less over the oceans. In areas of large sources, it can exceed 100 ppb. It has an atmospheric residence time ranging from hours to 10 days. Main sources include enzymatic decomposition of urea (from animal urine and excrement), production and application of fertilizers, biomass burning, and soil emissions. Industrial sources occur from paper manufacturing, wastewater treatment, petroleum refining, and combustion of fossil fuels. Its atmospheric concentration is especially high in agricultural areas. The major sinks include dry deposition as ammonia or in fine particle form (ammonium sulfate and ammonium nitrate), or wet deposition due to the high water solubility of ammonia and related compounds. It can absorb directly into surface water (lakes and oceans) or into water droplets in the atmosphere and then deposit with rain or snow. It acts as a nitrogen fertilizer source in water and can cause significant water degradation.

Table 10-2. U.S. Nitrous Oxide (N_2O) Emissions by Source (Tg CO_2 Equivalents), 2008.

Source Category	N_2O Emissions Tg CO_2 equivalent
Agricultural Soil Management	215.9
Mobile Combustion	26.1
Nitric Acid Production	19
Manure Management	17.1
Stationary Combustion	14.2
Forest Land	10.1
Wastewater Treatment	4.9
N_2O from Product Uses	4.4
Adipic Acid Production	2
Composting	1.8
Urban Areas	1.6
International Bunker Fuels	1.2
Field Burning of Agricultural Residues	0.5
Incineration of Waste	0.4
Total for U.S.	**318.2**

Note: 1 Tg = 10^{12} g; 1 g N_2O = 296 g CO_2 equiv.

10.2.4 Nitrogen Oxides (NO_x)

Nitrogen oxides consist of NO and NO_2, which occur together and are referred to as NO_x. They are inter-convertible in the atmosphere. They have a typical ambient concentration of approximately 15 ppb when measured as NO_2, though it can reach values of 60 ppb or even higher in urban environments [(Seetharam, et al., 2009)]. In remote environments the concentration can be just a few ppb. The main sources of NO_x in the atmosphere are biomass burning, fixation and denitrification reactions in soil, and as a byproduct of fossil fuel combustion. The most uncertain source is due to lightning (2 – 20 TgN/yr), which mostly produces NO_x in the upper troposphere, where it is difficult to measure. NO_x readily oxidizes in the atmosphere to form nitrates (NO_3-). Nitrates are very soluble in water and are removed from the atmosphere by wet or dry deposition. NO_x and HNO_3 can also be taken up directly by plants.

Nitrogen oxides are at the center of atmospheric chemistry. Most other air pollutants which oxidize or transform into other chemical species will directly or indirectly use NO or NO_2 during the process.

Table 10-3 quantifies the various human-caused sources of NO_x (measured as NO_2) in the USA (US-EPA, 2009). This table shows that the majority of NO_x emissions are due to combustion (first four entries in the table). Industrial process sources include nitric acid production, fertilizer production, and wastewater treatment.

Table 10-3. U.S. Anthropogenic Nitrogen Oxide Emissions by Source, 2008. Units are in 10^{12} g of N, rather than NO_x, as the mass of NO_x depends on the value of x which can be ½, 1, 2, 5/2, or 3

Sources	NO_x Emissions TgN/yr
On Road Vehicles	5.895
Non-Road Equipment	3.780
Electricity Generation	3.436
Fossil Fuel Combustion	2.165
Industrial Processes	1.057
Waste Disposal	0.141
Fires	0.086
Residential Wood Combustion	0.035
Solvent Use	0.006
Miscellaneous	0.003

10.2.4.1 Formation of NO_x from Combustion

There are three primary sources of NO_x in combustion processes:
- Thermal NO_x,
- Fuel NO_x, and
- Prompt NO_x.

Thermal NO$_x$ formation is highly temperature dependent and uses atmospheric N$_2$ as its nitrogen source. Fuel NO$_x$ forms during the combustion of fuels containing nitrogen, such as coal. Prompt NO$_x$ also forms from atmospheric N$_2$ but by a different mechanism than thermal NO$_x$. A fourth source, called feed NO$_x$, is associated with the combustion of nitrogen present in the feed material to rotary kilns used in producing cement, but this is considered a minor contributor.

Thermal NO$_x$ refers to NO$_x$ formed through high-temperature oxidation of the diatomic nitrogen found in combustion air. The formation rate is primarily a function of temperature and the residence time of nitrogen at that temperature. At high temperatures, usually above 1600°C (2900°F), oxygen (O$_2$) in the combustion air disassociates into its atomic state and participates in a series of reactions. The simplified, reversible reactions, referred to as the extended Zeldovich Mechanism, (Li, N; Thompson, S, 1996) producing thermal NO$_x$ are:

10.4 $\qquad N_2 + O \leftrightarrow NO + N$

10.5 $\qquad N + O_2 \leftrightarrow NO + O$

10.6 $\qquad N + OH \leftrightarrow NO + H$

The first two reactions are of primary interest in the formation of thermal NO$_x$ during controlled combustion. The third reaction only becomes important when reacting under very fuel rich conditions such as uncontrolled biomass burning.

An additional reversible reaction describing the formation of NO$_2$ is:

10.7 $\qquad 2\,NO + O_2 \leftrightarrow 2\,NO_2$

Equation (10.4) is the rate-limiting reaction. It is an endothermic reaction, so it is favored at high temperatures. Equation (10.7) is exothermic and is favored at low temperatures.

Fuel NO$_x$ refers to NO$_x$ production from nitrogen-bearing fuels such as coal and oil. It forms from the conversion of chemically bound nitrogen to NO$_x$ during combustion. During the combustion process, the nitrogen is released as a free radical that continues to react until it forms N$_2$ or NO. Fuel NO$_x$ can contribute as much as 50% of total NO$_x$ emissions from burning oil and up to 80% when burning coal. The amount of fuel N emitted as NO$_x$ depends mostly on the air to fuel (A/F) ratio. A fuel rich mixture (substoichiometric amounts of oxygen added) forms N$_2$ or NH$_3$ whereas a fuel lean mixture forms NO$_x$. In general, fuel NO$_x$ formation is oxygen sensitive and is not very temperature sensitive.

FIGURE 10-2. Fuel NO$_x$ Combustion Mechanism. Figure only shows the nitrogen-containing reactants and products. Species listed above arrows are possible reactants with the N compounds.

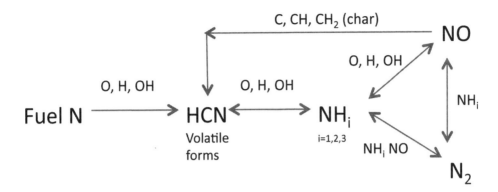

Although the complete mechanism is not fully understood, there are two primary paths of formation, see Figure 10-2. One path describes the oxidation of volatile nitrogen species during the initial stages of combustion. Fuel nitrogen oxidizes through several intermediaries into NO before and during the release of the volatile fraction of the fuel. If this initial stage of the combustion runs in a reducing atmosphere (sub-stoichiometric oxygen), the nitrogen evolves to form nitrogen gas, rather than NO$_x$. The second path describes the combustion of nitrogen contained in the residual fuel after the volatiles evolve (this material is called char). This reaction occurs more slowly than the volatile phase. Only around 20% of the char nitrogen forms NO$_x$. Char, which is nearly pure carbon, acts as a reducing agent to the NO$_x$ that has formed.

Prompt NO$_x$ refers to the reaction of atmospheric nitrogen, N$_2$, with radicals such as C, CH, and CH$_2$ fragments derived from fuel. Its formation is similar to fuel NO$_x$, except that the nitrogen source is from the air, not the fuel. It occurs in the earliest stage of combustion (hence the name prompt), and results in the formation of fixed species of nitrogen such as HCN (hydrogen cyanide), NH (nitrogen monohydride), H$_2$CN (dihydrogen cyanide) and CN$^-$ (cyano radical). These species can oxidize to NO or reduce to N$_2$. These reactions depend on both temperature and presence of oxygen.

10.2.4.2 Formation of NO$_x$ from Nitric Acid Manufacture and Uses

The principal end-use of nitric acid (HNO$_3$), also known as *aqua fortis* or spirit of nitre, is in the production of nitrogen fertilizers, explosive-grade ammonium nitrate, adipic acid - used for making nylon, and toluene di-isocyanate - used for manufacturing polyurethane.

Production of nitric acid is via the *Ostwald Process*, named after German chemist Wilhelm Ostwald [(Jones, et al., 1999)]. In this process, ammonia is converted to nitric acid. First, the ammonia is heated with oxygen in the presence of a platinum-rhodium catalyst to form nitric oxide and water. This step is strongly exothermic, making it a possible source of heat.

10.8 $\quad 4\,NH_3\,(g) + 5\,O_2\,(g) \rightarrow 4\,NO\,(g) + 6\,H_2O\,(g)$

The products are then reacted with oxygen from the air to form nitrogen dioxide.

10.9 $\quad 2\,NO\,(g) + O_2\,(g) \rightarrow 2\,NO_2\,(g)$

The product subsequently forms aqueous nitric acid and nitric oxide.

10.10 $\quad 3\,NO_2\,(g) + H_2O\,(l) \rightarrow 2\,HNO_3\,(aq) + NO\,(g)$

The nitric oxide can be recycled back for re-oxidation. The acid is concentrated to the required strength by distillation. The main waste from the manufacture of nitric acid is the release of NO in the final step. The release can be minimized by venting the exhaust gas through a water-scrubber absorption tower, but because there is an economic limit to the size of the tower, the efficiency achieved is in the range of 99%.

10.2.4.3 Formation of N_2O and NO_x from Inorganic Fertilizer (Synthetic Fertilizer)

There are two types of fertilizers, organic and inorganic. Organic fertilizers include naturally occurring organic materials, (e.g. manure, worm castings, compost, seaweed, guano), or naturally occurring mineral deposits (e.g. potassium nitrate also known as saltpeter). Inorganic fertilizer includes synthetic chemicals and/or minerals. Ammonia is usually a part of an inorganic fertilizer. It is usually synthesized from nitrogen in the air and natural gas (for hydrogen) using the Haber or Haber–Bosch process [(Hager, 2008), (Sittig, 1979)].

10.11 $\quad N_2 + 3H_2 \rightarrow 2NH_3$

The worldwide production of synthetic ammonia is about 125 million tons per year (2008), and it consumes about 5% of global natural gas production, which is slightly under 2% of world energy production (IFA, 2010). The ammonia may be used as is or as a feedstock for other nitrogen fertilizers, such as anhydrous ammonium nitrate or urea. Ammonia can be combined with rock phosphate [$Ca_3(PO_4)_2$] in the *Odda Process* to produce compound fertilizer (Steen, et al., 1986). The Odda process is a multistep reaction that can be summarized as:

10.12 $Ca_3(PO_4)_2 + 6HNO_3 + 10H_3PO_4 + 24NH_3 \rightarrow 3CaHPO_4 + 6NH_4NO_3 + (NH_4)HPO_4$

The compound fertilizer is often mixed with potassium fertilizer [KCl or K_2SO_4] to form an NPK fertilizer. The use of the ammonium-based fertilizer leads to several air pollution issues; increase methane emissions from crop fields, especially rice; release of excess ammonia to air or water; and increase nitrous oxide emissions – approximately 3 to 5% of industrially fixed nitrogen (i.e. Haber process ammonia) is converted to N_2O worldwide (Mosier, et al., 1998).

10.2.4.4 Formation of N_2O and NO_x from Wastewater Treatment

Human sewage often mixes with other household wastewater. The wastewater includes shower drains, sink drains, and washing machine effluent. It is transported by a collection system to either an on-site (e.g., septic system) or centralized wastewater treatment plant. NO_x and N_2O can generate during both nitrification and denitrification of the urea, ammonia, and proteins. These compounds convert to nitrate (nitrification) during aerobic (in the presence of oxygen) treatment. They also convert to nitrogen (denitrification) during anaerobic (in the absence of oxygen) treatment. NO_x and N_2O are intermediate products of both these processes (Corbitt, 1989).

Livestock manure management produces nitrous oxide as part of the nitrogen cycle through the nitrification and denitrification of the organic nitrogen in livestock manure and urine. The production of NO_x and N_2O from livestock manure depends on the composition of the manure and urine, the type of bacteria involved in the process, and the amount of oxygen and liquid in the manure system. NO_x emissions are most likely to occur in dry manure handling systems that have aerobic conditions, but that also contain pockets of anaerobic conditions due to water saturation (Corbitt, 1989).

10.3 Health and Welfare Effects of NO_x

Health and welfare effects from nitrogen oxides may potentially result from NO_2 itself or as it reacts and transforms to form secondary products including nitrates (NO_3^-), ozone (O_3) and particles. NO_x oxidizes to form nitrates, photo-reacts with volatile organic compounds to form ozone, and combines with ammonia, moisture, and other pollutants to create particulate matter (PM).

Figure 10-3 summarizes the main transformations and harmful effects. This section focuses only on NO_x as the other products are covered elsewhere in this book.

10.3.1 Effects on Public Health

Health effect assessments use short-term chamber studies with humans and animals and long-term epidemiological studies of populations exposed to elevated concentrations in their workplace or residence. The short-term studies show that direct effects of NO_2 alone on the lungs (or any other

FIGURE 10-3. Health and Welfare Effects of Nitrogen Emissions.

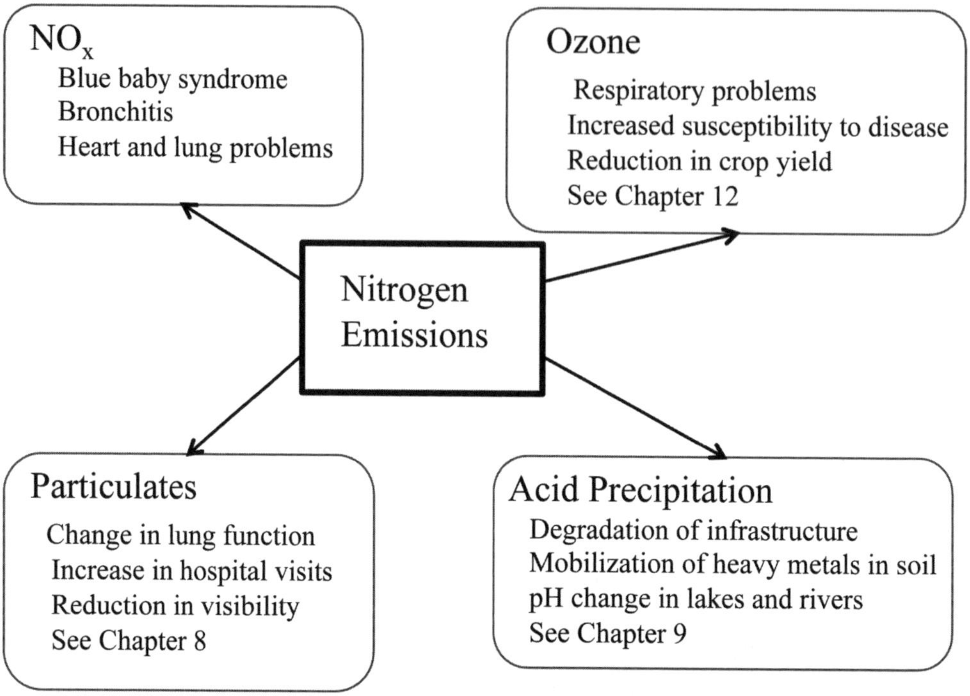

```
NOx                              Ozone
Blue baby syndrome               Respiratory problems
Bronchitis                       Increased susceptibility to disease
Heart and lung problems          Reduction in crop yield
                                 See Chapter 12

              Nitrogen Emissions

Particulates                     Acid Precipitation
Change in lung function          Degradation of infrastructure
Increase in hospital visits      Mobilization of heavy metals in soil
Reduction in visibility          pH change in lakes and rivers
See Chapter 8                    See Chapter 9
```

system) are minimal or undetectable at the levels normally encountered in the ambient outdoor air [(Chen, et al., 2008), (Hasselblad, et al., 1992)]. Mild airway inflammation may occur at NO_2 concentrations in excess of typical regulatory air levels (Goldstein, et al., 1987). Studies of people with asthma or allergies to pollen have shown that NO_2 can enhance these effects [(Panella, et al., 2000), (Kagawa, 1982)].

The National Institute for Occupational Safety and Health (NIOSH) and the Occupational Safety and Health Administration (OSHA) have set recommended exposure limits of 25 ppm for NO and 1ppm (NIOSH Short term) or 5 ppm (OSHA ceiling) for NO_2. The IDLH (immediately dangerous to life or health) concentrations are 100 ppm for NO and 20 ppm for NO_2 (NIOSH, 2005). Typical annual average concentrations in urban ambient air are less than 0.015 ppm in the USA, 0.022 ppm in Europe, and 0.015 ppm in Japan.

NO_2 concentrations in vehicles and near roadways are appreciably higher than concentrations measured at other locations, including industrial parks. In-vehicle concentrations can be 200% -300% higher than measured at nearby area-wide monitors. Near-roadway (within about 50 meters or 160 feet) ambient air concentrations of NO_2 have been measured at approximately 30 to 100% higher than away from roadways (Maruoa, et al., 2003), (Omatu, et al., 1988). NO_2 exposure concentrations near roadways are of particular concern for susceptible individuals. Approximately 16% of US housing (48 million people) are located within 90 meters (300 feet) of a major highway, railroad, or airport. This population likely includes a higher proportion of non-white and economically disadvantaged people, making this one of the major environmental justice issues in the US and the world. People that work near roadways (construction, maintenance, toll booths) also have higher incidences of respiratory problems.

Long-term studies show a correlation between ambient NO_2 concentrations with respiratory and cardiovascular health problems. A meta-analysis on mortality showed consistent associations with NO_2 [(Stieb, et al., 2002), (Stieb, et al., 2003)]. Hospital admissions for respiratory disease increase with increasing levels of NO_2 in urban areas [(Tecer, 2009), (Moolgavkar, 2000)]. Recent epidemiological studies have shown consistent associations between long-term exposure to NO_2 and decreased lung function in children as well as with decreased lung function and respiratory symptoms in adults. However, these studies do not show causality for NO_2 on health due to the many confounding pollutants (PM, SO_x, O_3) in ambient air.

NO_2 is a good indicator of fuel combustion and traffic-related air pollution. Its presence is easier to measure than the secondary pollutants (which require time before their formation), and its level correlates with the concentrations of more toxic pollutants suspected of producing adverse health effects. The potential of NO_2 to enhance the effects of other environmental pollutants, including allergens, suggests that regulations limiting NO_x would result in improved health. However, the World Health Organization could find no peer-reviewed scientific studies that clearly demonstrate that reduction of NO_2 leads to health benefits in the general population (WHO, 2003).

10.3.2 Effects on Public Welfare

Public welfare effects from NO_2 are mostly due to the secondary pollutants (particles, ozone, and acidic nitrates) associated with its emission, as shown in Figure 10-3. Effects of particles, acid precipitation, and ozone are described elsewhere (Chapter 8, Chapter 9, and Chapter 12, respectively). NO_2 absorbs visible light, causing a reddish-brown coloration of air, which reduces visibility. It is also a common component of smog and industrial haze. Acidic nitrate deposition, usually in the form of nitric acid, can change water and soil chemistry. These changes can lead to fish kills, altered plant communities and reduced tree growth. Nitrate deposition also results in excess nitrogen in ecosystems, which can cause changes in vegetation, loss of biodiversity, and increased greenhouse gas emissions. It can also increase the age of lakes through the process of eutrophication and hyper-eutrophication, which is an increase in the primary production of aquatic plants. The excess algae often cause water to appear bright green, and can cause significant reductions of oxygen in the water. There are areas, such as the Mississippi River Delta and entry into the Gulf of Mexico where the decaying algae use all the oxygen in the water, creating a condition known as hypoxia, or dead zone. It causes the death of most animals in the water (fish, crabs, mollusks).

10.4 History and Regulation

10.4.1 NH_3

Ammonia (NH_3) is not currently a regulated pollutant under the US EPA Clean Air Act and is not considered a Hazardous Air Pollutant (HAP). It is regarded as a precursor to fine particulate matter ($PM_{2.5}$) formation, and it may be regulated similarly to criteria pollutants. Many states and countries require the reporting of primary ammonia releases - ammonia released in the same chemical form as applied but not as a secondary emission due to a chemical reaction from some other chemical form. For example, if ammonia is directly used in a process, any escaping ammonia must be reported. However if nitrite is used in a process and some escapes to later form ammonia as a byproduct, the release would not be reported.

10.4.2 N_2O

The Kyoto Protocol ratified by 54 nations in 1997, and by 188 in 2009, classifies N_2O as an ozone depleting substance and as a greenhouse gas. It calls for substantial worldwide reductions in its emission (IPCC, 2007). Nitrous oxide is 298 times more potent than carbon dioxide in its ability to affect climate change. Also, results of a recent scientific study (Ravishankara, et al., 2009) indicate that nitrous oxide is currently the leading stratospheric ozone-depleting substance being emitted. Thus, legislation to restrict nitrous oxide emissions could contribute to both climate change protection and ozone recovery. The main proposed strategies to lower N_2O emissions is to increase the efficiency of fertilizer application, and to alter other agricultural practices. However, further work is needed to determine the feasibility of these activities as well as to develop techniques to measure and monitor the adoption rate and impact of N_2O emission reduction for various agricultural soil management practices. As of 2010, N_2O emissions are not regulated for ambient air, although there is discussion about the possibility (Perez-Ramirez, 2007).

10.4.3 Nitrates

Typically, nitrates are not primary air pollutants – the nitrogen is not emitted in the nitrate form, rather the nitrate forms from emissions of ammonia or nitrogen oxides. The main regulatory tool to control nitrates is to reduce the precursors. However, nitrates are collected, measured, and monitored in the environment since they are easy to measure; Also, nitrates are typically the final form in which the nitrogen emissions are removed from the atmosphere, see Figure 10-7.

10.4.4 NO_x

Table 10-4 shows the NO_x ambient air quality standard for several countries. Values are based on statistical evidence from epidemiological studies and short-term chamber experiments on humans and animals.

Table 10-4. National Ambient Air Quality Standards for NO₂ in Several Countries (2010).

Country / Time Period Average Units (unless noted)	Annual µg/m³	24-hour µg/m³	1-hr µg/m³
US	100 (53 ppb)	-	191 (100 ppb)
EU	40	-	200
Japan	-	76 - 114	-
India (Class I/II/III)*	15/60/80	30/80/120	-
China (Class I/II/III)*	40/40/80	80/80/120	120/120/240
Brazil	100	-	320
Mexico	-	-	395
South Africa	94	188	376
WHO	40	-	200

*Class I: Tourist, conservation area; Class II: Residential area; Class III: Industrial and Heavy

10.4.5 Regulatory Methodology – US Approach

Figure 10-4 provides a timeline of the regulatory tools used by the US-EPA to achieve NO_x reductions.

FIGURE 10-4. Timeline of US-EPA Regulatory Programs to Control NOx in the Ambient Air. * The 2009 Clean Air Interstate Rule (CAIR) has been vacated by the courts and EPA has been ordered to address flaws in the law while leaving CAIR in place temporarily. The first revision, the Cross-State Air Pollution Rule, stayed in August 2012, and CAIR remains in place. The stay was appealed to the Supreme Court by the US Government on December 10, 2013.

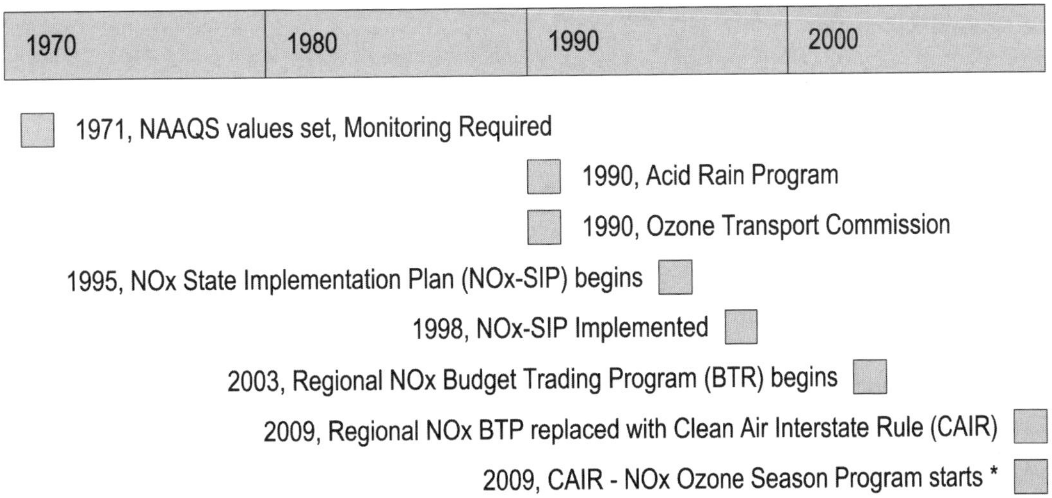

1970	1980	1990	2000

- 1971, NAAQS values set, Monitoring Required
- 1990, Acid Rain Program
- 1990, Ozone Transport Commission
- 1995, NOx State Implementation Plan (NOx-SIP) begins
- 1998, NOx-SIP Implemented
- 2003, Regional NOx Budget Trading Program (BTR) begins
- 2009, Regional NOx BTP replaced with Clean Air Interstate Rule (CAIR)
- 2009, CAIR - NOx Ozone Season Program starts *

Ambient air standards were set in 1971 as part of the Clean Air Act (CAA). Emission standards were revised in the 1990 Clean Air Act Amendments (CAAA), as part of the acid rain program, though they only included a few industries (chiefly coal-fired boilers) located in parts of the eastern US, and only for the summer months. The emissions standards were revised again in 1995. The revision includes most of the eastern US because evidence showed that NO_x travels across multi-state regions and causes areas with no NO_x emissions to exceed the ambient air quality standards for ozone (a photochemical produced by several atmospheric reactions involving NO_x). A trading program was developed from 2003 to 2009 to help states achieve their emission targets. Currently, NO_x emissions are regulated through the Clean Air Interstate Rule, which is designed to decrease NO_x, SO_x, Ozone, and particulate matter. However, the US Court of Appeals vacated the CAIR rules and has required the US-EPA to revise the rules. The first revision, known as the Cross-State Air Pollution Rule, has been stayed by the courts (stayed means that the regulation is not enforceable). The stay was challenged in the US Supreme Court (Dec. 10, 2013) court by the US Government. The CAIR program remains in place until the Court required revision occurs.

All areas in the U.S. presently meet the 1971 NO2 US-EPA NAAQS (less than 53 ppb). The current US annual average NO_2 concentrations range from approximately 10 to 25 ppb. Annual average ambient NO_2 concentrations have decreased by more than 40 percent since 1980, see Figure 10-5. NO_2 concentrations should continue decreasing as a number of mobile source regulations take effect. Tier 2 standards for light-duty vehicles began phasing in during 2004, and NO_x standards for heavy-duty engines began phasing in between the 2007 and 2010 model years. Current air quality monitoring data reflect only a few years of vehicles entering the fleet that meet these stricter NO_x tailpipe standards.

The US-EPA has used several programs to address the problems from nitrogen pollution, see Figure 10-4:

> The *Acid Rain Program* (ARP) was established by Congress through Title IV of the Clean Air Act Amendments (CAAA) of 1990. This program reduced sulfur dioxide (SO_2) from permitted sources through a cap and trade program across the lower 48 states. The ARP also reduced NO_x emissions from some of these units, but, unlike the SO_2 portion of the ARP, there was no cap on NO_x emissions or allowance trading. Instead, the ARP NO_x provisions applied boiler-specific NO_x emission limits on certain coal-fired boilers. Companies could use "emission averaging" plans across their units to provide flexibility with the rules. NO_x limits began on the largest boilers in 1996, while the second phase to reduce NO_x emissions from smaller coal-fired generating units was to begin in 2000. However, it was superseded by subsequent legislation.

The *Ozone Transport Commission* (OTC) was established under the 1990 CAAA to develop a NO_x reduction program. States in the Northeast and Mid-Atlantic collaborated to reduce summertime ground-level ozone in the region by achieving ozone season NO_x reductions in several phases. In 1995, Phase 1 required sources to reduce their annual NO_x emission rates to meet Reasonably Available Control Technology (RACT) requirements. In Phase II (1999–2002), states achieved reductions in NO_x from fossil fuel-fired units and large industrial boilers through an ozone season cap and trade program known as the OTC NO_x Budget Program. The third phase of the OTC NO_x Budget Program was replaced by EPA's NO_x SIP Call.

NO_x State Implementation Plan (SIP) Call: In 1995, EPA and the Environmental Council of the States formed the Ozone Transport Assessment Group to begin addressing the problem of ozone transport across the entire eastern United States. Based on the group's findings, EPA issued the NO_x SIP Call in 1998 to reduce the regional transport of ground-level ozone. This rule required states to reduce ozone season NO_x emissions that contribute to ozone nonattainment in other states. The NO_x SIP Call did not mandate which sources must reduce emissions; instead, it required states to meet emission budgets and gave them the flexibility to develop control strategies to meet those budgets.

NO_x Budget Trading Program (NBP): In 2003, the EPA began to administer the NBP under the NO_x State Implementation Plan. The NBP was a market-based cap and trade program implemented seasonally in the eastern United States. It eventually covered all or parts of 20 states. The NBP reduces NO_x emissions only during the warm summer months when ground-level ozone concentrations are highest. The states shared responsibility with EPA by allocating allowances, inspecting and auditing sources, and enforcing the program. However, regardless of the number of allowances a source held, it could not emit at levels that would violate other federal or state limits, e.g., for coal-fired units, New Source Performance Standards, ARP NO_x limits, Title V permit requirements, and Title I requirements of RACT for NO_x.

Cap and trade programs set a cap on overall regional emissions and give each source a set number of allowances based on previous emission records. Each allowance permits the release of one ton of pollutant emissions. This approach provides individual sources with flexibility in how they comply with emission limits. Sources may sell or bank (save) excess allowances if they reduce emissions and have more than they need, or purchase allowances if they are unable to keep emissions below their allocated budget. As a group, the participating sources cannot exceed the cap. The cap level protects public health and the environment. The cap also lends stability and predictability to the allowance trading market.

Results of this program showed that a cap and trade systems can work for regional and seasonal emission controls (NO_x), as well as national emissions (SO_2) when the program is well managed. This program also showed the importance of - reliable and continuous monitoring; having a simple nationally administered program adaptable by each state to include local concerns, and the utility of using facility level compliance rather than unit level compliance, where a facility consists of many units, some discharging from the same stack or nearby points (Napolitano, et al., 2007). In 2009, the Clean Air Interstate Rule (CAIR) NO_x ozone season program replaced NBP.

Clean Air Interstate Rule (CAIR): On March 10, 2005, EPA issued CAIR, to reduce criteria air pollution ($PM_{2.5}$, SO_2, and NO_x). One particular goal was to reduce ozone by reducing the precursors. CAIR accomplishes this by creating three separate trading programs: an annual NO_x program, an ozone season NO_x program, and an annual SO_2 program. However, as discussed above, this regulation has been vacated by the US Courts and is currently under appeal. This situation is currently unresolved (2014).

The US EPA added a new 1-hour NO_2 standard to NAAQS in 2010. It sets the maximum allowable concentration in ambient air at a level of 100 ppb. This value uses the 3-year average of the 98th percentile of the annual distribution of daily maximum 1-hour average concentrations. If an area exceeds this value, the state (or EPA) must develop and enact a plan to reduce the concentration as part of their State Implementation plan (SIP). SIPs are described in more detail in Chapter 3. The annual average remains the same (53 ppb) as set in the 1971 Clean Air Act. It also did not alter the secondary NO_2 standard (53 ppb) for protecting the public welfare. A new standard was set to protect public health from short-term exposure to NO_2, including the health of sensitive populations primarily near major roads, which often have higher NO_x concentrations than surrounding areas. The US EPA reviewed the NO_x standards two other times since 1971, but chose not to revise the standards during those reviews.

The US EPA started to identify or "designate" areas as attaining or not attaining the new standard beginning in January 2012 (US-EPA, 2013). Designations are based on the existing community-wide monitoring network. Areas with monitors recording concentrations below the standard are designated 'attainment'. Areas with monitors recording violations of the new standards are designated "nonattainment." Other areas of the country are labeled "unclassifiable" to reflect the fact that there is insufficient data available to determine if those areas are meeting the revised NAAQS. The unclassifiable areas are assumed to be in attainment.

The revised regulations included requirements for the placement of new NO_2 monitors in urban areas to begin operating no later than January 1, 2013.
- *Near Road Monitoring* – new monitors placed within 50 m of major roadways (those with high annual average daily traffic) in major urban areas (those with 500,000 people or more).

- *Community Wide Monitoring* – new or additional monitors placed in urban areas with a population greater than or equal to 1 million people to assess community-wide concentrations. Some NO_2 monitors already in operation may meet the community-wide monitor citing requirements.
- *Monitoring to Protect Susceptible and Vulnerable Populations* - site at least 40 additional NO_2 monitors to protect communities susceptible or vulnerable to NO_2 related health effects.

After the expanded network of NO_2 monitors are fully deployed, and collect three years of air quality data, EPA intends to re-designate areas in 2016 or 2017, as appropriate, based on the air quality data from the new monitoring network.

Figure 10-5 shows the US annual average concentration trend for NO_2 concentration. The average value has declined from an annual average of 28.2 ppb (1980) to 13.0 ppb (2010), a decrease of about 54%. During this same time, the US economy Gross Domestic Product (GDP) grew by 430%

FIGURE 10-5. US National Air Quality Trend for Annual Average NO_2 Concentration (ppm). The trend lines show the concentrations of the 10, 50 (annual average) and 90 percentile of measured values at air monitoring stations throughout the US. There were 81 stations operating in the 1980's, 150 in the 1990's, and 283 in the 2000's.

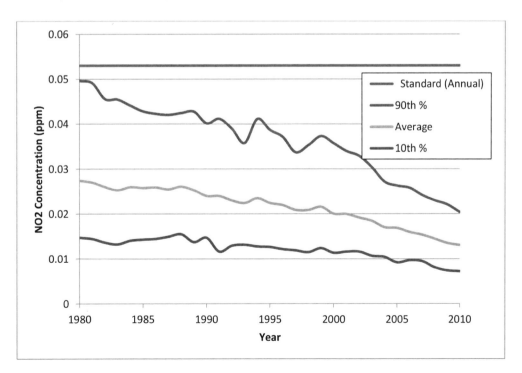

FIGURE 10-6. US National Air Quality Trend for One-Hour Average NO₂ Concentration (ppb). There were 29 stations measuring the hourly average in the 1980's, 98 in the 1990's, and 180 in the 2000's. Comparison from 1980 to 2013 show a 59% decline in hourly average values.

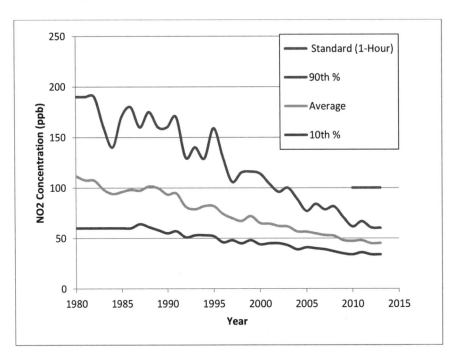

(Commerce, 2010), and traffic increased by 96% (FHA, 2010). As was shown in Chapter 9 -sulfur emissions, it required economy-wide regulation for 10 – 15 years to delink emissions from economic growth, and 10 – 30 years to reduce ambient concentrations by 50%. Newly industrializing economies may be able to shorten this time since much of the required regulatory and technological advancement has already occurred.

Figure 10-7 shows the change in wet deposition of nitrates and ammonia between 1994 and 2009. Comparison shows a small overall reduction in the amount. They also show a change in geographical distribution from the industrialized Ohio Valley to the agricultural Midwest. Dots show monitoring locations.

Figure 10-8 shows the quantity (million tons) of emissions of NO_2. The values only include sources with air emission permits, and do not include other known sources (as given in Table 10-3). This figure shows that the quantity of emissions did not change much until the implementation of the Acid Rain Program in 1990. Comparison with Figure 10-5, which shows ambient air concentrations, suggests that while emission rates did not decrease much until after 1995, the amount in any given area did decrease. The change is probably due to a shift in sources from power plants to agriculture and mobile sources.

FIGURE 10-7. Annual Average Inorganic Nitrogen Deposition (NO$_3$ and NH$_3$) Isopleths in Continental US.

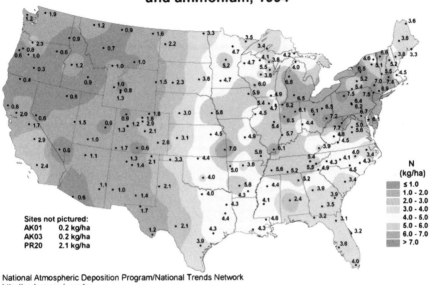

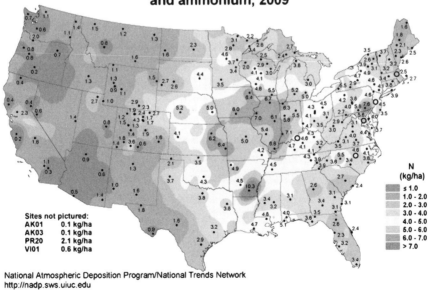

FIGURE 10-8. US National Air Quality Trend for Permitted, Stationary Sources of NO$_2$.

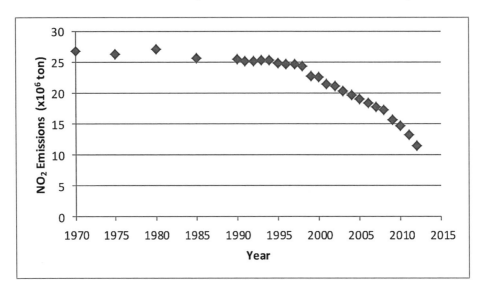

10.4.6 Regulatory Methodology – Japan

It is interesting to compare and contrast the US environmental programs with that of other economies such as Japan. Japan has a very advanced manufacturing and chemical industry. Like the US, Japan was an early adopter of environmental controls and the creation of regulatory policy and the technologies to implement pollution reduction. They have made significant strides in reducing emissions and improving ambient air quality.

Japan is a Pacific Ocean island country located east of China, Russia, and the Koreas. It consists of 6,852 islands, of which the four largest islands are Honshū, Hokkaidō, Kyūshū and Shikoku. These four islands account for 97% of Japan's land area. It has the world's tenth-largest population, about 128 million people, has the world's third-largest economy by nominal GDP, and the third largest in purchasing power parity. Japan currently is ranked 21st of 149 in the Environmental Sustainability Index (Yale, 2008), whereas the US is ranked 39th.

Japan has a parliamentary government with a constitutional monarchy. The country consists of forty-seven prefectures, each overseen by elected governors and legislatures, and administrative bureaucracy. Each prefecture includes cities, towns and villages. Japan's environmental history and current policies reflect a focus on economic development and environmental protection.

Environmental policies were downplayed by the government and industrial corporations during the rapid economic growth after World War II. As an inevitable consequence, several environmental catastrophes occurred in the 1950s and 1960s. These man-made catastrophes were the result of industrial pollution, though the groups of symptoms have disease names.

Minamata disease occurred because of organic mercury poisoning in Minamata Bay from the Chisso chemical factory (1932 – 1968). *Yokkaichi asthma* resulted from the uncontrolled release of particles, sulfur dioxide, and nitrogen dioxide from oil-based power generation in the city of Yokkaichi (1956 – 1972). Yokkaichi asthma has been identified in rapidly industrializing areas in the rest of the world, including Mexico City and mainland China. *Itai-Itai disease* results from the cadmium contaminated effluents of the Kamioka mine on the upper reaches of the Jintsu River in Toyama Prefecture (1912 – 1968). PCB (polychlorinated biphenyl) poisoning was discovered in the area in and around Kitakyushu through the ingestion of a contaminated commercial brand of rice oil (1968).

In the rising concern over these problems, the government introduced many environmental protection laws in 1970 and established the **Ministry of the Environment** (MoE) in 1971. The current environmental law and policy is known as Law No.91 of 1993 (Japan-MoEnv, 1993). The Law is a set of 46 articles that describe the overall mission, policies, and responsibilities of the state, local governments, corporations, and citizens. It has provisions for setting environmental quality standards, pollution controls, conservation policies, costs, and sets up an Environmental Council to study and discuss the requirements of the law.

The specific concerns for air pollution problems were codified in Law No. 97 of 1968 (Japan-MoEnv, 1968), with the latest amendment by Law No. 32 of 1996. This set of laws describes the general provisions, procedures, and responsibilities 'to protect the public health and preserve the living environment with respect to air pollution, by controlling emissions of soot, smoke and particle from the business activities of factories and business establishments; by controlling emissions of particle while buildings are being demolished; by promoting various measures concerning hazardous air pollutants; and, by setting maximum permissible limits for automobile exhaust gases, etc'. Its intent is to help victims of air pollution-related health damage.

The outdoor ambient air quality standards are listed in Table 10-4, which shows a NO_x standard of 76 to 114 µg/m³. The most recent data, from 2012, is provided in Figure 10-9. The data reports the average annual concentration from monitoring stations that operated at least 6,000 hours per year. The total number of air monitoring stations is approximately 1,400 across Japan, and the number of roadside monitoring stations is approximately 400. These results agree with the findings associating NO_x with vehicle emissions, and that NO_x is typically higher near roadways. The plot shows a quick decline from 1970 (when data was first recorded) through 1980, and then it stabilizes to a constant value until 2005, when the value begins to decline again. Compare this with Figure 10-5, which shows US annual averages, and note how both figures currently show similar ambient air concentrations.

These results show that Japan has been successful in identifying and regulating sources of NO_x in its economy. However, as seen in Figure 10-5 and Figure 10-9 and their discussion, it is difficult to achieve significant NO_x reductions before all the different types of sources in an economy can be controlled.

FIGURE 10-9. Annual Average Ambient Air NO2 concentration [ppm], Japan.
(Japan-MoEnv, 2012). Plot shows average value from all monitoring stations, and a subset of those monitoring stations located near major roadways.

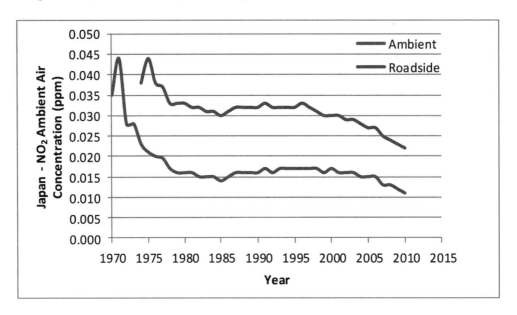

10.5 Control Technologies

Emissions standards are designed to limit the amount of NO_x present in the ambient air. There are many sources of nitrogen compounds (see Table 10-3) and in response, regulatory agencies typically focus on the large, stationary sources first and gradually include smaller sources and mobile sources over time. This section mainly discusses control of NO_x emissions from large, stationary combustion sources, but also includes a brief discussion on control from other sources.

■ Example 10-1.
Determine the uncontrolled hourly NO_x emissions for a 100 MW power generator (30% efficient) using the following combustion sources: a) Bituminous coal fired in a wall-fired, dry bottom boiler, b) No. 2 fuel oil, c) natural gas, and d) wet wood residue.

Solution:
Lookup the emission factors in Appendix 3 – Emission Factors. Use the given heating values to determine the amount of fuel needed, and the actual amount of emissions produced. When a range of values is provided, the average value was used.

a) Bituminous coal

$$x\frac{lbNO_x}{hr} = \frac{100\,MW}{0.3\,efficiency} \frac{3{,}413{,}000\,BTU}{MW \cdot hr} \frac{lb\,coal}{12{,}725\,BTU} \frac{ton}{2000\,lb} \frac{12\,lb\,NO_x}{ton\,coal} = 536.4 \frac{lbNO_x}{hr}$$

b) No. 2 Fuel Oil

$$x\frac{lbNO_x}{hr} = \frac{100\,MW}{0.3\,efficiency} \frac{3{,}413{,}000\,BTU}{MW \cdot hr} \frac{gal\,oil}{128{,}000\,BTU} \frac{24\,lb\,NOx}{1{,}000\,gal} = 213.3 \frac{lbNO_x}{hr}$$

c) Natural gas

$$x\frac{lbNO_x}{hr} = \frac{100\,MW}{0.3\,efficiency} \frac{3{,}413{,}000\,BTU}{MW \cdot hr} \frac{scf}{1000\,BTU} \frac{190\,lb\,NO_x}{10^6\,scf} = 216.2 \frac{lbNO_x}{hr}$$

d) Wet Wood Residue

$$x\frac{lbNO_x}{hr} = \frac{100\,MW}{0.3\,efficiency} \frac{3{,}413{,}000\,BTU}{MW \cdot hr} \frac{0.22\,lb\,NO_x}{10^6\,BTU} = 250.3 \frac{lbNO_x}{hr}$$

10.5.1 NO_x Control Technologies for Agriculture

Emissions result from the direct release of ammonia, urea, nitrates and other forms of nitrogen, and from the breakdown byproducts of fertilizer. These emissions are not currently regulated, even though it is one of the larger sources of anthropogenic nitrogen. The lack of regulation is probably due to the fact these emissions are individually small but widespread throughout agricultural areas. Also, there are few, if any, viable alternatives available to the use of fertilizers. The main control strategies are to minimize the over-application of fertilizer to cropland, to specify the time of application when plants are most likely to uptake it, and to use forms that are not readily emitted to the atmosphere. Irrigation practices can also affect the release of NO_x from cropland (Hall, et al., 1996), (Kurvits, et al., 1998).

10.5.2 NO_x Control Technologies for Combustion Processes

Combustion processes create NO_x emissions from several sources – fuel, thermal, and prompt (see section 10.2.4.1 *Formation of NO_x from Combustion*). It is important to understand the sources associated with a specific fuel and burning mechanism when designing and installing NO_x emission control equipment. For example, thermal NO_x emissions are dependent on the time and temperature profile of the combustion gases, whereas fuel NO_x depends on the air-to-fuel ratio. Prompt NO_x depends on both sets of variables, so thermal or fuel NO_x controls should also help limit prompt NO_x.

The following section highlights several boiler operating parameters and their effect on NO_x control. In general, *boilers* use combustion of fuel within a firebox to heat water. There are two main ways to design these systems. A *fire-tube boiler* is a type of boiler that boils water in a sealed container, or drum, and the gases pass around it. The gases may also flow through tubes that pass through the drum. A *water-tube boiler* is a type of boiler in which water circulates in tubes heated externally by the combustion gases. In both cases, the heat energy from the gas passes through the sides of the drum or tubes by thermal conduction, heating the water. The water may be used as hot water or as steam. The steam is then used to heat something else, or is sent through a turbine to generate electricity.

Figure 10-10 and Figure 10-11 show a water tube boiler. The hot combustion gases pass around the water containing tubes. The system may use several devices to improve efficiency. An economizer is used to preheat the incoming water before it is sent to the water tubes. The water in the tubes is heated to the boiling point, usually under pressure to increase this temperature. The water leaves the system as a saturated vapor (steam). Additional heat may be added in the superheater to further increase the temperature of the steam.

FIGURE 10-10. Schematic Plan View of Simplified Firebox with Water Tube Boiler.

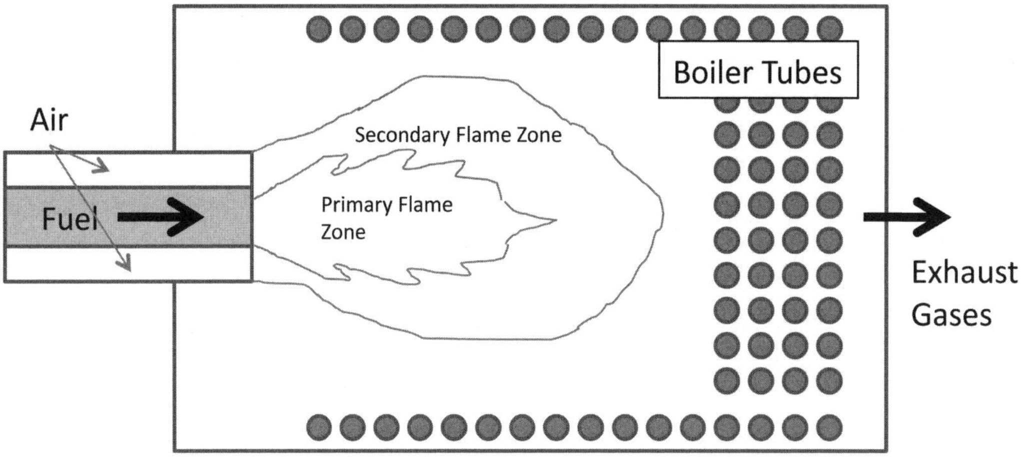

Boiler turndown is a ratio of capacity at full fire to the lowest firing point before shut-down. Old boilers may have only two firing positions, low and high. Newer boilers fire over a wider range of capacities. Depending on the controls, there may be fixed setting points or fully variable settings. If a 1 million BTU boiler can fire as low as 100,000 BTUs, then it has a 10:1 turndown ratio.

Turndown ratios are important for boilers that must operate over a wide range of capacities/demands. In general, from an efficiency standpoint it is best to have the boiler sized to match the load. Boilers that have a wider turn-down ratio are, therefore, typically more efficient at meeting

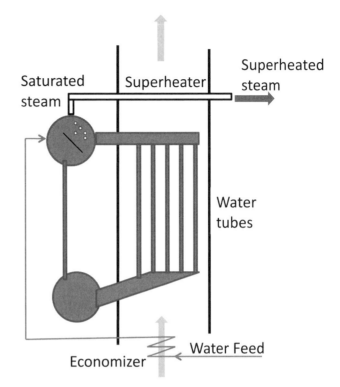

FIGURE 10-11. Schematic Showing the Heat Transfer Sections of a Water Tube Boiler.
The three heat transfer areas are the economizer, which preheats the incoming water; the water tubes, where the temperature and pressure of the water are increased to form steam; and the superheater, which further increases the temperature of the steam.

variable loads. However, there are exceptions, such as when it is possible to shut the boiler off for long periods of time, rather than run or cycle the boiler at very low fire conditions.

Boiler capacity is the amount of heat the boiler provides – often given in terms of BTUs or pounds of steam delivered per hour. The oldest method of rating boilers, still used to rate small boilers, is by horsepower (hp). One horsepower is defined as the ability to evaporate 34.5 lbs of water into steam at 212 °F (100°C). Larger boiler capacity is in lbs of steam evaporated per hour, under specified steam conditions. Maximum continuous rating is the hourly evaporation rate that can be maintained for 24 hours.

Boiler load is the horsepower, lbs of steam per hour, or BTU rating indicating the maximum capacity of the boiler. When a boiler operates at its maximum rated capacity, it is referred to as the maximum load. If the load varies from hour to hour, it operates at varying load. Load and load variation influence the operation and maintenance costs of the boiler.

Capacity and turndown should be considered together because some NO_x control technologies require boiler derating (reduction in the maximum load) in order to achieve guaranteed NO_x reductions. For example, flame shaping (primarily enlarging the flame to produce a lower flame temperature - thus lower NO_x levels) can require boiler derating because the shaped flame may expand so much that it impinges, or contacts, on the furnace walls at higher firing rates.

Boiler efficiency, in the simplest terms, represents the difference between the energy input from the fuel and energy output of the hot water or steam created. Some low NO_x controls reduce emissions by lowering flame temperature, particularly in boilers with inputs less than 100 MMBtu/hr. Reducing the flame temperature decreases the radiant heat transfer from the flame and lowers boiler efficiency. Depending upon the boiler design, this may be partly compensated by increased convective heat transfer.

Excess air means that more air (more oxygen) is added than required by stoichiometry. The excess air also contains nitrogen and water (humidity). All these excess gases absorb heat during combustion, thus lowering the maximum temperature of the combustion gases. The reduced temperatures cause a reduction in the formation of thermal NO_x.

Some excess oxygen is always needed because there are many ambient and atmospheric conditions that can affect oxygen supply. For example, colder air is denser and contains more oxygen than warm air by unit volume; wind speed affects every chimney/ flue/ stack differently, and barometric pressure further affects the air input. An excess oxygen setting allows there to be enough oxygen available for complete combustion even under unfavorable conditions. A boiler operated at the stoichiometric amount of oxygen or less cannot completely burn all the fuel and therefor emits some unburned or incompletely-burned fuel. Incomplete combustion causes air pollution and could potentially cause fires or explosions.

A boiler's excess air supply provides for safe operation. A typical burner is usually set up with 20-50% excess air (which yields a mole fraction of 3-7% O_2 in the exhaust). NO_x controls require higher excess air levels and result in the fuel's heat release being used to heat the extra air rather than transferring it to the boiler water as usable energy. Thus, excess air causes increased heat losses in the stack and reduced boiler efficiency.

■ Example 10-2.

Determine the amount of excess air needed to yield 7% oxygen in the exhaust gases from the combustion of natural gas (CH_4)

Solution:
First, determine the stoichiometric amount of O_2 needed. Next, determine the quantity of N_2 associated with the O_2 in the air. Assume the air is dry and has a composition of 79 mol% N_2 and 21 mol% O_2. The reaction is

$$CH_4 + 2O_2 \rightarrow CO_2 + 2H_2O$$

One mole of fuel has one mole of C and 4 moles of H, so 4 moles of O (or 2 moles of O_2) are needed. The amount of N_2 added is 79/21 of the amount of O_2 added.

Stoichiometric O_2 needed to burn one mole of CH_4:
   Atom Reaction 1: $C + O_2 \rightarrow CO_2$: one mole O_2 needed per mole of C in fuel
   Atom Reaction 2: $4H + O_2 \rightarrow 2H_2O$: one mole O_2 needed for four moles H in fuel

The actual amount of O_2 and N_2 added is increased by a factor of (1 + xs) to account for the excess air. The variable xs is the fraction of excess air added (50% excess would give xs=0.5).

Next, determine the species and quantities in the exhaust gas coming out. The atom reaction one shows one mole of CO_2 for each mole of C in the fuel. Atom reaction 2 shows ½ mole H_2O for each mole H reacted in the fuel.

Create a table to help keep the information organized:

IN			OUT	
	Stoich	Actual		Exhaust
Species	Moles	Moles	Species	Moles
C	1	1	CO_2	1
H	4	4	H_2O	2
O_2	2	2(1+xs)	O_2	2(1+xs)-2
N_2	79/21*moles of O_2	7.52(1+xs)	N_2	7.52(1+xs)

Note: xs is the fraction of excess oxygen added, and is the unknown quantity in the problem.

O_2 out is the difference between the actual O_2 in and the stoichiometric amount used. The N_2 remains constant since it does not participate in any of the reactions. If thermal NO_x formed, we would need to include another reaction, which would reduce the N_2 out and increase the stoichiometric O_2 needed. Similar extra reactions would be needed to account for the formation of fuel NO_x (see Example 10-3).

The mole fraction of O_2 (0.07) in the exhaust is equal to the ratio of O_2 to the sum of all species in the exhaust:

$$[O_2 \text{ exhaust}] = 0.07 = \frac{n_{O2}}{[n_{CO2} + n_{H2O} + n_{O2} + n_{N2}]} = \frac{\{2(1+xs)-2\}}{\left[\{1\}+\{2\}+\{(2(1+xs)-2\}+\{7.52(1+xs)\}\right]}$$

Solving for 'xs' yields 0.552. The solution shows that using 55.2 % excess air yields an exhaust gas that has 7% oxygen in it. A value of 7% is a typical target for boiler operation because it ensures there is enough oxygen present to complete combustion (also see Figure 10-13). The problem with too much excess air is that it reduces boiler efficiency. Boiler efficiency is directly proportional to the temperature of the gases. The extra mass from excess air absorbs some of the heat generated by combustion and acts to reduce the temperature.

Figure 10-12 highlights the four methods of NO_x reduction in a boiler, as well as several specific techniques in each method.

FIGURE 10-12. Major NO_x Reduction Techniques.

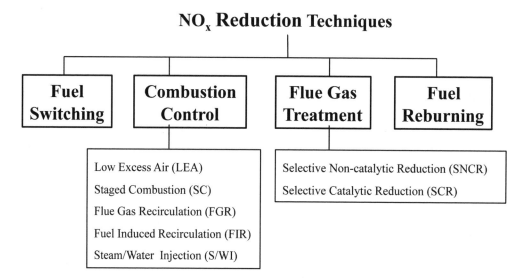

10.5.2.1 Fuel Switching

This method is used to reduce fuel NO_x emissions. The best fossil fuel for minimizing fuel NO_x is natural gas. Natural gas contains some nitrogen in the form of nitrogen gas, but not the organic compounds that lead to fuel NO_x. Refined oils, such as No. 2, contain less than 0.05% fuel-bound nitrogen. Residual oils can contain up to 0.6% fuel-bound nitrogen. NO_x formed by fuel-bound nitrogen from residual oils can account for 20-50% of the total NO_x level. Conversion from a fuel with high fuel bound nitrogen to a low level can significantly reduce total NO_x levels.

Nitrogen Emissions

However, it is not always easy to convert a boiler from one fuel to a different one. Considerations include boiler configuration, end use, and re-training of operators. In many cases, a new fuel requires the construction of an entirely new system.

■ Example 10-3.

Compare the NO_2 emissions in ppm from a) oil ($C_{12}H_{20}O_1N_{0.01}$) and b) coal (70.8 wt% C, 9 wt% H, 6 wt% O, 0.2 wt% N, and 14 wt% ash) burned with 50% excess air. In both cases assume that all nitrogen in the fuel is converted to fuel-NO_x and that no thermal NO_x forms.

Solution:
Write out balanced chemical reactions for each component of the fuel. Determine the amount of oxygen needed and apply the excess air. Finally, calculate the number of moles of each component in the exhaust.

Reaction 1: $C + O_2 \rightarrow CO_2$: one mole O_2 needed per mole of C in fuel
Reaction 2: $2H_2 + O_2 \rightarrow 2H_2O$: one mole O_2 needed for four moles H in fuel
Reaction 3: $N + O_2 \rightarrow NO_2$: one mole of O_2 needed per mole of N in fuel

a) Assume one mole of fuel reacts $C_{12}H_{20}O_1N_{0.01}$
$C_{12}H_{20}O_1N_{0.01} + (17.01 - 0.5) O_2 \rightarrow 12\ CO_2 + 10\ H_2O + 0.01\ NO_2$

One mole of this fuel contains 12 moles of C and 20 of H. We record the fuel O as negative because each mole of fuel O replaces 0.5 moles of air-O_2. Air N_2 is obtained by multiplying air O_2 by the ratio of their mole fractions in air (79/21).

The actual air O_2 and N_2 are calculated by multiplying each by the excess air ratio [(1 + xs) where xs is the excess fraction. xs = 0.5 in this example].

IN				OUT	
	Stoich	Actual			Exhaust
Species	Moles	Moles		Species	Moles
C	12	12		CO_2	12
H	20	20		H_2O	10
Fuel-O	-0.5	-0.5			
Fuel-N	0.01	0.01		NO_2	0.01
Air O_2	16.49	24.74		O_2	8.25
Air N_2	62.03	93.05		N_2	93.05

Finally, determine the amount of each species in the exhaust and calculate the concentration of NO_2 as the ratio of NO_2 present to the sum of all exhaust species. Multiply by 1,000,000 to obtain ppm:

$$\text{Wet basis: } [NO_2] = \frac{0.01}{[12+0.01+8.25+93.05]} \times 10^6 = 88.3 \, ppm$$

A wet basis is the actual composition of the emissions.

$$\text{Dry basis: } [NO_2] = \frac{0.01}{[12+10+0.01+8.25+93.05]} \times 10^6 = 81.1 \, ppm$$

A dry basis is used to make for easier comparison between different fuels, and to allow one to ignore the relative humidity of the incoming air.

b) This part is very similar to part a, except we must first convert the weights into moles. The following table shows the results from these calculations, where we assume 100 g of fuel reacts.

IN				OUT		
	Weight	Stoich	Actual			Exhaust
Species	g	Moles	Moles		Species	Moles
C	70.8	5.9	5.9		CO_2	5.9
H	9	9	9		H_2O	4.5
Fuel-O	-6	-0.375	-0.375			
Fuel-N	0.2	0.014	0.014		NO_2	0.014
Ash	14	-	-			
Air O_2		7.96	11.94		O_2	3.98
Air N_2		29.95	44.93		N_2	44.93

The concentration of NO2 in the exhaust in ppm is:

$$\text{Wet basis: } [NO_2] = \frac{0.014}{[5.9+4.5+0.014+3.98+44.93]} \times 10^6 = 236 \, ppm$$

$$\text{Dry basis: } [NO_2] = \frac{0.014}{[5.9+0.014+3.98+44.93]} \times 10^6 = 255 \, ppm$$

The ash plays no role in either calculation because it is a solid and does not contribute significantly to the exhaust gas volume.

10.5.2.2 Combustion Control Techniques

Combustion control techniques reduce thermal and fuel NO_x formation by controlling the flame temperature, the amount of time the gases are in the peak temperature zone, and/or the oxygen concentration in each part of the combustion chamber. These techniques are more economical than flue gas treatment (post-combustion) methods and are frequently utilized on industrial boilers. These methods can be used singly or in combination, depending on the system design, the required level of emissions, and cost. These techniques can be very difficult or impossible to retrofit into an existing system, unlike post-combustion methods.

10.5.2.2.1 Low Excess Air (LEA) Firing

Low excess air (LEA) reduces emissions by carefully controlling the amount and location of the air used to support combustion in a boiler. It is usually the simplest technique, and can be adapted to most boiler configurations which have controls for air flow rate and location. It works by reducing the amount of excess oxygen used during combustion. Figure 10-13 shows in general how NO_x emissions vary with the oxygen content.

FIGURE 10-13. Effect of Excess Oxygen in Emissions on NO_x. Adapted from (NGBBC, 2005)

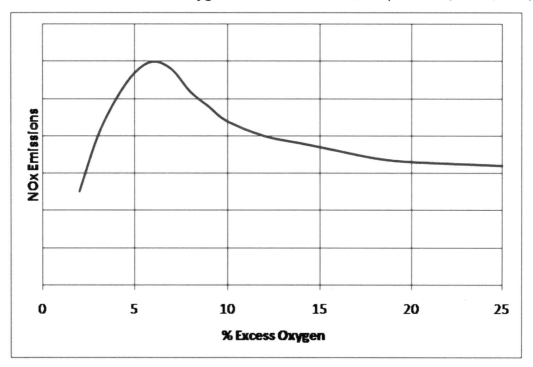

Operating with excess air is important because it ensures complete combustion in case of poor mixing, change in fuel quality over short times, or minor system upsets. It prevents or minimizes the formation of carbon monoxide, soot, or other unburned hydrocarbons. Incomplete combustion can also lead to the formation of combustible or explosive gas mixtures downstream from the boiler that creates the possibility of an explosion hazard.

The NO_x generation rate typically peaks at excess emissions oxygen levels of 5 – 7% (25% - 45% excess oxygen in the inlet) where the combination of high-combustion temperatures and the higher oxygen concentrations act together. At higher amounts of excess air, NO_x emissions decrease due to lower temperatures. At lower excess air amounts, which can create localized reducing conditions, NO_x emissions decrease due to a lack of oxygen.

LEA operation is a good combustion management practice because it maximizes boiler efficiency – which relates directly to boiler operating temperature. Therefore, most boilers operate at LEA regardless of whether NO_x reduction is an issue. Low excess air is achieved by changes in operating procedures, system controls, or both. LEA can result in a 10% reduction in NO_x emissions.

10.5.2.2.2 Staged Combustion (SC)

Staged combustion burners, the most common type of low-NO_x burners (LNB), achieve lower NOx emissions by staging the injection of either air or fuel in the combustion chamber, see Figure 10-14, where $\phi=1$ represents the stoichiometric air to fuel ratio. When $\phi > 1$, the mixture contains excess air. Staging increases the size of the flame zone, reduces the peak temperature and alters the availability of oxygen. Classifications of staged combustion burners are either air staged burners or fuel staged burners. Staged combustion can result in NO_x reductions of up to 60% for natural gas.

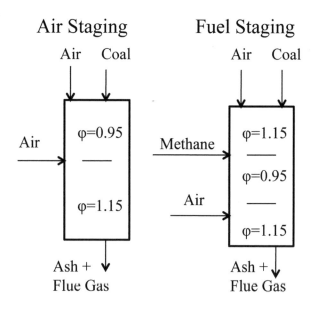

FIGURE 10-14. Feed Arrangements in Staged Combustion. $\phi=1$ represents the stoichiometric air to fuel ratio.

10.5.2.2.3 Air Staging

Staged air burning divides the combustion air to reduce the oxygen concentration in the primary burner zone, see Figure 10-14. Staged air lowers the NO_x formation and increases the amount of reducing agents. Secondary and tertiary air completes the combustion downstream of the primary zone, thus lowering the peak temperature and reducing efficiency and thermal NO_x formation.

Due to the staging effect of staged combustion air (SCA) burners, flame lengths tend to be longer than those of conventional burners. The extended length may be of particular concern when retrofitting packaged units because there is a possibility of flame impingement on the furnace walls, resulting in additional tube corrosion or failure.

Staged combustion can be accomplished external to the burner body by separate introduction of air. External air staging techniques commonly used for larger boilers include:

- **Burners-out-of-service (BOOS)** which is a staged combustion technique typically used for large boilers. Introducing additional gas through operational burners at the lower furnace zone creates fuel rich conditions which reduce NO_x. Additional air is supplied through the non-operating burners above the lower zone to complete combustion. Since burners are taken out-of-service, the unit capacity is reduced.
- **Over-fire-air (OFA)** which typically involves the injection of secondary air into the furnace through OFA ports above the top burner level, coupled with a reduction in primary airflow to the burners. The fuel-rich air-fuel mixture fed to the burners leads to reduced flame temperature and oxygen concentration. The over-fire-air helps finish the combustion when the gases are out of the primary flame zone.

10.5.2.2.4 Fuel Staging

In staged fuel burners, combustion air is introduced without splitting and instead the fuel is divided into primary and secondary streams. Despite the high-oxygen concentration in the primary combustion zone, thermal NO_x formation is limited by low-peak flame temperature, which results from the fuel-lean combustion. Quenching of the flame by the high excess air levels also occurs, further limiting the peak flame temperature and providing active reducing agents for NO_x reduction. Inerts (CO_2, H_2O, and N_2) entering the secondary combustion zone from the primary zone reduces peak flame temperatures and localized oxygen concentrations. These also reduce NO_x formation. An advantage of staged fuel burners over staged air burners is that they tend to have shorter flame lengths, decreasing the likelihood of flame impingement.

10.5.2.2.5 Cyclonic Low NO_x Burner

Cyclonic burners use high tangential velocities in the burner to create a swirling flame pattern in the furnace. The swirling, turbulent flow causes excellent internal mixing as well as recirculation of combustion gases, which reduces the temperature of the near-stoichiometric flame and lowers thermal NO_x formation. The tangential flame causes close contact between combustion gases and the furnace wall, adding a convective component to the radiant heat transfer within the furnace. The increased heat transfer and low excess air operation of the cyclonic burner results in increased boiler efficiency. A small quantity of low-pressure steam can be injected into the burner to achieve ultra-low NO_x levels and reduces the local flame temperature and NO_x formation.

10.5.2.2.6 Flue Gas Recirculation (FGR)

FGR mainly impacts thermal NO_x and has little effect on fuel NO_x emissions. FGR involves recycling a portion of the combustion gases from the emissions back into the boiler, see Figure 10-15. The extra gas reduces peak flame temperatures and allows a reduction of any recycled NO_x. The recycled combustion gas dilutes (not replaces) the incoming air and reduces the oxygen concentration (not amount) in the combustion zone. During the combustion, the recirculated flue gases absorb some of the heat and thereby reduce the peak combustion temperature.

For a given firing rate, FGR increases the throughput of gases in the burner and firebox. Increased FGR can increase the flame size and cause impingement on heat transfer surfaces thus limiting the maximum firing rate. Impingement occurs when the primary flame zone contacts a surface within the boiler. It creates hot spots on the surface leading to material degradation and corrosion. Reduced energy efficiency results from the reduced heat transfer due to the lower flame temperatures. Boiler thermal efficiency reductions resulting from FGR are limited to around 1% or lower. Natural gas boilers are usually not operated with more than 20 percent FGR due to flame stability considerations. Distillate-oil-fired boilers have even greater tendency for flame instability and higher emissions of unburned combustibles, which limits the amount of FGR even more than for natural gas.

FGR is not used for fuels with significant amounts of sulfur, due to the possible formation of acid gases. FGR is also not used when burning heavy fuel oil or coal, which can generate significant particulate matter (soot) that could cause a buildup on fans and ducts.

In order to retrofit a boiler with FGR, it is usually required to add a gas recirculation fan, dampers, ducting, and controls. A well-designed boiler can exploit the system fluid dynamics to induce flue gas recirculation without additional equipment. Overall NO_x reductions from FGR are typically in the 40 – 70% range.

10.5.2.2.7 Fuel Induced Recirculation

Fuel induced recirculation is a control technology for natural-gas-fired boilers. FIR involves recirculating a portion of the boiler flue gas and mixing it with the fuel at some point upstream of the burner, see Figure 10-15.

The primary difference between FIR and FGR is how the flue gas recirculates. FIR recirculates the flue gas with the fuel stream, whereas FGR recirculates the flue gas into the combustion air. FIR reduces the concentration of hydrocarbon radicals that produce prompt NO_x by diluting the fuel prior to combustion. This dilution also reduces the fuel volatility. Additionally, FIR reduces thermal NO_x in the same manner as FGR, by acting as a thermal diluent. Thus, the main benefit of FIR technology is that it impacts both prompt NO_x and thermal NO_x formation in gas-fired boilers.

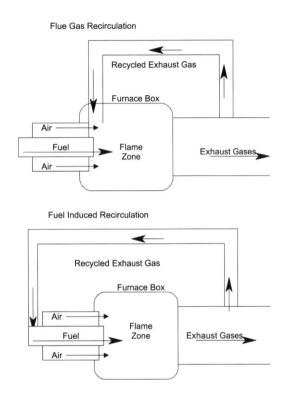

FIGURE 10-15. Schematic Showing Flue Gas Recirculation (FGR) and Fuel Induced Recirculation (FIR) Systems.

10.5.2.2.8 Steam / Water Injection

Steam or water injected into the flame reduces the peak flame temperature and the oxygen concentration. The reduction of the peak flame temperature reduces NO_x emissions by as much as 75 percent, depending on the amount of water or steam injected. Less water than steam is needed to achieve the same quenching effect because of the heat of vaporization required to change water into steam. Water/Steam can also be directly injected into the combustion air just prior to the flame. Advanced designs use the injected steam directly at the burner, where it helps to increase the mixing of fuel and air. This improved mixing can be used to reduce excess air requirements, and thereby reduce NO_x formation.

The amount of water injected normally varies between 25 and 75 percent of the natural gas feed rate, on a mass basis. However, the technique can create increased CO emissions due to the quenching effect of the water. It increases thermal losses in the stack due to the increased water content of the flue gas. Another concern is the potential for unsafe combustion conditions resulting from poor feed rate control.

10.5.2.3 Flue Gas Treatment Methods

Flue gas treatment is a post-combustion technology. It works by adding a reducing agent to the hot flue gases, which react with the NO_x to form N_2 and H_2O. Typical reducing agents are ammonia and urea. The reaction takes place non-catalytically at a high temperature (1,600 to 2,000°F) or catalytically at a lower temperature (575 to 800°F).

Retrofitting these technologies to boilers typically involves installation of reagent injection nozzles, reagent storage and control equipment, and catalytic reactors. NO_x reduction in the flue gas depends in large part on flue gas temperature, so injection nozzle placement is limited to where acceptable process temperatures are present. In packaged industrial boilers, available locations for reagent injection and catalyst placement are further limited by space considerations.

10.5.2.3.1 Selective Non-Catalytic Reduction (SNCR)

SNCR involves the injection of a NO_x reducing agent, such as ammonia (NH_3) or urea ($CO(NH_2)_2$), into the high-temperature boiler exhaust gases. The choice depends on local costs of the reactant. Figure 10-16 below illustrates the basic process for NO_x removal with ammonia.

FIGURE 10-16. Selective Non-Catalytic Reduction (SNCR).

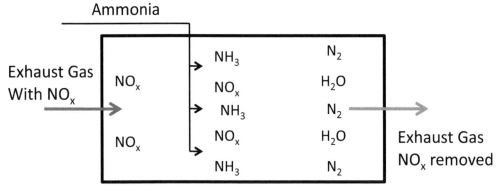

Similar NO_x reduction efficiencies occur with ammonia or urea. Reductions of 70% are possible under carefully controlled conditions, but reductions in the range of 40 to 50% are typical.

The technology can be difficult to apply to industrial boilers that modulate or cycle frequently. The difficulty arises because the ammonia (or urea) must be injected in the flue gases at a specific location that has the required temperature. However, the temperature profile changes when a boiler cycles, and thus the optimum location may not coincide with the location of the injectors.

10.5.2.3.1.1 Ammonia based SNCR

This method uses a gas phase homogeneous reaction between NO_x and ammonia to produce molecular nitrogen and water (equations (10.13) and (10.14)). The aqueous solution of ammonia or anhydrous ammonia is vaporized and injected into the flue gas through wall-mounted nozzles at a location selected for optimum reaction temperature and residence time. The optimum reaction temperature range for this process is 1,600 to 2,000°F, although this can be lowered to 1,300°F with additional injection of gaseous hydrogen.

10.13 $6\,NO + 4\,NH_3 \rightarrow 5\,N_2 + 6\,H_2O$

10.14 $6\,NO2 + 8\,NH_3 \rightarrow 7\,N_2 + 12\,H_2O$

The amount of ammonia added remains essentially constant during periods of normal operations. Unreacted and excess ammonia can be emitted and is called ammonia slip, breakthrough, or carryover. Ammonia slip is kept low because ammonia can react with other combustion constituents to form ammonia salts that result in additional particulate matter (PM) emissions. The amount added is adjusted during startup and shutdown to minimize slip: increasing loads use an ammonia-rich ratio, and decreasing loads use an ammonia-lean ratio.

Achievable NO_x reductions for individual boilers depend on the flue gas temperature, the residence time at that temperature, the initial NO_x concentration, the NH_3/NO_x ratio, the excess oxygen level, and the degree of ammonia/flue gas mixing. Also, stratification of both temperature and NO_x in the flue gas can affect the performance of the SNCR control. The optimum placement of SNCR injectors requires a detailed mapping of the temperature profile in the convective passes of the boiler, because of the narrow temperature window.

■ Example 10-4.

Determine the amount of ammonia required to remove 3.2 kg NO_2/min from an exhaust gas.

Solution:
Convert mass of NO_2 to moles, and then use equation (10.14) to perform the stoichiometric calculation. Assume a one-minute basis.

$$x\,NO_2\,mol = 3.2\frac{kg\,NO_2}{min} * \frac{1\,kg-mol}{46\,kg} = 0.07\frac{kg-mol\,NO_2}{min}$$

Equation (10.14) shows that six moles NO_2 require eight moles NH_3, so

$$x\,\frac{kg\text{-}mol\,NH_3}{min} = 0.07\,\frac{kg\text{-}mol\,NO_2}{min} * \frac{8\,kg-mol\,NH_3}{6\,kg-mol\,NO_2} = 0.093\,\frac{kg\text{-}mol\,NH_3}{min}$$

or 1.58 lb NH_3/min are needed for the removal.

10.5.2.3.1.2 Urea based SNCR

This process injects an aqueous solution containing urea and chemical enhancers into one or more locations, depending on the boiler type and size. The urea reacts with NO_x in the flue gas to produce nitrogen, carbon dioxide, and water, see equation (10.15). The main advantage of urea injection over ammonia injection is that urea is a non-toxic liquid that can be safely stored and handled.

10.15 $\quad CO(NH_2)_2 + 2\,NO + 1/2\,O_2 \rightarrow 2\,N_2 + CO_2 + 2\,H_2O$

Like ammonia injection, urea-based SNCR is effective only within a narrow temperature range. Without the use of chemical enhancers, urea injection effectively reduces NO_x at temperatures between 1,650 and 2,100°F. Residence time within this temperature range is the primary design consideration. By using chemical enhancers and adjusting concentrations, greater NO_x reduction efficiency can be achieved over a wider temperature window. Urea can oxidize to form NO_x if released at too high a temperature.

10.5.2.3.2 Selective Catalytic Reduction (SCR)

SCR involves the injection of ammonia in the boiler exhaust gases in the presence of a catalyst. The catalyst allows the ammonia to reduce NO_x levels at lower exhaust temperatures (500°F and 1,200°F), depending on the catalyst used. Selective catalytic reduction can result in NO_x reductions up to 90%.

The SCR process takes advantage of the selectivity of ammonia to reduce NO_x to nitrogen and water at lower temperature in the presence of a catalytic surface. The catalyst uses oxides of titanium, molybdenum, tungsten, vanadium, and/or zeolites - which are alumina-silicate-based. They may include other components that add structural stability. Catalysts come in various shapes and sizes, according to the particular application. Figure 10-17 shows the basic setup.

FIGURE 10-17. Selective Catalytic Reduction of NOx with Ammonia.

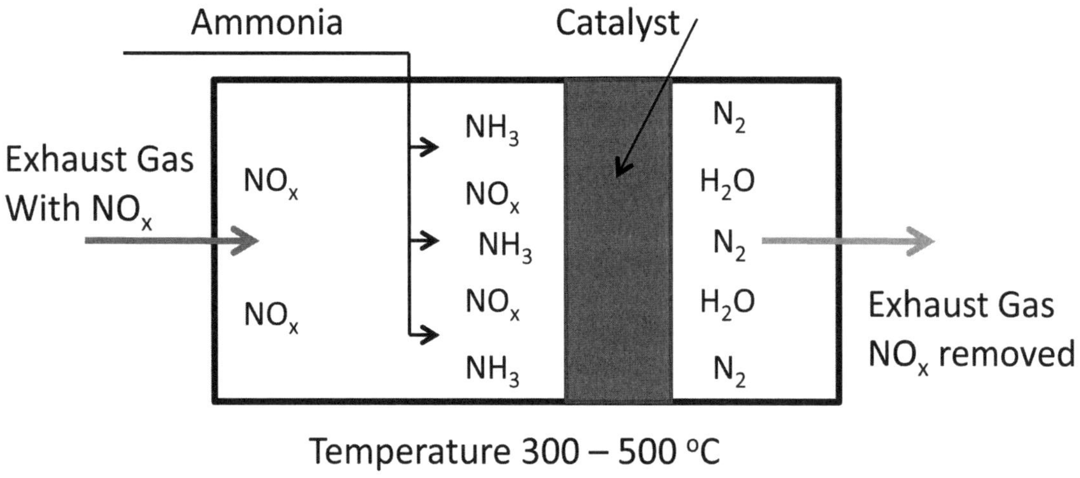

Gaseous ammonia is injected with a carrier gas, typically steam or compressed air, into the flue gas upstream of the catalyst. The ammonia/ flue gas mixture enters the catalyst, where it distributes through the catalytic bed, and reacts with the NO_x in the flue gas to form N_2 and water. The treated flue gas then leaves the catalytic reactor and continues to the exit stack or air pre-heater. Ammonia slip tends to be less with SCR than with SNCR.

SCR operates most efficiently at temperatures between 300 – 500°C (575 and 800°F) and when the flue gas is relatively free of particulate matter, which tends to contaminate or "poison" the catalytic surfaces. Operation above this temperature range can reduce the effectiveness of certain catalysts, and operation at lower temperatures increases the likelihood of ammonium sulfate particle formation within the catalyst.

10.5.2.3.3 Fuel Reburning

Fuel reburning is usually only used at large utility power plant boilers firing coal or residual oil. It uses an injection of natural gas after primary fuel combustion in a three-stage combustion process based on fuel staging. The main burner zone operates in a fuel-lean condition with about 90% of the required fuel input. The balance of the fuel (also known as reburn fuel) is injected at the point where primary combustion is complete. The location must be within the chemical reducing zone to decompose the NO_x and the chemical constituents likely to form NO_x. Tertiary air is blended in to complete the combustion process. Natural gas reburning can produce NO_x reductions in the 50 to 70 percent range. The reburn fuel makes up about 15 to 20 percent of the total fuel used.

10.5.3 Choosing a NO$_x$ Reduction Technology

NO$_x$ controls can worsen boiler performance while other controls can appreciably improve performance. Failure to account for all of the boiler operating parameters can lead to increased operating and maintenance costs, loss of efficiency, elevated CO levels, and shortening of the boiler's life. Selecting the best low-NO$_x$ control package should keep total boiler performance in mind. New systems designed for low-NO$_x$ technologies can provide NO$_x$ reductions without affecting total boiler performance. Table 10-5 shows a summary of expected NO$_x$ reductions for different techniques used on oil and gas fired boilers. Note the following abbreviations in the figure:

LNB – Low NO$_x$ Burner
FGR – Flue Gas Recirculation
SCA – Staged Combustion Air
SNCR – Selective Non-Catalytic Reduction
SCR – Selective Catalytic Reduction

Table 10-5. Performance Summary for NO$_x$ Control Techniques on Industrial Boilers. (Oland, 2002)

Boiler Type	NOx Control Technique	Natural Gas		Distillate Oil		Residual Oil	
		NO$_x$ Reduction [%]	NO$_x$ Emission lb/MMBTU	NO$_x$ Reduction [%]	NO$_x$ Emission lb/MMBTU	NO$_x$ Reduction [%]	NO$_x$ Emission lb/MMBTU
Firetube							
	LNB	20 - 78	0.02 - 0.08	10 - 20	0.15	10 - 60	0.09 - 0.25
	FGR	40 - 76	0.02 - 0.08	40 - 76	0.04 - 0.016	10 - 20	
	SCA	0 - 10	0.0 - 0.08	0 - 10		0 - 49	0.0 - 0.11
	LNB + FGR	30 - 80	0.02 - 0.04				
	SCR	53 - 91	0.006 - 0.05				
Water tube							
	LNB	0 - 71	0.03 - 0.20	10 - 40	0.08 - 0.33	10 - 60	0.09 - 0.60
	FGR	40 - 74	0.02 - 0.10	20 - 68	0.04 - 0.30	4 - 30	0.12 - 0.35
	SCA	0 - 60	0.06 - 0.24	0 - 40	0.09 - 0.30	0 - 60	0.20 - 0.74
	LNB + FGR	55 - 84	0.02 - 0.20	-	0.03 - 0.25	-	0.10 - 0.55
	LNB + SCA	30 - 80	0.10 - 0.20	-	0.10 - 0.30	-	0.20 - 0.40
	SNCR - ammonia	30 - 72	0.03 - 0.20				
	SNCR - urea	50 - 60	0.049 - 0.13				
	SCR	53 - 90	0.011 - 0.055				

● **Case Study 10-1. Nitrogen Emission Control in a Coal Fired Power Plant.**
In this case study we will design a nitrogen oxide control system for the coal fired power plant introduced in Case Study 2-1, and further developed in Case Study 3-3, 8-1, and 9-5. Summary of given information: Fuel is a bituminous coal (10,750 BTU/lb) of elemental analysis $C_7OH_{100}O_2N_{0.2}S_{0.3}Ash_{(7wt\%)}$. The fuel is burned with 50% excess air in a cyclone furnace of a 350 MW power plant with a 34.5% efficiency.

In case study 2-1 we assumed all the nitrogen in the coal, and only in the coal, oxidized and exited with the exhaust gas. This is not a very good assumption, as the nitrogen in the coal (fuel NO_x) and nitrogen in the air (thermal NO_x) can both contribute. We will revise this number by using the emission factor given in appendix 3. The material is bituminous coal, dry bottom, tangentially fired in the cyclone furnace. The emission factor from the table is 11 lb/ton of coal used.

The required removal depends on the NSPS limits for NO_x, as described in appendix 2. The NSPS for NO_x is 1.0 lb/MW-hr gross energy output

$$\frac{11\, lb_{NOx}}{1\, ton_{coal}} * \frac{ton}{2,000\, lb} * \frac{1\, lb\, coal}{11,750\, BTU_{in}} * \frac{1_{in}}{0.345_{out}} * \frac{3.413 \times 10^6\, BTU}{MW \cdot hr} = 4.63 \frac{lb_{NOx}}{MW_{out} \cdot hr}$$

The required removal efficiency to meet NSPS is:

$$\eta_{NO_x} = \frac{(4.63 - 1.0)}{4.63} * 100\% = 78.4\%$$

Next, we need to choose a method for control of the NO_x. Table 10-5 shows reduction strategies for gas and oil. The expected reductions for coal will be lower than those presented in the table. The three issues for the formation of NO_x; time, temperature, and turbulence, are generally more conducive to NO_x formation for coal combustion. The high sulfur and PM in the exhaust exclude the recirculation technologies. The combustion technologies work by changing the size and shape of the combustion zone. These can help, but are unlikely to achieve the required level of reduction. The remaining technologies focus on destruction of NO_x in the exhaust by adding ammonia to reduce the NO_x. This technology will work, but adding and storing ammonia is expensive, so minimization is important.

For this case study, a combination of low NO_x burners (LNB) and selective catalytic reduction (SCR) will be used. The LNB is expected to reduce NO_x by 30%. The remainder will be treated in the SCR using urea as the ammonia source. It is unlikely the LNB would be able to reduce NO_x to the emission limit, but the LNB will reduce the amount of NO_x reduction needed and so will also decrease the amount of chemical urea to be purchased and used.

The amount entering the SCR, after reduction due to LNB:

$$4.63 \frac{lb_{NOx}}{MW_{out} \cdot hr} * (1-0.3) * 350 MW = 1{,}134 \frac{lb_{NOx}}{hr}$$

The amount that can be discharged is

$$1.0 \frac{lb_{NOx}}{MW_{out} \cdot hr} * 350 MW = 350 \frac{lb_{NOx}}{hr}$$

Therefore, we must remove 784 $lbNO_x$/hr. The destruction reactions are given in equations (10.13) and (10.14). The amount of ammonia needed depends on the ratio of NO to NO_2. We will assume a mole ratio of 8:1, and that the reactions happen equally likely. Obviously, these two assumptions need verification with much more complex kinetics analysis of the particular system. It is well beyond the scope of this text to perform such calculations / simulations. It would be unwise to just assume the worst case (i.e. all NO_2), as too much ammonia will allow some ammonia to pass through the SCR and enter the environment, where it is likely to transform into NO_x. The number of moles of ammonia can be determined from the ratio (NO/NO_2) and the total mass (found using the molecular weights of each):

Let x = moles/hr of NO, y = moles/hr of NO_2. The MW_{NO} = 30 and MW_{NO2} = 46. Then:

$$\frac{x}{y} = 8$$

$$30x + 46y = 784 \frac{lb_{NOx}}{hr}$$

Solving yields x = 21.9 mol_{NO}/hr or 657.9 lb_{NO}/hr, and y = 2.74 mol_{NO2}/hr or 126.1 lb_{NO2}/hr.

Next, determine the amount of ammonia needed for each reaction:

$$21.9\frac{mol_{NO}}{hr} * \frac{4 mol_{NH3}}{6 mol_{NO}} = 14.6 mol_{NH3}$$

$$2.74\frac{mol_{NO2}}{hr} * \frac{8 mol_{NH3}}{6 mol_{NO2}} = 3.7 mol_{NH3}$$

Total required is 14.6 + 3.7 = 18.3 mol_{NH3}.

The final design would need to choose a catalyst and note the temperature range over which it is best to work. Simulations of the boiler system can show the temperature profile within it and help to place the catalyst and ammonia injection system.

10.6 Questions

* - Questions and problems may require additional information not included in the textbook.

1. Why is the ambient air concentration of ammonia higher over land than over the ocean?

2. Describe in your own words the differences between thermal NO_x and fuel NO_x.

3. Go to the US-EPA website (http://www.epa.gov/aircompare/) and find the ambient NO_2 in your hometown, where you work, or where you go to school.*

4. Go to Japan-Ministry of the Environment website and find the ambient NO_2 in Tokyo (http://www.env.go.jp/en/statistics/air/).*

5. Go to EU-EEA website (http://www.eea.europa.eu/themes/air/airbase) and find the ambient NO_2 in Paris, France (or another city of your choice).*

6. Why has NO_x been treated as a seasonal only problem in the eastern US?

7. Why didn't the US-EPA 'NO_x -SIP call' include the western states?

8. What effect would turndown have on SNCR performance?

9. Would you expect the open burning of wood in a campfire to generate NO_x?

Chapter 10

10. Describe the NO_x control strategy for a coal-fired boiler that generates mostly fuel NO_x

11. Describe the NO_x control strategy for an oil-fired boiler that generates both fuel and thermal NO_x

12. Describe the NO_x control strategy for a natural gas-fired boiler that generates only thermal NO_x

13. Nitrates are usually very soluble. Discuss NO_x removal by converting NO_2 to NO_3^- and then scrubbing the nitrate in a water spray tower. Would this be a good idea? *

14. How could the use of a catalyst reduce the need for ammonia in a NO_x reduction unit?

15. Make a list of three questions you have about this chapter or air pollution concerns you have.

10.7 Problems

1. Determine the NO_2 emissions rate (lb/hr) from a steam plant that uses sugar mill waste (3,300 BTU/lb) to generate 1,000 lb steam/ hr. The steam plant is 86% efficient in converting heat to steam. (Note: It takes about 1,100 BTU to generate 1 lb steam).

2. Determine the NO_2 emissions rate (lb/hr) from a steam plant that uses natural gas in a large wall fired boiler to generate 100,000 lb steam/hr. The steam plant is 85% efficient in converting heat to steam. (Note: It takes about 1,100 BTU to generate 1 lb steam).

3. A municipal waste incinerator burns the refuse from a small town (population 19,000). Determine the amount of NO_2 produced (lb/day) if the per capita production of solid waste refuse is 5 lb/person/day.

4. A medical waste incinerator burns the refuse from a 500-bed hospital. Determine the amount of NO_2 produced per day if the per capita production of solid waste refuse is 6 lb/bed/day. Assume that 85% of beds are in use on any given day.

5. Estimate the NO_2 emissions from a 500 MW electric power plant (32% efficient) that uses a) sub-bituminous coal, b) #4 fuel oil, and c) propane.

6. A boiler is fed with a distillate fuel oil of composition $C_9H_{16}O_2N_{0.03}$. Combustion occurs with 20% excess air. Assume all fuel nitrogen is converted to NO_2, and that no thermal NO_2 forms. Calculate the NO_2 concentration (ppm) in the exhaust gas on a dry basis.

Nitrogen Emissions

7. A 325 MW electric power plant (37.1% efficient) uses a distillate fuel oil ($C_{11}H_{18}O_1N_{0.025}$, 124,500 BTU/gal, and density 7 lb/gal). Combustion occurs with 50% excess air. Assume 65% of fuel nitrogen converts to NO, and that fuel NO is 40% of total NO (the remainder being thermal NO). Calculate a) the NO concentration (ppm) in the exhaust gas (dry basis) and b) the production of NO (lb/hr).

8. Repeat problem 10-7 but solve in terms of a) ppm of NO_2 (wet basis) and b) lb NO_2/hr (wet basis) rather than lb NO/hr (dry basis).

9. A furnace uses a coal with the following composition (C = 75.9 wt%, H = 5.7 wt %, O = 8.1 wt%, N = 0.3 wt%, and ash = 10 wt%). Combustion happens with 30% excess air to form CO_2, H_2O, and NO_2 gases. Assume all fuel nitrogen is converted to NO_2, and that no thermal NO_2 forms. Calculate the NO_2 concentration (ppm) in the dry exhaust gas.

10. A 825 MW electric power plant (36.2% efficient) uses a coal with the following composition (C = 74.8 wt%, H = 5.2 wt%, O = 7.3 wt%, N = 0.4 wt%, and ash = 12.3 wt %), with energy content 9,950 BTU/lb). Combustion occurs with 50% excess air to form CO_2, H_2O, and NO. Assume 35% of fuel nitrogen converts to NO, and that fuel NO is 80% of total NO (the remainder being thermal NO). Calculate a) the NO concentration (ppm) in the exhaust gas [dry basis] and b) the daily production of NO.

11. If the NO_2 emission standard for a steam plant is 225 ppm [dry basis], what is the maximum amount of N a fuel ($C_8H_{16}N_x$) may have, assuming all fuel N is converted to NO_2, no thermal NO_2 forms, and that 50% excess air is used.

12. If the emission standard for the power plant in problem 10-6 is 250 ppm NO_2, how much must the emissions be reduced? What retrofit technology(s) could be considered? Hint: If problem 10-6 was not done, have the student use a value of 500 ppm.

13. If the emissions of a facility are 1.7 kg NO/hr, what would the emissions be if expressed as NO_2?

14. If the emissions from a facility are 2.5 kg/min of NO_x (consisting of 70 wt% NO and 30 wt% NO_2), what would the emissions be if expressed as a) 100% NO and b) as 100% NO_2?

15. Calculate the stoichiometric amount of anhydrous ammonia (kg NH_3/day) needed to reduce 900 ppm NO to 150 ppm NO in a flue gas. The actual flow rate is 10,000 m³/min at 1,000°C and 1 atm.

16. Calculate the stoichiometric amount of anhydrous ammonia (kg NH_3 /day) needed to reduce 800 ppm NO_2 to 250 ppm NO_2 in a flue gas. The actual flow rate is 3,000 m^3/min at 700°C and 1 atm.

17. Calculate the stoichiometric amount of urea (kg urea /day) needed to reduce 1,200 ppm NO to 200 ppm NO in a flue gas. The actual flow rate is 525 m^3/min at 370 °C and 1 atm.

18. Consult Table 10-5 and determine which technologies will achieve NSPS for a new construction boiler [industrial steam generating unit] rated at 200 MW for the following fuels: a) natural gas (high heat release), b) residual oil (low heat release), and c) distillate oil (low heat release)

10.8 Group Project Ideas

For each project, the students should work in small groups and present their finding in either a short report (5-8 pages) or 15 minute presentation.

1. Are there any plans to modify the current NO_x ambient air quality standards? Write a short paper summarizing the reasons for the revision. Note that the US EPA is required to examine the NO_x NAAQS every five years.

2. There are several regions of the US that are in non-attainment for NO_x. Locate one such region and explore why. Next, determine what the state is proposing to achieve compliance. Finally, critique the state plan – do you believe it will work in a timely manner?

10.9 Bibliography

Asaf, D, et al. 2010. Long-term Measurements of NO3 Radical at a semiarid Urban Site:2. seasonal Trends and Loss Mechanism. *Environ. Sci. Technol.* 2010, Vol. In Press.

Bishop, G A, Peddle, A M and Stedman, D H. 2010. On-Road Emission Measurements of Reactive Nitrogen Compounds from Three California Cities. *Environ. Sci. Technol.* 2010, Vol. 44, 9, pp. 3616-3620.

Chen, Bingheng and Kan, Haidong. 2008. Air pollution and population health: a global challenge. *Environmental Health and Preventive Medicine*. 2008, Vol. 13, 2.

Commerce, US Department of. 2010. Gross Domestic Product. *National Economic Accounts*. [Online] 2010. [Cited: June 15, 2010.] http://www.bea.gov/national/nipaweb/Index.asp.

Corbitt, Robert A. 1989. 6. Wastewater Disposal. *Standard Handbook of Environmental Engineering*. New York : McGraw-Hill, Inc., 1989.

FHA, - Federal Highway Administration. 2010. Historical VMT Report. *Traffic Volume Trends*. [Online] March 22, 2010. [Cited: June 15, 2010.] http://www.fhwa.dot.gov/policyinformation/travel/tvt/history/historicvmt.pdf.

Gibson, E R, Hudson, P K and Grassian, V H. 2006. Physicochemical Properties of Nitrate Aerosols: Implications for the Atmosphere. *J. Phys. Chem. A*. 2006, Vol. 110, 42, pp. 11785-11799.

Goldstein, I F and Andrews, L R. 1987. PEAK EXPOSURES TO NITROGEN DIOXIDE AND STUDY DESIGN TO DETECT THEIR ACUTE HEALTH EFFECTS. *Environment International*. 1987, Vol. 13, 3, pp. 285-291.

Grassian, V H. 2002. Chemical Reactions of Nitrogen Oxides on the Surface of Oxide, Carbonate, Soot, and. *J. Phys. Chem. A*. 2002, Vol. 106, 6, pp. 860-877.

Hager, Thomas. 2008. *The Alchemy of Air*. Ney York : Harmony Books, 2008. ISBN 9780307351784.

Hall, A J and Matson, P A. 1996. NO_x EMISSIONS FROM SOIL: Implications for Air Quality Modeling in Agricultural Regions. *Annual Review of Energy and the Environment*. Nov 1996, Vol. 21, pp. 311-346.

Hasselblad, V and Kotchmar, D J. 1992. Synthesis of environmental evidence: Nitrogen dioxide epidemiology studies. *Journal of the Air and Waste Management Association*. 1992, Vol. 42, 5, pp. 662-671.

IFA. 2010. Statistics - Fertilizer Supply Statistics. *http://www.fertilizer.org/ifa/Home-Page/STATISTICS/Fertilizer-supply-statistics*. [Online] 2010. [Cited: 7 27, 2010.]

IPCC. 2007. *IPCC Fourth Assessment Report: Climate Change 2007*. New York, NY : Cambridge University Press, 2007.

—. 2001. Radiative Forcing of Climate Change. *Third Assessment Report*. Geneva : s.n., 2001, Chapter 6.

Japan-MoEnv. 1968. Air Pollution Control Law. *Ministry of the Environment, Government of Japan*. [Online] 1968. [Cited: July 29, 2010.] http://www.env.go.jp/en/laws/air/air/index.html. Law No. 97 of 1968.

—. 2012. Environmental Statistics 2012. *Annual Report on Environmental Statistics*. [Online] 2012. [Cited: July 1, 2013.] http://www.env.go.jp/en/statistics/contents/index_e.html#idouhasseigen.

—. 1993. The Basic Environmental Law. *Ministry of the Environment, Government of Japan*. [Online] 1993. http://www.env.go.jp/en/laws/policy/basic/index.html. Law No.91 of 1993.

Jones, Alan V, et al. 1999. *Access to Chemistry*. s.l. : Royal Society of Chemistry, 1999. ISBN 0854045643.

Kagawa, Jun. 1982. HEALTH EFFECTS OF AIR POLLUTANTS AND THEIR MANAGEMENT. *Atmospheric Environment*. 1982, Vol. 18, 3, pp. 613-620.

Konopinski, E. J., Marvin, C. and Teller, E. 1946. *Ignition of the Atmosphere with Nuclear Bombs*. s.l. : Los Alamos National Laboratory Research Library, LA-602, 1946.

Kurvits, T and Marta, T. 1998. Agricultural NH_3 and NO_x emissions in Canada. *International Nitrogen Conference*. 1998, Vol. 102, SUP1, pp. 187-194.

Li, N; Thompson, S. Boiler, A Simplified Non-Linear Model of NO_x Emissions in a Power Station. 1996. s.l. : IEE, 1996. UKACC International Conference on CONTROL '96.

Maruoa, Yasuko Yamada, et al. 2003. Measurement of local variations in atmospheric nitrogen dioxide levels in Sapporo, Japan, using a new method with high spatial and high temporal resolution. *Atmospheric Environment*. 2003, Vol. 37, 8, pp. 1065-1074.

Moolgavkar, S.H. 2000. Air pollution and hospital admissions for diseases of the circulatory system in three U.S. metropolitan areas. *Journal of the Air and Waste Management Association*. 2000, Vol. 50, 7, pp. 1199-1206.

Mosier, A and Kroeze, C. 1998. A New Approach to Estimate Emissions of Nitrous Oxide from Agriculture and Its Implications to the Global Nitrous Oxide Budget. *IGACtivities*. 1998, Vol. March, 12.

Napolitano, Sam, et al. 2007. The NOx Budget Trading: A Collaborative, Innovative Approach to Solving a Regional Air Pollution Problem. *The Electricity Journal*. November 2007, Vol. 20, 9, pp. 65 - 76.

NGBBC. 2005. Reduction and Control Technologies. *Low Excess Air*. [Online] 2005. [Cited: July 29, 2011.] http://cleanboiler.org/Workshop/RCTCombustion.htm.

NIOSH. 2005. *Pocket Guide to Chemical Hazards*. Washington, DC : US Department of Health and Human Services, 2005. NIOSH Publication:2005-149.

Oland, C B. 2002. *Guide to Low Emission Boiler and Combustion Equipment Selection*. s.l. : Oak Ridge NAtional Laboratory, 2002. ORNL/TM-2002/19.

Omatu, S, et al. 1988. ESTIMATION OF NITROGEN DIOXIDE CONCENTRATIONS IN THE VICINITY OF A ROADWAY BY OPTIMAL FILTERING THEORY. *Automatica*. 1988, Vol. 24, 1, pp. 19-29.

Panella, Massimiliano, et al. 2000. Monitoring nitrogen dioxide and its effects on asthmatic patients: Two different strategies compared. *Environmental Monitoring and Assessment*. 2000, Vol. 63, 3, pp. 447-458.

Park, S H, Lee, K M and Hwang, C H. 2009. Influences of Heat Loss on NO_x Formation in a Premixed CH4/Air-Fueled Combustor. *Energy Fuels*. 2009, Vol. 23, pp. 4378-4384.

Perez-Ramirez, Javier. 2007. Prospects of N_2O emission regulations in the European fertilizer industry. *Applied Catalysis B: Environmental*. January 2007, Vol. 70, 1-4, pp. 31-35.

Ravishankara, A R, Daniel, J S and Portmann, R W. 2009. Nitrous Oxide (N_2O): The Dominant Ozone-Depleting Substance Emitted in the 21st Century. Science. 2009, Vol. 326, October 2.

Russel, A R, et al. 2010. Space-Based Constraints on Spatial and Temporal Patterns of NO_x Emissions in California, 2005-2008. *Environ. Sci. Technol*. 2010, Vol. 44, 9, pp. 3608-3615.

Seetharam, A L and Simha, BL Udaya. 2009. Urban Air Pollution – Trend and Forecasting of Major Pollutants by Timeseries Analysis. *International Journal of Environmental Science and Engineering*. 2009, Vol. 1, 2.

Sittig, Marshall. 1979. *Fertilizer Industry: Processes, Pollution Control and Energy Conservation*. New Jersey : Noyes Data Corporation, 1979. ISBN 0815507348.

Skeen, S A, Kumfer, B M and Axelbaum, R L. 2010. Nitric Oxide Emissions during Coal and Coal/Biomass Combustion under Air-Fired and. *Energy Fuels*. In Press, 2010.

Steen, J, Aasum, H and Heggeboe, T. 1986. 15. The Norks Hydro Nitrophosphate Process. [book auth.] Nielsson. *Manual of Fertilizer Processing*. s.l. : CRC Press, 1986.

Stieb, David M., Judek, Stan and Burnett, Richard T. 2002. Meta-analysis of time-series studies of air pollution and mortality: Effects of gases and particles and the influence of cause of death, age, and season. *Journal of the Air and Waste Management Association*. 2002, Vol. 52, 4, pp. 470-484.

—. 2003. Meta-analysis of time-series studies of air pollution and mortality: Update in relation to the use of generalized additive models. *Journal of the Air and Waste Management Association*. 2003, Vol. 53, 3, pp. 258-261.

Tecer, L H. 2009. A factor analysis study: Air pollution, meteorology, and hospital admissions for respiratory diseases. *Toxicological and Environmental Chemistry*. 2009, Vol. 91, 7, pp. 1399-1411.

US-EPA. 1993. *Air Quality Criteria for Oxides of Nitrogen. Office of Health and Environmental Assessment*. Research Triangle Park, NC : Environmental Criteria and Assessment Office, 1993. EPA-600/8-91/049aF.

—. 2008. *Integrated Science Assessment for Oxides of Nitrogen — Health Criteria*. Research Triangle Park, NC : National Center for Environmental Assessment-RTP Division, 2008. EPA/600/R-08/071.

—. 2010. *INVENTORY OF U.S. GREENHOUSE GAS EMISSIONS AND SINKS: 1990-2008 (April 2010)*. Washington, DC : U.S. Environmental Protection Agency, 2010. EPA:430-R-10-006.

—. 2009. National Nitrogen Oxides Emissions by Source Sector. *Air Emission Sources*. [Online] 2009. [Cited: July 27, 2010.] http://www.epa.gov/air/emissions/nox.htm.

—. 2013. Nitrogen Dioxide Information. *Green Book*. [Online] Environmental Protection Agency, January 2013. [Cited: June 27, 2013.] http://www.epa.gov/airquality/greenbook/nindex.html.

WHO. 2003. *Health Aspects of Air Pollution with Particulate Matter, Ozone and Nitrogen Dioxide*. Bonn, Germany : World Health Organization, 2003. EUR/03/5042688.

Yale. 2008. 2008 Environmental Performance Index. [Online] 2008. [Cited: July 28, 2010.] http://epi.yale.edu.

CHAPTER **11**

Reactive Carbon Compounds

Carbon exists in many forms within the atmosphere. The forms include organic or inorganic, oxidized or reduced, and solid, liquid, vapor and gas. Most forms of carbon react in the atmosphere, and the most common product from these reactions is carbon dioxide (CO_2). This chapter focuses on the non CO_2 carbon compounds and labels these emissions as reactive carbon. Chapter 14 discusses carbon dioxide. The difference in regulatory history, chemistry, and removal technologies are significant between reactive carbon and CO_2. Also, the harm CO_2 causes in the ambient air is quite unlike the other compounds. Finally, large-scale CO_2 control technologies do not exist yet.

There are many natural and anthropogenic sources for the various forms of carbon emissions. The largest natural sources include volcanoes, fires, soil, and plant life. The most significant anthropogenic sources originate with the extraction and combustion of fuels. However, we note that almost all forms of economic activity have the potential to emit these compounds.

Carbon is transformed within the atmosphere into carbon dioxide by oxidation, and into particulate matter through complex reactions, often involving sunlight. The reactions are fast for some compounds (seconds to days) and very slow for others (methane may take 20 years to react in the atmosphere). Carbon compounds can be removed from the atmosphere by absorption in water, adsorption on soil, and biologically. Carbon is currently accumulating in the atmosphere and has been since the start of the industrial revolution in the 1800's. The primary sinks for carbon leaving the atmosphere are solubility and mineralization in the ocean, uptake by plants, and storage in the soil.

Regulation of carbon compounds focuses on the most harmful and toxic forms. Only one carbon compound, carbon monoxide, is considered a criteria pollutant and has ambient air standards. Some compounds, such as ethanol, are not measured in the ambient air but have limits placed upon their emissions from a source. Other compounds, such as methane, only require measurement and use of best practices to minimize releases. Most compounds are emitted in such small amounts that they are unregulated.

Reduction strategies for carbon compounds include absorption, adsorption, incineration, biodegradation, condensation, and product replacement and minimization. Results of emission controls are mixed, with much progress made in the reduction of volatile organics and carbon monoxide. Less success has occurred for other forms such as methane.

This chapter explores three categories of reactive carbon compounds: carbon monoxide (CO), methane (CH_4), and non-methane volatile organic compounds (NMVOC). Emissions of all three typically have similar origins in the use of organic based fuels and chemicals. However, their behaviors, interactions with the environment, and their control philosophies differ enough to consider them separately.

11.1 Carbon Monoxide (CO)

11.1.1 General Information

Carbon monoxide is a colorless, odorless gas formed during incomplete combustion of chemicals containing carbon. It has an ambient atmospheric concentration of 60-70 ppb in the Southern Hemisphere and 120-180 ppb in the Northern Hemisphere. Urban areas typically have much higher concentrations due to emissions from human activity. The high variability of CO concentration makes it difficult to estimate global trends, although the limited data suggest that the concentration may have doubled over the last century in the Northern Hemisphere. However, data suggest that the abundance of CO has decreased significantly in the past decade. The decrease is probably due to emission controls; unfortunately, there are very few large-scale, long-term data sources available to provide evidence as to the cause.

11.1.2 Sources, Sinks, and Ultimate Fate

Carbon monoxide (CO) forms when carbon-containing fuels burn incompletely. Incomplete combustion may be caused by using sub-stoichiometric amounts of oxygen (fuel-rich mixtures), poor mixing between the fuel and oxygen, or from quenching. Fuel rich mixtures lead to incomplete combustion because they do not have enough oxygen (from air usually) to completely burn the fuel. A fuel rich mixture provides short-term increases in power for an engine, such as when there is an increased load on the engine or boiler, and it is a transitory problem. Poor mixing can lead to oxygen-deficient regions inside the combustion zone, even if there is enough oxygen overall to complete combustion. Mixing can be improved by changing the flow pattern in a boiler or by using a large excess of inlet oxygen (typically 35 to 50%). Quenching happens when regions of the combustion zone are too cold to allow the reaction to occur in a timely manner. Typically, this occurs near boiler and engine walls during startup.

Table 11-1 lists the main sources of anthropogenic CO in the US (EPA, 2009). The largest fraction (~60%) derives from on-road motor vehicle exhaust. Other non-road engines and vehicles (such as construction equipment, planes, trains and ships) contribute another 25% of the CO emissions. Motor vehicle exhaust may contribute 85 to 95 percent of CO in urban areas. Woodstoves, gas stoves, cigarette smoke, and unvented gas and kerosene space heaters are the typical sources of CO indoors. Natural sources arise from forest fires, volcanoes, and emissions from decomposition of biomass in oxygen-poor environments.

Table 11-1. US Carbon Monoxide Emission Sources [tons/year], 2011.

Source Sector	Total Emissions	Percent of total
Highway Vehicles	26,807,738	35%
Miscellaneous	26,262,009	35%
Off-highway	14,960,378	20%
Fuel Conbustion - Other	2,920,978	4%
Waste Disposal / Recycling	1,113,593	1%
Fuel Combustion - Industrial	927,038	1%
Electricity Generation	784,244	1%
Metals Processing	766,380	1%
Petroleum	686,343	1%
Other Industry	336,462	0%
Chemical Production	167,097	0%
Storage and Transport	26,442	0%
Solvent Use	1,383	0%

The main removal mechanism, or sink, is oxidation with OH radicals (OH·) to form carbon dioxide.

11.1 $$CO + OH \cdot \rightarrow CO_2 + H \cdot$$

The average atmospheric lifetime of CO is approximately two months (Brasseur, et al., 1999), so it is stable enough to travel globally, but is not stable long enough to mix uniformly within the troposphere. In many areas, the concentration of CO controls the amount of OH radical in the atmosphere. A secondary removal mechanism involves consumption by microorganisms in soil or water.

11.1.3 Health and Welfare Effects

11.1.3.1 Human Health

The harmful effects of CO on human health are due to its interaction with hemoglobin in the blood. Hemoglobin absorbs oxygen through the lungs and transfers it throughout the body. When inhaled, CO is preferentially absorbed by hemoglobin by a factor of approximately 210:1. The equilibrium ratio of carboxyhemoglobin (HbCO) to oxyhemoglobin (HbO$_2$) is:

11.2 $$\frac{[HbCO]}{[HbO_2]} = 210 \frac{P_{CO}}{P_{O_2}}$$

Where: [HbCO] is the concentration of carboxyhemoglobin,
[HbO$_2$] is the concentration of oxyhemoglobin,
P$_{CO}$ is the partial pressure of CO in the air, and
PO$_2$ is the partial pressure of O$_2$ in the air.

Equilibrium typically takes six to eight hours when CO air concentrations are 50 ppm. CO is released from the hemoglobin slower than O$_2$, causing a toxic effect. As more CO is inhaled and adsorbed, the amount of oxygen that can be carried by the blood is reduced. The reduction in oxygen can cause (in increasing severity of exposure) headache, fatigue, dizziness, drowsiness, nausea, chest pain, vision problems, reduced ability to work or learn, reduced manual dexterity, vomiting, confusion, collapse, loss of consciousness, and death (Fierro, et al., 2001).

CO poisoning may occur faster and at lower concentrations in those most susceptible: young children, elderly people, people with lung or heart disease or blood problems (anemia), people at high altitudes, or those who already have elevated CO blood levels, such as smokers. Cigarette smokers are especially vulnerable because cigarette smoke contains 400 to 450 ppm CO. They can have levels of HbCO as high as 5 or 10%, as compared to 1 to 2% for non-smokers. It is interesting to note that carbon monoxide poisoning from motor vehicle exhaust was once a leading cause of suicides in many countries (Routley, 1998). However, regulations that reduce the allowable emissions of CO from motor vehicles have significantly reduced this public health issue.

Symptoms vary widely from person to person but begin around 10% HbCO. Significant impairments occur around 70%, and death happens at 90%. CO poisoning is reversible if detected in time. However, acute poisoning may result in permanent damage to parts of the body that require more oxygen such as the heart and brain.

Workers may be exposed to harmful levels of CO in indoor air in places such as boiler rooms, breweries, warehouses, petroleum refineries, pulp and paper production, steel production, around docks, blast furnaces, or coke ovens. Most deaths however occur in poorly ventilated homes that heat with gas or oil. Approximately 800 people die each year in the US due to CO poisoning. People in the following occupations typically have greater exposures:

■ Metal oxide reducer	■ Firefighter	■ Diesel engine operator
■ Longshore worker	■ Customs inspector	■ Garage mechanic
■ Welder	■ Police officer	■ Tollbooth or tunnel attendant
■ Forklift operator	■ Taxi driver	■ Organic chemical synthesizer

The OSHA permissible exposure limit (PEL) is 50 ppm (55 mg/m^3) in the air over an eight-hour average. The National Institute for Occupational Safety and Health (NIOSH) time-weighted average for a 40-hour workweek is 35 ppm (40 mg/m^3). The immediately dangerous to life or health (IDLH) value is 1,200 ppm (CDC, 2010).

■ Example 11-1.

Carbon Monoxide Exposure. Determine the % HbCO of a bartender exposed to 50 ppm for an eight-hour shift. Would you expect the person to show signs or symptoms of CO poisoning?

Solution:
Equation [11-2] allows us to calculate the ratio of HbCO to HbO$_2$ in an exposed person's blood.

$$\frac{[HbCO]}{[HbO_2]} = 210\frac{P_{CO}}{P_{O_2}} = 210\frac{50/10^6}{210,000/10^6}*100\% = 5\%$$

Symptoms begin at 10% HbCO, so there should not be any noticeable symptoms unless the bartender smokes, or has other lung problems.

11.1.3.2 Environmental Welfare

CO in combination with NO$_x$ contributes to the formation of smog and ground-level ozone, causing reduced visibility and deterioration of materials (e.g. stone, plastics, metals) in addition to harmful health effects (see Chapters 10 and 12).

11.1.4 History and Regulation

CO is one of the six criteria pollutants regulated through the US Clean Air Act (CAA). The CAA gives the US-EPA primary responsibility for determining the allowable concentration of pollutants in the ambient air. The EPA controls the concentration by regulating emissions from mobile sources such as automobiles, trucks, trains, and planes and large industrial sources such as power plants. The CAA gives state and local governments primary responsibility for regulating other stationary sources such as controlled burning of agricultural waste, and smaller municipal, commercial, and industrial sources. Federal, state, and local governments work together to reduce emissions of CO.

The US-EPA has set two National Ambient Air Quality Standards (NAAQS) for CO: a one-hour standard of 35 parts per million and an eight-hour standard of 9 parts per million. The concentrations are measured nationwide with a series of air quality monitoring stations. When a location has concentrations in excess of these values, the state is required to develop an action plan for the reduction of CO. The EPA classifies a region as serious when the CO concentration is above 16.5 ppm and moderate when the value is between 9.1 to 16.4 ppm. Regions with concentrations below 9.1 are in attainment. The US-EPA maintains a list of all areas that exceed the limits and requires the area to develop a plan for achieving compliance with NAAQS, (US-EPA, 2012).

Table 11-2. Milestones in Motor Vehicle Emissions Controls.

1970	Clean Air Act sets first auto emissions standards.
1974	EPA sets fuel economy standards.
1975	First catalytic converters for CO and hydrocarbons.
1975	First use of unleaded gas in catalyst-equipped cars.
1983	Vehicle inspection and maintenance programs established in 64 cities.
1990	Clean Air Act Amendments set new tailpipe standards.
1992	Oxyfuel introduced in cities with high CO levels.
1994	Phase-in of new vehicle standards and technologies begins.
2011	New CAFE Standards for Model Years 2012-2025

11.1.4.1 Control of Motor Vehicle Emissions

High levels of ambient CO mostly occur in urban areas with heavy traffic congestion. Starting in the early 1970's, the US-EPA set national standards to reduce emissions of CO and other pollutants from motor vehicles. The standards focus on tailpipe emissions, new vehicle technologies, and clean fuel programs. The US-Department of Transportation - National Highway Traffic Safety Administration (NHTSA) sets fuel use requirements for vehicles with the Corporate Average Fuel Efficiency (CAFE) standards (US-NHTSA, 2010). Table 11-2 lists a few of the important events in federal regulatory control of motor vehicle emission control.

The US-EPA and US-NHTSA regulate tailpipe emissions and fleet efficiencies for each vehicle manufacturer, vehicle type (passenger car, light duty truck), and model year. Figure 11-1 compares the actual passenger car and light truck fleet averages with the CAFE standards (NHTSA, 2014). Current law will increase the passenger car and light truck standard to 35.5 mpg in 2016 and 54.5 mpg for model year 2025.

Total CO emissions from on-road vehicles (which include cars, motorcycles, and light- and heavy-duty trucks) have been reduced by over 50 percent since 1970, despite large increases in the number of vehicles on the road and the number of miles they travel. Today's passenger cars are capable of emitting 90% less carbon monoxide over their lifetimes than their uncontrolled counterparts of the 1960's.

Vehicle technologies used to control CO emissions include catalytic converters, oxygen sensors, ambient air temperature and pressure monitors, and other engine sensors. These systems mostly work automatically and may warn owners when their emission control systems are not working properly. A second method of control allows regulators to require the addition of oxygen-containing compounds to gasoline. The additional oxygen has the effect of "leaning out" the air-to-fuel ratio, thereby promoting complete fuel combustion. The most common oxygen additives are alcohols or their derivatives. These fuel blends change with the seasons – lower volatility for summer and

FIGURE 11-1. US Average Fuel Efficiencies and Standards.

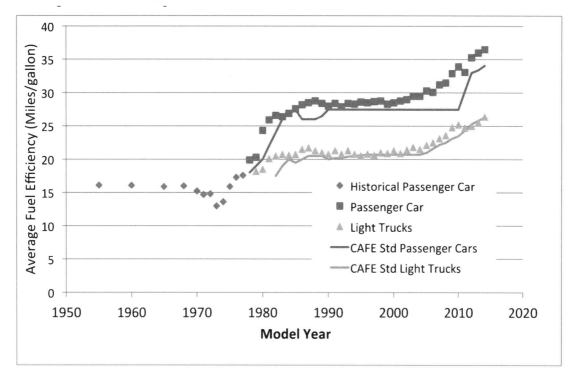

higher oxygen content for winter. The change for the winter season occurs because cars need more fuel to start at cold temperatures, and because some emission control devices (such as oxygen sensors and catalytic converters) operate less efficiently when cold.

■ Example 11-2.

Determine the expected yearly reduction in CO emissions from an average US passenger car if the fuel efficiency increases from 27.5 mpg to 35.7 mpg. Assume the car emits CO at a rate of 115 g CO/gallon and that the average car is driven for 15,000 miles in a year.

Solution:

First, determine how much CO is emitted at each efficiency, then use the difference and extrapolate to a one-year time:

$$CO_{27.5} = 115 \frac{g \cdot CO}{gallon} * \frac{gallon}{27.5 mile} = 4.18 \frac{g \cdot CO}{mile}$$

$$CO_{35.7} = 115 \frac{g \cdot CO}{gallon} * \frac{gallon}{35.7 mile} = 3.22 \frac{g \cdot CO}{mile}$$

$$CO_{reduction} = (4.18 - 3.22) \frac{g \cdot CO}{mile} * 15,000 \frac{mile}{year} = 14.4 \frac{kg \cdot CO}{year}$$

This calculation shows that CO is reduced $\frac{4.18 - 3.22}{4.18} * 100\% = 23\%$ by using more efficient automobiles.

11.1.4.2 CO Air Quality Trend

Figure 11-2 shows the US annual average CO concentration in the ambient air averaged over all the air monitors in the US, (US-EPA, 2010a). The result of the CAA regulations is that ambient carbon monoxide levels have dropped 80 percent between 1980 and 2010, exceeding the target of a 50% decrease over that timeframe. The figure also shows the average concentrations recorded at the lowest 10% and 90% of stations. These additional lines show the variation in reported values. It is interesting to note that approximately 50% of air monitoring stations reported concentrations above the NAAQS limit in 1980, and by 2010, less than 1% reported such high concentrations.

FIGURE 11-2. US National Air Quality Trend for CO Concentration (ppm).

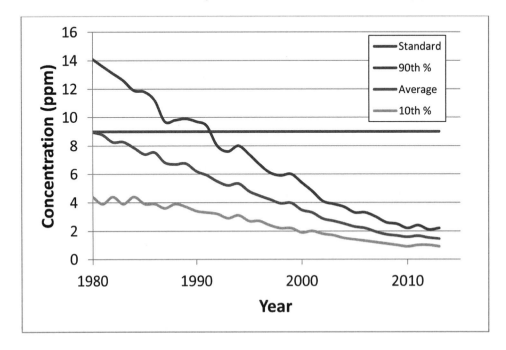

11.1.5 Control Strategies

The formation of CO is due to insufficient oxygen during combustion caused by poor air-fuel ratio control, poor mixing, or cold conditions at start-up. The primary control of CO emissions focuses on improved oxidation of the fuels. A secondary method of control oxidizes the CO in the exhaust gases with an additional oxygen source and a catalyst.

11.1.5.1 Good Combustion Practices

Excess air is the main strategy for reducing CO. Sub-stoichiometric amounts of oxygen guarantee generation of CO and other partial combustion products in the exhaust. Super-stoichiometric amounts of oxygen improves but does not eliminate CO emissions. When excess oxygen is present, the limiting factor to complete combustion becomes the mass transfer of oxygen to the fuel. Poor mixing and short residence times of fuel and oxygen in the flame limit mass transfer. See the discussion in chapter 10 for more information on combustion boiler operation. Note that the optimal conditions for controlling NO_x and CO are not the same. A design that minimizes NO_x may cause an increase in CO.

11.1.5.2 Catalytic Oxidation

When good combustion practices cannot reduce CO emissions to desired levels, an additional oxidation must take place during exhaust. Since the exhaust gas is no longer at a temperature to support the free combustion of CO, a catalyst is used. However, for the catalyst to work there must be an additional source of oxygen. The oxygen can originate in the excess air added for combustion, it can be added post-combustion, or through the reduction of other components in the exhaust, specifically NO_x, see section 16.6.2 for more information on catalytic converters.

11.2 Methane (CH_4)

11.2.1 General Information

Methane (CH_4) is a simple hydrocarbon, the first of the alkanes. It consists of a single carbon atom surrounded by four hydrogen atoms. It takes on the shape of a tetrahedron and is quite stable –much more stable than any other simple hydrocarbon. It has an estimated atmospheric lifetime of 9 to 15 years, depending on the contents of the air. Methane is produced all over the world by several processes, most of which involve the anaerobic bacterial decomposition of organic matter. It was first isolated in marshes and was originally called 'marsh gas.' Methane also occurs in the atmosphere of coalmines, where it is called 'fire damp' because it forms an explosive mixture with air. Methane is the primary constituent of natural gas and is one of the most important fossil fuel energy sources.

Methane is a greenhouse gas, which means that it is capable of absorbing infrared radiation. It is between 21 times (calculated over a period of 100 years) to 72 times (for a period of 20 years) more effective in trapping heat in the atmosphere than carbon dioxide (Forster, et al., 2007). Its concentration in the atmosphere is about 200 times less than CO_2, making it the third most important greenhouse gas, after water and carbon dioxide (see Chapter 14).

The current global average atmospheric concentration is about 1780 parts per billion on a volume basis (ppbv), a value that has held almost constant for the last ten years after a century of rapid rise. Historically, methane concentrations in the world's atmosphere have ranged between 300 and 400 ppbv during glacial periods (ice ages), and between 600 to 700 ppbv during the warm interglacial periods.

11.2.2 Sources, Sinks, and Ultimate Fate

A variety of natural and anthropogenic (human-related) sources emit methane, as listed in Table 11-3, (Denman, et al., 2007). Many of these sources are difficult to measure on a worldwide scale and the numbers in the table have been extrapolated from limited data sets, so the values should be considered as highly uncertain. Natural sources of methane include wetlands, plants and animals, soils, and geological activity. Human-related activities include fossil fuel extraction and use, animal husbandry (enteric fermentation in livestock and manure management), agriculture, and waste management. These activities release significant quantities of methane to the atmosphere. It is estimated that 50 to 65 percent of global methane emissions are due to human-related activities (US-EPA, 2010b). A brief description of each source is provided after the table.

Methane emission levels from a source can vary significantly from one country or region to another, depending on many factors such as climate, industrial and agricultural production characteristics, energy types and usage, and waste management practices. For example, temperature and moisture have a significant effect on the anaerobic digestion process, which is one of the key biological processes that cause methane emissions in both human-related and natural sources. Also, the implementation of technologies to capture and utilize methane from sources such as landfills, coal mines, and manure management systems affects the emission levels from these sources. The difference between methane emissions, methane removal processes, and methane lifetime in the atmosphere determines atmospheric methane concentrations.

Wetlands. Natural wetlands provide a habitat conducive to bacterial production of methane (methanogenic) during the anaerobic decomposition of organic material. These bacteria require environments with no oxygen (anaerobic) and abundant organic matter, both of which are often present in sediments and the lower water column of wetlands.

Termites. Termites produce methane as part of their normal digestive process, and the amount generated varies among different species. Ultimately, total emissions from termites depend on the

Table 11-3. Global Estimates for Methane in the Atmosphere [Tg CH$_4$/yr].

Natural Sources	Emission Range Tg/yr
Wetlands	145–231
Termites	20–29
Ocean	4–15
Hydrates	4–5
Geological Sources	4–14
Wild Life	15
Wildfires	2–5
Anthropogenic Sources	
Coal Mining	30-48
Gas, Oil Industry	36-68
Landfills and Waste	35-69
Ruminates	76-189
Rice Agriculture	31-112
Biomass Burning	14-88
Sinks	
Reaction with OH	428-511
Removal in Stratosphere	30-45
Removal in Soils	26-34
	Most Likely Values
Total Sources	582
Total Sinks	581
Imbalance / Unknown	1

1 Tg = 10^{12} g

population of these insects, which varies significantly around the world. They are very common in deforested tropical and temperate areas.

Oceans, Rivers, and Estuaries. The source of methane from oceans is not entirely clear, but two identified sources include the anaerobic digestion in marine zooplankton and fish, and from methanogenesis in sediments and drainage areas along coastal regions.

Hydrates. Methane hydrates are solid deposits composed of cages of water ice molecules that contain molecules of methane. The solids occur deep underground in polar regions and ocean sediments of the outer continental margin throughout the world. The average methane hydrate composition is one mole of methane for every 5.75 moles of water, though this is dependent on how many methane molecules "fit" into the various cage structures of the water ice lattice. The observed density is around 0.9 g/cm^3. One liter of methane hydrate solid contains, on average, 168 liters of methane gas at STP. Methane can be released from the hydrates with changes in temperature, pressure, salt concentrations, and other factors. Overall, the amount of methane stored in these hydrates globally is estimated to be very large with the potential for large releases of methane if there are significant breakdowns in the stability of the deposits. Because of this large potential for emissions, there is much ongoing scientific research related to analyzing and predicting how changes in the ocean environment affect the stability of hydrates. A 2008 measurement of methane in the Arctic atmosphere found that 8 million tons of methane is emitted annually from the permafrost below the seabed north of Russia (Shakhova, et al., 2010). It is expected that warming due to climate change will increase this rate of release, but currently no data is available to prove or disprove this claim.

Geologic. Geologic emissions are difficult to quantify because there are many small point sources all over the Earth. One of the dominant sources of geologic methane is mud volcanoes. Mud volcanos are not true volcanoes – they do not produce lava. Rather they emit heated water and gases, which lift soil with them and create a mud dome. These structures can be up to ten kilometers in diameter though most are much smaller, and often form on tectonic plate boundaries or near fossil fuel deposits. Over 1,000 such regions have been located worldwide on land or in shallow water. Mud volcanoes release methane gas from within the Earth, as well as smaller amounts of carbon dioxide, nitrogen and helium. Other structures that emit methane that would qualify as geologic sources include volcanoes, steam vents, and bubbling pools.

Wild Life. Another highly uncertain source of methane emissions is wild animals. Bison, elk, and moose are examples of animals that release methane. Estimates of methane emission from wild animals can be derived based on estimates of the population of these species and estimates of their methane production (Leng, 1993).

Wildfires. The incomplete combustion of organic material during fires generates methane. These fires typically are oxygen starved due to poor mixing with air. A large fraction of these emissions come from deforestation in tropical areas. However, the estimates in the table only include the emissions from natural forest fires. In addition to emissions from direct combustion, fires can lead to the release of large amounts of methane from soil, especially in high-latitude regions. Here, fires melt permafrost, which has trapped methane in the soil and the warmer soil temperatures after fire events lead to greater microbial activity.

Coal mining. Methane adsorbs on coal and can exist as a separate phase in the voids and pockets of a coal deposit. The methane trapped in coal deposits and the surrounding strata escapes during normal mining operations in both underground and surface mines. In addition, coal handling after mining results in methane emissions via desorption and diffusion.

Livestock enteric fermentation. Among domesticated livestock, ruminant animals (cattle, buffalo, sheep, goats, and camels) produce significant amounts of methane as part of their normal digestive processes. In the rumen, or large fore-stomach, of these animals microbial fermentation converts food into products that can be digested and utilized by the animal. This microbial fermentation process, referred to as *enteric fermentation*, produces methane as a by-product, which is exhaled by the animal. Methane is also produced in smaller quantities by the digestive processes of other animals, including humans, but emissions from these sources are insignificant. An adult cow emits 80-110 kg of methane per year, and there are about 100 million ruminants in the U.S. and 1 to 5 billion in the world, making ruminants one of the world's largest methane sources.

Landfills. Methane generation in landfills and open dumps originates from organic waste decomposition under anaerobic (without oxygen) conditions. The amount of methane created depends on the quantity and moisture content of the waste and the design and management practices at the site.

Natural gas and petroleum systems. Methane is the primary component of natural gas. Methane losses occur during the production, processing, storage, transmission, and distribution of natural gas. Because gas often occurs in conjunction with oil, the production, refinement, transportation, and storage of crude oil are also sources of methane emissions. Natural gas has seen a recent increase in use due to the development of new methods for extraction from underground reservoirs. There is currently a lot of debate about the climate benefits of switching from coal to natural gas as an energy source. In the end, the percentage of methane leaked by natural gas production and distribution is the key to whether switching yields a net climate benefit.

Livestock manure management. Methane is produced during the anaerobic decomposition of organic material in livestock manure management systems. Liquid manure management systems, such as lagoons and holding tanks, can cause significant methane production. Many of the larger swine and dairy operations use these systems. Manure deposited on fields and pastures, or otherwise handled in a dry form, usually produce insignificant amounts of methane.

Wastewater treatment. Wastewater from domestic municipal sewage and industrial sources is treated to remove soluble organic matter, suspended solids, pathogenic organisms, and chemical contaminants. These treatment processes can produce methane emissions if organic constituents in the wastewater are treated anaerobically and if the methane produced is allowed to escape to the atmosphere. Additionally, the sludge produced from some treatment processes is further biodegraded (digested) under anaerobic conditions, which also results in methane emissions.

Rice cultivation. Methane is produced during flooded rice cultivation by the anaerobic decomposition of organic matter in the soil. Flooded soils are ideal environments for methane production because of their high levels of organic substrates, oxygen-depleted conditions, and moisture. The level of emissions varies with soil conditions and production practices as well as climate.

Table 11-4 lists the anthropogenic methane emission sources in the US ((US-EPA, 2010b). The largest source is due to the large numbers of cows and related animals raised on farms and ranches.

Table 11-4. US Methane (2010) Emissions Reported by Source [Tg CH_4/yr].

Source Category	Methane Quantity [Tg/yr]
Enteric Fermentation	7.04
Landfills	6.32
Natural Gas Systems	4.82
Coal Mining	3.38
Manure Management	2.25
Petroleum Systems	1.46
Wastewater Treatment	1.22
Forest Land	0.60
Rice Cultivation	0.36
Stationary Combustion	0.34
Abandoned Coal Mines	0.30
Mobile Combustion	0.10
Composting	0.09
Agricultural Residue Burning	0.05
Petrochemical Production	0.05
Total for U.S.	**28.38**

There are three main methane sinks from the atmosphere. The largest sink is oxidation with hydroxyl radicals (OH) and oxygen (O_2) to form carbon dioxide and water, see equations [11.3] - [11.8]. These reaction sequences occur in both the troposphere and the stratosphere, and account for 90% of methane removal (Denman, et al., 2007). Note that carbon monoxide forms as an intermediary compound in this set of reactions. The two other known sinks include microbial uptake of methane in soils and methane's reaction with chlorine (Cl) atoms in the marine boundary layer. These sinks contribute 7% and 2% of total methane removal, respectively.

Reaction sequence in the absence of NO·:

11.3 $\quad CH_4 + OH\cdot \rightarrow CH_3\cdot + H_2O$

11.4 $\quad CH_3\cdot + O_2 \rightarrow CH_2O + OH\cdot$

11.5 $\quad CH_2O + OH\cdot \rightarrow HCO\cdot + H_2O$

11.6 $\quad HCO\cdot + O_2 \rightarrow CO + HO_2\cdot$

Finally, the CO is oxidized by an additional OH radical to form CO_2:

11.7 $\quad CO + OH\cdot + O_2 \rightarrow CO_2 + HO_2\cdot$

The overall reaction is:

11.8 $\quad CH_4 + 3O_2 + 2\ OH\cdot \rightarrow CO_2 + 2H_2O + 2\ HO_2\cdot$

The reaction sequence in the presence of NO·:

11.3 $\quad CH_4 + OH\cdot \rightarrow CH_3\cdot + H_2O$

11.9 $\quad CH_3\cdot + O_2 + M \rightarrow CH_3OO\cdot + M$

11.10 $\quad CH_3OO\cdot + NO\cdot \rightarrow CH_3O\cdot + NO_2\cdot$

11.11 $\quad CH_3O\cdot + O_2 \rightarrow CH_2O + HO_2\cdot$

11.12 $\quad CH_2O + h\nu \rightarrow H + HCO$

11.13 $\quad H + O_2 + M \rightarrow HO_2\cdot + M$

11.14 $\quad HCO + O_2 \rightarrow CO + HO_2\cdot$

11.15 $\quad CO + OH\cdot + O_2 \rightarrow CO_2 + HO_2\cdot$

The overall reaction is:

11.16 $\quad CH_4 + 5\ O_2 + NO\cdot + 2\ OH\cdot + h\nu \rightarrow CO_2 + H_2O + NO_2\cdot + 4\ HO_2\cdot$

Note that in both cases, formaldehyde (CH_2O) is a major intermediate compound during the degradation of methane (and most other hydrocarbons). Formaldehyde degradation by both the OH reaction [11.5] and the photolysis reaction (11.12) leads to the formation of carbon monoxide.

11.2.3 Health and Welfare Effects

11.2.3.1 Human Health

Methane is not considered toxic, however it is highly flammable and may form explosive mixtures with air (Air-Liquide, 2005). Methane is also an asphyxiant, causing harm when displacing oxygen to concentrations from 21% to below 19.5%. Methane has a lower explosive limit (LEL) of 5% and an upper explosive limit (UEL) of 15% by volume in air at 20°C and 1 atmosphere. If the concentration is between these limits, there is risk of burning or explosion when an ignition source is available. Potential ignition sources include open flames, electrical sparks, and static electricity discharges. Concentrations in storage areas are kept below 25% of the LEL for safety reasons.

11.2.3.2 Environmental Welfare

Methane on a global scale contributes to the increase in ozone formation in the troposphere (see section 12.1.2), the reduction of ozone formation in the stratosphere (see section 12.2.1), alters the oxidizing capacity of the atmosphere by influencing hydroxyl radical formation, and is a potent greenhouse gas (see section 11.2.3).

Methane was once considered irrelevant for addressing surface ozone pollution because its long atmospheric lifetime prevents it from contributing to the rapid photochemical production which leads to high-ozone episodes. However, methane, like other VOC's, can react in the troposphere with nitrous oxides (NO_x) in the presence of sunlight to form ozone. Increases in methane raise the baseline ozone level in air globally, including at the surface. Ozone episodes, fueled by the short-lived ozone precursors (nitrogen oxides and non-methane VOCs), then build on top of this baseline.

Methane is the only organic compound with a sufficiently long lifetime in the atmosphere to be transported (by convection and diffusion) into the stratosphere. Air crossing the tropopause contains CH_4 at about the average global concentration of 1780 ppb. Its destruction in the stratosphere is by reaction with the OH• radical and reaction with atomic oxygen or atomic chlorine as shown in the following reactions:

11.3 $\qquad CH_4 + OH\cdot \rightarrow H_2O + CH_3\cdot$

11.17 $\qquad CH_4 + O\cdot \rightarrow OH\cdot + CH_3\cdot$

11.18 $\qquad CH_4 + Cl\cdot \rightarrow HCl + CH_3\cdot$

The first reaction represents one of the main sources of water in the stratosphere. The second reaction uses free oxygen that could otherwise react with O_2 to form O_3 and is speculated to cause an overall reduction in stratospheric ozone. However, these reactions are quite slow in the lower stratosphere due to low temperatures and low concentrations of O and Cl.

Methane is also a greenhouse gas that contributes to the earth's overall greenhouse effect. The climate impact of methane differs from that of CO_2 in that methane is a transient gas (methane has about a 10 to 20 year lifetime), while CO_2 accumulates. The climate impact of methane depends on whether it is released quickly or slowly, relative to the methane lifetime. If released quickly, over just a few years or less, there would be a decade-timescale warming spike, followed by a recovery toward the lesser warming as the methane oxidizes into CO_2. The amount of methane trapped worldwide in methane hydrates and sealed in permafrost is conservatively estimated to twice the amount of carbon found in all known fossil fuels on Earth (Dillon, 1992). A release of 1% of the ocean methane hydrate reservoir over a five-year period would be the radiative equivalent of a doubling of the CO_2 content in the atmosphere.

FIGURE 11-3. Historical Trend in Atmospheric Methane Concentration.

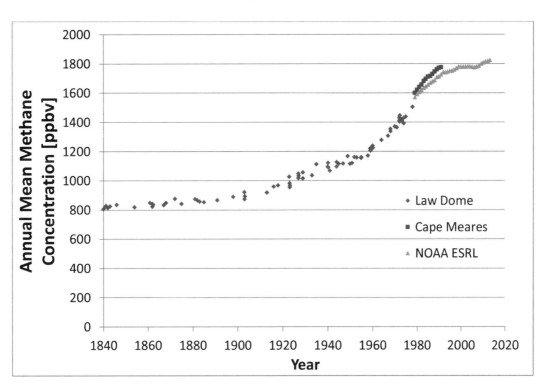

11.2.4 History and Regulation

The historical record, based on analysis of air bubbles trapped in ice sheets, indicates that methane is more abundant in the Earth's atmosphere now than at any time during the past 400,000 years (CSCC, 2001). Since 1750, global average atmospheric concentrations of methane have increased by 150 percent from approximately 700 to 1815 parts per billion by volume (ppbv) by 2012 ((Forster, et al., 2007), (Butler, 2012)). Figure 11-3 shows this 160-year trend in the annual average methane concentration in the atmosphere.

Law Dome is a large ice dome in Antarctica that was drilled for an ice core. The ice core was layered and samples of entrapped air were measured for several gases including methane and carbon dioxide. A layer forms each year, making dating of the gases relatively straight forward (similar to tree rings). Cape Meares is located on the Oregon, USA coast and the site is considered representative for atmospheric methane in the northern latitudes. National Oceanic and Atmospheric Administration Earth System Research Laboratory (NOAA ESRL) measure atmospheric methane at the Mona Loa Observatory, Hawaii, USA.

Methane is not considered a traditional air pollutant as it is neither toxic nor does it readily react in the atmosphere. However, many countries have realized that methane does have a negative influence in the atmosphere, and emissions represent a wasted resource. Note that reductions in methane are a quick route to reducing human contributions to global climate change. Unlike other pollutant gases, methane can be used to produce energy since it is the same hydrocarbon as natural gas. Consequently, for many methane sources, opportunities exist to reduce emissions cost-effectively or at low cost by capturing the methane and using it as fuel.

The US-EPA has taken a voluntary reduction path, where the government serves as a clearinghouse and provides expertise, low-cost loans, and other incentives to encourage the reduction of methane emissions. Several industrial programs have been developed to help decrease total methane emissions, see next section – 11.2.5 Control Strategies. These programs appear to have been somewhat successful in the US leading to a 10% decrease in emissions from 1990 to 2010 [(US-EPA, 2002a), (US-EPA, 2002b)]. These techniques are also used in many other countries.

11.2.5 Control Strategies

Control strategies for methane vary depending on the source, quantity and concentration generated. Large quantities of concentrated methane from a small source can be collected and used as a chemical feedstock or recovered for energy. Methane emitted in small quantities, low concentrations, or from multiple sources is typically managed by focusing on reduction and minimization. The cut off between the two depends on the cost of collection equipment, avoided energy costs, and the resources needed to run and maintain collection equipment. This section explores the control of methane from wastewater treatment, livestock, landfills, natural gas production, and coal production.

11.2.5.1 Wastewater Treatment.

Methane emissions can be avoided by direct capturing or by treating the wastewater and the associated sludge under aerobic conditions. Capturing methane requires all tanks and treatment units to be covered and connected in a way that allows off-gases to be collected and transported to a location where methane can be recovered or burned to generate power. The collection systems are very similar to those used to collect and control indoor air pollution, see Chapter 15. Aerobic conditions use oxygen to help bacteria break down and destroy the organic fraction of wastewater. These processes release carbon dioxide rather than methane.

11.2.5.2 Livestock

Cattle emit methane through a digestive process called enteric fermentation. Methane generation is a loss of carbon from the rumen and represents an unproductive use of food. Reducing methane emissions from livestock could improve the efficiency of livestock production, and lead to increased profitability. The general approach to improvement uses nutritional and genetic improvements. Current improvements have occurred mostly in dairy cows because of the greater control over feed. However, the cow-calf sector of the beef industry is the largest emitter of methane.

Emissions from beef cows are higher for a number of reasons. Beef cows are very large animals. Their diet consist mainly of forage of varying quality and this diet is poorer than in the dairy or feedlot sectors. Their level of management is typically not as good. Moreover, the beef cow population is very large (Worldwide estimates range from 1 to 5 billion animals). Some of the most effective management practices to improve efficiency and reduce emissions include:
- Soil testing of grazing areas, followed by the addition of proper amendments and fertilizers,
- Supplementing cattle diets with needed nutrients,
- Developing a preventative herd health program,
- Providing appropriate water sources and protecting water quality, and
- Improving genetics and reproductive efficiency.

The particular practices a livestock producer utilizes to improve production depends on the circumstances of their operation, including the producer's goals and the natural, financial, and labor resources available (US-EPA, 2007).

Another source of methane from livestock is from the collection and treatment of animal manure, see Table 11-5. Like wastewater treatment, the manure can be treated aerobically or anaerobically. Anaerobic systems can be less expensive as they do not require constant mixing or injection of oxygen, but they produce methane. The methane can be captured in an anaerobic digester with a biogas recovery system. The acid forming bacteria in the digester convert the volatile solids fraction into volatile acids, which then converts into methane and carbon dioxide by methane forming bacteria. The methane can be used to produce electricity, heat or hot water. Biogas recovery

Table 11-5. Daily Waste and Methane Production by Dairy, Beef, and Swine per 1000 Pounds of Animal Weight.

Animal (average animal weight)	Raw Waste lb/1000 lb animal	Total Solids lb/1000 lb animal	Volatile Solids lb/1000 lb animal	Methane Production ft^3/1000 lb animal
Dairy Cow (1000 lb/cow)	82	10.4	8.6	28.4
Beef Cow (2000 lb/cow)	60	6.9	5.9	19.4
Swine (200 lb/swine)	65	6	4.8	18.6

systems are effective at confined livestock facilities that handle manure as liquids and slurries, typically swine and dairy farms. Manure deposited on fields and pastures, or otherwise handled in a dry form, produces insignificant amounts of methane (AgSTAR, 2011).

11.2.5.3 Landfill Methane.

Landfills that accept organic waste materials generate landfill gas (LFG) from the anaerobic decomposition of these materials. LFG consists of about 50 percent methane, about 50 percent carbon dioxide (CO_2), and a small amount of non–methane organic compounds. The amounts and rates of production depend on the waste, water content, and temperature of the landfill. Unless a landfill captures this gas, it escapes into the atmosphere, where it can cause odor problems, explosion hazards, contribute to ozone formation, and global warming.

LFG can be captured, converted, and used as an energy source. It is extracted from landfills using a series of wells and a blower/flare (or vacuum) system. This system directs the collected gas to a central point for processing and treatment depending upon the ultimate use for the gas. Once collected the gas can be flared, used to generate heat and electricity, replace fossil fuels in industrial and manufacturing operations, or upgraded to pipeline–quality gas where the gas may be used directly or processed into an alternative fuel (US-EPA, 2011).

11.2.5.4 Natural Gas Production.

Methane emissions occur in all sectors of the natural gas industry: drilling, production, processing transmission, and final distribution. They result from normal operations, routine maintenance, fugitive emissions, and system upsets. Emissions occur as gas moves through the system due to intentional venting and unintentional leaks. Venting can occur through equipment design or operational practices, such as the continuous bleed of gas from pneumatic devices (that control gas flows, levels, temperatures, and pressures in the equipment), or venting from well completions during production. Methane losses can also occur from leaks (also referred to as fugitive emissions) in all parts of the infrastructure, from connections between pipes and vessels to valves and equipment. Methane emissions also occur in the oil industry via similar activities, although the methane composition is much lower in oil than natural gas (GasSTAR, 2010).

The different sources each require their own management plan to reduce emissions. Some of the reduction activities include:
- Installation of vapor recovery units (VRU) on storage and transmission equipment,
- Convert gas pneumatic controls to instrument air,
- Install and use aerial leak detection equipment,
- Eliminate unnecessary equipment and/or systems, and
- Survey and repair leaks.

11.2.5.5 Coal Production.

Coal contains many organic and inorganic compounds, including methane. The methane may escape the coal bed through natural fissures open to the atmosphere, or it may be stored or adsorbed in the coal matrix. When coal is excavated or mined, the coal mine methane (CMM) is released to the atmosphere. Coal mines that generate significant quantities of methane are called gassy. This source of methane can be dangerous during mining activities because it is an asphyxiant and an explosive hazard. Underground mines use ventilation systems to remove CMM (CMOP, 2011).

CMM is emitted to the atmosphere from a variety of sources:
- Degasification or drainage systems in underground coal mines. These systems use vertical and/or horizontal wells to recover methane in advance of mining (known as "pre-mine drainage") or after mining (called "gob" or "goaf" wells)
- Ventilation air from underground mines, which contains dilute concentrations of methane
- Abandoned or closed mines, from which methane may seep out through vent holes or fissures or cracks in the ground
- Surface mines, from which methane in the coal seams is directly exposed to the atmosphere
- Fugitive emissions from post-mining operations, in which coal continues to emit methane as it is stored in piles and transported.

CMM recovery methods depend on the gas quantity and quality, especially the concentration of methane and the presence of other contaminants. Nearly all CMM captured and used from active U.S. mines is injected into the natural gas pipeline system (US-EPA, 2010c). Worldwide, CMM is most often used for power generation, district heating, boiler fuel, or town gas, or it is sold to natural gas pipeline systems.

11.3 Volatile Organic Compounds (VOC)

11.3.1 General Information

There are three classifications of organic carbon: methane (CH_4) – discussed in section 11.2, volatile organic compounds (VOC) and non-volatile organic compounds. Volatile and non-volatile compounds can be grouped together as non-methane total organic carbon (NMTOC). In terms of ambient air pollution, methane and VOCs are of primary concern. The non-volatile fraction of NMTOCs are usually present in the atmosphere only as a component of particulate matter (e.g. droplets, adsorbed onto other PM, and as soot).

Volatile Organic Compound (VOC) pollutants include a wide range of carbon-based chemical species. There is no uniform definition for this group, but a few common characteristics of regulated VOC's include:
- High vapor pressures at ambient temperatures,
- Can affect human health and well-being, and
- Negatively interact with the environment.

Lists of regulated VOC chemicals vary between countries, and even within countries by state, region, and city. The lists also vary depending on the use of air – be it ambient, occupational, public, or private. This chapter explores VOCs in ambient air. Chapter 15 discusses VOCs and indoor air quality. Chapter 13 (Hazardous Air Pollutants) discusses air toxins such as benzene.

The lifetime of VOCs in the atmosphere is relatively short - on the order of hours to days – so they do not tend to accumulate to toxic levels. Concentrations are rarely large enough in ambient air to cause either acute or chronic toxic effects, although a few compounds like formaldehyde and benzene in some urban environments can occasionally create a hazard. The main concern is their interaction with sunlight and NO_x leads to the formation of harmful levels of ozone (O_3), as discussed in Chapter 12 - tropospheric ozone.

11.3.2 Sources, Sinks, and Ultimate Fate

Volatile organic compounds (VOC) can be natural or anthropogenic. It is difficult to derive a precise global emission rate of VOCs given their short atmospheric lifetimes and geographically varying sources. Estimates of the annual global natural VOC emissions range between 500 to 1,200 Tg per year; see Table 11-6 for comparison of two global estimates.

Tropical vegetation is the largest global source of natural emissions. Table 11-7 provides a short list of these natural chemicals, where isoprene and monoterpene are the largest component of the vegetation based emissions (Brasseur, et al., 1999). Oceans emit small amounts of NMVOC, although the precise source is unknown.

Table 11-6. Annual Global Average of Natural VOC Emissions.

Emission Sources				Emissions (TgC/yr)	
				Estimate 1[a]	Estimate 2[b]
Natural					
	Ocean			6 - 36	5
	Land				
		Microbial		6	
		Vegetation			655
			Isoprene	500	
			Monoterpene	125	
			Other	520	
Subtotal				1170	660
Anthropogenic					
	Fossil Fuel				56
		Transportation		22	
		Stationary		4	
		Industry		17	
		Organic Solvents		17	
	Biomass Burning			45	30
Subtotal				103	86
Total Emissions				1273	746

a) (Singh, et al., 1992) and (Guenther, et al., 1995)
b) (Ehhalt, 1999)
Tg = 1 Teragram = 10^{12} gram

Table 11-7. List of Major Biogenic VOC Species.

- Benzene
- Ethane
- Ethene
- Ethylene
- Isoprene
- Methylbutenol
- Monoterpenes
- Propane

Many factors affect natural VOC emissions. These factors include temperature, which determines rates of volatilization and growth, sunlight, which determines rates of biosynthesis, and precipitation. Biosynthesis describes the biological formation of these compounds. Emissions are also sensitive to climate-induced changes in plant species composition and biomass distributions. Emission occurs almost exclusively from the leaves, the stomata in particular. Isoprene, with the largest emission rate, is not stored in plant tissue and is only emitted during photosynthesis. Monoterpenes are stored in plant reservoirs and emitted throughout the day and night.

Providing a sense of scale – on a typical mid-summer day, a forest the area of Pennsylvania (about 62,000 km^2) emits approximately 3,400,000 kilograms of terpenes, which is equivalent to 1,200,000 gallons of gasoline - enough to fuel 2,000 average American cars for a year.

Anthropogenic VOC emissions generally originate with the production, refining, and use of fossil fuels (natural gas, oil, and coal), including the accidental release or incomplete combustion of these fuels. Table 11-8 lists several common VOCs and their anthropogenic sources. The unintentional emission due to leaks and spills, called *fugitive releases*, account for about 12% of industrial sources (Clearstone, 1994). Motivated by cost, environmental concerns, and regulation, industries are increasingly shifting toward aqueous based solutions to replace VOCs.

Table 11-8. Anthropogenic Sources of VOC Emissions.

Chemical	Source
Acetone	Paint, coatings, finishers, paint remover, thinner, caulking
Aliphatic hydrocarbons (octane, decane, undecane hexane, isodecane, mixtures)	Paint, adhesive, gasoline, combustion sources, liquid process photocopier, carpet, linoleum, caulking compound
Aromatic hydrocarbons (toluene, xylenes, ethylbenzene, benzene)	Combustion sources, paint, adhesive, gasoline, linoleum, wall coating
Chlorinated solvents (dichloromethane or methylene chloride, trichloroethane)	Upholstery and carpet cleaner or protector, paint, paint remover, lacquers, solvents, correction fluid, dry-cleaned clothes
n-Butyl acetate	Acoustic ceiling tile, linoleum, caulking compound
Dichlorobenzene	Carpet, moth crystals, air fresheners
4-Phenylcyclohexene (4-PC)	Carpet, paint
Terpenes (limonene, a-pinene)	Deodorizers, cleaning agents, polishes, fabrics, fabric softener, cigarettes

Ambient VOC concentrations are highest near their emission sources. The majority of natural emissions occur in the tropics (23°S to 23°N) with smaller amounts emitted in the mid-latitudes in the warmer seasons. Anthropogenic emissions occur mainly in urban regions (mostly in the

Northern Hemisphere between 40°N to 50°N). The biogenic emissions in urban areas are relatively low, so anthropogenic sources have significant impacts on regional atmospheric chemistry despite small global emissions.

Other forms of organic compounds in the atmosphere include NMVOCs, semi-volatile, and non-volatile organic compounds that form or adsorb to particles. Carbonaceous PM includes elemental carbon (EC), primary organic aerosols (POA) and secondary organic aerosols (SOA). Each has natural and anthropogenic sources. EC occurs during open burning, which is usually a fuel rich (insufficient oxygen) combustion processes. It appears as gray and black smoke and soot. POA occur directly during biomass burning, diesel and gasoline vehicle exhaust, and during meat cooking. Indeed some studies suggest up to 10% of POA originates from meat cooking emissions (Alves, et al., 2012). SOA are formed during photochemical reactions between trace organic gases and other particulates (e,g. nitrates) in the atmosphere. Partially oxidized organic compounds can condense onto pre-existing particles or form new SOA particles.

It has been observed that atmospheric organic particulate matter (PM) may contribute a significant fraction of total $PM_{2.5}$. They consist of a complex mixture of hundreds of organic compounds belonging to many different chemical classes. These classes and sources include:

- Polycyclic aromatic hydrocarbons (PAHs) are small in quantity but include known mutagenic and/or carcinogenic substances. PAHs originate from anthropogenic emissions such as industrial production, vehicle exhausts, waste incineration, and wood burning.
- Atmospheric aliphatic hydrocarbons (e.g. n-alkanes) are primary pollutants. They are highly resistant to biochemical degradation and are detectable in aerosols at relatively high concentrations. Ambient n-alkanes originate from two major sources: fossil fuel use and natural biogenic sources (e.g. plant waxes).
- Carbonyl compounds (e.g. n-alkan-2-ones) and methyl esters have primary and secondary sources. These compounds originate from oxidative processes, direct emission from biogenic sources, cooking smoke, wood smoke, and mobile sources.
- Numerous polar organic compounds such as acids, alcohols, sterols and anhydro-sugars in aerosols originate from plant waxes, bacterial activity, automobile exhaust, and incomplete biomass combustion.

US anthropogenic emissions are also dominated by fossil fuel derived sources. Table 11-9 lists the quantities reported by regulated users - those that emit more than 10 ton/yr of any one or 25 ton/yr of any combination (US-EPA, 2009).

VOC's emitted to the atmosphere are very reactive, and most have short residence times (hours to months). These species decompose to form a complex array of oxygenated compounds that continue to break down until they form carbon dioxide and water (and sulfuric acid if they contain sulfur). The following generalized set of reactions [11.19] - [11.26] describes one set of decomposition reactions using OH• and oxygen. Note that the VOC is represented as RH, where R is the carbon chain, H represents one hydrogen atom bound to the molecule, R' is an organic fragment having one less carbon than R, M is a catalyst - typically a metal on an atmospheric particle, and hv represents a unit of energy from a photon, typically provided by sunlight.

Table 11-9. US VOC Emissions Reported by Source, 2011.

Source Sector	Total VOC Emissions tons/yr
Biogenics - Vegetation and Soil	39,653,276
Fires	5,807,750
Industrial Processes	3,173,500
Solvent Use	2,815,288
Mobile - Onroad	2,413,026
Mobile - Nonroad	2,159,378
Gasoline (bulk and consumer)	796,522
Fuel Combustion	607,158
Miscellaneous	201,507
Waste Disposal	125,627
Agriculture - Livestock Waste	56,076

11.19 $RH + OH\cdot \rightarrow R\cdot + H_2O$

11.20 $R\cdot + O_2 + M \rightarrow RO_2\cdot + M$

11.21 $RO_2\cdot + NO\cdot \rightarrow RO\cdot + NO_2\cdot$

11.22 $RO\cdot + O_2 \rightarrow HO_2\cdot + R'CHO$

11.23 $HO_2\cdot + NO\cdot \rightarrow OH\cdot + NO_2\cdot$

11.24 $2[\ NO_2\cdot + h\nu \rightarrow NO\cdot + O\ ]$

11.25 $2[\ O + O_2 + M \rightarrow O_3 + M\ ]$

Overall:

11.26 $RH + 4O_2 + 2\ h\nu \rightarrow R'CHO + H_2O + 2O_3$

This set of reactions is very similar to the oxidation reactions for methane (see equations [11.3] - [11.8]). Note that this is not the only pathway for the oxidation of VOCs and many other pathways exist. Some of the pathways lead to combinations with NO_x to form PAN (peroxyacetyl nitrate) which provides a mechanism for permanent or temporary removal of NO_x from the atmosphere. Other oxidants include ozone and nitrate radical. The amount of time required to oxidize a particular organic compound depends on the oxidant (Atkinson, 2000) see Table 11-10. The oxidation reactions lead to the formation of many other intermediary species, some of which are hazardous air pollutants like formaldehyde, ketones, peroxides, hydroxycarbonyls and dicarbonyls, organic acids, and inorganic acids such as nitric acid and sulfuric acid. The formation of the aldehyde in reaction [11.22] also leads to additional formation of ozone as it undergoes photolysis.

Table 11-10. Calculated Lifetimes for Several NMVOC in the Atmosphere.

NMVOC	Lifetime due to			
	OH	NO_3	O_3	Photolysis
Propane	10 day	7 yr	>4500 yr	
n-Butane	4.7 day	2.8 yr	>4500 yr	
n-Octane	1.3 day	240 day		
Ethene	1.4 day	225 day	10 day	
Isoprene	1.4 hr	50 min	1.3 day	
a-Pinene	2.6 hr	5 min	4.6 hr	
Benzene	9.4 day	>4 yr	>4.5 yr	
Toluene	1.9 day	1.9 yr	> 4.5 yr	
m-Xylene	5.9 hr	200 day	>4.5 yr	
Phenol	5.3 hr	9 min		
Formaldehyde	1.2 day	80 day	>4.5 yr	4 hr
Acetaldehyde	8.8 hr	17 day	>4.5 yr	6 day
Acetone	53 day	>11 yr		60 day
Methanol	12 day	1 yr		
Ethanol	3.5 day	26 day		
2-Butanol	1.3 day	17 day		

Blank entries mean no data is available.

11.3.3 Health and Welfare Effects

11.3.3.1 Human Health

VOCs can cause various health effects, depending on the kind of compounds that are present, their concentrations, and on the individual response from the person exposed. Effects can vary from odor nuisance to decreases in lung capacity, cancer, and even death. Some compounds such as benzene, toluene, ethylbenzene, and xylene (collectively known as BTEX) are carcinogens and may cause leukemia. Hypersensitive individuals can have severe reactions to a variety of VOCs at very low concentrations. These reactions can occur following exposure to a single sensitizing dose or sequence of doses, after which time a lower dose can produce symptoms. Chronic exposure to low doses can also cause a response, however symptoms are usually non-specific and may be insufficient to permit identification of the offending compounds.

Concentrations in outdoor ambient air are very rarely at levels that cause immediate harm to humans. The low concentrations are due to low rates of emission and high rates of reaction with species such as OH•, O_2, and O_3. Hence, they do not tend to accumulate in the atmosphere to unhealthy concentrations. Outdoor air problems are usually of short duration and in a localized area, such as a spill site. Some VOCs, such as styrene, limonene, isoprene, and the monoterpenes, can react with NO_x and/or ozone to produce new oxidation products and secondary particles ($PM_{2.5}$), which create ambient air problems.

11.3.3.2 Environmental Welfare

The most serious regional and global environmental problem associated with the emissions of VOCs is due to their role in the formation of lower tropospheric ozone. Ozone directly affects ambient air quality through the production of smog and haze, as a facilitator in the production of particles, and its ability to oxidize and corrode metals, stone, polymers, and textiles. Chapter 12 provides a more detailed discussion of ozone formation and regulation.

11.3.4 History and Regulations:

The US EPA once listed these compounds as Reactive Organic Gases (ROG) but changed the terminology to Non-Methane Volatile Organic Compound (NMVOC) in order to match common usage by practitioners. Their definition of ROG included only compounds that are precursors of ozone. The list did not include VOCs that are non-reactive or of low-reactivity in ambient ozone production chemistry. The list would not be compatible with lists of species useful for monitoring indoor air or occupational exposure.

Identification and measurement of a particular NMVOC is expensive and time-consuming. In general, testing for individual compounds underestimate their total amount because some species present at low concentrations are difficult to identify or measure. A measurement of total VOCs (TVOC) to record all species present without distinguishing different chemicals was developed to deal with this situation. Many VOC air discharge permits require a TVOC test and additional tests for specific compounds known or suspected to be present in the air sample.

The main reasons for controlling VOCs in ambient air are to prevent the formation of hazardous levels of fine particles and ozone. Ozone formation in the lower troposphere is a function of the amount of ozone precursors (VOCs and NO_x) present, exposure to direct sunlight, and temperature. Generally, O_3 is a problem in the summer months in the US.

Some VOCs are known to be hazardous – acute or chronic exposure can lead to direct harm or death to humans. These species are regulated in the US under Hazardous Air Pollutant (HAP) laws. The Clean Air Act defines a *major stationary source* of HAPs as emissions that have the potential to emit 10 tons/year of one pollutant or a combination of 25 or more tons/year of any combination of HAPs. Sources emitting smaller amounts of HAPs are area sources and could also have their emissions regulated, although their permit may be very different than for major stationary sources. See chapter 13 for more information.

Indoor air is usually the most susceptible to VOC contamination, and many products are regulated to protect indoor air quality. Table 11-11 lists a few common consumer products with regulated VOC content (US-EPA, 1998). Table 11-12 lists several typical VOC emissions by industry. Typically, VOCs have the greatest impact on indoor air quality because users are close to the source, air exchange is often limited in comparison to outdoor air, and there is no energy source readily available to break down /oxidize them. See Chapter 15 for more information.

11.3.5 Control Strategies

There are many methods available for limiting and controlling the emissions of VOCs. The choice depends on overall concentration, vapor pressure, reactivity, flammability, and cost. High concentrations may allow for general recovery or direct combustion whereas low-concentration gases may need concentration before recovery or fuel addition before combustion or incineration. A material with a high vapor pressure may be recovered by cooling the gas stream, whereas if it has a low vapor pressure the cost of cooling may be prohibitive. A reactive compound may be destroyed in the exhaust stream by adding other chemicals (typically oxidants) and/or using a catalyst. Flammable materials may support or aid combustion and energy recovery. High-value materials allow for more costly and sophisticated recovery techniques since the cost of material replacement can be used to justify the equipment cost.

Table 11-11. VOC Limits in US Consumer Products.

Product	VOC max wt%	Product	VOC max wt%
Air fresheners:		Hairsprays	80
Single-phase	70	Hair mousses	16
Double-phase	30	Hair styling gels	6
Liquids/pump sprays	18	Household adhesives	
Solids/gels	3	Aerosols	75
Automotive windshield washer fluid	35	Contact	80
Bathroom and tile cleaners		Construction and panel	40
Aerosols	7	General purpose	10
All other forms	5	Structural waterproof	15
Cooking sprays—aerosol	18	Insecticides	
Dusting aids		Crawling bug	40
Aerosols	35	Flea and tick powder	25
All other forms	7	Flea and tick spray	35
Engine degreasers	75	Foggers	45
Fabric protectants	75	Lawn and Garden	20
Floor polishes/waxes		Laundry prewash	
Flexible flooring materials	7	Aerosols/solids	22
Nonresilient flooring	10	All other forms	5
Wood floor wax	90	Laundry starch products	5
Furniture maintenance products aerosol	25	Nail polish removers	85
General purpose cleaners	10	Oven cleaners	
Glass cleaners		Aerosols/pump sprays	8
Aerosols	12	Liquids	5
All other forms	8	Shaving creams	5

The objective of this section is to present various VOC reduction technologies, their strengths and weaknesses and the elements required to determine the preferred technology for a given application. Table 11-13 provides a list of these methods. Note that several of these methods also work with methane and/or carbon monoxide.

Table 11-12. Typical Industrial Emissions.

INDUSTRY	VOCS, TYPICAL SOLVENTS (S) AND OTHER OFF GASES
Alcohol Synthesis	C1, C2, C3, C6 Hydrocarbons
Automobile Coating	Ketones, Xylene, Toluene, Phenols
Bakery Ovens	Ethanol
Coffee Roasting	Heavy Oils from coffee beans
Electronic Components	Butyl Acetate, Xylene, Methylethyl Ketone
Formaldehyde	Formaldehyde, Methanol, Carbon monoxide
Metal Coating	Alcohols, Cellosolve Acetate, Phthalates
Paper Coating	High-Boiling Organics, Latexes
Pharmaceuticals	Isopropanol, Toluene, Hydrocarbons
Soil Remediation	Benzene, Toluene, Ethylene, Xylene
Sterilizers	Ethylene Oxide

Table 11-13. Unit Operations for Control of VOCs.

High Concentration	Low Concentration
Adsorption	Adsorption
Absorption	Absorption
Flaring	Biofiltration
Incineration	
Refrigerated Condensation	

11.3.5.1 Adsorption

An adsorption column is a mass transfer based unit operation in which the gas (adsorbate) bonds to the surface of a solid (adsorbent). The bond may be physical - due to electrostatic forces, or chemical - due to a reaction with the surface. The solid is usually in a particle form and designed to have a very large surface area per unit volume. The solids within the column are commonly referred to as a bed. Adsorption is carried out in fixed bed columns, which means the solids are not allowed to move about as they would in a fluidized bed column.

Typical solid adsorbents include activated carbon, silica gel, and activated alumina. Each of these are effective for VOC pollutants (as well as H_2S and SO_2). Most adsorbents also adsorb water, so gas streams need to be dry before treatment. In addition, most gases desorb at high temperatures, so gas streams must also be relatively cool before treatment. This desorption at high temperatures can be used to free the adsorbate in order to regenerate the solid after saturation. Regeneration allows multiple uses of the solid, until it physically degrades, or its adsorption capability declines too much.

The main design variables are solute concentration, temperature, flow rate, the choice of solid adsorbents and height of the column. Other important considerations are the *breakthrough time*, the cross-sectional area of the column, and the velocity of the adsorption front. Breakthrough is the time until the contaminant saturates the solid and begins emitting from the column. The adsorption front velocity relates the speed with which the solid is saturated to the column geometry. Figure 11-4 shows a plot of a contaminant breakthrough curve, with cartoons showing the adsorption wave in a column at several times. The dark color represents where the contaminant has adsorbed onto the solid.

FIGURE 11-4. Example of Breakthrough Curve and Adsorption Wave.

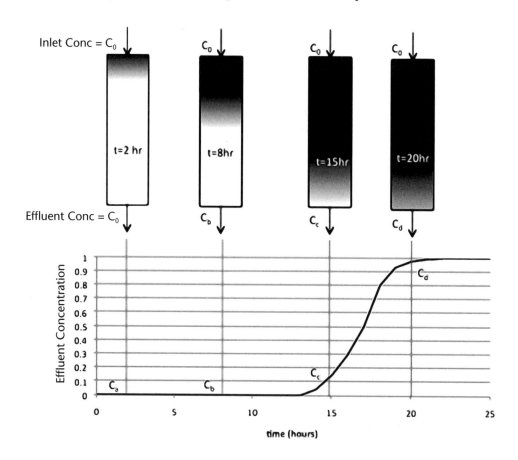

The amount (volume or mass) of a contaminant adsorbed on a solid depends on the temperature and the partial pressure of the contaminant. This relationship is called an equilibrium adsorption isotherm. There are several mathematical descriptions of isotherms. These empirical relationships are obtained by collecting equilibrium data (mass adsorbed at various partial pressures) at a given temperature. Figure 11-5 provides an example set of isotherm data.

FIGURE 11-5. Methane and Ethane on Activated Carbon Adsorbent (Isotherm at T=301.4 K).

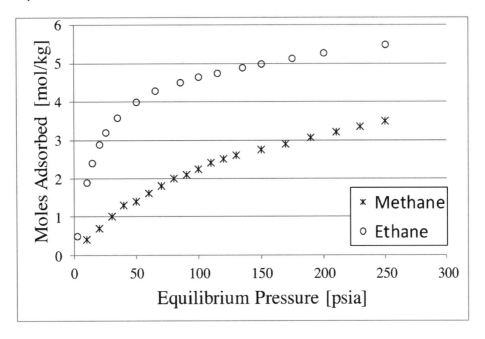

One common adsorption equation is the **Langmuir Isotherm**, given as:

11.27
$$W_i = \frac{aP_i^*}{1 + bP_i^*}$$

Where: W_i = moles of contaminant adsorbed per unit mass of adsorbent,
a,b = empirical constants (fitted from experimental observation), and
P_i^* = equilibrium partial pressure of contaminant (psia).

The moles of contaminant adsorbed (W_i) can be deduced from the total pressure of a gas before exposure to the solid, and the pressure after the gas is allowed to reach equilibrium adsorption with the solid, as:

11.28
$$W_i = \frac{(P^0 - P^*)V}{RT}$$

Where: P^0 = Total pressure before adsorption (vapor pressure) [psia],
 P^* = Equilibrium pressure after adsorption [psia],
 V = Volume of gas,
 R = Ideal gas constant, and
 T = Absolute temperature for the isotherm data [K].

The partial pressure of the gas is:

11.29
$$P_i^* = y_i P$$

Where: P_i^* = partial pressure of component i [psia],
 y_i = mole fraction of component i in the gas, and
 P = total pressure [psia].

A second isotherm model is the **Freundlich isotherm**, given as:

11.30
$$W_i = k(P_i^*)^n$$

Where: k, n = empirical constants, and n typically is between 0 and 1.

A third isotherm model is the **Brunauer-Emmett-Teller (BET) isotherm**:

11.31
$$W_i = \frac{cV_m P_i^*}{(P_{vapor} - P_i^*)\left[1 + (c-1)\frac{P_i^*}{P_{vapor}}\right]}$$

Where: W_i = moles of contaminant adsorbed per unit mass of adsorbent,
 c, V_m = empirical constants,
 P_i^* = partial pressure of component i [psia] (see equation [11.29]),
 P_{vapor} = vapor pressure of component i at isotherm temperature [psia].

The moles adsorbed can be determined using equation [11.28]. Vapor pressure is the pressure exerted by a vapor in equilibrium with its liquid form at a given temperature. It can also be thought of as the maximum volatility of a pure liquid in a closed container with no other liquids or gases in the system. Vapor pressures may be found using the Antoine equation (see problem 25) as found in Appendix VId or references such as (Lide, 2005) and (Perry, et al., 1997).

Adsorption columns run as batch systems. The collected contaminant remains on the solid within the column during a run. As long as there is enough capacity on the solids, no pollutants are emitted. Eventually, the surface becomes saturated, and no additional contaminant can be adsorbed. Once saturated, some contaminant begins to escape from the column; see point C_c in Figure 11-4. A formula for calculating breakthrough is:

11.32
$$t_B = \frac{Z_t - \delta}{v_f}$$

Where: Z_t = height of the bed,
  δ = depth of adsorption zone, and
  v_f = velocity of the adsorption front.

The height of the bed may be known, or it can be calculated based on a desired breakthrough time. Note that the isotherm data for a given material may change after multiple cycles of adsorption and desorption, and the time to breakthrough may change over time. The depth of the adsorption zone depends on the isotherm and the surface area per unit volume of the solid adsorbent. It is found experimentally and is small in comparison to the bed height. The velocity of the adsorption front [m/sec] may be calculated from:

11.33
$$v_f = \frac{Q_C}{(MW_i) W_i \rho_{bed} A_{bed}}$$

Where: Q_C = contaminant flow rate [g/sec],
  MW_i = molecular weight of contaminant i [g gas/mole gas],
  W_i = mole of gas adsorbed per unit mass solid [mole gas/kg solid],
  ρ_{bed} = bed density as packed (including packing voids) [kg solid/m³ bed], and
  A_c = cross-sectional area of the bed [m² bed].

The cross-sectional area of the column should be chosen to provide good mixing of the gas (turbulent flow), but not so small as to create large frictional losses in the gas stream nor so large as to be prohibitively expensive.

Regeneration of adsorption bed - The adsorbate must be removed from the column once breakthrough occurs. Removal can be done by replacing the saturated solids, or by regenerating the solids. Replacement is viable when an inexpensive and large quantity is readily available. Otherwise, the solids are regenerated. Regeneration occurs by taking the column off-line and desorbing the contaminant. Regeneration techniques include heating, reducing pressure, stripping with an inert gas, or a combination of these methods. A common method is to send steam through the column. Once the contaminant desorbs, it requires treatment to eliminate its potential hazard to human health and the environment. If the substance is valuable, it could be recovered using a refrigerated condenser (see section 11.3.5.5). Otherwise, it could be incinerated (section 11.3.5.3) or, if water-soluble, destroyed in a wastewater treatment bioreactor.

Regeneration may require several hours, so adsorption typically requires two or three parallel columns. One column is always available for adsorption while the others are undergoing desorption. Figure 11-6 presents a simplified process flow diagram showing a typical set up for adsorption with refrigerated condensation regeneration. In this figure, the left column is adsorbing, while the right column is being regenerated.

FIGURE 11-6. Schematic of Adsorption System with VOC Recovery.

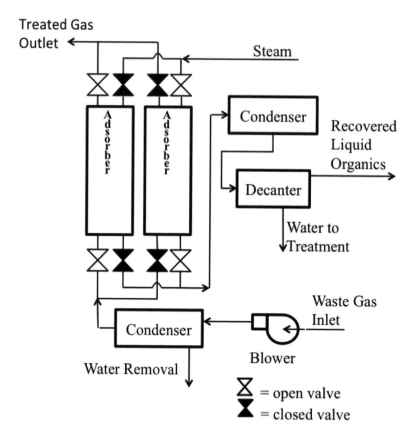

Example 11-3.

Generate a preliminary size determination for an adsorption column for removing ethane from air using an activated carbon adsorbent, see Figure 11-5 for isotherm data. The ethane mole fraction in the inlet is 10% and the inlet gas stream enters the column at 2.5 atm, 301.4 K, and volumetric flow rate of 0.65 m³/min. The bed density is 450 kg/m³. The column has an inside diameter of 1.5 meter, and the desired time until breakthrough is four hours. Neglect the depth of the adsorption zone.

Solution:
Steps in obtaining an answer for this problem:

A – Fit an isotherm to the provided experimental data.
B – Calculate the velocity of the adsorption front.
C – Determine the bed height.

We explore all three isotherm models to determine which best fits this set of data. The first step for each case is to linearize the isotherm equation so we can use a statistical line fit to determine the isotherm constants. We choose the best model using the linear regression term r^2. All three cases are plotted together with the data to confirm the choice. Linearize the equations into the form and use the data to find the best-fit line. The slope and intercept provide values for the constants.

The linearized form of the **Langmuir isotherm** model is:

11.34
$$\frac{P_i^*}{W_i} = \frac{1}{a} + \frac{b}{a} P_i^*$$

Where: $y = P_i^*/W_i$,
$x = P_i^*$,
m_0, the intercept, $= 1/a$, and
m_1, the slope, is b/a.

The linearized form of the **Freundlich isotherm** model is:

11.35
$$\ln(W_i) = \ln(k) + n \ln(P_i^*)$$

Where: $y = \ln(W_i)$
$x = \ln(P_i^*)$
$m_0 = \ln(k)$
$m_1 = n$

The linearized form of the **BET isotherm** model is:

11.36
$$\frac{P_i^*}{W_i(P_0 - P_i^*)} = \frac{1}{cV_m} + \frac{c-1}{cV_m}\left(\frac{P_i^*}{P_0}\right)$$

Where: y = $P_i^*/(W_i(P_0-P_i^*))$
x = P_i^*/P_0
m_0 = $1/cV_m$
m_1 = $(c-1)/cV_m$

A vapor pressure of 670 psia was used for ethane at 301.4 K (Perry, et al., 1997). Values for the best-fit slope, intercept, and r^2 value are summarized in Table 11-14. Values for the model constants are given in Table 11-15. All three of the model predictions and the experimental data are plotted in Figure 11-7.

Table 11-14. Example 3 Best Fit Line Parameters.

Model	Langmuir	Freundlich	BET
m_0 (intercept)	3.9111	-0.4606	0.00142
m_1 (slope)	0.1700	0.4330	0.2630
r^2	0.9986	0.8867	0.9884

Table 11-15. Example 3 Isotherm Model Parameters.

Model	Parameter	Value
Langmuir	a	0.2557
	b	0.04346
Freundlich	k	0.6309
	n	0.4330
BET	c	186.1
	Vm	3.781

FIGURE 11-7. Example 11-3 Model fits of Ethane Isotherm Data at Temperature of 301.4 K.

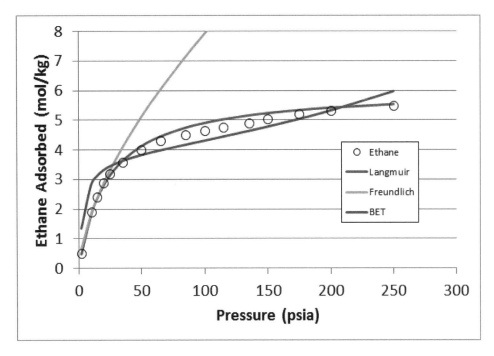

Based on these results, the Langmuir isotherm is the best model of the three explored for this particular data set. Note that other, more complex models exist, and one of those could be a better fit although they will remain unexplored in this work. With this model, we can estimate the amount of gas adsorbed onto the solid over the pressure range of the data.

B. Next, find the velocity of the adsorption front with equation [11-33]. For this calculation, use each term as follows, recalling $Q_g = (G_s)(\varrho_{gas})(y_i)(MolWt_i)$:

$$Q_g = \left(0.65 \frac{m^3}{min}\right)\left(\frac{2.5 atm}{0.082 \frac{l \cdot atm}{mol \cdot K} 301.4K} \frac{1000 l}{m^3}\right)(0.1)\left(30 \frac{g}{mol}\right) = 197.2 \frac{g}{min}$$

$$W_i = \frac{0.2557 \frac{mol}{kg_{solid} psia}(3.675 psia)}{1+(0.04346 psia^{-1})*(3.675 psia)} = 0.810 \frac{mol}{kg_{solid}}$$

Where $P_i^* = y_i * P_{total} = 0.1 * 2.5 atm * 14.7 psia/atm = 3.675 psia$. Solving for v_f leads to:

$$v_f = \frac{197.2 \frac{g}{min}}{\left(30 \frac{g}{mol}\right)\left(0.81 \frac{mol}{kg_{solid}}\right)\left(450 \frac{kg_{solid}}{m^3_{bed}}\right)\left(\frac{\pi}{4} 1.5^2 m^2_{bed}\right)} = 0.0102 \frac{m}{min}$$

C: Finally use equation [11.32] to obtain the bed height:

$$Z_t = v_f \tau_B = \left(0.0102 \frac{m}{min}\right)\left(4 hour * 60 \frac{min}{hour}\right) = 2.45 m$$

11.3.5.2 Absorption

An absorption column is a mass transfer based unit operation in which gas (absorbate) dissolves into a liquid (absorbent). The amount of gas that can dissolve into the liquid controls the flow rates of liquid and gas. The concentration gradient of the contaminant between the gas and liquid phases control the rate of absorption. These factors control the size of the column.

Absorption can be carried out in a spray tower (good when also removing particulate matter) or in a column containing plates or solid packing. The column design uses the solid to help increase the liquid surface area to volume ratio (creating thin liquid films) and to improve the mixing between the gas and liquid. The choice of liquid requires that the contaminant in the gas phase is highly soluble but that the rest of the gas phase is insoluble. It is also useful if the liquid has a low vapor pressure, such that very little of it evaporates into the gas phase within the tower or column.

The main design variables for absorption are the gas flow rate, the liquid flow rate, and the column height. These variables relate through the mole balance and the rate of mass transfer between the gas and liquid phases. Figure 11-8 shows the basic design and system variables for a counter-current absorption column. Other flow arrangements are possible (co-current, cross flow) but they are not as efficient.

The operating mole balance (in = out) on this system (assuming T and P constant) is:

11.37
$$G_{in} y_{in} + L_{in} x_{in} = G_{out} y_{out} + L_{out} x_{out}$$

Where: G_{in}, G_{out} = total gas flow in or out [mole/time],
L_{in}, L_{out} = total liquid flow in or out [mole/time],

FIGURE 11-8. Schematic System Design for Absorption System.

x_{in}, x_{out} = mole fraction of contaminant in the liquid phase in or out, and
y_{in}, y_{out} = mole fraction of contaminant in the gas phase in or out.

This unit operation works when the contaminant is not in equilibrium between the gas and liquid phases. Absorption occurs when the gas phase contains more contaminant than when in equilibrium with the liquid. The difference causes the contaminant to move from the gas phase into the liquid phase. The opposite case, stripping, is used when the contaminant moves from the liquid into the gas phase. Henry's law is used to describe the equilibrium between the gas and liquid phase for dilute solutions (where there is very little of the gas phase component in the liquid):

11.38
$$P_i^* = H_i C_i^*$$

Where: P_i^* = Equilibrium partial pressure for component i, with partial pressure defined in equation 11.29. The * superscript denotes equilibrium conditions.
H_i = Henry's Law constant for component i
C_i^* = Equilibrium liquid concentration for component i (may be represented as the mole fraction in the liquid phase – check units of H).

Appendix VIb provides Henry's law constants for several gases in water.

Any of the following equations can describe the rate of mass transfer between the gas and liquid phases:

11.39
$$N = K_y(y - y^*) = K_x(x - x^*)$$

11.40
$$N = K_P(P_i - P_i^*) = K_C(C_i - C_i^*)$$

Where: N = molar flux [moles/ area/ time]
K are the mass transfer coefficients [units vary]
y = gas phase mole fraction
x = liquid phase mole fraction
P_i = partial pressure of component i [atm]
C_i = liquid concentration of component i [moles/volume]
* = superscript denotes equilibrium value.

The choice depends on what information is available, (Henley, et al., 1981). Note that the equilibrium values are just a helpful reference point and do not represent any actual condition in the column. The difference-term in each form represents the driving force that causes preferential transfer of the contaminant from one phase to the other. The larger this term is, the greater the transfer.

Equation [11.37] has multiple terms to represent the gas and liquid flow rates. As written, there is no easy way to use the equation to design a column. To help reduce the number of variables, we redefine the gas and liquid flow rates. The gas stream excludes the contaminant and is called the carrier gas, G_c. The liquid stream excludes the contaminant and is called the liquid solvent, L_s. The redefinition allows the flow of liquid and gas to have the same value between the inlet and outlet. In order to exclude the contaminant this way, we must replace mole fractions with mole ratios:

11.41
$$Y = \frac{y}{1-y} \quad X = \frac{x}{1-x}$$

11.42
$$y = \frac{Y}{1+Y} \quad x = \frac{X}{1+X}$$

Where: Y = mole ratio of contaminant in the gas phase,
X = mole ratio of contaminant in the liquid phase,
y = mole fraction of contaminant in the gas phase, and
x = mole fraction of contaminant in the liquid phase.

Problem 15 in the homework asks you to show that this transformation is valid and necessary. Rewriting equation [11.37] with these variables and rearranging leads to:

11.43
$$G_C(Y_{out} - Y_{in}) = L_S(X_{out} - X_{in})$$

Where: G_C = flow of the carrier gas (the insoluble species of the entering gas), and
L_S = flow of the solvent liquid.

This operating line is a linear equation in X-Y coordinates. It has a slope of L_S/G_C. Figure 11-9 plots the operating and equilibrium lines. The equilibrium line was generated using Henry's Law data for CO_2 at 10°C and 25 atm. An absorption operating line is always above the equilibrium line on this type of plot. An operating line that is below the equilibrium line would represent a stripping column, where the contaminant is moving from the liquid into the gas. The vertical distance between the equilibrium line and the operating line represents the driving force, or concentration gradient, for absorption at those conditions in the column. This driving force is what causes the contaminant to move from the gas phase into the liquid phase.

The end points of the operating line represent the conditions at the top and bottom of the column. The bottom of the column has coordinates (X_{out}, Y_{in}) and the top of the column has coordinates (X_{in}, Y_{out}).

In a typical design, the known variables include the gas phase compositions (Y_{in} and Y_{out}) and the liquid inlet composition (X_{in}). If the gas flow rate is chosen as the basis, then just the liquid flow rate (L_S) and the liquid exit composition (X_{out}) remain to be determined. The problem has two unknowns and one equation, making the design problem open-ended. If one of the values is set, then the other can be solved for; otherwise, there are multiple valid solutions.

Constraints may be placed on the solution set. The first is that there is a minimum liquid flow rate, below which the column cannot remove enough of the contaminant from the gas to reach the desired removal efficiency (Y_{out}). The other limit is a maximum liquid flow rate determined from the cost of associated equipment to move and treat the liquid – pumps, piping, and regeneration equipment.

The minimum flow rate (L_{min}) can be deduced from the operating line – equilibrium line plot. The point (X_{in}, Y_{out}) is a known point of the operation line, and the line $Y = Y_{in}$ constrains the other end. The slope of the operating line is L_S/G_C, and since G_C is known, the minimum value of L_S can be determined from the minimum slope of the operating line. This minimum occurs when the operating line touches the equilibrium line. The minimum usually occurs at the location where $Y=Y_{in}$ intersects the equilibrium line, but it could happen at a different value. The reasoning here is that when the operating line and the equilibrium line touch, there is no longer any driving force to move the contaminant from the gas into the liquid, so it would take an infinitely large column to achieve the desired amount of transfer. If the operating line crossed the equilibrium line, it would lead to a region with the contaminant moving from the liquid into the gas! This minimum flow rate is undesirable, so the actual design flow rate must be larger. As a rule of thumb, the flow rate is typically 1.3 to 1.7 times larger than the minimum.

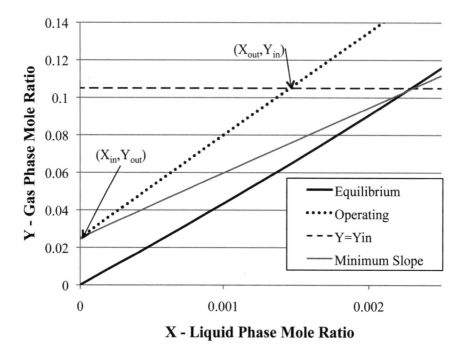

FIGURE 11-9. Absorption Design Diagram - CO_2 in Water at 10 °C and 25 atm.

Calculate the column height from:

11.44 $$Z_t = H_{og} N_{og}$$

Where: H_{og} = the height of a transfer unit, and
N_{og} = the number of transfer units.

A transfer unit is a concept that relates how much liquid is needed to achieve equilibrium with the contaminated gas. Equilibrium means that the concentration of a contaminant in the liquid is as high as it can be for a gas of constant composition as related by Henry's law for dilute systems, equation [11.38]. It does not mean that the liquid and gas have the same concentrations. The concentration of a contaminant in an absorption column is always changing, so the system moves beyond this local equilibrium point. In fact, it may move beyond it several times – and that several times is the number of transfer units. The height of a transfer unit is often constant for dilute solutions. The value depends on the mass transfer coefficients for the contaminant moving through the gas, across the gas-liquid interface, and into the liquid. It is beyond the scope of this book to make the calculation; details for the interested can be found in (Fair, et al., 1997). The number of transfer units is given as:

11.45
$$N_{og} = \int_{y_2}^{y_1} \frac{y_{LM}^* dy}{(1-y)(y-y^*)}$$

Where: y_{LM}^* is the log-mean value of the equilibrium vapor phase mole fraction at the gas-liquid interface. It can be calculated using the equation below:

11.46
$$y_{LM}^* = \frac{(1-y)-(1-y^*)}{\ln\left[\dfrac{1-y}{1-y^*}\right]}$$

Where: y = bulk vapor phase mole fraction, and
y* = equilibrium vapor phase mole fraction.

Alternatively, a simple graphical technique can be used to estimate the number of transfer units (N_{og}). Draw a horizontal line starting at (X_{in}, Y_{out}), connecting the operating line and the equilibrium line. Where the line intersects the equilibrium line, draw a vertical line to the operating line. Then draw another horizontal line to the equilibrium line. The horizontal lines form a series of steps, see Figure 11-10. N_{og} is equal to the number of horizontal steps to reach the point (X_{out}, Y_{in}). A fractional step may be needed to complete the determination. Example 11-4 includes a determination of N_{og}. The value of N_{og} increases as L_S approaches its minimum flow rate.

Other design considerations include the column diameter and pressure drop. The diameter directly controls the gas velocity:

11.47
$$v_g = \frac{G}{\dfrac{\pi}{4}(Dia^2)}$$

Where: G = total gas flow rate, and
Dia = column diameter.

The gas velocity, v_g, is a superficial velocity – it does not account for the fraction of the area excluded by the flowing liquid or any solid plates or packing that may be present. A smaller diameter causes faster velocities. At very fast velocity, the gas slows the descent of the liquid, which causes liquid holdup. Even faster velocities can cause the column to flood because too much liquid is held up. There is a much larger pressure drop in the gas flow during holdup. The large pressure drop can lead to other problems and is expensive for gas handling after the column.

Additional discussion of absorption may be found in references such as (Henley, et al., 1981) or (Wankat, 1988).

■ Example 11-4.

How many equilibrium transfer units are required to absorb benzene (C_6H_6) from air using water at 10°C and 1 atm. pressure. The incoming gas flow rate is 3 mol/min, contains 6.5 mol-% benzene in air and must be reduced to 0.5 mol-%. The liquid flow rate is 1.5 times the minimum needed. Neglect the solubility of N_2 and O_2 in the water.

Solution:
  A- Obtain and plot X-Y data for equilibrium line.
  B – Add the point (X_{in}, Y_{out}) and the line Y=Yin to determine the minimum liquid flow rate.
  C – Plot the operating line with LS using design rate (Ls = 1.5 $L_{S\,min}$), Find X_{out}.
  D – Graphically solve for N_{og}.

Henry's Law constant, from Appendix 6b, for benzene at 10°C is 156 atm/mole fraction in liquid. These values need to be converted to mole ratios as shown in table below.

x	y= 156 x	X=x/(1-x)	Y=y/(1-y)
0	0	0.00000	0.00000
0.00005	0.0078	0.00005	0.00786
0.00010	0.0156	0.00010	0.01585
0.00020	0.0312	0.00020	0.03220
0.00030	0.0468	0.00030	0.04910
0.00040	0.0624	0.00040	0.06655
0.00050	0.078	0.00050	0.08460

Figure 11-10 plots these last two columns as the equilibrium line.

The inlet and outlet gas mole ratio and the carrier gas flow rate are:

$$Y_{in} = \frac{0.065}{1-0.065} = 0.0695$$

$$Y_{out} = \frac{0.005}{1-0.005} = 0.00503$$

$$G_C = 3\frac{mol}{hr}(1-y_{in}) = 3(1-0.065) = 2.805\frac{mol_{benzene}}{hr}$$

The point (X_{in}, Y_{out}) = (0, 0.00503) is shown as a hollow circle, and the line Y=Y_{in} (Y = 0.065) is shown as a dashed horizontal line on the figure.

Reactive Carbon Compounds

Next, a line is sketched for the minimum conditions connecting (X_{in}, Y_{out}) to the point where the line $(Y=Y_{in})$ crosses the equilibrium line. This point is (X_{out}^{min}, Y_{in}). $(L/G)_{min}$ is the slope of this line. The actual flow rate can be calculated by multiplying the minimum by the design factor stated in the problem:

$$\left(\frac{L_S}{G_C}\right)_{min} = slope = \frac{Y_{in} - Y_{out}}{X_{out}^{min} - X_{in}} = \frac{0.0695 - 0.00503}{0.000446 - 0.0} = 144.5$$

$$(L_S)_{min} = 144.5(G_C) = 405.3 \frac{mol}{hr}$$

$$(L_S)_{actual} = 1.5(L_S)_{min} = 608.0 \frac{mol}{hr}$$

FIGURE 11-10. Absorption Design Diagram for Example 11-4 Benzene Absorption in Water.

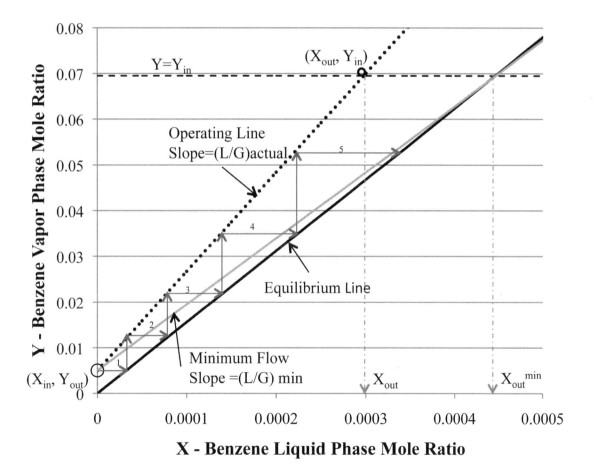

Next, calculate the slope for the operating line, and plot it as the actual operating line on the figure:

$$(slope)_{actual} = \frac{L_S}{G_C} = \frac{608.0 \frac{mol}{hr}}{2.805 \frac{mol}{hr}} = 216.8$$

The mole ratio of benzene in the outlet water is calculated from this slope (216.8) and the intercept (Y=0.00503 at X =0) to yield X_{out} = 0.000297.

The number of transfer units can be determined graphically by stepping off the stages on the column, starting at (X_{in}, Y_{out}). There are about 4.7 steps, or stages, for this system.

We can check this solution with the mass balance (equation [11.43])

$(3mol)(0.065) + (1470mol)(0.00) = (2.805mol)(0.00503) + (608.2mol)(0.000297)$

$0.195 = 0.195$ Answer checks.

11.3.5.3 Thermal Destruction

Some emissions cannot be reduced by adsorption or absorption, either a suitable sorbent is not available, or it is too costly to construct such a facility. Another possible way to reduce these emissions is direct oxidation. Combustion is useful when the emissions are capable of supporting a flame. Combustion may be carried out in the open air (flaring) or a combustion chamber. Incineration is useful when fuel must be added to complete the oxidation. In each case, the process involves three types of materials in the input: fuel, oxidant, and diluent. Fuel generates the combustion reaction. The oxidant supports the combustion by breaking the chemical bonds in the fuel and releasing heat. The diluent(s) are any other substances that do not take part in the combustion, but is present. An example of a diluent is nitrogen when air is the source of oxygen.

11.3.5.3.1 Flaring

Flaring is used to control almost any VOC stream and can handle large fluctuations in flow rate, heating value, and VOC concentration. Flaring is performed in the open air, so there is not the safety concern of keeping the gas mixture under the lower explosion limit. Flares may be located at ground level or elevated. Elevating the flare moves the flame from the ground to prevent dangerous conditions of an open flame near other process equipment. It also allows the combustion effects, noise, heat, smoke, and odors to occur away from work areas.

Flaring occurs in either continuous or batch mode. The batch can be cyclic or variable. The petroleum and petrochemical industries use flares as safety devices to control large volumes of VOC emissions resulting from system upsets. They are designed to relieve emergency process upsets that release large volumes of gases, but can also be used to control vent streams from various unit operations. Gases from refineries and the chemical industry contain VOCs that have high heating values (methane and ethane). Emission streams in other industries, such as bakery oven emissions, may not have a high heat content and require additional fuel to complete combustion.

FIGURE 11-11. Elevated Flare in Operation. The image is of a clustered flare, with multiple gas risers and flare tips. The stack is self-supporting, with both steam and non-assisted flare tips. Image courtesy of Flaregas Corporation.

The most important design variables are flow rate, temperature, heat content of the emission gas stream, and mixing. Flow rates can vary from near zero to about 500 scm/sec (1,000,000 scf/min), where scm is standard cubic meters and scf is standard cubic feet. The discharge temperature is usually in the range of 500 to 1,100°C. The gas stream needs to have a heating value of at least 11 MJ/scm (300 BTU/scf) to support combustion. Otherwise, an auxiliary fuel must be added. Mixing can be enhanced by the injection of steam or air, or by pressurizing the vent stream feeding the flare.

Advantages of flares include the low capital and operating costs, ability to handle large flows, and they work well with intermittent and highly fluctuating flows. Disadvantages include the open-air flame produces noise, smoke, heat and light pollution; they can be a source of SO_x, NO_x, and CO; they cannot be used with halogen containing waste streams, and the heat released during combustion is lost to the atmosphere.

The Autoignition Temperature is the minimum temperature required to ignite a gas or vapor in the air without a spark or flame being present. Appendix VI f lists the autoignition temperature for several common fuels.

11.3.5.3.2 Incineration

This technology reduces emissions by raising the temperature of the contaminant above its autoignition point in the presence of oxygen, and keeping it at that temperature for a time needed to complete oxidation to carbon dioxide and water. The important design variables (as discussed in section 10.5.2.2) are temperature, time, oxygen, and mixing.

FIGURE 11-12. Interior of Combustion Chamber. The burners in this chamber are steam assisted for olefins service, and the image shows the modular ceramic fiber block used for insulation of the inside walls of the unit. Also note the pilots (used to ignite the gases) located adjacent to the burners. Image courtesy of Flaregas Corporation.

A thermal incinerator includes a fuel source, the inlet emission stream, an ignition chamber, and an outlet emission stream – see Figure 11-13 A. A recuperative incinerator also includes a heat exchanger to recover some heat energy from the system, see Figure 11-13 B.

FIGURE 11-13. Schematic of A) an Incinerator and B) an Incinerator with Recuperative Heat Exchanger.

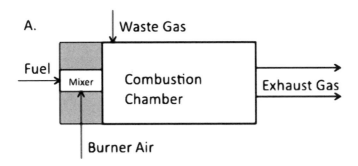

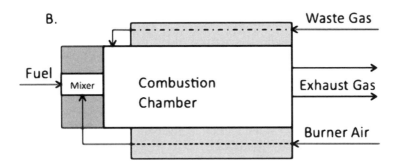

The main component in combustion chambers is a nozzle-stabilized flame maintained by a combination of waste gas compounds, auxiliary fuel, and supplemental air. Extra fuel and air are added on an as-needed basis to ensure complete destruction of the waste gas contaminants. The waste gas is heated to its ignition temperature as it passes through the flame. This temperature depends on the gas composition, and must be determined by experiment. The required level of VOC destruction determines the time and temperature needed in the combustion chamber. A shorter time is needed at higher temperatures. A nominal residence time, τ, can be calculated using the following equation:

11.48
$$\tau = \frac{Vol}{G}$$

Where: Vol = combustion chamber volume [m³], and
G = volumetric flow rate of the gas [m³/sec].

Most thermal units provide no more than one second of residence time, and the operating temperature is adjusted to complete oxidation in this amount of time. If the gas stream characteristics change over time, only the temperature can be easily changed, so it becomes the main operating variable.

An energy balance on the inlet and outlet streams, and the thermodynamic properties of the combustion gases allows calculation of the oxidation temperature:

11.49
$$\frac{dE_{system}}{dt} = Q - W + \sum_{i=1}^{n} F_i^{in} E_i^{in} - \sum_{i=1}^{n} F_i^{out} E_i^{out}$$

Where: dE_{system}/dt = the rate of accumulation of energy in the system,
Q = rate of energy transfer across the system boundary not associated with mass,
W = rate of energy transfer associated with work done across the system boundary, and
$F_i E_i$ = the rate of energy (E_i) transferred in and out of the system by mass flow (F_i) across the system boundary for multiple species (i=1, n).

The first term (dE/dt) represents accumulation in the system. At steady state, the combustion chamber reaches its operation temperature, and no additional accumulation occurs. It may take minutes to hours after startup before steady state operation occurs.

The second term (Q) represents the energy transfer across the system boundary not associated with mass, such as heat exchanged by conduction, convection, or radiation. Under adiabatic conditions (Q = 0) the oxidation temperature is the *adiabatic flame temperature*.

The work term (W) can be separated into two terms – flow work and shaft work:

11.50
$$W = W_s - \sum_{i=1}^{n} F_i^{in} PV_i^{in} + \sum_{i=1}^{n} F_i^{out} PV_i^{out}$$

Flow work is the energy associated with adding and removing mass from the system. Shaft work originates with equipment (likes pumps, mixers, and turbines) that transfers energy across the system boundary. We assume that shaft work (Ws) is zero for the calculation of the flame temperature.

The remaining two terms in equation [11.49] represent the energy transferred into and out of the system with the flowing mass (F_i). The energy terms (Ei) represent the sum of all the energy forms – kinetic, potential, internal, and other that the mass may have. The kinetic, potential, and other terms are insignificant for combustion. The internal energy is important as it represents the energy stored in the molecules as vibrations, rotations, and translations, and is often thought of as thermal energy:

11.51
$$E_i = \frac{u_i^2}{2} + gz_i + U_i + other \cong U_i$$

Enthalpy, H_i, is defined as the sum of the internal energy (U_i) and the flow work (PV_i). It is a function of temperature:

11.52
$$H_i = U_i + PV_i = H_i^{ref}(T_{ref}) + \int_{T_{ref}}^{T} Cp_i dT$$

Where: H_i^{ref} = the enthalpy at some reference temperature (typically 298 K), and
Cp$_i$ = the species heat capacity.

Heat capacity is a function of temperature, generally of the form:

11.53
$$Cp = \alpha + \beta T + \gamma T^2 + \frac{\varepsilon}{T^2}$$

Appendix VIc lists values for the constants for several species. Additional values may be found in other sources, such as (Weast, 2003).

Applying the steady-state, adiabatic, no shaft work assumptions to equation [11.49] and with some algebraic rearrangement, yields the following relationship:

11.54
$$\sum_{i=1}^{n} F_i^{in} H_i^{in} - \sum_{i=1}^{n} F_i^{out} H_i^{out} = 0$$

We can develop a relationship between the flow of mass in and out of each species for the system by using the stoichiometric relationships from the combustion reaction, assuming the reaction goes to completion. A general form of a reaction, written on a per mole of species A basis, is given as follows:

11.55
$$A + \frac{b}{a}B \rightarrow \frac{c}{a}C + \frac{d}{a}D$$

Table 11-16. Stoichiometric Relationships between Molar Flows In and Out of Combustion Chamber.

Species	Amount Entering	Amount Exiting
A	F_A^{in}	$F_A^{in} - \frac{a}{a} F_A^{in}$
B	F_B^{in}	$F_B^{in} - \frac{b}{a} F_A^{in}$
C	F_C^{in}	$F_C^{in} + \frac{c}{a} F_A^{in}$
D	F_D^{in}	$F_D^{in} + \frac{d}{a} F_A^{in}$
Inerts	F_{Inert}^{in}	F_{Inert}^{in}

For a typical combustion reaction A is the fuel molecule, B is oxygen, C is carbon dioxide, and D is water. Species A is assumed to be the limiting reagent – that is it is completely used up in the reaction. Species B is in excess. There may also be some inert gas(es) such as nitrogen when air is the source of oxygen. There may be multiple inert species to carry along, and each must be included in the calculations. Table 11-16 lists the amounts of each species entering and leaving.

Note that the amount of species A exiting is zero when the reaction goes to completion. Expanding the sum in equation [11.54] (i=A, B, C, D, Inert) and substituting in each term from Table 11-16 and rearranging yields the following equation:

11.56
$$\sum_{i=1}^{n} F_i^{in} H_i^{in} - \sum_{i=1}^{n} F_i^{out} H_i^{out} = F_A^{in}\left(H_A^{in} - H_A^{out}\right) + F_B^{in}\left(H_B^{in} - H_B^{out}\right) + F_C^{in}\left(H_C^{in} - H_C^{out}\right)$$
$$+ F_D^{in}\left(H_D^{in} - H_D^{out}\right) + F_{Inert}^{in}\left(H_{Inert}^{in} - H_{Inert}^{out}\right) - F_A^{in}\left(\frac{d}{a}H_D^{out} + \frac{c}{a}H_C^{out} - \frac{b}{a}H_B^{out} - H_A^{out}\right) = 0$$

The last term is defined as the heat of reaction on a per mole A basis at the reaction temperature, written as:

11.57
$$\Delta H_{rxn} = \left(\frac{d}{a}H_D^{out} + \frac{c}{a}H_C^{out} - \frac{b}{a}H_B^{out} - H_A^{out}\right)$$

The heat of reaction is known at a reference temperature (typically 298 K), so use equation [11.52] to modify to the standard reference temperature, and rewrite it as:

11.58
$$\Delta H_{rxn} = \left[\frac{d}{a}H_D^{ref} + \frac{c}{a}H_C^{ref} - \frac{b}{a}H_B^{ref} - H_A^{ref}\right] + \left[\frac{d}{a}Cp_D + \frac{c}{a}Cp_C - \frac{b}{a}Cp_B - Cp_A\right](T - T_{ref})$$

The first term in brackets is the heat of reaction at the reference temperature ΔH_{rxn}^{ref}, and the second term is the overall change in heat capacity per mole A, ΔCp. Thus, equation [11.58] is the sum of the heat of reaction at the reference temperature plus the reaction heat capacity:

11.59
$$\Delta H_{rxn} = \Delta H_{rxn}^{ref} + \Delta Cp(T - T_{ref})$$

Substituting equation [11.59] into [11.56] and reverting to summation notation yields:

11.60
$$\sum_{i=1}^{n} F_i^{in} H_i^{in} - \sum_{i=1}^{n} F_i^{out} H_i^{out} = \sum_{i=1}^{n} F_i^{in}\left(H_i^{in} - H_i^{out}\right) - \left[\Delta H_{rxn}^{ref} + \Delta Cp(T - T_{ref})\right]F_A^{in}$$

The change in enthalpy between inlet and outlet conditions can also be rewritten in terms of heat capacity by again applying equation [11.52] to yield:

11.61
$$H_i^{out} - H_i^{in} = \left[H_i^{ref}(T_{ref}) + \int_{T_{ref}}^{T_{out}} Cp_i dT\right] - \left[H_i^{ref}(T_{ref}) + \int_{T_{ref}}^{T_{in}} Cp_i dT\right]$$
$$= \int_{T_{in}}^{T_{out}} Cp_i dT = \overline{Cp}_i(T_{out} - T_{in})$$

Where $\overline{Cp}_i$ is the average heat capacity over the given temperature range. It is calculated with the following equation, using the constants ($\alpha, \beta, \gamma, \varepsilon$) from appendix 6c:

11.62
$$\overline{Cp}_i = \frac{\int_{T_{in}}^{T_{out}} Cp_i dT}{(T_{out}-T_{in})} = \frac{\alpha_i(T_{out}-T_{in}) + \frac{\beta_i}{2}(T_{out}^2-T_{in}^2) + \frac{\gamma_i}{3}(T_{out}^3-T_{in}^3) - \varepsilon_i\left(\frac{1}{T_{out}}-\frac{1}{T_{in}}\right)}{(T_{out}-T_{in})}$$

Substituting equation [11.61] into [11.60] gives the adiabatic flame temperature equation:

11.63
$$\sum_{i=1}^{n} F_i^{in} \overline{Cp}_i (T_{out}-T_i^{in}) + F_A^{in}\left[\Delta H_{rxn}^{ref} + \Delta \overline{Cp}(T_{out}-T_{ref})\right] = 0$$

The usual unknown in the equation, Tout, is solved for with a trial and error technique, see Example 11-5.

Incinerators can achieve 98 – 99.9999% destruction of VOCs. They work best on concentrated streams of the contaminant (the waste gas is above 20% of the lower explosive limit (LEL). Incineration is a poor choice for low-concentration high-flow waste stream. Also, highly variable flow may create operational problems due to highly changing residence times in the combustion chamber.

Incinerators have a relatively high operating cost when they require supplemental fuel. The problem may be overcome by using regenerative incineration to recover heat. The heat from the combustion reaction is used to pre-heat the incoming streams, thus lowering the amount of fuel needed to achieve destruction temperature. Regeneration may reduce fuel use as much as 70% for a well-designed system, and less for system retro-fits. However, the required heat exchanger adds to the capital cost of the equipment.

Finally, incineration is not recommended for waste streams containing halogens or sulfur because they create corrosive acid gases in the exhaust and require special materials and additional removal equipment before the gas can be emitted to the atmosphere.

■ Example 11-5.

A bakery generates a waste gas stream containing nitrogen with 30,000 ppmv carbon dioxide and 1000 ppmv ethanol at a flow rate of 4.4 m³/min and conditions of 25°C and 1 atm. They are required to reduce the ethanol by 98%. One option for treatment is incineration, which needs to be carried out at 650°C for 1.0 seconds to achieve the required reduction. Can this stream support incineration at 650°C if oxygen is supplied from air and fed at 20% excess? If not, how much methane is required to support the incineration, again with air supplying the oxygen, but in this situation using 50% excess oxygen? How large must the combustion chamber be to carry out the destruction?

Solution:

This problem asks to find the temperature obtained from the complete oxidation of the VOC (ethanol) in the waste gas. If it is not enough to reach the destruction temperature, how much methane must be added? First, we determine the adiabatic flame temperature from the combustion of ethanol in the waste gas stream.

Balanced Chemical Reaction:

$$4C_2H_5OH + 13O_2 \rightarrow 8CO_2 + 10H_2O$$

We choose a time basis of one second and determine the number of moles of each inlet species using the ideal gas law:

$$n = \frac{PV}{RT} = \frac{(1\,atm)(4.4\,\frac{m^3}{min})}{\left(0.082\,\frac{l\cdot atm}{mol\cdot K}\right)(298K)} \cdot \frac{1000\,l}{m^3} \cdot \frac{min}{60\,sec} = 3.00\,\frac{mol}{sec}$$

Next, determine the actual molar flow rates (Fi) for each species (inlet and outlet). These values are in the following table. Note that the amount of oxygen is increased by a factor of 1.2 due to the excess requirement in the problem statement. Also note that the amount of nitrogen is increased due to the addition of air, and it is split into a sum in the table to make this clear.

Table 11-17. Flow Rates for Example 11-5.

Species	Amount Entering (mol)	Amount Exiting (mol)
C_2H_5OH	$3*\frac{1000}{10^6} = 0.003$	0
O_2	$0.003*\frac{13}{4}(1.2) = 0.0117$	$0.0117 - \frac{13}{4}*0.003 = 0.00195$
CO_2	$3*\frac{30,000}{10^6} = 0.09$	$0.09 + \frac{8}{4}*0.003 = 0.096$
H_2O	0	$0 + \frac{10}{4}*0.003 = 0.0075$
N_2	$3*\frac{969,000}{10^6} + \frac{79}{21}*0.0117 = 2.951$	2.951

Next, we must find the average heat capacities for each species with data from appendix 6c and using equation [11.62]. The following table lists the data and the calculated value of the average. The temperature range was chosen as 25°C to 650°C (298 K to 923 K) since these are the required inlet and outlet temperatures.

Table 11-18. Mean Heat Capacities for Example 11-5.

Species	α	β	γ	ε	$T_{in}(K)$	$T_{out}(K)$	$\bar{C}p\left(\dfrac{cal}{mol \cdot K}\right)$
C₂H₅OH	6.99	39.741	-11.926	0	298	923	26.42
O₂	7.16	1.00	0	-0.4	298	923	7.63
CO₂	10.57	2.1	0	-2.06	298	923	11.10
H₂O	7.3	2.46	0	0	298	923	8.80
N₂	6.83	0.9	0	-0.12	298	923	7.34

Finally, the heat of reaction can be looked up from the table in appendix 6c. Ethanol has a value of -66,200 cal/mol (lower heating value since the produced water is in the gas phase; use the higher heating value when the product water is in the liquid phase). The reaction heat capacity is:

$$\Delta \bar{C}p = \left[\frac{d}{a}\bar{C}p_D + \frac{c}{a}\bar{C}p_C - \frac{b}{a}\bar{C}p_B - \bar{C}p_A\right]$$

$$= \left[\frac{10}{4}(8.8) + \frac{8}{4}(11.1) - \frac{13}{4}7.63 - 26.42\right] = -7.02 \frac{cal}{mol \cdot K}$$

Substituting the flow rates, the average heat capacities, and the heat of reaction data into equation [11.63] and solving for T_{out} gives a value of 333 K. This is not hot enough and would be unlikely to support a flame so additional fuel is required.

B. How much methane is needed to increase the temperature to 923 K (using 50% excess air)?

An additional reaction is added

$$CH_4 + 2O_2 \rightarrow CO_2 + 2H_2O$$

Table 11-19. Flow Rates with Additional Reaction for Example 11-5.

Species	Amount Entering	Amount Exiting
C₂H₅OH	0.003	0
CH₄	Unknown = Fm	0
O₂	$0.003 * \dfrac{13}{4}(1.5) + F_m * \dfrac{2}{1}(1.5) = 0.0146 + 3F_m$	$(0.0146 + 3F_m) - \left(\dfrac{13}{4}.003 + \dfrac{2}{1}Fm\right)$
CO₂	0.09	$0.09 + \left(\dfrac{8}{4}.003 + F_m\right)$
H₂O	0	$\dfrac{10}{4}.003 + \dfrac{2}{1}F_m$
N₂	$2.907 + (0.0146 + 3F_m)\dfrac{79}{21} = 2.962 + 11.29F_m$	$2.962 + 11.29F_m$

Methane has a heat of reaction of -191,760 cal/mol. Again, create a data set like Table 11-17 and Table 11-18. Calculations must include the new reaction. The other values are carried over from the previous table, since the waste gas remains the same, and the ethanol in it still adds to the heat generated.

Only one line needs to be added to the heat capacity table.

Table 11-20. Additional Heat Capacities Needed in Example 11-5.

Species	α	β	γ	ε	$T_{in}(K)$	$T_{out}(K)$	$\bar{C}p \left(\dfrac{cal}{mol \cdot K} \right)$
CH_4	3.381	18.044	-4.3	0	298	923	12.65

The reaction heat capacity for this reaction is

$$\Delta \bar{C}p = \left[\frac{2}{1}(8.8) + \frac{1}{1}(11.1) - \frac{2}{1}7.63 - 12.65 \right] = 0.79$$

Substituting these new values into equation [11.63], the adiabatic flame temperature equation, and solving for F_m, the molar flow rate of methane yields (after a bit of algebra) F_m=0.52 mol methane /second.

■ Example 11-6.

Rework Example 11-5, incineration of bakery waste, but use a heat exchanger to preheat all the inlet streams (waste gas, fuel, and air) to 500°C. How much will this reduce the fuel use?

Solution:
Recalculate, noting that all the calculations are the same except the average heat capacities and reaction heat capacities use T_{in} = 500°C. Solving for F_m yields 0.19 mol methane/second. Therefore, the fuel use reduction is:

$$\% reduction = \frac{0.52 - 0.19}{0.52} * 100\% = 63\%$$

The amount of fuel saved is significant. A decision to do this, or not, depends on the cost of the equipment needed to move the heat from the exhaust to the feed. It typically will require heat exchanges and some piping to help move the gases through the heat exchanger.

11.3.5.4 Biological Control

Bioremoval processes use microbes in a bioreactor to consume contaminants from an air stream. The microbes use the contaminants as a food source. The bioreactor is a specialized process that provides contact between the gas phase contaminant and a suitable population of microbes. The process requires contacting the contaminated gas stream with a liquid or wetted bed of solids that contain a healthy population of biomass. There are three types of systems used as bioreactors:
- Biofilter– uses a fixed bed of solids to grow the microbe community, water does not flow through the bed, but there is a layer of water around the solids/ microbes.
- Biotrickling filter – uses a fixed bed of solids to grow the microbe community, water flows through the system.
- Bioscrubber – biomass suspended in the water, which flows through the system. There is no bed of solids required.

Biofiltration is the most common biological method for removal of gas phase contaminants, and it is the focus of this section.

A biofilter is simply a large box containing an air inlet chamber at the bottom (also called a plenum), a support rack above the plenum, a bed of solids used to grow the biofilm, and a device to add water (as needed) to the top of the bed – see Figure 11-14. The bed is usually several feet deep. The bed can be made of many materials such as wood chips, peat, composted yard waste, bark, soil, gravel or even plastic shapes. Other materials may be added to aid the growth of the biomass such as fertilizer or to reduce acid build-up such as oyster shells.

FIGURE 11-14. Schematic of a Basic Biofilter.

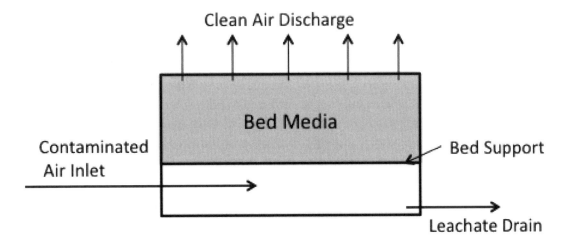

Biofilters work well for high gas flow and low contaminant concentration conditions. The important design and operating variables include:

Design:
    Identifying the type and amount of contaminants,
    Gas flow rate,
    Microbe population, and
    Bed material and size.

Operation:
    Temperature,
    Moisture,
    Macronutrients, and
    Control of acidity.

Good design of a biofilter creates a process unit that meets the emission limits of the contaminant, is easy to use and maintain, and is capable of handling flow variations. It is, therefore, necessary to specify the contaminant to determine the amount of treatment. The contaminant must be soluble in water in order to be available to the microbes. Knowledge of the contaminant and its solubility should allow determination of the residence time for the system (using a laboratory or pilot scale model of the system). Next, the average and maximum gas flow rates should be determined, and the bed size calculated.

The most important variable affecting operation of a biofilter is temperature. A hot gas stream can kill the biomass faster than any other factor. Microbes, in general, flourish at temperatures between 30 and 40°C (60 – 105°F). If the gas stream is too hot, it can usually be cooled with a humidifier, which cools the incoming stream by using the heat to evaporate water. The humidified air helps to maintain bed moisture. Cold air is not usually lethal to microbes, but it may slow their activity or even cause them to go into a state of suspended animation. It can take hours to days to re-animate them.

Microbes need moisture to survive, and moisture helps create the biofilm that can absorb and destroy the contaminant. Low moisture can be corrected by adding moisture directly to the bed or by humidifying the inlet gas stream. High moisture may drown the microbes or prevent the flow of gas through the bed. A drain system can help reduce the probability of flooding within the biofilter. The drain may be built into the plenum located below the bed. The drained liquid, called leachate, probably requires treatment before release to the environment. Many of the reactions performed by the microbes form water, and it may not be necessary to add much water to the system.

The microbes also need additional nutrients besides the contaminant to survive and thrive. These macronutrients – nitrogen, phosphorus, potassium, sulfur, magnesium, calcium, sodium, and iron can be from an organic bed medium (i.e., wood chips). They may also originate from a fertilizer added directly to the bed or with the makeup water added to maintain bed moisture. The nutrient content of the bed should be checked periodically by submitting samples to a soils lab for analysis.

Microbes can be obtained directly from a population already existing in the bed media (soil, wood chips, or compost) or added from any variety of sources, such as activated sludge from a wastewater treatment plant. Once added, it takes a short time (days to weeks) for the population to acclimate to the biofilter and achieve its maximum efficiency. Beds of organic materials can last 2 – 5 years, and they are easy to replace.

The removal efficiency of a biofilter is related to the inlet and outlet concentrations of the contaminant in the gas stream:

11.64
$$\eta = \frac{C_{in} - C_{out}}{C_{in}}$$

Theoretical determination of removal efficiency based on considerations of mass transfer with a chemical reaction yields (Heinsohn, et al., 1999):

11.65
$$\eta = 1 - \exp\left[\frac{-K_0 Z}{U_G}\right]$$

Where: K_0 = empirical reaction and mass transfer constant of the bioreactor,
Z = Height of the biofilter bed, and
U_G = superficial gas velocity in biofilter (assumes no solids present).

Advantages:
- Efficient removal of contaminants
- Little or no by-product pollutants (only CO_2 and H_2O for well-run system)
- Simple and inexpensive installation

Disadvantages:
- Not good with high concentration contaminants (unless highly soluble)
- Often requires a large area to achieve needed residence time
- Difficult to control temperature, pH, or nutrient levels, since there is normally no continuous flow of liquid into the system.
- Can be clogged by particles in the gas stream or excessive biomass growth.

■ Example 11-7.

Determine the outlet concentration of VOCs from a biofilter bed. The bed is 8 ft by 16 ft in cross section and 6 ft deep. The incoming emissions flow at 1,000 ft³/min and contains 210 ppm VOCs. The reaction and mass transfer coefficient of the bed is 4.8 min⁻¹.

Solution:

Apply equation [11.65] to find the overall efficiency, then use equation [11.64] to determine the outlet concentration, C_{out}.

Where:
$K_0 = 4.8$ min-1
$Z = 6$ ft as given

$$U_G = \frac{Q}{Area} = \frac{1,000 \frac{ft^3}{min}}{(8ft)(16ft)} = 7.8 \frac{ft}{min}$$

Equation [11.65]

$$\eta = 1 - \exp\left[\frac{-K_0 Z}{U_G}\right] = 1 - \exp\left[\frac{-4.8 \min^{-1}(6ft)}{7.8 \frac{ft}{\min}}\right] = 1 - .025 = 0.975$$

Equation [11.64]

$$\eta = \frac{C_{in} - C_{out}}{C_{in}} \text{ rearrange to solve for } C_{out}: C_{out} = C_{in}(1-\eta) = 210\,ppm(1-0.975) = 5.2\,ppm$$

11.3.5.5 Condensers

Refrigerated condensers control VOCs and hazardous air pollutants (HAP) by cooling an emission stream in order to condense the organic vapors into liquids or solids. The condensed material is recovered and possibly reused. Pre-cooling of the emission stream is sometimes used to remove the moisture. Pre-cooling can use either a different refrigerant, the refrigerant leaving the cooling coils, or the cold exhaust stream from the condenser.

Refrigeration works by removing heat from a heat exchange surface inside the cooling region and discharging the heat somewhere else. An intermediary fluid such as a chilled brine solution, ammonia, sulfur dioxide, chlorofluorocarbons, hydrofluorocarbons, liquid nitrogen, or air may remove the heat. Each has advantages and disadvantages relating to the toxicity of the fluid, the ability to absorb and move heat, and the cost of equipment to move the fluid.

The condenser works best when there is excellent contact between the waste gas stream and the cooled heat exchanger surface. Contact improves with turbulent flow profiles in the gas stream, and by generating a concentration gradient at the condensation surface. Turbulence in the gas stream increases mixing, which causes more of the gas stream to contact the heat exchanger surface. Condensation of organic vapor causes the gas stream to lose volume, which causes a lower concentration of vapor near the heat exchanger surface, thus producing a concentration gradient that increases the flow of the contaminant toward the surface. Usually, only a negligible fraction of a percent escapes this combined action when the condenser is large enough, and the refrigerant temperature is low enough.

The removal efficiency of refrigerated condensers depends on the temperature of the refrigerant fluid, see Table 11-21, and on the cooling capacity. The refrigeration side working fluid and its pressure control the temperature (US-EPA, 2001). Cooling capacity relates the ability of the working fluid to pick up heat in the condenser and to shed this heat elsewhere. The removal efficiency is then limited by the vapor pressure of the VOC at the condenser temperature. All substances have a finite (greater than zero) vapor pressure even at the coldest temperatures. This means that there is always some amount of the VOC in the vapor phase. The amount can be estimated by determining the vapor pressure of the contaminant at the condenser operating temperature, which is usually several degrees warmer than the lowest temperature listed in Table 11-21. The Antoine equation is used to model vapor pressure (P_{sat}) as:

11.66
$$\log_{10} P^{sat} = A - \frac{B}{T+C}$$

Where: P_{sat} = the saturation or vapor pressure in mmHg,
T = temperature in °C.
A, B, C = constants (See Appendix VI d).

Table 11-21. Refrigerated Condenser Operating Fluids.

Working Fluid	Lowest Temperature °C	Removal Efficiency %
Brine	-14 (0 °F)	50 – 90
CFC, HFC	-53 (-65)	90+
Air (reverse Brayton Cycle)	-73 (-100)	98
Liquid Nitrogen	-195 (-352)	99+

Efficiency can be increased by:
- Decreasing the amount of non-condensable gases (e.g. air),
- Maximizing the portion of the organic vapor that is affected by the cold surface in the condenser (i.e., make sure the residence time is large enough and the flow sufficiently turbulent),
- Minimizing the partial pressure of the VOC vapor by decreasing the temperature of the condenser, and
- Increasing the residence time of the vapor in contact with the condenser.

It is difficult to recover a single organic compound with high purity from a mixture of organic vapors by condensing it because all of the organic compounds are condensed and collected together. All condensers have this limitation. Subsequent purification by another process such as fractional distillation, skimming, or membrane separation is required before recycling or selling recovered organic compounds.

A major use for refrigerated condensers today is in dry cleaners to capture either the petroleum or the perchloroethylene dry cleaning fluid vapors. Another use is the recovery of VOCs that are removed from an adsorption column during regeneration.

■ Example 11-8.
Determine the maximum removal efficiency of ethanol in a condenser using chilled brine. The incoming emissions contain 530 ppm ethanol at 35°C and 1 atm (760 mmHg).

Solution:
Determine the vapor pressure of ethanol at -14°C, the operating fluid temperature for brine. The vapor pressure can be calculated using the data in appendix 6d.

$$\log_{10} P^{sat} = A - \frac{B}{T+C}$$

$$\log_{10}(P^{sat}) = 8.1122 - \frac{1592.864}{(-14)+226.184} = 0.6052$$

$$P^{sat} = 10^{0.6052} = 4.029 \, mmHg$$

$$P_{in} = 0.032 * 760 \, mmHg = 24.32 \, mmHg$$

$$\eta = \frac{C_{in} - C_{out}}{C_{in}} = \frac{P_{in} - P_{out}}{P_{in}} = \frac{24.32 - 4.03}{24.32} = 0.83 \text{ or } 83\% \text{ removal.}$$

We used the fact that partial pressures are directly related to concentration through the ideal gas law in order to simplify the efficiency equation.

11.4 Questions

* - Questions and problems may require additional information not included in the textbook.

1. Visit a government environmental website (US, EU, or other) and determine if any regions within their jurisdiction have ambient air that is not currently in compliance for CO.*

2. Find the current level of CO in the air where you live.

3. Figure 11-1 shows passenger cars have exceeded CAFE standards since the year 2000. Why? and Why haven't the CAFE standards risen with the actual value?*

4. How is a) CO b) CH_4 c) TVOC d) benzene measured in the ambient air? Is there more than one way to determine the concentration?*

5. Determine if you live or work near any local emission sources for a) CO b) CH_4 or c) VOCs. If so, detail one source in terms of amount released and the process by which it is released. Also, does the site have any emission control equipment?*

6. Consult Figure 11-3. Speculate on why methane concentrations have leveled off over the last decade. Would you expect the concentration to remain the same, increase, or decrease in the next decade?

7. How might you determine methane generation from termites? What do you think the important variables are?*

8. How do uncontrolled fires and biomass burning cause VOC emissions?

9. How would you expect the VOC emissions to vary over time (day, week, year) in a) an urban area, b) a rural farming area, and c) a tropical forest.

10. One criticism of the Freundlich isotherm model is that it only provides a good fit over a limited range of pressures. Examine Figure 11-7 and comment on this criticism. When would the Freundlich model be acceptable for ethane adsorption on activated carbon?

11. Compare equation 11.45 and the graphical technique for determining N_{og}. Is it coincidence that the two methods yield similar results?

12. Good operation of a biofilter requires good solubility of the contaminant in water. Why must this be so?

13. Would a biofilter be a good choice for controlling VOCs from a low flow, high concentration waste gas? If not, what technology would be better?

14. Why will the operating temperature of a condenser be warmer than the temperature of the working fluid?

11.5 Problems

1. Calculate the concentration of CO in $\mu g/acm$ for air with 0.01 ppm CO at 115°F and 0.98 atm.

2. Find the partial pressure of CO in air needed to cause CO poisoning symptoms in a) a healthy adult, b) a healthy child (equilibrium ratio = 290) and, c) an adult smoker that has a current $HbCO/HbO_2$ of 5%. Assume equilibrium contact between the person and the contaminated air (8 hours of exposure).

3. What would be the $HbCO/HbO_2$ ratio of someone exposed to the immediately dangerous to life or health (IDLH) concentration of CO (assume equilibrium)?

4. Determine the amount of methane produced from a herd of 75 dairy cows. Assume each cow weighs an average of 1,300 lbs and that all their waste is captured for digestion.

5. Determine the concentrations of methane and carbon dioxide from the anaerobic digestion of landfill waste, assume a chemical formula of $C_6H_{10}O_4$, and that it chemically reacts with water to form only these two products.

6. Determine the methane production (m^3/day) from a landfill that accepts household waste for a town of 1,500 people. Assume that the waste generation is 1250 kg/person/year. Also assume that the landfill generates methane at a rate of 43 m^3/ 1000 kg.

7. Determine the amount of energy (watts) that could be produced from the emissions from the landfill in problem 6. Assume the heat content of landfill gas is 540 kJ/m^3, and that this is used to generate electricity (25% efficiency). The USA currently generates about 1 GW electricity from landfill gas.

8. Determine the best fit isotherm model for the methane on activated carbon data in Figure 11 5, between the Langmuir, Freundlich, and BET models.

9. Find the adsorption front velocity for 12 m³/min of air containing 500-ppm propyl alcohol (C_3H_7OH) flowing through an activated carbon packed bed of density 540 kg/m³ and area 1.5 m². The isotherm can be modeled with the BET equation where V_m=6.1 liter and c=4.8. The vapor pressure for propyl alcohol is 15 mm Hg at the system entrance conditions of 298 K and 1 atm.

10. Find the breakthrough time (day) for 80-ppm vinyl chloride (C_2H_3Cl) adsorption on activated alumina for a bed 3 m in height. The total gas flow rate is 8 m³/min, the bed density is 480 kg/m³, the bed cross sectional area is 1 m², and the isotherm is given by a Freundlich model with k=0.86 and n=0.45. Neglect the mixing front depth. Assume T = 298 and P = 1 atm.

11. Generate a mole-ratio plot showing the Henry's law equilibrium for H_2S at 8 atm and temperatures of a) 0°C, b) 10°C, c) 20°C, d) 30°C, and e) 40°C. Let the H_2S in water mole fraction range from 0 - 0.005 mole %.

12. Generate a mole-ratio plot showing the Henry's law equilibrium for SO_2 at a temperature of 10°C and at a) 0.5 atm, b) 1 atm, c) 2 atm, d) 5 atm, and e) 10 atm. Let the liquid mole fraction range from 0 - 0.005 mole %.

13. Find the rate of mass transfer of ethene from a gas that contains 105 ppm C_2H_4 at 25 atm and 10 °C into water with 0.00015 mol % C_2H_4. The mass transfer coefficient is 18.4 [mol/(vapor mol%•m2 •min)].

14. Find the rate of mass transfer of ethane from a gas that contains 105 ppm C_2H_4 at 25 atm and 10°C into water with 0.00015 mol % C_2H_4. The mass transfer coefficient is 5,650 [mol/(liquid mol% m² min)].

15. Derive equation [11.43] starting from [11.37] and using [11.41] and [11.42]. Why is this transformation needed? When is it not necessary?

16. Calculate the number of transfer units needed to adsorb 85% of chloroform ($CHCl_3$) from a gas stream at 10°C and 2 atm using water. The gas inlet contains 15 mol% $CHCl_3$ and 85% inert, insoluble other gas (N_2). The inlet water contains no $CHCl_3$. The water flow rate is 1.7 times the minimum flow rate. Henry's Law constant for chloroform at 10°C is 109.

17. Determine the column height for removing 80% of the benzene from an insoluble gas at 20°C and 1 atm using water. The inlet water contains 0 mol% of benzene and the inlet gas contains 5.1 mol%. The water flow rate is 1.5 times the minimum and the height of a transfer unit is known to be 0.40 m for this unit. Henry's Law constant for benzene at 20°C is 247.

18. Estimate the combustion chamber volume needed to treat 1.00 kg/sec of an offgas (MW= 30.5) at 600°C and 1.5 atm for 1 second.

19. Calculate the mean heat capacities [Cp] for a) CO_2, b) H_2O, and c) N_2 over the temperature range 100 – 700°C.

20. Find the adiabatic flame temperature for the combustion of methane in 25% excess air entering a combustion chamber at 1 atm and a) 25°C or b) 100°C.

21. A waste gas stream at 1 atm and 298 K contains 10,000-ppm benzene, 5 mol% water, and the remainder is nitrogen. Determine the maximum temperature obtained from combusting this gas with 35% excess air.

22. A waste gas stream contains 1,000-ppm benzene, 10% carbon dioxide, and the remainder is nitrogen. A 99% destruction of benzene can be achieved by incineration at 750°C for 0.5 seconds. How much methane must be added per mole of waste gas to achieve this destruction? Air is added at 25% excess. [0.044 mol CH_4/mol waste gas]

23. Find the reaction and mass transfer constant, K_0, of a 2.7 m deep bioreactor removing acetaldehyde from a waste gas. The reactor operation conditions were recorded as:

C_{in} [g/m^3]	C_{out} [g/m^3]	U_G [m/s]
1.9	0.005	0.3
2.0	0.05	0.5
2.2	0.32	1.0
1.0	0.40	2.0
2.1	0.80	2.0

24. Determine the biofilter bed depth needed to remove 85% of the acetaldehyde as described in problem 23. The bed is 4 m long by 2 m wide and has a void volume of 42%. The gas stream flows at a rate of 10 m³/sec.

25. The vapor pressure of a substance can be modeled with the Antoine equation, see Appendix 6d. The constants for carbon tetrachloride (CCl_4) are A= 6.841, B=1178, and C = 220.6. If a gas stream at 760-mmHg pressure contains 10 mol% CCl_4, which of the condenser operating fluids listed in Table 11-21 could be used to achieve at least 95% removal?

11.6 Group Project Ideas

For each project, the students should work in small groups and present their finding in either a short report (5-8 pages) or 15 minute presentation.

1. Are there any plans to modify the current CO ambient air quality standards? Write a short paper summarizing the reasons for the revision or lack thereof. Note that the US EPA is required to examine the CO NAAQS every five years.

2. Methane is currently (2012) regulated by voluntary compliance in the US. Why has the US-EPA chosen this method? How well is it working? Are there any plans for creating regulations either as an ambient air quality issue or from a particular source? If so, elaborate; if not, explore methane emissions from natural gas production via fracking and comment on the technical or political feasibility of regulating these sources.

11.7 Bibliography

AgSTAR, US-EPA. 2011. AgSTAR Program. *An EPA Partnership Program.* [Online] January 27, 2011. [Cited: February 3, 2011.] http://www.epa.gov/agstar/.

Air-Liquide. 2005. MSDS - *Methane.* Cambridge, MD, USA : Air Liquide, 2005. http://alemis.us.airliquide.com/ChemSafe/MSDS/MSDSpdf.asp?pdf=117068_1.PDF.

Alves, Celia, et al. 2012. Organic Compounds in Aerosols from Selected European Sites - Biogenic versus Anthropogenic Sources. *Atmospheric Environment.* 2012, Vol. 59, pp. 243-255.

Atkinson, R. 2000. Atmospheric chemistry of VOCs and NO_x. *Atmospheric Environment.* 2000, Vol. 34, 12, pp. 2063-2101.

Brasseur, Guy P, Orlando, John J and Tyndall, Geoffrey S. 1999. *Atmospheric Chemistry and Global Change- Chapter 5.* New York, NY, USA : Oxford University Press, 1999. ISBN 0-19-510521-4.

Butler, James. 2012. THE NOAA ANNUAL GREENHOUSE GAS INDEX (AGGI). *U.S. Department of Commerce / National Oceanic & Atmospheric Administration / NOAA Research.* [Online] September 24, 2012. [Cited: July 3, 2013.] http://www.esrl.noaa.gov/gmd/aggi/.

CDC, US-. 2010. *NIOSH Pocket Guide to Chemical Hazards.* Washington, DC, USA : US Department of Health and Human Services, 2010.

Clearstone. 1994. *A National Inventory of Greenhouse Gas (GHG), Criteria Air Contaminant (CAC) and Hydrogen Sulphide (H2S) Emissions by the Upstream Oil and Gas Industry, Volume 1, Overview of the GHG Emissions Inventory.* s.l. : Canadian Association of Petroleum Producers, 1994. http://www.capp.ca/default.asp?V_DOC_ID=763&PubID=86220.

CMOP, US-EPA. 2011. Coalbed Methane Outreach Program (CMOP). *EPA Home Climate Change Methane Coalbed Methane Outreach Program (CMOP).* [Online] February 2, 2011. [Cited: February 3, 2011.] http://www.epa.gov/cmop/.

CSCC, Committee on the Science of Climate Change, National Research Council. 2001. *Climate Change Science: An Analysis of Some Key Questions*. Washington DC, USA : National Academies Press, 2001. ISBN 0-309-07574-2.

Denman, Kenneth, et al. 2007. Chapter 7 Couplings Between Changes in the Climate System and Biogeochemistry. [book auth.] IPCC. *Fourth Assessment Report*. Cambridge, UK : Cambridge University Press, 2007.

Dillon, William. 1992. *Gas (Methane) Hydrates -- A New Frontier*. Woods Hole, MA, USA : US Geological Survey, marine and Coastal Geology Program., 1992. http://marine.usgs.gov/fact-sheets/gas-hydrates/title.html.

Ehhalt, D H. 1999. Gas Phase Chemistry of the Troposphere. [book auth.] Darmstadt Steinkopff. *Chapter 2 in "Global Aspects of Atmospheric Chemistry"*. Ney York, NY, USA : Springer, 1999.

EPA, US -. 2009. National Summary of Carbon Monoxide Emissions. *Air Emission Sources - Carbon Monoxide*. [Online] November 2009. [Cited: February 2, 2011.] http://www.epa.gov/air/emissions/co.htm.

Fair, James R, et al. 1997. Section 14. Gas Absorption and Gas-Liquid System Design. [book auth.] Robert H Perry and D W Green. *Perry's Chemical Engineers' Handbook, 7th Ed*. New York, NY, USA : McGraw Hill, 1997.

Fierro, Maria A, O'Rourke, Mary Kay and Burgess, Jefferey L. 2001. *ADVERSE HEALTH EFFECTS OF EXPOSURE TO AMBIENT CARBON MONOXIDE*. s.l. : University of Arizona, College of Public Health, 2001.

Forster, Piere, et al. 2007. Chapter 2 Changes in Atmosphereic Constituents and in Radiative Forcing. [book auth.] IPCC. *Fourth Assessment Report*. Cambridge, UK : Cambridge University Press, 2007, Table 2.14, p. 212.

GasSTAR, US-EPA. 2010. Natural Gas STAR Program. *EPA Home Climate Change Methane Natural Gas STAR Program*. [Online] December 14, 2010. [Cited: February 3, 2011.] http://www.epa.gov/gasstar/.

Guenther, A, et al. 1995. A Global Model of Natural Volatile Organic Compound Emissions. *Journal Geophysical Research*. 1995, 100, p. 8873.

Heinsohn, Robert J and Kabel, Robert L. 1999. *Sources and Control of Air Pollution*. Upper Saddle River, NJ, USA : Prentice Hall, 1999. ISBN 0-13-624834-9.

Henley, Ernest J and Seader, J D. 1981. *Equilibrium-Stage Separation Operations in Chemical Engineering*. New York, NY, USA : J Wiley & Sons, 1981. ISBN 0-471-37108-4.

Leng, R A. 1993. Quantitative ruminant nutrition - A green science. *Australian Journal of Agricultural Research*. 1993, Vol. 44, pp. 363-380.

Lide, David. 2005. *CRC Handbook of Chemistry and Physics, 85th Edition*. Boca Raton, FL, USA : CRC Press, 2005.

NHTSA. 2014. CAFE - Fuel Economy. *National Highway Traffic Safety Administration*. [Online] June 2014. [Cited: December 17, 2014.] http://www.nhtsa.gov/fuel-economy.

Perry, Robert H and Green, Don W. 1997. *Perry's Chemical Engineering Handbook, 7th ed*. New York, NY, USA : McGraw Hill, 1997.

Routley, V. 1998. *Motor Vehicle Exhaust Gassing Suicides in Australia: Epidemiology and Prevention*. Monash, Austrailia : Monash Injury Research Institute, 1998.

Shakhova, Natalia, et al. 2010. Extensive Methane Venting tothe Atmosphere from Sediments of the East Siberian Artic Shelf. *Science*. 327, 2010, Vol. 5970, p. 1246.

Singh, H B and Zimmerman, P R. 1992. Atmospheric Distribution and Sources of Nonmethane Hydrocarbons. [book auth.] J O Nriagu. *Gaseous Pollutants: Characterization and Cycling*. New York, NY, USA : Wiley and Sons, 1992.

US-EPA. 2010c. *2010 US Greenhouse Gas Inventory Report*. Washington DC, USA : US Government Printing Office, 2010c. EPA # 430-R-10-006.

—. 2009. Air Emission Sources. *EPA Home Air & Radiation Common Air Pollutants Air Emission Sources Volatile Organic Compounds*. [Online] November 4, 2009. [Cited: February 4, 2011.] http://www.epa.gov/air/emissions/voc.htm.

—. 2012. Carbon Monoxide Information. *EPA Home/Green Book*. [Online] December 14, 2012. [Cited: July 2, 2013.] http://www.epa.gov/airquality/greenbook/cindex.html.

—. 2007. EPA Home > Climate Change > Methane > Ruminant Livestock > Frequent Questions. *Ruminant Livestock*. [Online] March 21, 2007. [Cited: February 3, 2011.] http://www.epa.gov/methane/rlep/faq.html.

—. 2002a. *INVENTORY OF U.S. GREENHOUSE GAS EMISSIONS AND SINKS: 1990-2000*. Washington DC, USA : US Government Printing Office, 2002a. EPA 430-R-02-003.

—. 2011. Landfill Methane Outreach Program. [Online] February 2, 2011. [Cited: February 3, 2011.] http://www.epa.gov/lmop/.

—. 2010b. *Methane and Nitrous Oxide Emissions From Natural Sources*. Washington, DC, USA : US Government Printing Office, 2010b. http://www.epa.gov/methane/sources.html. EPA 430-R-10-001.

—. 2010a. National Trends in CO Levels. *US-Epa Home/ Air & Radiation/ Air Trends*. [Online] Decenber 17, 2010a. [Cited: February 2, 2011.] http://www.epa.gov/airtrends/carbon.html.

—. 1998. *National Volatile Organic Compound Emission Standards for Consumer Products*. Washington DC, USA : US Government Printing Office, 1998. EPA-453/R-98-008B.

—. 2001. *Refrigerated Condensers for Control of Organic Air Emissions*. Washington DC, USA : US Government Printing Office, 2001. EPA-456/R-01-004.

—. 2002b. *US Climate Action Report 2002*. Washington DC, USA : National Service Center for Environmental Publications, US Government Printing Office, 2002b. 430R02016 EPA.

US-NHTSA. 2010. CAFE - Fuel Economy. *National Highway Traffic Safety Administration*. [Online] march 5, 2010. [Cited: December 1, 2010.] http://www.nhtsa.gov/fuel-economy.

Wankat, Phillip C. 1988. *Equilibrium Staged Separations*. Englewood Cliffs, NJ, USA : Prentice Hall, 1988. ISBN 0-13-500968-5.

Weast. 2003. *CRC Handbook of Chemistry and Physics*. Boca Raton, FL, USA : CRC Press, 2003.

CHAPTER **12**

Ozone

Ozone is both good and bad in the atmosphere. When it is located far above the earth's surface in the stratosphere, it blocks out harmful ultraviolet radiation. It acts like a shield to protect all life on earth from radiation damage. When located at the earth's surface in the troposphere, its highly reactive nature causes harm to most living things. Unfortunately, it is not possible to move the ozone from the troposphere, where it causes harm, to the stratosphere, where it protects. Even more troublesome, human activity is destroying the ozone in the stratosphere and increasing it in the troposphere.

Ozone (O^3) is a highly reactive trace gas composed of three atoms of oxygen arranged in a bent linear structure. It is very reactive and oxidizes most substances it encounters. It has a lifetime in the troposphere (0 – 12 km elevation) of 5 - 30 days and the stratosphere (12 – 60 km elevation) of 30- 300 days, depending on the season (shorter in the summer). It is pale blue in color and has a distinct odor often encountered during and after lightning storms (which can create ozone). It absorbs light in the ultraviolet part of the spectrum (270 – 400 nm) and, therefore, plays a major role in the earth's energy balance and the structure of the atmosphere. It is found throughout the earth's atmosphere (see Figure 12-1). Its concentration is highest in the mid stratosphere due to the role UV radiation plays in its formation.

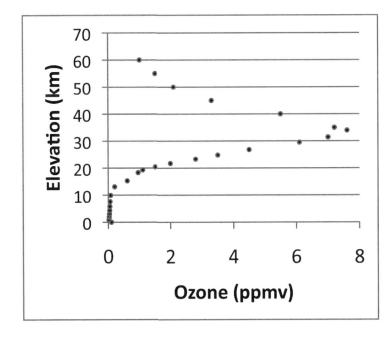

FIGURE 12-1. Ozone Distribution in the Atmosphere. Abstracted from (Brasseur, et al., 1999).

The standard way to express total ozone level measurements (the amount of ozone in a vertical column) in the atmosphere is by using Dobson units (DU). The Dobson unit is named after a researcher from the University of Oxford - Gordon Dobson. He built the first spectrophotometer for ozone in 1924. The measurement correlates the amount of ozone in a column of the atmosphere above the measurement device with the amount of UV radiation striking the device. The greater the amount of UV, the lower the amount of ozone in the column of air between the device and the sun. This measurement approximates an integration of the ozone concentration over the thickness of the atmosphere. One Dobson unit refers to a layer of pure ozone that would be 10 μm thick under standard temperature and pressure. For example, 300 DU of ozone at 273 K and 1 atm would occupy a layer 3 mm thick. One DU is equivalent to 2.69×10^{20} ozone molecules per square meter.

The average concentration of ozone at a point is measured in parts per million by volume (ppmv) or in $\mu g/m^3$. Methods to determine oxone concentration include - *Electrochemical Concentration Cells*, which measure current produced by chemical reactions with ozone; *Photospectroscopy*, which uses film or electronic sensors sensitive to UV light to measure wavelengths affected by ozone; and *Laser in Situ Sensors,* which measure absorption of laser light between a projector and a sensor some distance away. Measurements at different altitudes in the atmosphere are collected by balloons (ozonesondes), aircraft, rockets, and/or satellites.

Figure 12-2 shows a map of the total ozone above the northern hemisphere during the record ozone loss in spring 2011 (Canada, 2013). The amount over the poles will change significantly between seasons, whereas over the equator it is relatively constant in concentration.

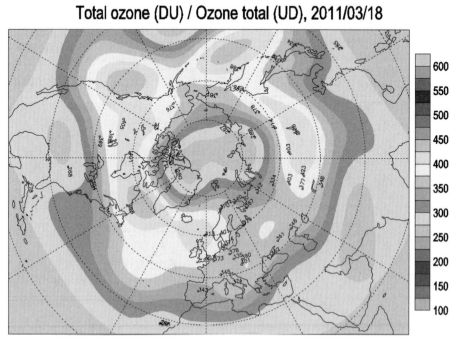

FIGURE 12-2. Global Total Ozone Distribution for March 18, 2011. http://exp-studies.tor.ec.gc.ca/ozone/images/graphs/gl/current.gif

12.1 Tropospheric Ozone

12.1.1 General Information

Ozone is ubiquitous throughout the troposphere; the concentration is highest in urban areas, but it is also present in remote areas. The global distribution is due to atmospheric transport processes (winds, mixing, convection) as well as the presence of its chemical precursors, NO_x and VOCs.

12.1.2 Sources and Sinks

Figure 12-3 shows the tropospheric sources and sinks for ozone. There are two sources for tropospheric ozone – 1) transport from the stratosphere (400 Tg/year) and 2) photo-chemical reactions involving NO_x and reactive carbon (CO, CH_4, and NMVOC) in the presence of sunlight (4,300 Tg/year). Ozone has no significant anthropogenic emission sources, although it is useful as a water treatment disinfectant. The sinks for ozone include 1) photochemical degradation and reaction with water to form hydroxyl radicals (4,000 Tg/year) and deposition on the earth's surface (700 Tg/year). Note that 1 Tg = 10^{12} grams.

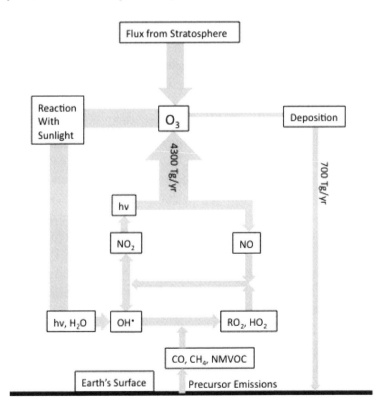

FIGURE 12-3. Sources and Sinks of Ozone (O_3) in the Troposphere. Number values represent ozone flux in Teragrams (10^{12} g) per year.

Ozone is generated in the stratosphere through the interaction of oxygen and sunlight (see section 12.2). Normally, there is little mixing between the stratosphere and troposphere. However, there are certain conditions that can lead to stratospheric air being pulled into the troposphere and then mixing. This weather phenomenon is called *tropopause folding*. It occurs when a large cold front passes beneath a jet stream (large river of air moving much faster than the air around it), which causes a downward motion of the tropopause and the stratosphere on the windward (upwind) side of the jet-stream. The mixed region is large: 100 - 200 km long, 100 – 300 km wide, and 1 – 4 km thick. It typically only brings air from the stratosphere to the upper and middle regions of the troposphere. It is rare for this air mass to be transported directly to ground level. Once in the troposphere, the ozone (and other constituents) can migrate to ground level by diffusion and convection. This mechanism accounts for 40 – 60% of the background level of ozone in the troposphere. Prior to studies in the 1960's and 1970's, this was believed to be the only source of ozone in the troposphere.

The other source of O_3 in the troposphere is due to chemical reactions between ozone precursors: nitrogen oxides (NO_x), reactive carbon (CO, CH_4, and NMVOC), and sunlight. These reactions are why ozone is a secondary air pollutant. Secondary means not directly emitted but generated from other pollutants. The formation of O_3 and other oxidants and oxidation products from these precursors is a complicated and nonlinear relationship between many factors: the concentrations and ratios of the precursors in ambient air; the intensity and spectral distribution of sunlight; atmospheric mixing; temperature; presence of catalytic particles; and the rates of chemical reactions between ozone and its many precursors. These processes can also lead to the formation of other photochemical products, such as peroxyacetyl nitrate (PAN), nitric acid (HNO_3), sulfuric acid (H_2SO_4), formaldehyde (CH_2O), and other carbonyl compounds, such as aldehydes and ketones.

The most important variables are the concentrations of NO_x and VOC, and the VOC/NO_x ratio. NO_x concentrations are highest in urban areas due to emissions from automobiles and fossil fuel combustion. VOCs have many natural and human sources. VOC concentrations in urban areas are usually high due to the use of fossil fuels, but may also be high near forested areas (see Chapter 11).

Regions with low NO_x concentrations, such as rural, suburban, and remote continental areas have an increase in O_3 production with increasing NO_x. At higher concentrations found in urban areas, especially near busy streets and highways and in power plant plumes, there is a decrease in O_3 production due to the reaction with NO. In between these two regimes, O_3 production shows only a weak dependence on NO_x concentrations. The rate of O_3 production per NO_x oxidized is highest in areas where NO_x concentrations are lowest and decreases with increasing NO_x concentration. In regions with low VOC concentrations, the NO_x competes with VOC for available hydroxyl radicals, decreasing the rate of VOC oxidation and subsequently the rate of ozone production.

Classes of organic compounds important for the photochemical formation of O_3 include alkanes, alkenes, aromatic hydrocarbons, carbonyl compounds, alcohols, organic peroxides, halogenated organic compounds, and carbon monoxide. These compounds have a wide range of chemical properties and lifetimes: the atmospheric lifetime of isoprene is about one hour, and methane is about a decade. In urban areas, all classes of reactive carbon are important for O_3 formation. In nonurban vegetated areas, biogenic VOCs emitted from vegetation (isoprene and monoterpenes) tend to be the most important. In the upper and middle troposphere, CH_4 and CO are the main carbon-containing precursors to O_3 formation.

The photochemical formation of O_3 results from the photolysis of nitrogen dioxide (NO_2) to nitric oxide (NO) and a ground-state oxygen atom, (O). This free oxygen then reacts with molecular oxygen in the presence of a catalyst (M) to form O_3:

12.1 $NO_2 + h\nu \rightarrow NO + O$ for $\lambda < 430$ nm

12.2 $O + O_2 + M \rightarrow O_3 + M$

12.3 $NO + O_3 \rightarrow NO_2 + O_2$

Where: O is a free oxygen atom in its ground state
   $h\nu$ = energy from an absorbed photon (J)
   Where h = Planck's constant (6.63×10^{-34} J s)
      ν = frequency of photon (= c/λ, where $c = 3 \times 10^8$ m/s)
      λ = wavelength of the photon (measured in nm)

The reaction of NO to NO_2 is enhanced by the availability of organic (RO_2) or hydroperoxy (HO_2) radicals, see equations 12.6 and 12.18. These radicals form during the oxidation of reactive carbon. When reaction 12.3 occurs, it leads to additional NO_2, which increases the formation of ozone by the above equations (12.1 and 12.2). The presence of reactive carbon enhances the formation of ozone from the photolysis of NO_2. However, without NO_x, these compounds would not produce ozone.

The following sets of reactions provide one possible reaction scheme for the oxidation of reactive carbon in the presence of NO_x leading to the formation of ozone. Note that RH represents a VOC, where R is the carbon chain, and H represents one hydrogen atom bound to the molecule, R' is an organic fragment having one less carbon than R, M is a third body that transfers energy to/from the reaction, the dot superscript denotes that the molecule is in its radical form (missing an electron), and $h\nu$ represents a unit of energy from a photon (see equation 12.1) typically provided by sunlight:

12.4 $RH + OH\cdot \rightarrow R\cdot + H_2O$

12.5 $R\cdot + O_2 + M \rightarrow RO_2\cdot + M$

12.6 $RO_2\cdot + NO \rightarrow RO\cdot + NO_2$

12.7 $\quad RO\cdot + O_2 \rightarrow HO_2\cdot + R'CHO$

12.8 $\quad HO_2\cdot + NO \rightarrow OH\cdot + NO_2$

12.1 $\quad 2\,[NO_2 + h\nu \rightarrow NO + O]$

12.2 $\quad 2\,[O + O_2 + M \rightarrow O_3 + M]$

Overall:
12.9 $\quad RH + 4\,O_2 + 2\,h\nu \rightarrow R'CHO + H_2O + 2\,O_3$

Oxidation of methane dominates ozone production in air where the NMVOC concentrations are small:

12.10 $\quad CH_4 + OH\cdot + \rightarrow CH_3\cdot + H_2O$

12.11 $\quad CH_3\cdot + O_2 + M \rightarrow CH_3O_2\cdot + M$

12.12 $\quad CH_3O_2\cdot + NO \rightarrow CH_3O\cdot + NO_2$

12.13 $\quad CH_3O\cdot + O_2 \rightarrow HO_2\cdot + CH_2O$

12.14 $\quad HO_2\cdot + NO \rightarrow OH\cdot + NO_2$

12.1 $\quad 2\,[NO_2 + h\nu \rightarrow NO + O]$

12.2 $\quad 2\,[O + O_2 + M \rightarrow O_3 + M]$

Overall:
12.15 $\quad CH_4 + 4\,O_2 + h\nu \rightarrow CH_2O + H_2O + 2\,O_3$

A similar mechanism for carbon monoxide:

12.16 $\quad CO + OH\cdot \rightarrow CO_2 + H\cdot$

12.17 $\quad H\cdot + O_2 + M \rightarrow HO_2\cdot + M$

12.18 $\quad HO_2\cdot + NO \rightarrow OH\cdot + NO_2$

12.1 $NO_2 + h\nu \rightarrow NO + O$

12.2 $O + O_2 + M \rightarrow O_3 + M$

Overall:

12.19 $CO + 2\,O_2 + h\nu \rightarrow CO_2 + O_3$

Note that oxidation products from methane and NMVOC are themselves further oxidized until they form carbon dioxide and water.

The main sinks for ozone include deposition onto the earth's surface and photochemical reaction with water to form hydroxide and molecular oxygen. Surface deposition occurs when the ozone reacts with any oxidizable material, and it is greatest above growing vegetation. The photochemical loss is due to the following reactions:

12.20 $O_3 + h\nu \rightarrow O_2 + O\cdot$

12.21 $O\cdot + H_2O \rightarrow 2\,OH\cdot$

12.22 $HO_2\cdot + O_3 \rightarrow OH\cdot + 2\,O_2$

While this mechanism suggests a route of photochemical loss of tropospheric ozone, it generates a significant amount of OH radical. Recall that the hydroxyl radical (OH) plays a significant role in the photochemical oxidation of hydrocarbons and the formation of ozone. It is not clear that this reaction leads to a net loss in ozone.

The photochemical lifetime of ozone in the troposphere depends on the season, latitude and elevation of a location. Lifetimes range from a few days to a few weeks in the summer, whereas it can be three months in the winter. Table 12-1 shows a model calculated photochemical lifetime of O_3 at various altitudes, seasons, and latitudes (Brasseur, et al., 1999).

Table 12-1. Photochemical Lifetime (days) of Ozone in the Troposphere.

Latitude →	40 °N		20 °N	
Altitude (km)	Summer	Winter	Summer	Winter
0	8	100	5	17
5	15	160	10	35
10	40	300	30	90

The ambient concentration of ozone depends on the time of day, season, temperature, and location. The reactions creating tropospheric ozone depend on the availability of sunlight as well as the chemical precursors. Ambient O_3 concentrations are higher during warmer seasons and during the weekday, peaking during the later portion of the day.

12.1.2.1 Daily Variations

Ozone concentrations vary over the course of a day. Ozone formation depends on the interaction between solar energy and its chemical precursors. At night, ozone concentration declines due to its participation in oxidation reactions and lack of sunlight. The maximum rate of formation coincides with maximum solar radiation – typically occurring between 10 am and 2 pm. The maximum concentration usually lags this time because the rate of formation is greater than the rate of its destruction. Ozone concentrations tend to peak early- to mid-afternoon in areas where there is strong photochemical activity and later in the day in areas where transport from upwind regions is the main source. Also, nighttime automobile emissions of NO in urban areas can reduce O_3 to low levels overnight.

12.1.2.2 Seasonal Variations

Ozone is typically a summertime pollutant. Summer is the time when the most sunlight is available; temperatures are higher, and the largest emissions of natural sources of VOCs occur. Monthly maxima can occur any time from June through August. However, springtime maxima are observed in national parks, mainly in the western United States and at a number of other relatively unpolluted monitoring sites throughout the Northern Hemisphere. For example, the highest O_3 concentrations at Yellowstone National Park (Wyoming, Montana, and Idaho, USA) tend to occur during April and May. Monthly minima O_3 concentrations tend to occur from November through February at polluted sites and during the fall at relatively remote sites. The springtime peaks may be the result of stratospheric ozone intrusion events (see tropopause folding, section 12.1.2).

12.1.2.3 Temperature Dependency

The reactions that generate ozone are temperature dependent; as temperature rises, so do the rates of reaction. The photochemical process responsible for the formation of tropospheric ozone typically becomes important when the ambient temperature exceeds 15°C (Alley, et al., 1962).This means that, in the mid-latitudes, ozone is rarely a concern during winter months. Many regions do not require testing or collecting data for ozone during the winter months. However, high concentrations of wintertime ozone can occur in valleys. These occur because snow cover enhances solar insolation (through reflection) into a shallow boundary layer where local VOC and NO_x emissions are trapped.

12.1.2.4 Location Dependency

In western North America, the presence of mountain barriers limits atmospheric mixing (as in Los Angeles, Salt Lake City, and Mexico City) and causes a higher frequency and duration of days with high O_3 concentrations. In eastern North America, high O_3 concentrations are associated with summer weather and high-pressure weather systems. Such events can extend over hundreds of thousands of square kilometers for several days. Ozone concentrations in southern urban areas (such as Houston, TX and Atlanta, GA) tend to decrease with increasing wind speed. In northern cities (such as Chicago, IL; New York, NY; Boston, MA; and Portland, ME), the average O_3 concentrations over metropolitan areas increase with wind speed, indicating that transport of O_3 and its precursors from upwind areas is important [(Husar, et al., 1998); (Schichtel, et al., 2001)]. Ozone in city centers tends to be lower than in regions either upwind or downwind because of destruction by NO emitted by motor vehicles, see equation 12.28.

An additional location dependency is associated with precursor pollutant sources – the eastern US has a higher population density, and greater automobile emissions than the western US. Also, there is a general west to east direction of surface winds. Other cities, such as Houston, TX, have large sources of VOCs from local petroleum based industries. In general, the eastern US has regional (multiple state) ozone problems and the western US ozone problems are more confined to urban areas.

12.1.3 Health and Welfare

12.1.3.1 Effect on Human Health

Numerous scientific studies have linked ground-level ozone exposure to a variety of problems, including:
- Shortness of breath;
- Wheezing and breathing difficulties during exercise or outdoor activities;
- Coughing and sore or scratchy throat;
- Pain when taking a deep breath;
- Increase the frequency of asthma attacks;
- Aggravate and increase susceptibility to respiratory illnesses like pneumonia, asthma, emphysema, and bronchitis;
- Inflame and damage the lung lining;
- Reduce lung function; and
- Permanent lung damage with repeated exposures.

These effects may lead to increased work and school absences, visits to doctors and emergency rooms, and hospital admissions. Research also indicates that ozone exposure may increase the risk of premature death from heart or lung disease. A recent report suggests that 4 to 6% of heart attacks in the US can be linked to air pollution (Brook, et al., 2004). These affects begin to appear in healthy populations at air concentrations between 40 - 80 ppbv of ozone [(Adams, 2002), (Adams, 2003), (Adams, 2006)]. The symptoms become stronger and affect larger percentages of the population as the concentration increases. The OSHA exposure limit is 100 ppbv time weighted average for a 10 hour workday during a 40 hour workweek. The IDLH (immediately dangerous to life or health) level is 5,000 ppbv (NIOSH, 2011).

Some people are more sensitive to ozone than others. Sensitive groups include - children; healthy adults who are active outdoors; people with lung disease, such as asthma, emphysema, or chronic bronchitis; and older adults. Direct evidence of human health effects due to O_3 exposure are obtained through controlled human exposure studies of volunteers or field and epidemiologic studies of populations exposed to ambient O_3. Controlled human exposure studies typically use fixed concentrations of O_3 under carefully regulated environmental conditions and subject activity levels. The majority of controlled human studies have investigated the effects of exposure to O_3 in young nonsmoking healthy adults (18 to 35 years of age) performing continuous exercise (CE) or intermittent exercise (IE). These studies use various combinations of O_3 concentration, exercise routine, and exposure duration. The most salient observations from studies the US-EPA reviewed were that: young healthy adults exposed to O_3 concentrations of 60 ppbv develop significant reversible, harmful effects (as listed above), and higher concentrations increased the negative effects [(US-EPA, 1996), (US-EPA, 2006a).

It is interesting to note that some of the early 20th century measurements of ground level atmospheric ozone were done at European alpine sites (e.g., Arosa, Switzerland) because people were encouraged to visit these sites to take the ozone rich air for their health.

12.1.3.2 Effect on Human Welfare

Ground-level ozone can have detrimental effects on plants and ecosystems. These effects include:
- interference with the ability of sensitive plants to produce and store energy,
- increased susceptibility to diseases, insects, other pollutants, and weather;
- damage to the leaves of trees and other plants negatively impacting their appearance; and
- reduced forest growth and crop yields.

Sensitive plant species that are potentially at increased risk from ozone exposure include trees such as black cherry, quaking aspen, ponderosa pine, and cottonwood. These trees are found across the United States, including in protected parks and wilderness areas. Not all species are equally sensitive to ozone.

(Murphy, et al., 1999) estimated the yearly damage from motor vehicle ozone precursor emissions (NO_x and VOCs) caused losses of $2.8 billion to $5.8 billion (1990 dollars) to the eight largest production crops. The US-EPA estimates the loss in crop production due to just ozone at $500 million each year (US-EPA, 2010a). Air pollutant effects on tree species and forest habitats have resulted in measurable changes in total biomass, changes in composition of forest species (biodiversity), and forest health [(Kurczynska, et al., 1997), (Bringmark, et al., 1995), (McLaughlin, et al., 1999), (Vacek, et al., 1999)]. No loss estimate is available for changes in forest biomass and composition because this is more difficult to quantify than marketable products, and there is little agreement in how to value these natural services.

Ozone and other photochemical oxidants also react with many economically important man-made materials, decreasing their useful life and aesthetic appearance. Some susceptible materials include elastomers, fibers, dyes, and paints [Chapter 11 from (US-EPA, 2006a)].

Elastomers such as natural rubber, synthetic butadiene, isoprene, and styrene polymers and copolymers are particularly susceptible to even low concentrations of O_3. Ozone damages these compounds by breaking the molecular chain at carbon double bonds; a chain of three oxygen atoms is added directly across the double bond. The change in structure promotes the characteristic cracking of stressed/ stretched rubber called "weathering." Tensile strain produces cracks on the surface of the rubber that increase in number with increased strain. The rate of crack growth is dependent on the degree of stress, the type of rubber compound, O_3 concentration, time of exposure, and temperature. After initial cracking, there is further O_3 penetration, resulting in additional cracking, loss of strength, decrease in ductility and eventually failure. One of the first tests for ozone was to stretch a rubber band around a jar and see how long it took to break (AQMD, 1997). High levels of ozone could cause the rubber band to snap in less than 15 minutes.

Ozone can damage textiles and fabrics by methods similar to those associated with elastomers. Synthetic fibers are less affected by O_3 than natural fibers; however, O_3 contribution to the degradation of textiles and fabrics is not very significant, resulting only in slight decreases in fiber strength.

Ozone causes color fading in textile dyes by reacting with the dye molecules. Ozone molecules break the aromatic ring portion of the dye molecule, oxidizing the dye. The rate of fading depends upon the rate of diffusion of the dye to the fiber surface. The type of textile fiber and the manner of dye application influence the rate and severity of the O_3 attack. Several artists' pigments are also sensitive to fading and oxidation by O_3 when exposed to concentrations found in urban areas. Because of the potential of O_3 to damage works of art, recommended limits on O_3 concentrations in museums, libraries, and archives are relatively low, ranging from 10 to 15 ppbv. These facilities use air de-ozonators as part of their air conditioning systems when outside air exceeds these concentrations.

Ozone acts to erode some surface coatings (paints, varnishes, and lacquers). However, many of the available studies on O_3 degradation of surface coatings do not separate the effects of O_3 from those of other pollutants or environmental factors such as weather, humidity, and temperature. In addition, many manufacturers alter the materials or add additional components to reduce the harmful effects of environmental factors.

12.1.4 History and Regulation

Table 12-2 shows the O$_3$ ambient air quality standard for several countries. The values are based on statistical evidence from epidemiological studies and short-term chamber experiments.

Table 12-2. Ozone Ambient Air Quality Standards in Several Countries.

Country /	Time Period Average Units (unless noted)	8-hr µg/m^3	1-hr µg/m^3
US		150	
EU		120	
Japan			120[A]
India (Class I/II/III)		100/100/100	180/180/180
China (Class I/II/III)			120/160/200
Brazil			160
Mexico		150[A]	
South Africa			235
WHO		100	

Notes:
Class I: Tourist, conservation; Class II: Residential; Class III: Industrial and Heavy Traffic.
A - Includes all photochemical oxidants (e.g. O$_3$, PAN)

The US-EPA establishes National Ambient Air Quality Standards (NAAQS) for ground-level ozone (O$_3$) and other criteria pollutants based on human health and welfare standards. The 1970 Clean Air Act (CAA) and the 1990 Clean Air Act Amendments (CAAA) authorized the process for determining and implementing these standards. Table 12-3 provides a brief timeline of the milestones in the development and implementation of ambient air ozone standards in the US (US-EPA, 2010b).

Under the CAA and CAAA each state must develop a plan describing how it will attain and maintain the NAAQS. The plan is called the State Implementation Plan (SIP). A SIP is a collection of programs (e.g. monitoring, modeling, emission inventories, control strategies) and documents (policies, rules, and enforcements) that the state uses to attain and maintain the NAAQS for all air contaminants, see section 3.2. A state must engage the public in approving its plan prior to sending it to EPA for approval. In some cases where the EPA fails to approve a SIP, the Agency can issue and enforce a Federal Implementation Plan (FIP) to ensure attainment and maintenance of the NAAQS. The SIP is a working document revised as often as needed.

Table 12-3. History of US-EPA Ground-level Ozone Standards.

Year	Action
1971	EPA established a 1-hour NAAQS ozone standard of 0.08 ppm.
1979	EPA revised the 1-hour standard to 0.12 ppm.
1991	The number of counties designated for non-attainment reached 371. Concerned about the new science indicating adverse effects at levels allowed by the NAAQS, the American Lung Association went to court to compel EPA to act.
1994	EPA obtained a voluntary remand based on a promise to consider the newer studies.
1995	EPA and 37 eastern states form the Ozone Transport Assessment Group - work with stakeholders to study ozone transport for two years.
1996	EPA issued a three-volume criteria document encompassing hundreds of new scientific studies, finding "strong" scientific evidence of adverse health effects from ozone at levels allowed by the 1979 NAAQS (US-EPA, 1996).
1997	(July) EPA revised the air quality standards for ozone replacing the 1979 standard with an 8-hour standard set at 0.08 ppm. Three states and dozens of industry plaintiffs quickly challenged the new standards. (October) EPA acts on the work of the Ozone Transport Assessment Group and proposes NO_x regional reductions in the eastern US.
1998	EPA issued a final rule on regional NO_x reductions, known as the NO_x SIP Call (see chapter 10).
1999	The DC Circuit Court of Appeals sent the 1997 standards back to EPA for further study. EPA appealed.
2001	The U.S. Supreme Court unanimously upheld the constitutionality of the Clean Air Act as EPA had interpreted it in setting the 1997 health-protective air quality standards. ***The Supreme Court also reaffirmed EPA's long-standing interpretation that it must set these standards based solely on public health considerations without consideration of costs.***
2002	EPA began the process by which states (governors) and tribes submit recommendations for what areas would be designated non-attainment (failing to meet the 1997 standard).
2003	(June) EPA proposed the clean air ozone implementation rule with options for how areas would transition from the 1-hour ozone standard to the 8-hour ozone standard. (July) States and tribes recommended designations - 412 counties included. (December) EPA responded to states and tribes describing intended modifications to their recommended designations - 506 counties included. (December) EPA proposed the Clean Air Interstate Rule (CAIR) to help areas in the US meet the 8-hour ozone standard.
2004	EPA finalized the Clean Air Ozone designations and basic implementation rule.
2008	EPA reduced the National Ambient Air Quality Standards for ground-level ozone, from 0.08 ppmv to 0.075 ppmv.

2010	EPA announces plans to reconsider its 2008 decision setting national standards for ground-level ozone – the 2008 level conflicted with the CASAC report showing adverse effects at lower concentrations. The 2009 Clean Air Interstate Rule (CAIR), which is currently in effect, set the values lower. However, CAIR has been vacated by the courts, and EPA has been ordered to address flaws in the law, while leaving CAIR in place temporarily. The first revision, the Cross-State Air Pollution Rule, was stayed in August 2012, and CAIR remains in place. The stay was appealed to the Supreme Court by the US Government on December 10, 2013.
2014	U.S. Supreme Court granted the Environmental Protection Agency's (EPA) motion to lift the stay of the Cross-State Air Pollution Rule (CSAPR). EPA begins implementation of CSAPR.

The CAA requires the US-EPA to review the latest scientific information and standards every five years for each criteria pollutant. New standards and policy decisions undergo review by the scientific community, industry, public interest groups, the general public, and the Clean Air Scientific Advisory Committee (CASAC) before becoming established. Table 12-2 lists the current US primary standard. The secondary (human welfare based) standard is the same as the primary standard. The O_3 concentration is calculated and reported from the 3-year average of the fourth-highest daily maximum 8-hour average ozone concentrations measured at each monitor within an area over each year. If this level does not exceed 0.075 ppm (as of May 27, 2008), the area is designated as in 'attainment.' If the average concentration exceeds this value, the area becomes a 'nonattainment' area. Once 'nonattainment' designations take effect, the state and local governments have three years to develop implementation plans outlining how the area will attain the ambient air standard and how it will maintain the standard. Typical plans include reducing the emissions contributing to ground-level ozone concentrations. The actual methods to bring this about are allowed to vary to the specific circumstances of the area. Most plans include regulations to control stationary sources of NO_x and VOCs. Some include reductions on mobile sources or the fuels used in mobile sources.

The US-EPA has created a set of national and regional rules to reduce emissions of pollutants that form ground-level ozone. The rules affect two main emission sources that release the ozone precursor chemicals (NO_x and VOCs) – stationary fossil-fuel powered sources (power plants) and mobile sources (cars, trucks, planes, ships, and trains). The rules for stationary sources include:
- Clean Air Interstate Rule (2005 and 2009) designed to reduce ground-level ozone in the east by permanently capping emissions of SO_2 and NO_x.
- Clean Air Visibility Rule (2005) amended EPA's 1999 Regional Haze Rule to require emission controls for industrial facilities emitting air pollutants that reduce visibility.
- Regional Transport Rule (1998) reduces regional emissions of NO_x in 22 eastern states (and DC) in order to reduce the regional transport of ozone.

- NO_x SIP Call (1998) reduces the regional transport of NO_x in the eastern US, see chapter 10.
- Acid Rain Program (1990) uses a combination of traditional requirements and a market-based cap and trade program to reduce power plant emissions of NO_x and SO_2, see Chapter 9.

The rules for mobile sources include (see Chapter 15 for more detailed information):
- Clean Air Non-Road Diesel Rule (2004) set emission standards for the engines used in construction, agricultural, and industrial equipment, and reduced the amount of sulfur allowed in the fuel they use.
- Clean Diesel Trucks and Buses Rule (2007) issued in December 2000, requires a 95% reduction in emissions of heavy-duty trucks and buses – comparable to achievements made for automobiles.
- Tier 2 Vehicle Emission Standards and Gasoline Sulfur Program, sets tailpipe emissions standards for all passenger vehicles, including SUVs, pickups, vans, and large personal passenger vehicles beginning with the 2004 model year to the same national emission standards as cars. It also requires reduced levels of sulfur in gasoline.
- Emissions standards for highway motorcycles (2005) sets limits on hydrocarbon (HC) and carbon monoxide (CO) emissions for motorcycles based on engine size, see Table 12 4.
- Emission standards for engines (2010) that power forklifts, electric generators, recreational boat engines, snowmobiles, all-terrain vehicles and off-road motorbikes to reduce HC, CO, and NO_x emissions between 35 – 95% depending on engine type (US-EPA, 2001), (US-EPA, 2011a).
- Emission standards for locomotives and marine diesel engines (2004) to reduce sulfur in fuel by 99% and reduce PM and NO_x emissions by as much as 90% and 80% respectively (US-EPA, 2011b), (US-EPA, 2011c).

Table 12-4. US Motorcycle Emission Standards.

Class	Model Year	Emission Standard (g/km)	
		HC	CO
I	2006 and later	1	12
II	2006 and later	1	12
III	2006 - 2009	1.4	12
III	2010 and later	0.8	12

Notes: Class I - Engine Size of 50 to 169 cc (3.1 to 10.4 cu. in.).
Class II - Engine Size of 170 to 279 cc (10.4 to 17.1 cu. in.).
Class III - Engine Size of 280 cc and over (17.1 cu. in. and over).

This list describes the regulatory requirements of manufacturers. Only a few areas in the US require users to have their equipment tested and repaired if necessary. Additional tests may be necessary for areas in ozone non-attainment– that is they have trouble meeting the NAAQS standards.

The US-EPA has also created voluntary programs to help reduce ground-level ozone formation. These programs include the National Clean Diesel Campaign, Voluntary Diesel Retrofit Program, the SmartWay Transport Partnership, and Clean School Buses USA. These programs provide scientific and engineering technical assistance to help states, local governments, and public and private concerns.

Nationwide, the average daily maximum 8-hr O_3 concentrations decreased about 30% from 1980 to 2010, see Figure 12-4. The decrease is larger at the higher end (90th percentile) than at the lower end (10th percentile). The difference in distribution of concentrations is becoming narrower and that the worst locations have improved the most over time. The change is probably due to reductions of NO_x ozone precursors in the most impacted areas, but less in other areas. A look at the trends for NO_2 concentrations (Figure 10-5) shows a similar trend. Most of the US is NO_x limited for ozone formation, hence reductions in NO_x have the most impact.

FIGURE 12-4. US National Air Quality Trend for Ambient O_3 Concentration (ppm).

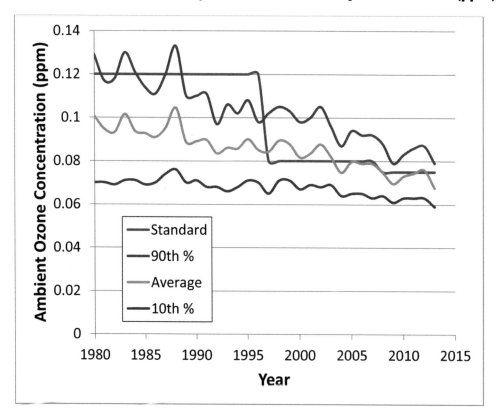

● **Case Study 12-1. World Health Organization Guidelines for Ozone**
The World Health Organization (WHO) is an international non-government organization (NGO) that acts as the directing and coordinating authority for health within the United Nations system. WHO's constitution, originating on April 7, 1948, states that its objective "is the attainment by all people of the highest possible level of health." It is responsible for providing leadership on global health matters, shaping the health research agenda, setting norms and standards, articulating evidence-based policy options, providing technical support to countries and monitoring and assessing health trends. Its mission is to combat disease, especially key infectious diseases and to promote the general health of the people of the world. The WHO flag features the Rod of Asclepius as a symbol for healing.

One primary focus area for WHO is air pollution, both indoor and outdoor. Air pollution is a major environmental health problem affecting everyone in developed and developing countries alike. WHO has published several reports on Air Quality Guidelines (AQG) [(WHO-Europe, 1987), (WHO-Europe, 2000), (WHO, 2006). These documents offer global guidance on reducing the health impacts of air pollution. The AQG applies worldwide and provides recommendations based on expert evaluation of current scientific evidence. The AQG - Global Update 2005 report revised limits from earlier reports for the concentration of particulate matter, ozone, nitrogen dioxide, and sulfur dioxide, applicable across all WHO regions (193 member states). The guidelines are not legal limits or regulatory, rather they provide the common background information to help nations and states to develop health and science based laws. No country is required to implement the WHO guidelines.

Key findings in the 2005 Air Quality Guidelines:
- *There are serious risks to health from exposure to PM and O_3 in many cities of the world.*
- *It is possible to derive a quantitative relationship between pollution levels and specific health outcomes (increased mortality or morbidity). The relationship allows insight into health improvements expected from a reduction in air pollution.*
- *Even relatively low concentrations of air pollutants correlate to adverse health effects.*

- *Poor indoor air quality may pose a risk to the health of over half of the world's population, including citizens in every member state.*
- *Significant reductions in exposure to air pollution due to lowering the concentrations of several of the most common air pollutants emitted during the combustion of fossil fuels. Such measures also reduce greenhouse gases and contribute to the mitigation of global warming.*

In addition to guideline values, the AQGs give interim targets for each outdoor air pollutant aimed at promoting a gradual shift from high to lower concentrations. Achieving these targets leads to significant reductions in risks for acute and chronic health effects from air pollution can be expected. Progress towards the guideline values, however, is the ultimate objective.

The WHO guidelines for Ozone (O_3) are 100 µg/m^3 for an 8-hour average. The previously recommended limit, 120 µg/m^3 8-hour mean, was reduced based on new peer-reviewed scientific evidence showing conclusive associations between daily mortality and ozone levels occurring at ozone concentrations below 120 µg/m^3.

This evidence showed that the excessive ozone in the air can have a marked effect on human health causing breathing problems, triggering asthma, reducing lung function and causing lung diseases. The conclusion showed that the daily mortality rises by 0.3% and heart diseases by 0.4 %, per 10 µg/m^3 increase in ozone exposure.

● Case Study 12-2. Ozone Action Days

An Ozone Action Day, alternately called "Ozone Alert" or "Clean Air Alert," is called when the Air Quality Index (see chapter 1) gets into the unhealthy ranges (40 – 80 ppbv) for ozone. Different air pollution control agencies call them at different levels or set points. An action day is declared when the AQI is Moderate, or Code Yellow, if the levels are expected to approach Code Orange levels. In most places an action day is called when the AQI is forecast to be Code Orange - Unhealthy for Sensitive Groups (children, adults who are active outdoors, and people with lung disease, such as asthma). In other places, action days are called when the AQI is forecast to be Unhealthy, or Code Red. They can be declared by a local municipality, county or state, depending on the size of the affected region. The alert can be declared up to 24 hours before the ozone concentration is expected to exceed the action level. Local air quality experts (usually meteorologists) use air quality computer models, weather data, measurements of pollution levels, and local experience to generate daily air pollution forecasts.

An ozone incident typically occurs during the summer when warm temperatures, sunny days, and the buildup of ozone precursors combine with a high-pressure system over an urban area. High temperatures enhance the rate of the ozone formation reactions. Sunshine increases the photo-chemical activity involving the precursors (NO_x and reactive carbon). High-pressure systems usually limit vertical mixing and have low-wind speeds that minimize the dispersal of pollutants and allow the concentrations of pollutants to build up.

An Ozone Action Day is a public notification designed to allow citizens to make informed choices about limiting their exposure to potentially harmful air pollution conditions. There are no laws requiring anyone to do or not do particular activities in the event of an action day. The declaring group typically provides a list of voluntary activities that can help limit or reduce the impact of the pollution incident.

Ozone Action Day Tips
- *Conserve electricity and set your air conditioner to a higher temperature.*
- *Choose a cleaner commute—share a ride to work or use public transportation. Bicycle or walk to errands when possible.*
- *Defer use of gasoline-powered lawn and garden equipment.*
- *Refuel cars and trucks after dusk.*
- *Combine errands and reduce trips.*
- *Limit engine idling.*
- *Use household, workshop, and garden chemicals in ways that keep evaporation to a minimum, or try to delay using them when expecting poor air quality.*

A daily US map showing the states and regions that have declared an Ozone Action Day is available at the US EPA AirNow website - http://www.airnow.gov/. Many other countries use similar systems to reduce harm from ozone air pollution.

12.1.5 Control

Tropospheric ozone is a secondary pollutant whose ambient concentration depends on the precursor concentrations of NO_x and reactive carbon (CO, CH_4, and NMVOCs). Since it is a secondary pollutant, control strategies focus on reducing emissions of the precursor pollutants. Chapter 10 discusses reductions of NO_x and Chapter 11 discusses reductions of reactive carbon. However, the photochemical production of ozone is non-linear with respect to the concentrations of the precursors, so a simple reduction in one may not produce an associated reduction in ozone and could even cause an increase.

The current US-EPA modeling tool for forecasting ozone concentrations for Air Quality Index (AQI) decisions is the Community Multi-scale Air Quality (CMAQ) modeling system. This tool is useful for examining the effect of policy decisions on NOx and reactive carbon emissions linked to ambient ozone concentrations. It combines current peer-reviewed atmospheric science and air quality modeling with multi-processor computing techniques in an open-source framework to generate estimates of ozone, particles, toxics, and acid deposition. The latest information and a copy of the current CMAQ model is available for download from the Community Modeling and Analysis System (CMAS) Website – [http://www.cmascenter.org/].

The US-EPA's first-generation atmospheric models (1970's – 1990's) were created to predict the concentrations of single pollutants. The EPA had a regional acid deposition model, a regional model for predicting ozone, and a different regional model for particle pollution. These models tended to work well for a limited time and spatial scales. However, it was found that air pollution problems are dependent on multiple pollutants and that a better approach would require a single atmospheric model that incorporated all the pollutants together.

Starting in 1994 EPA began work on a "super model" framework that used data from other related models, in order to apply, combine, and interpret the previously separate information. The point of this model was to help scientists and policy makers at EPA determine how changes to proposed or existing air emission regulations would impact future air quality and provide benefit to public health. This super model simulates the atmospheric processes associated with air pollution by calculating solutions to mathematical model equations over a three-dimensional grid using millions of data points (e.g. position, temperature, pressure, humidity, wind speeds, concentrations and emissions of pollutants, availability of sunlight, and the effects of clouds and aerosols). It was first released for public use in 1998 after a concentrated development effort by researchers in EPA, NOAA, and the academic and private sectors.

The CMAQ modeling system simulates the chemical and physical processes that occur in the atmosphere for all pollutants simultaneously. It combines three modeling components: meteorology, emissions, and chemistry-transport. The meteorology model is used to describe the state (e.g. temperature, pressure, humidity) and motions (horizontal and vertical wind speed and diffusion) of the atmosphere. The emissions model estimates the man-made and natural emissions injected into the atmosphere, including point sources (e.g. smoke stacks), and area sources (e.g. forests and crops). The chemical-transport model simulates the chemical transformations, including photolytic reactions, and ultimate fates of primary and secondary pollutants.

The grid resolutions and domain sizes for CMAQ range spatially and temporally over several orders of magnitude, and simulations can be performed to evaluate long-term (annual to multi-year) and short term (weeks to months) issues. CMAQ can be used for urban and regional scale model simulations because of its ability to handle a large range of spatial scales.

This model is extremely useful, but its complexity is beyond the scope of this text to use. However, a common relationship between the concentrations of NOx, and reactive carbon and ozone can be graphically represented using an Empirical Kinetic Modeling Approach (EKMA) plot (which originates from the first-generation models). It relates changes in the precursors to changes in the maximum ozone concentration and provides a simplified representation of possible effects from changes in the precursor pollutants. Figure 12-5 shows an EKMA ozone isopleth (ppb) as a function of a constant emission rate for NO_x and VOC (10^{12} molecules/($cm^2 \cdot s$)) in a zero-dimensional box model calculation (US-EPA, 2006b). The isopleths (solid lines) represent the peak ozone concentration during the three-day simulation period. The ridge line, shown by solid circles, shows the transition from NO_x-saturated (above) to NO_x-limited (below) conditions.

A zero-dimensional box model means that the calculations do not consider any spatial variations (calculations occur at a single point), the emissions of the precursors are held constant over the simulation, and that time increments are through three days. From the solution sets (NO_x and VOC emission rate variables) the maximum value of O_3 was chosen, even if it did not correspond to the same time between sets.

FIGURE 12-5. Calculated Ozone Isopleths Resulting from Constant Emission Rates of NO_x and VOCs.

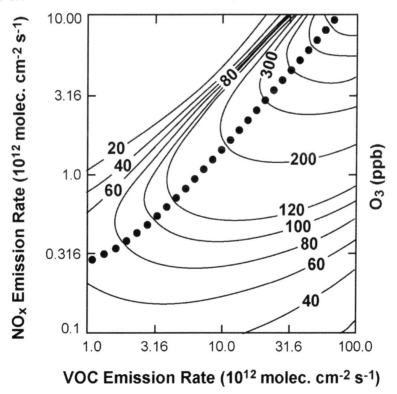

This figure shows that reducing ozone concentrations requires different strategies depending on whether the atmosphere is VOC-limited or NO_x-limited. A VOC limited atmosphere (above the dotted line) requires reductions in VOC emissions to reduce ozone concentrations. A NO_x limited atmosphere (below the dotted line) requires reductions in NO_x emissions to reduce ozone concentrations. If VOCs were reduced in the NO_x-limited region it may have no effect, or it could cause an increase in the ozone concentration. Similarly, for the VOC limited region a reduction of the NO_x emission rate may not lead to a reduction in the formation of ozone. These results show how important it is to have good measurements of the actual concentrations of the precursors and to know their emission rates. It is also worth remembering that the atmospheric concentration of each precursor change with weather conditions and time of year. Therefore, there is no single strategy that works everywhere and at all times.

● **Case Study 12-3. Ozone Removal by BASF Catalyst's PremAir® Technology.**
Most air pollution control technology focuses on controlling emissions. Once emitted, people have relied on natural processes to remove or destroy the pollutants. The control of ground level ozone is very difficult because it is not an emitted pollutant; rather, it is the result of photochemical reactions involving NO_x and VOCs. Both have natural and anthropogenic emission sources and can travel hundreds of miles before reacting to form ozone. It has been the most difficult of the criteria air pollutants to control.

BASF Catalyst has developed a commercial product that removes ozone from ground level air. The PremAir® catalyst can be applied to any heat exchange surface - such as car, bus and truck radiators, or stationary HVAC (Heating, Ventilation, and Air Conditioning) condensers. The catalyst uses the waste heat available from a radiator or condenser and uses it in the catalyzed destruction of ozone to form molecular oxygen.

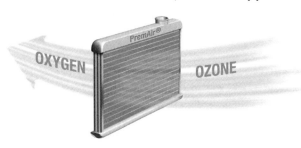

The technology takes advantage of the huge volumes of air that are processed daily by both mobile and stationary heat exchange devices.

The treated automotive radiator is capable of removing up to 75% of the ozone in the air passing over it. The coating has no significant (less than 3%) impact on the perfor-

mance of the radiator. Durability tests from actual vehicle on-road fleet testing show that the catalyst retains up to 80% of its ability to destroy ozone after 100,000 miles.

PremAir catalysts have been successfully installed on over three million automotive radiators throughout the world. Manufacturers of Ultra Low Emission Vehicles (ULEV) and Super Ultra Low Emission Vehicles (SULEV) frequently use this technology. These vehicles receive an emission credit for an equivalent amount of VOCs, thus allowing greater flexibility in vehicle design. Air conditioners are also good candidates for the technology since they are often used in densely populated areas and when the outdoor temperature is high (both conditions correlate with high levels of ground level ozone).

The technology may be able to provide a positive benefit to the environment while being passive in the end user application. It does not require maintenance, upgrades, or repairs in normal use and lasts for the lifetime of the heat exchanger. However, no data exists to show that installation of this technology leads to an actual reduction in ambient ozone concentrations.

12.2 Stratospheric Ozone

12.2.1 Sources and Sinks

Ozone (O_3) is formed in the stratosphere by a reaction between molecular oxygen (O_2) and high-energy photons from the sun, a process called photolysis. This reaction occurs mainly over the tropics and mid-latitudes during summer when solar radiation is the most intense. Chapman first proposed the set of reactions that lead to the formation of stratospheric ozone (Chapman, 1930):

12.23 $\quad O_2 + h\nu \rightarrow O + O \qquad\qquad \lambda < 242$ nm

12.24 $\quad O + O_2 + M \rightarrow O_3 + M$

12.25 $\quad O + O_3 \rightarrow 2 O_2$

12.26 $\quad O_3 + h\nu \rightarrow O_2 + O \qquad\qquad 200 < \lambda < 360$ nm

Where: O is a free oxygen atom,
M is a catalyst particle,
hν = energy from an absorbed photon (J),

h = Planck's constant (6.63×10^{-34} J s),
ν = frequency of photon (= c/λ, where $c = 3 \times 10^8$ m/s), and
λ = wavelength of the photon (measured in nm).

The type of energy needed to drive this reaction comes from solar radiation. Specifically it is the ultraviolet photons that are capable of breaking the molecular oxygen bonds. Table 12-5 provides a list of photons and their wavelengths. The most energetic photons, those with the shortest wavelengths, are absorbed in the highest layers of the atmosphere. The longer wavelength photons travel further into the atmosphere.

Table 12-5. Photons Characterized by their Wavelength (λ).

NAME	Wavelength (nm)
X-rays	0.01 - 10
Extreme Ultraviolet	10 - 120
Ultraviolet	
UV-C	100 - 280
UV-B	280 - 315
UV-A	315 - 400
Visible Light	
Violet	425
Blue	475
Green	525
Yellow	575
Orange	625
Red	675
Infrared	700 - 10^6
Microwaves	10^6 - 10^8

The amount of ozone predicted by the Chapman model is much larger than is observed. In the 1970's, it was shown that ozone levels also depend on the presence of odd-nitrogen compounds. Additional work has shown that the amount of stratospheric ozone depends on the presence of hydroxyl radical (OH·), nitrous oxide (N_2O), atomic chlorine (Cl), and atomic bromine (Br). Each contributes to a net reduction in ozone.

Hydroxyl radicals are produced in the stratosphere by the oxidation of water, methane, and hydrogen gas. An additional source is transport from photolysis of water in the mesosphere (atmospheric layer above the stratosphere, starting at elevations around 70 km) into the stratosphere.

The main source of nitrous oxide (N_2O) is denitrification occurring in soils, of which about 1/3 is due to modern agricultural practices, see Chapter 10. The primary sink for N_2O is the reaction with free oxygen in the stratosphere to form nitric oxide (NO), see equation 12.27. Photolysis of molecular oxygen (equation 12.23) generates the free oxygen.

12.27 $N_2O + O \rightarrow 2\,NO$

The NO can catalytically convert ozone to molecular oxygen:

12.28 $NO + O_3 \rightarrow NO_2 + O_2$

12.29 $NO_2 + O \rightarrow NO + O_2$

NO may also be formed from the direct photolysis of N_2O in the higher regions of the stratosphere.

Sources of stratospheric chlorine originate in both natural and synthetic compounds. The chief natural source is from the photo disassociation of methyl chloride (CH_3Cl). The most important man-made source materials are the chlorofluorocarbon (CFC) class of compounds. Examples include CCl_3F (Freon-11) and CCl_2F_2 (Freon-12). The reaction freeing the chlorine requires sunlight:

12.30 $CH_3Cl + h\nu \rightarrow CH_3\cdot + Cl\cdot$

Chlorine participates in a similar set of reactions, catalytically destroying ozone:

12.31 $Cl + O_3 \rightarrow ClO + O_2$

12.32 $ClO + O \rightarrow Cl + O_2$

Bromine acts similar to chlorine and has similar sources, the chief one being methyl bromide (CH_3Br) which occurs naturally and from anthropogenic sources. The main sink is surface deposition of hydrobromic acid. The effectiveness of bromine for destroying ozone is enhanced because: its radical is not reacting with hydrocarbons; it is more readily photolyzed than chlorine, and by reactions coupling chlorine and bromine:

12.33 $BrO + ClO \rightarrow Br + Cl + O_2$

12.34 $Br + O_3 \rightarrow BrO + O_2$

12.35 $Cl + O_3 \rightarrow ClO + O_2$

These reactions are not the only pathways for these compounds to cause ozone destruction; rather, they show the general catalytic nature of the materials. The actual ozone concentration depends on these and similar reactions, their kinetics, the concentrations of each species as well as the concentration ratios of the species.

It may be useful to consider the atmospheric chemistry of ozone as two intersecting cycles: the fast primary photolytic cycle, see equations 12.23 – 12.26, and a slower VOC oxidation cycle that acts to also convert NO to NO_2, see equations 11.19 – 11.26.

The number of catalytic reactions each atom can undergo depends on the conversion of the catalytic species into more stable compounds and their removal from the stratosphere. The OH radicals can form water, or oxidize other compounds. NO_x can convert to nitric acid (HNO_3) and chlorine converts to hydrochloric acid (HCl), both of which are removed by wet or dry deposition. Bromine can form HBr or $BrONO_2$, which are removed by deposition, but they are also easily destroyed by photolysis, so it is more difficult for them to diffuse out of the stratosphere. Bromine is a more efficient catalyst for ozone destruction than nitrogen or chlorine.

12.2.2 The Ozone Hole

FIGURE 12-6. South Pole Stratospheric Ozone Hole: Size and Minimum Measured Concentration, 1979 to 2009. Area of Ozone hole (circles) use left y-axis and Ozone Concentration (squares) use right y-axis.

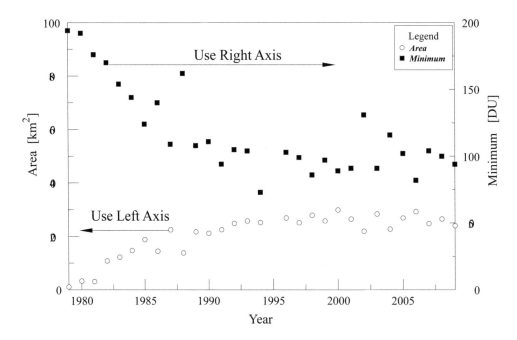

Each spring, since 1985, the catalytic destruction reactions have overwhelmed the ozone formation reaction over the North and South Pole. The North Pole has seen an average maximum reduction of 30% and the South Pole has seen an average reduction of more than 50%. Figure 12-6 shows a time plot of the minimum ozone measurement (in Dobson Units or DUs) over the South Pole (squares using the right y-axis for values), and the ozone hole area (circles using the left y-axis values). Data for the plot is from NASA (US-NASA, 2013). The area of the ozone hole is determined from a map of total column ozone. It is calculated from the area on the Earth that is enclosed by a line with a constant value of 220 Dobson Units. This number was chosen because this value was not found in Antarctic observations prior to 1979. No data are available for 1995 due to budget constraints. The actual dates for the minimum value occur between September 17 and November 2, suggesting the yearly variation of conditions.

The conditions that create these large reductions begin with the onset of winter when the polar stratosphere cools and descends closer to the surface. The Coriolis Effect sets up a strong circulation pattern around the pole forming an oblong vortex, which varies in size and shape each year. The strong wind pattern is called the circumpolar vortex. It prevents polar air from mixing with mid-latitude air, providing an effective barrier to horizontal transport into or out of the polar region. The barrier prevents the air from being warmed by mass transport from other regions. When temperatures cool to -80°C (-110°F) *Polar Stratospheric Clouds* (PSC) begin to form. PSC are liquid or solid condensates of water, sulfuric acid, and nitric acid. Other constituents may also be present in smaller quantities. Particle size ranges from 0.5 to 10 μm diameter. In the darkness of autumn and winter, heterogeneous reactions on the surface of the PSC convert NO_x into HNO_3 and chlorine from its destructive forms (ClO and Cl_2) into reservoir forms ($ClONO_2$ and HCl). The PSC reservoirs keep these compounds in the stratosphere and prevent them from diffusing into the troposphere and being removed from the atmosphere. They also serve as heterogeneous catalyst supports for springtime ozone destruction.

The return of sunlight in the spring begins the process of photolyzing these compounds from the surface of the PSC back into their ozone-destroying form. The return of sunlight also starts to heat the polar atmosphere. Within a few weeks, the circumpolar vortex breaks up and allows horizontal mixing to occur between the polar air and midlatitude air. During this mixing, the ozone hole weakens and spreads out – spreading away from Antarctica to the Falkland Islands, South Sandwich Islands, the southern tip of South America, New Zealand, and Australia.

This phenomenon causes similar, though smaller, decreases in ozone concentrations over the Artic. The reason it is smaller is that the Arctic air does not get as cold because the circumpolar vortex is usually not as strong due to the varied topography around the North Pole. These surface features make it easier for mid-latitude air to mix with the polar air all year long.

The enhanced ozone destruction process requires the availability of Cl (as well as OH, NO, and Br) and weather conditions cold enough for the stratospheric clouds to form. Ozone depletion has also been observed in the mid-latitudes where measurements show ozone levels fall about 10% during the winter and 5% in the summer. Depletion worsens at higher latitudes.

The main source of additional chlorine and bromine in the stratosphere are the man-made chlorofluorocarbon (CFC) compounds, commonly called Freon, and the bromofluorocarbon compounds, known as Halons. The first CFCs were discovered by Thomas Midgley Jr. in 1928 as non-toxic, non-flammable, water-insoluble, non-carcinogenic refrigerants replacing hazardous materials like ammonia. DuPont setup the first commercial production in 1930. Their usage grew enormously over the years as aerosol-spray propellants, refrigerants, solvents, and foam-blowing agents. Ironically, their safe, non-reactive properties are what cause them to become a problem in the stratosphere. They do not interact with living organisms, nor do they degrade in the troposphere, so they have long atmospheric lifetimes (decades to centuries). This persistence allows them to mix throughout the troposphere. Eventually, after a few years, they reach the tropopause and diffuse across it. Once in the stratosphere, the availability of more energetic photons allows the chlorine and bromine compounds to photo-disassociate, thus enhancing the ozone depletion reactions (equations 12.31 - 12.35).

CFC compounds are also very strong greenhouse gases. They range in strength from 5 to 10,000 times as effective as carbon dioxide at trapping heat within the atmosphere, see Table 12-7. This means that even small amounts of these compounds in the atmosphere will create strong responses. Chapter 14 discusses how the greenhouse effect works. CFCs remain in the atmosphere between 20 to 1,000 years before their destruction. The release of these compounds in the mid 1900's will remain an important contributor to global climate change for several centuries.

12.2.3 Health and Welfare

Stratospheric molecular oxygen and ozone absorb 97-99% of the sun's ultraviolet light that reaches Earth. Ozone is the principal absorber of solar radiation in the 250 – 300 nm wavelength regions. Ozone depletion leads to increased levels of UV-B radiation at the Earth's surface. Ultraviolet light is known to damage DNA, causes skin cancer and eye damage, and it may also cause harm to plants and animal life.

DNA damage occurs when DNA absorbs UV-B light and the absorbed energy breaks chemical bonds in the DNA. Proteins within the cell's nucleus repair most of this damage, but unrepaired genetic damage can lead to cancers. In fact, one method used to analyze amounts of genetically damaging UV-B is to expose samples of DNA to the light and then count the number of breaks in the DNA. One researcher used human DNA to find that genetically significant doses of solar radiation could penetrate as far as 9 feet into non-turbulent ocean water (Regan, et al., 1995).

Skin cancer mostly occurs to light-skinned peoples. A 1% decrease in the ozone layer causes an estimated 1-2% increase in UV-B irradiation. The decrease in ozone leads to an estimated 2-4% increase in basal-cell cancer, and 3-6% increase in squamous-cell cancer. 90% of these skin cancers can be attributed to UV-B exposure. These cancers are relatively easy to treat if detected in time, and are rarely fatal. The more dangerous malignant (cancerous) melanoma is not as well understood, although there is a correlation between brief, high-intensity exposures to UV and the eventual appearance (as long as 10-20 years later) of melanoma. Twice as many deaths due to melanomas occur in the southern states of Texas and Florida, as in the northern states of Wisconsin and Montana, but there could be other factors involved. Another effect of long-term exposure to UV-A, UV-B, and UV-C is the premature aging of the skin. Even careful tanning kills skin cells, damages DNA and causes permanent changes in skin connective tissue, which leads to skin wrinkling later in life. There is no such thing as a safe tan.

Eye damage results from high doses of UV light. The cornea is particularly susceptible because it is a high absorber of UV light. High doses of UV light can cause temporary clouding of the cornea, called 'snow-blindness', and there appears to be a link between chronic doses of UV-B and the formation of cataracts. Higher incidences of cataracts are found in populations that live at high elevations and lower latitudes (approaching the equator).

Ultraviolet levels are over 1,000 times higher at the equator than in the polar regions, so life at the equator is better adapted to higher UV light than organisms in the polar regions. However, plant and animal damage varies widely due to differences in sensitivity to UV-B. Some plants may show sensitivity to UV-B light, others show negligible effects, and a few may increase in vigor. Even between cultivars within a species, there may be large differences. An increase in UV-B would probably cause a shift in population rather than a large die-off of plants.

The World Health Organization estimates that as many as 60,000 people a year worldwide die from too much sun, mostly from malignant skin cancer. Of these deaths, 48,000 are from melanoma, and 12,000 are from other skin cancers. An estimated 90 percent of these cancers are caused by ultraviolet light from the sun.

12.2.4 History and Regulation

"Perhaps the single most successful international agreement to date has been the Montreal Protocol." — **Kofi Annan, Former Secretary-General of the United Nations**

M.J. Molina and F.S. Rowland published the first scientific paper describing the threat of CFC's on the stratospheric ozone (Molina, et al., 1974). This laboratory study demonstrated the ability of CFC's to catalytically breakdown ozone in the presence of UV light. They calculated that if CFC production continued to increase at the going rate of 10%/year until 1990, then remain steady, CFCs would cause a global 5 to 7 percent ozone loss by 1995 and 30-50% loss by 2050. Based on such studies the US and several European countries banned CFC's in aerosol sprays starting in 1978.

Little else was done after this due to a successful media campaign from the makers and users of CFCs, portraying the scientific community as 'Chicken Littles screaming that the sky is falling.'

Irving S. Shapiro, Chairman and CEO of DuPont at the time, commented that the ozone depletion theory was "a science fiction tale...a load of rubbish...utter nonsense." (Chemical Week, 16 July 1975).

However, the discovery of the Antarctic ozone hole in 1985 proved the skeptics wrong. The Earth's protective ozone layer was being destroyed by CFCs, and the ozone depletion was worse than predicted (Farman, et al., 1985). The 1985 measurements showed a 10 – 20% loss of ozone and 1987 measurements showed a 50% loss over the Antarctic. None of the scientists studying the ozone layer anticipated the eventual ozone losses reaching 70% in the Antarctic spring, 50% in the Arctic spring, 8% in Australia in summer, 10-15% in New Zealand in summer, and 3% globally year-round (WMO, 2002). This was a case where the uncertainty of the measurements and projections erred toward the best case scenario, and the worst case scenario was not as pessimistic as reality.

In response to these unprecedented measurements confirming a global threat to all living things, the Montreal Protocol, an international agreement to phase out ozone-destroying chemicals, was approved by the United Nations in 1987 to address the threat (eventually 191-member states ratified the Protocol and are referred to as signatory countries). As ozone depletion worsened globally throughout the 1990's, additional amendments to the Montreal Protocol were ratified to promote a faster phase out of ozone-destroying chemicals:

> **The Montreal Protocol on Substances that Deplete the Ozone Layer** (1987) states that developed countries were required to begin phasing out CFCs in 1993 and achieve a 50% reduction relative to 1986 consumption levels by 1998. Under this agreement, CFCs were the only Ozone Destroying Substance (ODS) addressed.
>
> **The London Amendment** (1990) changed the ODS emission schedule by requiring the complete phase-out of CFCs, halons, and carbon tetrachloride by 2000 in developed countries, and by 2010 in developing countries. Methyl chloroform was also added to the list of controlled ODSs, with phase-out in developed countries targeted in 2005, and in 2015 for developing countries.
>
> **The Copenhagen Amendment** (1992) significantly accelerated the phase-out of ODSs and incorporated a hydrochlorofluorocarbon (HCFC) phase-out for developed countries, beginning in 2004. Under this agreement, CFCs, halons, carbon tetrachloride, and methyl chloroform were targeted for complete phase-out in 1996 in developed countries. In addition, consumption of methyl bromide was capped at 1991 levels.
>
> **The Montreal Amendment** (1997) included the phase-out of HCFCs in developing countries, as well as the phase-out of methyl bromide in developed and developing countries in 2005 and 2015, respectively.

The Beijing Amendment (1999) included tightened controls on the production and trade of HCFCs. Bromochloromethane was also added to the list of controlled substances with phase-out targeted for 2004.

Figure 12-7 shows a projection used by the United Nations Environment Program to show the results of the Montreal Protocol and amendments on the concentration of stratospheric chlorine and the number of expected additional cases of skin cancer. The results from the adoption of the Beijing amendment in 1999 are expected to allow the eventual recovery of the ozone layer.

The Montreal Protocol and Amendments is one of the first international environmental agreements to use trade sanctions and offer major incentives for non-signatory nations to sign the agreement. The treaty negotiators justified the sanctions because depletion of the ozone layer is an environmental problem most effectively addressed on a global level. Furthermore, without the trade sanctions, there would be economic incentives for non-signatories to increase production, damaging the competitiveness of the industries in the signatory nations as well as decreasing the motivation to search for less damaging CFC alternatives.

As of 2011 it appears that the concentrations of ODS have decreased in the troposphere, and the concentrations of chlorine and bromine have stopped increasing in the stratosphere. The most recent measurements show that the 2006 Antarctic ozone hole was the largest in surface area and had the greatest mass deficit. The 2011 Arctic ozone hole had the largest decrease in measured ozone (45 – 50%). Damage to the stratospheric ozone layer will continue well into the 21st century until the excess chlorine and bromine are removed from the stratosphere and deposited onto the Earth's surface. The ozone layer is projected to return to pre-ozone hole values (1980) around 2050 for midlatitudes and 2065 for polar regions. The exact date of recovery depends on the effectiveness of present and future regulations on the emission of CFCs and their replacements. It also depends on the level of climate change, which is expected to cause long-term cooling in the stratosphere even as the earth's surface temperature increases (see Chapter 14). This additional cooling would increase ozone loss due to the easier formation of PSC and would prolong recovery of the ozone layer. Another potential problem chemical is N_2O, which currently has increasing emissions (see Chapter 10) and is a known ozone depleting chemical.

Molina and Rowland were awarded the Nobel Prize in 1995. The citation from the Nobel committee credited them with helping to deliver the Earth from a potential environmental disaster.

FIGURE 12-7. United Nations Environment Program Projection of the Effect of the Montreal Protocol Amendments on Stratospheric Chlorine Concentration and the Resulting Excess Number of Skin Cancers. (UNEP/GRID-Arendal, 2007)

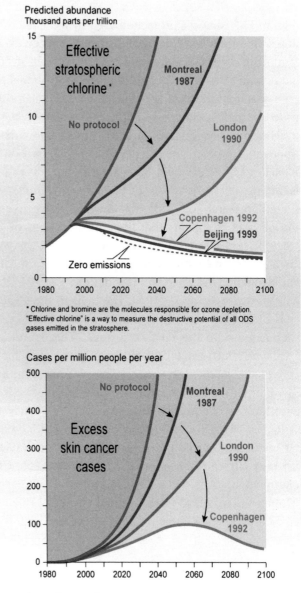

THE EFFECTS OF THE MONTREAL PROTOCOL AMENDMENTS AND THEIR PHASE-OUT SCHEDULES

*Chlorine and bromine are the molecules responsible for ozone depletion. "Effective chlorine" is a way to measure the destructive potential of all ODS gases emitted in the stratosphere.

Source: *Twenty Questions and Answers about the Ozone Layer: 2006 Update*, Lead Author: D.W. Fahey, Panel Review Meeting for the 2006 ozone assessment.

● Case Study 12-4. Ozone and UV Radiation in Australia

Australians suffer the highest rates of skin cancer in the world. Each year, around 1,200 Australians die from skin cancer (Population of Australia in 2010 was 22 million). UV radiation levels in Australia are higher than in Europe or North America. During the summer, asymmetries in the earth's orbit bring Australia closer to the sun than the Northern Hemisphere during its summer, resulting in an additional seven percent solar UV intensity. This increase coupled with the generally clearer atmospheric conditions means Australians are exposed to up to 15 per cent more UV than people at similar latitudes in the northern hemisphere. Also, Australia's proximity to Antarctica means that when the circum-polar vortex breaks up in mid-spring, the ozone depleted air can spread to the surrounding regions, causing ground level UV radiation to be even higher.

One response by the Australian government was a massive public education campaign to raise awareness of skin cancer and to help people prevent it. A cartoon seagull (Sid Seagull) singing 'Slip! Slop! Slap!' appeared on TV screens. The message was that anyone can get skin cancer, and sunburn is a leading indicator. The advice given is to 'slip' on a shirt, 'slop' on sunscreen, and 'slap' on a hat.

SunSmart (Victoria, 2011) is another Australian initiative; this one is funded by the Victorian Health Promotion Foundation, and began in 1988. Its goal was also to combat spiraling skin cancer incidence and mortality rates. Since then, research by The Cancer Council Victoria shows SunSmart's messages have reached a majority of Australians and had a strong impact on their behavior. Public attitudes towards tanning and sun protection have changed dramatically. Australians are taking more preventative measures to reduce their risk of sun damage and skin cancer. Consequently, skin cancer deaths have slowed and for females, have even started to decline.

This case study is an example of using public education to reduce the risk associated with an environmental problem. It is just one tool to help people cope with such problems.

The US National Weather Service and the US-EPA have developed a UV index to indicate the strength of solar UV radiation on a scale from 1(low) to 11 (extremely high). They have also developed general guidelines how to protect oneself from overexposure to UV radiation. Table 12-6 lists the categories. The local conditions are available from local weather services or the US-EPA SunWise website http://www.epa.gov/sunwise/uvindex.html.

Table 12-6. US-EPA UV Index.

Exposure Category	UV Index	Protective Actions
Very Low	0, 1, 2	Apply skin protection factor (SPF) 15 sunscreen.
Low	3, 4	SPF 15 & protective clothing and hat.
Moderate	5, 6	SPF 15, protective clothing, and UV-A&B sunglasses.
High	7, 8, 9	SPF 15, protective clothing, sunglasses and make attempts to avoid the sun between 10am to 4pm.
Very High	10+	SPF 15, protective clothing, sunglasses and avoid being in the sun between 10am to 4pm.

12.2.5 Control

The only way to control stratospheric ozone depletion is to reduce emissions of the *Ozone Depleting Substances (ODS)* that cause it. The Montreal Protocol and Amendments provide a legal framework to reduce production of such compounds but do not force any particular solution. Hence, the control issue becomes a replacement issue. The services provided by the banned substances are essential services no one wants to live without - refrigeration, painting, and fire extinguishment to list a few. The criteria for replacement include reduction in ozone depletion potential, reduction in global warming potential, and that the alternatives are functional, economical, and have global availability.

Ozone Depletion Potential (ODP) is a number that relates the amount of ozone depletion caused by a substance to a similar mass of CFC-11. Thus, the ODP of CFC-11 is 1.0. Other CFCs and HCFCs have ODPs that range from 0.01 to 1.0. Class I ODS have an ODP of 0.2 or greater. Class II ODS have an ODP less than 0.2. HFCs have zero ODP because they do not contain chlorine. Table 12-7 lists several of the known ODS and provides their atmospheric lifetime, ODP, and GWP values. A list of all ozone-depleting substances (ODS) recognized by the Montreal Protocol is available (US-EPA, 2010c).

● **Case Study 12-5. How is an Ozone Depletion Potential Calculated?**
The ozone depletion potential (ODP) of a compound is a measure of its ability to destroy stratospheric ozone. It is a relative measure: the ODP of CFC-11 is defined to be 1.0, and the ODP's of other compounds are calculated with respect to this reference point. Thus a compound with an ODP of 0.2 is, roughly speaking, one-fifth as "bad" as CFC-11.

More precisely, the ODP of a compound "x" is defined as the ratio of the total amount of ozone destroyed by a fixed mass of compound x to the amount of ozone destroyed by the same mass of CFC-11:

12.36
$$ODP_x = \frac{\text{Loss of Ozone due to Compound x}}{\text{Loss of Ozone due to CFC-11}}$$

The ODP of CFC-11 is 1.0 by definition. The two loss terms are measured experimentally from laboratory and field measurements combined with various atmospheric chemistry and transport models. The actual calculation is beyond the scope of this text and the interested reader should explore (US-EPA, 2010). Note that ODP is a relative measure, not an absolute, so it is insensitive to many of the details in the model (since they are included in both the numerator and denominator.

ODP depends on several factors:
- *The type of halogen in the molecule – Bromine containing halocarbons have higher ODP than chlorine containing compounds. Fluorine does not contribute to ODP.*
- *The number of bromine and chlorine atoms in the molecule*
- *Molecular weight. ODP depends on mass of the substance rather than moles.*
- *Atmospheric lifetime. Compounds with hydrogens can be destroyed in the troposphere and will have reduced impact.*

Global Warming Potential (GWP) is a number that relates the amount of global warming caused by a substance to a similar mass of CO_2. Thus, the GWP of CO_2 is defined to be 1.0. Water, a substitute in numerous end-uses, is given a value of 0. Table 12-7 lists the GWP values for several ODS. These classes of chemicals are all strong greenhouse gases because they have the ability to absorb infrared radiation. Chapter 14 provides more information about global warming.

Substitutes and alternative compounds must provide a similar function to the chemicals they are replacing. The important properties are the thermal (heat capacity and latent heats), physical (vapor pressure), chemical (low reactivity and low toxicity), and economical (modest cost and worldwide availability). Multiple replacement chemicals have been identified because there is no single material that optimizes all these properties for all applications.

Table 12-7. Ozone Depleting Substances with their associated atmospheric lifetime (years), Ozone Depletion Potential (ODP), and Global Warming Potential (GWP).

Class I				Atm Mixing Ratio	Growth
Compound Abbreviation and Name	Atm Life (yr)	ODP	GWP	ppt_v	ppt_v/yr
CFC-11 (CCl_3F) Trichlorofluoromethane	45	1	4680	243.4	-2
CFC-12 (CCl_2F_2) Dichlorodifluoromethane	100	1	10720	537.4	-2.2
CFC-113 ($C_2F_3Cl_3$) 1,1,2-Trichlorotrifluoroethane	85	0.8	6030	76.5	-0.6
CFC-114 ($C_2F_4Cl_2$) Dichlorotetrafluoroethane	300	1	9880	16.4	-0.04
CFC-115 (C_2F_5Cl) Monochloropentafluoroethane	1700	0.6	7250	8.4	0.02
Halon 1211 (CF_2ClBr) Bromochlorodifluoromethane	16	3	1860		
Halon 1301 (CF_3Br) Bromotrifluoromethane	65	10	7030		
Halon 2402 ($C_2F_4Br_2$) Dibromotetrafluoroethane	20	6	1620		
CFC-13 (CF_3Cl) Chlorotrifluoromethane	640	1	14190		
CCl_4 Carbon tetrachloride	26	1.1	1380	88.7	-1.1
Methyl Bromide (CH_3Br)	0.8	0.6	5	7.5	-0.2
Class II					
Compound Abbreviation and Name	Atm Life (yr)	ODP	GWP		
HCFC-21 ($CHFCl_2$) Dichlorofluoromethane	1.7	0.04	148		
HCFC-22 (CHF_2Cl) Monochlorodifluoromethane	11.9	0.055	1780	192.1	8.6
HCFC-123 ($C_2HF_3Cl_2$) Dichlorotrifluoroethane	1.3	0.02	76		
HCFC-124 (C_2HF_4Cl) Monochlorotetrafluoroethane	2.9	0.022	600	1.47	-0.01
HCFC-141b ($C_2H_3FCl_2$) Dichlorofluoroethane	9.2	0.11	710	19.5	0.7
HCFC-142b ($C_2H_3F_2Cl$) Monochlorodifluoroethane	17.2	0.065	2270	20.2	1.3
HCFC-225ca ($C_3HF_5Cl_2$) Dichloropentafluoropropane	1.9	0.025	120		
HCFC-225cb ($C_3HF_5Cl_2$) Dichloropentafluoropropane	5.9	0.033	586		

Table Notes:
- Mixing Ratio and Growth are from 2008. Data in the "ODP" column come from World Meteorological Organization's Scientific Assessment of Ozone Depletion: 2010 (WMO, 2011).
- Atmospheric life refers to the time needed to remove 63% of the compound from the atmosphere.
- ODP and GWP values listed are semi-empirical. The GWP values listed are for direct radiative forcing. All GWP values represent global warming potential over a 100-year time horizon.
- All of the listed chemicals have seen economic use.

The Montreal Protocol required quick action on phase-outs – between five and fifteen years. Phase-out means a ban on consumption, production, and trade. Class I ODS followed a very short time phase-out schedule in the United States and other developed countries:
  CFCs – 75% by 1994 and 100% by 1996,
  Halons – 100% by 1994,
  CCl4 – 50% by 1994, 100% by 1996, and
  CH3Br – 25% by 1999, 100% by 2005.

Developing countries had a longer timeline since they produce and use much less, and they may not have the required capital or infrastructure for quick action. Some exemptions are allowed for critical and essential uses in medical and laboratory equipment, and quarantines of agricultural materials.

Class II ODS (mostly HCFCs) reductions have a longer phase-out for consumption and production: 75% in 2010, 90% in 2015, 99.5% in 2020, and 100% in 2030. The longer phase-out allows these substances, which have less ODP, to be used to reduce the ozone depletion problem while allowing continued use of older equipment. These time lines may seem long, but the time needed to identify replacements, redesign equipment, and then to distribute across an economy, is comparable. This work requires a tremendous worldwide effort by engineers and scientists.

■ *Example 12-1.*
Determine the atmospheric lifetime of CFC-114.

Solution:
Use data from Table 12-7 and equation 2.1.

$$\tau_{total} = \frac{M_{global}}{F_{total}} = \frac{16.4\,ppt}{0.04\,ppt/yr} = 410\,yr$$

CHAPTER 12

The reported value is 300 years. The reported value represents the average over many years, and accounts for differences in rates in different parts of the atmosphere (latitude, elevation, and time of year).

■ Example 12-2.

Use the atmospheric lifetime from Table 12 7 and create a plot of the expected atmospheric concentration of CFC-11 for the next 50 years. Assume that the starting concentration is 243 pptv.

Solution:
Atmospheric lifetime represents the time required for removal or chemical transformation of 63% (i.e. $1 - 1/e$) of the total global atmospheric burden. This is modeled as

12.37
$$C = C_0 * \left[\exp\left(-\frac{t}{\tau}\right) \right]$$

Where C is the time dependent concentration, C_0 is the initial concentration, t is the time of interest, and τ is the atmospheric lifetime of the substance. The units if t and τ must match.

Data for Example 12-2

Year	Conc.
0	243.0
10	194.6
20	155.8
30	124.8
40	99.9
50	80.0

Air Pollution: Engineering, Science, and Policy

FIGURE 12-8. Projection of Global Atmospheric Concentration of CFC-11 over Next 50 Years, Solution to Example 12-2.

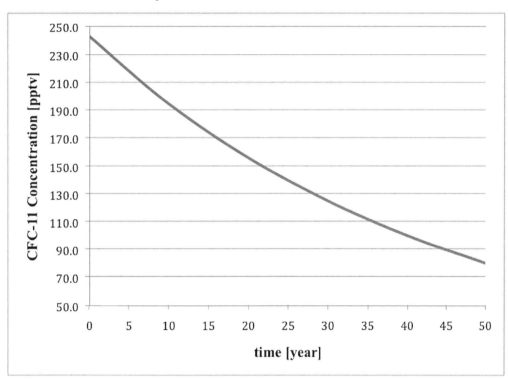

12.2.6 Alternatives to Chlorofluorocarbons.

Attempts have been made to develop compounds which have low-ozone depletion potential (ODP) to replace CFCs as refrigerants, aerosol propellants and solvents. Possible alternatives for CFCs include:

Hydrochlorofluorocarbons (HCFC) - These compounds are chemically similar to the CFCs, but have less chlorine overall. They also have shorter atmospheric lifetimes, so less of the material reaches the stratosphere. HCFCs are less stable than CFCs because they contain carbon-hydrogen bonds, which readily react with the hydroxyl radical (OH·) in the troposphere. The freed chlorine typically forms hydrochloric acid and is removed by wet or dry deposition onto the Earth's surface. However, because they still contain chlorine and have the potential to destroy stratospheric ozone, they are viewed as temporary replacements for the CFCs.

Hydrofluorocarbons (HFC) – These compounds contain carbon, hydrogen, and fluorine, but no chlorine or bromine. They have shorter atmospheric lifetimes than CFCs or HCFCs and readily react in the troposphere. All HFCs have an ozone depletion potential of 0. Industry and the scientific community view certain chemicals within this class of compounds as acceptable alternatives to CFCs and HCFCs on a long-term basis. Like HCFCs, the HFCs contain hydrogen that is susceptible to attack by the hydroxyl radical. Oxidation of HFCs by OH is believed to be the major destruction pathway in the atmosphere, leading to atmospheric lifetimes of less than 12 years. Although HFCs do not deplete the ozone within the stratosphere, these compounds are greenhouse gases (see Chapter 14). Concern over their greenhouse effects may make it necessary to regulate production and use of these compounds at some point in the future.

Hydrocarbons, such as butane and propane are cheap, readily available, and contain no chlorine, so they are not ODSs. However, they are flammable, toxic, poisonous, and their atmospheric release contributes to the formation of tropospheric ozone (see section 12.1.2).

Water and steam are effective for some cleaning applications and thus can replace some CFCs as solvents used in cleaning.

Carbon dioxide, although a greenhouse gas, is non-toxic, non-flammable, cheap and abundant. It can be used as the working fluid in refrigeration but must be run at extremely high pressures - up to several thousand psi. Such high pressures pose a potential danger to users and producers of the equipment. It can also serve as a solvent and cleaning agent in its supercritical state (T $\geq$ 31.1°C, P $\geq$ 72.9 atm) – although the current cost of such systems prohibits large-scale use.

The US-EPA maintains a data set of currently acceptable alternatives to CFCs and other ODSs in their Significant New Alternatives Policy (SNAP) Program (US-EPA, 2011d).

12.3 Questions

* - Questions and problems may require additional information not included in the textbook.

1. One proposal concerning too much ozone in the troposphere and too little in the stratosphere is to pump the ozone from the troposphere into the stratosphere. After consulting Figure 12-1, comment on the utility of this solution (assuming it was possible to move large volumes of air around the atmosphere).

2. Explain in your own words why the lifetime of ozone in the troposphere – 1) increases during winter; 2) increases with elevation; and 3) increases as you move away from the equator and toward the poles.

3. Visit the US-EPA AirNow website http://www.airnow.gov/, or the EU Air Quality Now website - http://www.airqualitynow.eu/. What is today's ozone forecast for where you live? For Washington DC - USA, London – UK, and Paris – France?*

4. Examination of Table 12-2 shows a large variation between the ozone ambient air limits. How do these levels compare with the observed level of known health effects on healthy, non-smoking, young adults? Why is there such a difference?

5. Describe how a 'super-model' for air quality might work. Include the input information you would need and what sort of output information it should provide. Explore the CMAQ website for more information.*

6. During discussions on the Montreal Protocol, skeptics raised the argument that CFCs are heavy molecules (average MW = 90 - 120), and since they are heavier than air (MW = 29), they would pool at the bottom of the atmosphere and therefore could not be causing problems in the stratosphere. Why is this argument false?

7. Why are HCFCs considered a temporary (or stop gap) replacement for CFCs?

8. Find the latest spring's size and minimum concentration of stratospheric ozone at the North and South Pole.*

9. Make a list of three questions you have about this chapter or air pollution concerns you have.

12.4 Problems

1. Consider a typical summer sunny day in an urban area. Sketch the relative concentrations of NO_x, VOCs, and O_3 for a 24 hour period. Assume sunrise occurs at 6 am, sunset occurs at 8 pm, traffic maximums occur around 8 am and 6 pm, traffic minimum occurs between 2 am and 5 am, and that industrial emissions occur mainly between 6 am and 6 pm.

2. Calculate the amount of hydrocarbons and carbon monoxide emitted by a class III motorcycle over one year (8,000 miles). Compare the value to a car (model 2005, bin 4) as shown in chapter 16 Table 16-7.

3. The environmental impact statement for a proposed coal fired power plant shows that it would increase ambient amounts of NO_x and CO by 4 ppb and 7 ppb in the nearby urban area. Neither value would cause the area to exceed NAAQS, nor would they exceed PSD

levels for these pollutants (see chapter 3). However, the area is in non-attainment for ozone (8 hour standard). Would this facility be allowed to be built?

4. Determine the number of automobiles with PremAir® catalyst treated radiators needed to reduce the ozone concentration in an urban area by 10% in 1 day. Assume the urban area covers 500 square miles, that there is little wind to add or remove ozone, that there is an inversion height of 2000 ft, that the air is well mixed so there is no spatial variation in concentrations. The initial concentration of ozone is 90 ppbv. Each car can treat 750,000 ft^3/hr of air removing 70% of the ozone. If there are 300,000 cars operating at any given time, what percentage need to have the coating? In your opinion, what would be an acceptable cost for coating a radiator?

5. Calculate the emission rates in units of 10^{12} molecules/(cm^2·sec) for NO$_x$ and VOC (or NMOG) for a car that just meets the emission standards (NO$_x$ standard is 0.07 g/mile, NMOG standard is 0.075 g/mile, see Chapter 16 Table 16-7 for more detail). Assume the car is traveling at a highway speed of 60 miles per hour and emits over an area the width of the car and distanced traveled.

6. An urban area has a NO$_x$ emission rate that is 0.3 x 10^{12} molecules/(cm^2·sec) and the VOC emission rate is 30 x 10^{12} molecules/(cm^2·sec). Use Figure 12-5 to estimate the O$_3$ concentration. What would happen to the O$_3$ if a) the NO$_x$ emission rate was reduced by 25%, b) the VOC emission rate was reduced by 25%, c) both were reduced by 25%?

7. Assume Figure 12-5 represents the actual O$_3$ concentrations of an urban area. If the NO$_x$ emission rate is 0.4 x 10^{12} molecules/(cm^2·sec) and the VOC emission rate is 9 x 10^{12} molecules/(cm^2·sec), suggest a strategy to reduce O$_3$ to below ambient air quality standards.

8. How far would a polar stratospheric cloud particle fall during winter (100 days)? Assume the particulate diameter is 1 um, and that the settling or terminal velocity is approximated in figure 7-8. Would this be enough for a mid-stratospheric particulate to reach the tropopause?

12.5 Group Project Ideas

For each project, the students should work in small groups and present their finding in either a short report (5-8 pages) or 15 minute presentation.

1. There have been several attempts by the US-EPA to regulate ozone, including the Clean Air Interstate Rule (2009) and the Cross-State Air Pollution Rule (2011). However, both

have been found inadequate by the US Court of Appeals. These are currently (2013) being appealed by the US Government. Summarize the problems with these rules and provide an update on the latest updates to this process.

2. There are several regions of the US that are in non-attainment for O_3. Locate one such region and explore why. Next, determine what the state is proposing to do to achieve compliance. Finally, critique the state plan – do you believe it will work in a timely manner?

12.6 Bibliography

Adams, W C. 2006. Comparison of chamber 6.6-h exposures to 0.04–0.08 PPM ozone via square-wave and triangular profiles on pulmonary responses. *Inhalation Toxicol*. 2006, 18, pp. 127-136.

—. 2003. Comparison of chamber and face mask 6.6-hour exposure to 0.08 ppm ozone via square-wave and triangular profiles on pulmonary responses. *Inhalation Toxicol*. 2003, 15, pp. 265-281.

—. 2002. Comparison of chamber and face-mask 6.6-hour exposures to ozone on pulmonary function and symptoms responses. *Inhalation Toxicol*. 2002, 14, pp. 745-764.

Alley, F C and Ripperton, L A. 1962. The Efect of Temperature on Photochemical Oxidant Production. *Journal of the Air Pollution Control Association*. 1962, Vol. 12, p. 464.

AQMD. 1997. The Southland's War on Smog: Fifity Years of Progess Toward Clean Air. *South Coast Air Quality Management District*. [Online] 1997. http://www.aqmd.gov/home/library/public-information/publications/50-years-of-progress.

Brasseur, Guy P, et al. 1999. *Atmospheric Chemistry and Global Change - Chapter 13*. New York, NY, USA : Oxford University Press, 1999. ISBN 0-19-510521-4.

Bringmark, E and Bringmark, L. 1995. Disappearance of spatial variability and structure in forest floors - a distinct effect of air pollution? *Water Air Soil Pollution*. 1995, 85, pp. 761-766.

Brook, R D, et al. 2004. Air Pollution and Cardiovascular Disease: A Statement for Healthcare Professionals From the Expert Panel on Population and Prevention Science of the American Heart Association. *Circulation*. 2004, Vol. 109, pp. 2655-2671.

Canada, Enviroment. 2013. Global Ozone Maps. *Enviroment Canada Ozone and UV Monitoring*. [Online] July 5, 2013. [Cited: July 5, 2013.] http://exp-studies.tor.ec.gc.ca/ozone/images/graphs/gl/current.gif.

Chapman, S. 1930. A Theory of Upper-Atmosphere Ozone`. *Memoirs of the Royal Meteorological Society*. 3, 1930, Vol. 26, pp. 103-125.

Farman, J C, Gardiner, B G and Shanklin, J D. 1985. Large losses of total ozone in Antarctica reveal seasonal ClO_x/NO_x interaction. *Nature*. 1985, Vol. 315, pp. 207-210.

Husar, R B and Renard, W P. 1998. Ozone as a function of local wind speed and direction: Evidence of local and regional transport. *91st annual meeting and exhibition of the Air & Waste Management*. June 1998.

Kurczynska, E U, et al. 1997. The influence of air pollutants on needles and stems of scots pine (Pinus Sylvestris L.) trees. *Environ. Pollut*. 1997, 98, pp. 325-334.

McLaughlin, S and Percy, J. 1999. Forest health in North America: some perspectives on actual and potential roles of climate and air pollution. *Water Air Soil Pollution*. 1999, 116, pp. 151-197.

Molina, M J and Rowland, F S. 1974. Stratospheric Sink for Chlorofluoromethanes: Chlorine Atom-Catalysed Destruction of Ozone. *Nature*. 1974, Vol. 249, pp. 810-812.

Murphy, J J, et al. 1999. The cost of crop damage caused by ozone air pollution from motor vehicles. J. Environ. Manage. 1999, Vol. 55, pp. 273-289.

NIOSH. 2011. *Pocket Guide to Chemical Hazards*. Washington, DC, USA : US-Center for Disease Control, 2011. http://www.cdc.gov/niosh/npg/.

Regan, J D and Yoshida, H. 1995. DNA UVB Dosimeters. *Journal of Photochemistry and Photobiology B: Biology*. November, 1995, Vol. 31, 1-2, pp. 57-61.

Schichtel, B A and Husar, R B. 2001. Eastern North American transport climatology during high- and low-ozone days. *Atmos. Environ.* 35, 2001, pp. 1029-1038.

UNEP/GRID-Arendal. 2007. Effects of the Montreal Protocol amendment and their phase-out schedules. [Online] UNEP/GRID-Arendal Maps and Graphics Library, 2007. [Cited: March 24, 2011.] http://maps.grida.no/go/graphic/effects-of-the-montreal-protocol-amendment-and-their-phase-out-schedules.

US-EPA. 2006a. *Air Quality Criteria for Ozone and Related Photochemical Oxidants (2006 Final)*. Research Triangle Park, NC, USA : National Center for Environmental Assessment-RTP Office, 2006a. EPA 600/R-05/004aF.

—. 1996. *Air quality criteria for ozone and related photochemical oxidants*. Research Triangle Park, NC, USA : Office of Research and Development, 1996. EPA 600/AP-93/004aF-cF. 3v..

—. 2010a. Basic Information. *Ground-level Ozone*. [Online] US EPA, January 7, 2010a. [Cited: March 16, 2011.] http://www.epa.gov/air/ozonepollution/basic.html.

—. 2011c. Diesel Boats and Ships. *Nonroad Engines, Equipment, and Vehicles*. [Online] EPA's Office of Transportation and Air Quality, February 9, 2011c. [Cited: March 16, 2011.] http://www.epa.gov/otaq/marine.htm.

—. 2001. *Emission Standards for New Nonroad Engines*. Washington, DC, USA : US-EPA Office of Transportation and Air Quality, 2001. EPA 420-F-01-026.

—. 2006b. *Figure AX2-22 in Air Quality Criteria for Ozone and Related Photochemical Oxidants Volume II of III*. Washington DC, USA : US Government Printing Office, 2006b. pp. AX2-135. EPA 600/R-05/004bF.

—. 2011a. Forklifts, Generators, and Compressors (gasoline/propane). *Nonroad Engines, Equipment, and Vehicles*. [Online] EPA Office of Transportation and Air Quality, January 12, 2011a. [Cited: March 16, 2011a.] http://www.epa.gov/otaq/largesi.htm.

—. 2010b. History of Ground-level Ozone Standards. *Ground-level Ozone*. [Online] US-EPA, September 16, 2010b. [Cited: March 16, 2011.] http://www.epa.gov/glo/history.html.

—. 2011b. Locomotives. *Nonroad Engines, Equipment, and Vehicles*. [Online] EPA Office of Transportation and Air Quality, January 6, 2011b. [Cited: March 16, 2011.] http://www.epa.gov/otaq/locomotives.htm.

—. 2010. Ozone Layer Protection. [Online] US Environmental Protection Agency, December 15, 2010. http://www.epa.gov/ozone/defns.html.

—. 2010c. Ozone-depleting Substances. *Ozone Layer Protection - Science*. [Online] August 19, 2010c. [Cited: March 28, 2011.] http://www.epa.gov/ozone/science/ods/index.html.

—. 2011d. Significant New Alternatives Policy (SNAP) Program. *Ozone Layer Protection - Alternatives / SNAP*. [Online] US-EPA, MArch 23, 2011d. [Cited: March 28, 2011.] http://www.epa.gov/ozone/snap/.

US-NASA. 2013. Annual Records. *Ozone Hole Watch*. [Online] National Aeronautics and Space Administration, May 31, 2013. [Cited: July 8, 2013.] http://ozonewatch.gsfc.nasa.gov/meteorology/annual_data.html.

Vacek, S, Bastl, M and Leps, J. 1999. Vegetation changes in forests of the Krkonose Mountains over a period of air pollution stress (1980-1995). *Plant Ecology*. 1999, 143, pp. 1-11.

Victoria, SunSmart. 2011. SunSmart. [Online] February 2011. [Cited: March 28, 2011.] http://www.sunsmart.com.au/.

WHO. 2006. *Air quality guidelines - global update 2005*. Geneva, Switzerland : WHO Press, 2006. WHO SDE/PHE/OEH/06.02.

WHO-Europe. 2000. *Air quality guidelines for Europe, 2nd ed*. Copenhagen, Netherlands : WHO Regional Publications, 2000. European Series, No. 91.

—. 1987. *Air quality guidelines for Europe*. Copenhagen, Netherlands : WHO Regional Publications, 1987. European Series, No. 23.

WMO. 2011. *Scientific Assessment of Ozone Depletion: 2010, Global Ozone Research and Monitoring Project*. Geneva, Switzerland : World Meteorological Organization, 2011. Report No. 52.

CHAPTER **13**

Hazardous Air Pollutants

Air toxics, also known as hazardous air pollutants (HAPs), are those pollutants known to or suspected of causing cancer or other serious health problems. These health concerns are associated with both short- and/or long-term exposures. While HAPs are in many ways similar to the criteria pollutants, they are not discharged in large volumes, and they are limited to just a few locations. It also has not been possible to determine a safe level in ambient air, either because of very limited health data or the exact composition of chemicals within the emissions are unknown (such as oil refineries and waste incinerators). Because of these issues, *regulations focus on emissions sources rather than ambient air quality* and may include specific technology, specific work practices, or levels of emission control.

The US-EPA identified 187 substances plus diesel exhaust as HAPs as part of the 1990 Clean Air Act Amendments (CAAA), see appendix 5 for the complete list of these substances. The US-EPA also began identifying and regulating categories of major industrial sources for these HAPs in 1992. The source category list is constantly updated and expanded by the US EPA – see the Air Toxics web site (EPA, 2010) for the latest lists.

The US EPA excludes five of the six criteria pollutants from the list of HAPs because the Clean Air Act addresses them specifically. These criteria pollutants are particulate matter (PM), nitrogen oxides (NO_x), sulfur oxides (SO_x), ozone (O_3), and carbon monoxide (CO). The only substance listed as both a criteria pollutant and HAP is lead (Pb). Lead and two other HAPs (benzene and mercury) are the subjects of focused study in this chapter.

13.1 Hazardous Air Pollutants - General Information

13.1.1 Emission Sources

Categories of air toxics include stationary, mobile, background, and secondary formation. *Stationary sources* include large and small industrial facilities, business and commercial enterprises, municipal works, and residential areas. *Mobile sources* include cars, trucks, trains, planes, ships, and construction equipment. *Background sources* result from the long-range transport from other parts of the world and may include natural sources like fires, forests, and volcanoes. Secondary formation sources result from other pollutants that have been emitted to the air and then chemically react in the environment. Secondary formation generates the largest and the most harmful fraction of these substances.

Stationary source emissions are either major or area sources.
- "Major" sources are defined as sources that emit 10 tons per year of any of the listed toxic air pollutants, or 25 tons per year of a mixture of air toxics. These sources may release air toxics from equipment leaks, when materials are transferred from one location to another, or during discharge through emission stacks or vents
- "Area" sources consist of smaller-size facilities that release less than 10 tons per year of a single air toxic, or less than 25 tons per year of a combination of air toxics. Though emissions from individual area sources are often relatively small, collectively their emissions can be of concern - particularly where large numbers of sources are located in heavily populated areas.

Mobile source air toxics are compounds emitted from on-road and off-road equipment that are known or suspected to cause cancer or other serious health and environmental effects. They include both directly emitted compounds and the secondary pollutants that result from their release. Examples of mobile source air toxics include benzene, 1,3-butadiene, formaldehyde, acetaldehyde, acrolein, polycyclic organic matter (POM), naphthalene, diesel exhaust organic gases, and diesel particulate matter.

Background sources are compounds that exist in air that is remote from the impact of mobile or stationary sources and is considered clean. Several HAPs have been identified in these locations. Their concentrations are assumed to be constant across the entire US, mostly because there is insufficient data to do otherwise. Table 13-1 lists the background concentrations for the 12 HAPs for which this information is available (US-EPA, 2012a). These background levels are assumed to be from natural sources, generally microbial activity or from plants.

Secondary formation sources, also called atmospheric transformation, is a process by which chemicals in the environment are transformed into hazardous air pollutants through a series of chemical and biological reactions. HAPs can form as the original material reacts although other chemical species which are not hazardous may also form. These products may subsequently react to form other compounds. The level of hazard depends on the entire chain of transformation compounds and their atmospheric removal processes. The US-EPA maintains a data repository of volatile organic gas and particulate matter (PM) speciation profiles of air pollution sources. It provides information on the compositions of source materials and their transformation products (SPECIATE, 2012). Over 40 percent of the listed HAPs have atmospheric lifetimes of less than one day. The atmospheric transformation products of these HAPs are more likely material for human and environmental exposure, especially in cases where the transformation products are more toxic than the HAP itself.

Table 13-1. US Background Concentrations of Several Hazardous Air Pollutants (HAPs).

HAP	Background Concentration [μg/m3]
Benzene	0.48
Carbon tetrachloride	0.88
Chloroform	0.083
Ethylene Dibromide	0.0077
Ethylene Dichloride	0.061
Formaldehyde	0.25
Hexachlorobenzene	0.00009
Methylene Chloride	0.15
Polychlorinated Biphenyls (PCBs)	0.00038
Tetrachloroethylene	0.14
Trichloroethylene	0.081
Xylene	0.17

Examples of secondary HAPs are aromatic, nitrogenated, and oxygenated compounds. Almost every hydrocarbon photo-oxidation reaction produces formaldehyde and acetaldehyde (see Chapter 12). Estimates in urban areas show in situ formation contributes as much as 85 percent of the ambient levels of formaldehyde and 95 percent of acetaldehyde, (Kao, 2012).

The amount of HAPs released into the atmosphere in the US has been steadily declining since detailed recordkeeping began after passage of the 1990 Clean Air Act Amendments. Table 13-2 shows the reported releases of HAPs from the largest sources for the 1990-93 to 2002 timeframe. An overall reduction of approximately 34% was reported (US-EPA, 2008).

Table 13-2. Quantities of HAPs released in the US.

Source	1990-1993 10^6 tons	2002 10^6 tons	Reduction %
Major	2.6	0.9	65
Area	1.3	1.6	-23
On-road	2.6	1.4	46
Off-road	0.8	0.9	-13
Total	7.3	4.8	34

13.1.2 Health and Welfare

People exposed to toxic air pollutants at sufficient concentrations and durations have an increased chance of cancer, immune system damage, neurological, reproductive (e.g., reduced fertility), developmental, respiratory, or other health problems. Exposure to these substances at sufficient concentrations and durations may also cause health problems for plants and animals. The harmful level may be very different than the concentrations that cause harm to humans.

The US EPA publishes a National Air Toxic Assessment (NATA) report every three years, (US-EPA, 2012a). The 2005 NATA, originally published in 2011, assessed 178 of the 187 HAPs (plus diesel particulate matter), the remaining compounds had insufficient information to include in the report. The 2005 report is the latest report as of 2015. The assessment assigns a risk for each substance in terms of health effects (cancer and non-cancer), number of citizens exposed, and geographic extent of exposure (national or regional). The risk assessment includes stationary, mobile, and secondary sources, and excludes background sources. The risks estimated do not consider ingestion exposure or indoor sources. The assessment only includes estimates of chronic cancer risks for those air toxics that are known to cause cancer and whose risk the US EPA can quantify with available dose-response data. Therefore, the risk estimates represent only a subset of the total potential cancer risk associated with air toxics. The assessment is intended to be used as one tool in prioritizing actions for minimizing the health effects from air toxics.

The assessment calculates the probability (risk) that a person could contract cancer from breathing air toxics when exposed to 2005 emission levels for 70 years. A risk level of 1 in a million implies a likelihood that one person, out of one million equally exposed people, would contract cancer if exposed continuously (24 hours per day) to that specific concentration over 70 years (an

Table 13-3. 2005 NATA Health Effects Drivers and Contributors Risk Characterization.

Risk Characterization Category	Risk Exceeds (per million)[1]	HI > 1.0[2]	Millions of People Exposed (minimum)
National Cancer Driver	10		25
Regional Cancer Driver	10		1
Regional Cancer Driver	100		0.01
National Cancer Contributor	1		25
Regional Cancer Contributor	1		1
National Noncancer Driver		1.0	25
Regional Noncancer Driver		1.0	0.01

1. Cancer risks are upper-bound lifetime cancer risks (i.e., a plausible upper limit to the true probability that an individual will contract cancer over a 70-year lifetime, as a result of a given hazard (such as exposure to a toxic chemical)). This risk can be measured or estimated in numerical terms (e.g., one chance in a million).
2. HI = the sum of hazard quotients for substances that affect the same target organ or organ system. Because different pollutants may cause similar adverse health effects, it is often appropriate to combine hazard quotients associated with different substances to understand the potential health risks associated with aggregate exposures to multiple pollutants.

assumed lifetime). This risk would be an excess cancer risk that is in addition to any cancer risk borne by a person not exposed to these air toxics. Table 13-3 shows the definitions of the various risk categories assigned to each substance based on the assessment.

The assessment identified a few air toxic substances as having national or regional health effects, see Table 13-4, (US-EPA, 2012a). The assessment also showed that:
- Urban areas tend to have higher cancer and non-cancer risk estimates than rural areas. The increased risk is most likely caused by the greater number and quantity of sources. Urban air also tends to have higher levels of secondary formation because of the greater amounts and types of emitted pollutants.
- The average additional national cancer risk due to HAPs is estimated to be 50 per million people.
- About 5% (3,100/66,000) of census tracts had enhanced cancer risks of greater than 100 per million.

Table 13-4. 2005 National Air Toxics Assessment Results.

National cancer risk driver	Formaldehyde
Regional cancer risk drivers	Benzene
	PAHs
	Naphthalene
National cancer risk contributors	1,3-Butadiene
	Arsenic compounds
	Chromium compounds
	Coke oven emissions
	Acetaldehyde
	Acrylonitrile
	Carbon tetrachloride
	Ethylene Oxide
	Tetrachloroethylene
	1,4-Dichlorobenzene
	Ethylbenzene
Regional cancer risk contributors	Nickel compounds
	1,3-Dichloropropene
	Methylene chloride
National noncancer hazard drivers	Acrolein
Regional noncancer hazard drivers	2,4-Toluene diisocyanate
	Chlorine
	Diesel PM
	Hexamethylene diisocyanate
	Hydrochloric acid
	Manganese compounds

A sense of perspective of this risk can be obtained by comparison with historical data that shows approximately 1 out of every 3 Americans (or 336,000 in a million) contracts cancer during their lifetime when all causes are taken into account, see Table 13-5. This table shows the leading causes of death in the US for 2009, (US-CDC, 2009). These values will be different for different countries or for the global population. The table also shows the proportion of all deaths for a given cause as well as the risk of any person dying from a given cause in any given year for every million people in a population. For example a 10^{-6} increased cancer risk represents an increase risk of 1 in 1,000,000 for developing cancer and would put the risk at 1,850 cancer deaths per million people per year, rather than the 1,849 deaths that would be expected in an unexposed population. The reported risk assumes that the odds of a particular cause are equally distributed throughout the population. This last assumption ignores factors such as gender, race, age, and pre-existing conditions, which are known to alter individual risk levels. It is interesting to note that none of the leading causes are directly attributed to environmental pollution, yet such pollution is a known contributing factor. The World Health Organization has estimated that 1 in 8 deaths worldwide can be attributed to air pollution, either indoor or ambient (WHO, 2014).

There are approximately 1.5 million diagnosed cancer cases per year in the US, of which about one-third die from their disease, (Xu, et al., 2010). Of these cancer deaths, almost one-third can be attributed to tobacco use alone, and another third can be related to lifestyle factors such as poor nutrition, physical inactivity, and obesity. The inhalation of air toxics contributes less than 0.1% of the risk of contracting cancer. For comparison purposes, the national risk of contracting cancer from radon exposure (a naturally occurring air toxic, see Chapter 15) is about 2,000 in a million. Cancers from air toxics are generated, on average, at a rate of about 215 per year nationwide, (and lead to about 70 deaths/yr). As this is a statistical trend, the identities of these cases will never be known.

Table 13-5. Causes of Death and Associated Risk in US, 2009.

Causes (2009)	Total Number	% of all deaths	Risk/10^6/year
All Causes	2,437,163	100.00	7,938
Heart Disease	599,413	24.59	1,952
Cancer	567,628	23.29	1,849
Respiratory	137,353	5.64	447
Stroke	128,842	5.29	420
Accidents	118,021	4.84	384
Alzheimer's	79,003	3.24	257
Diabetes	68,705	2.82	224
Influenza	53,692	2.20	175
All Other	684,506	28.09	2,229

Besides direct exposure from breathing, ingesting, or absorption through the skin, some toxic air pollutants are deposited onto soils or surface waters, where they are taken up by plants or ingested by animals. These materials may be destroyed or altered by these organisms, and some substances, like mercury, are stored in the organism's tissue. If this organism is consumed, the consumer also takes in the stored contamination. The contaminant can accumulate to greater concentrations than in the contaminated prey. This *bio-accumulation* process provides a pathway for these consumers to have much greater exposures than expected from the concentrations in the water, air, or soil.

Bioaccumulative chemicals are not easily metabolized and are more soluble in fatty tissue than in water. This preferential solubility allows these substances to accumulate in human or ecological food chains through consumption and uptake. The six EPA listed bioaccumulative species include dioxins, furans, chlorinated solvents (trichloroethane), polycyclic aromatic hydrocarbons (anthracene and naphthalene), and mercury.

13.1.3 History and Regulations

The original 1970 CAA controlled toxic air pollutants on a species basis. However, from the time the act was passed through 1990, only seven substances had been reviewed and regulated (Arsenic, Asbestos, Benzene, Beryllium, Mercury, Radionuclides, and Vinyl Chloride). The amount of effort to make a determination of the safe levels of each contaminant was found to be much more challenging than originally estimated, both in terms of time and expense.

Throughout the 1970's and 80's it was found that the centralized command and control structure design used for the criteria pollutants was unworkable and ineffective for HAPs. The nature of the sources varied tremendously between pollutants and geographic regions. While one region may have a problem with HAP from a particular source category and have need for regulation, it was determined to be unfair to regulate those sources in other regions where that HAP was not a problem. After it was determined that national air quality standards were inappropriate for HAPs, Congress significantly altered US-EPA's authority to regulate hazardous air pollutants. The 1990 Clean Air Act Amendments directed the US-EPA to develop regulations for all industries that emit one or more of the listed HAPs in significant quantities and to regulate emissions of air toxics from a published list of industrial sources (source categories).

Section 112 of the CAAA lists the HAPs to be regulated by source category. The rules focus on controlling emissions rather than setting ambient air quality guidelines. A "National Emissions Standards for Hazardous Air Pollutants" (NESHAPs) is issued to limit the release for specific industrial sectors. The emission control levels required by NESHAPs are technology-based, meaning that they require the best available control technology (BACT) an industrial sector could afford. The levels are not based on health risk considerations because the safe level of allowable releases and ambient air concentrations are unknown or cannot be determined with an acceptable degree of certainty. This significant difference between NESHAPs and NAAQS means that

emissions regulation is by each industrial source category, which can have very different limits than another category. Examples of source categories include:
- Electric utility steam generating units,
- Steel production and manufacture,
- Chlorine production,
- Goldmine ore processing and production,
- Rubber tire manufacturing,
- Dry cleaning,
- Plywood and composite wood products,
- Manufacturing of nutritional yeast,
- Paint stripping operations, and
- Rocket testing facilities.

The decision of what emission level is acceptable involves four provisions:
1) development of Maximum Achievable Control Technology (MACT) standards for each of the listed 188 pollutants for all specified categories of sources,
2) health-based standards, when available,
3) standards for stationary area sources, and
4) prevention of catastrophic releases.

MACT

MACT standards are developed for each source category. It is done on a source by source basis because each pollutant in each industry has its own unique conditions. MACT is set by assessing the current controls used in an industrial source and looking at the best performing control technologies (usually the top 12% in terms of reduction efficiency). These emissions are then used to set a baseline called the MACT floor. The actual MACT standard must, at a minimum, achieve this baseline level. Existing sources are given three years after publication of standards to achieve compliance, with a possible one-year extension; additional extensions may be available for special circumstances or certain categories of sources. Existing sources that achieve voluntary early emissions reductions receive a six-year extension for compliance with MACT.

More stringent standards may be set when it makes sense in terms of economic, environmental, and public health benefit. EPA is required to revise the standards periodically, at least every eight years. EPA can, on its initiative or in response to a petition, add or delete substances or source categories from the lists. Section 112 establishes a presumption in favor of regulation for the designated chemicals. It requires regulation of a designated pollutant unless EPA or a petitioner can show *"that there is adequate data on the health and environmental effects of the substance to determine that emissions, ambient concentrations, bioaccumulation or deposition of the substance may not reasonably be anticipated to cause any adverse effects to human health or adverse environmental effects."*

Health-Based Standards

The second provision requires the EPA to address situations in which a significant health risk remains after implementing MACT. The EPA must report to Congress on these residual risks, and recommend legislation. If Congress does not legislate in response to these recommendations, the EPA is required to issue standards to protect public health. A residual risk standard is required for any cancer-causing pollutant that poses an added risk to the most exposed person of more than one-in-a-million. Residual risk standards are required eight years after publishing final MACT rules for a source category. In general, residual risk standards do not apply to area sources.

Stationary Area Sources

The third provision requires standards for stationary "area sources" defined as small, but numerous sources emitting one or more HAPs. Examples of area sources include gas stations, dry cleaners, and bakeries, which collectively emit significant quantities of hazardous pollutants. The provision requires the EPA to regulate the stationary area sources together. The standards often take the form of 'best practices.' These sources are responsible for up to 90% of the emissions of the 30 most hazardous air pollutants in urban areas. In setting such standards, the EPA can impose less stringent "generally available" control technologies, rather than requiring MACT. An example of such control technology is that all storage containers must have a lid.

Prevention of Catastrophic Releases

Finally, Section 112 addresses prevention of sudden, catastrophic releases of air toxics by establishing an independent Chemical Safety and Hazard Investigation Board. The Board is responsible for investigating accidents involving releases of hazardous substances, conducting studies, and preparing reports on the handling of toxic materials and measures to reduce the risk of accidents. The EPA is also directed to issue prevention, detection, and correction requirements for catastrophic releases of air toxics by major sources. The law requires owners and operators to prepare risk management plans which include hazard assessments, measures to prevent releases, and a response program.

The regulatory responsibility for controlling air toxics is shared among federal, state, local and tribal air programs. The US-EPA sets the national standards for air toxic emissions. The state, local, and tribal programs are responsible for implementing these rules through the State Implementation plan (SIP) program. Additional rules at the local level may exist in some areas.

The federal standards are the minimum set for the country. They cover over 96 categories of major stationary sources. The US EPA provided notice on March 21, 2011, that they have completed the emission standards required by the CAA (Jackson, 2011). The rules provide standards for 90 to 100 percent of the emissions from urban air toxic pollutants, as well as 90 percent of the six potentially bio-accumulative toxic pollutants. These standards are projected to reduce annual air toxics emissions by about 1.7 million tons.

The federal standards also cover mobile source air toxics (MSAT). These regulations are applied only to new vehicles. Improvements in air quality may take several years from the beginning of their implementation and occur as older vehicles are replaced with cleaner ones that meet new emission standards. Most of the reductions in air toxics occur as a secondary benefit from the regulations to control particulate matter, carbon monoxide, and ozone (which is controlled by reducing volatile organics and nitrogen oxides). EPA estimates that these programs will result in a 65 percent reduction in gaseous air toxics from highway mobile sources between 1999 and 2030, despite large increases in vehicle miles traveled. An example of a program aimed specifically at air toxics is the 'Control of Hazardous Air Pollutants from Mobile Sources' (US-EPA, 2007). This program is designed to lower emissions of benzene and other air toxics in three ways: (1) by lowering the benzene content of gasoline, beginning in 2011; (2) by reducing exhaust emissions from passenger vehicles operating at cold temperatures (under 75 degrees), beginning in 2010; and (3) by reducing emissions that evaporate from, and permeate through, portable fuel containers, beginning in 2009.

13.1.4 Control Strategies

Actual control strategies depend on the type, amount, and concentration of air toxics and/or their precursors in the emissions. There are many methods available for the reduction of emissions.

Before use: Product substitution, product replacement, product minimization, and source blending (see chapter 17).
During use: Product minimization, reduced concentration, product reuse, and improved maintenance (see chapter 17).
After use: Product reuse, material recovery, and recycling.
Emission Treatments:
    PM: Cyclones, Filtration, ESP (see Chapter 8)
    Organics: Adsorption, Absorption, Incineration, Condensation (see Chapter 11)
    Minerals: Wet scrubbing, Adsorption (see Chapters 8, 9 and 11)

The following three sections explore three air toxics and their treatment in the US: benzene, lead, and mercury. These three have been the subject of much study (in the US and internationally) and have been shown to have significant effects on humans and the environment. Benzene and mercury are hazardous air pollutants. Lead is regulated as a criteria pollutant and as a hazardous air pollutant and is the only pollutant treated as both in the US.

13.2 Benzene

13.2.1 General Information

Benzene (C_6H_6) is a volatile aromatic organic chemical compound. The six carbons are in a ring pattern, where each carbon atom has one carbon-carbon double bond, one carbon-carbon single bond, and one bond to a hydrogen atom. The shape is much like a hexagon with hydrogen atoms attached to each corner. It is a colorless, sweet smelling, highly flammable liquid. It was first isolated and identified in 1825 by Michael Faraday, who gave it the name bicarburet of hydrogen. He produced it by collecting the lightest fraction during distillation of coal tar.

13.2.2 Sources, Sinks, and Ultimate Fate

Benzene is a natural component of petroleum and gasoline (3 to 9%), and crude oil is the main source worldwide. It is also emitted to the atmosphere from volcanoes and forest fires. It is a constituent in tobacco smoke (30 – 60 µg/cig). Benzene is widely used in the United States and ranks in the top 20 chemicals for production volume. Industrial production began in 1849, and current production is 8 billion gallons (54 million tons, density 0.8765 g/cm^3) worldwide and 1.8 billion gallons in the US during 2010.

While benzene is a component in oil, there is not enough to satisfy the demand. Other components in crude oil can convert into benzene. Current production from crude oil is by three pathways. The first is cracking (breaking up molecules into smaller ones) of the naptha (C5-C10) and gas oil (C14-C20) fractions into a raw pyrolysis gasoline, which is rich in benzene and other aromatics. The second pathway is from reforming (rearrangement of atoms in a molecule) of the naptha fraction. The third pathway is the disproportionation of methylbenzene (toluene) which is converted to xylene and benzene. None of these production processes are solely dedicated to benzene production and the method chosen depends on how the producer refines the crude oil that is available. Major US benzene producers include BP, ExxonMobil, Flint Hills resources, LyondellBasell, Shell, and Total (ICIS, 2013).

Benzene has several industrial uses including as a solvent for chemical and pharmaceutical industries and as a raw material or an intermediate in the synthesis of other chemicals. It is part of the synthesis of *ethylbenzene* which is used to make styrene and polystyrene; *cumene* which is used to produce phenol which is another chemical intermediary compound; *cyclohexane* which is used to form adipic acid which is used to make nylon; *nitrobenzene* which is used to produce aniline and explosives (TNT); *detergent alkylates* which act as surfactants in detergent production; and *chlorobenzene* which is used to manufacture insecticides. Its use as a solvent has decreased as less hazardous solvents are identified. Benzene was added to gasoline as an octane enhancer when tetra-ethyl lead was first phased out. It is now limited in gasoline to concentrations of less than 0.62 vol%.

When released into the environment, it quickly enters and mixes with air. Its vapors are easily transported in the atmosphere where it can be destroyed or be absorbed into water and then deposited onto the ground. Destruction reactions typically occur by reaction with hydroxyl radicals (OH·). The reaction speed is increased in the presence of nitrogen oxides as part of the smog formation cycle. Benzene is somewhat soluble in water allowing some of it to be washed out of the atmosphere during precipitation events. The washed out benzene deposits on plants and soil where it may be sorbed and destroyed, re-volatilized, or it may remain in the water phase and get carried into surface or ground waters. Benzene can be destroyed by some microorganisms in the soil or water, but the typical destruction mechanism is oxidation in the atmosphere. The expected lifetime in the atmosphere is on the order of several days and in water and soils of weeks. It can be readily absorbed by the lipid phase (fatty parts) of aquatic organisms. However, it is not known to buildup concentration levels (bioaccumulate) in plants or animals.

13.2.3 Health and Welfare

13.2.3.1 Human Health

Benzene is a known human carcinogen (causes cancer). It also causes other, non-cancer related health concerns in humans and animals. It has a human odor threshold of 1.5 ppm (5,000 $\mu g/m^3$).

Humans are exposed to benzene from inhalation of contaminated air, ingestion of contaminated water and food, and tobacco smoke. Inhalation exposure is highest near heavily traveled roadways, gasoline service stations, and oil refineries. Ingestion may occur anywhere that water or food has been contaminated and can occur in urban and rural areas. Tobacco cigarettes contain 30 – 60 µg benzene each. Smoking accounts for half of the exposure in the US. Inhalation from other sources accounts for 20% of exposures, with water and food ingestion accounting for the remainder.

Once benzene enters the body by inhalation about half leaves during exhalation. The other half passes through the lung lining and enters the bloodstream. Benzene exposure through ingestion and skin contact also enters the bloodstream. Once in the blood it quickly travels throughout the body and is temporarily stored in bone marrow and fat. The liver metabolizes it, and these products leave the body through the urine, typically within 48 hours.

Acute
Acute effects (single or short-term exposures) include drowsiness, dizziness, headaches, and unconsciousness. Ingestion may cause vomiting, dizziness, and convulsions. Animal studies of lethal

dosages (LD50), where 50% of the exposed population dies from the exposure, have been used to determine effects of different types of exposure. Tests involving rats, mice, rabbits, and guinea pigs show benzene has low acute toxicity from inhalation (LD50 > 20,000 mg-benzene/kg-body-mass), moderate acute toxicity from ingestion (500-5000 mg/kg), and low or moderate acute toxicity from dermal exposure (2,000 – 20,000+ mg/kg). (Duarte-Davidson, et al., 2001).

Chronic

Chronic (daily or long-term) non-cancer effects from inhalation cause blood disorders, bone marrow disorders, aplastic anemia, excessive bleeding, and damage to the immune system. It also causes reproductive and developmental toxicity, with some evidence suggesting structural and numerical chromosomal aberrations. Adverse effects on the fetus include low-birth weight, delayed bone formation, and bone marrow damage based on observation of animals exposed to benzene by inhalation (ATSDR, 2007).

Cancer risk from benzene exposure increases the incidence of leukemia (cancers in the blood forming organs). The US-EPA has calculated the increase in risk to an individual who is continuously exposed to benzene in the air for different concentrations, see Table 13-6.

Table 13-6. US-EPA Calculated Cancer Risk from Inhalation of Benzene.

Concentration of Benzene in Air $\mu g/m^3$	Increased Risk of Cancer over Lifetime
0.13 to 0.45	1×10^{-6}
1.0	2.2×10^{-6} to 7.8×10^{-6}
1.4 to 4.5	1×10^{-5}
13.0 to 45.0	1×10^{-4}

The inhalation reference concentration is 0.03 mg/m^3. The oral reference dose is 0.004 mg/kg/day. These reference concentrations and dosages represent an estimate of the exposure at or below which adverse health risks are unlikely to occur over a lifetime. These estimates are not direct estimators of the risk but are intended to serve as a reference point to gauge the potential effects.

13.2.3.2 Environmental Welfare

The environmental effects of benzene in air and water are similar to human effects. Land animals have similar exposure patterns from ambient air and water, and it causes similar reproductive problems, blood disorders, and cancer, at least among the studied mammals. Aquatic life, like fish and shellfish, can receive much higher exposures in contaminated water. It appears to cause reproductive problems, changes in their appearance, and shorten their lives. Plants exposed to benzene in the soil may have their growth slowed and have reduced reproductive rates (DEWHA, 2009).

13.2.4 History and Regulation

Benzene is not a criteria pollutant like particulate matter and sulfur dioxide, but is a hazardous air pollutant (HAP). It is regulated by OSHA for occupational air, by the US-EPA for water, and by the US-FDA for food, but has no outdoor air regulatory level (US-CDC, 2005)

Indoor Air - The indoor and occupational air regulations are NIOSH Recommended Exposure Limits (REL) is 0.1 ppm[1] 10 hour-time weighted average (TWA). This level indicates the maximum average concentration for up to a 10-hour work day during a 40-hour work week. OSHA Permissible Exposure Limits (PEL) is 1 ppm 8-hour TWA. This level must not be exceeded during any 8-hour shift of a 40-hour workweek.

Ambient Air -The US-EPA is allowed to restrict benzene because it is a hazardous air pollutant under section 112 of the Clean Air Act. The US-EPA does not have set national ambient air quality goals for benzene, but it has developed regulations designed to lower industrial emissions of benzene. In addition, regulations have been proposed that would control benzene emissions from industrial solvent use, waste operations, transfer operations, and gasoline marketing. Proposed rules would require gas stations to install new equipment restricting benzene emissions while storage tanks are being filled (Stage I controls, see section 16.6.3). The use of clean ("oxygenated") fuels was mandated as a means of reducing motor vehicle emission-related air pollutants, and the EPA predicts that this clean fuels program will decrease ambient benzene levels by 33%.

The US-EPA has developed a program specifically targeted at Mobile Source Air Toxics or (MSAT) which includes benzene. This program, called the Control of Hazardous Air Pollutants from Mobile Sources Final Rule, was promulgated in 2007. The rule lowers emissions of benzene and other air toxics in three ways: (1) lowering the benzene content of gasoline to be less than 0.62% by volume, begun in 2011; (2) reducing exhaust emissions from passenger vehicles operating at cold temperatures (under 75 degrees), begun in 2010; and (3) reducing emissions from portable fuel containers (begun in 2009). These standards are expected to reduce emissions of benzene by 61,000 ton per year when fully implemented. EPA estimates that these programs will result in a 65% reduction in benzene emissions from highway mobile sources between 1999 and 2030, despite large increases in vehicle miles traveled. In addition, as a result of non-road equipment emission controls, EPA estimates that benzene emissions from non-road equipment will be reduced by over 60% between 1999 and 2030, despite significant increases in activity (US-EPA, 2012b).

The US-EPA has monitored the ambient concentration of benzene since 1994. Figure 13-1 shows monitored concentration levels at 22 monitoring sites nationwide that have measurements since 1994, (US-EPA, 2010). There are currently 339 HAP monitoring sites within the US.

---

1. 1 ppm benzene = 3.19 mg/m^3

FIGURE 13-1. US National Air Quality Trend for Benzene Concentration [μg/m³].

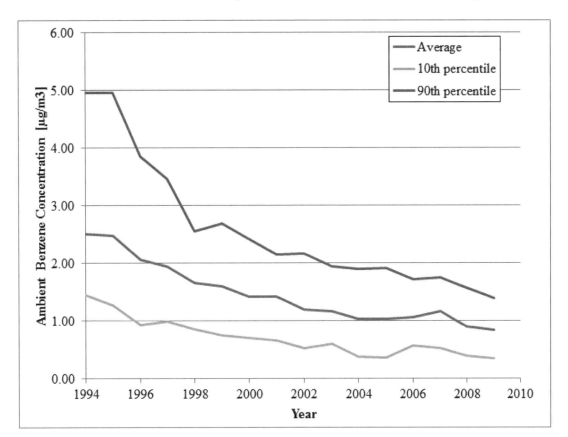

13.2.5 Control Strategies

Control of benzene emissions is typically approached by reducing the allowed emissions from large sources, producers, and users. The largest source of benzene occurs in petroleum refineries.

The US-EPA has focused on the reduction of benzene from gasoline. By controlling the amount of benzene in the fuel, less escapes while filling and less is present in the exhaust from automobiles. Benzene is a natural component of gasoline and is added as an octane enhancer. Most gasoline sold in the US has an octane rating between 85 and 91. Benzene has an octane rating of 106, whereas cyclohexane has an octane rating of 83. In most developed countries, the benzene content of gasoline is regulated to 1.0 vol% or less. The US-EPA's Mobil Source Air Toxics Phase 2 rule requires refiners and importers to the U.S. to achieve a benzene content of conventional, as well as reformulated gasoline to a corporate annual average of 0.62 vol% as of 2011.

Chapter 13

Reducing the content of benzene in gasoline requires most refiners to add benzene reduction facilities. While benzene is a natural component of crude oil, the refining process typically generates more during the stages of changing the heavier fractions into lighter ones or lighter fractions into heavier ones. Catalytic reforming is the source of about 70-85 vol% of gasoline benzene while fluid catalytic cracking accounts for another 10-25 vol%.

There are several methods that can be used to reduce the formation and overall concentration of benzene. These include:
- Removing benzene precursors from the oil refinery feed to the reformer. The precursors are cyclohexane, methyl cyclopentane, and n-hexane.
- Saturating benzene's carbon-carbon double bonds in various blends/ locations. Saturation requires the addition of hydrogen, which is attached to the ring while splitting the double bonds. Cyclohexane forms by addition of six hydrogen atoms.
- Solvent extraction of benzene from the reformate. Extraction removes much of the benzene that gets generated in the reformer. This benzene can be recovered and sold as a chemical feedstock rather than used in gasoline. The process is a liquid-liquid extraction using a glycol or proprietary solvent.
- Alkylating benzene to ethylbenzene or propylbenzene. These can both be used to enhance the octane of the fuel blend.

Another potential source of benzene emissions is from chemical industries that use it as a feedstock or chemical intermediary in their production of other materials. Benzene in such exhaust streams can be removed using gas-solid adsorption or incineration, both of which are described in detail in section 11.3.5 VOC Control Strategies. The choice of method depends on the actual concentration, variations in concentrations, and what other compounds makeup the exhaust.

■ Example 13-1.

Benzene removal from an air stream. Estimate the dimensions of an adsorption column needed to remove 125 ppm benzene from 25 m3/min air (at 30°C and 1.5 atm) using activated carbon (bed density = 450 kg/m^3). The Langmuir isotherm constants are a = 64.45 and b = 12.96 when pressure is in kPa, and the loading is in mol-benzene per kg of activated carbon. Assume the column runs for eight hours per day, and is then regenerated. The column height should be twice the diameter.

Solution:
Following the procedure from *section 11.3.5.1 Adsorption*, first calculate the flow rate of benzene.

$$Q_{Bz} = \left(25\frac{m^3}{min}\right)\left(\frac{1.5atm}{0.082\frac{l \cdot atm}{mol \cdot K}303.15K}\right)\left(\frac{1000l}{m^3}\right)\left(\frac{125}{10^6}\right)\left(78\frac{g}{mol}\right) = 14.7\frac{g}{min}$$

The partial pressure of benzene:

$$P_i^* = y_i * P_{total} = \frac{125}{10^6} * 1.5atm * 101.33\frac{kPa}{atm} = 0.019kPa$$

The loading of benzene on activated carbon:

$$W_{Benzene} = \frac{64.45\frac{mol}{kg_{carbon}kPa}(0.019kPa)}{1+(12.96kPa^{-1})*(0.019kPa)} = 0.983\frac{mol_{Benzene}}{kg_{carbon}}$$

The required volume of activated carbon for eight hours of operation

$$Vol_{bed} = \frac{14.7\frac{g}{min}(8hour)\left(\frac{60\min}{hour}\right)}{\left(78\frac{g}{mol}\right)\left(0.983\frac{mol}{kg_{carbon}}\right)\left(450\frac{kg_{carbon}}{m^3_{bed}}\right)} = 0.205m^3$$

Finally, calculate the bed geometry.

$$Vol = H*\frac{\pi}{4}D^2 = 2D*\frac{\pi}{4}D^2 = \frac{\pi}{2}D^3 \text{ and } D = \sqrt[3]{\frac{2Vol}{\pi}} = 0.507m$$

and $H = 1.014m$

The actual column must be somewhat bigger so that no benzene escapes due to bypass or poor packing, as well as to account for the inlet and outlet components. At least two columns are needed so the benzene can continue to be removed while one column is being cleaned.

13.3 Lead

13.3.1 General Information

Lead is a metallic chemical element (Atomic Weight = 207.2) in the carbon group (IVA). Its symbol is Pb, from the Latin name plumbum for soft metal. It has a high density (11.3 g/cm³), is ductile and malleable, and can react with organic chemicals. When it reacts, it typically forms the lead II ion

(Pb2+) but may also form a lead IV ion (Pb^{4+}). Lead has been used for thousands of years because it is available worldwide, easy to extract, and easy to work. The Romans had an estimated 80,000 ton annual production, mostly as a byproduct of silver smelting. It has been recognized as a poison since before 100 BC and lead poisoning is probably the oldest recorded occupational disease.

13.3.2 Sources, Sinks, and Ultimate Fate

Worldwide production of lead was 3.7×10^6 tons in 2007 (USGS, 2013). The top producers are China (1.4×10^6 tons), Australia (6.4×10^5) and the US (4.4×10^5). Lead is obtained from mined ores (primary production) or from recycled materials (secondary production). Worldwide annual use was 9×10^6 tons, of which nearly 60% was from recycled sources. The largest US source is located in the state of Missouri.

Primary production is from three minerals: galena (PbS) is the most common, cerussite (PbCO$_3$), and anglesite (PbSO4). It can be found in its metallic state very rarely. Production involves concentrating the mineral form, heating it, and then oxidizing it to form lead oxide, and finally it is reduced to form the metal, see Case Study 13-1.

The main uses in the US are in lead-acid batteries (88%), ammunition (4%), additives to glass and ceramics (3.3%), sheet lead (2%), and casting metal (1.3%). Lead was more common throughout the economy before environmental regulations resulted in substitutions with less harmful materials. Some of those uses included additives to gasoline, paints, and solder, and as piping for water systems. Each of these uses resulted in actual or potential exposure of humans to toxic levels of lead.

Lead is released to the environment as either a vapor or particle. The atmospheric residence time is on the order of weeks – long enough for worldwide tropospheric circulation. It is removed from the atmosphere by gravity settling (for particles) or washout with any of the water precipitation forms. Once it is deposited on the ground it may adhere to the soil (most likely), it may absorb into plant biomass, or it may transport into surface or ground water. Lead is an element so it cannot be broken down in the environment. Once removed from the atmosphere it remains in the biosphere until it is geologically sequestered by burial in sediments. There are no other known processes for lead removal from the environment. Lead that is absorbed by plants either return to the soil or get eaten. The consumer may absorb some fraction (15 – 85%) of the eaten lead which may be taken into the bones or be slowly excreted.

Source categories in the US are divided into on-road vehicles, metals processing, fuel combustion, non-road vehicles and engines, and other. Figure 13-2 shows the reported or estimated emissions from each of these categories since 1970. This figure demonstrates the excellent progress achieved in getting the lead out of the US economy.

FIGURE 13-2. US Reported Annual Lead Emissions [x10³ ton/year].

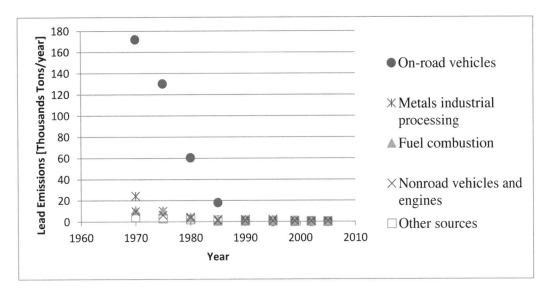

● **Case Study 13-1. Primary Lead Production.**
Primary lead production begins with sintering. Concentrated lead ore is fed into a sintering machine with iron, silica, limestone fluxes, coke, soda ash, pyrite, zinc, caustics and/ or pollution control particles. Sintering is used to join all these powdered materials into larger particle mixtures. The hot air is also used to burn off any sulfur in the lead ore. These larger particles (clinkers) are then sent to the smelter.

The smelter is a blast furnace that heats the clinkers to melting. The melt forms several layers within the furnace. The molten lead layer forms at the bottom of the furnace. Lighter minerals and elements, (e.g. arsenic and antimony) float to the top. This topmost layer is called speiss. A matte layer also forms from the copper and metal sulfides. Finally, a layer of blast furnace slag (mostly silicates) forms just above the lead. The speiss and the matte are usually sold to copper smelters and refined for the copper. The slag is stored and partially recycled if the metal content is sufficient or used as fill or aggregate for concrete or road asphalt. However, it may need to be placed into a controlled landfill if it contains leachable toxic materials.

The lead from the blast furnace, called lead bullion, then undergoes the drossing process. The bullion is agitated in kettles and cooled. This process results in molten lead and dross. Dross refers to the lead oxides, copper, antimony and other elements that float to the top of the lead. Dross is usually skimmed off and sent to a dross furnace to recover the non-lead components which are sold to other metal manufactures.

Finally, the molten lead is refined. Pyrometallurgical methods are used to remove the remaining non-lead components of the mixture that are then sold to other metal processing plants. The refined lead is either directly cast or it is made into alloys.

Primary lead production produces air emissions, process wastes, and solid wastes. Air emissions consist primarily of sulfur oxides (SO_x) and particulate matter (PM). Particle emissions from blast furnaces include lead oxides, quartz, limestone, iron pyrites, iron-limestone-silicate slag, arsenic and other metallic compounds. The exact type and amount of particles depends on the input material. The particulate matter emissions are usually controlled using a filter bag-house. This waste is combined with the process and other solid wastes. (US-EPA, 1995).

13.3.3 Health and Welfare

Environmental levels of lead have increased more than 1,000-fold over the past three centuries as a result of human activity. The greatest increase took place between 1950 and 2000 and reflected the use of leaded gasoline worldwide (ASTDR, 2011).

13.3.3.1 Human Health

Today, lead exposure in the US general population is mainly through eating and drinking of contaminated foodstuffs. Produce, meats, grains, seafood, soft drinks, and wine all may contain lead. Meat from hunting may contain unusually high levels if lead-containing ammunition was used. Drinking water usually has very low amounts, but these can be elevated if the water supply is acidic and leaches it from distribution equipment that has lead pipes, lead solder, or brass fixtures. Another source of lead is contaminated consumer products. In 2007, several million toys worldwide were recalled because the manufacturer used lead containing paint (Story, 2007).

Workers may become exposed to lead in their workplace, especially lead smelting and processing, rubber products, plastic manufacturing, steel welding and cutting, battery manufacturing, construction/demolition work, incinerators, pottery and ceramics workers, and anyone removing lead-based paints. The main pathway of exposure is inhalation of vapors, fumes, or dust. Lead is almost completely absorbed if it makes its way into the lower respiratory tract. Absorption through ingestion is usually very low, except for people who are fasting or dieting, and it can be very high for children in any case.

The occupational limits for airborne lead exposure are:
- NIOSH Recommended Exposure Limit for an 8 to 10 hour time-weighted-average exposure = 0.10 mg/m3.
- NIOSH Immediately Dangerous to Life or Health = 100 mg/m^3.
- OSHA Permissible Exposure Limit for an 8-hour work day = 0.5 mg/m^3.

The US Consumer Products Safety Commission (CPSC), under the Federal Hazardous Substance Act, has banned most "paint and surface coatings containing more than 0.06% lead, and furniture, toys and other articles intended for use by children that are coated with such paint." Lead in paint causes it to have a sweet flavor, encouraging children to eat the paint flakes and become poisoned with it. It is very important for parents to keep children away from flaking painted surfaces.

It appears that 1 µg/m^3 lead in air directly contributes approximately 19 µg lead per liter blood in children and about 16 µg per liter blood in adults, although the relative contribution from air is less significant in children than in adults. These values are approximations, recognizing that the relationships are curvilinear in nature and apply principally at lower blood lead levels.

There is no beneficial purpose for lead in the human body. Lead has been shown to affect virtually every major organ in the body. The most sensitive organs appear to be the nervous system (particularly in children), the circulatory system and the cardiovascular system. In addition, colic is a consistent early symptom of lead poisoning. Indeed, lead poisoning has been called painter's colic. In adults, symptoms of neurological effects include dullness, irritability, poor attention span, headaches, muscular tremors, loss of memory, and hallucinations. The condition may then worsen, sometimes abruptly, to delirium, convulsions, paralysis, coma and death. In children, many of the same symptoms occur along with hyperirritability and convulsions. There is a greater incidence of permanent neurological and cognitive impairments in children. Even at lower levels without the severe symptoms described above, there may be permanent damage. Estimates of blood-lead levels for children under five years old indicate that nearly nine percent, or approximately two million children in the US have blood-lead levels of 100 µg/liter or higher, which the Center for Disease Control considers to be the level indicating lead poisoning.

High levels of lead cause adverse effects on both male and female human reproductive functions. Lead is a teratogen that can cause fetal malformation, a mutagen that can affect both sperm and eggs, and a reproductive toxin that can impair fertility. Women who are exposed during pregnancy may experience miscarriages and stillbirths. Moderate blood-lead levels (400-500 µg/l) may affect sperm development and production in males.

13.3.3.2 Environmental Welfare

Once removed from buried minerals and ores, lead remains within the biosphere for very long times. The only removal mechanism is sequestration in sedimentary deposits. Lead in the biosphere resides primarily within and upon particles of soil. These may become airborne as dust, washed into surface waters with rain, leach into groundwater (especially in areas susceptible to acid rain), or they may be taken up by plants. Lead poisoning in plants causes slower growth and reduced viability. The lead within plants may be ingested by animals, where many of the same problems associated with humans may be observed. Lead harms aquatic fowl at blood levels near 0.2 ppm and doses above 1.2 ppm cause death.

13.3.4 History and Regulation

The major sources of lead emissions have historically been from fuels in on-road motor vehicles (such as cars and trucks) and industrial sources. Emissions from on-road vehicles decreased 99% in the US between 1976 and 1995 due primarily to the ban on adding tetra-ethyl lead to gasoline. Use of leaded gasoline in highway vehicles was phased out between 1976 and 1986, when manufacturers could no longer offer leaded gasoline for sale. It was outright prohibited from use on December 31, 1995, when users could no longer purchase it to add to gasoline on their own. The major sources of lead emissions to the air today are ore and metals processing and piston-engine aircraft operating on leaded aviation gasoline.

● Case Study 13-2. Lead in Gasoline.

Lead was added to gasoline in the form of tetraethyl lead (TEL) starting in the 1920's (Kitman, 2000). One of the major research questions for internal combustion engines at that time was how to increase engine efficiency. The main focus was to increase the amount of compression within the engine cylinders (see section 16.2). When compression was increased, the engine used less fuel and had increased power, but the air-fuel mixture could combust before the optimally timed spark ignited the mixture. This early ignition could be heard as a knocking or pinging sound, and it caused increased wear and tear on the engine, and reduced the engine's lifespan. TEL was found to have anti-knock capabilities in 1921 by Thomas Midgley Jr., a General Motors researcher who invented another environmentally disastrous compound – Freon (see section 12.2). TEL was first discovered in 1854 by a German chemist. It was noted as only a technical curiosity on account of its known deadliness. It is highly poisonous and even casual contact is known to cause hallucinations, difficulty in breathing, madness, spasms, *palsies, asphyxiation, and death. The poisoning may occur by inhalation, ingestion, or absorption through the skin of the liquid or its vapors. Indeed, each of the first production facilities of the 1920's saw several worker deaths from lead poisoning within a year of starting operation. Its chief rival for improving fuel combustion was ethanol, however a well-funded marketing campaign by GM, Standard Oil, and E.I. du Pont de Nemours & Company (which controlled a significant amount of GM's stock and owned several lead mines) saw TEL become the primary octane enhancer for gasoline by 1936 when 90% of gasoline sold in the US included TEL. Because of patents, GM and du Pont realized royalties of $0.03 per gallon of gasoline leading to annual profits of over $300 million from the use of leaded fuels. The US-EPA began a lead phase-out in 1973, but was sued by the makers of TEL. The standards were upheld by the US court of Appeals and the Supreme Court refused to hear the case in 1976. The primary phase-out began in 1976 and was completed by 1986. As an interesting side-note, President Reagan's administration proposed to relax or eliminate the lead phase-out in 1981, but reversed their opposition in 1982 due to public pressure and allowed the phase-out to continue. The European Union banned leaded gas in 2000.*

The highest levels of lead in air are near lead smelters. Other stationary sources are iron and steel manufacturers, waste incinerators, utilities, and lead-acid battery manufacturers. Combustion and smelting processes operate at high temperatures and emit submicron lead PM. Material handling and mechanical operations emit larger particles of lead as fugitive dust, which settles from the air near the source.

Lead is a criteria pollutant in the US. The US-EPA has set a National Ambient Air Quality Standard (NAAQS) for lead throughout the entire US. Table 13-7 lists the ambient air quality goals for several countries. The US level was changed from 1.5 to 0.15 in 2008 based on human health studies conducted during the previous decade. Many countries use the WHO guidelines for setting standards because they do not have the regulatory infrastructure or capabilities to conduct the necessary studies. It is interesting to note that Japan has not set an ambient level for lead. They instead regulate it as a hazardous air pollutant and set emission limits for producers. The US has also started to regulate lead more like a HAP rather than criteria pollutant because of the great successes in reducing it as an air pollutant for the majority of the country.

The US-EPA's primary tool for controlling lead emissions is the State Implementation Plan (SIP). Each state SIP must contain an enforceable control strategy to ensure attainment and maintenance of the NAAQS limits. A control strategy must be selected "that provides the degree of emissions reductions necessary for attainment and maintenance of the national ambient air quality standards. The emission reductions must be sufficient to offset any increases in air quality concentrations that are expected to result from emission increases due to projected growth of the

Table 13-7. Ambient Air Quality Standards for Pb in Several Countries.

Country	Standard [μg/m^3]	Time Period Average
US	0.15	3-month
EU	0.5	1-year
Japan	No Standard	
China (I/II/III)[*1]	1.0/1.0/1.0	1-year
WHO[*2]	0.5	1-year

*1 = Class I: Tourist, conservation area; Class II: Residential area;
Class III: Industrial and Heavy Traffic area.
*2 = Recommendation, No Regulatory Authority.

population, industrial activity, motor vehicle traffic, or other factors" [40CFR51.110a]. In general, a control strategy must consist of emission limitations applicable to all sources within specified categories. The selected sets of emission limitations are established based on the judgment that they are adequate to bring about the NAAQS attainment.

Figure 13-3 presents the results of ambient air testing in the US at over 30 sites that have quality data since 1980 (US-EPA, 2012g). The achievement in lead reduction is obvious and remarkable – over 89% reduction between 1980 and 2010. The majority of this occurred because of the ban on adding lead to gasoline as of 1986. No national data set exists before this time, or from the beginning of the lead ban in 1976. It could be supposed that ambient levels were even greater in the pre-phase-out time period. Non-attainment areas in the US, as of December 2012, include the East Helena area of Montana, and Herculaneum, Missouri. Both locations have or had lead-smelting operations in the direct vicinity. Additional areas may be added to this list in 2013 due to the new standards adopted in 2008, see (US-EPA, 2012c).

The US-EPA finalized the National Emission Standards for Hazardous Air Pollutant (NESHAP) Emissions for Primary Lead Processing on November 15, 2011. The standards set stack emission limits for lead smelters, work practice standards to minimize fugitive dust emissions, and codified recordkeeping and reporting requirements. There is only one US facility in this source category - the Doe Run Company in Herculaneum, Missouri. The US-EPA finalized the NESHAP Emissions for Secondary Lead Processing on January 5, 2012. The standards set emission limits for lead compounds, fugitive dust standards, and added limits on total hydrocarbons, dioxin, and furan emissions. Other industries that have lead emissions also have NESHAP emission limits. These industries, typically iron/steel mills and foundries, include emission limits on particulate matter which strongly correlates with HAP metal emissions.

FIGURE 13-3. US National Air Quality Trend for Pb Concentration, as Total Suspended Particles [μg/m³].

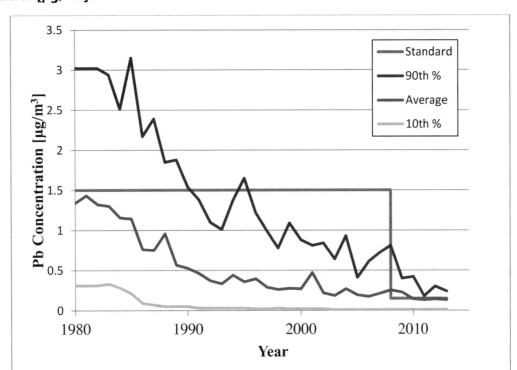

13.3.5 Control Strategies

There are two basic methods for reducing the amount of lead in air emissions. The first, source reduction, removes it from use such that it cannot enter the air. Source reduction was shown to be successful in the US after the addition of tetraethyl lead to gasoline was banned starting in 1976. The second method, emission reduction, separates lead from emissions before release to the environment. The emission reduction methods and technologies are becoming more important as most source reduction programs have been implemented.

13.3.5.1 Source Reduction:

Success in eliminating lead from source materials depends on identifying alternatives that meet or exceed the qualities that lead has and that can do so economically. Table 13-8 lists several former uses of lead and the material or technology used in replacement.

Table 13-8. Replacement for Lead.

Former Uses of lead	Replacements
Octane enhancer in gasoline	Benzene, MTBE, Ethanol
Paint	Organic pigments
Ammunition	Copper-Steel Alloys, Tungsten, Stainless Steel
Weights	Polymers, Tungsten,
Radiation Shield	Tungsten and polymer composites
Ceramic Glazes	Lead-free glazes
Solder	Tin, Silver, Bismuth, Copper and their alloys
Batteries	Lithium, Cadmium, Nickel, others
Fishing Sinkers	Bismuth, Tin, Stainless Steel, Glass, Stone, others

In each case, lead was identified as a problem and alternatives found. In some cases, the alternatives were already competitive with the lead-containing material, and in other cases the drive to remove lead caused discovery of alternatives. One of the few markets where no economical alternative has been found is automobile batteries. There is significant research underway to find an alternative, especially if it can be made to weigh less. A successful alternative in this market will enable many new commercial applications such as electric cars.

13.3.5.2 Emission Reduction:

Lead emissions to the air occur from two main sources – stack emissions and fugitive dust emissions.

Stack Emissions: Lead in stack emissions is almost always associated with particulate matter (PM), and control is based on the same technologies and equipment as other PM. See Chapter 8 for detailed information on the control of PM from stack emissions.

Fugitive Dust Control: One of the largest sources of lead in the atmosphere is from dust that becomes entrained by wind and is swept away from a contaminated site. Most such dust deposits near the source, but smaller dust particles can be carried hundreds or thousands of miles away. These emissions can be minimized with many techniques. Not all are appropriate for every site, and some sites may need to use several techniques (CABQ, 2005).

Lead containing dust may originate from many sources, including:
- Minerals and ores used in primary lead production (delivery, storage, waste);
- Collection and processing during secondary production;

- Construction and demolition of structures;
- Waste piles that contain lead materials – production residuals, paint, toys, ceramics;
- Welding and flame-torch cutting of steels or other lead-containing materials;
- Sanding, stripping, scraping, grinding, heating to remove leaded paint;
- Abrasive blasting of steel structures; and
- Unpaved roads and shoulders where soil was contaminated during the leaded gas era

Various dust suppression techniques include:
- *Watering* – Spraying or applying water with spray bars, hoses, sprinklers, and water trucks. The dust particles interact with the water droplets, which removes them from the air and holds them onto the ground surface, where they can be cleaned up or added to the soil.
- *Wind Sheltering/Barrier* – Provide a minimum of three sided barriers that extend to at least the protected materials height. The barriers reduce the wind speed and create low-wind zones where particles can re-deposit.
- *Wind Break Fencing and Vegetation* – The windbreak needs to be 3- 5 feet above top of the protected surface. The material should have a low porosity, so wind must go around rather than through it. This method would be used near piles and along roads to reduce the surface speed of the wind, hence reducing its ability to entrain any particles.
- *Coverings* – Temporary covers (tarps, plastic sheeting) are anchored to prevent wind or rain access to surface of protected material. Longer term storage should consider roofed areas.
- *Site Access Improvements* – Control traffic to a few well established routes maintained in a manner that prevents wind erosion. Control includes clearly marking travel routes and preventing access to other others.
- *Chemical Stabilization* – adds a bonding element to the top of the soil or pile, making the surface particles larger and thus unlikely to be entrained by wind. It works best on areas that remain undisturbed for extended times. It becomes expensive if done frequently or in large areas.
- *High Wind Protections* – Additional temporary protections may be needed during periods of high winds.

13.4 Mercury

13.4.1 General Information

Mercury (Hg) is an elemental metal (atomic weight 200.59) that is liquid at room temperature. Upon heating it becomes a colorless, odorless gas. It is also called quicksilver and liquid silver, which in Latin is hydrargyrum, the name from which its elemental symbol is derived. Production is from the mineral cinnabar (HgS). Elemental mercury has a very high density (13.5 g/cm^3), good thermal expansion, poor thermal conductivity, and good electrical conductivity. These properties have made it useful for electrical switches, float valves, thermometers, pressure measurements (barometers, manometers, sphygmomanometers), and as an essential component in fluorescent lighting. It is capable of forming a mixture (amalgam) with gold, silver, aluminum, and zinc. This ability to dissolve other metals makes mercury very useful in collecting trace amounts of those metals from ores or soils, and it is used extensively in the gold and silver mining industries. Mercury and its salts also have antiseptic properties. These compounds were used in the past to treat syphilis, and today as an antiseptic and preservative (Mercurochrome and Thimerosal) for vaccines, cosmetics, and eye drops.

13.4.2 Sources, Sinks, and Ultimate Fate

There are many natural and anthropogenic sources of mercury in the atmosphere. Natural sources include volcanoes and the weathering of rocks. Anthropogenic sources include mining, chlorine production, waste incineration and the combustion of coal. These sources include primary and secondary sources. Primary sources are those that directly emit mercury into the atmosphere. Secondary sources include re-emission from sources that have been disposed of or deposited on land, in water, or were adsorbed from the atmosphere by plants and soil. Forest fires and discarded fluorescent lights are examples of secondary sources.

The main forms of mercury for air pollution are elemental, reactive gas mercury, and particulate mercury. *Elemental mercury* (Hg^0) makes up 95% of the total amount of mercury in the atmosphere. It has low solubility in water, which allows it to have an atmospheric residence time on the order of one year. The long residence time means that it can be globally redistributed and that it has a global impact. *Reactive gas mercury* (Hg^{+2}) includes the oxide (HgO), chloride ($HgCl_2$), and organic forms such as methyl-mercury (CH_3Hg^+). These forms are more water soluble, which causes them to have atmospheric residence times on the order of a week. *Particulate mercury* (Hg^p) consists of particulate matter that has mercury in any form sorbed in or on it. The particles are not pure mercury and/or its compounds; rather, they are some other material (organic soot, minerals) that has some mercury associated with it. The actual chemical form is not well known, but it may consist mainly of the oxide. It has an atmospheric residence time distribution similar to other PM, on the order of a few weeks. The denser or larger size PM have shorter atmospheric residence times and have greater local impacts, smaller sizes have longer residence times and have more of a regional or global impact.

Natural sources of mercury, such as volcanic eruptions and emissions from the ocean, contribute about a third of current worldwide mercury air emissions. Anthropogenic sources account for the remaining two-thirds, although these estimates are highly uncertain. Land, water, and other surfaces can repeatedly re-emit mercury into the atmosphere after its initial release into the environment. Much of the mercury circulating through today's environment is mercury released years ago. The anthropogenic sources are roughly split between these re-emitted emissions from previous human activity and direct emissions from current activity. The US-EPA has estimated that about one-third of US direct emissions deposit within the country, and the remainder enters the global cycle. US anthropogenic mercury emissions are estimated to account for roughly 3 percent of the total global emissions, and the US power sector is estimated to account for about 1 percent of the total global emissions (US-EPA, 2006).

Total global mercury emissions from all sources (natural and anthropogenic) are estimated to be between 4,400 to 7,500 Mg (1 Mg = 10^6 gram = 1 metric ton) emitted per year. Table 13-9 and Table 13-10 show estimates of the amount of mercury released globally and in the US, respectively (Selin, 2009). Table 13-11 lists the total anthropogenic emissions from 2012 by region (UNEP, 2013).

Table 13-9. Global Mercury Sources and Sinks.

Global Sources and Sinks			
	Ref 1	Ref 2	Ref 3
Sources	Mg/yr	Mg/yr	Mg/yr
Natural			
Land	810	500	
Ocean	1300	400	
Secondary			
Land (re-emission)	790	1500	
Ocean (re-emission)	1300	2400	
Total Natural	4200	4800	
Anthropogenic			
Fossil Fuel Combustion			878
Metal Production (inc gold)			661
Cement Production			189
Waste Incineration			125
Dental Amalgams (cremation)			26
Other			280
Total Anthropogenic	2400	2200	2200
Total Sources	6600	7000	
Sinks			
Land Deposition	3520		
Ocean Deposition	3080		
Total Sinks	6600		
Amount in Atmosphere (Mg)	5000	5360	
Atmosphere residence time (yr)	0.76	0.79	

Ref. 1 (Mason, et al., 2002), Ref. 2 (Selin, et al., 2007), Ref. 3 (UNEP, 2008).

Table 13-10. US Anthropogenic Sources of Mercury (1999). (Cohen, 2005)

1999 Sources	Mg/yr
Coal Combustion (Electricity)	43.4
Metallurgical Processes	13.6
Chlor-Alkali Process	6.3
Hazardous Waste Incineration	6
Mobil Sources	6
Other Manufacturing	5.7
Municipal Waste Incineration	5.3
Medical Waste Incineration	3.9
Other Chemical Manufacturing	3
Oil Combustion (non-mobile)	2.7
Cement and Concrete	2.1
Other Fuel Combustion	1.9
Industrial Waste Incineration	1.7
Pulp and Paper	1.4
Mining	1.4
Fluorescent Lamp Manufacture and Breakage	0.8
Coal Combustion (other)	0.2
All Other	0.1
Total	105.6

Table 13-11. Global Anthropogenic Mercury Emissions by World Region, 2012.

Region	Emission ton/yr	Global percentage
Australia, New Zeeland, Oceania	22.3	1.1
Central America, Caribbean	47.2	2.4
CIS and other non-EU European	115	5.9
East and Southeast Asia	777	39.7
EU	87.5	4.5
Middle East	37	1.9
North Africa	13.6	0.7
North America	60.7	3.1
South America	245	12.5
South Asia	154	7.9
Sub-Saharan Africa	316	16.1
Other (unknown)	82.5	4.2
Total	1957.8	100

Figure 13-4 shows the historical trend in atmospheric concentration deduced from ice cores. The pre-industrial values represent the assumed natural background level. The peaks in 1815, 1883, and 1980 represent the addition of mercury to the atmosphere from major volcanic eruptions. The broader peak from 1850 – 1884 represents the mercury released during the US gold rush. The mercury was used to dissolve trace amounts of gold from soil and rocks. This mixture was later separated by distillation in which the mercury boils off leaving gold and other precious metals behind. Larger scale implementation may have included some mercury recovery, though most of the mercury fumes were allowed to escape into the atmosphere. The growth after 1900 represents the mercury released from modern industry –of which the largest component is from coal powered electricity generating units.

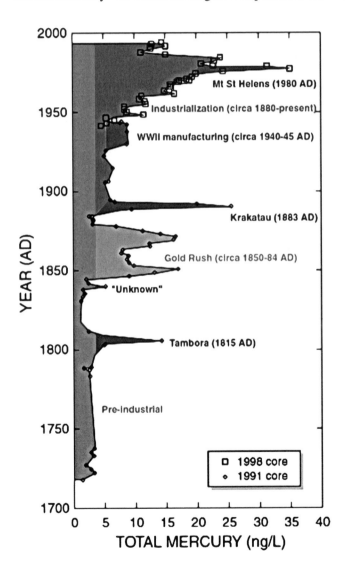

FIGURE 13-4. Historical Atmospheric Concentration of Mercury from Ice-Cores. (USGS, 2002). Ice cores are used to measure historical atmospheric contents. An ice core is drilled using a hollow bit which carves out a cylinder of ice. The cylinder of ice is extracted from the site and removed to a refrigerated lab for study. Ice cores layers are similar to tree rings where the ice layers are created by snow in the winter, which traps atmospheric gases. The layers build from year to year as more snow falls in the winter making dating of the layer relatively straight forward. Samples of entrapped air can be measured for gases like carbon dioxide and other constituents like mercury or pollen.

The fate of mercury depends on several factors, including:
- the form of mercury emitted (elemental, reactive gas, or particle),
- the location of the emission source,
- how high above the landscape the mercury is released (e.g., the height of the stack),
- the surrounding terrain, and
- the weather.

Depending on these factors, atmospheric mercury can be transported locally, regionally, continentally and/or globally before deposition. Mercury that remains in the air for long periods of time (months) and travels across continents is considered part of the global cycle.

Figure 13-5 is a schematic of the global cycle of mercury, adapted from (Pirrone, et al., 2001). Elemental mercury is the main type of emission that enters the global cycle because it has low water solubility. Typical emission sources for the elemental form include volcanoes and coal fired power plants. Once emitted, this form has a residence time of about one year, which is enough time to travel through the atmosphere around the world several times. It may retain its elemental form or be transformed to the reactive gas form or particulate matter form during this transport. Eventually, the mercury is deposited onto the earth's surface through either a wet or dry process. The deposition may be on land or on water. Land deposition may fall directly onto plants or onto the ground. The mercury on plants may be absorbed by the plant, or be ingested by an animal eating the plant, which may cause harm to these organisms. Ground surface deposition may be washed into surface or ground water by later precipitation events. Mercury in any form in water can be readily transformed into either elemental mercury (vapor) or reactive gas mercury (chiefly alkyl-mercury such as methyl-mercury) by microorganisms, phytoplankton, or by abiotic processes, especially under anaerobic conditions. These transformations can make the mercury more volatile or can deposit it in locations where it can re-transform into a volatile form. The deposited and transformed mercury can thus be re-emitted to the atmosphere to rejoin the global cycle. This mercury re-emission rate is temperature dependent – with lower rates at colder temperatures, which may lead to higher net deposition in colder regions and higher altitudes.

The major long term (on a human lifespan scale) sink for mercury occurs with deposition into sediments. Mercury is deposited into ocean sediments at a rate of less than 200 Mg/yr, and into land sediments at a rate of 5,000 Mg/yr. Land sediments wash into the oceans and then enter the ocean sediments. During the journey from land to ocean, the mercury in these sediments may be re-volatilized into the atmosphere. The current amount held in these reservoirs is estimated at 10^6 Mg.

13.4.3 Health and Welfare Effects

Mercury generates the most concern of any of the heavy-metal pollutants due to its high level of toxicity, the high level of emissions, its ability for re-emission, and its ability to bioaccumulate. It is

FIGURE 13-5. Schematic of Global Mercury Cycle.

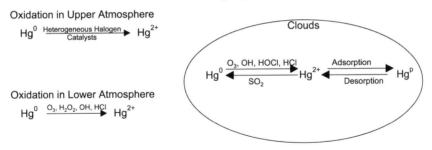
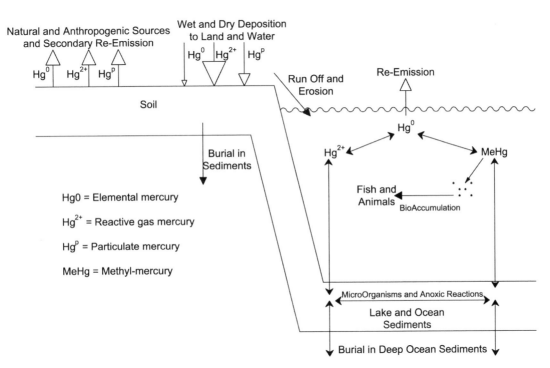

toxic to humans at very low doses (elemental IDLH[2] = 10 mg/m^3 and alkyl-mercury IDLH = 2 mg/m^3 [NIOSH]). The US-EPA's Reference Dose (RfD) for methyl-mercury is 0.1 μg/kg body weight/day, which is defined as the exposure level without recognized adverse effects. It is calculated as a level not to be exceeded in order to protect the nervous system.

When mercury is deposited or transported to water, it can be transformed biologically by microorganisms and phytoplankton or chemically in anaerobic conditions. The products include elemental mercury and reactive gas compounds like methyl-mercury. Elemental mercury has low water

---

2. IDLH is the Immediately Dangerous to Life and Health concentration.

solubility and is easily re-emitted to the atmosphere. Methyl-mercury is soluble in water and is even more soluble in organic matrices, such as the fatty tissue within the organisms that live in the water. The high solubility allows it to be absorbed and buildup (accumulate) within aquatic plants and animals. Fish and animals that eat these organisms buildup methyl-mercury in their bodies. The methyl-mercury is concentrated further up the food chain as bigger fish and animals eat smaller ones, a process called *bioaccumulation*. Methyl-mercury concentrations in fish depend on many factors, including the concentration of mercury in the water, water pH and temperature, the amount of dissolved solids and organic matter in the water, and what organisms live in the water. Methyl-mercury concentrations in fish may also be affected by the presence of sulfur and other chemicals in the water. Because of these variables, and because food webs are very complex, bioaccumulation is hard to predict and can vary from one water body to another. However, in a given water body, the highest concentrations of methyl-mercury are found in large fish that eat other fishes. The concentrations of methyl-mercury in large fish can be over a million-fold larger than in the surrounding water.

13.4.3.1 Human Health

Human exposure to mercury compounds can result from misuse or overuse of mercury-containing products. Exposure to mercury compounds is primarily through ingestion, but can occur through other pathways such as inhalation of the vapor. Ingested organic mercury compounds are more readily absorbed through the gastrointestinal tract than are inorganic compounds. The US-EPA has determined that eating mercury-contaminated fish is the primary route of exposure to mercury for most people in the US. Another less common exposure to mercury is breathing mercury vapor. These exposures can occur when elemental mercury or products that contain elemental mercury break and release mercury to the air, particularly in warm or poorly-ventilated indoor spaces (see chapter 15).

Symptoms of mercury poisoning can include a variety of effects. The effects include mood swings, emotional changes, memory loss, changes in nerve responses, performance deficits on tests of cognitive function, mental disturbances, muscle weakness, impairment of the peripheral vision, disturbances in sensations ("pins and needles" feelings, usually in the hands, feet, and around the mouth), tremors, lack of coordination of movement, and impairment of speech, hearing, and walking. At higher exposures, there may be kidney effects, respiratory failure and death. None of these symptoms or groups of symptoms is unique to mercury poisoning. No human data indicate that the exposure to any form of mercury causes cancer, but the human data currently available are very limited (US-EPA, 1997). Mercury's health effects can be very subtle and the effects can begin long after the exposure occurred.

The greatest risk from mercury is for fetuses and young children because their nervous systems are still developing. They are four or five times more sensitive to mercury than adults. Damage occurring before birth or in infancy can cause a child to be late in beginning to walk and talk and may cause lifelong learning problems. Unborn children can be seriously affected even though the methyl-mercury causes no symptoms in their mothers.

Almost all people worldwide have trace amounts of methyl-mercury in their tissues, reflecting mercury's widespread presence in the environment. Whether exposure to the various forms of mercury harms a person's health depends on a number of factors:
- chemical form of mercury,
- dosage,
- age of the person exposed (the fetus is the most susceptible),
- duration of exposure,
- route of exposure—inhalation, ingestion, dermal contact, and
- health of the person exposed.

All forms of mercury can produce adverse health effects at sufficiently high doses. The primary exposure route for elemental (Hg^0) mercury is inhalation of the vapor and subsequent absorption through the lungs and into the bloodstream. Once in the blood it can be converted into other forms which are soluble in fatty tissue (especially the brain and nervous system). The primary exposure route for inorganic and organic reactive mercury (Hg^{++}) is ingestion. It is absorbed through the gastrointestinal tract, which it may damage and then moves to the bloodstream. It may also affect the nervous system and kidneys. Organic forms are especially soluble in the fatty tissue of the brain and can be readily transported through the placenta towards a fetus. Mothers who have been exposed to methyl-mercury and later breast-feed their babies may expose their children through their milk. The main exposure pathways for particulate mercury (Hg^p) is inhalation and ingestion.

Recent human biological monitoring show that most people have blood mercury levels below the level (5.8 μg/L of whole blood) associated with possible health effects (US-CDC, 1999), (US-CDC, 2009). However, the US-EPA estimated (US-EPA, 1997), based on patterns of fish and shellfish consumption and methyl-mercury concentrations present in fish and shellfish, that 7% of women of childbearing age would have blood mercury concentrations greater than 5.8 μg/L. Actual blood mercury analyses in the 1999-2000 National Health and Nutrition Examination Survey (US-CDC, 2000) for 16-to-49 year old women showed that approximately 8% of women in the survey had blood mercury concentrations greater than 5.8 ug/L. Based on these values and the overall U.S. population of women of reproductive age and the number of U.S. births each year, it is estimated that more than 300,000 newborns each year may have increased risk of learning disabilities associated with in utero exposure to methyl-mercury.

The US-EPA and US-FDA studied these exposure analyses and in 2004 issued joint consumer advice about methyl-mercury in fish and shellfish (US-FDA, 2004). The advice is for women who may want to become pregnant, women who are pregnant, nursing mothers, and young children. The advisory provides three recommendations for selecting and eating fish or shellfish to ensure that these consumers will receive the benefits of eating fish and shellfish and be confident that they have reduced their exposure to the harmful effects of methyl-mercury.

1. Do not eat shark, swordfish, king mackerel, or tilefish because they contain high levels of mercury.
2. Eat up to 12 ounces (two average meals) a week of a variety of fish and shellfish that are lower in mercury.
   a. Five of the most commonly eaten fish that are low in mercury are shrimp, canned light tuna, salmon, pollock, and catfish.
   b. Another commonly eaten fish, albacore ("white") tuna has more mercury than canned light tuna. So, when choosing your two meals of fish and shellfish, consume only up to six ounces (one average meal) of albacore tuna per week.
3. Check local advisories about the safety of fish caught by family and friends in your local lakes, rivers, and coastal areas. If no advice is available, eat up to 6 ounces (one average meal) per week of fish you catch from local waters, but don't consume any other fish during that week.

Follow these same recommendations when feeding fish and shellfish to young children, but serve smaller portions. The US-EPA hosts a web-based compilation of fish advisories issued by states, tribes, territories and local governments [http://www.epa.gov/hg/advisories.htm.]

13.4.3.2 Environmental Welfare

Fish are the main source of food for many birds and other animals, and mercury can seriously damage the health of these species. Loons, eagles, otters, mink, kingfishers, ospreys, and bear eat large quantities of fish. Because these predators rely on speed and coordination to obtain food, mercury may be particularly hazardous to them. Studies suggest that some highly-exposed wildlife species are being harmed by methyl-mercury (US-EPA, 1997). Effects on wildlife can include mortality (death), reduced fertility, slower growth and development, and abnormal behavior that affect survival, depending on the level of exposure. In addition, the report indicates that the endocrine system of fish, which plays an important role in fish development and reproduction, may be altered by the levels of methyl-mercury found in the environment.

13.4.4 History and Regulation

The Clean Air Act regulates air toxics, known as hazardous air pollutants (HAP). Mercury is one of these air toxics. The Act directs the US-EPA to establish technology-based standards for sources that emit air toxics. Those sources also are required to obtain Clean Air Act operating permits and to comply with all applicable emission standards. The law includes special provisions for dealing with air toxics emitted from utilities, giving the US-EPA the authority to regulate power plant mercury emissions by establishing "performance standards" or "maximum achievable control technology" (MACT), whichever the Agency deems most appropriate.

13.4.4.1 US Federal

The original CAA provided for control of HAPs based on chemical emission basis. The review proved to be extremely challenging and by 1990, only seven substances (of the original list of 188) had been reviewed. One of these substances was mercury, and the following regulations were applied to two industrial sources [NESHAPS title 40 part 61 (pre-1990 CAAA lists)]:

(a) Emissions to the atmosphere from mercury ore processing facilities and mercury cell chloralkali plants shall not exceed 2.3 kg (5.1 lb) of mercury per 24-hour period.

(b) Emissions to the atmosphere from sludge incineration plants, sludge drying plants, or a combination of these that process wastewater treatment plant sludges shall not exceed 3.2 kg (7.1 lb) of mercury per 24-hour period.

The 1990 Clean Air Act Amendments gave the US-EPA authority to control mercury (as well as the other listed HAPs) from major sources rather than on chemical emissions basis. Emission controls could now be based on the category of emitters that would typically have mercury emissions, rather than requiring each emission source to test for mercury. The initial activity determined the major source categories. Next, the US-EPA determined whether regulation of a major source category was appropriate and necessary. The US-EPA is required by the US Congress to issue standards of control on a major source if it is found to be appropriate and necessary. The major source categories found to require emission standards are:

- Medical waste incinerators,
- Hazardous waste incinerators,
- Solid waste incinerators,
- Portland cement plants,
- Iron and steel foundries,
- Electric arc furnace steelmaking facilities,
- Mercury cell chloralkali plants,
- Industrial, commercial, and institutional boilers and process heaters, and
- Coal powered electricity generating units.

Final standards have been approved and implemented for all, including electricity generating units with the issuing of the Mercury and Air Toxics Standards (MATS), in December 2011 (40CFR part 63 for source categories). These standards regulate mercury and a number of other toxic air pollutants, such as arsenic, acid gases, cadmium, chromium, nickel, selenium, and cyanide. Each standard is developed for a particular industrial process and carefully considers any special circumstances peculiar to that industry (US-EPA, 2012d). For some industries it was found that the mercury is emitted in their waste water or solid waste streams, which are not covered by the Clean

Air Act, but are covered by other regulations such as the clean water act and the resource conservation and recovery act. These waste streams are regulated in different parts of the Code of Federal Regulations.

The following set of tables (Table 13-12 to Table 13-21) summarize the emission limits for mercury in the New Source Performance Standards (40CFR part 60):

Table 13-12. Hospital/Medical/Infectious Waste Incinerators.

Type	Start	End	Small	Medium	Large	units	
Construction	6/20/1996	12/1/2008	0.55	0.55	0.55	mg/dscm^a	or 85% reduction
Modification	3/16/1998	4/6/2010	0.55	0.55	0.55	mg/dscm^a	or 85% reduction
Construction	12/1/2008	to present	0.014	0.0035	0.0013	mg/dscm^a	
Modification	4/6/2010	to present	0.014	0.0035	0.0013	mg/dscm^a	

Table 13-13. Large Municipal Waste Combustors.

Type	Start Date	End Date	μg/dscm^a	% reduction	
Construction	9/20/1994	12/19/2005	80	85	whichever is less stringent.
Reconstruction	6/19/1996	12/19/2005	80	85	whichever is less stringent.
New	12/19/2005	to present	50	85	whichever is less stringent.

Table 13-14. Small Municipal Waste Combustion.

Type	Start Date	End Date	mg/dscm^a	or % redux	
Construction	-	8/30/1999	0.08	85	whichever is less stringent.
Construction	8/30/1999	to present	0.47		
Reconstruction	6/6/2001	to present	0.47		

Table 13-15. Other Solid Waste Incineration Units.

Type	Start Date	End Date	μg/dscm^a
Construction	12/9/2004	to present	74
Reconstruction	6/16/2006	to present	74

Table 13-16. Sewage Sludge Incineration Units.

Type	Start Date	End Date	Fluidized Bed mg/dscm^a	Multi Hearth mg/dscm^a
New	3/21/2011^b	to present	0.001	0.15
Existing	-	3/21/2011^b	0.037	0.28

Table 13-17. Portland Cement Plants.

Type	Start Date	End Date	Average^c	Upper Limit	Units
Existing	-	9/9/2010	19.51	32.8	lb/MM ton feed
			31.7	54.1	lb/MM ton clinker
New	9/9/2010	to present	8.49	12.3	lb/MM ton feed
			14	20.3	lb/MM ton clinker

The following set of tables summarizes the emission limits for mercury in the National Emission Standards for Hazardous Air Pollutants - Maximum Achievable Control Technology Rules (40CFR part 63):

Table 13-18. Mercury Cell Chlor-Alkali Plants MACT.

Type	Start Date	End Date	Value	Units	Notes
Existing	-	12/19/2002^b	0.076	g Hg/Mg Cl$_2$	w/ end-box ventilation
			0.033	g Hg/Mg Cl$_2$	w/o end box ventilation
			23	mg/dscm^a	w/ oven type thermal recovery
			4	mg/dscm^a	w/ non-oven type thermal recovery
New	12/19/2002^b	to present	zero		

Table 13-19. Gold Mine Ore Processing MACT.

Type	Start Date	End Date	lb/million tons ore
New	2/17/2011^b	to present	84
Existing	-	2/17/2011^b	127

Table 13-20. Boilers and Process Heaters MACT.

Type	Start Date	End Date	lb Hg /MMBTU heat input	Notes
New	12/23/2011[b]	to present	8.60×10^{-7}	Solid Fuel
			4.90×10^{-7}	Liquid Fuel
			7.90×10^{-6}	Gas
Existing	-	12/23/2011[b]	3.10×10^{-6}	Solid Fuel
			2.60×10^{-6}	Liquid Fuel
			7.90×10^{-6}	Gas

Table 13-21. Coal and Oil Electric Generating Units (EGU) MACT.

Type	Start Date	End Date	lb/GWh	Fuel
Existing	-	2/16/2012[b]	0.0130	not low-rank virgin coal
			0.1200	Low-rank virgin coal
			0.0300	IGCC
			0.0020	Solid- Oil derived
			0.0020	Liquid Oil
New	2/16/2012[b]	to present	0.0002	not low-rank virgin coal
			0.0400	Low-rank virgin coal
			0.0030	IGCC
			0.0020	Solid- Oil derived
			0.0001	Liquid Oil

Electric Arc Furnace Steelmaking Facilities.
These MACT standards are based on pollution prevention guidelines. They require that any scrap metal obtained from motor vehicles must originate from a provider that participates in an EPA approved program for the removal of mercury from vehicles. The approved programs must have a goal of removing 80% of all mercury switches from the vehicles before scrapping.

Notes for above tables:
    a = Corrected to 7 mol% oxygen in exhaust.
    b = Date of the final rule.
    c = An average and upper limit are used due to the high variability in limestone feed.
    Units: dscm = dry standard cubic meter, MMBTU = 10^6 BTU, GWh = GigaWatt-hours

Example 13-2.

Determine the required mercury removal efficiency from a small medical waste incinerator constructed in 2010. Use the emission factor from appendix 3.

Solution:

Find the emission factor from Appendix 3, compare with the limits in Table 13-12.

Emission factor is 0.107 lb Hg/ ton of waste.
Performance Standard is 0.014 mg/dscm

To compare these units we need to estimate the volume of gaseous emissions from a ton of waste. To convert from mass of waste to gas emissions, we need to know the composition of the waste and we will assume that 50% excess air is used. An article was located using an internet search engine searching for the terms 'medical waste elemental composition,' (Li, et al., 1993). The article reports an analysis yielding 58% solids, 38% water, and 4 % ash. The solids were composed of carbon (34%), oxygen (15%), Hydrogen (5%), plus a small amount of other elements, nitrogen (1%), sulfur (0.3%), and chlorine (2.7%), which we will ignore for this calculation. All values are weight per-cents. The emissions can now be calculated by estimating the combustion products.

Basis: 1 ton medical waste				
Inlet	In Waste	mw	moles	O2 required
Species	[lb]	[lb/lb-mol]	[lb-mol]	[lb-mol]
carbon	680	12	56.7	56.7
oxygen	300	16	18.8	-9.4
hydrogen	100	1	100.0	25
water	760	18	42.2	0

Next, determine the amount of oxygen required. Multiply by [1 + excess%/100] or 1.5 to find the actual amount of oxygen and nitrogen added from the air for combustion.

Air used for combustion at 50% excess oxygen		
Species	[lb-mol]	
Required O$_2$	72.3	from sum for each species above.
Actual O$_2$	108.4375	from multiplying required O2 by 1.5
N$_2$	407.9315	multiply actual O2 by 79/21 mole ratio of air

Next, we can determine the moles of exhaust gases

Emissions	
Species	[lb-mol]
CO_2	56.7
H_2O waste	42.2
H_2O burn	50
O_2 excess	36.1
N_2	407.9
Total	593.0

Finally, we can convert from moles to volume at standard conditions:

$$V = \frac{nRT}{P} = \frac{(593 \text{ lb-mol})\left(0.7302 \frac{ft^3 \cdot atm}{lb-mol \cdot ^\circ R}\right)(536^\circ R)}{(1 atm)} = 232,270 scf * \frac{1 m^3}{35.315 ft^3} = 6,580 scm$$

Now we can convert the known emission factor into the same units as the emissions standard and determine the require deficiency.

$$0.107 \frac{lb-Hg}{ton_{waste}} * \frac{ton_{waste}}{6,580 scm} * \frac{454 g}{lb} * \frac{1,000 mg}{g} = 7.39 \frac{mg}{scm}$$

$$\eta = \frac{7.39 - 0.014}{7.36} * 100\% = 99.81\%$$

This is a very difficult to achieve efficiency for a single post-combustion technology. A more realistic plan would involve source separation of mercury containing items to remove mercury before combustion. An efficient product management system with proactive employee participation is capable of removing over 90% of mercury from the medical waste stream.

13.4.4.2 US States

The US-EPA works with state, local and tribal governments to implement a variety of programs designed to reduce mercury pollution and impacts through the State Implementation Plan (SIP) program. The states implement most environmental regulations and programs. They may adopt local environmental laws and regulations that are more stringent than federal requirements, or to act in areas where the US-EPA has not acted. Twenty- two states had implemented or developed overall state-based mercury action plans before the US-EPA began developing the federal regulations. Many of the state plans include pollution reduction elements that exceed current federal requirements. In particular, the states in the Great Lakes basin and northeast region have efforts to identify and pursue ways to reduce and prevent mercury releases to the environment, both as individual states and in multi-state collaborations.

The US-EPA has developed a state budget for mercury emissions from electric generating units (US-EPA, 2007). Each state is provided an allowance of emissions based on past reported emissions of mercury. The total allowances amount to 38 tons per year for the years 2011 – 2017, and 15 tons per year for 2018 and thereafter. The states with the largest budgets (Texas, Indiana, Ohio, and Pennsylvania) are those with many coal-fired power plants. Some states were given no allowance (Vermont, Rhode Island) because they have not historically emitted mercury. Each state must describe how these limits will be achieved in their State Implementation Plan (see section 3.2.3 Environmental Laws and Regulations – Delegation to States and Tribes). The plan must contain emission standards and compliance schedules, or prescribe allowable rates of emissions. The unit of allowances is to be given in ounces of mercury emissions per year.

13.4.4.3 International

The U.S. engages with international partners, bilaterally and multilaterally, to address key mercury issues including data collection and inventory development, source characterization, and best practices for emissions and use reduction. Some of the main partnerships include the United Nations through the Environment Program Governing Council and the Industrial Development Organization (UNIDO); Russia and Canada through the Arctic inventory and emission reduction projects; Brazil, Russia, India, China (BRIC) along with several other South American governments through assessment and industrial sector improvement planning; and Canada for the Great Lakes Bi-National Toxics Strategy. All of these programs center on better understanding of mercury emissions and reduction strategies.

- The US has been a catalyst in a number of global mercury partnerships designed to achieve early reductions in use and emissions of mercury globally as called for by United Nations Environment Program Governing Council Decision 23/9 IV of February 2005 (UNEP, 2010).
- The US provides support for the development of Russia's mercury action plan and inventory, as well as support for a regional Arctic inventory and emissions reductions projects (Rackley, et al.).
- The Canada-United States Strategy for the Virtual Elimination of Persistent Toxic Substances in the Great Lakes Basin, known as the Great Lakes Bi-national Toxics Strategy, provides a framework for actions to reduce or eliminate mercury and other persistent toxic substances from the Great Lakes Basin (US-EPA, 2012e).

● **Case Study 13-3. The Long Path to Mercury Emission Controls at Electric Generating Units (EGUs) in the US.**

The US Congress ordered the US-EPA to identify and regulate emission sources of mercury in the 1990 Clean Air Act Amendments (CAAA). The US-EPA issued its first list of industrial categories for which it would develop HAP standards in 1992. The electrical power plant category was not included in this list. The National Resources Defense Council (NRDC) filed suit (sued) the US-EPA for allegedly unlawfully omitting power plants from the list. The case was based on a reading of the CAAA that the US-EPA was supposed to conduct a health effects study addressing toxic air pollution from fossil fuel-fired power plants and issueing regulations if the US-EPA finds it is appropriate and necessary, and if so to issue Maximum Achievable Control Technology standards (MACT). Many health studies had shown that power plants were one of the leading sources nationwide of HAPS. As a result of the original lawsuit, NRDC and US-EPA entered into a settlement agreement in 1994, under which the US-EPA was required to complete the study and report to Congress by November 1995. Following several delays, the US-EPA submitted the report to Congress (US-EPA, 2012f) in February 1998—but still without making a determination about the appropriateness and necessity of MACT standards. Following notice of intent to file an "unreasonable delay" lawsuit by NRDC and the Sierra Club later in 1998, the original NRDC settlement agreement was modified twice more to require the US-EPA to make the necessary regulatory determination by Dec. 15, 2000. Then the US-EPA Administrator Carol Browner did so and determined that it was "appropriate and necessary" to reduce toxic air pollutants from fossil fuel-fired power plants using MACT standards.

The US-EPA continued to study the issue and in 2004 issued a set of rulemaking proposals for public comment. In 2005, the head of the US-EPA retracted (canceled) the 2004 rule by issuing a rescission rule. Simultaneously, the US-EPA issued a mercury "cap-and-trade" rule that would delay any significant reductions in power plant mercury emissions for 13 years. This rule also disclaimed any need to reduce the remaining 70 or so toxic air pollutants from power plants leaving known hazardous contaminants like arsenic, lead, dioxins, acid gases, and heavy metals completely unregulated in EGU emissions. Some cynics at the time suggested that this change was done to mostly further delay action because the cap and trade would be found to violate the CAAA by the courts, and the new rule would be thrown out forcing the US-EPA to start again once the cases were heard by the federal courts.

In February 2008, the U.S. Court of Appeals for the D.C. Circuit ruled (Appeals, 2008) that the US-EPA had illegally evaded the Clean Air Act's protective safeguards (MACT standards) that should have required deep and timely reductions in toxic air pollution, including mercury, from the nation's coal-fired power plants. The court further ruled that the US-EPA had illegally substituted a mercury pollution trading scheme for the protections required by the Clean Air Act. The court vacated the US-EPA rules and made clear that the US-EPA now had a firm legal obligation to adopt MACT standards to reduce all toxic air pollutants from power plants. The unanimous court ruling even went so far as to mock the US-EPA's defiance of the plain language of the law. The court compared the US-EPA's actions to the capricious Queen of Hearts in Alice in Wonderland, since the US-EPA had -- in the court's words -- "substituted [its] desires for the plain text" of the law.

New MACT standards were opened for public comment in May 2011 (over 900,000 comments were received on this topic), and the final rules were adopted in December 2011, see Table 13 21. The US-EPA estimates that the new standards will prevent as many as 11,000 premature deaths and 4,700 heart attacks a year. The standards will also prevent 130,000 cases of childhood asthma symptoms and about 6,300 fewer cases of acute bronchitis among children each year. The US-EPA estimates that manufacturing, engineering, installing and maintaining the pollution controls to meet these standards will provide employment for thousands, potentially including 46,000 short-term construction jobs and 8,000 long-term utility jobs. Also, as part of the commitment to maximize flexibilities under the law, the standards are accompanied by a Presidential Memorandum that directs the US-EPA to use the tools in the Clean Air Act to implement the rules in a cost-effective manner that ensures electric reliability. For example, under these standards, the US-EPA is not only providing the standard three years for compliance, but is also encouraging permitting authorities to make a fourth year broadly available for technology installations, and if still more time is needed, providing a well-defined pathway to address any localized reliability problems should they arise.

13.4.5 Control Technologies

Control of mercury emissions is based upon eliminating mercury from use when possible or on reduction of the mercury concentrations in emissions and other releases into the atmosphere. Some emissions result from a stack release, but a larger amount occurs from non-point specific locations such as dust from material handling, evaporation from ash or sludge ponds, or from previously spilled mercury which slowly evaporates from the spill site. A location may have a monitor on its stack to measure mercury in its emissions, but fail to measure most of the mercury leaving the location.

Each industry varies in the ways in which it may emit mercury. Table 13-22 lists the mercury emission factors for several industries. These emission factors are obtained by averaging the results from many sources, and these values could change over time as an industry alters their input materials. Also note that these emission factors represent the amount emitted if there were no controls on the source. The numbers provide a useful baseline to compare various reduction strategies.

Table 13-22. Mercury Emission Factors for Several Industrial Processes.

Industry	Emission Factor	Units
Coal Fired Power Plant	0.1 to 0.3	g/Mg coal
Coal Fired Residential and Commercial Boiler	0.3	g/Mg coal
Oil Combustion	0.001	g/Mg oil
Municipal Waste Incineration	1	g/Mg waste burned
Sewage Sludge Incineration	5	g/Mg waste burned
Cement Production	0.1	g/Mg Cement produced
Steel and Pig-Iron	0.1	g/Mg steel produced
Primary Mercury Production	200	g/Mg ore mined
Gold Production (Industrial scale)	25000	g/Mg gold produced
Copper Production	5	g/Mg Cu produced
Lead Production	3	g/Mg Pb produced
Zinc Production	7	g/Mg Zn produced

Note: $1 \text{ Mg} = 10^6 \text{ g}$

The emission sources associated with the emission factors in Table 13-22 can be categorized into three groups – fuel sources, direct mercury usage, and material contaminant. The major fuel source is coal, and any industry that uses coal has mercury emissions. The direct uses of mercury include primary mercury production and gold production, where it is used to dissolve the gold and is later removed by distilling the gold-mercury amalgam. Finally, there are many other industries that use material contaminated by mercury – ores that contain copper, lead, and zinc also contain trace amounts of mercury; waste streams from hospitals and municipalities contain mercury from discarded items.

Strategies for controlling mercury include pollution prevention measures such as product substitution, process modification, work-practice standards, and materials separation; flue gas treatment technologies; coal cleaning (which includes mercury control); and alternative strategies.

13.4.5.1 Product Substitution

Fuel Sources - The largest quantities of mercury stack emissions are from combustion processes - coal-fired power plants and waste incineration, see Table 13-10. The US Geological Survey examined 7,000 coal samples and found that they contained between 0 and 1.8 ppm (mass basis) of mercury, with an average concentration of 0.17 ppm and standard deviation of 0.17 ppm (USGS, 2001). Natural gas, the current favored replacement fuel, generally contains much lower levels of mercury, generally less than 0.010 ppm and with an average of 0.002 ppm.

Fuel replacement would be one way to achieve reductions. The switch from coal to natural gas is already occurring within the US because of the lower cost of natural gas and the greatly simplified pollution control requirements. The switch reduces mercury emissions from power generation by about 90%. However, it is not a simple matter to switch fuels unless the boiler has been specifically designed to use multiple fuels. The switchover may require an entirely new boiler and requires retraining of the facilities boiler operators. Further, if many facilities choose to switch fuels, the market price of coal will be reduced while the price of natural gas may increase – thus complicating the economics of such a change.

Direct Usage - The second largest direct source of mercury emissions is in the production of metals – chiefly mercury, gold, and silver. Reductions from mercury primary production occur as the demand for mercury is reduced. The reduction happens with material replacement or improvements in reclaiming and recycling. The chief replacement material for recovering gold and silver is cyanide, which is also very toxic and a hazardous air pollutant. However, cyanide can be destroyed before it is released to the environment.

Materials Contamination - The other large category of mercury emissions is from incineration of medical and municipal waste, including the solids removed from wastewater. The mercury from these sources could be reduced/eliminated if the discarded products that use mercury were switched to non-mercury products. Eliminating these sources prevents the subsequent release of mercury from incinerator emissions.

Many states and local governments have enacted bans on consumer goods that contain mercury; light switches in car doors, trunks, and hoods, light-up shoes, thermometers, pressure measurements (barometers and sphygmomanometers), and thermostats. The concept is simple – eliminate the sale of such products, and there is a corresponding decrease in mercury in the waste streams. The half-life of mercury-containing devices in hospitals is on the order of 80 days. Any replacement with non-mercury medical devices shows up as a reduction in hospital waste streams soon afterwards.

13.4.5.2 Process Modification, Work-Practice Standards, and Material Separation.

These changes are based on separating wastes that contain mercury from those that do not. They reduce the total amount of contaminated waste and increases the mercury concentration, which may allow for easier separation or control. The ideas can be applied to specific point sources, area sources, and fugitive emissions. Practices include:
- Covering tanks that process mercury-containing materials and collecting their off gases for treatment;
- Altering solids handling to reduce dust emissions – fewer movements, different equipment, protection from wind;
- Implementing clearly defined processes for use and disposal of objects that contain mercury; and
- Establishing excellent house-keeping procedures that minimize re-entrainment of dust to the atmosphere.

13.4.5.3 Flue Gas Treatment Technologies

When mercury reduction or replacement methods are insufficient to reduce mercury emissions, then emission gas treatments must be considered. Mercury removal from flue gas occurs by the chemical interactions between mercury, acid gases, and unburned carbon in ash. The variety of composition of materials burned in incinerators or boilers results in a wide range of compositions of the flue gas. There is no one-size-fits-all technology for mercury emission control and solutions must be developed for each system. Also, these technologies are still in various phases of research and development.

During combustion and incineration, the mercury (Hg) in the fuel is volatilized and converted to elemental mercury (Hg^0) vapor in the high-temperature regions of the boiler. As the flue gas cools, a series of reactions convert Hg^0 to Hg^{+2} and Hg^p. These oxidation reactions are slow (kinetically limited) so the flue gas entering a cleaning device(s) will be a mixture of Hg^0, Hg^{+2} and Hg^p and this process is called mercury speciation partitioning. These various species exit the boiler in gas or solid form, and may be attached or adsorbed onto other solid particles. The partitioning of Hg has considerable influence on the selection of mercury control approaches. In general, the majority of gaseous mercury in bituminous coal-fired boilers is known to be Hg^{+2}, whereas the majority of gaseous mercury in subbituminous- and lignite-fired boilers is known to be Hg^0.

There are two general approaches under development to control mercury in flue gases – (1) powdered activated carbon (PAC) injection, and (2) multi-pollutant control, in which Hg capture is enhanced in existing and new control devices for sulfur dioxide (SO_2), nitrogen oxides (NO_x), and particulate matter (PM). A third option that is currently in research is coal cleaning - a process that pretreats a coal to remove harmful contaminants before the coal is burned, thereby eliminating the need to treat the emission stream.

The first two approaches improve collection efficiency by chemically changing elemental mercury into the oxidized and particulate mercury forms. Oxidized mercury is more soluble in water and more prone to adsorption onto particle surfaces. Particle removal can be highly efficient. Oxidation of mercury in flue gas can be promoted in two ways – addition of oxidation chemicals and fixed bed oxidation catalysts. The oxidized mercury can then be removed in the wet scrubber used in sulfur removal. Additional treatment may be required to prevent the re-emission of captured mercury but is subsequently chemically reduced back to the elemental form.

13.4.5.4 Powdered Activated Carbon

Powdered Activated Carbon (PAC) is the most widely used mercury adsorbent, because it can usually achieve 90% removal efficiencies. It has been used on facilities that burn municipal solid waste since the 2000's. Particles of activated carbon are injected into the exit gas flow, downstream of the boiler. Activated carbon can be made from any organic carbon-containing material and can be enhanced by including halogens (chlorine, bromine, iodine) or sulfur onto the surface. The mercury attaches to the carbon particles and is removed in a traditional particle control device (cyclone, electrostatic precipitator [ESP], fabric filter). Use of PAC may not achieve 90% removals in systems that have high levels of sulfate (SO_3^-), that use hot-side ESPs, or that burn low-rank coals (sub-bituminous or lignite).

■ Example 13-3.

Estimate the rate of addition for the adsorbent x-7,000 to remove 90% of mercury from the 150 °C flue gas of a 200 MW bituminous coal (11,500 BTU/lb) fired boiler (35% efficient). The Hg concentration in the coal is 0.17 ppm(mass) and the concentration in the exhaust gas is 6.46 x 10^{-5} mg Hg/L. Assume adsorption can be modeled using the Langmuir isotherm as shown in Table 13-23 from (Yuan, et al., 2004).

Table 13-23. Two Models for Estimating the Adsorption of Mercury Chloride on Adsorbent x-7000.

Adsoption Isotherm Model	Equation	Parameter Values		
			30 °C	150 °C
Langmuir	$q = \dfrac{q_m k_L C}{1 + k_L C}$	$q_m =$	3.3	8.623
		$k_L =$	13.05	21.833
Freundlich	$q = AC^n$	$A =$	18.288	32.443
		$n =$	0.247	1.2

Where,

q = amount of mercury adsorbed by x-7000 adsorbent [mg/g]
C = concentration of mercury in gas stream [mg/L]

Solution:
First, find the amount of coal needed to generate 200 MW.

$$X\frac{Mg_{coal}}{day} = \frac{200MW}{0.35 eff}\frac{3.413 \times 10^6 BTU}{MW-hr}\frac{24hr}{day}\frac{1lb_{coal}}{11,500BTU}\frac{454g}{lb}\frac{Mg}{10^6 g} = 1,848\frac{Mg_{coal}}{day}$$

Next, determine the amount of mercury to be removed from the coal.

$$1,848\frac{Mg_{coal}}{day}\frac{0.17 g_{Hg}}{1 Mg_{coal}} \times (0.9 eff) = 282.7 \frac{g_{Hg}}{day}$$

Now solve for q, the amount of mercury that can be adsorbed by the additive x-7000

$$q_{150°C} = \frac{q_m k_L C}{(1+k_L C)} = \frac{(8.623)(21.833)(6.46 \times 10^{-5})}{(1+(21.833)(6.46 \times 10^{-5}))} = 0.0121 \frac{mg_{HG}}{g_{X7000}}$$

Finally, using the quantity of mercury to be removed and the amount adsorbed, solve for the amount of x-7000 required.

$$Xgx7000 = 282.7\frac{g_{Hg}}{day}\frac{g_{X7000}}{0.0121 mg_{Hg}}\frac{1000 mg_{Hg}}{1 g_{Hg}} = 23 \times 10^6 g_{X-7000}$$

This facility would use 23 tons of the adsorbent each day - roughly two to three rail-car loads per week of the material. It may need to be separated from other collected solids before disposal.

13.4.5.5 Multi-Pollutant Control

Multi-pollutant control can be an inexpensive option if a facility already uses pollution control technologies for the capture of sulfur, particulate matter, or nitrogen oxides. It has been observed that these control systems remove some of the mercury before the stack can release it. Some systems only need to add mercury monitoring equipment. An example of such a system is the combination of selective catalytic reduction and wet flue gas desulfurization on bituminous coal-fired boilers. The effectiveness of these technologies for mercury removal varies, depending on characteristics of the fuel, other burned materials and the configuration of the boiler/combustor. Table 13-24 shows average capture efficiencies for a variety of systems that burn pulverized coal (PC) (Srivastave, et al., 2003). The table splits results based on the type of control equipment and on the type of coal used. In almost all cases the efficiencies are greatest for bituminous (high-rank) coals.

Table 13-24. Average Mercury Capture Efficiency by Existing Post-Combustion Control Equipment for Pulverized Coal Fired Boilers, by Coal Type.

Post-Combustion Control Method	Post-Combustion Emission Control Device	Bituminous	Subbituminous	Lignite
PM Only	CS-ESP	36	3	0
	HS-ESP	9	6	NT
	FF	90	72	NT
	PS	NT	9	NT
PM and Spray Dryer	SD + CS-ESP	NT	35	NT
	SD + FF	98	24	0
	SD + FF + SCR	98	NT	NT
PM and Wet FGD	PS + FGD	12	0	33
	CS-ESP + FGD	75	29	44
	HS-ESP + FGD	49	29	NT
	FF + FGD	98	NT	NT

Where: CS-ESP = Cold Side Electro-Static Precipitator,
HS-ESP = Hot Side Electro-Static Precipitator,
FF = Fabric Filter,
PS = Particle Scrubber,
SD = Spray Dryer,
SCR = Selective Catalytic Reduction,
FGD = Flue Gas Desulfurization, and
NT = Not tested.

13.4.5.6 Alternative Strategies

The main alternative strategy involves what is called a clean coal option. The coal is treated to remove mercury before combustion. Pyrolysis is used to preheat the coal in two stages – the first heats the coal to just above the boiling point of water (100°C) to drive off any trapped moisture. Then the coal is heated again to remove mercury, as well as other volatile substances (such as H_2S). This coal cleaning technology increases the heat content of the coal and reduces the emissions of mercury and sulfur. Additional steps have been proposed that may significantly reduce the formation of particulate matter. However, no commercial power plants currently use this experimental treatment due to cost concerns.

13.5 Questions

* - Questions and problems may require additional information not included in the textbook.
NOTE: Appendix V provides the complete list of HAPs.

1. Look at a map of emission sources in your state or a nearby state. Can you find a location that may be used to determine the background concentration of any one or more HAPs? Develop your own criteria, such as must be 250 miles from any major source, 100 miles from any area source, and at least 25 miles from any major road or railway.*

2. Identify 3 HAPs that may be expected to originate from:
    a) stationary sources only,
    b) mobile sources only, and
    c) secondary formation sources only.

3. Visit the US-EPA website and locate the most recent NATA (www.epa.gov/nata/). Identify any national cancer risk drivers, national cancer risk contributors, national non-cancer hazard drivers, and any substances of regional concern for where you live.*

4. What is MACT and how is it determined? Choose a HAP and determine if a MACT has been developed for that substance for a source category.*

5. What is a bio-accumulative air toxic? List the six HAPs that are thus categorized.

6. Plot the data in table 13-5 and determine the increased risk of cancer over a person's lifetime if they are exposed to an atmospheric concentration of 10 $\mu g/m^3$. Critique the assumptions used in determining these values. How could you improve the estimate?

7. What areas of the US are in non-attainment for lead (Pb)? Choose one and determine the source/s which cause the excessive values.*

8. Suggest several ways to reduce lead in nearby air and soil from:
    a) Open pit lead mine,
    b) Auto battery recycler,
    c) Military ammunition reclamation facility,

d) Wetland dredging (assuming lead is from gunshot due to duck hunting)

e) Renovation/demolition of building with lead plumbing, leaded glass windows, and electrical work.

9. Compare and contrast the various alternatives to lead shot in ammunition.*

10. Mercury has been used in many consumer objects such as electrical switches, float valves, thermometers, pressure measuring barometers, manometers, and sphygmomanometers, and in fluorescent lighting. Choose one of these applications and discuss why mercury is/was used and what has been done to minimize or replace it.

11. What is the chloralkali process and why does it use mercury?*

12. Why does mercury seem to have higher atmospheric deposition rates in colder regions?

13. Is there an active fish consumption advisory in waters near where you live?*

14. Make a list of three questions you have about this chapter or air pollution concerns you have.

13.6 Problems

1. Determine if the following emission sources would be classified as a major source, an area source, or not a significant source of hazardous air pollutants.
    a. 22 tons/yr of methylene chloride.
    b. 7 ton/yr benzene, 5 tons/yr of toluene, 2 tons/yr of a mix of o-, m-. and p- xylene.
    c. 12 tons/yr of ethanol.
    d. Mercury from combustion of 10^6 tons/yr of bituminous coal with uncontrolled emissions.

2. Would the average person be able to detect (by smell) benzene at the concentration considered harmful?

3*. Does the benzene in cigarettes pose an increase cancer risk for the smoker? Use the data in Table 13-6 and make any necessary assumptions about the consumption of a cigarette.

4. A typical American spends 20 hours per day indoors, 3 hours per day outdoors, and 1 hour a day in a car. Assume the benzene concentrations in a particular region are 6.7 $\mu g/m^3$ indoors, 4.5 $\mu g/m^3$ outdoors, and 50 $\mu g/m^3$ in a car.

a. Calculate the time weighted average benzene concentration ($\mu g/m^3$) encountered from all the sources during a day?

b. How does the average change if 8.5 hours of the day is spent working in a place where the indoor air concentration of benzene is 13.5 $\mu g/m^3$.

c. According to the NIOSH Recommended Exposure Limit of 0.1ppm weighted 10-hour average, would taking a 10 hour drive be considered hazardous?

5. Determine if the activated carbon adsorption column in example 13-1 could remove 200 ppm benzene from 20 m^3/min of air at 28 °C and 1.0 atm. Assume the column must run for 8 hours.

6. Plot the data in table 13-6 and determine the increased risk of cancer over a person's lifetime if they are exposed to an atmospheric benzene concentration of 10 $\mu g/m^3$. Critique the assumptions used in determining these values. How could you improve the estimate?

7. Determine the impact on global primary production of lead if an alternative to the lead-acid battery is found and reduces this lead demand by 90%. Assume the ratio of recycled lead decreases to 50% due to this change, and that the US use percentages are similar to the global use values.

8. Calculate the blood lead concentration in an adult if they are exposed to the NIOSH REL threshold (0.10 mg/m^3) in all the air they breathe. Would this person exceed the level of blood lead indicating lead poisoning?

9*. Make a simple estimate of the efficiency of one of the fugitive dust suppression methods in section 13.3.5.2.

10. Determine the mercury composition (dry wt%) in the emissions from burning coal (assume to be $C_{75}H_{100}$) with 50% excess air. The coal has a Hg emission factor of 0.3g/Mg coal.

11. Determine the amount of Hg released [lb/hr] from a 500 MW (34% efficiency) coal EGU. The coal is considered low-rank and has a Hg concentration of 4.8 lb/10^{12} BTU. Does this emission rate comply with the 2012 EGU-MACT limit (Table 13-20)?

12*. Determine the amount of mercury an average size adult would need to consume to reach the level of possible health effect (5.8 $\mu g/l$ whole blood).

13. Would the amount of mercury in a compact fluorescent light bulb (3 mg) be enough to cause a 30 ft by 30 ft room with a 7 ft ceiling to exceed the IDLH if all the mercury was released

instantaneously upon breaking and mixed perfectly within the closed room? In reality, only about 20% will evaporate, it will mix poorly, and this will take several days.

14. Determine the amount of fish a person could consume [ounces] to reach the level of possible health effect (5.8 µg/l whole blood). Assuming all mercury ingested goes instantly into the blood and is very slowly removed. Assume a fish has a concentration of
    a. Catfish – 0.05 ppm on a weight basis.
    b. Tuna (canned chunk light) – 0.20 ppm
    c. Tuna (canned albacore) – 0.40 ppm
    d. Shark – 0.75 ppm.

15. Calculate the concentration of mercury in a 10 year old, 15 oz. Minnesota yellow perch if the water body it was caught in has a concentration of 0.65 ngHg/L and a bioaccumulation factor (BAF) of 350,000 L/kg.

$$BAF = \frac{\text{mercury in biota}\left[mg_{Hg}/kg_{fish}\right]}{\text{mercury in water}\left[mg_{Hg}/L\right]}$$

16. Estimate the safe dose of fish consumption per week for an adult. Assume that 50% of all ingested mercury is absorbed into the bloodstream, an adult has 5.5 L of blood, mercury is removed from the body at a rate as given in the equation below (it is based on the assumption that mercury has a 300 day half-life in the bloodstream before being excreted), the safe concentration of mercury is 5.8 µg/l blood, and the fish has a concentration of 0.05 ppm by weight.

17*. Estimate the 2010 annual US emissions [tons/year] from one of the industrial processes listed in Table 13-22. You will need to find the annual production numbers from an outside source (not in this text).

13.7 Group Project Ideas

For each project, the students should work in small groups and present their finding in either a short report (5-8 pages) or 15 minute presentation.

1. Explore the list of HAPs and choose one. Write a short paper discussing its sources and sinks, health effects, and reduction strategies.

2. Write a short paper to address the question of what is the acceptable risk level from a particular HAP. Be sure to discuss the benefits from the process that emits the HAP or its precursors, an estimate of the cost to reduce the emissions, and identify who bears the most risk (who, where, when, and why).

3. Compare and contrast the various alternatives to lead shot in ammunition.*

4. Mercury has been used in many consumer objects such as electrical switches, float valves, thermometers, pressure measuring barometers, manometers, and sphygmomanometers, and in fluorescent lighting. Choose one of these applications and discuss why mercury is/was used and what has been done to minimize or replace it.*

13.8 Bibliography

Appeals, US-Court of. 2008. *State of NJ, et al., v. US-EPA*. Washington DC, USA : District of Columbia Circuit, 2008. No. 05-1097.

ASTDR. 2011. Lead. *Toxic Substances* Portal. [Online] United States Agency for Toxic Substances and Disease Registry, March 11, 2011. [Cited: March 3, 2013.] http://www.atsdr.cdc.gov/substances/.

ATSDR. 2007. *Toxicological Profile for Benzene*. Agency for Toxic Substances and Disease Registry, US Department of Health and Human Services. Atlanta, GA, USA : US Public Health Service, 2007.

CABQ. 2005. *Fugitive Dust Control Methods*. Environmental Health Department, Air Quality Division, City of Albuquerque, NM, USA. 2005. http://www.cabq.gov/airquality/pdf/dustcontrolmethods.pdf.

Cohen, M. 2005. *Source-attribution for atmospheric mercury deposition: Where does the mercury in mercury deposition come from?* Air Resources Laboratory, NOAA. 2005. Power Point Presentation. http://www.arl.noaa.gov/ss/transport/cohen.html.

DEWHA. 2009. *Benzene*. Department of the Environmant, Water, Heritage and the Arts. s.l. : Australian Government, 2009. pp. 1-2.

Duarte-Davidson, R, et al. 2001. Benzene in the Environment: An Assessment of the Potential Risks to the Health of the Population. *Occupational and Environmental Medicine*. 58, 2001, Vol. 1.

EPA, US. 2010. Air Toxics Web Site. *Technology Transfer Network*. [Online] US EPA, May 20, 2010. [Cited: September 23, 2011.] http://www.epa.gov/ttn/atw/socatlst/socatpg.html.

ICIS. 2013. Benzene Market Overview. *Chemicals*. [Online] February 2013. http://www.icis.com/chemicals/benzene/.

Jackson, Lisa P. 2011. *Completion of the Requirement To Promulgate Emission Standards*. Washington DC, USA : US Federal Register, 2011. FR Doc. 2011-4489.

Kao, Alan S. 2012. Formation and Removal Reactions of Hazardous Air Pollutants. *Journal of the Air and Waste Management Association*. 44:5, 2012, pp. 683-696.

Kitman, J.L. 2000. The Secret History of Lead. *The Nation*. March 20, 2000, Vol. 270, 11.

Li, Chih-Shan and Jenq, Fu-Tien. 1993. Physical and Chemical Composition of Hospital Waste. *Infection Control and Hospital Epidemiology*. 1993, pp. 145-150.

Mason, R P and Sheu, G R. 2002. Role of the Ocean in the Global Mercury Cycle. *Global Biogeochemical Cycles*. 2002, Vol. 16, p. 1093.

Pirrone, N, et al. 2001. Atmospheric Mercury Emissions from Anthropogenic and Natural Cources in the Mediterranean Region. *Atmospheric Environment*. 35, 2001, pp. 2997-3006.

Rackley, K, et al. *The U.S. Mercury Emission Inventory for the Arctic Council Action Plan*. Research Triangle Park, NC, USA : US Environmental Protection Agency. Office of Air Quality Planning and Standards.

Selin, N E, et al. 2007. Chemical cycling and deposition of atmospheric mercury: Global constraints from observations. *Journal of Geophysical Research Atmospheres*. 2007, Vol. 112.

Selin, Noelle E. 2009. Global Biogeochemical Cycling of Mercury: A Review. *Annual Review of Environment and Resources*. 34, 2009, DOI: 10.1146/annurev.environ.051308.084314.

SPECIATE. 2012. SPECIATE Version 4.3 . *Technology Transfer Network - Clearinghouse for Inventories & Emissions Factors*. [Online] US-EPA, September 2012. http://www.epa.gov/ttn/chief/software/speciate/index.html.

Srivastave, RK, Staudt, JE and Jozewicz, W. 2003. Preliminary Estimates of Performance and Cost of Mercury Emission Control Technology Applications on Electric Utility Boilers: An Update. [Online] 2003. [Cited: Nov 23, 2012.] http://www.andovertechnology.com/dOCUMENTS/MEGA_Paper_59.pdf.

Story, L. 2007. *Lead Paint Prompts Mattel to Recall 967,000 Toys*. New York, NY, USA : The New York Times, 2007.

UNEP. 2013. *Global Mercury Assessment 2013. Sources, Emissions, Releases and Environmental Transport*. Geneva, Switzerland : United Nations Environmental Program, 2013.

—. 2010. Mercury Programme Establishment of Partnerships. *United Nations Environment Programme - Chemicals*. [Online] January 12, 2010. [Cited: November 24, 2012.] http://www.chem.unep.ch/mercury/partnerships/.

—. 2008. *United Nations Environment Program - Global Atmospheric Mercury Assessment: Sources, Emissions, and Transport*. Geneva, Switzerland : s.n., 2008. http://www.unep.org/hazardoussubstances/Mercury/MercuryPublications/.

US-CDC. 2005. *Facts About Benzene*. Emergency Preparedness and Response. Atlanta, GA, USA : s.n., 2005.

—. 2009. *Fourth National Report on Human Exposure to Environmental Chemicals*. Department of HEalth and Human Services. Atlanta, GA, USA : s.n., 2009. http://www.cdc.gov/exposurereport/.

—. 2000. *National Health and Nutrition Examination Survey*. Atlanta, GA, USA : NAtional Center for Health Statistics, 2000. http://www.cdc.gov/nchs/nhanes/nhanes1999-2000/nhanes99_00.htm.

—. 1999. *Toxicological Profile for Mercury*. s.l. : Agency for Toxic Substances and Disease Registry, 1999. CAS# 7439-97-6.

US-EPA. 2012a. 2005 National-Scale Air Toxics Assessment. *Technology Transfer Network Air Toxics*. [Online] US-EPA, May 2012a. http://www.epa.gov/ttn/atw/nata2005/.

—. 2010. Ambient Concentrations of Benzene. *Report on the Environment*. [Online] December 2010. [Cited: March 2, 2013.] http://www.epa.gov/ncea/roe/index.htm.

—. 2012e. Binational Toxics Strategy. *Great Lakes*. [Online] April 24, 2012e. [Cited: November 24, 2012.] http://www.epa.gov/glnpo/bns/index.html.

—. 2007. *Control of Hazardous Air Pollutants from Mobile Sources*. Washngton, DC, USA : Office of Transportation and Air Quality, 2007. EPA420-F-07-017.

—. 2012f. Controlling Power Plant Emissions: Chronology. *Mercury*. [Online] February 7, 2012f. [Cited: November 24, 2012.] http://www.epa.gov/hg/control_emissions/decision.htm.

—. 2006. *EPA's Roadmap for Mercury*. Washington, DC, USA : US Environmental Protection Agency, 2006. EPA-HQ-OPPT-2005-0013.

—. 2008. *Latest Findings on National Air Quality (2006)*. Research Triangle Park, North Carolina, USA : Office of Air Quality Planning and Standards, 2008. EPA-454/R-07-007.

—. 2012g. Lead. *Air Trends*. [Online] November 28, 2012g. [Cited: July 12, 2013.] http://www.epa.gov/airtrends/lead.html.

—. 2012c. Lead Nonattainment Areas. *Green Book*. [Online] December 2012c. http://www.epa.gov/oaqps001/greenbk/lindex.html.

—. 2012d. Mercury and Air Toxics Standards. [Online] US Environmental Protection Agency, March 27, 2012d. [Cited: March 2, 2013.] http://www.epa.gov/mats/.

—. 1997. *Mercury Study Report to Congress. Volume 5: Health Effects of Mercury and Mercury Compounds*. Washington DC, USA : Office of Air Quality Planning and Standards, 1997. EPA-452/R-97-007.

—. 2012b. *Mobile Source Air Toxics*. 2012b. http://www.epa.gov/otaq/toxics.htm.

—. 1995. *Profile of the Nonferrous Metals Industry*. Washington, DC, USA : s.n., 1995. EPA 310-R-95-010.

—. 2007. *Revisions to Definition of Cogeneration Unit in Clean Air Interstate Rule (CAIR), CAIR Federal Implementation Plans, Clean AirMercury Rule (CAMR); and Technical Corrections to CAIR, CAIR FIPs,CAMR, and Acid Rain Program Rules*. Washungton, DC, USA : Government Printing Office, 2007. October 19. 72 FR 59204.

US-FDA, US-EPA and. 2004. *What You Need to Know about Mercury in Fish and Shellfish*. 2004. http://water.epa.gov/scitech/swguidance/fishshellfish/outreach/advice_index.cfm. EPA-823-F-04-009.

USGS. 2002. *Glacial Ice Cores Reveal A Record of Natural and Anthropogenic Atmospheric Mercury Deposition for the Last 270 Years*. s.l. : United States Geological Survey, 2002. USGS FS-051-02.

—. 2013. Lead: Statistics and Information. *Mineral Resources Program*. [Online] United States Geological Survey, February 25, 2013. [Cited: March 3, 2013.] http://minerals.usgs.gov/minerals/pubs/commodity/lead/.

—. 2001. *Mercury in US Coal - Abundance, Distribution, and Modes of Occurance*. s.l. : United States Department of the Interior, 2001. FS-095-01.

WHO. 2014. *7 Million Premature Deaths Annually Linked to Air Pollution*. Geneva, SW : World Health Organization, 2014. http://www.who.int/mediacentre/news/releases/2014/air-pollution/en/.

Xu, Jiaquan, et al. 2010. *National Vital Statictics Report*. s.l. : US Department of Health and Human Services, CDC, 2010. Vol. 58, No 19.

Yuan, Chung-Shin, Lin, Hsun-Yu and Chen, Wei-Chin. 2004. *Adsorption Isotherm of Vapor-Phase Mercury Chloride onto Spherical Activated Carbon via Thermogravimetric Analysis*. London, UK : 13th World Clean Air and Environmental Protection Congress, 2004. 987654321 / 12717.

CHAPTER **14**

Carbon Dioxide and Climate Change

Carbon dioxide (CO_2) is unlike other air pollutants. At current atmospheric concentrations it is not toxic, it is relatively unreactive, and it does not directly cause the death of plants or animals. It causes no harm at the local or regional scales, and it does not cause problems at short time scales. It remains unregulated in most of the world because of these traits. However, it does cause harm on a global scale and over long time scales by changing the earth's climate.

This pollution problem is often referred to as global climate change (GCC) or global warming. The discussion linking carbon dioxide to GCC can be summarized by four basic ideas:
1. The greenhouse effect is real, and it plays a central role in the earth-sun energy balance which determines the earth's temperature.
2. CO_2 is a greenhouse gas.
3. CO_2 is increasing in the atmosphere due to human activity.
4. Changing the energy balance changes the climate.

Observation, direct measurement, and theoretical understanding provide strong evidence for each idea, and there is no scientific evidence countering these ideas. This chapter explores the evidence for each, although a detailed description of climate processes is beyond the scope of this book. The chapter concludes with a discussion of carbon dioxide control measures.

14.1 General Information and Basic CO_2 Chemistry

Carbon dioxide (CO_2) at normal atmospheric conditions is a colorless, odorless gas. It is a linear molecule (O=C=O) that contains symmetric double bonds between the central carbon atom and each oxygen atom. It has no electrical dipole when in its rest (lowest energy) state. It is non-flammable but is moderately reactive and supports the combustion of metals, such as magnesium. It was first discovered in the 17th century by Jan Baptist van Helmont, a Flemish chemist (Wikipedia, 2011a). Many of its physical properties were measured in the 1750's by Joseph Black, a Scottish physician.

Carbon dioxide is an essential component of the earth's carbon cycle and is of central importance to all life on earth. Plants, algae, and photosynthetic bacteria convert CO_2 and water in the presence of sunlight to create carbohydrates and oxygen, see equation (14.1). Oxygen is a waste gas, and the carbohydrates provide energy storage.

14.1 $$n(CO_2) + n(H_2O) + h\upsilon_{sunlight} \rightarrow (CH_2O)_n + n(O_2)$$

During times when there is no sunlight, the organisms combine the carbohydrates with oxygen to form energy, CO_2 and water, see equation (14.2). Consumers of these organisms may also use these carbohydrates for their own energy needs. Note that the energy created is thermal energy rather than the sunlight used in the equation (14.1).

14.2
$$(CH_2O)_n + n(O_2) \rightarrow n(CO_2) + n(H_2O) + energy$$

All molecular oxygen present in the atmosphere is due to an imbalance between the formation of these carbon molecules and their use. The imbalance happens when these carbon compounds are removed from the carbon cycle for long periods of time (on the order of 10 million years or longer). Examples of this long-term storage include reservoirs of natural gas or oil and coal beds.

CO_2 in the atmosphere can readily dissolve in water. Once in solution it can react with water to form carbonic acid (H_2CO_3) and acidify the water.

14.3
$$CO_2 + H_2O \rightarrow H_2CO_3$$

Depending on the pH of the water, the carbonic acid may dissociate to form bicarbonate ion (HCO_3^-) at pH between 6.3 to 10.3 or carbonate ion (CO_3^{2-}) at pH greater than 10.3.

14.4
$$CO_3^{2-} + 2H_2O \rightleftarrows HCO_3^- + H_2O + OH^- \rightleftarrows H_2CO_3 + 2OH^-$$

The presence of cations may allow the formation of bicarbonate or carbonate salt. Some of these salts are soluble ($NaHCO_3$) and some are insoluble. A chemical reaction between calcium ion and carbonate ion forms an insoluble solid precipitate known as limestone or chalk:

14.5
$$Ca^{2+} + CO_3^{2-} \rightarrow CaCO_3$$

If this limestone is later exposed to acid, it reacts to reform carbonic acid:

14.6
$$2HCl + CaCO_3 \rightarrow CaCl_2 + H_2CO_3$$

Heating limestone to approximately 850 °C produces lime. The reaction also releases carbon dioxide as a byproduct. This reaction is common in the creation of cement and concrete:

14.7
$$CaCO_3 + heat \rightarrow CaO + CO_2$$

Another important reaction producing CO_2 emissions is the combustion of carbon-containing materials (e.g. natural gas, oil, coal, peat, wood, bagasse). The combustion reaction for methane is:

14.8
$$CH_4 + 2O_2 \rightarrow CO_2 + 2H_2O + heat$$

Similarly, yeast metabolizes sugar to produce carbon dioxide and ethanol, also known as alcohol, in the production of bakery goods, wine, beer, and other spirits:

14.9
$$C_6H_{12}O_6 \rightarrow 2CO_2 + 2C_2H_5OH$$

14.2 Sources and Sinks

The amount of atmospheric carbon dioxide is increasing. An increase in sources of CO_2 (e.g. the extraction and release of carbon from fossil fuels and limestone) and the lack of a similar size sink causes accumulation in the atmosphere. The long-term atmospheric concentration over the previous 800,000 years has fluctuated between 180 - 280 ppmv, making it the fourth most common substance in dry air. However, since the 1800's, the concentration in the atmosphere has been increasing above this long-term average, currently by approximately 1%/yr. Figure 14-1 shows atmospheric CO_2 concentrations from 1744 to 1953 as measured in ice core samples from Antarctica, (Neftel, 1994).

Figure 14-2 shows more recent data from an observatory atop Mauna Loa, Hawaii, USA (Tans, et al., 2011). This observatory started to measure CO_2 in March 1958 and has observed an annual increase every year. The location was chosen for its remote location with minimal influence from vegetation or human activity. Mauna Loa is not an active volcano.

FIGURE 14-1. Historical Values of CO_2 Concentration in Atmosphere from the Siple Station Ice Core in Antarctica (1744 – 1953).

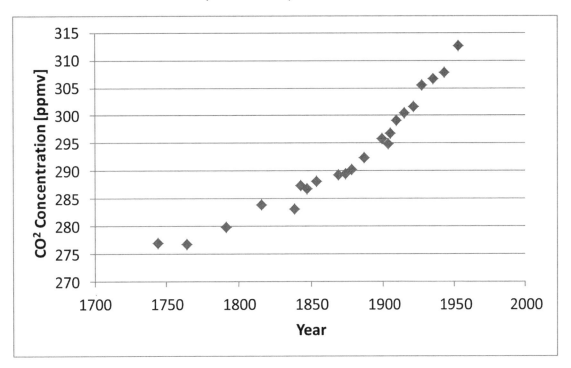

FIGURE 14-2. Concentration of CO_2 in the Atmosphere.

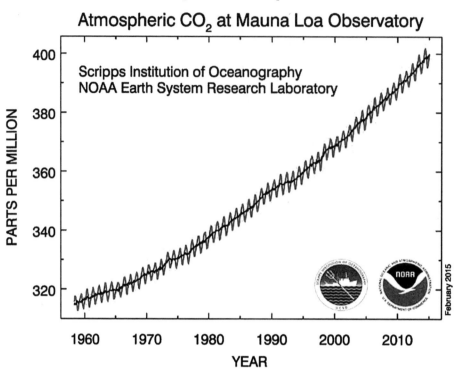

14.2.1 Sources

There are many natural and anthropogenic sources for CO_2 emissions. The typical unit for atmospheric CO_2 is 'GtC', which means gigaton of carbon (1Gt=10^9 ton =10^{12} kg = 10^{15} gram=1Pg). One GtC is equal to 44/12 or 3.667 $GtCO_2$.

Natural sources of CO_2 emission to the atmosphere balanced with similar size sink processes until the start of the Industrial Revolution. The land exchanges carbon with the atmosphere at an approximate rate of 120 GtC/yr. Land uptake occurs through the reduction reactions occurring during primary production (a process called photosynthesis). Atmospheric return occurs through the emissions (respiration) occurring from the oxidation of these photosynthetic products. The ocean exchanges carbon with the atmosphere at an approximate rate of 90 GtC/yr. Accumulation is due to both biological (photosynthesis) and chemical uptake (carbonate chemistry – see reactions (14.3) to (14.7). One other major natural source is volcanoes, which contributes approximately 0.13 - 0.23 GtC/yr.

Anthropogenic sources are due to the extraction and use of fossil fuels and limestone. These sources add carbon to the atmosphere at a rate of approximately 9 GtC/yr (IEA, 2012). There currently are no human activities to balance these emissions. This imbalance causes a similar size increase of carbon in the atmosphere, in the oceans, and on land, with an approximate partition of 50% into the atmosphere, 30% into the oceans, and 20% stored on land. It is believed that the amount of uptake by the oceans will decrease over time due to the decrease in solubility as water warms and as the surface water becomes saturated with carbonates. It is very uncertain how the land uptake will change as it is exposed to increasing quantities of carbon dioxide, although it is currently increasing its rate of uptake. Total emissions for these global sources are listed in Table 14-1, (Boden, et al., 2011).

Table 14-1. Cumulative Worldwide Anthropogenic Emissions of Carbon to the Atmosphere (1750 – 2008).

Source	Total Emissions [Gigaton C]
Coal	168
Oil	122.7
Nat Gas	44.8
Cement	8
Flaring	3.3
Total	346.8

Figure 14-3 provides the global annual total emissions from fossil fuels and cement production (Boden, et al., 2011). The estimates of emissions prior to 1990's are determined by assuming that all the fuel produced in a given year was used in that year. It is a fair assumption because these fuels are rarely stockpiled in any large quantity or for long periods of time.

Table 14-2 provides a more detailed list of the breakdown of US anthropogenic emissions (US-EPA, 2011). The largest anthropogenic source (US and global) of carbon emissions to the atmosphere is from the use of fossil fuels. These materials are extracted from buried reservoirs and release carbon to the atmosphere during or after their use. The US estimates are based on reported production values. The remaining categories are based on the following considerations.

Other fossil fuel uses include CO_2 released from activities where the fossil fuel serves as a chemical feedstock, such as for polymer and plastic production.

Metal production includes the purification and refinement of iron, steel, aluminum, lead, and zinc. The CO_2 is released from carbonate ores and fossil fuels used in production.

Natural gas production includes leaks, ventings, and flarings during production and transportation. Natural gas releases are reported as a CO_2 equivalent value since most of the release converts to CO_2 in the atmosphere, see equation (14.8).

FIGURE 14-3. Annual Anthropogenic Emissions of Carbon to Atmosphere (1800 – 2008).

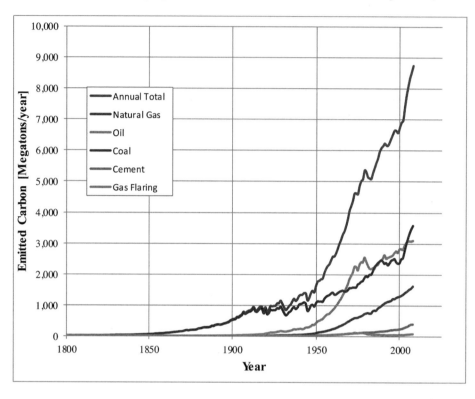

Land use is the net change due to cropland and wetland alterations. A positive value reflects that the carbon is moving from the soil to the atmosphere.
Cement production releases CO_2 due to the chemical reaction to form lime from limestone, as well as from the heat needed to drive the reaction, see equation (14.7).

Note that forests have a negative value, which means that US forests are accumulating carbon rather than emitting it. This value is tracked to provide an offset if carbon emissions in the US become regulated. Globally, deforestation is occurring faster than reforestation and adds about 1 GtC/yr to the atmosphere.

The values in these tables and figures exclude carbon sources from short term sinks, such as the use of biomass, combustion of wood, decomposition of biomass from seasonal and annual changes in growth, and respiration of humans and animals. These short-term fluctuations count as part of the 120 GtC/yr respiration part of the land exchange. It is wrong to compare them with fossil fuel emissions because the source of the carbon is different – the CO_2 humans and all living things expire originated with carbon that is currently part of the short term atmosphere - ocean - land exchange.

Table 14-2. US Anthropogenic Emissions of Carbon to Atmosphere by Source (1990 – 2009) [Megaton C/year].

Source	1990	1999	2009
Fossil Fuel Combustion	1292.29	1470.3	1420.64
Other Fossil Fuel uses	32.35	45.79	33.65
Lime/Limestone/Soda Ash	38.4	36.71	22.69
Metal Production	12.33	10.85	10.5
Natural Gas Systems	9.08	10.72	7.91
Land use (net)	2.56	3.38	3.85
Cement	4.58	4.8	3.22
Ammonia/ Urea	3.14	3.9	3.05
Waste Incineration	1.94	2.05	2.13
Petro/Petrochemical	1.06	1.39	0.87
Other	0.9	0.87	0.79
Total Source	1398.63	1590.76	1509.3
Off-sets:			
Forests	-861.5	-534.1	-1015.1
Total (Sources - Offsets)	537.13	1056.66	494.2

14.2.2 Sinks

Figure 14-4 shows the global carbon cycle, based on data from Figure 7-3 of the fourth-assessment report of the Intergovernmental Panel on Climate Change (IPCC, 2007). The values are a bit out of date already – the 2013 value for carbon in the atmosphere is 840 GtC, with an accumulation rate of approximately 4 GtC/yr. Other values are more difficult to quantify on a yearly basis and no update in their values is currently available.

There are three main carbon sinks, or processes for removing carbon from the atmosphere - land storage, ocean storage, and sedimentation:

Land biota and soils storage values are relatively stable except for northern hemisphere reforestation, which removes approximately 0.6 GtC/yr from the atmosphere and tropical deforestation, which currently adds approximately 1.6 GtC/yr to the atmosphere. Land biota and soil storage is considered a short to medium term sink (1 – 1,000 years). Most land plants live only a year or just a few years. While alive they take in and store carbon which returns to the atmosphere very soon after they die (1 to 5 years). The residence time is somewhat different for forests, where the storage time

FIGURE 14-4. Global Carbon Cycle, 2005 Data.

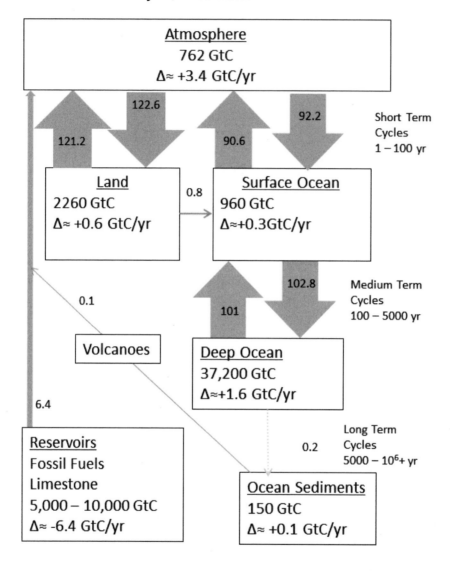

of the carbon depends on the lifespan of the trees (25 – 300 years) and what happens when they die. Specifically, carbon from trees is returned the atmosphere immediately during a fire, after about 10 years if allowed to decompose on the forest floor, and after approximately 1000 years if the tree is buried, such as in a peat bog.

The second sink arises from the accumulation of dissolved organic carbon and inorganic carbon in the ocean (surface and deep regions), which currently accumulates approximately

1.9 GtC/yr. Carbon is transported to the deep ocean by diffusion or advection. Diffusion occurs everywhere but is slow and accounts for a small fraction of the total transport. Advection between layers occurs in only a few locations where currents move vertically. These locations occur where the surface water becomes cooler and saltier and thus denser, so that it sinks. These down-welling areas are in the Arctic and Antarctic regions. Upwelling occurs in equatorial locations and the Antarctic. Deep ocean storage is a medium to long-term sink (100 – 5,000 years). Particle organic carbon returns to the atmosphere within ten years of formation, dissolved inorganic carbon within 1,000 years, and dissolved organic carbon within 5,000 years. Variations occur due to specific patterns of ocean circulation, temperatures, pressures, and presence of marine biota.

The third sink arises from the storage of carbon in land and ocean sediments with subsequent burial, which currently accumulates approximately 0.2 GtC/yr. This storage is a long term sink (1,000 – 10^6 or more years). Land sediments may last 1,000 years during which it may move to the oceans through erosion. Methane clathrates, kerogen, and sedimentary formations of limestone may remain buried for millions of years (Reeburgh, 1997). The rate of this value, the only geologically long term sink, is very similar to the rate at which volcanoes release carbon into the atmosphere. This balance is why CO_2 concentrations in the atmosphere had been so stable over the last million years.

■ *Example 14-1.*

Use the data from Figure 14-4 to determine how long it may take for the carbon removed from the long term geological reservoirs to be naturally sequestered in the ocean sediments for long term storage.

Solution:

Figure 14-4 shows that, in 2005, 6.4 GtC was added to the atmosphere from the long term geological reservoirs (mainly fossil fuels). It also shows that approximately 0.1 GtC/yr is sequestered into the ocean sediments.

$$time = \frac{6.4 GtC}{0.1 GtC/yr} = 64 \, years$$

Current information suggests that the fossil fuel removal is around 10 GtC (see homework problem 14.2). Each year of current production adds 100 years of time before atmospheric CO_2 levels can return to the current concentration.

14.3 Health Effects

CO_2 is not a traditional air pollutant. It is not toxic at typical atmospheric concentrations. It does not create problems near the source of emissions, nor does it directly cause harm to humans, plants, or animals. Health effects of very high concentrations of CO_2 are discussed in section 15.2.1, Indoor Air Quality. Deaths from CO_2 are rare, and associate with large, quick releases which displace local oxygen and cause suffocation.

The concern arises from the role CO_2 and a few other gases (CH_4, N_2O, CFCs, SF_6) have in the global energy balance. The global energy balance requires that the amount of energy added to the earth from the sun must be equal to the amount of energy leaving the earth. The energy enters as sunlight and leaves as infrared radiation. Carbon dioxide is a greenhouse gas, this means that it absorbs infrared and near-infrared light emitted from the earth's surface and re-emits it later in a random direction, much like the panes of glass in a greenhouse. Because the re-emission is in a random direction, some of it returns to the earth, thus causing warming to occur. The warming, in itself, is not a bad thing because the effect from all the greenhouse gases currently warms the planet approximately 38°C and prevents the earth from freezing. The problem arises because, as we saw in the previous section, CO_2 is accumulating in the atmosphere. With this increase comes a change in the greenhouse effect. This change will cause an increase in the earth's average global temperature. Any change in average temperature will alter climate, though in very unpredictable ways. It is this effect on the global energy balance that causes CO_2 to be an air pollutant.

14.3.1 CO₂ and Climate Change Science

The most important consideration in understanding climate and climate change is determining how energy enters, leaves, and distributes over the Earth and its atmosphere. The first law of thermodynamics provides the mathematical tool to describe these relationships. The first-law states that the total energy of a system and its surroundings is constant. Any change in one is balanced by a change in the other:

14.10
$$\Delta E_{system} + \Delta E_{surroundings} = 0$$

In this equation, the system is defined as the Earth and its atmosphere, and the surroundings is the Sun and the rest of space. The most important source of energy entering the system is from the Sun. We can quantify this amount using the Stefan-Boltzmann equation

14.11
$$I = \varepsilon \sigma T^4$$

Where: I = intensity [W/m²],
  ε = emissivity [unitless],
  σ = Stefan-Boltzmann constant [5.67 x10-8 W/m²K⁴], and
  T = absolute temperature [K]

Carbon Dioxide and Climate Change

Intensity represents energy transferred from/to an object per unit area. Emissivity is a number between zero and one, representing how well an object radiates energy. We assume a value of one (perfect blackbody) for all our calculations. The most important consideration here is that the amount of radiation an object emits is directly related to its temperature. A warmer object emits more radiation than a cold object.

14.3.1.1 Black Body Model

The simplest model for a global energy balance of the Earth assumes a transparent atmosphere and models the Earth as a spherical black body. Applying an energy balance on this system gives the average global temperature. This obvious oversimplification provides the average temperature of an object in space with no atmosphere, such as the moon or an asteroid, see Figure 14-5. Equation (14.10) can be applied to this system:

14.12 $$E_{in} - E_{out} = 0$$

Where: E_{in} = the energy from the Sun, and
E_{out} = the energy radiated away from the surface.

The incoming energy term can be determined from the intensity of the energy received from the sun reduced by the fraction that is reflected from the earth (see Table 14-3), and multiplied by the Earth's cross-sectional area (πr^2).

14.13 $$E_{in} = I_{in}(1-\alpha) A_{Earth} = I_{in}(1-\alpha)\pi r^2_{Earth}$$

Where: I_{in} = energy intensity into the Earth system,
α = Earth's surface albedo, average value of 0.33, and
A_{Earth} = Earth's cross-sectional area.

Intensity is determined from the Sun's energy flux, and accounting for the spreading out of that energy as it leaves the Sun's surface (r_{Sun} = 6.95 x10^8 m) and travels to the Earth ($d_{Sun-Earth}$ = 1.49 x 10^{11}m) which decreases the intensity by the square of the ratio of these distances:

14.14
$$I_{In} = \varepsilon_{Sun} \sigma T^4_{Sun} \left(\frac{r_{Sun}}{d_{Sun-Earth}}\right)^2$$

$$= 1 * 5.67 \times 10^{-8} \frac{W}{m^2 K^4} * (5800K)^4 * \left(\frac{6.95 \times 10^8 m}{1.50 \times 10^{11} m}\right)^2 = 1377 \frac{W}{m^2}$$

The actual measured value of this quantity is I_{in} = 1361 W/m^2, with a range of 1321 – 1413 W/m^2 to account for the earth's elliptic orbit and spherical surface.

FIGURE 14-5. Global Energy Flux Models.

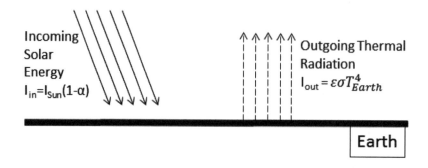

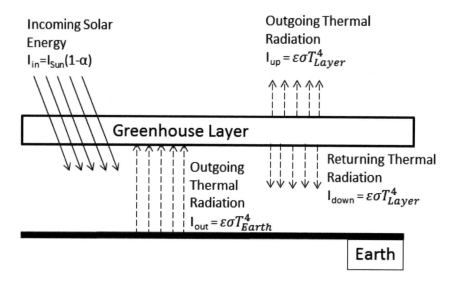

Albedo represents the reflective nature of the earth's surface, see Table 14-3. Ice, water, and some atmospheric particles act like mirrors and reflect much of the incoming energy back into space without it ever interacting with the Earth system. This value varies over different surfaces and varies seasonally. Changes in climate and land use can affect albedo. It takes on values between 0 and 1 and in general dark colored objects have a low albedo and light colored have a high value.

Table 14-3. Albedo of Several Materials.

Surface	Albedo, α
Perfect Absorber	0.00
Fresh Asphalt	0.04
Open ocean	0.06
Worn Asphalt	0.12
Conifer Forest	0.12
Deciduous Forest	0.16
Bare Soil	0.17
Green Grass	0.25
Earth (average)	0.33
Desert Sand	0.40
Concrete	0.55
Ocean Ice	0.60
Fresh Snow	0.85
Perfect Reflector	1.00

Other atmospheric constituents, such as aerosols, may interact with incoming solar radiation. Some compounds reflect the sunlight and cause a cooling effect. Examples include sulfuric acid, nitrates, and organic carbon compounds. Some compounds can scatter/adsorb the light and cause localized heating. Examples include elemental carbon particles and mineral dust. In addition to these direct effects, some substances have indirect effects. Materials that act as cloud condensation nuclei (CCN) can alter the location and size distribution of water droplets. The change in cloud cover and type of cloud will alter the incoming solar radiation and the outgoing thermal radiation. The overall global impact of the direct and indirect radiative forcing of aerosols is a net cooling of the Earth, although there is some uncertainty about this.

The energy out (E_{out}) term arises from the radiation of heat from the Earth's surface to its surroundings (outer space). This energy is in the infrared part of the light spectrum, so it is of much lower quality than the sunlight entering the Earth system. It is invisible to the human eye. Also, this energy leaves from the Earth's entire surface, including the lit and unlit sides, and it leaves in equal proportion in every direction. The energy out term can be rewritten using the Stephan-Boltzmann equation as:

14.15
$$E_{out} = I_{out} SA_{Earth} = \varepsilon \sigma T_{Earth}^4 \left(4\pi r_{Earth}^2\right)$$

Equating (14.13) and (14.15) then solving for the Earth temperature yields:

$$14.16 \quad T_{Earth} = \left[\frac{I_{in}(1-\alpha)}{4\varepsilon\sigma}\right]^{1/4} = \left[\frac{1361\frac{W}{m^2}(1-0.33)}{4*1*5.67x10^{-8}\frac{W}{m^2K^4}}\right]^{1/4} = 251K$$

The black-body temperature is lower than the actual Earth's average temperature of 289K. This model is too cold because it lacks the greenhouse effect of the atmosphere. The next section explores how the greenhouse effect works with another simplified system called the one-layer model.

14.3.1.2 One-Layer Model

The one-layer model simplifies the actual greenhouse effect by replacing a complex atmosphere with a layer equivalent to a greenhouse pane of glass. The layer is assumed to be completely transparent to the Sun's energy and completely absorbs the thermal energy leaving the Earth's surface, see

Figure 14-5. The energy leaving this layer is either directed up and escapes, or is directed down and returns to the Earth's surface.

$$14.17 \quad \begin{aligned} E_{in} + E_{down} &= E_{out} \quad &\text{Earth Energy Balance} \\ E_{out} &= E_{up} + E_{down} \quad &\text{Layer Energy Balance} \\ E_{in} &= E_{up} \quad &\text{Overall Energy Balance} \end{aligned}$$

Where: E_{in} and E_{out} are given in equations (14.13) and (14.15),

$$14.18 \quad E_{up} = I_{up}SA_{Earth} = \varepsilon\sigma T_{layer}^4 \left(4\pi r_{Earth}^2\right), \text{ and}$$

$$14.19 \quad E_{down} = I_{down}SA_{Earth} = \varepsilon\sigma T_{layer}^4 \left(4\pi r_{Earth}^2\right)$$

Substitution of these expressions into the equations (14.17) and canceling of common quantities yields:

$$14.20 \quad \begin{aligned} I_{in}(1-\alpha) + 4\varepsilon\sigma T_{layer}^4 &= 4\varepsilon\sigma T_{Earth}^4 \\ T_{Earth}^4 &= 2T_{layer}^4 \\ I_{in}(1-\alpha) &= 4\varepsilon\sigma T_{layer}^4 \end{aligned}$$

This set of equations contains only two unknowns, so only two of the three equations need to be solved. Solving the third equation, and then the second is probably the easiest way, and yields:

T_{layer} = 251 K (same as the black body model), and T_{Earth} = 298 K (25°C or 77°F).

Comparison of these two models, black body and the one-layer, show how the greenhouse effect works. Both models are too simplistic to predict the actual global temperature of 289K (61°F), although the one-layer provides a closer prediction. More realistic models include complexities that account for air motion, thermal radiation absorption that depends on the wavelength of the radiation, the chemical species in the atmosphere, their concentrations, variations in solar energy, and the latitude of thermal emission.

14.3.1.3 Greenhouse Gases

There are several important greenhouse gases in the earth's atmosphere. These gases are listed in Table 14-4 along with their concentration in the atmosphere, an estimate of their current contribution to the earth's greenhouse effect, the average time spent in the atmosphere, and their global warming potential (GWP). GWP is similar to carbon dioxide equivalents, in that it relates the global warming potential of each gas as compared to the same amount of carbon dioxide. All of these species are greenhouse gases because they have the ability to interact with thermal radiation (they are infrared active). Other atmospheric gases, such as the symmetrical diatomic molecules of nitrogen and oxygen, do not interact with the visible or IR portions of the light spectrum, and so they are not greenhouse gases.

Table 14-4. The Major Greenhouse Gases, 2011.

Species	Formula	Atmospheric Concentration* ppm	Greenhouse Effect Contribution %	Atmospheric Lifetime
Water	H_2O	9000**	36 - 66 excluding clouds 66 - 85 including clouds	9 days
Carbon Dioxide	CO_2	390	9 - 25	50 years***
Methane	CH_4	1.8	4 - 9	12 years
Nitrous Oxide	N_2O	0.02	1 - 2	120 years

Notes: * Dry basis except for water
** Water in atmosphere varies depending on temperature and humidity. Reported value assumes an average global temperature of 289K and 50% relative humidity.
*** This is the average time a molecule remains in the atmosphere before it cycles to land or ocean. It can reenter the atmosphere as part of these cycles, which if included would change this value to approximately 100,000 years.

IR radiation is a form of energy associated with photons that have wavenumbers between 100 cm^{-1} (far IR) and 14,000 cm^{-1} (near IR). The energy of a photon depends on its frequency (or wavelength or wavenumber) as:

14.21
$$E = h\nu = \frac{hc}{\lambda} = hc\sigma$$

Where: h = Planck's constant = 6.63 x 10^{-27} erg sec,
ν = frequency of the photon [Hz or sec^{-1}]
c = velocity of light, 3 x 10^8 m/sec
λ = wavelength of the photon [nm]
σ = wavenumber of the photon [cm^{-1}]

■ **Example 14-2:**
CO_2 absorbs photons with an infrared energy at a wavenumber of 667 cm^{-1}. Calculate the a) wavelength [nm] and b) frequency of these photons [THz].

Solution:
Apply equation (14.21) and make sure units are considered. We are given λ=667 cm^{-1}

a) $\frac{hc}{\lambda} = hc\sigma$ which simplifies to $\lambda = \frac{1}{\sigma}$. Wavelength is reported in nanometers, so convert the wavenumber from cm^{-1} to nm^{-1} before applying the equation. Recall 1 cm = 10^7 nm.

$$\lambda = \frac{1 cm}{667.3} * \frac{10^7 nm}{1 cm} = 14,986 nm$$

The interaction between IR radiation and a molecule occurs when a molecule absorbs a photon. This absorbance causes the molecule to become excited. The excitement lasts 10-9 to 10-6 seconds, after which the molecule relaxes and emits a photon. This emitted photon is often at the same wavenumber as the absorbed photon, but this does not always happen and depends on the energy absorbed and the final relaxed state. However, the total energy emitted must be equal to the total energy absorbed for a molecule moving from its ground state to an excited state and then back to its ground state. Note that at normal earth temperatures, molecules spend almost all of their time in their lowest energy (ground) state.

Photons emit from the molecule in any direction. The one-layer model accounted for this by assuming that half the thermal energy leaving the layer was emitted up, and half was emitted down. Any sideways motion would eventually either return to the Earth or escape.

The particular values of energy a molecule absorb depend on the types of atoms, types of chemical bonds, and the actual geometric orientations between the atoms. A three atom molecule has four possible vibrational states:

symmetric stretch – the two outside atoms move further away and closer to the center atom at the same time and rate,

up/down bending – the two outside atoms move up and down while the center atom moves down and up, out of sync with the others;

in/out bending – similar to the up/down except the atoms are moving horizontally rather than vertically; and

asymmetric stretch – the outside atoms move further away and closer to the center atom at different times and rates.

Note that the up/down and in/out bending have the same energy values because they are only different when considering the overall orientation of the molecule. Such similarities are called degeneracies. A linear orientation (where all three atoms connect in a line) do not have an IR active symmetric stretch because the change in molecular energy from each outer atom's motion is exactly canceled by an equal but opposite in sign motion of the other outer atom.

Note that any integer multiple or sum of the energies associated with the excited states also are allowable energy changes (degeneracies), but values in between are not allowed. Another way of saying this is the excitation energies are discrete and quantized. Table 14-5 lists the first excited states for three greenhouse gases (Atkins, et al., 2010). Methane, a symmetric five atom molecule in the shape of a tetrahedron has two distinct IR excited states – 3020 cm^{-1} and 1300 cm^{-1}, each of which has several degeneracies associated with it. Ozone is also a three atom molecule that is arranged in a triangle shape with excited states at 528 cm^{-1}, 1033 cm^{-1}, and 1355 cm^{-1}.

Table 14-5. IR Wavenumber (cm-1) of Various Excited States for Three Greenhouse Gases.

Molecule	CO_2	H_2O	N_2O
Orientation	linear	non-linear	non-linear
symmetric stretch	NA	3651.7	1285
up/down bending	667.3	1595	588.8
in/out bending	667.3	1595	588.8
asymmetric stretch	2349.3	3755.8	2223.5

Figure 14-6 shows a modeled thermal spectrum [adapted from (Archer, 2007)] of the radiation emanating from Earth (the squiggly line) and several additional spectrums associated with black-body radiation for several temperatures (smooth lines). The model calculates the energy intensity

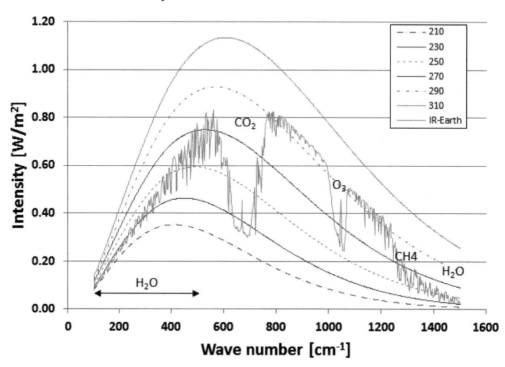

FIGURE 14-6. Model IR Spectrum Emitted from Earth Compared with Blackbody Radiation from Various Temperature Surfaces.

for each value of wavenumber as would be measured by a sensor at 70 km height, assuming a constant surface temperature, no clouds or rain, and a 1976 standard US atmosphere to provide concentrations of the atmospheric constituents.

The Earth's modeled spectrum would look like a blackbody if it had no atmosphere. Instead, it ranges between two of the black-body spectra – one just above the 290 K line (surface temperature) and one just below the 230 K (tropopause temperature) line. The upper-line values correspond to wave numbers that do not interact with any atmospheric constituents and thus emit at the Earth's surface temperature. The lower line corresponds to wave numbers that interact with the IR active constituents of the atmosphere. When this radiation is finally emitted to the sensor, it originates from molecules at higher elevations, where the atmospheric temperature is colder (see section 5.3). Note how these dips correspond to the wave number values in Table 14-5 and its preceding paragraph. The labels on the graph help distinguish which absorption region corresponds with the various compounds.

When thermal radiation is absorbed and re-emitted, the intensity of emissions corresponds with the temperature of the emitting molecule. Figure 14-7 demonstrates how this may happen. A photon at 667.3 cm^{-1} is emitted from the ground (T=298 K) and after some distance (1000m) it is absorbed by and excites a CO_2 molecule. This molecule then emits a photon at 667.3 cm^{-1} but with an intensity associated with a temperature of 292 K. The photon can emit up or down, and repeats this process multiple times until an emission occurs that allows the photon to escape the atmosphere without another interaction. The most likely elevation for such a photon to escape is near the tropopause (elevation approximately 12 km and a temperature of 225 K in this example).

FIGURE 14-7. Simplified Model of a Path a Photon (wavenumber 667.3 cm^{-1}) may Take between Emission from the Earth's Surface until it Escapes into Space. An actual path includes many interactions and more variety in directions and distances.

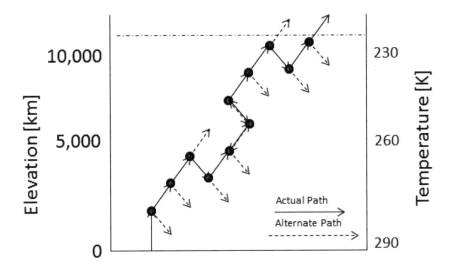

The width of the absorbed region, called a band, is due to the concentration of the absorbing molecule in the atmosphere. Higher concentrations lead to deeper and wider bands. Bands only go as deep as the blackbody temperature profile associated with their final emission temperature. The change in width may seem contradictory to the fact that each molecule only absorbs at a single wavenumber. There are several mechanisms that cause these lines to widen: Doppler shift, lifetime broadening, and band saturation (Atkins, et al., 2010). Doppler shift arises from the velocity difference between the molecule and the observer and causes a change of (1 + v/c) to the wavelength, where c is the speed of light, and v, the molecules velocity, is negative if it is approaching the observer and is positive if it is receding. Lifetime broadening is due to the time required to enter / leave the excited state. Since the energy is spread out during this time, the wavelength becomes less certain as $\Delta E \approx h/\tau$. Short lived excited states are characterized by broad lines and long-lived states have narrow lines. For example if the excited state has a lifetime of 10^{-10} seconds, the line will expand by about 0.05 cm^{-1}. Neither of these mechanisms account for the width of the observed bands. Almost all of the width is attributed to band saturation.

Whenever a molecule absorbs or emits a photon, it requires some amount of time – it is not instantaneous. If, during the absorption/relaxation time it collides with another molecule, the total energy of the photon plus the collision must be equal to the energy of the excited state. For example, if a particular molecule of CO_2 collided with an N_2 molecule (which causes a change in velocity or acceleration of the CO_2 molecule during the time interval of the collision) and were hit by a photon, it would only accept that photon for excitation if the total energy (photon plus collision) was identical to the energy of a photon of 667.3 cm^{-1} wavenumber. In this case, the photon would need a lower energy (higher wavenumber). If the CO_2 had transferred energy to the N_2, then a higher energy photon would be needed to make up the difference. As the concentration of CO_2 increases, such three-way collisions are more common and cause the band to expand.

Recall that the global energy balance requires that the incoming solar radiation must balance the outgoing earthlight radiation. Figure 14-6 can be used to determine the total outgoing radiation by summing the intensity across the wavenumbers. Table 14-6 lists this value for several different CO_2 concentrations and surface temperatures, data adapted from (Archer, 2007). It shows that the amount of outgoing thermal radiation declines as the concentration of CO_2 increases, and increases as the surface temperature increases. This is because as CO_2 concentration increases the saturation band expands, and so less energy is emitted within that band. Another way to present this information is given in Figure 14-8 where the relationship between $T_{surface}$ and CO_2 concentration is given for the case of $I_{earth} = I_{sun} = 260$ W/m². This calculation assumes that the Earth's temperature is constant everywhere and there are no clouds. Such simplifications can be overcome with more sophisticated models. However, the point is that the Earth's surface temperature must increase when CO_2 increases in order to satisfy the energy balance.

Table 14-6. Intensity of Earthlight, I_{earth} [W/m²], for Surface Various Temperatures and Atmospheric CO_2 Concentrations.

[CO$_2$] ppmv	Temperature			
	278 K	288 K	298 K	308 K
0	248.9	286.2	326.2	369.6
10	238.6	273.8	311.9	352.9
100	231	264.1	299.9	338.8
280	227.9	260	294.7	332.5
400	226.8	258.5	292.9	330.3
560	225.6	257.2	291.2	328.1
1000	224	254.8	288.1	324.4
10000	217.1	244.9	274.9	307.4

FIGURE 14-8. Effect of CO_2 Concentration on the Required Surface Temperature of Earth to Achieve $I_{sun} = I_{earth} = 260$ W/m². Figure assumes 1976 US standard atmosphere, no rain or clouds, and constant surface temperature. Model ignores any feedback mechanisms like increased water content as the atmosphere warms.

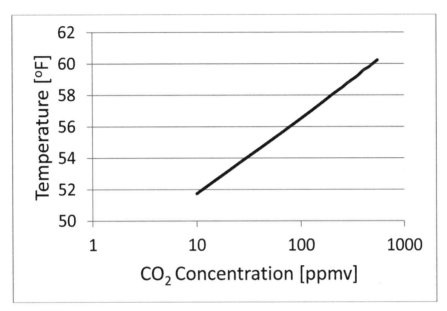

Another key point is that this analysis does not depend on temperature measurements. The rising temperature is an effect of the increase of CO_2. There is much discussion in the news about the temperature record and its historical values. Most such analysis show that there has been an increase in average temperature and that the rate of temperature change is increasing. However, even with increasing CO_2 concentrations there may be years or even decades where the average atmospheric temperature does not increase by much or even decreases. The time-lag is not inconsistent with the energy balance model, as there are many variables and processes which determine the temperature. Such an observation of no or negative temperature change for a year or more is not significant and does not indicate that there is no problem. More or less energy may be transferred to the ocean or soil in a particular year, and this would show a decrease or increase in the atmospheric temperature. The natural variation in global temperature is between 1 and 3°C each year. The natural yearly variation is of the same order of magnitude as the total expected increase due to the current increase in concentration of CO_2.

The impact of this energy difference (or temperature increase) is important to understand. Figure 14-8 shows approximately one degree Kelvin increase (or 2°F) for an increase in CO_2 from 280 ppmv to 560 ppmv. More sophisticated models, with less static assumptions, estimate an increase of 2 to 3 K (4 – 8°F) for such a doubling. Svente Arrhenius, in 1890's, estimated that a doubling of CO_2 would lead to an increase in temperature of 3°C (Arrhenius, 1896), a value that has not changed much since. This is because the physics and chemistry of these processes, at a global scale, are well understood and have been well understood for at least a century. While this value seems small, especially compared to daily and seasonal temperature fluctuations, it is enough to alter the Earth's climate and weather.

How can such a small change alter the Earth's climate? First – the main drivers of climate are temperature and water distribution. Many climate processes and regions exist in pseudo-equilibrium. When any parameter changes, the equilibrium points move. An example is the size and extent of a glacier. If it is slightly colder and/or wetter, the glacier grows, when it is slightly warmer and/or drier, it shrinks. The edge exists at the place where the average climate is balanced between growth and decay. Any change in climate causes this location to move. Actual observation of glacier extent shows that global glacier mass loss has been happening every year for the last 20 years and that the rate of loss is accelerating (Blunden, et al., 2011).

Other projections for a warming climate include:
- Increase in warm weather events such as record high temperatures, earlier day of the year that the spring growing season begins and later last day of frost-free weather in the fall, increase in precipitation extremes (both wet and dry periods), and an increase in intensity of temperature-driven weather phenomena such as cyclones and hurricanes.
- Reductions in areas covered by ice (sea and land) and permafrost in both the northern and southern hemispheres.
- Increases in ocean surface temperature, increases in ocean acidity (from the absorption of CO_2), and increases in sea level due to thermal expansion and melting land ice.
- More rapid increase in temperature in the Arctic and Antarctic and less rapid increase in tropical areas.

All of these projections have been directly observed as of 2007 (IPCC, 2007). These trends are continuing today, and are expected to become larger and more common as the concentration of carbon dioxide continues to increase.

● Case Study 14-1. Debunking the Mythology of Denialism.

There is a great deal of confused information in the public knowledge sphere concerning global climate change. It is based on skeptical inquiry but is more closely described as denialism. Skepticism of climate change is in no longer justified because the unknowns and uncertainties of the science and projections no longer leave sufficient uncertainty to justify a position defending doing nothing (aka business as usual). However, it is wise to remain skeptical of proposed solutions, because there is no clear choice for how to change, see section 14.5. Several of the myths of the denialists are discussed below:

CO_2 is not caused by human activity. Actually it is, see Figure 14-2 and Figure 14-3. Combustion of fossil fuels alone accounts for 220% of the observed increase of CO_2 in the atmosphere. Human activity is adding more than the amount of atmospheric increase. Additional sinks on land and ocean are currently removing some of these fossil fuel caused additions, but they are not enough to balance the additional input, and we may expect these sinks to decrease in the future, especially the ocean.

Volcanoes are the main source of CO_2. This belief was true before the advent of the Industrial Age. Volcanoes add about 0.2 GtC/yr (with a variation on the order of 50%). However, this addition is now dwarfed by the fossil fuel contribution of 8 - 10 GtC/yr.

The amount of human-caused CO_2 is insignificant compared to the total amount. One percent per year may seem trivial, but over decades it adds up. There are approximately 850 GtC in the atmosphere; humans are currently adding about 8 GtC/yr, and atmospheric accumulation is less than 4 GtC/yr, which is less than 0.5 % of the total. However, this has been ongoing for 150 years and has increased CO2 levels by about 40%, which is significant.

Temperature is not changing, or has stopped changing. This belief is simply untrue. There are many temperature records, and obtaining an annual average for the earth is difficult because the measured values are spaced heterogeneously, and few places record for more than a few decades. Any creative subset can show any trend – increasing, decreasing, or neutral. Several good estimates can be seen at RealClimate.org (Schmidt, et al., 2007) or the Berkeley Earth Temperature work at berkeleyearth.org (Rohde, et al., 2011). While there may be time periods of a year to a decade when the change is stagnant or even reverses, the long term trend (which is what climate is) is an increase. Even in a year that sees no change in the atmosphere global temperature, the average temperature still exceeds the long term average.

Climate is too complex to predict; weather predictions are always wrong. Climate is less chaotic than weather, since climate represents long-term averages and excludes most of the daily extremes. It is too complex to make projections at small scales (cities and counties). The models have improved enough to make some *projections at regional and national levels and are excellent at the global scale. They are better at determining temperature than water-related measures (humidity, cloud cover, and total precipitation). The models also are able to predict today's climate from historical data, as well as the effects from perturbations like volcanic eruptions, the North Atlantic Oscillation, and El Nino – La Nina in the equatorial Pacific.*

It is too expensive to change. *A value judgment, but consider that the cleanup after a weather-related disaster can cost $10's of billions, and all projections show an increase in weather extremes, meaning more weather-related disasters. It is possible to consider that it is too expensive to not change.*

It is too hard to change. *Again, this is another value judgment. Many of the required changes make life better and reduce conditions leading to global conflicts. However, all future scenarios, including doing nothing, require large-scale changes in human organizations. It is a matter of choice and control. Waiting until a choice is forced reduces the number of options.*

A new technology will save us. *It does not exist yet, and it is naive to not, at least, invest in such technologies if the solution depends on them, see section 14.5.3.1.*

I do not care about the future, or I will be dead within 50 years so why should I care. *This belief is a value judgment. Many people do care, especially about the type of future our generation leaves for the next, and the one after that.*

The climate is always changing, even without human interference. *This belief is true, but past climate changes are no cause for complacency; indeed, they tell us that the Earth's climate is very sensitive to change in the energy balance. Two main conclusions can be drawn from climate history:*
- *Climate has always responded strongly if the radiation balance of the earth was disturbed. That suggests the same will happen again, now that humans are altering the radiation balance by increasing greenhouse gas concentrations. In fact, data from climate changes in the Earth's history have been used to quantify how strongly a given change in the radiation balance alters the global temperature (i.e., to determine the climate sensitivity). The data confirm that our climate system is as sensitive as the climate models suggest, perhaps even more so.*
- *Impacts of past climate changes have been severe. The last great ice age, when it was globally 4-7 °C colder than now, completely transformed the earth's surface and its ecosystems, and sea level was 120 meters lower. When the earth last was*

2-3°C warmer than now, during the Pliocene 3 million years ago, sea level was 25-35 meters higher due to the smaller ice sheets present in the warmer climate. US cities such as Los Angeles, New Orleans, Miami, New York, and Washington DC will be very significantly impacted by a 5 m rise in sea level.

Past climate changes show that the temperature causes CO_2 to change, not the other way around. *Actually, it works both ways: CO_2 changes affect temperature due to the greenhouse effect while temperature changes affect CO_2 concentrations due to the carbon cycle response. This response is called a feedback loop.*

If global temperatures change, the carbon cycle responds (typically with a delay of decades to centuries). The phenomenon has been observed in the geological records from the ice age cycles of the past 3 million years, which were caused by variations in the earth's orbit (the Milankovitch cycles). The CO_2 feedback amplified and globalized these orbital based climate changes: without the lowered CO_2 concentrations and reduced greenhouse effect, the full extent of ice ages cannot be explained, nor can the fact that the ice ages occurred simultaneously in both hemispheres. The details of the lag-relationship of temperature and CO_2 in Antarctic records have recently been reproduced in climate model experiments (Ganopolski, 2009), and they are entirely consistent with the major role of CO_2 in climate change. During the warming at the end of ice ages, CO_2 was released from the oceans – just the opposite of what we observe today, where CO_2 is increasing in both the ocean and the atmosphere.

If the CO_2 concentration in the atmosphere changes, then the temperature follows because of the greenhouse effect. The CO_2 concentration change is what is happening now from fossil fuel sources. However, this has also happened many times in earth's history. CO_2 concentrations have changed over millions of years due to natural carbon cycle changes associated with plate tectonics (continental drift), and climate has tracked those CO_2 changes such as the gradual cooling into ice-age climates over the past 50 million years.

A rapid carbon release, not unlike what humans are causing today, also occurred at least once in climate history, as sediment data from 55 million years ago show. This "Paleocene-Eocene thermal maximum" brought a major global warming of ~ 5°C, which caused a detrimental ocean acidification and a mass extinction event (Shellito, et al., 2003). However, this event took place over twenty thousand years not hundreds. It serves as a warning to us today.

14.4 Regulatory Considerations

As stated above, CO_2 is not toxic, poisonous, or deadly to people at current atmospheric levels. Levels would need to increase 10,000% to the improbable amount of 40,000 ppm or more before such dangers occurred. As such, CO_2 emissions from human-caused sources pose no short-term problem at local, regional, state or national levels. However, CO_2 and other greenhouse gases (GHGs) do pose a tremendous global scale problem with their ability to trap heat and cause the Earth's climate to change in many ways – both obvious (increase in global average temperature) and unknowable (weather, agriculture, spread of diseases). There has only ever been one other global only air pollution problem – CFC's and stratospheric ozone destruction (Section 12.2).

The emission of CO_2 is lumped together with a few other greenhouse gases. They are looked at together because they share some common properties: infrared activity, emitted by humans in significant amounts (even if they also have natural sources), are well mixed into the atmosphere, and have long lifetimes in the atmosphere. Table 14-7 lists these gases as well as their carbon dioxide equivalent value (CO_2equivalent) (US-EPA, 2011). The equivalence value allows the six to be reported as a single value using the fact that they all interact with thermal radiation. The equivalent number represents the level of activity of each molecule compared to a molecule of CO_2. Thus a single molecule of CH_4 yields as much greenhouse gas effect as 25 molecules of CO_2, as averaged over a 100 year timespan.

Table 14-7. The US-EPA Six Major Greenhouse Gases.

Name	Formula	CO_2 Equivalent
Carbon dioxide	CO_2	1.0
Methane	CH_4	25
Nitrous oxide	N_2O	298
Hydrofluorocarbons	HFC	675 – 14,800
Perfluorocarbons	PFC	7,400 – 13,300
Sulfur hexafluoride	SF_6	22,800

Water, the most important greenhouse gas, is not included in this list because it is not added to the atmosphere by humans at a significant rate, and it has an atmospheric lifetime of only a few days.

14.4.1 US Regulatory Considerations

The US has recently started to address climate change by beginning the process to establish greenhouse gas standards as part of the Clean Air Act. The US-EPA was forced to do this after several states, citizen groups, and NGOs sued to force that federal agency to regulate carbon dioxide and other greenhouse gases as pollutants.

The petitioners 'asked' the EPA to issue a decision concerning the development of regulations of GHGs from motor vehicles. The petitioners specifically referred to section 202(a)(1) of the Clean Air Act (CAA), 42 U.S.C. § 7521(a)(1), which requires the US-EPA Administrator to set emission standards for "any air pollutant" from motor vehicles or motor vehicle engines "which in his judgment cause[s], or contribute[s] to, air pollution which may reasonably be anticipated to endanger public health or welfare." The case began in 2003 when the EPA made two determinations,

1. The EPA lacked authority under the Clean Air Act to regulate carbon dioxide and other greenhouse gases (GHGs) for climate change purposes.
2. Even if the EPA did have such authority, it would decline to set GHG emissions standards for vehicles, believing the US Congress should do so.

At this point, the petitioners appealed the decision, believing it violated the CAA. The U.S. Court of Appeals for the District of Columbia Circuit decided on September 13, 2005, to uphold the decision of the EPA. The case was appealed to the Supreme Court, which agreed to hear the case on June 26, 2006. The specific issues decided were:

1. Whether the petitioners had standing.
2. Whether CO_2 causes air pollution as defined by the CAA. If CO_2 is not an air pollutant, then the EPA has no authority under the CAA to regulate emissions. If CO_2 is an air pollutant, then the EPA Administrator could decide not to regulate carbon dioxide, but only consistent with the terms of the CAA.
3. Whether the EPA Administrator may decline to issue emission standards for motor vehicles on the basis of policy considerations not enumerated in section 202(a)(1).

The consensus judgment found [Massachusetts v. Environmental Protection Agency, April 2, 2007, (SCotUS, 2007)]:

1. The petitioners were found to have standing[1]. Justice Stevens reasoned that the states had a particularly strong interest in the standing analysis.

    "The case has been argued largely as if it were one between two private parties; but it is not. The very elements that would be relied upon in a suit between fellow-citizens as a ground for equitable relief are wanting here. The State owns very little of the territory alleged to be affected, and the damage to it capable of estimate in money, possibly, at least, is small. This is a suit by a State for an injury to it in its capacity of quasi-sovereign. In that capacity the State has an

---

1. Standing refers to the idea that a party must have some material interest or potential for harm from the actions of the other party.

interest independent of and behind the titles of its citizens, in all the earth and air within its domain. It has the last word as to whether its mountains shall be stripped of their forests and its inhabitants shall breathe pure air."

2. The Court held that the Clean Air Act gives the United States Environmental Protection Agency (EPA) the authority to regulate tailpipe emissions of greenhouse gases. The Clean Air Act provides:

"The Administrator shall by regulation prescribe (and from time to time revise) in accordance with the provisions of this section, standards applicable to the emission of any air pollutant from any class or classes of new motor vehicles or new motor vehicle engines, which in his judgment cause, or contribute to, air pollution which may reasonably be anticipated to endanger public health or welfare."

The Act defines "air pollutant" as "any air pollution agent or combination of such agents, including any physical, chemical, biological, radioactive . . . substance or matter which is emitted into or otherwise enters the ambient air." The majority report commented that "greenhouse gases fit well within the Clean Air Act's capacious definition of air pollutant."

3. The Court remanded[2] the case to the EPA, requiring the agency to review its contention that it has discretion in regulating carbon dioxide and other greenhouse gas emissions. The Court found the current rationale for not regulating to be inadequate and required the agency to articulate a reasonable basis to avoid regulation.

Two members of the nine-member court, Chief Justice Roberts and Justice Scalia issued dissenting opinions [(SCotUS, 2007)].

On December 7, 2009, the US-EPA Administrator signed two distinct findings regarding greenhouse gases under section 202(a) of the Clean Air Act as published in the *Federal Register* under Docket ID No. EPA-HQ-OAR-2009-0171:

1. Endangerment Finding. The current and projected concentrations of the six key well-mixed greenhouse gases in the atmosphere threaten the public health and welfare of current and future generations.
2. Cause or Contribute Finding. The combined emissions of these GHG from new motor vehicles and engines contribute to the greenhouse gas pollution which threatens public health and welfare.

---

2. Remand means to send back to. In other words, the Court required the EPA to reconsider its former decision and to make a new determination based on the Court findings.

This final rule became effective January 14, 2010. These findings do not themselves impose any requirements on industry or other entities. However, this action is a prerequisite for implementing greenhouse gas emissions standards for vehicles. In collaboration with the National Highway Traffic Safety Administration, EPA finalized emission standards for light-duty vehicles (2012-2016 model years) in May of 2010 and heavy-duty vehicles (2014-2018 model years) in August of 2011. The new CAFE (corporate averaged fleet efficiency) standards will increase the passenger car and light truck standard to 35.5 mpg in 2016 and 54.5 mpg for model year 2025.

After this case, the US-EPA began two additional programs to address GHGs and climate change:

- On May 13, 2010, EPA issued a final rule (Greenhouse Gas Tailoring Rule) that establishes thresholds for GHG emissions that define when permits under the New Source Review Prevention of Significant Deterioration (PSD) and Title V Operating Permit programs are required for new and existing industrial facilities. Many state plans require permits at pollutant emission levels of 100 to 250 tons per year. However, this value is very small for GHGs. This rule tailors the requirements of the CAA permitting programs to limit which facilities will be required to obtain PSD and title V permits so only the largest sources—power plants, refineries, and cement production facilities – would be required to include GHGs in their permits (construction and operating). These programs do not cover emissions from small farms, restaurants, and all but the very largest commercial facilities at this time. This 'Tailoring Rule' requires permits (construction and operating) for major sources that currently are part of the CAA permitting process and that have GHG emissions of 100,000 tons per year (tpy) or more of carbon dioxide equivalents (CO_{2e}). These permits have the emitter report their emissions and describe any practices at their facility to control the release GHGs.
- On December 23, 2010, the EPA issued a proposed schedule for establishing greenhouse gas (GHG) standards under the Clean Air Act for fossil fuel-fired power plants and petroleum refineries. This schedule provides a path forward that allows the agency to address pollution from sources that makeup nearly 40 percent of the nation's greenhouse gas emissions.

EPA is currently (2014) in the process of creating carbon pollution standards for new power plants, and is engaging with states, stakeholders, and the public to propose carbon pollution standards for existing power plants.

14.4.1.1 Actions for New Power Plants

The US-EPA issued a proposal for carbon pollution from new power plants in March 2012. They received more than 2.5 million comments from the public about the proposal. The US-EPA considered these comments and modified their work. They released a new proposal in September 2013, which sets separate standards for natural gas-fired turbines and coal-fired units. Under the proposed standards, carbon dioxide emissions are controlled by requiring high-efficiency technologies for gas units and requiring both high efficiencies with carbon capture and storage for coal units. The proposed standards are:

Natural Gas Units:
- Use combined cycle process (efficiency of 40-50%), and
- 1,000 lb CO_2/MW-hr (>850 MMBTU/hr), or
  1,100 lb CO_2 / MW-hr ($\leq$ 850 MMBTU/hr)

Coal Units
- 1,000 lb CO_2/MW-hr over 12-month period, or
  1,000 – 1,050 lb CO_2/MW-hr over 84-month period.

■ **Example 14-3:**
Determine the uncontrolled emission rate of CO_2/MW-hr for a) natural gas combined cycle unit (45% efficient), and b) a coal unit (35% efficient). Assume the natural gas is methane with a heat content of 960 BTU/SCF (22,800 BTU/lb), and coal is 75 wt% carbon

Solution:
a) Natural gas

$$\frac{1 lb_{NG}}{22,800 BTU_{in}} * \frac{12 lb_C}{16 lb_{NG}} * \frac{44 lb_{CO2}}{12 lb_C} * \frac{3.413 \times 10^6 BTU}{MW \cdot hr} * \frac{e_{in}}{0.45 e_{out}} = 915 \frac{lb_{CO2}}{MW \cdot hr_{out}}$$

The unit does not need carbon capture as long as the unit achieves a minimum of a 41% efficiency. Such efficiencies are possible with combined cycle technology. More traditional gas plants would have only a 35% efficiency. The proposed standards could be used to justify the additional cost for the NGCC technology if its total cost is less than capturing carbon.

b) Coal

$$\frac{1 lb_{coal}}{10,000 BTU_{in}} * \frac{0.70 lb_C}{1 lb_{coal}} * \frac{44 lb_{CO2}}{12 lb_C} * \frac{3.413 \times 10^6 BTU}{MW \cdot hr} * \frac{e_{in}}{0.35 e_{out}} = 2,500 \frac{lb_{CO2}}{MW \cdot hr_{out}}$$

This unit would require the capture of 60% of its carbon emissions. A higher efficiency unit would somewhat reduce the need for capture.

■ Example 14-4:

Use the data from Example 14-3 to determine the amount of natural gas leakage that would negate the benefit from reduced carbon dioxide emissions. Recall that methane is a more powerful greenhouse gas, see Table 14-7.

Solution:

Table 14-7 lists the $CO_{2\,equivalent}$ of methane as 25. This number can vary somewhat depending on the time scale of concern. The value in the table assumes a 100-year time scale. If the time scale were 20 years, then the value would be 72.

From example 1 we see the reduction in CO_2 per unit of energy is

$$(2,500-915)\frac{lb_{CO2}}{MW \cdot hr_{out}} = 1,585\frac{lb_{CO2}}{MW \cdot hr_{out}}$$

The equivalent amount of methane is $63.4\frac{lb_{CH4}}{MW \cdot hr_{out}}$

The amount of methane used is:

$$\frac{1\,lb_{CH4}}{22,800\,BTU_{in}} * \frac{3.413 \times 10^6\,BTU}{MW \cdot hr} * \frac{e_{in}}{0.45\,e_{out}} = 333\frac{lb_{CH4}}{MW \cdot hr_{out}}$$

The results imply that if approximately 15% of the methane escapes during production, transportation, or use that the benefit of reduced carbon release from using natural gas will be lost. This level of loss is probably on the low side of actual conditions. There are no good estimates of leakage during production; transportation is assumed to have a 3% leakage rate; the end user (consumer) is variable but may be another 3% leakage rate. Note that most current natural gas facilities do not use the combined cycle process and will have 35-38% efficiency. The lower efficiency reduces the allowable leakage rate to 5-6% over the entire life cycle from production to final consumption.

14.4.1.2 Actions for Existing Power Plants

The US-EPA has not yet (2014) issued standards that would apply to existing power plants. The Clean Air Act recognizes that the ability to build emissions controls into a source's design is greater for new sources than for existing sources, so it allows different approaches to set standards for each case. A guiding Presidential Memorandum specifically directs EPA to build on state leadership, provide flexibility, and take advantage of a wide range of energy sources and technologies towards addressing carbon emissions. It is possible that the proposal may not include quantitative carbon emission limits, and instead focus on improving efficiency, standardized operation and maintenance procedures, and use of best practices. However, until the release of such regulation, this discussion is mere speculation.

14.4.2 International Regulatory Considerations

The first international agreement to limit GHG emissions is the Kyoto Protocol (UN, 2012). It is an international agreement linked to the United Nations Framework Convention on Climate Change (UNFCCC) that sets binding targets for 37 industrialized countries and the European community for reducing greenhouse gas (GHG) emissions. The agreed to reductions amount to an average of five per cent against 1990 levels over the five-year period 2008-2012.

This agreement recognized that developed countries are principally responsible for the current high levels of GHG emissions in the atmosphere. So, the Protocol places a heavier burden on developed nations under the principle of "common but differentiated responsibilities."

The Protocol was initially adopted on December 11, 1997, in Kyoto, Japan, and entered into force on 16 February 2005, when the government of Russia adopted the protocol. As of September 2011, 191 states have signed and ratified the protocol. The only industrialized country not to have ratified the protocol is the United States. Other states that did not ratify Kyoto include Afghanistan, Andorra and South Sudan. The reluctance of the US to ratify the treaty is centered on the issues of differences in CO_2 reduction expectations between developed economies (like the US and EU) and developing economies (like China and India).

Under the Treaty, countries meet their targets primarily through national measures. However, the Kyoto Protocol offers additional means of meeting targets by way of three market-based mechanisms.
1. Emissions trading – known as "the carbon market,"
2. Clean development mechanism (CDM), and
3. Joint implementation (JI).

These mechanisms help stimulate green investment and help each country meet their emission targets in a cost-effective way.

Under the Protocol, countries' actual emissions have to be monitored, and precise records have to be kept of the trades carried out. A registry system tracks and records any transactions. The UN Climate Change Secretariat, based in Bonn, Germany, keeps an international transaction log to verify that transactions are accounted for by all parties and that each is consistent with the rules. Reporting is done through submitting annual emission inventories and national reports at regular intervals. A compliance system ensures that each state is meeting its commitments or helps them if they have problems doing so.

The Kyoto Protocol is seen, except in the US, as an important first step towards a global emission reduction regime that stabilizes GHG emissions. However, like the Montreal Protocol for reducing ozone depleting substances (see section 12.2.5), it will not completely solve the problem, but rather provide the tools and ideas for future international agreements that are necessary for a long term solution on climate change.

14.5 Control Technologies

There currently are no large scale technologies to control greenhouse gas emissions or to control the effects of higher atmospheric concentrations of these gases. The problem is exacerbated by the fact that CO_2 is emitted in huge quantities on a daily basis from thousands of large sources and that it is a byproduct from creation of essential resources – energy and power. The 'obvious' solution of stopping production and use of fossil fuels to generate power is unlikely to occur rapidly or soon. The transition away from fossil fuels will occur only when an alternative source of power can replace it because humans will not voluntarily give up power. Alternative sources of power must be inexpensive, abundant worldwide, be easy to transport, have no or few byproducts, and be usable any time of day and any day of the year.

14.5.1 Scale of the Required Yearly Carbon Emission Changes

Engineers and scientists considering solutions to the problem of global climate change suggest carbon emissions from fossil fuels need to be reduced by 80 – 95% from current values to minimize the impact on the earth's climate. Figure 14-9 shows several possible pathways to achieve this reduction (Allison, et al., 2009). Each pathway would reduce emissions such that the impact will be limited to a 2°C rise from the enhanced greenhouse effect. The plot shows that the annual change is less if begun sooner, and increases to harsh levels if begun later. To place this in perspective, when the USSR broke up in the 1990's, the new countries underwent a GDP and energy decline of approximately 5%/yr for the following decade, resulting in rather harsh economic hardship for the majority of citizens. Perhaps this experience can guide world leaders in finding alternative ways to change besides the austerity programs used by the Russian and other governments.

FIGURE 14-9. Various Pathways for Reducing Fossil Fuel Carbon Emissions to Minimize Harm to Human Health and the Environment.

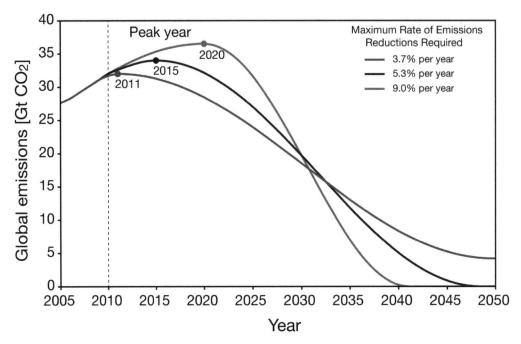

■ Example 14-5:

Human use of fossil fuels in 2011 emitted approximately 10 GtC to the atmosphere. Determine the volume and dimensions of this carbon if it could be stored in one place as a pile of plain carbon. Assume the density is the same as coke (a nearly pure carbon material used industrially for making steel), and that the pile has an angle of repose of 40°.

FIGURE 14-10. Geometry of a cone-shaped pile.

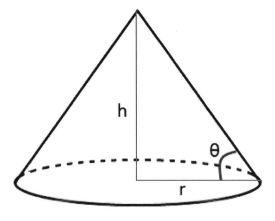

Solution:

To solve this problem we first need to determine the volume of the mass using the density of coke (1500 kg/m³ as per Table 14-8), and then find the dimensions of a pile of this volume.

14.22 $$10 GtC \frac{10^{12} kgC}{1 GtC} \frac{1 m^3}{1500 kgC} = 6.67 \times 10^9 m^3$$

The pile forms a cone shape, and the dimensions depend on the material's angle of repose, see Figure 14-10. This angle, θ, is between the pile's edge and the ground. Also shown in the figure are the height of the cone and radius of the base. The height is related to the base as h = r (tan θ). The volume of a cone follows in equation (14.23). Use the volume obtained above in order to determine the height and area of the base of the pile:

14.23 $$V = \frac{\pi}{3} r^2 h = \frac{\pi}{3} \frac{h^3}{(\tan \theta)^2}$$

14.24 $$h = \left[V \frac{3}{\pi} (\tan \theta)^2 \right]^{1/3} = \left[6.67 \times 10^9 m^3 \frac{3}{\pi} (\tan 40)^2 \right]^{1/3} = 1,649 m = 5,410 ft$$

The area of the base is:

14.25 $$BaseArea = \pi r^2 = \pi \left(\frac{h}{\tan \theta} \right)^2 = \pi \left(\frac{1649 m}{\tan 40} \right)^2 = 1.22 \times 10^7 m^2 = 4.68 mile^2$$

The dimensions are similar to a very large hill the size of a small city! Obviously, there is no way to collect all the excess emitted carbon dioxide in one spot and convert it all into carbon, rather this example is intended to place the magnitude of the problem into a perspective that is comprehendible.

Table 14-8. Storage Considerations for a Few Forms of Carbon.

Substance	Density	Angle of Repose
Carbon (as coke)	1,500 kg/m³	40°
Diamond	3,200 kg/m³	Cubic Crystal
Limestone (CaCO₃)	2,700 kg/m³	37°
Solution in water	1 - 3 g CO₂ / kg H₂O	fluid
Liquid CO₂	1000 kg/m³	fluid
Super-Critical CO₂	600 kg/m³	fluid

14.5.2 Scale of the Required Energy Production Changes.

Fossil fuels are a relatively inexpensive and dense form of energy. Any solution to the global climate change problem requires finding and using non-fossil fuels for energy production.

Figure 14-11 shows the sources of fuel for energy consumption in the US, (US-EIA, 2011c). It is very similar to Figure 14-3 which showed sources of C into the atmosphere. This figure however includes additional, non-carbon emitting sources such as hydroelectric, nuclear, and other renewable energy sources (solar, wind, geothermal, waste, and non-wood biofuels). Fossil-fuels account for about 83% of this energy, down from a high of 94% in the 1960's.

FIGURE 14-11. US Primary Energy Consumption by Fuel Source 1800 – 2010.

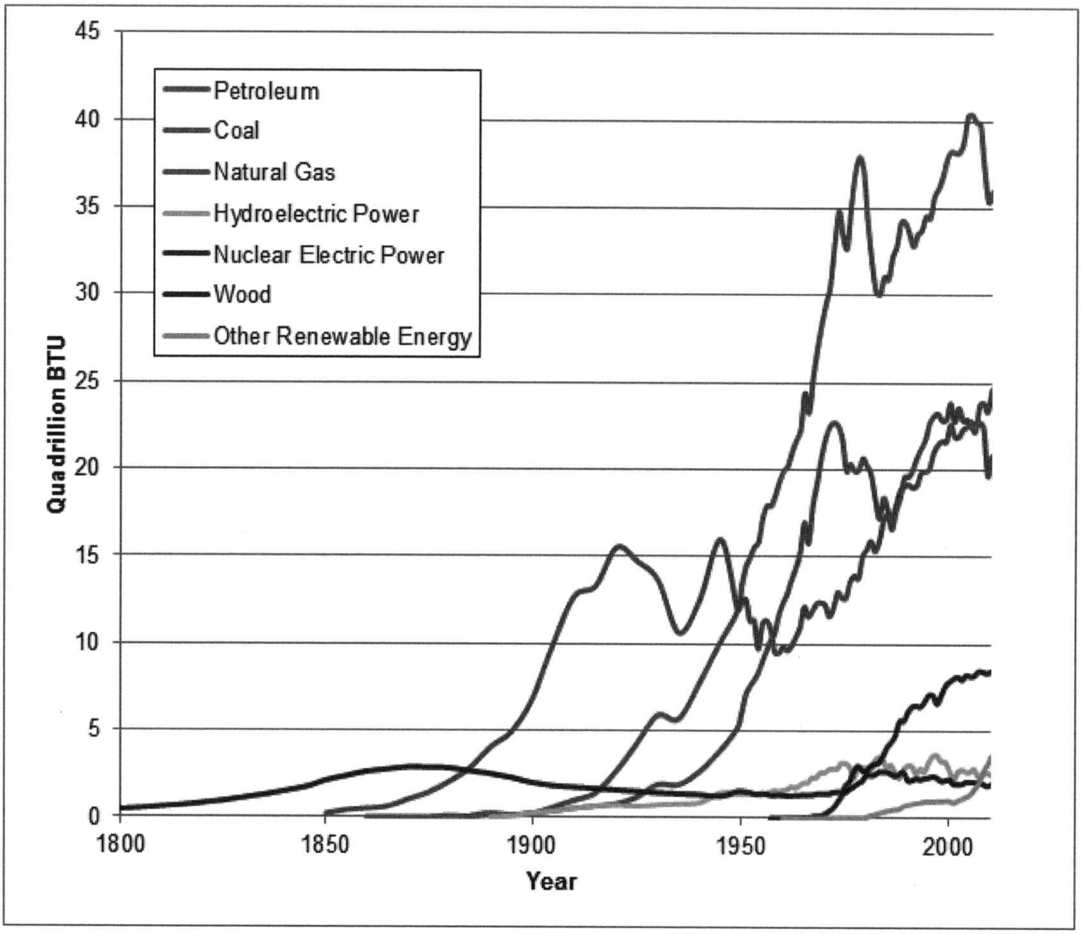

Table 14-9 lists the energy density, carbon content, and carbon dioxide emissions of several fuels (Wikipedia, 2011c). The table uses approximate or average values for more complex fuels such as coal, wood, and oil. The data does not include the energy lost from externalities in extracting the energy nor in transportation of the fuel to location of energy use.

Table 14-9. Energy Density and Carbon Content of Several Fuels.

Fuel	Energy Density MJ/kg$_{fuel}$	C Content kgC/kg$_{fuel}$	CO_2 Emissions kgCO$_2$/kg$_{fuel}$	CO_2 Emissions gCO$_2$/kWh
Uranium	20,000,000	0	0	0
Hydrogen	123	0	0	0
Natural Gas	53.6	0.75	2.75	184.5
Propane	49.6	0.82	3.01	218.3
Butane	49.1	0.83	3.04	222.7
Crude Oil	46.3	0.85	3.12	242.4
Fat (animal or vegetable)	37	0.75	2.75	267.4
Ethanol	31.1	0.52	1.91	220.9
Coal, bituminous	24	0.7	2.57	385.2
Wood	16.2	0.5	1.83	406.3
Coal, lignite	14	0.3	1.1	282.6
Hydro - 100 m elevation	0.001	0	0	0

1 MJ = 0.278 kWh   1 kWhr = 3413 BTU

The total US energy consumed in 2010 was about 98 QUADs (one quad = one quadrillion BTU = 10^{15} BTU) not including imports and exports. Prior to 1850, almost all energy was derived from wood, and it produced a maximum of 3 QUADs in the late 1800's. It continues to provide similar amounts today. Table 14-10 shows a breakdown of the users of this energy, (US-EIA, 2011c). These energy source and consumption distributions are similar worldwide. US energy consumption accounts for about 20% of the world's total and is expected to grow about 20% over the next 25 years. Energy consumption worldwide is expected to grow 50% from 505 QUADs in 2008 to 770 QUADs in 2035.

Table 14-10. US Energy Consumption by Economic Sector, 2010.

Sector	Energy Use [QUAD]
Residential	22.15
Commercial	18.21
Industrial	30.14
Transportation	27.51
Export (Not included in total or Fig 11)	8.17
Import (Non-petroleum)	4.50

The top three energy sources in Figure 14-11 are fossil fuels. The carbon released from these fuels removes it from the long-term geological storage and puts it into the atmosphere. As stated earlier (section 14.2.2) it takes natural processes on the order of 100,000 years to remove this much carbon from the atmosphere-land-surface ocean cycles and return it to long term storage reservoirs. Carbon released from biofuels (wood, alcohol, animal and vegetable oils) originates with carbon already in the atmosphere-land-surface ocean cycles; therefore, it does not contribute to an increase in atmospheric concentrations of carbon.

Note that natural gas has the lowest release of CO_2 per kWh of the fossil fuels. The reduced value is because a large fraction of the heat from burning natural gas is from hydrogen, and natural gas has the most hydrogen of all the fossil fuels. However, unlike the other fossil fuels, natural gas has a significant lifetime if released into the atmosphere (approximately 15 years, but may be up to 100). If it is accidently released during any step between reservoir extraction to the actual combustion, it may be worse than the other forms, since the methane molecule is a much stronger greenhouse gas than carbon dioxide.

The fourth largest energy source in the US is nuclear energy. It generates heat from the radioactive decay of uranium and daughter products. It does not use or release carbon during energy production. Expansion of nuclear energy production is one way to decrease fossil fuel use. However, it has several environmental and financial concerns keeping it a small part (approximately. 8%) of the global energy picture. Environmental concerns include ore mining, waste handling, waste storage, and the risk of catastrophic releases of radioactive materials during accidents. The waste is radioactive at levels that cause human health concerns for tens of thousands of years after use. There is no known method to keep the waste and humans apart for such lengths of time. Very few people are willing to live near such storage areas or to even allow such wastes to pass near them on the way to a storage area, irrespective of the safety assurances of experts. Financial problems include the high cost of construction, insurance, and maintenance. No nuclear plant has ever been able to afford private insurance; they all have made plans with local governments (taxpayers) to cover the costs of accidents. In the US, the national Price-Anderson Act provides compensation up to $300 million covered by a self-insurance fund supported by each reactor and also by a reactor-operator pool from the US reactors, which provides $10.4 billion. The lack of private insurance, required of almost all other industrial concerns, is because the costs of an accident are difficult to estimate and thought to be much larger than the amount of capital that any (or even all) insurance companies have access to. In the world's two major nuclear plant accidents the costs were indeed very large - Chernobyl, Ukraine, at $235 billion and Fukushima Daiichi, Japan, at $300 billion as of 2011. Finally, the cost of maintaining a nuclear plant is larger than conventional fuel plants because of the increased need of vigilance against any errors, mistakes, or accidents. The nuclear plant fleet is much safer and better run than any other human enterprise, and this costs more to do.

Renewable energy includes all the various forms of energy derived from incident solar energy – hydroelectric, wind, biofuels (e.g. wood, ethanol) and solar. Hydroelectric converts the stored potential energy of water moving from a high elevation to a lower elevation through use of a turbine and generator. Large plants are difficult to site and often require flooding a great deal of land to form a sufficiently deep reservoir. Table 14-11 lists examples of several of the world's largest hydroelectric facilities. Smaller plants (<10MW) are common in any location with fast, high-flow rivers and modest elevation changes. They also have a significant local environmental footprint because they alter the normal flow of surface water.

Table 14-11. Large Hydroelectric Electricity Generating Units.

Name	Country	Electricity Produced (MW)	Reservoir Area (km^2)
Three Gorges Dam	China	18,000	1,045
Churchill Falls	Canada	5,400	5,700
Tehri Dam	India	2,400	52
Aswan	Egypt	2.100	5,250
Hoover Dam	US	2,000	640

Wind power converts the kinetic energy of moving air to electricity through use of a turbine and generator. It currently generates about 1- 3% of worldwide electricity demand (US-EIA, 2011a) and is growing quickly. It is expected to take on 4 – 6% of world electricity demand by 2035 (US-EIA, 2011b). It is unlikely to grow beyond a level of 20% unless better ways to store energy are discovered. Wind power is not always available at the time it is most needed. Wind often moves strongest at night, when demand for electricity is weak. Conversely, when demand is large during the day, wind is at its least powerful. Wind also varies with time of year, in both direction and magnitude. Possible energy storage ideas include hydrogen generation from the electrolysis of water, or pumped hydro. Hydrogen generated from off peak electricity production could be used to generate electricity on demand, either in a combustion process like natural gas, or from fuel cells (which can achieve efficiencies over 50%). Pumped hydro uses the off-peak electricity to pump water into a raised storage area. An elevation increase of at least 300 feet is typically sought. This water is then allowed to run downhill through turbines (could be the same pumps used to transport uphill) to create electricity on demand. These systems are thought to be able to achieve 70% efficiencies. They also require large storage basins for the pumped water.

Biofuel originates from the solar energy stored by plants during photosynthesis, see equation (14.1). Typically used forms include wood and certain crops that store their energy as sugars, starches, and/or oils in especially high concentrations. Sugars and starches (including cellulose from wood and agricultural residue) can be converted to ethanol. Plant oils, generally from seeds, are converted to biodiesel. All agricultural materials (plant and animal) can also be converted into methane in anaerobic biodigesters. Table 14-12 provides a comparison of several energy crops, excerpted from (Wikipedia, 2011b).

Table 14-12. Energy Yields of Various Energy Crops.

Energy Crop	Yield of Oil US Gal / acre-yr	Yield of Biofuel kg/hectare-yr
Corn	18	145
Oats	23	183
Cotton	35	273
Hemp	39	305
Soybean	48	375
Sesame	74	585
Sunflower	102	800
Rapeseed	127	1,000
Coconut	287	2,260
Palm Oil	635	5,000
Algae	10,000	50,000 - 100,000 *

* Note that claims of potential yield from microalgae vary widely by source.

The final energy source, solar power, converts the suns energy into electricity either directly in a solar cell or indirectly by boiling a working fluid that passes through a turbine/ generator to create electricity (similar in many ways to conventional electricity production in a coal or nuclear plant). Commercially available solar cells (2010) have efficiencies around 10 – 20%, although research proto-types have achieved efficiencies of 40% conversion of sunlight to electricity.

14.5.3 Control Strategies

"Necessity is the mother of invention."

As mentioned before there, is currently no existing commercially available technology to control CO_2 emissions at the required scale. However, this has been the case for every air pollutant; no technology for control exists before there is a need to do so. In every case so far, once it is decided that the problem must be solved, a solution is found, and cost efficiencies follow quickly. Most likely many different solutions will be found, since there are many different sources of CO_2, and it is emitted in many different ways.

Control strategies for CO_2 can be organized as to whether they reduce CO_2 emissions (mitigation) or reduce the harm caused by the emissions (adaptation). Mitigation is a preferred strategy for the long-term solution and to minimize overall harm. Adaptation is required when mitigation occurs too slowly or in too small a focus. Both strategies require large changes in the way humans organize their personal and public time, culture, and ways of living. Both strategies also offer many opportunities for new ideas, new technologies, new regulatory apparatus, and leadership. Anyone trained in engineering or the sciences can make important and lasting contributions by helping solve these problems.

14.5.3.1 Mitigation

There are three basic ways to reduce emissions of carbon to the atmosphere: supply reduction, demand reduction, and emission controls. Judicial use of each could greatly affect the release of carbon at moderate costs.

SUPPLY REDUCTION is a method to reduce carbon emissions through changes in energy supplier equipment and habits. Ideas include increasing the conversion efficiency from fuel energy content into electricity, switching to fuels that release less carbon per amount of energy produced, and use of fuels that do not release carbon from the long-term reservoirs,

Energy conversion efficiency. Power plant efficiencies are typically defined as either the percentage of the heat content in the fuel that is converted to usable electricity output or as a heat rate (BTU/kWh) which compares the energy content of the fuel (BTU) to the electric energy output (kWh). The efficiency of an existing plant is difficult to change by more than a few percent. The efficiency of a new power plant is largely a function of economic choice. Current worldwide heat rate is about 9,000 BTU/kWh, while the US has an average operating heat rate of 10,400 BTU/kWh.

■ *Example 14-6:*
Determine the % efficiency of fuel with a heat rate of 10,400 BTU/kWh.

Solution:
Perform a unit conversion, recalling from Appendix 1 that 3413 BTU = 1 kWh

14.26
$$\%Eff = \frac{3,413 \frac{BTU}{kWh}}{10,400 \frac{BTU}{kWh}} * 100\% = 32.8\%$$

Factors affecting Electricity Generating Units (EGU) efficiency include design choices, operational practices, fuel, pollution control methods, and ambient conditions.

- Design choices: Construction of natural gas and coal-fired power plants represent a tradeoff between capital cost, efficiency, operational requirements, and availability. For example, a steam turbine system that operates at a higher temperature and pressure can achieve a higher efficiency. The higher temperatures and pressures, however, require more exotic materials of construction for both the boiler and turbine, thus the capital cost increases.
- Operational Practices: Efficiency can be improved by reducing over-fire air (see chap 10) to a minimum, fully utilizing heat integration systems, staying after steam leaks and exchanger fouling, and a large number of other practices. Operating at full load capacity enhances efficiency. However, the reality is that load is ever changing, and the requirements of market-based systems focus on reliability leads to an inability to run at full load.
- Fuel: Among coals the higher ranking (more BTUs per unit mass) coals enable higher efficiency because they contain less ash and less water. However, most new coal production yields sub-bituminous coal, which is of a lower rank.
- Pollutant control methods: The level of pollutant emission control effects efficiency. NO_x reduction units (Selective Catalytic Reduction) and SO_x scrubbers represent side loads that decrease net generation and thus reduce efficiency. SCR's typically require about 1% load, and scrubbers require 2% of load. By 2020, up to 90% of all coal-fueled plants will use at least one of these control technologies, and most gas units will use SCR. The capture of CO_2 is expected to have a 5 – 30% side load.
- Ambient conditions: The thermodynamic efficiency of EGUs using steam as a working fluid increases when cooling water and air are colder. Additionally, higher altitudes have lower ambient pressure and efficiency is reduced due to changes in the expansion and compression of the steam.

EGUs use the heat content of a fuel to change the pressure and temperature of a gas (working fluid). The change is accomplished by transferring the chemical heat of combustion of the fuel to the working fluid. The working fluid gains energy (pressure and temperature) that can be extracted mechanically in a turbine, to expand the working fluid (thus lowering the pressure and temperature) to spin a turbine. This spinning mechanical energy is then transferred to a generator that converts the spinning into electricity. Figure 14-12 shows a simplified schematic of a coal-fired power plant. The thermodynamic efficiency is a function of the pressure and temperature of the steam leaving the boiler and of the water entering the boiler (which depends on a cooling tower to remove heat). Higher pressure and temperature steam and cooler water increase efficiency.

Table 14-13 shows a range of efficiencies for several types of coal and gas fueled EGUs (US-EIA, 2011b). Pulverized coal (PC) plants burn coal in a furnace and transfer as much heat as possible to the boiler-water to produce steam. Supercritical, advanced super-critical and ultra-supercritical systems are used by PC plants that operate at higher temperatures and pressures than the traditional

FIGURE 14-12. Simplified Schematic of a Steam Turbine Electricity Generating Unit.

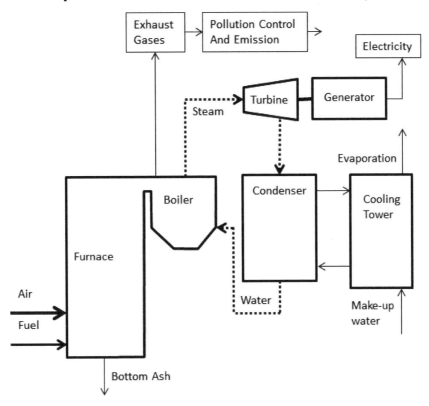

sub-critical PC plant. The critical point for water is 647 K (374°C) and 218 atm (3206 psia), at which point the liquid and vapor phases become indistinguishable. Operating at temperatures and pressures above this point allows greater efficiencies but requires advanced alloys for construction.

Combustion turbine (CT) plants are designed to start and stop quickly to meet fluctuating demand loads. They run with natural gas or low-sulfur fuel oil. The combustion gases are run through a turbine directly and operate much like a jet engine – they draw in air at the inlet, compress it, mix in fuel, and then ignite the mixture, see Figure 14-13. The hot gases expand through the turbine which is connected to a generator to produce electricity. It does not use a boiler or steam system. These units are the most flexible systems because they can start and shutdown very quickly (minutes as compared to hours for a steam turbine system).

Integrated gasification combined cycle (IGCC) plants use either natural gas or gasified solid fuel. The combustion products first pass through a gas turbine. The gas turbine runs much like a CT system, but the exhaust gases are sent to a boiler to run a steam turbine rather than sent to discharge. This combined system extracts much more of the chemical energy of the fuel, and it has a higher construction cost.

FIGURE 14-13. Simplified Schematic of Combustion Turbine Electricity Generating Unit.

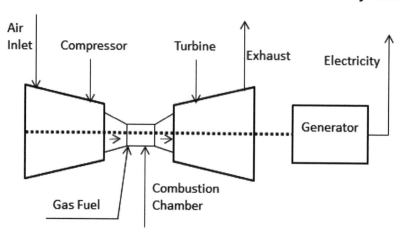

Table 14-13. Power Plant Efficiencies.

Power Plant Type		Pressure psig	Temperature °F	Heat Rate BTU/kWh	Efficiency %
Coal					
	PC, ultra super-critical	5500	1300	7760	44
	PC, advanced super-critical	4700	1130	8130	42
	PC, super-critical	3500	1100	8710	39
	PC, sub-critical	2400	1050	9280	37
	IGCC			8920	38
	PC, Average			10,400	33
Natural Gas					
	CC, advanced			6330	54
	CC, conventional			6800	50
	Conventional CT			10840	31

Note: PC = Pulverized Coal, CC = Combined Cycle, CT = Combustion Turbine

■ *Example 14-7:*

Determine the carbon emission reduction if a 500 MW, 35% efficient coal plant is replaced by a 40% efficient coal plant. Assume the coal has a heat content of 9,800 BTU/lb and is 72% carbon.

Carbon Dioxide and Climate Change

Solution:
Determine the carbon emission from the 35% efficient plant, then repeat for the 40% plant. The difference is the reduction.

35% eff: $\dfrac{500\,MW}{0.35\,efficiency} * \dfrac{3{,}413{,}000\,BTU}{MW\,hr} * \dfrac{lb-coal}{9{,}800\,BTU} * \dfrac{0.72\,lb-C}{lb-coal} = 3.58 \times 10^5 \dfrac{lb-C}{hr}$

40% eff: $\dfrac{500\,MW}{0.40\,efficiency} * \dfrac{3{,}413{,}000\,BTU}{MW\,hr} * \dfrac{lb-coal}{9{,}800\,BTU} * \dfrac{0.72\,lb-C}{lb-coal} = 3.13 \times 10^5 \dfrac{lb-C}{hr}$

The difference is 45,000 lb-C/hr or 197,000 ton-C/year.

Fuel conversion – Convert plants from high CO_2 production fuels (coal) to use a fuel that produces less CO_2 per unit of energy such as natural gas, see Table 14-9. The number of natural gas production plants would need to triple to replace ½ of current coal plants (Bellman, 2007). This could lead to increased prices for gas and reduced prices for coal, making coal a more attractive fuel and preventing additional fuel conversion projects.

■ Example 14-8:
Determine the reduction in CO_2 emissions when a 500 MW electricity generating unit using coal (as described in Example 14-7) is replaced with natural gas (1000 BTU/scf). Assume both fuels convert energy to electricity at 35% efficiency.

Solution:
Solve this problem using the result from Example 14-7.

Coal emissions: $3.58 \times 10^5 \dfrac{lb-C}{hr}$

Natural gas emissions (assuming 0°C, one atm as standard conditions):

$\dfrac{500\,MW}{0.35\,efficiency} * \dfrac{3{,}413{,}000\,BTU}{MW\,hr} * \dfrac{scf}{1{,}000\,BTU} * \dfrac{0.00298\,lb-mol}{scf} * \dfrac{16\,lb-C}{1\,lbmol} = 2.32 \times 10^5 \dfrac{lb-C}{hr}$

The difference shows a reduction of 35% from the coal unit, which is 126,000 lb-C/hr or 550,000 ton-C/year

Fuel Replacement – This method replaces a fossil fuel with non-fossil fuels such as hydro, geothermal, nuclear, biofuels, wind, or solar. If the energy is produced without releasing carbon from a long-term reservoir, it does not contribute to carbon induced climate change. Section 14.5.2 discussed many of these systems. Replacement is the ideal case for supply reduction, but it requires large capital costs to replace existing fossil fuel plants, as well as meeting the anticipated growth in energy demand. However, the wide variety of systems and design choices allows the flexibility needed to succeed with such replacement.

DEMAND REDUCTION is a method to reduce carbon emissions through changes in end-user equipment and habits. When the end user (consumer, business, industry, or public organization) reduces their energy/power needs, it reduces the need for energy suppliers to produce it.

Conservation. This method reduces emissions by using energy only when it is needed and in just the minimum required amount. It can be applied at large scales and small. A factory might consider heat integration, which uses heat exchangers to move heat from where it is unneeded (in the waste stream) to where it is useful (preheat a feed stream). A government building or shopping center can reduce the amount of heating in winter and cooling in summer. Auto manufacturers have found that drivers are much more likely to drive conservatively if fuel efficiency is displayed on the control panel. Conservation can also be applied at a small scale - turning off lights when they are not being used, walking or biking rather than driving, and combining chores when driving. Changing consumer habits requires public education campaigns with frequent reminders as well as an incentive program (high-energy costs work well).

Efficiency. This method works by using devices such as refrigerators, furnaces, air conditioners, water heaters, washers and dryers that use less energy to achieve a given task. It also includes replacing incandescent lights lighting with alternatives such as compact fluorescent lights (CFL) and light emitting diodes (LED). These alternatives typically require just 10 – 20% of the energy needed by incandescent lights.

■ Example 14-9:

Calculate the one-year energy reduction (kW-hour/year) in conversion of a home to compact fluorescent light (CFL) light bulbs from incandescent light bulbs.

Solution:
To solve we need to make a few assumptions: how many bulbs will be replaced and the difference in bulb energy demand. Our typical home has five bulbs that can be easily replaced. They will each be 100 W incandescent bulbs replaced with 13 W compact fluorescent bulbs. Finally, we assume that each bulb is used for six hours per day.

Before replacement case:

14.27
$$5 \text{ bulbs} * \frac{100W}{bulb} * \frac{1kW}{1000W} * 6\frac{hours}{day} * 365\frac{day}{year} = 1,095 \frac{kW \cdot hour}{year}$$

After replacement case:

14.28
$$5 \text{ bulbs} * \frac{13W}{bulb} * \frac{1kW}{1000W} * 6\frac{hours}{day} * 365\frac{day}{year} = 142 \frac{kW \cdot hour}{year}$$

The difference is 953 kWh, which if the electricity is from coal would represent a reduction of 137 kg CO_2 in one year (conversion from Table 14-9). A typical cost of electricity is $0.09 /kWh, so this change would save a typical household $86/year, which is much more than the cost of these five new bulbs.

The number of households in the US is approximately 100 million. If every US house converted five bulbs, it would create an emission reduction of approximately four million tons of carbon per year or 0.2% of the US annual emissions. This modest change requires almost no significant change in lifestyle or consumption patterns. By itself, it does not solve the problem, but combined with many other such changes it would be part of a cost-effective strategy.

Thoughtful consumption. An idea that is in opposition to that of some societies and cultures that value people based on their worth as consumers, with no thought to creativity, art, or happiness. Thoughtful consumption is a modification of these values that realizes the greatest happiness for the greatest number of people occurs when all needs are met, and only some wants are satisfied. Needs are identified as activities that can be satiated, such as water, adequate food, clothing, housing, and health-care. Once a person has had enough they do not accept more, until the need returns. (i.e. a thirsty person needs a glass of water, but after having two or three glasses they do not continue to drink more even if it is free). A want is never or rarely ever satiated (i.e. if someone was handing out free $100 bills, who would stop taking one every time they were handed one? – money is a want). When thoughtful consumption happens, people have fewer want things, yet have greater happiness. This happiness manifests as more time for friends and family, and spending one's time in pursuit of activities that make them happy, rather than making more money to increase their worth as a consumer. Note that thoughtful consumption does not mean returning to prehistoric societies or living in poverty, rather it is a way of living that does not require more-more-more, or the latest fad or fashion (though it allows for each person to have some of those things). This radical idea strikes incredible fear into the people who control the economies and societal structure of our current world. It appears that these 'leaders' would prefer the harsh effects of climate change over a change in consumption habits of consumers. This fear is probably because they have sufficient wealth to adapt to almost any problems that emerge – unlike most of the population. These fears appear to be the foundation of the extreme onslaught of global change denialism. The idea of a restructured civilization can have tremendous positive impact on most people's lives, but requires large-scale collaborative efforts to implement.

EMISSION CONTROLS. This process is called carbon capture and sequestration (CCS). It follows the traditional pollution control methods. It works by collecting and altering carbon emissions before or after they are released to the atmosphere. The captured carbon is turned into a form that does not reenter the atmosphere for long times comparable to the geological reservoirs for fossil fuels. Three technologies need to be developed at large scales for the process to work: capture, transportation, and storage. These technologies, once developed, will create an energy penalty or a side load to energy production, because each requires energy. Additional uncertainties relate to proving the technologies, anticipating environmental impacts, and how governments should incentivize the process. Application of CCS for biomass sources (such as when co-fired with coal) could result in the net removal of CO_2 from the atmosphere.

Capture - There currently is no known or proven method to capture CO_2 from point sources on a large scale. Methods currently being explored include pre-combustion, post-combustion, and source separation for non-combustion industrial processes. These technologies have been shown to be economically feasible under certain conditions (IPCC, 2005), but are not currently deployed in any significant amount (under 0.1 $MtCO_2$/yr or fewer than five installations worldwide).

Pre-combustion capture could be used in IGCC power plants where the carbon in the fuel is converted to CO in a gasification step, and the CO is then converted to CO_2 in the water-gas-shift reaction, which also generates hydrogen for use in the combustion step.

14.29 $$CO + H_2O \leftrightarrow H_2 + CO_2$$

A single reactor can convert about 80% of the CO; additional reactors would be needed to achieve higher conversion. The CO_2 produced could be scrubbed from the reactor product stream before adding air for combustion, when it is most concentrated.

Post-combustion and source separation capture of CO_2 can best be applied to large carbon point sources that have emission streams with high concentrations of CO_2. Potential sources include coal, gas, or biomass-fired power generation, major energy-using industries, synthetic fuel plants, natural gas fields and chemical facilities for producing hydrogen, ammonia, cement and coke.

All capture methods rely on using an adsorption or absorption step to remove the CO_2 from other gases. Section 11.3.5 described these unit operations. Three such capture methods are currently being explored – wet scrubbing, dry adsorption, and biogenic capture.

Wet scrubbing sprays a water-amine solution from the top of a scrubbing tower and introduces the gas to the bottom for a counter-current contact between the two fluids. The liquid solution is chosen such that the CO_2 is highly soluble in it, and that other components of the gas stream have low solubility. It also requires that the liquid stream does not evaporate into the gas stream. Under these circumstances, the CO_2 transfers from the gas to the liquid phase. The systems may run at atmospheric pressure but may benefit from a higher operating pressure. The amines used for capturing CO_2 include monoethanolamine (MEA), diethanolamine (DEA), and methyldiethylanolamine (MDEA). Once it dissolves into the water-amine solution, it must be removed so that the amines can be reused. This operation is performed in a similar type of column called a stripper. In the stripper, energy is added to the solution to raise its temperature and cause the CO_2 to degas from the liquid. The CO_2 is high purity since it is the only gas dissolved into the solution.

Dry adsorption is similar to the above process except the contact is between a porous solid material and the CO_2 gas stream. The choice of solid is important to selectively adsorb CO_2. Solid carbon materials such as coal and activated carbon are good choices. Once the solid sorbs its limit of CO_2, it must be removed from the gas stream. Typically, the solid can be regenerated while creating a high-purity CO_2 stream. Regeneration occurs by raising the temperature of the solid, very much like the wet scrubbing process. Typical regeneration methods allow the solid to be reused dozens to hundreds of times.

Biogenic capture uses photosynthesis to remove CO_2 from a gas stream and to fix it into a solid form, see equation (14.1). The organisms used to perform the photosynthesis can be either land or water based. The solid forms generated can be used as fuels or placed into long term storage. Land based systems include organisms such as crops and trees. Water based organisms include algae or bacterial solutions. These processes use solar energy to alter the carbon dioxide chemically. These systems may use special greenhouses to provide contact between the organisms and the emission stream, or simply use atmospheric air.

■ Example 14-10:
How large of a land area is needed to control the CO_2 emissions from a 500 MW coal power plant.

Solution:
Find the emissions of CO_2 from the plant. Then calculate the surface area needed to consume it by some biomass. Assume that the biomass usage of CO_2 remains constant throughout the year.

$$\frac{500\,MW}{0.35\,efficiency} * \frac{3{,}413{,}000\,BTU}{MW\,hr} * \frac{lb-coal}{9{,}800\,BTU} * \frac{0.72\,lb-C}{lb-coal} * \frac{44\,lb-CO_2}{12\,lb-C} = 1.31 \times 10^6 \frac{lb-CO_2}{hr}$$

Soil has an average uptake rate of $5.24 \times 10^{-6} \frac{lb-CO_2}{ft^2 hr}$. This will require

$$\frac{1.31 \times 10^6 \frac{lb-CO_2}{hr}}{5.24 \times 10^{-6} \frac{lb-CO_2}{ft^2 hr}} = 2.5 \times 10^{11}\,ft^2 = 9{,}000\,mi^2$$

This is impractical, other forms of biomass should be explored. Microalgae have much higher uptake rates when exposed to strong light and limited to shallow bodies of water (1 m deep). Microalgae have been shown to utilize CO_2 at a rate that yields an area of around 183 mi^2. Better, but still impractical considering the number of power plants.

Transport – Once the CO_2 has been captured and processed to produce a concentrated stream, it must then be transported to long-term storage. Dry CO_2 is not corrosive to pipelines even if it contains contaminants, but it becomes corrosive when moisture is present. Any moisture, therefore, needs to be removed to prevent corrosion and avoid the high cost of constructing pipes made from corrosion-resistant material. Risks in CO_2 transportation include rupture or leaking of pipelines, possibly leading to the accumulation of a dangerous level of CO_2 in localized air.

There are two basic transportation methods for moving large quantities of CO_2 – pipelines and tanker ships. It is not economical to transport CO_2 in its gas phase due to its low density. Also, both transportation methods require an increase in the density of the gas stream. The increase can be done by converting it to a liquid (LCO_2), supercritical fluid (SCO_2), or solid form. The solid form (dry ice) would require very large amounts of energy for conversion and will not be considered here.

Figure 14-14 shows the phase diagram for CO_2, highlighting the phase behavior at different temperatures and pressures, data obtained from (Weast, 1986). It relates the phase behavior of CO_2 as temperature and pressure are varied. Super-critical CO_2 (600 kg/m^3) exists above the critical point (305 K, 73 atm) where the liquid and gas become indistinguishable. This material has the nice characteristic of existing near room temperature (although at much higher pressures). SCO_2 can be obtained by compressing the gas to a pressure above 73 atm and it does not require much cooling. Liquid CO_2 has a density near that of water (1000 kg/m^3) and would most likely be used near the triple point (218 K, 5.2 atm). The triple point is where all three phases coexist. With this much higher density, nearly twice as much CO_2 can be transported per unit volume. LCO_2 requires cold temperatures (-50 °C) but does not require as high of pressures as SCO_2. Transportation as LCO_2 would require refrigeration of the ship or at periodic distances along a well-insulated pipeline.

Storage - Potential storage methods include injection into underground geological formations, into the deep ocean, or industrial fixation as inorganic carbonates.

Geological formation storage - Injection of CO_2 into geological reservoirs could lead to permanent storage of CO_2. Geological storage is the most mature of the storage methods, with a number of commercial projects in operation. Storage of CO_2 can be in deep underground water-saline formations, oil and gas reservoirs, and deep un-minable coal seams using injection and monitoring techniques similar to those currently utilized in the oil and gas industry. Of the different types of potential storage formations, storage in coal formations is the least developed. If injected into

FIGURE 14-14. CO_2 Phase Diagram.

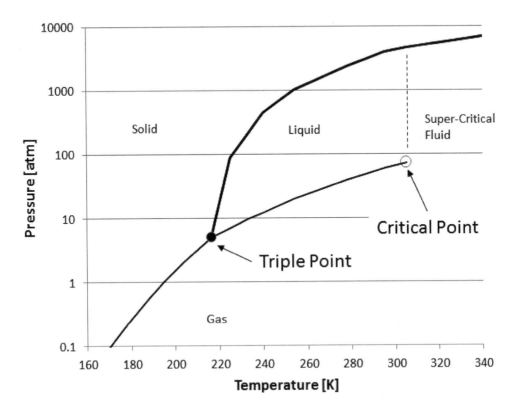

suitable saline formations or oil and gas fields at depths below 800 m, various physical and geo-chemical trapping mechanisms prevent the CO_2 from migrating to the surface. Storage capacity in oil and gas fields, saline formations, and coal beds is uncertain. The IPCC reported a storage capacity of between 675 to 900 GtCO_2 for the relatively well-characterized gas and oil fields, more than 1000 GtCO_2 (possibly up to an order of magnitude higher) for saline formations, and up to 200 GtCO_2 for coal beds. Concerns around geological storage include rapid release of CO_2 as a consequence of seismic activity and the impact of old and poorly sealed well bores on the storage integrity of depleted oil and gas fields (IPCC, 2005).

Ocean storage – Injection of CO_2 into the ocean at great depths via pipelines or ships is not expected to retain CO_2 permanently as the CO_2 will re-equilibrate with the atmosphere over the course of several centuries. CO_2 in the ocean would be unable to interact with the atmosphere. Significant research is needed into potential biological impacts to clarify the nature and scope of environmental consequences, especially in the longer term (IPCC, 2005).

Over millennia, carbon dioxide injected into the oceans would approach the same equilibrium as if it were released to the atmosphere. Sustained atmospheric CO_2 concentrations in the range of 350 to 1000 ppmv imply that 2,300 to 10,700 Gt of anthropogenic CO_2 will eventually reside in the ocean. Effects of elevated CO_2 levels have mostly been studied on time scales up to several months on individual organisms that live near the ocean surface. Observed phenomena include reduced rates of calcification, reproduction, growth, circulatory oxygen supply and mobility as well as increased mortality over time. In some organisms, these effects are seen in response to small additions of CO_2. Immediate death of these organisms will occur at the injection site.

Injection of carbon dioxide into the ocean could also affect the water's pH. When carbon dioxide reacts with water, it forms carbonic acid, a weak acid, see equation (14.3). The change could be offset by dissolving mineral carbonates (such as limestone) along with the CO_2 injection. However, large amounts of limestone would be required.

Mineralization - Mineral storage, also known as mineral carbonation, is the process by which carbon dioxide is reacted with a metal oxide to form an insoluble carbonate. The following reaction shows the process by which calcium oxide (lime) reacts to form calcium carbonate (limestone):

14.30 $$CaO + CO_2 \rightarrow CaCO_3 + heat$$

The most promising metal oxides considered for these processes are the alkali-earth metals like calcium and magnesium because they are very insoluble in water. However, they are rare in nature, making silicate rock or industrial residues ideal alternatives because of their abundance and availability. The minerals must be pretreated by means of crushing, grinding, milling, mechanical separation, and chemical processing before mineral carbonation. The carbon dioxide requires little processing and can be used in its pipeline state.

The main environmental concern with mineral storage is large-scale mining for the minerals required for the process, the mineral preparation, and waste disposal. Impacts of such a project include, but are not limited to land clearing, habitat destruction, water, soil and air pollution. This option of carbon storage can result in increased production of carbon dioxide due to the energy needs of the many processes and treatments required. These problems make it seem like an infeasible option until better methods are identified.

Cost estimates of the components of a CCS system vary widely depending on the base case and the wide range of source, transport and storage options Table 14-14, (IPCC, 2005). In most systems, the cost of capture (including compression) is the largest component, but this could be reduced by 20–30% over the next few decades using technologies still in the research phase as well as by upscaling and learning from experience. The extra energy required is a further cost consideration. CO_2 storage is economically feasible under conditions specific to enhanced oil recovery (EOR), and in saline formations, where it is used to avoid carbon tax charges for offshore gas fields in Norway. Pipeline transport of CO_2 operates as a mature market technology costing 1–5 US$/$tCO_2$ per 100 km (high end for very large volumes) (US-EIA, 2011b). Several thousand kilometers of pipelines already transport 40 Mt/yr of CO_2 to EOR projects. The costs of transport and storage of CO_2 could decrease slowly as technology matures further and the scale increases.

Table 14-14. Cost Ranges for Components of a CCS System Applied to a Given Source, 2011.

CCS system components	Cost range	Remarks
Capture from a coal- or gas-fired power plant	15-75 US$/$tCO_2$ net captured	Net costs of captured CO_2 compared to the same plant without capture
Capture from hydrogen and ammonia production or gas processing	5-55 US$/$tCO_2$ net captured	Applies to high-purity sources requiring simple drying and compression
Capture from other industrial sources	25-115 US$/$tCO_2$ net captured	Range reflects use of a number of different technologies and fuels
Transport	1-8 US$/$tCO_2$ transported	Per 250 km pipeline or shipping for mass flow rates of 5 (high-end cost) to 40 (low-end cost) $MtCO_2$/yr.
Geological storage*	0.5-8 US$/$tCO_2$ net injected	Excluding potential revenues from EOR or ECBM.
Geological storage: monitoring and verification	0.1-0.3 US$/$tCO_2$ injected	Covers pre-injection, injection, and post-injection monitoring, and depends on the regulatory requirements
Ocean storage	5-30 US$/$tCO_2$ net injected	Including offshore transportation of 100–500 km, excluding monitoring and verification
Mineral carbonation	50-100 US$/$tCO_2$ net mineralized	Range for the best case studied. Includes additional energy use for carbonation

Note: * Over the long term, there may be additional costs for remediation and liabilities

Estimates of the role CCS will play over the course of the century to reduce GHG emissions vary. It has been seen as a 'transitional technology,' with deployment anticipated from 2015 onwards, peaking after 2050 as existing heat and power-plant stock is turned over, and declining thereafter as the de-carbonization of energy sources progresses (IEA, 2011). Other studies show a more rapid deployment starting around the same time, but with continuous expansion even towards the end of the century. Yet other studies show no significant use of CCS until 2050, relying more on energy efficiency and renewable energy (IPCC, 2005). Long-term analyses by use of integrated assessment models using a simplified carbon cycle indicated that a combination of bioenergy technologies together with CCS could decrease costs and increase attainability of low stabilization levels (below atmospheric levels of 450 ppmv).

14.5.3.2 Adaptation

The IPCC projections suggest that the best response to this global threat is to reduce (mitigate) carbon emissions by 80% by the year 2050, see Figure 14-9. These scenarios are intended to limit the change in global average temperature to no more than 2°C by keeping the atmospheric level of CO_2 below 450 ppm. These solutions require major changes in the global economy, global energy infrastructure, as well as lifestyle and cultural changes for nearly all societies. Even if this succeeds, many problems may still happen due to the carbon that has already been emitted and accumulated in the atmosphere. All of the adaptations are expensive and require well trained engineers and scientists to carry out. The most likely and foreseeable problems include the following:

Changes in weather patterns. These will increase the quantity and magnitude of severe weather problems such as floods and droughts. Timing of seasons will be altered, such as the first frost-free day and last frost free-day. It may also impact agriculture requiring changes in types of crops grown, changes in planting times, and altering availability of surface water. Storm water and irrigation systems may require large-scale changes.

Ocean sea level increase. Many coastal communities will be impacted. They will require adaptations such as construction of coastal/storm water defenses, relocation and/or protection of groundwater from salt water intrusion, and relocation of infrastructure.

Alteration of habitats. Most biomes will move towards poles or increased elevations. Tropical species will become more common in midlatitudes. Many species will need to migrate, and some specialist species may not be able to find new habitats. These alterations may cause problems for plants dependent on other species for pollination. The changes will most likely lead to human population movement and migration.

While these problems are not directly related to air pollution, they are the expected outcome of an air pollution problem and should be included in determining the cost (or avoided cost) of any actions related to control of carbon emissions.

14.5.3.3 Cost

A final word concerns cost and how to account for the financial burden of reducing fossil fuel carbon emissions. The current cost associated with these activities is externalized onto the global environment. Externalized costs are paid for with the health and well-being of all life on earth. The consumers do not pay the true cost of using this resource, but it is difficult to account for the value of an acidified ocean or the growth of a forest. There are those who say an economic system is fairest when no costs are externalized. There are others that say the easiest route to wealth is to externalize your costs.

● Case Study 14-2. How to Solve the Climate Problem.

The following is an excerpt from James Hansen's book "Storms of My Grandchildren," (Hansen, 2009). He is a member of the US National Academy of Sciences, an adjunct professor in the Department of Earth and Environmental Sciences at Columbia University and at Columbia's Earth Institute, and retired director of the NASA Goddard Institute for Space Studies.

> *"Our goal is a global phase-out of fossil fuel carbon dioxide emissions. We have shown, quantitatively, that the only practical way to achieve an acceptable carbon dioxide level is to disallow the use of coal and unconventional fossil fuels (such as tar sands and oil shale), unless the resulting carbon is captured and stored. We realize that remaining, readily available pools of oil and gas will be used during the transition to a post-fossil-fuel world. But a rising carbon price surely will make it economically senseless to go after every last drop of oil and gas--even though use of those fuels with carbon capture and storage may be technically feasible and permissible.*
>
> *...*
>
> *Also, remember that the solution to the climate problem requires a phase-down of carbon emissions, not necessarily a phasedown of energy use. We will need to slow the energy growth rate and decarbonize our energy sources to solve the problem.*
>
> *Why do fossil fuels continue to provide most of our energy? The reason is simple. Fossil fuels are the cheapest energy. This is in part due to their marvelous energy density and the intricate energy-use infrastructure that has grown up around fossil fuels. But there is another reason: Fossil fuels*

are cheapest because we do not take into account their true cost to society. Effects of air and water pollution on human health are borne by the public. Damages from climate change are also falling on the public, but they will be borne especially by our children and grandchildren.

How can we fix the problem? The solution necessarily will increase the price of fossil fuel energy. We must admit that. In the end, energy efficiency and carbon-free energy can surely be made less expensive than fossil fuels, if fossil fuels' cost to society is included. The difficult part is that we must make the transition with extraordinary speed if we are to avert climate disaster. Rather than immediately defining a proposed framework for a solution, which may appear to be arbitrary without further information, we need to first explore the problem and its practical difficulties.

Two alternative legislative actions have been proposed in the United States: "fee-and-dividend" and "cap-and-trade." Let's begin by looking at the simpler approach, fee-and-dividend. In this method, a fee is collected at the mine or port of entry for each fossil fuel (coal, oil and gas), i.e., at its first sale in the country. The fee is uniform, a single number, in dollars per ton of carbon dioxide in the fuel. The public does not directly pay any fee or tax, but the price of the goods they buy increases in proportion to how much fossil fuel is used in their production. Fuels such as gasoline or heating oil, along with electricity made from coal, oil or gas, are affected directly by the carbon fee, which is set to increase over time. The carbon fee will rise gradually so that the public will have time to adjust their lifestyle, choice of vehicle, home insulation, etc., so as to minimize their carbon footprint.

Under fee-and-dividend, 100 percent of the money collected from the fossil fuel companies at the mine or well is distributed uniformly to the public. Thus those who do better than average in reducing their carbon footprint will receive more in the dividend than they will pay in the added costs of the products they buy.

The fee-and-dividend approach is straightforward. It does not require a large bureaucracy. The total amount collected each month is divided equally among all legal adult residents of the country, with half shares for children, up to two children per family. This dividend is sent electronically to bank accounts, or for people without a bank account, to their debit card.

A rising carbon price does not eliminate the need for efficiency regulations, but it makes them work much better. The best enforcement is carbon price--as the fuel price rises, people pay attention to waste.

Let's discuss cap-and-trade explicitly. Then I will provide a bottom-line proof that it cannot work. In cap-and-trade, the amount of a fossil fuel for sale is supposedly "capped." A nominal cap is defined by selling a limited number of certificates that allow a business or speculator to buy the fuel. So the fuel costs more because you must pay for the certificate and the fuel. Congress thinks this will reduce the amount of fuel you buy--which may be true, because it will cost you more. Congress likes cap-and-trade because it thinks the public will not figure out that a cap is a tax. How does the "trade" part factor in? Well, you don't have to use the certificate; you can trade it or sell it to somebody else. There will be markets for these certificates on Wall Street and such places. And markets for derivatives. The biggest player is expected to be Goldman Sachs. What is the advantage of cap-and-trade over fee-and-dividend, with the fee distributed to the public in equal shares? There is an advantage to cap-and-trade only for energy companies with strong lobbyists and for Congress, which would get to dole out the money collected in certificate selling, or just give away some certificates to special interests.

Okay, I will try to be more specific about why cap-and-trade will be necessarily ineffectual. Most of these arguments are relevant to other nations as well as the United States. First, Congress is pretending that the cap is not a tax, so it must try to keep the cap's impact on fuel costs small. Therefore, the impact of cap-and-trade on people's spending decisions will be small, so necessarily it will have little effect on carbon emissions. Of course that defeats the whole purpose, which is to drive out fossil fuels by raising their price, replacing them with efficiency and carbon-free energy. The impact of cap-and-trade is made even smaller by the fact that the cap is usually not across the board at the mine. In the fee-and-dividend system, a single number, dollars per ton of carbon dioxide, is applied at the mine or port of entry. No exceptions, no freebies for anyone, all fossil fuels covered for everybody. In cap-and-trade, things are usually done in a more complicated way, which allows lobbyists and special interests to get their fingers in the pie. If the cap is not applied across the board, covering everything equally, any sector not covered will be able to lower its price. Sectors not covered then increase their fuel use."

14.6 Questions

* - Questions and problems may require additional information not included in the textbook.

1. Visit the Moana Loa Carbon dioxide website and update Figure 14-2 http://www.esrl.noaa.gov/gmd/ccgg/trends/*

2. Fit your own trend line to the data in Figure 14-2 and estimate the projected level of CO_2 in the atmosphere in 10 years and in 50 years.

3. What is kerogen? What is a methane clathrate? *

4. Where is the closest limestone mine to where you live? What is its annual production?*

5. Where is the closest active fossil fuel source to where you live? What is its annual production?*

6. What is today's maximum solar intensity where you live (account for your latitude)?*

7. Describe in your own words what CO_{2e} (equivalent carbon dioxide) means.

8. Sketch the motions of the excited states of CO_2 and H_2O.

9. Explore one of the climate change denialist websites. Who owns and pays for the site? Read their arguments as to why climate change is not important. Does the argument make logical sense? Are their facts provable using the scientific method? Do they include the necessary and sufficient data to support their claims? Is there insufficient data to reject the claims?*

10. What are the Milankovich cycles? make a time series sketch that includes all three.*

11. Are there any current or proposed US-EPA regulations to control emissions of carbon dioxide for stationary sources? Summarize the regulatory ideas as to whom the regulations apply to, the expected impact, and when must the regulations be met.*

12. Why is water not included in the list of US greenhouse gases?

13. What is the current distribution of the US energy portfolio (create an update for Figure 14-11).*

14. Discuss ways to pay for carbon capture and sequestration (CCS). Who pays? How much? To whom does it get paid? How should that money be used? Who decides? Make any reasonable assumptions as needed.*

15. Make a list of three questions you have about this chapter or air pollution concerns you have.

14.7 Problems

1. How much would 1 Gt of dollar bills be worth? Compare to world economy of $70 trillion. How tall would a stack of these bills be? Can you compare it to something of the same magnitude e.g. distance to New York. [Assume: a dollar weighs 1 gram and has a thickness of 0.11 mm].

2. How much oil, natural gas, and coal were produced in the last year for which the data is available (report in industry standard units – bbl, scf, ton respectively)? What mass [Gt] of carbon will be released to the atmosphere from burning these products?

3. Determine the length of a train needed to haul one years' worth of US coal production. Compare to something of the same magnitude, i.e. distance to New York.

4. Determine the mass [Gt] of CO_2 released by human respiration in one year. How does this compare to the one year increase in the atmosphere? Is this a reasonable comparison? Note that humans breathe at a rate of 10 liter/min per person and $[CO_2]$ = 4% by volume.

5. Calculate the Earth's surface temperature using a two layer model, similar to section 14.3.1.2. Does this provide any insight into the temperature structure of the troposphere (see Figure 5.1)?

6. Calculate the frequency and wavelength of the excited states of a) CO_2, b) H_2O, or c) O_3.

7. Calculate the blackbody and one layer surface temperatures of Mars and Venus. Compare these temperatures to the observed value. Does this suggest anything about the respective atmospheres of these planets?

	Albedo	Distance from Sun (10^6 km)	$T_{observed}$ (K)
Venus	71	108	700
Mars	17	228	240

8. Explore ways to quantify the scale of one year of fossil fuel carbon emissions (approximately 10 GtC):
   a) Size of a cubic diamond.
   b) Size of a water body needed to dissolve it as CO_2.
   c) Size of a storage basin to contain it as liquid CO_2.
   d) Size of a storage basin to contain it as super-critical CO_2.
   e) Pile of limestone ($CaCO_3$).
   f) Surface area of a 1 meter deep photo-bioreactor for growing algae (35 g-CO_2/ m²/ day) used to convert CO_2 into oil.
   g) Your own creative idea.*

9. Calculate the efficiency of growing corn for food energy. Make whatever assumptions are needed (Fermi type problem, see Chapter 2, example 2-4).

10. The US annual production of methane from enteric fermentation (digestion system of cows and other animals) is 5.6 Tg CH_4. Determine the 100 year - CO_2 equivalent of this emission value.

11. What mass (ton) of CO_2 is released from a 950 MW coal power plant (33% efficiency) in one year? Assume the coal provides 9,980 BTU/lb and contains 72 wt% carbon.

12. Calculate the CO_2 emission reduction (tons) if the US car fleet (250 million cars, SUVs and pickup trucks) improves its fuel mileage from 22 miles/gallon to 50 miles/gallon. Assume an average car drives 15,000 miles per year and gasoline (ρ=0.720 ton/m³) is approximated by C_8H_{18}.

13. Calculate the direct carbon emission reduction (ton/year) of converting a 200 MW power plant (33% efficiency) from coal (70 wt%C) to wood (50 wt%C). Assume the coal has an energy content of 9,500 BTU/lb and the wood 7,000 BTU/lb. Would this conversion alter the release of PM, SO_x, and NO_x?

14. Calculate the direct carbon emission reduction (ton/year) of increasing the efficiency of a 1000 MW coal fired power plant from 33% to 39%. The coal is 74% C and has a heat content of 9800 BTU/lb.

15. Calculate the direct carbon emission reduction (ton/year) of converting a 1000 MW coal fired power plant (33% efficient) with a natural gas plant (39% efficient). The coal is 74% C and has a heat content of 9800 BTU/lb. The gas is 75 wt% carbon and has a heat content of 980 BTU/scf.

16. What would be the expected decrease in direct carbon emissions (ton/year) for converting ½ of the US coal powered electricity generating units to using natural gas by using the data from problem 15? What would be the change in demand for natural gas? The US currently has 350 GW of coal power plants, and currently uses 24 trillion scf of natural gas.

17. Assume a carbon capture and sequestration (CCS) unit on a power plant requires a 20% side load to capture 80% of carbon emissions. How much carbon will be released from a 700 MW (output) natural gas plant (35% efficiency excluding CCS) compared to one without CCS?

18. Calculate the CO_2 concentration (mole fraction – dry basis) in the exhaust gas of a natural gas power plant. Assume 50% excess air is used. Ignore water in inlet air.

19. Calculate the CO_2 concentration (mole fraction) in the exhaust gas of coal power plant assuming 50% excess air is used and that the coal has a molecular form of $C_{61}H_{225}O_5$. Ignore water in inlet air.

20. Determine the size (length in meters) for an adsorption column for removing carbon dioxide from the exhaust stream of a power plant using a molecular sieve-13x (MS-13x) as the adsorbent. The exhaust gas flow rate is 9.0 mol/sec and contains 6.5 mol% CO_2 at 50 psi and 25°C. The bed density is 625 kg/m³, the column has an inside diameter of 1.5 meter, and the desired breakthrough time is 8 hours. Neglect the depth of the adsorption zone. See Section 11.3.5.1 Adsorption for additional help. Isotherm data (Siriwardane, et al., 2001) is given as:

$$W_i = \frac{aP_i^*}{1+bP_i^*}$$

Where W_i = moles of CO_2 adsorbed per kg MS-13x, a = 1.33 [psi-1], b = 0.214 [psi-1], and = equilibrium partial pressure of CO_2 [psi].

21. A water-amine solution is used in an absorption column to remove carbon dioxide from the exhaust gas of a 500 MW natural gas power plant (36% efficient). The exhaust gas contains 7.5 mol% CO_2. The system is designed to capture 80% of the CO_2. The column is run at 10°C and 2 atm pressure, and the Henrys law constant for this system is 420 atm/ CO_2 mol fraction. The liquid flow rate is 1.5 times the minimum needed. Find the actual liquid flow rate in moles solution per minute. Neglect the solubility of other gases and assume no liquid evaporates. See Section 11.3.5.2 Absorption for additional help.

22. Calculate the diameter of a pipeline needed to transport the CO_2 emissions (as pure CO_2) from a 700 MW natural gas power plant (36% efficiency), as a) liquid CO_2 and b) as super-critical CO_2. Assume a pipeline fluid velocity of 3 m/sec.

23. Determine the additional cost if CCS is used on an 800 MW coal power plant (35% efficiency without CCS). Include capture, 500 km transport, storage, and monitoring/verification. Report the cost as a) total cost [$/year], b) per ton coal [$/ton], c) per unit electricity [$/kWh], while assuming steady production at 98% of rating and operation time of 8500 hours/year. Consult Table 14-14 for cost estimates, and that the CCS side load is 15% and it is capable of capturing 80% of CO_2.

24. Rework problem 23, except use natural gas and change units of b) to [$/MMscf].

14.8 Group Project Ideas

1. The Kyoto Protocols expired in 2012. What is replacing them? Did it achieve its objectives? Why did the US choose not to sign the Kyoto Protocol?

2. Explore the works of Svente Arrhenius as related to CO_2, spectroscopy, and industrialization. What did he think of CO_2 induced climate change?

3. Explore additional sources of information on non-fossil fuel energy sources. Prepare a projection of how this will fit the US/ World energy portfolio in 10 years and 50 years. Only explore one such source.

4. Choose a topic that relates to global climate change and determine where your elected representative stands on the issue. Write a letter to your representative to express your opinion of their stand. Send a letter to your representative and forward a copy to your instructor. For help with the mechanics of writing such a letter see: https://writerep.house.gov/writerep/welcome.shtml or http://usgovinfo.about.com/od/uscongress/a/letterscongress.htm.

14.9 Bibliography

Allison, Ian, et al. 2009. *The Copenhagen Diagnosis, 2009: Updating the World on the Latest Climate Science*. Sydney, Australia : The University of New South Wales, 2009.

Archer, David. 2007. *Global Warming: Understanding the Forecast*. Malden, MA, USA : Blackwell Publishing, 2007. ISBN: 9781405140393.

Arrhenius, Svante. 1896. On the Influence of Carbonic Acid in the Air Upon the Temperature of the Ground. *Philosophical Magazine*. 1896, Vol. 41, pp. 237-276.

Atkins, P W and Depaula, J. 2010. *Physical Chemistry, 9th Ed*. Oxford, UK : Oxford University press, 2010. ISBN: 9780199543373.

Bellman, D.K. 2007. *Topic Paper #4: Electric Generation Efficiency*. s.l. : National Petroluem Council, 2007.

Blunden, J, Arndt, D S and Baringer, M O, Eds. 2011. State of the Climate in 2010. *Bulletin of the American Meteorlogical Society*. 2011, Vol. 92, 6, pp. S1-S266.

Boden, Tom, Marland, Gregg and Andres, Bob. 2011. *Global CO2 Emissions from Fossil-Fuel Burning, Cement Manufacture, and Gas Flaring: 1751-2008*. Oak Ridge, TN, USA : Oak Ridge National Laboratory, Carbon Dioxide Information Analysis Center, 2011.

Ganopolski, A. and D.M. Roche. 2009. On the nature of lead–lag relationships during glacial–interglacial climate transitions. *Quaternary Science Reviews*. 2009, Vol. 28, pp. 3361-3378.

Hansen, J. 2009. *Storms of My Grandchildren*. Ney York, NY, USA : Bloomsbury USA, 2009. ISBN-13: 978-1608195022.

IEA. 2011. *World Energy Outlook*. Paris, France : International Energy Agency, 2011.

—. 2012. *World Energy Outlook*. http://www.iea.org/ : International Energy Agency, 2012.

IPCC. 2005. *Carbon Dioxide Capture and Storage*. Cambridge, UK : Cambridge University Press, 2005.

IPCC, Solomon, S., D. Qin, M. Manning, Z. Chen, M. Marquis, K.B. Averyt, M. Tignor, and H.L. Miller (eds.). 2007. *Contribution of Working Group I to the Fourth Assessment Report of the Intergovernmental Panel on Climate Change*. Cambridge, UK : Cambridge University Press, 2007.

Neftel, A., H. Friedli, E. Moor, H. Lötscher, H. Oeschger, U. Siegenthaler, and B. Stauffer. 1994. *Historical CO2 record from the Siple Station ice core*. Oak Ridge, Tenn., U.S.A. : U.S. Department of Energy, 1994. In Trends: A Compendium of Data on Global Change. Carbon Dioxide Information Analysis Center, Oak Ridge National Laboratory.

Reeburgh, W S. 1997. Figures Summarizing the Global Cycles of Biologically Active Elements. *Dr. William S. Reeburgh, Professor Marine and Terrestrial Biogeochemistry*. [Online] 1997. [Cited: Jan 9, 2012.] http://www.ess.uci.edu/~reeburgh/figures.html.

Rohde, Robert, et al. 2011. Berkeley Earth Temperature Averaging Process. *Berkeley Earth Surface Temperature*. [Online] 2011. [Cited: Jan 18, 2012.] http://www.berkeleyearth.org.

Schmidt, Gavin, et al. 2007. Start Here. *Real Climate*. [Online] May 22, 2007. [Cited: Jan 18, 2012.] http://www.realclimate.org/index.php/archives/2007/05/start-here/.

SCotUS. 2007. *Massachusetts ET AL. vs. Environmental Protection Agency ET AL. 05–1120*, Washington DC, USA : US Supreme Court, April 2, 2007. http://www.supremecourt.gov/opinions/06pdf/05-1120.pdf.

Shellito, Cindy J., Sloan, Lisa C. and Huber, Matthew. 2003. Climate model sensitivity to atmospheric CO2 levels in the Early-Middle Paleogene. *Palaeogeography, Palaeoclimatology, Palaeoecology*. 2003, Vol. 193, pp. 113-123.

Siriwardane, Ranjani, et al. 2001. *Adsorption and Desorption of CO2 on Solid Sorbents*. Morgantown, WV, USA : US Department of Energy, 2001.

Tans, Pieter and Keeling, Ralph. 2011. Atmospheric CO2 at Mauna Loa Observatory. *Trends in Atmospheric Carbon Dioxide*. [Online] Dec 2011. [Cited: Jan 10, 2012.] http://www.esrl.noaa.gov/gmd/ccgg/trends/.

UN. 2012. Kyoto Protocol. *United Nations Framework Convention on Climate Change*. [Online] United Nations, January 2012. [Cited: January 19, 2012.] http://unfccc.int/kyoto_protocol/items/2830.php.

US-EIA. 2011a. How Much of the World's Electricity Supply is Generated from Wind and Who are the Leading Generators? *US Energy Information Administration - Energy in Brief*. [Online] Aug 30, 2011a. [Cited: Jan 9, 2012.] http://www.eia.gov/energy_in_brief/wind_power.cfm.

—. 2011b. *Projections: EIA Annual Energy Outlook*. Washington, DC, USA : s.n., 2011b. www.eia.gov/ies. DOE/EIA-0383(2011).

—. 2011c. US Primary Energy Consumption Estimates by Source, 1775-2010. *US Energy Information Administration*. [Online] Oct 19, 2011c. [Cited: Jan 9, 2012.] http://www.eia.gov/totalenergy/data/annual/perspectives.cfm.

US-EPA. 2011. *Inventory of US Greenhouse Gas Emissions and Sinks: 1990 - 2009*. Washington, DC, USA : US Environmental Protection Agency, 2011. EPA 430-R-11-005.

Weast, Ed. 1986. *Handbook of Chemistry and Physics, 67th*. Boca Raton, FL, USA : CRC Press, 1986. ISBN 0849304679.

Wikipedia. 2011a. Carbon Dioxide - History. *Wikipedia*. [Online] Nov 2011a. [Cited: Jan 16, 2012.] http://en.wikipedia.org/wiki/Carbon_dioxide.

—. 2011b. Energy Content of Biofuel. *Yields of Common Crops Associated with Biofuels Production*. [Online] September 6, 2011b. [Cited: January 19, 2012.] http://en.wikipedia.org/wiki/Energy_content_of_biofuel.

—. 2011c. Energy Density. *Common Energy Densities*. [Online] December 22, 2011c. [Cited: January 19, 2012.] http://en.wikipedia.org/wiki/Energy_density.

CHAPTER **15**

Indoor Air Quality

15.1 History and General Information

Indoor air quality (IAQ) is a product of the interaction between a building, its location, climate, uses, and occupants. Quality may range from much better than outdoor air to much worse. Great effort must be applied to create excellent air, whereas neglect typically causes significant problems.

Problems associated with indoor air quality range from discomfort and reduced productivity, to serious health problems, including death. Typically, indoor air problems are not the result of a single source or factor, but to a multitude of factors interacting in unforeseen ways. Potential problems can arise due to outdoor air quality, work practices and procedures, maintenance, furnishings, temperature, humidity, and moisture sources.

The causes of air quality concerns can be simple or very complex, and they may be difficult to identify. They can involve chemical, microbiological, physical, and psychological mechanisms. Non-IAQ related factors, such as stress, poorly designed work areas, lighting levels, and seasonal illnesses or colds may cause concern about air quality, even though they may not be caused by poor air quality.

The control of indoor air pollutants can be placed into two categories – direct removal and material avoidance. Direct removal requires air-handling equipment to move the air from inside to the outside. Material avoidance eliminates problems by not allowing a substance to discharge into the indoor air.

A growing body of evidence suggests that poor indoor air quality contributes to increased health care costs and productivity loss [(CPSC, et al., 1994), (US-EPA, et al., 1991), (US-EPA, 1993), (Morrison, et al., 2008), (ALA, 2009), (US-EPA, 2009b), (US-CDC, 2009a)]. The US Occupational Safety and Health Administration (OSHA) estimates, based on limited sampling, that 30 percent of the workers in non-industrial buildings - including offices, schools, and hospitals - are exposed to poor indoor air quality. The US-EPA estimates that medical costs, lost productivity, and increased sick leave due to poor indoor air quality, cost businesses and government more than $60 billion per year.

The average American adult spends 90 percent of their time indoors. Children spend less total time indoors, but appear to be spending an increasing amount of time indoors. Pollutants affect children more strongly because their smaller, developing bodies require more food, water, and air on a per weight basis. Schools tend to have particle problems because of renovations and retrofitting to the buildings so they can accommodate more students. Typically, ventilation rates are not adapted to meet these demands. In addition, limited budgets often do not provide for proper maintenance of ventilation systems. Schools can contain a variety of pollutant sources including art, biology and chemistry supplies, shop areas, gymnasiums, cleaning solvents and pesticides. Nearly 50 percent of schools tested had problems linked to poor indoor air quality (ELI, 2002). The US-EPA and many state governments have focused on improving indoor air quality in schools since children spend much of their day in the classroom [(HSN, 2009), (PA-Department of Health, 2002)].

Concerns about health effects and related costs have prompted some states to enact legislation to improve indoor air quality. Banning tobacco smoking inside public buildings (schools, government offices, airports, train stations) and in private buildings where the public may gather (offices, restaurants, movie theatres, enclosed shopping malls) is the most commonly used legislative tool to improve indoor air quality. There are no national regulations or laws directed at public exposure to harmful indoor air, and there is no federal level group monitoring indoor air quality within public or private buildings. However, property owners are required to provide information to potential buyers concerning indoor air quality problems (such as lead paint, flooding, moisture problems, mold, and radon).

Nearly all regulatory guidance for indoor air quality originates at the local level through city / county/state building codes. Building codes set minimum guidelines for what is acceptable for new construction, renovation, and rebuilding. Building codes may be set by national organizations such as, in the US, the National Electric Code, National Fire Code, Seismic Code, and Leadership in Energy & Environmental Design (LEED). However, it must be noted that these are based on private trade association works and are not federal laws. These codes can be adopted by a region or state. Insurance rates may reflect adaptation of these, and other similar codes.

● Case Study 15-1. Origin of an indoor air quality problem.

An office employee was working inside on a warm spring day and opened a window to allow some fresh air into the office (which sometimes smelled funny due to the presence of a printer). The employee did not completely close the window in the evening. Later that night a storm occurred and the wind blew the window open and rain entered. The next day the window was found open, and water was on the carpeted floor. No one thought anything about it. A few days later, the room smelled musty, and several employees were complaining of cold-like symptoms. Within the next week, several employees had to take time off due to the sickness. Some people that had taken time off to recover noticed that as soon as they returned, the symptoms returned.

The recurring symptoms were informally reported to the maintenance department, which sent a person to check the air quality. They discovered a very high level of mold. Problem solving discovered mold growing at the wall base under the window. It was then reported that rain had soaked the carpet a few weeks earlier. It was found to have absorbed into the wallboard material and created an opportunity for mold to grow. Removal of the flooring and wallboard material showed several square feet of moldy surfaces releasing spores into the nearby air. The carpeting and wallboard material had to be replaced. Repairs required about a week and resulted in several additional hours of lost work time.

15.1.1 Occupational Indoor Air Quality

Although there are no laws for general exposure to pollutants in indoor air in the US, there are extensive guidelines for indoor occupational air. This type of air is encountered at a person's workplace. The owner of the work place is required to maintain a quality of air that is not harmful to the employees. However, these standards would not necessarily apply to non-employees in the same space.

● Case Study 15-2. History of OSHA and NIOSH.

In 1968, the US Secretary of Labor, W. Wirtz, testified before Congress that 'Each day there will be 55 dead, 8,500 disabled, and 27,200 hurt in America's workplaces.' He was trying to gain support for a comprehensive occupational safety and health bill as proposed by President L. B. Johnson's administration. In 1970, Congress passed the Occupational Safety and Health (OSH) Act with a large majority, and on December 29, 1970, President Richard M. Nixon signed the bill into law. It became effective 120 days later.

These regulations were strongly opposed by the business community (especially small businesses). They instead favored a system of voluntary employer efforts; state-based activities; and federally funded research rather than a 'system of federal penalties and other attributes of overwhelming federal authority reaching into hundreds of thousands large and small business operations.' They were worried that the OSH Act would become 'the most extensive federal intervention into the day-to-day operation of American business in history' (Teplow, 1971). Prior to 1970, the country had been maintaining workplace safety on a purely voluntary basis. However, during the 1960's the government began to see that such practices led to rising accident rates, an increasing number of workplace illnesses, and an under-funded worker compensation system that did not sufficiently motivate employers to improve workplace safety and health.

The Occupational Safety and Health Act of 1970 (29 CFR § 671) [(OSHA, 1970)] created both the National Institute for Occupational Safety and Health (NIOSH) and the Occupational Safety and Health Administration (OSHA). This act requires employers to provide a safe and healthy workplace for their employees.

OSHA, a part of the U.S. Department of Labor, is responsible for developing and enforcing workplace safety and health regulations. OSHA's role is to promote the safety and health of America's working people by setting and enforcing standards; providing training, outreach and education; establishing partnerships; and encouraging continual process improvement in workplace safety and health. OSHA and its state partners maintain a staff of inspectors, complaint discrimination investigators, engineers, physicians, educators, standards writers, and other technical and support personnel in offices throughout the country.

NIOSH is in the U.S. Department of Health and Human Services and is an agency established to help ensure safe and healthful working conditions. They provide research, information, education, and training in occupational safety and health. The Institute does the following:

* Develops recommendations for occupational safety and health standards
* Conducts research on worker safety and health
* Conducts training and employee education
* Develops information on safe levels of exposure to toxic materials and harmful physical agents and substances
* Conducts research on new safety and health problems
* Conducts on-site investigations (Health Hazard Evaluations) to determine the toxicity of materials used in workplaces
* Funds research by other agencies or private organizations through grants, contracts, and other arrangements.

These two agencies' work is not specific only to indoor air quality, but cover all the ways a worker may be exposed to safety and health problems. Because inhalation is one of the primary methods of exposure to a contaminant, these agencies determine workplace-air exposure standards.

15.2 Sources and Health Effects

There are hundreds to thousands of different chemicals in most indoor air. Table 15-1 lists a few of the most common pollutants. Unfortunately, no single test can measure for all contaminants. Detection methods depend on the actual chemical, and only the contaminants tested for are found. Measurement types include direct air sampling, exposed surface wipe tests, air filtration and adsorption, and spectroscopy methods. The particular techniques are detailed in publications such as the National Institute for Occupational Safety and Health (NIOSH) Pocket Guide to Chemical Hazards (US-CDC, 2008).

Table 15-1. Typical Indoor Air Quality Materials of Concern.

Carbon Monoxide
Carbon Dioxide
Ozone
Radon
Asbestos
Volatile Organic Compounds
Inorganic Metals – Mercury, Lead
Biologically Active Materials – Molds, Allergens, Bacteria
Tobacco Smoke

The following sections discuss each of these materials and include information on sources, health effects, and controls.

15.2.1 Carbon Monoxide

Carbon monoxide (CO) is a colorless, odorless gas that forms whenever there is combustion of carbon-containing fuels (natural gas, propane, oil, wood, coal). It forms when the combustion process does not go to completion and is caused by poor air/fuel mixing or poor burning characteristics (time, temperature, technique). Section 11.1 discusses CO in ambient air.

CO interferes with the delivery of oxygen throughout the body. It binds to hemoglobin molecules in the blood, which transfer oxygen throughout the body. If a person is exposed to high concentrations, it can cause unconsciousness and even death. Exposure to lower concentrations can cause a range of symptoms in otherwise healthy people, including headaches, dizziness, weakness, nausea, confusion, disorientation, and fatigue. People who have chronic heart disease may experience episodes of increased chest pain. The symptoms of carbon monoxide poisoning are sometimes confused with the flu or food poisoning. Fetuses, infants, elderly people, and people with anemia or a history of heart or respiratory disease can be especially sensitive to carbon monoxide exposures.

CO can be controlled by proper cleaning and maintenance of fuel burning equipment, proper ventilation, and careful consideration of fuel choice. The detection of CO has become automated, and CO detectors are recommended for all buildings, including homes, that have combustion based energy sources (furnaces, hot-water tanks, and cooking).

15.2.2 Carbon Dioxide

Human activity is the primary cause of carbon dioxide (CO_2) in indoor air. CO_2 concentration is considered a marker for indoor air pollutants generated by humans and it correlates with human metabolic activity. It is also caused by fossil fuel combustion, fermentation, and wastewater and solid waste destruction, where it is the desired end product. Chapter 14 discusses CO_2 in ambient air.

Indoor concentrations of CO_2 above 1000 ppm suggest that the building has inadequate ventilation (NIOSH, 2009). Unusually high levels (10,000 ppm) of carbon dioxide indoors may cause occupants to grow drowsy, get headaches, or function at lower activity levels. At extremely high values (100,000 ppm even if normal oxygen levels are present) it can cause loss of consciousness or death by suffocation. Indoor levels are an indicator of the adequacy of outdoor air ventilation relative to indoor occupant density and metabolic activity.

CO_2 can be controlled with proper design of venting and/or increased ventilation. The UK standards for schools say that carbon dioxide in all teaching and learning spaces, when measured at seated head height and averaged over the whole day should not exceed 1,500 ppm. Canadian standards limit carbon dioxide to 3,500 ppm. US-OSHA limits carbon dioxide concentration in the workplace to 5,000 ppm for prolonged periods, and 30,000 ppm for 15 minutes (US-CDC, 2009b).

15.2.3 Ozone

Indoor ozone may originate from outside air used in ventilation, from air and water purification or disinfection, and unintentionally as a byproduct from the use of all types of electrical equipment (computers, printers, fans, motors). In addition, any electrical discharge (using an electric switch, static electricity) that causes a spark can create ozone, although these sources are usually insignificant.

Ozone is irritating to lung tissue and harmful to human health (See chapter 12 for a more detailed discussion of the effects of ozone). Indoor ozone reacts with many substances including common everyday substances such as skin oils, cosmetics, and hair care products to create other harmful chemicals such as aldehydes and ketones. Many of these reactant products are known sensory irritants, and some are known carcinogens (Apte, et al., 2007). Products containing citrus or terpene extracts will react very quickly with ozone to form toxic and irritating chemicals such as formaldehyde. These reaction byproducts will also lead to the formation of fine and ultra-fine particles.

If outside ozone concentrations are high, ventilation air can reduce it, or it may be necessary to use an ozone filter. Typical filters use activated carbon or specific ozone catalysts. Indoor generation can be difficult to eliminate or reduce, although clustering of electrical equipment allows for focused ventilation or air filtration.

● **Case Study 15-3. Ozone**
An acceptable level for ozone byproduct for household devices has been set in the US-Code of Federal Regulations (CFR) at a maximum 50 parts per billion (ppb). Most air filter manufacturers and makers of other household electronics have also voluntarily adopted this maximum. However, machines called ozone generators directly produce ozone (O_3) molecules—not as a byproduct, but as a direct product—and blow it into the room to clean the air. The idea is that the ozone is very reactive and chemically reacts with other indoor air pollutants to oxidize them. It is also an excellent disinfectant and can be used to kill pathogens in the air. Unfortunately, these machines can produce ozone up to 10-times more than the acceptable standard. Therefore, it has been recommended that you do not use ozone generator machines in your home (AAFA, 2009).

15.2.4 Radon

Radon is a naturally occurring, colorless, odorless gas that is generated underground during the decay of radionuclides such as uranium. Uranium is found in small quantities throughout the earth's soils, and it naturally decays to release alpha, beta, and gamma particles and to form new elements. These new elements may also undergo spontaneous radioactive decay in a chain of reactions through a series of intermediate elements that ultimately form the non-radioactive isotopes of lead. One intermediate element along the decay chain for uranium is radon gas.

Radon is a radioactive element with a half-life of 3.8 days. All of its known isotopes are radioactive. When radon decays, it produces an alpha particle and new radioactive elements called radon daughters or decay products, the first being Polonium, which is also radioactive. Unlike the gaseous radon, its daughters are solids and stick to surfaces, such as dust particles in the air. When inhaled, these particles can stick to the airways of the lung and increase the risk of developing lung cancer

Radon, once created, travels upward through rock and soil until it escapes from the ground. This general upward motion allows it to enter buildings from below, where it can accumulate in basements. It may also dissolve into groundwater and can be brought inside the building from well water. Any building can collect radon – new or old, with or without a basement.

Radon appears throughout the world. Every state in the US has identified buildings with elevated radon levels (above four picocuries per liter of air in a yearly average), see Figure 15-1. It is difficult to predict where radon will show up. It may appear at a high concentration in one building, and not at all in a nearby building. The only way to know if radon is a problem in a specific building is to measure it, typically with either a short term (3 – 90 days) or long-term (one-year) measurement kit.

Radon is a known carcinogen and is estimated to be the second leading cause for lung cancer (after tobacco smoking). Radon enters the body when breathing, and with drinking water. It causes between 7,000 and 30,000 deaths per year in the United States (US-EPA, 2009a). This risk is elevated for smokers and those exposed to second hand smoke (CPSC, et al., 1994), (ALA, 2006). Reduction of radon concentration falls into two categories – reduced infiltration or increased

FIGURE 15-1. US-EPA Map of Radon Zones. Regions in Zone 1 (dark) have a predicted average indoor radon screening level greater than 4 pCi/L [pico-Curies per liter], Zone 2 (intermediate) has a predicted average between 2 and 4 pCi/L, and Zone 3 (light) has a predicted average less than 2 pCi/L, (US-EPA, 2011).

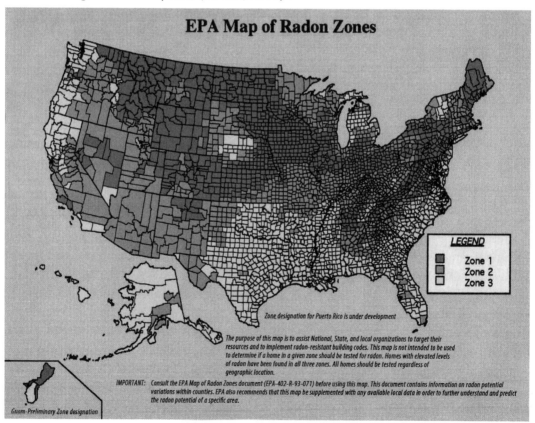

ventilation. Infiltration can be reduced by sealing cracks in floors and walls and changing the flow of air into the home with a system known as sub-slab depressurization, which uses pipes and fans to remove radon gas from beneath the concrete floor and foundation before it can enter the building. The collected air with radon is vented above the roof, where it safely disperses. Ventilation should focus on increasing circulation and mixing in below grade areas, as well as increasing the air exchange rate of all building air.

15.2.5 Asbestos

Asbestos is a mineral fiber that has been used in a variety of building construction materials for insulation and as a fire-retardant. The US-EPA and the Consumer Product Safety Commission (CPSC) banned several asbestos products in 1977 but does not require removal. Manufacturers also have voluntarily limited use of asbestos. Today, asbestos can be found in older buildings, in pipe and furnace insulation materials, asbestos shingles, millboard, textured paints and other coating materials, and floor tiles. It can be positively identified only with a special type of microscope. There are several types of asbestos fibers, some more harmful than others. Elevated concentrations of airborne asbestos can occur after asbestos-containing materials are disturbed by cutting, sanding, or other remodeling activities. Improper attempts to remove these materials can release asbestos fibers into the building's air, increasing asbestos levels and endangering the occupants.

The most dangerous asbestos fibers are too small to be visible. After inhalation, they can be captured and accumulate in the lungs. Asbestos can cause lung cancer, mesothelioma (a cancer of the chest and abdominal linings), and asbestosis (irreversible lung scarring that can be fatal). Symptoms of these diseases do not show up until many years after exposure begins. Most people with asbestos-related diseases were exposed to elevated concentrations while working around asbestos, some develop the disease from exposure to clothing and equipment brought home from job sites. While hazards correlate with chronic exposure, even limited (single-event) exposure may cause problems for some people.

Usually, it is best to leave asbestos material that is in good condition alone. Material in good condition does not release asbestos fiber. There is no danger unless the fibers are released and inhaled into the lungs. Removal of asbestos from a building should only occur when necessary — when the building is to be extensively remodeled for example. Procedures for removal are covered in the Toxic Substances Control Act (TSCA).

The US-EPA requires removal only in order to prevent significant exposure. A location with known intact asbestos-containing materials should develop a plan of action to help decide how to protect the material, and under what conditions it must be removed (GETF, 2003). Most states require the use of a licensed asbestos abatement contractor for removal work.

15.2.6 Volatile Organic Compounds

Volatile Organic Compounds (VOCs) found in indoor air may originate from fuels, in products such as glues, adhesives, carpets, furniture, paints, coatings, varnishes, and waxes, as well as cleaning, disinfecting, cosmetic, and degreasing products. VOCs are released whenever these products are used, and some may even be released during storage. VOCs may also be introduced into indoor air when products manufactured with them give off vapor inside a building. Section 11.3 discusses VOCs in ambient air.

The US-EPA found levels of a dozen common organic pollutants to be 2 to 5 times higher inside homes than outside, regardless of rural or industrial area. They also found that when people are using products containing organic chemicals, they can expose themselves and others to very high pollutant levels. These elevated concentrations can persist in the air long after the activity is completed (US-EPA, 1991). Another study by the US-EPA, covering six communities in various parts of the United States, found indoor levels up to ten times higher than those outdoors—even in locations with significant outdoor air pollution sources, such as petrochemical plants (US-EPA, 1987).

Typical health effects depend on the particular compounds, the exposure time and the airborne concentration of the material. Effects may range from no effect to immediate effects like eye, nose, and throat irritation, headaches, dizziness, visual disorders, and memory impairment. Long-term effects include increased risk of some cancers, although there is very little information about the long-term, low level and episodic exposures to these compounds. Many VOCs, such as benzene and formaldehyde are known to cause cancer in animals; some are suspected of or are known to cause cancer in humans.

Control methods for VOCs include increasing ventilation during usage, replacing the product with less hazardous alternatives (e.g. water based latex paint rather than oil based paint, non-petroleum based cleaning solvents), minimizing amounts in storage, and designing excellent ventilation in storage areas. It is recommended to keep on hand only the amount needed for short times – purchase frequent small amounts rather than storing large amounts. The rate at which most VOCs evaporate increases at higher temperatures, and the rate may depend somewhat on the humidity level.

● Case Study 15-4. Formaldehyde in the Home.

The most significant sources of formaldehyde are pressed-wood products made using adhesives that contain urea-formaldehyde (UF) resins. Pressed wood products made for indoor use include particleboard (used as subflooring and shelving and in cabinetry and furniture); hardwood plywood paneling (used for decorative wall covering and used in cabinets and furniture), and medium density fiberboard (used for drawer fronts, cabinets, and furniture tops). All of these products off-gas formaldehyde for the first few years after manufacture. Medium density fiberboard contains a higher resin-to-wood ratio than any other UF pressed wood product and is recognized as being the highest formaldehyde-emitting pressed wood product [(US-EPA, 2009b), (Alert, 2008)].

Formaldehyde is a colorless, pungent-smelling gas that can cause watery eyes, burning sensations in the eyes and throat, nausea, and difficulty in breathing in some humans exposed at elevated levels (concentrations above 0.1 ppm). High concentrations may trigger attacks in people with asthma. There is evidence that some people can develop sensitivity to formaldehyde. It causes cancer in animals and humans.

During the 1970s, many homeowners had urea-formaldehyde foam insulation (UFFI) installed in the wall cavities of their homes as an energy conservation measure. However, many of these homes had relatively high indoor concentrations of formaldehyde soon after the UFFI installation. Few homes now use this product. Studies show that formaldehyde emissions from UFFI decline with time; therefore, homes in which UFFI was installed many years ago are unlikely to have high levels of formaldehyde now.

15.2.7 Metals

15.2.7.1 Mercury

Mercury is a very versatile product. The elemental form is a liquid at room temperature, and its molecular forms have found uses in many products such as rodenticides because of its toxic effects. As a liquid it has seen use in many measurement instruments (pressure manometers, barometers, thermometers, electrical switches, thermostats) because of its high density and consistent thermal expansion properties. Section 13.4 discusses mercury in ambient air.

Elemental mercury evaporates at a rate of 7 µg/cm^2/hr at 20°C [(Andren, et al., 1979), (Garetano, et al., 2006)]. Up to 80% of inhaled mercury is absorbed and can be stored within the body (Cherian, et al., 1978). The primary health concern associated with inhaled mercury vapor is its neurotoxicity, and infants are considered particularly vulnerable. The Agency for Toxic Substances and Disease Registry (ATSDR) and the US-EPA, respectively, have established a minimal risk level (MRL) of 300 ng/m^3 and a reference concentration (RfC) of 200 ng/m^3 for elemental mercury vapor in residential quarters. The release of elemental mercury in a household may pose some health risk for those who are exposed. For example, broken clinical thermometers typically contain 600–675 mg elemental mercury but can generate mercury vapor concentrations an order of magnitude above both the ATSDR MRL and the US-EPA RfC (Smart, 1986). It should be noted that mercury thermometers are rarely used in the US anymore. Significant levels of mercury vapor have been found in buildings decades after spillage, resulting in the significant exposure of subsequent building occupants without their knowledge. Twelve residential and commercial sites were surveyed in the New York metropolitan area without prior knowledge of the sites mercury contamination. Eleven of these locations were found to have mercury vapor concentrations significantly elevated over outdoor concentrations. Prior breakage of clinical fever thermometers was subsequently identified as the probable mercury source in two of the locations (Orloff, et al., 1997).

● Case Study 15-5. Mercury in Compact Fluorescent Lamps.

Compact Fluorescent Lamps (CFL) produce light directly from electricity. They require only about 25% of the energy used by standard incandescent bulbs for the same amount of light. They also have much longer life spans, typically 2 to 10 times longer, depending on the application. Each CFL bulb contains about 4 mg of mercury. No mercury is released as long as the bulb is intact. However, if broken the mercury is released to the local environment where it may evaporate and contaminate the local air or water. See problem 13 at end of this chapter. Most retail sellers of CFLs collect them at no charge for proper disposal.

Note that even though fluorescent bulbs contain mercury, using them contributes less mercury to the environment than using regular incandescent bulbs. The reduction is because they use less electricity, and coal-fired power plants are the biggest source of mercury emissions into the air, due to the mercury content in coal.

15.2.7.2 Lead

There are many ways in which humans are exposed to lead: through air, drinking water, food, contaminated soil, deteriorating paint, and dust. It has been used in paint, gasoline, water pipes, lipstick, and many other common products [(CPSC, et al., 1994), (US-EPA, 1993), (US-CDC, 2008)]. Lead in ambient air is discussed in section 13.3.

Airborne lead enters the body when an individual inhales vapor phase lead compounds or breathes or ingests lead particles or dust. Lead affects practically all systems within the body. At high levels, it can cause convulsions, coma, and even death. Lower levels of lead can adversely affect the brain, central nervous system, blood cells, and kidneys. The effects of lead exposure on fetuses and young children can be severe. They include delays in physical and mental development, lower IQ levels, shortened attention spans, and increased behavioral problems. Fetuses, infants, and children are more vulnerable to lead exposure than adults since lead is more easily absorbed into growing bodies, and the tissues of small children are more sensitive to the damaging effects of lead. Children have higher exposures since they are more likely to get lead dust on their hands and then put their fingers or other lead-contaminated objects into their mouths. Signs and symptoms in children may include irritability, abdominal pain, vomiting, marked ataxia, and seizures or loss of consciousness. Symptoms in adults may include headaches, nausea, anorexia (and weight loss), constipation, fatigue, personality changes, and hearing loss. Obviously, these symptoms are shared with many other problems that have no connection to lead poisoning.

Lead exposure can be minimized by good housekeeping to remove dust and dirt, good ventilation, and avoiding the use of lead-containing materials (paint, toys, vinyl window blinds, plasticizer in many plastics) inside a building.

15.2.8 Biological Air Pollutants – Molds, Allergens, Bacteria

Biological air pollutants are found to some degree in every building. Sources include outdoor air, water reservoirs such as humidifiers, dehumidifiers, air conditioners and refrigerators, animals such as insects and rodents, and human occupants. The most important factor is a high relative humidity, which encourages the growth of molds and fungi on damp surfaces and allows dust mite populations to increase.

Components of mechanical heating, ventilating, and air conditioning (HVAC) systems can act to introduce and distribute biological air pollutants. They may introduce material from outside air if the air intake is near a source such as standing water (ponds, puddles from rainstorms) or organic debris (landscaping, grass clippings, bird droppings). The sub-units of such equipment can also provide the conditions necessary to grow these organisms especially in the humidification, ductwork, and mixing boxes, where either water or dust may collect.

Biological air pollutants may cause harmful effects to humans through infections, toxicity, and hypersensitivity. Infections occur when pathogenic organisms invade the human body. Toxicity occurs when the human body is exposed to toxins produced by another organism or its waste products. Hypersensitivity occurs when the organisms or their byproducts cause a specific activation of the immune system – this may be expressed as a cough, runny nose and eyes, difficulty breathing, asthma attack, or even taking the form of a disease.

The best control method for biological air pollutants is prevention. The major preventive measures include:

- Adequate design and maintenance of HVAC equipment to prevent the buildup or collection of water and dust,
- Adequate disinfection of water reservoirs and potable water systems,
- Prevention of standing water accumulation inside or near air intake,
- Rapid repair of leaks and seepage,
- Thoroughly clean and dry water-damaged carpets and building materials within 24 hours of damage, or consider removal and replacement,
- Maintain humidifiers and dehumidifiers to keep relative humidity below 50 percent,
- Use of dedicated exhaust fans to remove excess water vapor, and
- Control rodent and insect populations.

● **Case Study 15-6. Toxic Mold Threatening Health and Insurance Policies.**
One indoor air pollutant that has recently found notoriety is mold. Time, CNN, Newsweek, FOX, and other newspapers have run stories on toxic mold and its threat to human health and housing. One interesting story focused on a couple who became ill with headaches, respiratory problems and memory loss after purchasing a new $6.5 million mansion in Dripping Springs, Texas. Investigators found that much of the house was contaminated with toxic mold, and the house was condemned and bulldozed. The owners subsequently won a $32 million lawsuit against the insurance company, which was accused of not fixing the plumbing leaks that caused the damage. This case and others have raised public consciousness, leading to a dramatic increase in the number of mold-related claims submitted to insurers over the past few years. Insurers are scrambling to address the problem, with some stating they cannot address the increase in claims without raising rates or excluding mold coverage.

Nationwide, states have introduced legislation to deal with the issue. Approaches include setting up advisory and research panels to discover more about the risks of mold, creating public health standards for mold, and requiring that sellers and lessees disclose known mold hazards to buyers and lessors.

15.2.9 Tobacco Smoke

This hazard includes the gas and particle mixture released from the burning of tobacco (pipes, cigars, cigarettes). It also includes the mixture emitted from a smoker known as second-hand smoke. Those who smoke may generate it inside a building; it may enter the building from outside when smoking occurs near a building entrance or open window, or when smoking occurs near an air intake. Some may also enter on the clothes and equipment of people exposed to smoke. It has recently been noted that secondhand marijuana smoke may also harm similar to tobacco smoke (Springer, 2014).

Cigarette smoke contains over 4,800 chemicals including nicotine, formaldehyde, and carbon monoxide, 69 of which are known to cause cancer. Smoking is directly responsible for approximately 90 percent of lung cancer deaths and approximately 80-90 percent of emphysema and chronic bronchitis deaths [(US-CDC, 2009c), (ALA, 2008), (US-CDC, 2007)]. Secondhand smoke has been classified as a Group A carcinogen by the U.S. Environmental Protection Agency, a rating used only for substances proven to cause cancer in humans. Second hand smoke is estimated to be responsible annually for approximately 3,000 lung cancer deaths, 35,000 - 62,000 cardiovascular disease deaths, and 2,300 deaths from Sudden Infant Death Syndrome. Secondhand smoke is a direct health threat to people who already have heart and lung disease, and increases the risk of serious respiratory disease during the first two years of a child's life. It also contributes to 150,000-300,000 lower respiratory tract infections annually in children less than 18 months of age, resulting in 7,500-15,000 hospitalizations (ALA, 2009).

Exposure can be reduced or eliminated by imposing an indoor smoking ban (including outdoor areas where air may normally enter a building), or providing a separate and well-ventilated area for smoking. Typical mechanical ventilation equipment cannot remove these pollutants as quickly as they are generated.

15.3 Regulation

Relatively few indoor air quality regulations exist at the Federal or State level for a number of reasons: indoor air pollution is mostly invisible, it comes from an extremely large variety of sources, and air quality problems vary from building to building and by location. Most of the regulatory guidance originates at the local level using building codes (e.g. LEED, National Electric Code).

Many of the symptoms caused by indoor air pollution are not easy to pinpoint and may occur at extremely low pollutant levels that are difficult to measure, making it hard to determine what contaminant to regulate and at what level. In addition, many symptoms of low-level exposure to chemicals may take years to become apparent, making it extremely difficult to determine a single cause. Typical symptoms are the same as those for the common cold, flu, and seasonal allergies.

15.3.1 National Laws

OSHA has no indoor air quality (IAQ) standards but it does provide guidelines addressing the most common workplace complaints about indoor air quality. These complaints are usually due to temperature, humidity, lack of outside air ventilation or smoking. Only when workplace problems become potentially hazardous conditions leading to serious physical harm or death, do OSHA standards become applicable. Such standards may include those for specific air quality contaminants, ventilation systems, or the General Duty Clause of the OSH Act (OSHA, 2008a).

Growing consciousness about indoor air quality problems in non-industrial settings has prompted the federal government to take a variety of actions. OSHA's April 1994 proposal to adopt standards on indoor air quality generated the largest public response the agency had ever seen at that time; more than 100,000 comments were received when the comment period closed in August 1995. The proposed regulations (OSHA, 2008b) were removed in 2001. OSHA continues to study the issue, but no target date has been established for a reintroduction of a proposed set of standards for indoor air quality.

The US-EPA has dealt with the problem by publishing a series of voluntary guidance documents to assist building owners and others in maintaining good indoor air quality (US-EPA, 2008a). It also developed a program called "Tools for Schools," which helps schools set up programs to identify and correct indoor air quality problems (US-EPA, 2008b).

15.3.2 State Laws

States are central players in the development of policies addressing indoor air quality. The absence of a general federal regulatory scheme in this area allows states the freedom to develop local and regional approaches to managing indoor air issues. Since the late 1980's, there has been considerable activity in state legislatures on the subject of indoor air quality. State policies reflect the multi-faceted nature of IAQ issues. Laws and regulations address a variety of individual pollutants, practices and building types (CARB, July, 2005).

State indoor air quality legislation varies. Examples include: setting up commissions to study and assess health risks of indoor air quality; educating residents about the hazards of various indoor air pollutants; notification of parents before pesticides are used in schools; requiring sellers of property to disclose environmental hazards such as radon or asbestos; improving school indoor air quality; and establishing state programs or offices to deal with indoor air quality issues. A few states, such as New Hampshire, have set standards for specific pollutants such as carbon dioxide and formaldehyde. Minnesota authorized the Commissioner of Health to provide assessment and evaluation of indoor environments of public and non-profit buildings and facilities in disaster-affected communities (ELI, 2009).

One of the most common topics of legislation has involved indoor air quality in schools. The reasons for this involve the previously discussed susceptibility of schools to indoor air problems, children are considered an 'at-risk' population, and the fact that schools are usually government-owned buildings, making them easier to regulate than private homes or businesses.

As of October 2008, 49 states and DC have clean indoor air provisions restricting smoking in public places, such as offices, restaurants, and government buildings. Such laws have improved indoor air quality by reducing exposure to secondhand smoke, which contains many harmful and irritating chemicals.

Indoor air quality is a complex and constantly evolving issue. However, it appears that the research base is not deep enough to support exposure standards for many indoor air pollutants. Nonetheless, states have a variety of reasons for requiring standards and continue to set their own standards as the need arises.

● **Case Study 15-7. What is LEED?**
Leadership in Energy & Environmental Design (LEED) is a green building certification program that recognizes best-in-class building strategies and practices. To receive LEED certification, building projects must satisfy prerequisites and earn points to achieve different levels of certification. Prerequisites and credits differ for each rating system, and teams choose the best fit for their project.

There are five rating systems that address multiple project types:
Building Design and Construction *applies to new construction and major renovation. Interior Design and Construction applies to interiors of buildings.* **Building Operations and Maintenance** *applies to existing buildings that are undergoing improvements or little to no construction. Neighborhood Development applies to new land development or redevelopment containing residential, nonresidential uses, or a mix.* **Homes** *applies to single family housing, low-rise multi-family (1 – 3 floors), and mid-rise multifamily (4 – 6 floor) units.*

Each rating system consists of a combination of credit categories. The categories include; **Location and Transportation** *with credits for projects in areas with certain population densities, near diverse uses, and with multiple transportation options.* **Materials and Resources** *with credits for using sustainable building materials and minimizing waste. Indoor* **Environmental Quality** *gives credits for projects that promote good indoor air quality, access to daylight, and good views.* **Water Efficiency** *gives credits for reducing potable water consumption for both inside and outside uses.* **Energy and Atmosphere** *gives credits for innovative and efficient energy performance.* **Sustainable Sites** *gives credits for design strategies that minimize the project's impact on ecosystems and water resources.*

There are specific prerequisites that a project must satisfy and a variety of optional credits that earn points. The number of points determines the level of LEED certification. The levels include: certified (40-49 points), silver (50-59 points), gold (60-79 points), and platinum (80+ points).

Examples of required and optional credits include:
storage and collection of recyclables (required), tobacco smoke control (required), use of low emitting materials (up to 3 points), conducting indoor air quality assessment (up to 2 points), daylight (up to 3 points), sustainable wastewater management (1 point), walkable project site (1 point), bird collision deterrence (1 point). See the US Green Building Council for the complete list (USGBC, 2014).

15.4 Control Technologies

The most common methods used to reduce the risk and improve indoor air quality from pollutants include ventilation to remove the contaminants, filtration/ adsorption to remove the contaminants, proactive maintenance and housekeeping, attentive initial and retrofit building design, and source control. Table 15-2 lists several IAQ control methods and briefly describes their advantages and disadvantages. Most systems combine some or all of these techniques.

Table 15-2. Typical Methods to Control Indoor Air Quality.

Method	Advantages	Disadvantages
Ventilation.	Removes all contaminants, widely used.	Requires extensive electrical/ mechanical equipment. Emits contaminant to outside air.
Filtration/ Adsorption.	Removes a contaminant from the air, so it is not discharged outside.	Removes only some contaminants, requires periodic cleaning/ replacement of media.
Proactive maintenance, good housekeeping.	Reduces amounts discharged to air, Eliminates problems before they occur.	Constant effort required, additional workers.
Well thought out design and redesign.	Minimizes problems and exposures.	Difficult to adapt to existing situations, redesign can be very expensive.
Contaminant source control.	Reduces the amount released from storage, minimizes spill hazards.	Requires excellent inventory control to ensure supplies are always on hand.

Techniques for analyzing IAQ include collection of air samples, collection of samples on building surfaces, and computer modeling of airflow inside buildings. The resulting samples can be analyzed for mold, bacteria, chemicals, or other stressors. These investigations can lead to an understanding of the sources of the contaminants and ultimately to strategies for removing the unwanted elements from the air.

15.4.1 Ventilation and Equipment

The OSH Act established performance and design standards for specific substances and specific operations. Included are standards for ventilation equipment. Most systems consist of hoods, ductwork, fans, and vent stacks. This equipment may require additional filters, scrubbers, or adsorption equipment to remove contaminants before release to the outside air.

15.4.1.1 Air Exchange Rate

One of the key concepts in ventilation is air exchange rate. The exchange rate describes how long it takes for the amount of air equal in volume to the room or building to move through the ventilation equipment. The air exchange rate, τ, is calculated as:

15.1 $$\tau = \frac{V}{Q}$$

Where: V is the volume of the vented air space, and
Q is the air volumetric flow rate.

The air exchange rate has units of time. There are two ways to consider how the air may move through the ventilated space. First, *no mixing* - the new air completely replaces the old air in one-exchange rate. Therefore, it takes the same amount of time as calculated from equation one. Second, *complete mixing* - the vented air has the same concentration as the air inside the vented space at any given time (this value changes as new air enters). This first case is quite unrealistic, except in a duct with one-entrance and one-exit, as most airflow is turbulent and, therefore, has significant mixing. The second case is also idealized, but more closely approximates actual ventilation where there are many places for air to enter and leave the building. This condition leads to the interesting problem of how much time is required to vent a room or building. The problem can be solved by performing a mass balance on the space.

15.2   Rate of Mass − Rate of Mass + Generation in = Rate of Mass
        Flowing in   Flowing out   Building   Accumulation

This can be rewritten mathematically as:

$$15.3 \qquad Q_{in} C^a_{in} - Q_{out} C^a_{out} + \text{Gen}^a = \frac{d(V C^a_{out})}{dt}$$

Where:
Q_{in} = the volumetric flow of gas entering,
C^a_{in} = the concentration of 'a' in the entering flow,
Q_{out} = the volumetric flow of gas exiting,
C^a_{out} = the concentration of 'a' in the exiting flow,
t = time, and
Gen = the amount of 'a' generated within the building, such as particles from an oil-fueled heater or formaldehyde from tobacco smoke.

This equation is always written in terms of one particular mass species (usually the contaminant) which we call 'a'. Typically, for a room or building the flow in and out is the same, so Q_{in} is equal to Q_{out}, and we can drop the subscripts. We also note that the volume, V, does not change so this constant can be taken out of the derivative. The only remaining variables are Cout and time. Rearranging equation (15.3) we get:

$$15.4 \qquad Q\left[C^a_{in} - C^a_{out}\right] = V \frac{dC^a_{out}}{dt}$$

Next, solve this first-order differential equation by separation and integration, using the integration limits of $C^a_{out} = C^a_0$ at time = 0 and $C^a_{out} = C^a_{t'}$ at some later time, t':

$$15.5 \qquad \int_0^{t'} \frac{Q}{V} dt = \int_{C^a_0}^{C^a_{t'}} \frac{-1}{\left[C^a_{in} - C^a_{out}\right]} dC^a_{out}$$

Solving this equation for t', and substituting τ for V/Q, yields:

$$15.6 \qquad t' = \tau \left[\ln\left(\frac{C^a_{in} - C^a_0}{C^a_{in} - C^a_{t'}}\right)\right]$$

The time needed to reduce a contaminant is proportional to the air exchange rate and to the natural log of the ratio of the initial to later time concentration differences.

Indoor Air Quality

■ Example 15-1.

Determine the concentration of ozone in a 500 m³ room 8 hours after a ventilation fan starts. The room houses electrical equipment that generates ozone at a rate of 100 mg/hr. The fan moves 1800 m³/hr of air out of the room. The outside make-up air has an ozone concentration of 25 ppb. The initial concentration inside the room is 65 ppb.

Solution:
Since there is flow in, flow out, generation, and accumulation (decreasing) term, we must start at the general form of the mass balance:

$$15.3 \quad Q_{in}C_{in}^a - Q_{out}C_{out}^a + \text{Generation} = \frac{d(VC_{out}^a)}{dt}$$

The room volume is constant, so we can factor it out of the derivative, and then divide the entire equation by V to get,

$$15.7 \quad \frac{Q_{in}C_{in}^a}{V} - \frac{Q_{out}C_{out}^a}{V} + \frac{\text{Generation}}{V} = \frac{d(C_{out}^a)}{dt}$$

Identifying terms and working out units:

$Q_{in} = Q_{out} = 1800$ m³/hr
$V = 500$ m³
$C_{in} = 25$ ppb
$C_{out}(t=0) = 65$ ppb
Generation = 100 mg/hr

Terms 1, 2 and 4 in the above equation all have units of ppb/hr, term 3 must be converted to these same units. We use the ideal gas law to convert from units of mass to volume:

$$\frac{\text{Generation}}{V} = \frac{100 \frac{\text{mg O}_3}{\text{hr}}}{500 \text{ m}^3} \frac{\text{mg} - \text{mol O}_3}{48 \text{ mg O}_3} \frac{\text{kg} - \text{mol}}{10^6 \text{mg} - \text{mol}} \left[\frac{0.082 \text{ m}^3\text{atm}}{\text{kg} - \text{mol K}} \right] \frac{298 \text{ K}}{1 \text{ atm}} 10^9 = 101.8 \frac{\text{ppb}}{\text{hr}}$$

The final multiplication by 10^9 changes (volume of O_3 / total volume) into ppb units.

Substituting these values:

$$\frac{d(C_{out})}{dt} = \frac{1800 \frac{\text{m}^3}{\text{hr}}(25\text{ppb})}{500 \text{ m}^3} - \frac{1800 \frac{\text{m}^3}{\text{hr}} C_{out}}{500 \text{ m}^3} + 101.8 \frac{\text{ppb}}{\text{hr}} = (191.8 - 3.6 C_{out}) \frac{\text{ppb}}{\text{hr}}$$

Separate and integrate:

$$\int_{C_0^a}^{C_i^a} \frac{d(C_{out})}{(191.8 - 3.6 C_{out})\frac{ppb}{hr}} = \int_0^{8\,hr} dt$$

$$\frac{1}{3.6} \ln\left[\frac{191.8 - 3.6(65)}{191.8 - 3.6 C_{8hr}}\right] = 8hr$$

Solving for C_{8hr} yields 53.3 ppb. Even with 3.6 room air exchanges each hour, the concentration of ozone has decreased by only 18%.

15.4.1.2 Hoods

Hoods isolate and capture indoor contaminants before they can enter the general building air space. They also capture nearby ambient air, which mixes with the contaminant. The amount of ambient air captured increases with the distance between the hood and the contaminant sources. Good hood design minimizes airflow requirements, protect worker's breathing air, and provide worker's space for their work.

There are three main types of hood – enclosure, capturing, and canopy, see Figure 15-2. Enclosure hoods contain the work or process such that all contaminants are released inside the device. They typically have only one open side or face through which a worker may interact with the process. Air requirements typically require an air velocity of 100 to 500 ft/min (0.5 to 2.5 m/sec) through the open face. Volumetric flows are obtained by multiplying the opening area by this velocity:

15.8 $\quad Q = u_{face} Area_{face}$

Where: $\quad u_{face}$ = velocity of the gas passing through the open face of the hood, and
$\quad\quad\quad\quad Area_{face}$ = the cross-sectional area of the open face of the hood.

Capturing hoods create a strong airflow around the work area in order to capture any contaminants. A capture velocity of 50 to 100 ft/min should exist at the most distant point of contaminant source from the hood. This value increases if the contaminants are released with a high velocity, if crosscurrents exist in the workspace, or if the capture zone is expected to have turbulent air currents. The relationship between the capture velocity and the distance from the hood opening is:

15.9 $\quad u_{capture} = \dfrac{0.1\, Q}{x^2 + 0.1\, Area_{face}}$

Where: Q is calculated with equation (15.8).

x = the distance from the source to the center of the hood face.

Canopy hoods are used to remove humidity from the air (it does not matter what air it captures) or above heated equipment to capture fumes rising from a tank or other process. A common example is the vent used over a kitchen stove. They use low airflows and allow the escape of some fumes. They should never be used to vent hazardous materials. Flow rates can be calculated as

15.10 $$Q = (1.4) H u_{control} 2(L+W)$$

Where: L = the length of the canopy,
W = the width of the canopy,
H = the height of the canopy above the source, and
$u_{control}$ = the control velocity.

The control velocity typically has a value between 0.5 to 2.5 m/s.

Pressure drop from a hood depends on the hood shape, size, and air velocity at the duct leaving the hood.

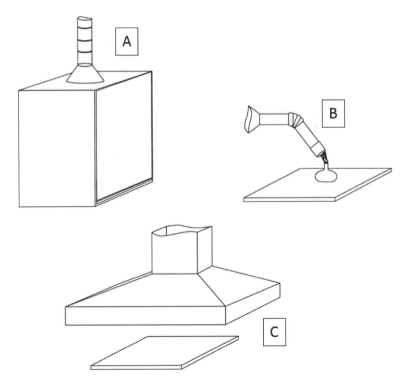

FIGURE 15-2. Three Basic Types of Hoods. A is an example of an enclosed hood – all work is contained within the device. B is a movable capture hood that can be placed near a temporary source. C is a canopy hood, located above a heated source.

15.4.1.3 Ductwork

Ducts control flowing air from hoods to fans, control equipment, and exit ports or vent stacks. Ducts for air ventilation are typically constructed of carbon steel or stainless steel. Carbon steel is easy to work with, form to shape, and is relatively inexpensive. Stainless steel may be required if the air is hot or contains corrosive contaminants. Other materials may be used, depending on the gas properties, available materials, and cost.

Air velocity in the duct is between 2 and 6 m/sec. Higher values may be used in certain industrial applications that expect high-particle loadings in the air. Higher velocity prevents particle buildup in the ductwork. Figure 15-3 provides the basic relationship between air velocity, duct diameter (or effective diameter), and volumetric flow.

FIGURE 15-3. Basic Sizing Relationships for Airflow in a Duct.

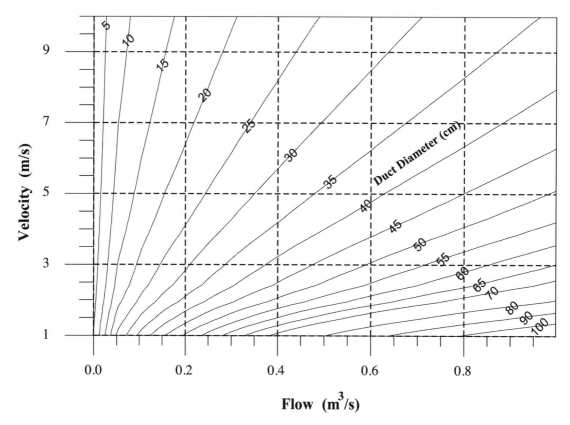

Indoor Air Quality

■ **Example 15-2.**
Find the duct velocity when air flows at a rate of 350 cfm through a circular 6-inch diameter duct.

Solution:
Rearrange equation (15.8) and solve for the velocity:

$$u_{duct} = \frac{Q}{A_{duct}} = \frac{350 \frac{ft^3}{min}}{\frac{\pi}{4}\left(6 \, in \, \frac{ft}{12 \, in}\right)^2} = 1783 \frac{ft}{min} = 29.7 \frac{ft}{sec} = 9.06 \frac{m}{sec}$$

This problem could also be solved using Figure 15-3. First, convert to metric units

Flow: $350 \frac{ft^3}{min} \frac{min}{60 \, sec} \frac{m^3}{35.315 \, ft^3} = 0.165 \frac{m^3}{s}$

Diameter: $6 \, inch \frac{2.54 \, cm}{inch} = 15.24 \, cm$

Then locate this flow and diameter and read velocity from the figure: velocity = 9 m/s.

■ **Example 15-3.**
Find the effective diameter of a 10 cm by 12 cm rectangular duct.

Solution:
Apply equation 4.32 to calculate the effective diameter:

Area = 12 cm * 10 cm = 120 cm²

Perimeter = (10 cm + 12 cm + 10 cm + 12 cm) = 44 cm

$$D_{eff} = \frac{4 \, Area}{Perimeter} = \frac{4 \, (120 \, cm^2)}{44 \, cm} = 10.9 \, cm$$

■ **Example 15-4.**
Find the effective duct diameter needed for moving 25 m³/min of air at a velocity of 500 m/min.

Solution:
Solve equation (15.8) for the duct area and then determine the effective diameter using the geometric formula for the area of a circle.

$$Area_{duct} = \frac{Q}{u_{duct}} = \frac{25 \frac{m^3}{min}}{500 \frac{m}{min}} = 0.05 \, m^2$$

$$D_{eff} = \sqrt{\frac{4 \, Area_{duct}}{\pi}} = \sqrt{\frac{4(0.05 \, m^2)}{\pi}} = 0.252 \, m = 25.2 \, cm = 10 \, inch$$

Checking with Figure 15-3 yields an answer of about 25 cm.

■ Example 15-5.

Find the pressure drop in 20 m of a 25 cm diameter (commercial steel pipe) duct when air flows at a rate of 30 m³/min. The air is at 1 atm and 20°C.

Solution:

This problem requires the use of equations described in Chapter 4 - Fluid Mechanics Review (FMR) section. Begin by sketching the system and placing all known information on the diagram.

FIGURE 15-4. System Sketch for Example 15-5.

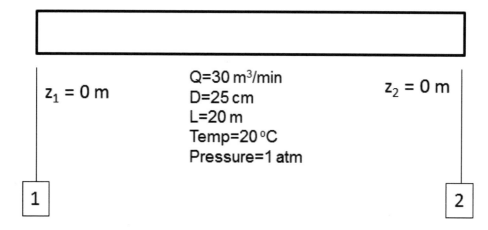

Next, determine the flow regime (laminar or turbulent) by calculating the Reynolds Number (equation 4.28). Re is necessary to determine the friction term. To determine it, we need the velocity of the air in the duct:

$$u_{duct} = \frac{Q}{Area_{duct}} = \frac{30\frac{m^3}{min}}{\frac{\pi}{4}(0.25m)^2} \frac{min}{60\,sec} = 10.2\frac{m}{sec}$$

We also need the air density and viscosity. Determine the air density using the ideal gas law:

$$\rho_{air} = \frac{mass}{V} = \frac{nMW_{air}}{V} = \frac{PMW_{air}}{RT} = \frac{(1atm)(29\frac{g}{gmol})}{\left(0.082\frac{l\,atm}{mol\,K}\right)(273+20)K} = 1.207\frac{g}{l}$$

Viscosity is found using the data and equation in appendix 6a:

$$\mu_{air} = \mu_0 \left(\frac{T}{T_0}\right)^n = 0.0171cP\left(\frac{293K}{273K}\right)^{0.768} = 0.01805cP$$

$$= 1.805\times10^{-5}\frac{kg}{m\,sec}$$

Now determine the Reynolds number:

$$Re = \frac{\rho_{air}u_{duct}D}{\mu_{air}} = \frac{\left(1.207\frac{kg}{m^3}\right)\left(10.2\frac{m}{sec}\right)(0.25m)}{1.805\times10^{-5}\frac{kg}{m\,sec}} = 170,520$$

The value of Re implies that the flow is turbulent. Next, use equation 4.36 to determine the friction factor. Use a roughness factor of 0.07 mm from Table 4-2.

$$f_M = \frac{0.25}{\left[\log_{10}\left(0.27\frac{\varepsilon}{D} + \frac{5.74}{Re^{0.9}}\right)\right]^2} = \frac{0.25}{\left[\log_{10}\left(0.27\frac{0.00007}{0.25} + \frac{5.74}{170,520^{0.9}}\right)\right]^2} = 0.018$$

Next, compute the pressure drop from equation 4-22:

$$\Delta P = -\rho_{air}\left(\frac{\Delta u_{duct}^2}{2} + g\Delta h + W + F\right)$$

Where: Δu_{duct} = zero, since the diameter does not change, and we assume that the air density is constant,

Δh = zero, as shown in the problem,

W = zero, as no work is done between the system (the ductwork) and the surroundings, and from equation 4.29:

$$F = \frac{1}{2} f_M u_{duct}^2 \frac{L}{D_{eff}} = \frac{1}{2}(0.018)(10.2\frac{m}{sec})^2 \left(\frac{20m}{0.25m}\right) = 74.9\frac{m^2}{sec^2}$$

so:

$$\Delta P = -\rho F = -1.207\frac{kg}{m^3} 74.9\frac{m^2}{sec^2} = -90.4\,Pa = -0.36\,inch\,H_2O$$

The result is a modest pressure loss (0.1% of atmospheric pressure), which confirms our assumption for constant density. Note that the pressure drop directly correlates to the cost of operation, so its' value should be minimized, but this must be balanced against the cost of the ductwork, where a larger diameter is a greater cost. Ways to reduce the pressure loss include reducing the flow rate, or specifying a larger diameter during the duct design phase. See the next example.

■ **Example 15-6.**

Determine the duct diameter needed to reduce the pressure loss in the above example to 70 Pa.

Solution:

This problem uses the information from Example 15-5, except that the duct diameter (D) is now the unknown, and the pressure drop (ΔP) is 70 Pa.

Without the diameter, we cannot determine the air duct velocity. That also means that we cannot determine the Re or the Friction factor. It may appear that we are stuck; however, we may guess a value of the diameter, calculate the unknown factors, and then use the mechanical energy balance to calculate the velocity needed to yield the given pressure drop. If this value is close (±10 %) to the value from our guessed diameter, then that is a good answer. If not, re-guess the diameter using the new value of velocity. Typically, only two or three iterations are needed before the solution converges on a good answer.

INDOOR AIR QUALITY

FIGURE 15-5. System Sketch for Example 15-6.

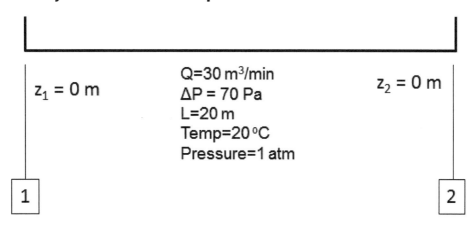

For this problem, we need to reduce the pressure drop, so we might guess a slightly larger diameter than in Example 15-5 – try 28 cm. It might also be useful to realize that a velocity of 5 m/s is typical, and you could use that fact to determine a good first guess. With this guessed diameter, we can proceed. Again, we first calculate the velocity:

$$u_{duct} = \frac{Q}{Area_{duct}} = \frac{30 \frac{m^3}{sec}}{\frac{\pi}{4}(0.28m)^2} \cdot \frac{min}{60 sec} = 8.12 \frac{m}{sec}$$

Next, calculate the Reynolds number, for which we need the air density and viscosity – which we determined in example 4:

$$Re = \frac{\rho_{air} u_{duct} D}{\mu_{air}} = \frac{\left(1.207 \frac{kg}{m^3}\right)\left(8.12 \frac{m}{sec}\right)(0.28m)}{1.805 \times 10^{-5} \frac{kg}{m\,sec}} = 152{,}000$$

The value of the Re implies that the flow is turbulent. Next, use equation 4-37 to determine the friction factor. Use a roughness factor of 0.07 mm from Table 4-2:

$$f_M = \frac{0.25}{\left[\log_{10}\left(0.27 \frac{\varepsilon}{D} + \frac{5.74}{Re^{0.9}}\right)\right]^2} = \frac{0.25}{\left[\log_{10}\left(0.27 \frac{0.00007}{0.28} + \frac{5.74}{152{,}000^{0.9}}\right)\right]^2} = 0.0181$$

Next, solve 4.23 for the duct velocity:

$$\Delta P = -\rho_{air}(F) = -\rho_{air}\left(\frac{1}{2}f_M u_{duct}^2 \frac{L}{D_{eff}}\right)$$

Or

$$u_{duct} = \sqrt{\frac{-\Delta P D_{eff}}{(\rho_{air})(\frac{1}{2}f_M L)}} = \sqrt{\frac{-(0-70\,Pa)(0.28\,m)}{1.207\frac{kg}{m^3}(0.5)(0.0181)(20\,m)}} = 9.47\frac{m}{s}$$

Which is more than 10% different from 8.12. So we recalculate by first determining the diameter needed to yield a velocity of 9.47 m/s

$$D_{eff} = \sqrt{\frac{4Q}{\pi u_{duct}}} = \sqrt{\frac{4(30\frac{m^3}{min})}{\pi(9.47\frac{m}{s})}\frac{min}{60\,sec}} = 0.26\,m$$

$$Re = \frac{\rho_{air} u_{duct} D}{\mu_{air}} = \frac{\left(1.207\frac{kg}{m^3}\right)\left(9.47\frac{m}{sec}\right)(0.26\,m)}{1.805x10^{-5}\frac{kg}{m\,sec}} = 165,000$$

for which $f_M = 0.0180$, and

$$u_{duct} = \sqrt{\frac{-\Delta P D_{eff}}{(\rho_{air})(\frac{1}{2}f_M L)}} = \sqrt{\frac{-(0-70\,Pa)(0.26\,m)}{1.207\frac{kg}{m^3}(0.5)(0.018)(20\,m)}} = 9.15\frac{m}{s}$$

The result is within 5% of the previous answer (9.47) so no additional iteration is required. This solution yields a duct diameter of 26 cm or 10.2 inches. Typically, ducts are made at just a few standard sizes, and not made to exact dimensions. If the calculated size is different than the standard sizes, the next larger standard size is used, possibly 12-inch diameter for this example.

15.4.1.4 Fans

Fans are used to move air from one location to another. The motion is caused by creating areas of high and low pressure, with airflow being in the direction of high pressure to low pressure. Fans work by imparting energy to the air with a rotating blade, which accelerates the air away from the blades and increases its pressure.

A fan is selected based on the required volumetric airflow, the pressure needed to move the required volume of air through the ventilation system, and the density of the air at the fan (the temperature and pressure must be known). Additional information that may help in the selection include the type of contaminant (solid or liquid particles, corrosive properties, flammability or combustibility), the available space for the fan, the fan efficiency, and the building occupants' sensitivity to noise.

Volumetric airflow is determined by summing all the airflow needs from hoods and any other sources. Air density can be calculated from the ideal gas law using known pressure and temperature. It may also be looked up in tables provided by the fan manufacturer. The required pressure generated by the fan must be able to overcome any frictional losses due to movement through the ductwork and control devices, and to provide sufficient velocity at the vent stack. This pressure is usually called the fan static pressure (FSP) and its calculation is:

15.11 $\quad FSP = \Delta P_{suction} + \Delta P_{discharge} - VP$

Where: $\Delta P_{suction}$ = pressure drop due to friction for the air as it moves from the hood (sources) through the ducts to the fan,

$\Delta P_{discharge}$ = pressure drop due to friction for the air as it moves from the fan, through any control equipment, and out of the building.

Velocity Pressure (VP) is calculated to determine the additional pressure needed at the hood inlets to accelerate the air from non-moving to the required inlet velocity. It must have the same units as the pressure terms.

Fan curves are used to describe how a particular fan operates. The curve is a plot of FSP (y-axis), and volumetric airflow rate (x-axis) see the dashed line in Figure 15-6. The flow rate must be referenced to a particular air density (or temperature and pressure) to allow corrections for an application where T and P are not the same. The operation of a fan is typically described with a fan performance chart or a fan characteristic curve, which relates the pressure increase due to the fan as a function of the air flow rate. A simplified model equation can be used to describe this curve, given as,

15.12 $\quad \Delta P_{fan} = a - bQ^2$

where a and b are constants for a given fan.

The operation of the entire system (fan and ductwork) can be explored with a system-operating curve. The plot is created by superimposing the system curve onto the fan curve. The two curves intersect at the system operating point. This point shows how the given fan operates in a particular system. Example 15-7 demonstrates these calculations.

■ Example 15-7.

Determine the system operating point for the system described in Example 15 5. The fan characteristic curve can be described with equation (15.12) where a = 120 Pa and b = 0.05 Pa min²/m⁶.

Solution:
Generate a table of pressure change for various flow rates for the fan (column 2) and the system (column 3). Note that the system curve data is generated using the equations developed in Example 15-5 and varying the flow rate (which in turn alters the velocity).

Fan pressure change example calculation, with Q =5 m³/min

$$\Delta P_{fan} = 120 \, Pa - 0.05 \frac{Pa \, min^2}{m^6} Q^2 = 118.75 \, Pa$$

Table 15-3. Pressure Change as a Function of Flow for Example 15-7.

Col. 1	Col. 2	Col. 3
Q	ΔP Fan	$-\Delta P$ System
0	120.0	0.0
5	118.8	3.9
10	115.0	14.3
15	108.8	31.1
20	100.0	54.3
25	88.8	83.7
30	75.0	120.5
35	58.8	160.0

Finally, superimpose the two curves onto a single plot and locate their intersection to obtain the system operating point.

FIGURE 15-6. Example 15-7, System Operating Point.

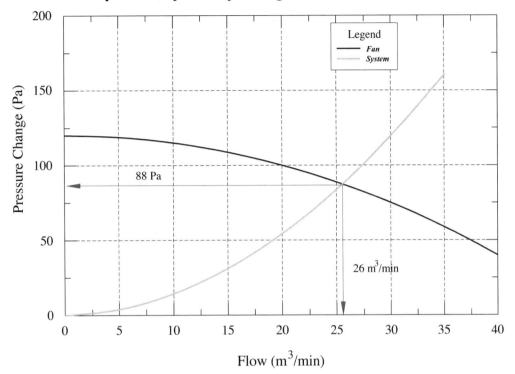

From the plot, we determine that this system and fan operate together at a flow of 26 m³/min and a pressure change of 88 Pa.

Actual fan curves do not extend through the entire range of flows. At the low and high end the fan becomes very inefficient. Manufacturers of fans provide a set of fan curves for their equipment. If their fan curves do not intersect your system curve, that particular fan is not recommended. Note that fan curve lines should never be extrapolated in order to find a system operating point, rather a different fan with an appropriate fan curve must be selected.

15.4.1.5 Venting Stacks

Proper design of venting equipment disperses the contaminated air into the outside air such that it does not reenter the building, nor enter nearby buildings. It requires information on the volumetric flow rate, temperature, humidity, and velocity of the released air. It also requires knowledge of nearby buildings (especially air intake and discharge locations) and the local weather patterns. Chapter 6.6 discusses stack design.

15.4.2 Filtration and Adsorption

Filtration and adsorption subsystems can remove contaminants from the air stream before release into the building or out to the venting stack. These systems are sized to accommodate the volumetric airflow with minimal pressure drop and to handle contaminant loading. Chapters 8 and 11 discuss these unit operations.

15.4.3 Maintenance and Housekeeping

Proper maintenance requires an active, organized, and proactive management plan. It includes daily, weekly, biweekly, monthly, quarterly, and yearly activities on every piece of equipment and every location within a facility. The primary objectives should be to provide periodic examinations to discover problems early and to shutdown equipment for repair in an orderly manner. Provisions include for cleaning, inspection, replacement of worn parts, and general repair. Regularly scheduled maintenance reduces or eliminates unexpected shutdowns and generally saves on costs over the long term.

Good housekeeping requires work areas are kept free of dust and dirt, and spilled liquids are cleaned up in a timely manner. Remove dust from equipment and building surfaces and floors before it can become airborne by traffic, vibrations, and air currents. Immediate cleanup of liquid spills can prevent the creation of an environment for mold and bacteria growth. A regular cleanup schedule is very effective in removing dust from work areas and for spotting problems before they get out of control. A good program is essential where using, handling, or storing solvents.

Any airspace that may become dangerously contaminated should have automated instrumentation including detectors and alarms that notify workers when there is a problem.

15.4.4 Green Engineering and Indoor Air Quality

Chapter 17 discusses the basic ideas and concepts of green engineering and pollution prevention. Green engineering is the design of chemical products and processes that reduce or eliminate the use and/or generation of hazardous substances.

Many indoor air quality issues can be reduced by careful consideration of the materials brought inside the building. A simple technology is the door mat, where shoes can be cleaned before entering a building, thus eliminating shoe-carried fumes, odors, or particulates from entering the building. Adequate design of operations that generate solids may prevent the release of particles into building air. Preventive maintenance and good housekeeping can prevent a small problem from becoming larger by remedying the issue in a timely manner. Reduction of storage quantities can minimize risks from spills and leaks. Always ask if a particular chemical is needed, what quantity is needed, and if any less harmful products exist to replace it. Many VOC's now have water-based counterparts that can greatly reduce volatile emissions.

● **Case Study 15-8. Green Engineering — Water based solvents used to replace volatile organic solvents (US-EPA, 1999).**
Nalco Company - The Development and Commercialization of ULTIMER®: The First of a New Family of Water-Soluble Polymer Dispersions.

Innovation and Benefits: *The Nalco Company developed a novel way to synthesize the polymers used to treat water in a variety of industrial and municipal operations. Nalco now manufactures these polymers in water, replacing the traditional water-in-oil mixtures and preventing the release of organic solvents and other chemicals into the environment.*

High-molecular-weight polyacrylamides are commonly used as process aids and water treatment agents in various industrial and municipal operations. Annually, at least 200 million pounds of water-soluble, acrylamide-based polymers are used to condition and purify water. These water-soluble polymers assist in removing suspended solids and contaminants and effecting separations. Traditionally, these polymers are produced as water-in-oil emulsions. Emulsions are prepared by combining the monomer, water, and hydrocarbon oil–surfactant mixture in approximately equal parts. Although the oil and surfactant are required for processing, they do not contribute to the performance of the polymer. Consequently, approximately 90 million pounds of oil and surfactant are released to the environment each year. Nalco has developed a new technology that permits the production of the polymers as stable colloids in water, eliminating the introduction of oil and surfactants into the environment.

The Nalco process uses a homogeneous dispersion polymerization technique. The water-soluble monomers are dissolved in an aqueous salt solution of ammonium sulfate then polymerized using a water-soluble, free-radical initiator. A low-molecular-weight dispersant polymer is added to prevent aggregation of the growing polymer chains. For end-use applications, the dispersion is simply added to water, thereby diluting the salt and allowing the polymer to dissolve into a clear, homogeneous, polymer solution. This technology has been successfully demonstrated with cationic copolymers of acrylamide, anionic copolymers of acrylamide, and non-ionic polymers.

Development of water-based dispersion polymers provides three important environmental benefits. First, the new process eliminates the use of hydrocarbon solvents and surfactants required in the manufacture of emulsion polymers. Dispersion polymers produce no VOCs and exhibit lower biological oxygen demand (BOD) and chemical oxygen demand (COD) than do emulsion polymers. Second, the salt used, ammonium sulfate, is a waste byproduct from another industrial process, the production of caprolactam. Caprolactam is the precursor in the manufacture of nylon; 2.5–4.5 million pounds of ammonium sulfate are produced for every million pounds of caprolactam, providing a ready supply of feedstock. Finally, dispersion polymers eliminate the need for costly equipment and inverter surfactants needed for mixing emulsion polymers. This technological advantage makes wastewater treatment more affordable for small- and medium-sized operations.

Nalco's dispersion polymers contain the same active polymer component as traditional emulsion polymers without employing oil and surfactant carrier systems. The polymers are produced as stable colloids in water, retaining ease and safety of handling while eliminating the release of oil and surfactants into the environment. By adopting this new technology, Nalco has conserved over one million pounds of hydrocarbon solvent and surfactants since 1997 on two polymers alone. In 1998, the water-based dispersions used 3.2 million pounds of ammonium sulfate, a byproduct from caprolactam synthesis that would otherwise be treated as waste. Additional environmental benefits will be realized as the dispersion polymerization process is extended to the manufacture of other polymers.

15.5 Questions

* - Questions and problems may require additional information not included in the textbook.

1. How would you know if there is an IAQ problem in your home or workplace?

2. Find a newspaper editorial from the early 1970's concerning the creation of the OSHA? Summarize the findings and compare the editorial's main point with current concerns of the issue.*

3. Compare and contrast the health effects of carbon monoxide and carbon dioxide.

4. Determine how to obtain a free or low-cost radon test kit for your home or workplace.*

5. Reread Case Study 15-1. What was the true cause of the air problem? How could the situation have been prevented?

6. Compile an inventory of three possible sources of VOCs in your home or workplace. Suggest possible alternatives that reduce the overall hazard.*

7. Reread Case Study 15-3. How would you know if the level of ozone in your home or workplace exceeds the standards?

8. Visit your national or state environmental quality website to determine if any lakes or rivers in your area are affected by mercury contamination. What do you think of this?*

9. Find a MSDS (Material Safety Data Sheet) for one of the materials listed in Table 2. Prepare a one-paragraph summary of the information and share with a friend and/or classmate. What are their reactions to the information?*

10. Is your national or state government considering IAQ regulation? Write a letter to your representative concerning your viewpoint on the regulation.*

11. Explore a national or state government website for IAQ. Write a one paragraph summary of an issue you found interesting.*

12. Make a list of three questions you have about this chapter or air pollution concerns you have.

15.6 Problems

1. Determine the air flow rate needed in a laboratory (8 m by 10 m with a 3.5 m ceiling) if the average air exchange rate must be 20 minutes or less.

2. Find the average air exchange rate in a 3000 m^3 building if the exhaust operates at a flow of 20 m^3/min. How does this compare with a standard that requires a minimum of one exchange per hour? Should the exchange flow rate be altered?

3. Find the flow rate when air travels at 500 ft/min through a 1.5 ft by 2.5 ft rectangular duct.

4. Calculate the mass (in kg) of inlet air (or makeup air) needed (at 20°C and 1 atm) per day for a 10,000 m³ building with an exchange rate of 60 min. How will this differ in summertime and wintertime?

5. The design of a duct system requires a 50 cm diameter duct; however, the space available for the duct is only 100 cm by 35 cm. What dimensions must the duct have in order to have the same flow and pressure drop as the circular duct?

6. What diameter circular duct is needed to transport room temperature air at a speed of 5.0 m/s and to have a Reynolds number of 350,000?

7. Find the Reynolds number for air at 1 atm and 5°C traveling at 5.3 m/s in a 10 cm diameter duct. Would the flow be classified as laminar or turbulent?

8. Calculate the expected frictional loss (m²/min² and inchH$_2$O) when moving 150 m³/min of standard air (dry, 1 atm, 298 K) through 400 m of horizontal 60 cm diameter, commercial steel duct. Hint: See chapter 4 for discussion of terms, especially equation 4.22.

9. Size a 2:1 rectangular duct system made of galvanized steel to move 1200 m³/m of chilled air (10°C and 1 atm) for a length 500 m, while having a pressure drop of no more than 1.0 cm H$_2$O.

10. Generate an operating curve by plotting pressure drop as a function of flow rate for the system in problem 15-8 above. Now superimpose the following fan curve (FSP in Pa = 700 - 30 Q² when Q is in [m³/min]). Determine the system operation point from the intersection of the two lines.

11. Determine the concentration (mg/m³ and ppm) of benzene in a closed room (300 m³) if 3.20 grams are spilled. Assume that it all evaporates instantaneously, and that the air in the room is well mixed. Next, allow the room's air to change once in 20 minutes. How long will it take until the concentration in the room is reduced by 50%? How long until the concentration is below the recommended NIOSH limit (1ppm)? Note that 1 ppm is equivalent to 3.25 mg/m³ for benzene.

12. In Example 15-1, we saw that the given air exchange rate only reduced the ozone concentration by 18% after 8 hours. a) How much time would be needed at the original example conditions to get the 50% reduction? b) What air exchange rate would be needed to reduce the ozone concentration in the room by 50% after 8 hours?

13. How long will it take for the mercury from a broken thermometer (about 0.7 gram with a density of 13.5 g/cm^3) to evaporate completely? Assume mercury evaporates from a spherical drop at a constant rate of 7 μg/cm^2/hr. If the spill occurred in a bathroom (12 m^3 with no air exchange) what would you expect the room's air concentration to be after 1 hour? How does this value compare with the NIOSH limit for mercury in air (0.01 mg/m^3)? Hint to simplify: use an average diameter rather than a continuously changing diameter for the evaporating drop.

14. Formaldehyde is a chemical found in second hand smoke caused by the combustion of tobacco. A smoker will emit up to 1 mg from a single cigarette smoked in an average manner. Determine the amount of this chemical in the air of a club which permits smoking. You will need to estimate the number of smokers and the amount each smokes over a given time period. You may assume the club can hold 200 people and has a volume of 4000 m^3. How much formadehyde is released in one hour? Assume the club's ventilation will exchanges the club air by 1/3 in one hour – what will the concentration of formaldehyde be after one hour? How does this compare to the OSHA maximum allowed concentration (0.12 mg/m^3)?

15.7 Group Project Ideas

For each project, the students should work in small groups and present their finding in either a short report (5-8 pages) or 15 minute presentation:

1. Compare and contrast a current IAQ law or regulation in your country and a foreign country. Do both countries address the needs of their citizens in an adequate manner?

2. Consider the creation of a federal regulation covering IAQ. How would you determine which materials and concentrations to cover? How to pay for it? Who should monitor it? What exceptions would be allowed and not allowed? Would you include different levels of protection for different buildings (i.e. public, private, schools, hospitals, government offices, sports arenas)?

15.8 Bibliography

AAFA. 2009. Air Filters: What Do I Need to Know About Air Filters? *Asthma and Allergy Foundation of America*. [Online] 2009. http://www.aafa.org/display.cfm?id=8&sub=16&cont=37.

ALA. 2006. Radon. [Online] March 2006. http://www.lungusa.org/site/pp.asp?c=dvLUK9O0E&b=35395.

—. 2009. Second Hand Smoke. Americal Lung Association. [Online] 2009. http://www.lungusa.org/site/pp.asp?c=dvLUK9O0E&b=35421.

—. 2008. Smoking 101 Fact Sheet. american Lung Association. [Online] August 2008. http://www.lungusa.org/site/c.dvLUK9O0E/b.39853/k.5D05/Smoking_101_Fact_Sheet.htm#two#two.

Alert, Business. 2008. EPA to Assess Risks of Formaldehyde in Pressed Wood Products. *Business Alert - US*. [Online] Dec 18, 2008. http://info.hktdc.com/alert/us0825d.htm.

Andren, A. W. and Nriagu, J. O. 1979. *The global cycle of mercury. In: The Biogeochemistry of Mercury in the Environment.* New York : Elsevier, 1979.

Apte, M G, Buchanan, I S.H. and Mendell, M J. 2007. Outdoor Ozone and Building Related Symptoms in the Base Study. *Indoor Air.* Feb 21, 2007.

CARB, California Air Resources Board. July, 2005. *Report to the California Legislature - INDOOR AIR POLLUTION IN CALIFORNIA.* Sacramento, CA : California Environmental Protection Agency, Air Resources Board, July, 2005.

Cherian, M. G., Hursh, J. G. and Clarkson, T. W. 1978. Radioactive mercury distribution in biological fluids and excretion in human subjects after inhalation of mercury vapor. . *Arch Environ Health.* 1978, Vol. 33, pp. 190-214.

CPSC, et al. 1994. *Indoor Air Pollution.* Washington DC : U.S. Government Printing Office, 1994. CPSC #455.

ELI. 2009. *Environmental law Institute Database of State Indoor Air Quality Laws Complete Database.* Washington DC : Environmental law Institute, 2009.

—. 2002. Healthier Schools: A Review of State Policies For Improving Indoor Air Quality. [Online] Environmental Law Institute, 2002. http://www.eli.org.

Garetano, Gary, Gochfeld, Michael and Stern, Alan H. 2006. Comparison of Indoor Mercury Vapor in Common Areas of Residential Buildings with Outdoor Levels in a Community Where Mercury Is Used for Cultural Purposes. *Environmental Health Perspectives.* 2006, Vol. 114, 1, pp. 59-62.

GETF. 2003. *Report of Findings and Recommendations On the Use and Management of Asbestos.* Annandale, Virginia : Global Environmental & Technology Foundation, 2003.

HSN, Inc. 2009. What You Can Do. Healthy Schools Network, Inc. [Online] 2009. [Cited: July 31, 2009.] http://www.healthyschools.org/index.html..

Morrison, Blake and Heath, Brad. 2008. Thousands of School Kids Routinely Exposed to Toxic Chemicals. *USA Today.* Dec. 8, 2008.

NIOSH. 2009. Chemical. [Online] 2009. http://www.cdc.gov/niosh/topics/chemical.html.

Orloff, K G, et al. 1997. Human exposure to elemental mercury in a contaminated residential building. *Arch Env Health.* 1997, Vol. 52, 3, pp. 169-172.

OSHA. 2008b. IAQ Standards. [Online] 2008b. http://www.osha.gov/SLTC/indoorairquality/standards.html.

—. 2008a. Indoor Air Quality. [Online] 2008a. http://www.osha.gov/SLTC/indoorairquality/index.html.

—. 1970. Occupational Safety and Health Act of 1970. [Online] 1970. [http://www.osha.gov/pls/oshaweb/owasrch.search_form?p_doc_type=OSHACT. 29 CFR 671.

PA-Department of Health. 2002. *Indoor Air Quality Guidelines for Pennsylvania Schools.* 2002.

Smart, E. R. 1986. Mercury vapour levels in a domestic environment following breakage of a clinical thermometer. *Science Total Environment.* 1986, Vol. 57, pp. 99-103.

Springer, M. 2014. *Secondhand marijuana smoke may damage blood vessels as much as tobacco smoke.* s.l. : American Heart Association, 2014.

Teplow, Leo. 1971. *Safety Expert, steel industry.* [interv.] Wall Streel Journal. Dec 1, 1971.

TSCA. Toxic Substances Control Act. *15 U.S.C. § 2601 et seq.*

US-CDC. 2009a. *Air Pollution.* [Online] 2009a. http://www.cdc.gov/nceh/airpollution/.

—. 2009b. Carbon Dioxide. CDC Documentation for IDLH Concentration. [Online] 2009b. http://www.cdc.gov/niosh/idlh/124389.html..

—. 2008. NIOSH Pocket Guide to Chemical Haxards. *National Institute for Occupational Safety and Health.* [Online] 2008. http://www.cdc.goc/niosh/default.html.

—. 2007. Tobacco Information and Prevention Source (TIPS). Adult Cigarette Smoking in the United States. *U.S. Center for Disease Control and Prevention.* [Online] Nov 2007. http://www.cdc.gov/tobacco.

—. 2009c. Tobacco Use: Targeting the Nation's Leading Killer. *National Center for Chronic Disease Prevention and Health Promotion.* [Online] Centers for Disease Control and Prevention, 2009c. http://www.cdc.gov/nccdphp/publications/aag/osh.htm.

US-EPA. 2009b. An Introduction to Indoor Air Quality. *Formaldehyde.* [Online] 2009b. http://www.epa.gov/iaq/formalde.html.

US-EPA and CDC-NIOSH, US. 1991. *Building Air Quality: A Guide for Building Owners and Facility Managers.* Wasington DC : US Government Printing Office, 1991.

US-EPA. 2011. EPA Map of Radon Zones. *EPA Indoor Air - Radon.* [Online] February 11, 2011. [Cited: August 9, 2011.] http://www.epa.gov/radon/zonemap.html. EPA-402-R-93-071.

—. 2008a. Green Indoor Environments. [Online] 2008a. http://www.epa.gov/iaq/greenbuilding/index.html.

—. 2008b. IAQ TfS Program. [Online] U.S. Environmental Protection Agency, 2008b. http://www.epa.gov/iaq/schools/.

—. 1999. Nalco Company. *Green Chemistry Award Winners.* [Online] 1999. http://www.epa.gov/oppt/greenchemistry/pubs/pgcc/winners/grca99.html.

—. 1987. *Project Summary: The Total Exposure Assessment Methodology (TEAM) Study.* Washington DC : US Government Printing Office, 1987. EPA-600-S6-87-002.

—. 2009a. Radon: The Health Hazard with a Simple Solution. *Radon.* [Online] 2009a. http://www.epa.gov/radon/index.html.

—. 1991. *Sick Building Syndrome.* Washington DC : US Government Printing Office, 1991. EPA 402-F-94-004.

—. 1993. The Inside Story: A Guide to Indoor Air Quality. [book auth.] U.S. Environmental Protection Agency Office of Radiation and Indoor Air. Washington DC : US Government Publishing Office, 1993.

USGBC. 2014. LEED Credits. *LEED.* [Online] 2014. [Cited: January 9, 2015.] http://www.usgbc.org/credits.

CHAPTER **16**

Mobile Sources

16.1 Introduction

Mobile sources of air pollution originate from on-road and off-road engines and movable equipment. On-road engines include devices like motorcycles, cars, light trucks, heavy trucks, and buses. Off-road engines and movable equipment include trains, ships, aircraft, construction vehicles and portable power generators. Devices such as recreation vehicles, golf carts, snowmobiles, lawnmowers, and chainsaws are also considered to be off-road engines. While there is a great variety in the nature of mobile sources, they share certain characteristics that suggest similar treatment for pollution reduction strategies:
- small individual sources,
- large total emissions,
- emission at ground level, near potential human receptors,
- combustion of hydrocarbon based fuels, and
- ubiquitous throughout the economy.

Each source adds only a small amount to an area's pollution emissions. However, there are large numbers of these sources, causing the small individual amounts to sum to a significant fraction of a region's total pollution (20% to 50% of some pollutants). Table 16-1 lists estimates for the number of mobile sources worldwide and in the US (US-BTS, 2012). It also provides some information on the number of new sources manufactured each year.

Pollutants from mobile sources include carbon monoxide (CO), hydrocarbons or volatile organic compounds (abbreviated as HC, VOC, or THC, NMOG), particulate matter (PM), nitrous oxides (NO_x), sulfur oxides (SO_x), hazardous organic and inorganic compounds (chiefly lead), and carbon dioxide (CO_2). Figure 16-1 compares the quantities of reported emissions from mobile sources (highway and off highway) in the US since 1970 (US-EPA, 2012a) Additional data can be found at 'http://www.epa.gov/ttn/chief/trends/index.html'. Off-road sources, in general, contribute similar total amounts of each pollutant as compared to on-road sources. Mobile sources also contribute significantly to the creation of ozone and smog, but do not directly emit it. Ozone and smog result from a chemical reaction between VOCs, NO_x, and sunlight in the atmosphere (see Chapter 12).

Table 16-1. Numbers of Mobile Sources (2010).

Type of Vehicle	Number		Production	
	Global	US	Global	US
Highway Vehicles	1.015×10^9	250×10^6	77.7×10^6	7.8×10^6
Locomotives		23,893		259
Marine Ships	50,054	196[a]	3.3×10^8 GT[b]	0
Commercial Aircraft	46,300	7,431	1977[c]	1334
Military Aircraft	65,800			
General Aviation	223,370			
Small SI Engines				
Class I				10×10^6
Class II				4×10^6
Classes III, IV, V				12×10^6
Marine SI				500,000

a = Includes only ocean-going vessels above 1000 GT
b = gross tonnage of ship, not number of ships
c = Number of commercial aviation aircraft (960 from the top producers Boeing and Airbus) (CAPA, 2011).

■ *Example 16-1.*
Determine the emissions of each mobile source pollutant for the year 2000.

Solution:
Read the data from Figure 16-1 and apply the appropriate scaling factor. An additional column is added setting all the values to the same scale.

CO: 90×10^6 ton/yr = 90×10^6 ton/yr
NO_x: 125×10^5 ton/yr = 12.5×10^6 ton/yr
SO_x: 70×10^4 ton/yr = 0.7×10^6 ton/yr
PM_{10}: 60×10^4 ton/yr = 0.6×10^6 ton/yr
$PM_{2.5}$: 45×10^4 ton/yr = 0.45×10^6 ton/yr
VOC: 80×10^5 ton/yr = $8. \times 10^6$ ton/yr
NH_3: 30×10^4 ton/yr = 0.3×10^6 ton/yr

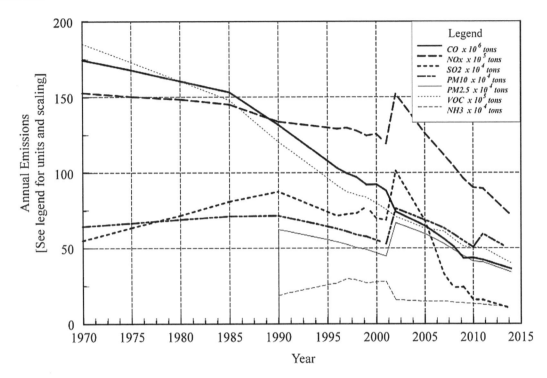

FIGURE 16-1. Mobile Source Pollutants Reported to US-EPA, 2015. Values on the chart are scaled, to find the actual emissions [ton/year] multiply the chart value by the scaling factor given in the legend. Data reporting requirements were modified starting in 2002 to include more sources.

Mobile source emissions can be categorized into three types: exhaust emissions, evaporative emissions, and refueling losses. Exhaust emissions originate with the combustion products and byproducts (CO_2, H_2O, CO, VOC, NO_x) from burning fuel. Evaporative emissions originate from fuel that escapes from the vehicle by diffusion through the storage tank and fuel lines, from small leaks in the fuel system, and the daily heating and cooling due to temperature changes between day and night. Refueling losses occur when liquid fuel displaces the fuel vapors within the fuel storage tank and by spilling. Effective strategies for reducing emissions from mobile sources address all three areas.

The US-EPA began mobile source emission cleanup by focusing on on-road engines in the early 1970's and off-road equipment in the mid-1990's. Table 16-2 lists the US-EPA's organization of mobile sources. Each of the categories listed in the table may have multiple subcategories based on engine type, engine size, fuel type, and date of manufacture.

Table 16-2. US-EPA Mobile Source Categories.
- Highway Vehicles
    - Light Duty Vehicles (LDV) [motorcycles, automobiles, pickup trucks, SUVs, minivans]
    - Heavy Duty Vehicles (HDV) [Freight trucks, buses]
- Non-road Engines
    - Land-based Diesel
    - Land-based Spark-Ignition
        - Small Spark Ignition (≤25 hp)
        - Large Spark Ignition (>25 hp)
        - Recreation Vehicles
- Marine Engines and Vessels
    - Marine Spark Ignition
    - Marine Diesel
- Locomotives
- Aircraft

This chapter describes the mobile source equipment, mobile source fuels, and then explores the specific vehicles and regulations. The chapter concludes with a discussion of the technologies and practices used to reduce emissions.

16.2 Engines

There are two main types of engines used in mobile sources: reciprocating and gas turbine. Both are considered to be internal combustion engines. Reciprocating engines use a cylinder and piston arrangement whereby the combustion of fuel and air within a cylinder generates linear motion of the piston. This linear motion is mechanically converted to rotation within the rest of the engine. A gas turbine engine uses the energy from combustion to directly spin a turbine, which transforms the chemical energy into either mechanical energy or thrust.

16.2.1 Internal Combustion Reciprocating Engines

This section explores the three most common reciprocating engines; the two-stroke, the Otto, and the Diesel engines. The latter two types are four-stroke engines.

16.2.1.1   Two-Stroke Reciprocating Engine

Two-stroke engines are the simplest, lightest, and least expensive to manufacture of the reciprocating engines. They are useful when low-weight is important and when engine orientation may vary. Common uses include lawn-mowers, chainsaws, and small boat engines.

Figure 16-2 shows a schematic of a two-stroke engine. It consists of a crankcase, which contains the cylinder and piston, an air/fuel/engine-oil inlet, and an exhaust port. The cylinder contains the combustion chamber and spark plug that is used to ignite the combustion.

FIGURE 16-2. Schematic of Two-Stroke Engine.

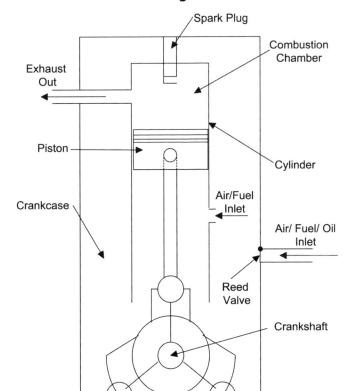

A cycle of the engine starts with the combustion stroke. This stroke begins when the piston is at the top of its cycle, and the spark plug ignites the air-fuel mixture. The explosion of the combustion gases pushes the piston down. This linear downward motion is converted to angular motion by the crankshaft. The downward motion of the piston also acts to compress the air/fuel mixture in the crankcase. While the piston is moving downward, it first uncovers the exhaust port, allowing the combustion products to exit from the cylinder. It then uncovers the intake port and the air/fuel mixture, which has just been compressed in the crankcase, enters the cylinder and displaces the remaining exhaust. Typically, the piston is designed in such a way as to prevent the direct release of the air/fuel mixture out the exhaust port before combustion.

The momentum of the crankshaft drives the piston back up toward the sparkplug for the compression stroke. As the piston moves up in the cylinder, the air/fuel mixture is compressed, and a vacuum is created in the crankcase. The vacuum opens the reed valve and pulls the air/ fuel/ engine-oil mixture from the carburetor into the crankcase. The compression stroke ends when the piston returns to the top of the cylinder and the spark plug fires again to repeat the cycle.

Two-stroke engines have three important advantages over four-stroke engines:
- Two-stroke engines do not have inlet and exhaust valves, which simplifies their construction and lowers their weight.
- Two-stroke engines fire once every revolution, four-stroke engines fire once every other revolution. This allows two-stroke engines to gain a significant power boost.
- Two-stroke engines can work in any orientation, which can be important in something like a chainsaw. A standard four-stroke engine may have problems with oil flow unless it is upright, and solving this problem can add complexity to the engine.

The disadvantages of the two-stroke engine include:
- Performance and pollution problems occur because the crankcase is used for compression of the air-fuel mixture leading to mixing between the intake and exhaust within the cylinder.
- The crankcase cannot be used for storage and distribution of engine oil, so oil must be mixed into the fuel itself. The oil does not burn well in the engine and causes the exhaust to be oily and to have a high concentration of unburned hydrocarbons and particles.
- The overlap of inlet and exhaust phases allows some of the fuel to be vented unburned from the cylinder.

These problems can be overcome by using a four-stroke engine, which completely separates the inlet and exhaust phases. It also allows the air/ fuel compression to occur solely within the cylinder, thus freeing the crankcase to store and distribute engine oil onto the moving parts.

16.2.1.2 Four Stroke Reciprocating Engines

Otto Engine or Spark Ignition Engine:
The four-stroke approach is known as the Otto cycle, in honor of Nikolaus Otto, who invented it in 1867. The four strokes are intake, compression, combustion, and exhaust. Figure 16-3 illustrates each. The various strokes refer to what is occurring within the cylinder. Figure 16-4 labels these over a plot of the piston position within the cylinder.

A description of the complete cycle begins with the piston located at its topmost position within the cylinder (called top dead center, TDC). Note that the piston does not travel to the top of the cylinder; rather, it leaves a *headspace* between the piston at the top of its travel and the top of the cylinder. The intake stroke begins when the intake valve opens, and the piston moves downward. The cylinder fills with air and the fuel injector adds an appropriate amount of fuel. The fuel can be added either directly to the cylinder or into the air stream rushing into the cylinder. This fuel mixes with the air for the remainder of the intake stroke, which ends when the piston reaches the bottom of its travel (called bottom dead center, BDC). Next, the piston travels up again and compresses the air/fuel mixture increasing the temperature and density of the mixture[1], which helps increase the power of the engine. The combustion stroke ends when the piston returns to its topmost position. At this point, the spark plug emits a spark and ignites the compressed air/fuel mixture starting an explosive combustion reaction within the cylinder. The explosion forces the piston to travel downward again, which provides the power to the engine. Finally, after the piston reaches BDC and starts to travel upwards, the exhaust valve opens, and the exhaust stroke begins. Once it reaches the top, the exhaust valve closes, and the cycle repeats. Note that the timing of these events coincides with the piston position as summarized in Figure 16-4, which shows the process through 720 degrees (4π) of travel. The actual timing of events may vary by several degrees in either direction depending on operating conditions.

There is a short period during the end of the exhaust stroke when both the exhaust and inlet valves are partially open. This period is called *valve overlap*. The amount of overlap can affect engine performance and is an important part of the engine's design. A well designed and controlled valve overlap can be used to reduce NO_x emissions, and this method is referred to as internal exhaust gas recirculation (EGR, see section 16.6.2.5). Excessive valve overlap increases the emissions of unburned and partially burned fuel.

The up/down travel or linear motion of the piston is converted to rotational motion by a mechanical connection (piston rod) between the piston and the crankshaft. The spinning of the crankshaft is measured in revolutions per minute (rpm). A typical value is 600 rpm for an idling engine and 2400 rpm for highway driving.

Car engines have multiple sets of cylinders and pistons. Each is connected to the same crankshaft and provides power to the engine. The number of cylinders is typically four, six, or eight. They are arranged in one of three ways: inline, V, or flat. An inline engine arranges all the cylinders linearly, and the pistons move vertically. A V-engine arranges the cylinders in two banks, alternately offset such that the view along the crankshaft would show the cylinders in a V alignment. A flat engine arranges the cylinders in two banks on opposite sides of the engine, and the pistons move in the horizontal direction. There are specialty engines that use more cylinders and different arrangements, but the basic principles of operation remain the same.

---

1. See homework problem 16-14 for a method to calculate the change in temperature and pressure during the compression stroke.

FIGURE 16-3. Diagram Showing the Stages of a Four-Stroke Engine in Operation.

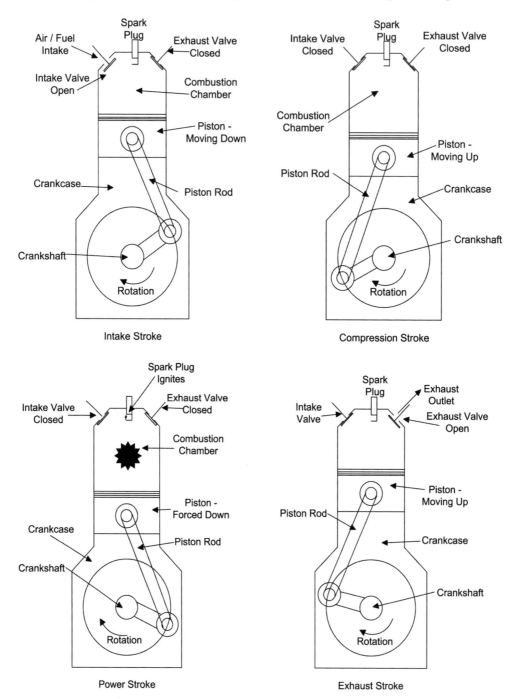

FIGURE 16-4. Timing of Major Events of a 4-Stroke, Spark-Ignition Engine.

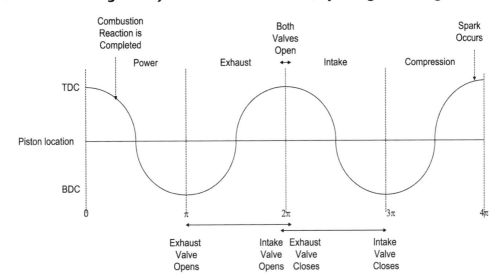

Diesel Engine:

Rudolf Diesel invented the diesel engine in 1892. He wanted to improve the efficiency of the four-stroke engine. The diesel engine is very similar to the Otto engine. Both use a four-stroke mechanism to combust fuel to generate engine power and create rotational energy. The main difference is that diesel engines do not use a spark plug to ignite the air-fuel mixture. They instead increase the amount of compression during the compression stroke to increase the temperature of the air and then inject the fuel, which spontaneously combusts when it makes contact with the compression heated air. The diesel engine always adds fuel directly into the cylinder chamber (direct fuel injection) near the end of the compression stroke, rather than during the intake stroke. This timing allows the air to be compressed and heated, which gives good control over the timing of the combustion reaction. The other main difference is the fuel. Diesel is a heavier, oilier blend with higher energy content. Section 16.3 discusses engine fuels in more detail.

Diesel fuel and automobile gasoline are not interchangeable. Adding diesel fuel to a gasoline powered car may cause severe problems. The fuel pump will have a difficult time getting the wrong fuel to the engine, and the burning characteristics of the fuel would cause many problems within the engine. It is very likely to destroy the engine.

16.2.1.3 Turbine Engines

Gas turbine engines convert internal (chemical) energy in the fuel to kinetic energy in an exhaust gas. The kinetic energy is used to produce power in a turbine and / or is used to create thrust. The engine consists of three basic unit operations or systems: compressor, combustor, and turbine, see Figure 16-5. Additional systems (fan, mixer, afterburner, and nozzle) may be used to increase the efficiency of the engine for certain applications.

FIGURE 16-5. Schematic of a Turbofan Engine.

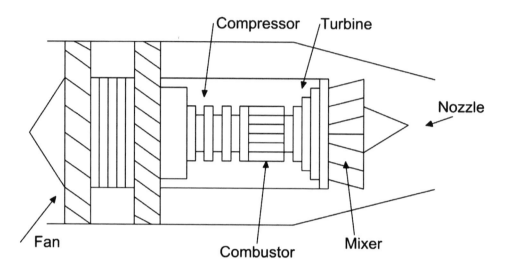

The fuel is typically a hydrocarbon, like propane, natural gas, kerosene, or jet fuel. The heat from combustion expands the reaction product gases which are then used to spin a turbine. The turbine generates power which can be used to spin a mechanical device and/or to generate thrust. The mechanical device may be used to drive the rotor of a helicopter or spin the shaft of a generator to produce electricity. Thrust is used to move the engine and whatever it is attached to, like an airplane. Gas turbine engines have greater power to weight ratios than reciprocating engines, and can be considerably simpler and smaller than a reciprocating engine of the same power. Turbine engines are expensive because the moving parts spin faster and operate at higher temperatures, so they use high-performance materials. They operate best at constant or slowly varying loads.

The main components and subsystems in gas turbine engines include:
- **Compressor** - The compressor is the first component in the engine core. The compressor consists of fans with many blades attached to a shaft. The compressor increases the incoming air pressure between 3 and 20 times the original pressure by reducing its volume.
- **Combustor** - The combustor mixes the compressed air with fuel and then ignites it, typically with a spark. Multiple spray nozzles add the fuel. The combustion products produce a hot expanding gas stream that can reach temperatures of 1,000 – 2,700 °F (540 - 1,500 °C).

- **Turbine** - The turbine receives the high-energy gas stream coming out of the combustor. The turbine, like the compressor, is made of a series of fan blades attached to the shaft. Unlike a compressor, the turbine expands the gas and extracts energy as the expanding gas causes the fan blades to rotate. The turbine is linked by a shaft to turn the compressor blades. The shaft may also spin some auxiliary equipment like a fan, removing some of energy from the turbine. The remainder is extracted as work, either to operate some mechanical device or to provide thrust.
- **Fan** - The fan pulls excess air into the engine, increases its velocity, and splits the air into two streams. One stream, the core, continues into the engine where it is reacts with the fuel. The other part, the bypass, moves around the engine and is remixed with the combustion products leaving the turbine. The bypass air increases the amount of thrust an engine can produce and also helps quiet the engine.
- **Mixer** – A mixer is used to rejoin the bypass air and combustor products before they exit the engine.
- **Afterburner** – An afterburner is used to increase the thrust provided by a turbine engine. It is not shown in Figure 16 5. The afterburner is a secondary combustor used between the turbine and the exit-nozzle. It is used to increase the temperature of the gas ahead of the nozzle. The result of this increase in temperature is an increase of 40 to 50 percent in thrust.
- **Nozzle** - The nozzle is the exhaust duct of the engine. It is the engine part that produces thrust. The energy depleted airflow that passed the turbine, in addition to the colder air that bypassed the engine core, produces a force when exiting the nozzle that acts to propel the engine, and whatever it is attached to. If the turbine and compressor are efficient, the pressure at the turbine discharge is nearly twice atmospheric pressure. The majority of the thrust comes from the bypass air, which flows at much larger flow rates than the combustion air.

There are several variations of the gas turbine engine used for mobile sources: turbojets, turboprops, turbofans, turboshafts, and ramjets.
- **Turbojets** – Turbojets consist of just the basic systems (compressor, combustor, and turbine). All the air ingested by the inlet is passed through the engine, and the exhaust is used for thrust.
- **Turboprops** – A turboprop engine includes a propeller, which is driven by the turbine. The propeller provides additional thrust by moving more air. Turboprops generate the majority of their thrust with the propeller. The turboprop, compared to a turbojet engine, has better propulsion efficiency at flight speeds below about 500 miles per hour. They become increasingly noisy and inefficient at high speeds
- **Turbofans** – A turbofan engine uses a large fan at the air intake to increase the air flow through the engine, as shown in Figure 16-5. The fan receives power from the turbine section of the engine. Most modern jet aircraft are powered by turbofans. The fan increases

the total air through the engine, although some of the air bypasses the main engine and is ejected directly as a "cold" jet or mixed with the gas-generator exhaust as a "hot" jet. The bypass system increases thrust without increasing fuel consumption by increasing the total air-mass flow and reducing the velocity with the same total energy supply.
- **Turboshaft** – These engines are very similar to turboprops, differing only in that nearly all energy in the exhaust is extracted by the turbine to spin a rotating shaft, and they generate little to no jet thrust. This shaft work is used to power machinery rather than a propeller. They are often used to power helicopters.
- **Ramjets** – These engines are the simplest jet engine because it has no moving parts. The speed of the engine "rams" or forces air into the intake. It is essentially a turbojet in which rotating machinery has been omitted. Its application is restricted by the fact that the inlet air compression depends wholly on forward speed. The ramjet develops no static thrust and very little thrust in general below the speed of sound. As a consequence, a ramjet vehicle requires some form of assist to reach operating speed. It has been used primarily in guided-missile systems and space vehicles.

16.3 Fuel Characteristics

Most mobile sources are powered by fossil fuels, although electricity and biofuels are increasingly used. Electric powered mobil sources typically obtain their charge of electricity from fossil-fuel power plants. Some mobile sources have been developed to use liquid fuels derived from the oils, starches, and sugars in plant biomass.

Fossil fuel or petroleum is a derived from ancient fossilized organic materials, such as zooplankton and algae. It is created when large quantities of their remains settle to the sea or lake bottoms, mix with other sediments (clay, sand, soil) and get buried under anoxic conditions. Once the layers become thick and heavy enough, the bottom portions become pressurized from the weight above and begin to heat up from the earth's interior heat and the anoxic decomposition reactions of the biomass. This process causes the organic matter to change, first into a waxy material known as kerogen and eventually, as more heat and pressure is added, into liquid and gaseous hydrocarbons via a process known as catagenesis.

Three conditions must be present for oil reservoirs to form: a source rock rich in organic material buried deep enough for subterranean heat to cook it into oil; a reservoir rock that is porous and permeable enough for it to accumulate in; and a cap rock (seal) or another mechanism that prevents it from escaping to the surface. Within these reservoirs, fluids typically organize themselves into three layers: gas, oil, and water, from top to bottom. The actual thickness of the layers depends on the source material and the time, temperature, and pressure conditions. Most hydrocarbons are less dense than rock or water, they tend to migrate upward through adjacent rock layers until either reaching the surface or becoming trapped within porous rocks (known as reservoirs) or under impermeable rocks. This process is influenced by underground water flows which can transport the oil hundreds of kilometers horizontally or even short distances downward before becoming trapped in the reservoir. When hydrocarbons collect in a trap, an oil field forms, from which the liquid can be extracted by drilling and pumping.

The extraction of petroleum (crude oil) from a reservoir is called production. Crude oil is classified by the geographic location of the production (e.g. West Texas Intermediate, Brent, Oman), density (API gravity), and sulfur content. Low-density crude is termed light, and high-density crude is termed heavy. Low sulfur crude (< 0.5 wt%) is sweet and high-sulfur crude (> 0.5 wt%) is sour. Each reservoir oil has unique characteristics which make it more or less valuable. The value is associated with the ability to take the crude form of the oil and convert it into more useful forms.

Once produced, the petroleum is transported to a refinery where various fractions are separated and classified, typically by vapor pressure and boiling point. Table 16-3 lists these fractions from lightest to heaviest along with the general carbon chain size.

Crude oil does not contain the desired amounts of the various fractions. It typically contains more heavy components than can be profitably sold and fewer light components than demanded. Refineries change the distribution of each fraction, as well as some of the other properties with specialty equipment. *Distillation towers* are used to separate the fractions. *Thermal and catalytic crackers* are used to break large molecules into smaller ones. *Catalytic reformers* are used to generate hydrogen and larger molecules from the lighter fractions. *Hydro-desulfurization* is used to remove sulfur atoms from molecules. *Claus units* are used to recover the removed sulfur as elemental sulfur. With these processes, a refinery can maximize the crude oil utility.

Table 16-3. Fractions of Crude Oil (with range of carbon chain sizes) [and other names].
- Light Distillates
    - LPG (C3 – C4's, primarily propane and butane)
    - Gasoline (C4 – C12)
    - Naptha (C5 – C12)
- Middle Distillates
    - Jet Fuel (C5 – C16)
    - Kerosene (C6 – C16)
    - Diesel (C8 – C21)
- Heavy Distillates
    - Fuel Oils
        - No. 1 (C9 – C16)   [also known as coal oil, stove oil and range oil]
        - No. 2 (C10 – C20)   [distillate home heating oil]
        - No. 3 (C10 – C50)   [outdated, now merged with No. 2]

- No. 4 (C12 – C70)   [commercial heating oil]
- No. 5 (C12 – C70)   [industrial heating oil, Bunker B, requires heating]
- No. 6 (C20 – C70)   [residual fuel oil, Bunker C, requires heating]
  o Lubricating Oils (C18 – C34)
  o Wax (C20 – C40)
- Residuum
  o Asphalt (solid, C28+)
  o Petroleum Coke (solid, C40+)

The amount of refining needed varies for each market. Markets may be as large as a country or as small as an individual city. Most countries have content and property standards for each fuel. Many have different standards for different regions, different times of the year (i.e. summer and winter), and control the allowable additives. The most commonly regulated properties are vapor pressure, oxygen content, and sulfur content. Vapor pressure directly affects how readily the fluid vaporizes. A higher vapor pressure allows the fuel to burn easier, but also means more of the fuel can escape (evaporate) during refueling, or during engine startup and shutdown. The evaporation contributes to regional VOC emissions and the formation of smog and ozone. Oxygen content is especially important in cold conditions – because having some oxygen in the fuel helps it burn more completely. Sulfur content generates SO_x emissions leading to acid precipitation (see Chapter 9) and poisons the catalysts used to reduce other emissions in the exhaust (see section 16.6.2.1). Section 16.5 provides additional discussion of fuel content as it relates to emissions.

16.3.1 Air to Fuel Ratio

The most important mobile source engine performance parameter is the ratio of air to fuel (A/F) used in the combustion process. The A/F ratio is a mass ratio describing how much air is used to combust the fuel.

16.1 $$A/F = \frac{\text{mass of air}}{\text{mass of fuel}}$$

This ratio is stoichiometric when just enough air is added to burn all the fuel completely. The stoichiometric A/F for gasoline is 14.7:1 and for methanol it is 6.5:1. The comparison between the actual amount added and the stoichiometric amount can be used to classify engine designs and emission control systems.

16.2 $$\lambda = \frac{(A/F)_{actual}}{(A/F)_{stoich}}$$

The system is fuel rich when λ (lambda) < 1 and it is fuel lean when λ > 1. An engine typically gives the highest power when running fuel rich. It has the best fuel economy when run fuel lean. Emissions are lowest when run stoichiometric (Wu, et al., 2004).

■ Example 16-2.
Determine the stoichiometric A/F ratio for octane (C_8H_{18}).

Solution:
Begin by writing the chemical reaction for complete combustion.

16.3 $$2C_8H_{18} + 25O_2 \rightarrow 16CO_2 + 18H_2O$$

Next, determine the mass of fuel and oxygen

16.4 $$m_{fuel} = 2\,mole * \left[114\frac{gram}{mole}\right] = 228g$$

16.5 $$m_{oxygen} = 25\,mole * \left(32\frac{gram}{mole}\right) = 800g$$

Next, determine the mass of air associated with the required amount of oxygen. Assume that air is 21 mol% oxygen and the MWair is 28.85 (see Example 2-8).

16.6 $$m_{air} = 25\,molO_2 * \frac{1\,mol_{air}}{0.21\,mol_{O_2}} * \frac{28.85\,g_{air}}{1\,mol_{air}} = 3434.5\,g_{air}$$

Finally, take the ratio of air to fuel

16.7 $$A/F = \frac{m_{air}}{m_{fuel}} = \frac{3{,}434.5\,g_{air}}{228\,g_{fuel}} = 15.1$$

If more air is added, the mixture is fuel lean (λ > 1), and if less air is added, the mixture is fuel rich (λ < 1)

16.4 US Regulation of Mobile Sources

Emission control solutions utilize a variety of approaches: better engine and vehicle design, improved fuel formulations, enhanced education for owners and operators, and alternative transit options. Regulations and voluntary programs have reduced emissions from individual automobiles 75 to 98% since the 1970's (US-EPA, 2012b). The US-EPA regulates mobile sources based on the type of mobile source, as noted in Table 16-2.

The regulatory requirements are typically outcome-based and do not force a particular technology or device. Auto manufacturers and petroleum refiners have responded to the regulations with engine improvements, use of additional systems, and improvements in fuel formulations. Urban planners have responded with the creation of better road systems, creation of transportation choices (mass transit, bicycle lanes, carpooling incentives, and pedestrian friendly designs), and vehicle maintenance requirements.

16.4.1 Highway Vehicles

Highway Light Duty Vehicle (LDV) standards cover new engines used for LDVs independent of fuel used (gas, diesel, or alternative fuel). LDVs include passenger cars, light-duty trucks, sport utility vehicles (SUV), minivans and pickup trucks. The gross vehicle weight rating (GVWR) is used to further categorize these vehicles as shown in Table 16-4.

Table 16-4. US-EPA Vehicle Categories for Highway Light Duty Vehicles.

Light-Duty Vehicle	LDV	< 8,500 lb GVWR
Light-Duty Truck	LDT	<8,500 lb GVWR, and
		<6,000 lb curb weight, and
		<45 ft² frontal area
Light Light-Duty Truck	LLDT	<6,000 lb GVWR
Light-Duty Truck 1	LDT1	<3,750 lb LVW[a]
Light-Duty Truck 2	LDT2	>3,750 lb LVW[a]
Heavy Light-Duty Truck	HLDT	6,000 – 8,500 lb GVWR
Light-Duty Truck 3	LDT3	<5,750 lb ALVW[b]
Light-Duty Truck 4	LDT4	>5,750 lb ALVW[b]
Medium-Duty Passenger Vehicle[c]	HLDT	< 10,000 lb GVWR

Notes: a) LVW (loaded vehicle weight) = curb weight + 300 lb.
b) ALVW (adjusted loaded vehicle weight) = average of GVWR and curb weight.
c) Manufacturers may alternatively certify engines for diesel-fueled MDPVs through the heavy-duty diesel engine regulations.

16.4.1.1 Emission Factors

There is no single emission factor value for automobiles because the engine control technology is constantly changing, sometimes more than once per year. Rather emission factors are developed for each model year. Table 16-5 lists estimated emission factors for automobiles for model years 1990 to 2020. The data was calculated from the US-EPA MOVES (MOtor Vehicle Emission Simulator) database (US-EPA, 2010a). This tool allows users to estimate emissions based on manufacturer data, expected fleet makeup, and regional variations. The fleet makeup includes age of vehicles or model year, their expected mileage or age of engine, expected usage, an estimate of user maintenance, and an estimate of how well their pollution control equipment is working. Table 16-6 shows the effect of engine wear on the emission factors of each pollutant between a pre-control era model and a 1995 model.

Table 16-5. LDV Emissions 1990 - 2020, US-EPA MOVES Estimate.

Year	CO g/mile	NOx g/mile	SO2 g/mile	PM10 g/mile	PM25 g/mile	Distance miles
1990	42.79	4.70	0.145	0.182	0.174	2.14×10^{12}
2000	16.73	3.79	0.089	0.104	0.099	2.74×10^{12}
2005	11.10	2.67	0.053	0.084	0.080	2.99×10^{12}
2010	7.89	1.79	0.009	0.054	0.052	2.96×10^{12}
2015	5.67	1.13	0.006	0.034	0.032	3.19×10^{12}
2020	4.44	0.70	0.006	0.020	0.019	3.50×10^{12}

Table 16-6. Emissions Factors for LDVs (Gas Powered).

Pollutant	Model year	Emissions at engine miles [g/mile]		
		0 [New]	50,000	100,000
HC	pre-1968	7.250	8.150	9.050
	1995	0.233	0.598	1.973
CO	pre-1968	78.270	89.520	100.770
	1995	2.147	9.387	26.557
NO_x	pre-1968	3.440	3.440	3.440
	1995	0.240	0.655	1.620

■ *Example 16-3.*
Compare the total CO generated from LDVs in 1990 and in 2015.

Solution:
Use the data from Table 16-5.

CO_{1990}: $\dfrac{42.79 \, g}{mile} * 2.14 \times 10^{12} \, mile * \dfrac{ton}{10^6 \, g} = 91.6 \times 10^6 \, ton$

CO_{2015}: $\dfrac{5.67 \, g}{mile} * 3.5 \times 10^{12} \, mile * \dfrac{ton}{10^6 \, g} = 19.8 \times 10^6 \, ton$

This is an almost 80% decrease, even though total miles increased by 60%.

16.4.1.2 Emission Standards

The US-EPA developed emission standards for LDVs starting with the 1973 model year. The standards have changed considerably over time in terms of numerical values, pollutants, and applicable vehicles. The 1990 CAAA altered the standards from a one-size fits all (Tier 0) to standards based on the vehicle size (Tier 1) or manufacturer chosen bins (Tier 2). All the classifications place limits on emissions of carbon monoxide (CO), oxides of nitrogen (NO_x), particulate matter (PM), formaldehyde (HCHO), and non-methane organic gases (NMOG) or non-methane hydrocarbons (NMHC).

Each bin in the Tier 2 system has different emission standards, and manufacturers may choose into which bin to place each vehicle. They are constrained by the requirement that their overall fleet of vehicles must attain a certain average. The use of bins allows manufacturers flexibility in vehicle design and greatly simplifies the determination of their fleet's average emissions.

Table 16-7. US-EPA Tier 1 and 2 Emission Standards for LDV at 100,000 miles [g/mile].

Program	Model Year	Vehicles	NOx	NMOG	CO	PM	HCHO
Tier 1							
	1994 - 2003	LDV	0.6	0.31	4.2	0.1	
	1994 - 2003	LDT1	0.6	0.31	4.2	0.1	0.8
	1994 - 2003	LDT2	0.97	0.4	5.5	0.1	0.8
	1994 - 2003	LDT3	0.98	0.46	6.4	0.1	0.8
	1994 - 2003	LDT4	1.53	0.56	7.3	0.12	0.8
	1994 - 2003	LDV Diesel	1.25	0.31	4.2	0.1	
	1994 - 2003	LDT1 Diesel	1.25	0.31	4.2	0.1	0.8
Tier 2							
Bin 1	2004 +	b	0	0	0	0	0
Bin 2	2004 +	b	0.02	0.01	2.1	0.01	0.004
Bin 3	2004 +	b	0.03	0.055	2.1	0.01	0.011
Bin 4	2004 +	b	0.04	0.07	2.1	0.01	0.011
Bin 5	2004 +	b	0.07	0.09	4.2	0.01	0.018
Bin 6	2004 +	b	0.1	0.09	4.2	0.01	0.018
Bin 7	2004 +	b	0.15	0.09	4.2	0.02	0.018
Bin 8	2004 +	b	0.2	0.125	4.2	0.02	0.018

b LDV, LLDT, HLDT, and MDPV

Table 16-7 lists the recent emission standards. It shows the tier 1 standards by vehicle type and the tier 2 standards by bin. In the original published standards, there were some additional bins to help manufacturers phase into the new regulations, but these bins have been retired. All the listed standards have units of grams of pollutant per driven mile. The standards account for the aging of a vehicle by providing separate values for new, 50,000 miles, and 100,000 miles. These tables only show one set of values in order to allow the reader to see how they have evolved. The complete record is available at (US-EPA, 2012c). It should also be noted that a vehicle is only required to meet the emission standards for its model year, an older vehicle does not need to meet newer standards; they are grandfathered in at their engine's model year. Appendix 16-1 lists the historical emissions standards (1970 – 1993) and provides the standards for LDVs and LDTs separately.

■ Example 16-4.

Estimate the improved efficiency in NO_x emissions between a 1994 model year LDV, and a tier 2 bin 4 vehicle built in 2006.

Solution:
Use data from Table 16-7.

$$\eta = \frac{0.6 - 0.04}{0.6} * 100\% = 93.3\%$$

The tier 2 regulations also added new requirements for fuel quality. Refiners and importers must meet a corporate average gasoline sulfur standard of 30 ppm with an 80 ppm sulfur cap, as of 2006. Several alternative and voluntary standards have also been set as part of the Clean Fuel Fleet with standards for low-emission vehicles (LEV), inherently low emission vehicles (ILEV), ultra-low emission vehicles (ULEV), and zero-emission vehicles (ZEV), see (US-EPA, 2012d).

16.4.2 Heavy Duty Vehicles

Highway Heavy Duty Vehicles (HDV) have a gross vehicle weight rating (GVWR) above 8,500 lbs. Diesel heavy-duty vehicles are further divided into service classes as follows:
- Light heavy-duty diesel engines: 8,500 - 19,500 lbs.
- Medium heavy-duty diesel engines: 19,500 - 33,000 lbs.
- Heavy heavy-duty diesel engines (including urban bus): > 33,000 lbs.

US-EPA standards only cover new HDV diesel engines and do not apply to older model vehicles. The state of California has slightly different definitions and somewhat different standards,

(CARB, 2011). A default set of diesel emission factors used by the California Air Resources Board (CARB) is given in Table 16-8.

Table 16-8. California Diesel Engine Emission Factors.

		Default Emission Factors (g/hp-hr)			
Horsepower	Model Year	PM	HC	CO	NO$_x$
120 or less	all years	0.84	1.44	4.8	13
121 - 250	pre - 1970	0.77	1.32	4.4	14
	1970 and newer	0.66	1.1	4.4	13
251+	pre - 1970	0.74	1.26	4.2	14
	1970 and newer	0.63	1.05	4.2	13

Under the federal light-duty Tier 2 regulation (phased-in beginning 2004) vehicles of GVWR up to 10,000 lbs used for personal transportation were re-classified as "medium-duty passenger vehicles" (MDPV - primarily larger SUVs and passenger vans) and are subject to the light-duty vehicle standards. The regulation allows a diesel engine model used for the 8,500 - 10,000 lbs vehicle category to be classified as either light- or heavy-duty and certified to different standards depending on the application.

The US-EPA standards for diesel engines were first set in the 1988 model year and have been revised several times. Appendix 16-2 lists the standards by year and vehicle type (US-EPA, 2012e). Standards for the 2004 to 2007 model years gave flexibility to manufacturers by providing two options to certify their engines. Similar to the LDVs, several alternative and voluntary standards have been provided for LEV, ILEV, ULEV, ZEV diesel vehicles (US-EPA, 2012f).

The 2007 regulations required diesel fuel production to achieve lower sulfur levels, 15 ppm-wt from the previous limit of 500 ppm-wt. This standard is a technology enabler, allowing engine manufacturers to use control technologies that would be damaged or destroyed by sulfur, such as catalytic particle filters and NO$_x$ reducing catalysts.

16.4.3 Non-Road Land-Based Diesel

US-EPA standards cover new mobile non-road diesel engines used in construction, agriculture, municipal, and industrial applications. The definition of a non-road engine uses the idea of mobility or portability. It includes engines that are self-propelled, on equipment that is propelled while performing its function, or on equipment that is portable (has wheels, skids, carrying handles …). It excludes highway engines, railway locomotives, marine engines, aircraft engines, engines used in underground mining, and stationary engines (in place for more than 12 months) all of which have their own standards.

The first US standards for new non-road diesel engines (tier 1) were adopted in 1994 and phased in between 1996 and 2000. Appendix 16-3 lists these standards and the subsequently updated standards (tier 1 to tier 4) for engines below 560 kW (US-EPA, 2012g). The goal of the standards is to reduce the emissions from these sources by more than 90% from their non-regulated ancestors. The standards were developed between the US-EPA, California Air Resources Board (CARB) and many engine makers (including Caterpillar, Cummins, Deere, Detroit Diesel, Deutz, Isuzu, Komatsu, Kubota, Mitsubishi, Navistar, New Holland, Wis-Con, and Yanmar). The emission reductions are achieved by changes in engine designs, operation methodology, exhaust-gas after-treatments, and alterations in diesel fuel formulations.

Table 16-9 lists the standards for large diesel engines (above 560 kW) also known as the Tier 4 emission standards. The 2011 standards only apply to engines above 900 kW, and were sometimes referred to as 'transitional Tier 4.' The 2015 limits represent the final Tier 4 standards. Engine sizes between 560 and 900 kW use the 2006 tier 2 standards (shown in Appendix 16-3) until 2015. Engines of all sizes must also meet smoke standards of 20/15/50% opacity at acceleration /lug /peak modes, respectively. Opacity is a measure of the ability of light to pass through a sample. The modes refer to different ways to operate an engine – acceleration occurs when the engine rpms increase. Lug mode refers running the engine at very high load and slow speed (run engine slow in a high gear – it is bad to do this and it can seriously harm the engine). Peak mode refers to running the engine at its top speed.

Table 16-9. Tier 4 Emission Standards—Engines Above 560 kW [g/kWh].

Year	Category	CO	NMHC	NO_x	PM
2011	Generator sets > 900 kW	3.5	0.4	0.67	0.1
	All other engines > 900 kW	3.5	0.4	3.5	0.1
2015	Generator sets	3.5	0.19	0.67	0.03
	All other engines	3.5	0.19	3.5	0.04

The tier 4 standards require the use of exhaust control equipment that would be harmed by high levels of sulfur in the fuel. The US-EPA mandated reductions in sulfur content from pre-tier 4 standards of 0.3 wt% (3,000 ppm-wt) to 500 ppm in 2007, and to the same level as highway ultra-low sulfur diesel of 15 ppm in 2010.

16.4.4 Non-Road Spark Ignition Engines

Regulations for non-road spark ignition (SI) engines started in 1995 and are frequently updated (US-EPA, 2010b). Multiple categories and phases have been implemented since this time, and each additional regulation changes or adds a few new values. It is a very active research and regulatory area and methods of emission reduction are constantly changing. The interested reader is encouraged to explore the latest rules at http://www.epa.gov/otaq/standards/index.htm.

Small SI Engine regulations apply to engines with power below or equal to 25 hp. They are placed into five classes, based on use and size. The majority of these engines are in lawn and garden equipment. Table 16-10 lists these classifications. Class I and II include the larger non-handheld devices such as push mowers, pressure washers, home-use generators, and riding mowers. Classes III, IV, and V include handheld devices such as chainsaws, leaf blowers, edgers, and weed trimmers (US-EPA, 2012h).

Table 16-10. Small Spark Ignition (SI) Engine Classifications.

Class	Engine size [cm^3]	Use
I	<225	Non Handheld
II	≥225	Non Handheld
III	<20	Handheld
IV	20-50	Handheld
V	≥50	Handheld

Large spark ignition engines above 25 hp are found in industrial equipment and used in a wide variety of applications, including forklifts, airport ground-service equipment, terminal tractors, generators, compressors, welders, aerial lifts, and ice grooming machines. These engines may operate on gasoline, liquefied petroleum gas, or compressed natural gas.

Appendix 16-4 list the regulatory emission limits for non-road SI engines for the three classifications: small, large, and marine. Emission standards for SI engines >25 hp were finalized in November 2002. Marine SI engines apply to all manner of watercraft and equipment used in them (generators). Appendix 16-5 compares the uncontrolled emission factors with the phase 1 and 2 regulated values. Uncontrolled emission values show the expected emissions from the engine without any pollution control equipment. The phase 1 and phase 2 values show the expected emissions for an engine that meets current regulatory limits.

Evaporative emissions from these engines were first regulated in 2008. The limits for fuel systems (tanks and lines) use the permeability limits of the as-used equipment. The limit for fuel lines is 15 g/m²/day. The limit for fuel tanks is 1.5 g/m²/day. Additional restrictions to reduce evaporation losses during storage (diurnal losses) and filling require certain design methods, such as the fuel storage tank must be closed with a tethered cap and with physical indication of the seal (i.e. click when tight). The regulations do not require use of a particular piece of equipment; rather, they describe the characteristics of an appropriate system.

16.4.5 Marine Diesel Engines.

16.4.5.1 Background:

Marine Diesel Engines typically refer to the engines used in the Great Lakes and Ocean Shipping merchant vessels. The United Nations Conference on Trade and Development (UNCTAD) estimates that the operation of merchant ships contributes about US$380 billion in freight rates within the global economy, equivalent to about 5% of total world trade.

Ocean-going ships carry around 90% of international trade. Shipping trade estimates are often calculated in ton-miles, as a way of measuring the volume of trade (or transportation work). In 2008, it was estimated that the industry transported over 7.7×10^9 tons of cargo, equivalent to a total volume of world trade by sea of 3.2×10^{19} ton-miles. Shipping by ocean freight adds 1 – 5 % to the total shelf price of items. Table 16-11 lists the number of international ocean ships. The numbers (and the international regulations) do not include other ocean going ships such as private yachts, military equipment, or fishing vessels. Figure 16-6 shows the amount of cargo shipped, by cargo type from 1970 – 2010 (MARISEC, 2010). The current largest category 'other' includes manufactured items such as durable consumer goods, automobiles, and parts.

Table 16-11. Number of Ocean Going Ships Subject to International Air Pollution Regulations, 2008.

Type of Service	Quantity
General Cargo Ships	16,224
Bulk Carriers	8,687
Container ships	4,831
Tankers	13,175
Ferries, Cruise, and Passenger ships	7,137
Total	50,054

16.4.5.2 Regulations:

Marine engines are divided into three categories based on cylinder volume (the displacement volume of the piston), see Table 16-12. Categories 1 and 2 range in size from 500 to 8,000 kW and are subdivided into tiers based on the displacement and engine power output. Category 3 marine diesel engines are the largest mobile engines (2,500 to 70,000 kW). Emission controls for category 3 engines are limited due to their use of residual fuels. Residual fuel is the by-product of distilling crude oil to produce lighter petroleum products. It has high viscosity, high density, and it typically

FIGURE 16-6. Annual World Shipping by Marine Vessels.

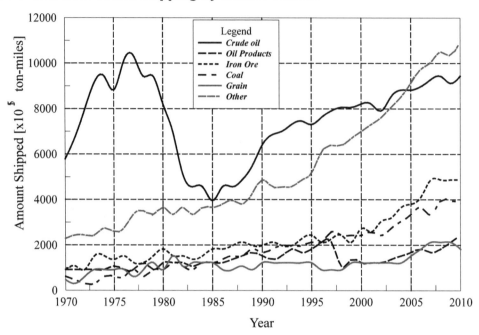

has high ash, sulfur, and nitrogen content in comparison to other fuels. Also, these properties are highly variable due to the left-over nature of this fuel (i.e. it is the bottom of the barrel). These properties make it very inexpensive, cause ignition quality problems, and usually require heating before use. The US-EPA estimated that residual fuel can increase engine NOx emissions from 20-50% and PM from 750% to 1250% (typically as sulfate particles) when compared to distillate fuel.

Table 16-12. Marine Engine Categories Based on Engine Cylinder Displacement Volume.

Category	Tier 1 -2	Tier 3 - 4	Basic Engine Technology
1	D < 5 Liter [a]	D < 7 Liter	Land-Based Non-Road Diesel
2	5 Liter < D < 30 Liter	7 Liter < D < 30 Liter	Locomotive Engines
3	D ≥ 30 Liter	D ≥ 30 Liter	Unique Marine Engine Designs

Where D is the displacement per cylinder in liters.
a and Power ≥ 37 kW

The US-EPA began to regulate emissions from marine engines in 1999. The Category 1 and 2 regulations are similar to those for land-based mobile engines. The rules closely follow the International Maritime Organization's (IMO) "International Convention on the Prevention of Pollution from Ships" rules, also known as MARPOL 73/78 and its updates called Annex (IMO, 2011). The IMO is an agency of the United Nations missioned to promote maritime safety. The current rules, MARPOL Annex VI "Regulations for the Prevention of Air Pollution from Ships," sets limits on NO_x and SO_x emissions and prohibits the intentional emission of ozone depleting substances.

One interesting aspect of marine vessels is that they commonly move between the waters in different countries and international waters (which are not governed by any single nation). The operation would be confusing if every nation had different standards and if no rules applied in international waters. Therefore, most countries (167 as of 2008) have agreed to achieve the IMO MARPOL regulations. The IMO Annex VI has established four Emission Control Areas (ECA) around the world in which specific regional pollution problems (typically NO_x and/or SO_x) must be addressed by vessels in that area. The ECAs ensure that foreign flagged vessels comply with IMO Tier III NO_x limits while within any nations waters (i.e. the IMO Tier III standards are only applicable within ECAs). The ECAs also trigger low-sulfur fuel requirements for all vessels within any nation's waters. These regulations only apply to ships operating within the specially designated ECAs. The US has obtained designation for two areas: the North American ECA (as of August 2012), and the Caribbean ECA (January 2014). The other two regions that have been adopted as air pollution ECAs are the Baltic Sea (May 2006) and the North Sea (November 2007), both of which are for SO_x only.

Category 1 and 2:
Emission standards for Category 1 and 2 engines use the land-based standard for non-road and locomotive engines. The tier 2 standards apply to engines built in 2004 or later, and limit CO to 5.0 g/kWh for all size engines. NO_x + HC are limited to between 7.2 and 11.0 g/kWh based on engine displacement, and PM to between 0.20 and 0.50 g/kWh. Tier 3 phases in between 2009 – 2014, and reduces limits on NO_x+ HC to between 4.7 and 11.0 g/kWh, PM to between 0.11 and 0.40 g/kWh, and CO to between 5.0 and 8.0 g/kWh. The higher limit values apply to smaller engines (as small as 8 kW). Tier 4 limits phase in from 2014 to 2017 and reduce limits for NO_x to 1.8 g/kWh, HC to 0.19 g/kWh, and PM to between 0.04 and 0.12 g/kWh, for engines ranging in size from 600 kW to greater than 3,700 kW.

Category 3:
NOx emission limits for diesel engines depend on the engine maximum operating speed, n [rpm] as shown in Table 16-13. Tier I and Tier II limits are global, while the Tier III standards apply only in NO_x Emission Control Areas.

Tier I limits are for engines built between 2004 and 2010. Tier II limits apply for engines built between 2011 and 2015 and require the use of engine-based emission controls. Tier 3 limits apply to engines built beginning in 2016. They require engine based controls and add-on equipment. The US-EPA also added a CO standard of 5.0 g/kWh and HC limits of 2.0 g/kWh. No standards were added for PM, but manufacturers are required to measure and report PM emissions.

Table 16-13. MARPOL Annex VI NOx Emission Limits.

Tier	Date Year	NO$_x$ Limits n < 130 g/kWh	130 ≤ n < 2000 g/kWh	n ≥ 2000 g/kWh
I	2000	17	$45n^{-0.2}$	9.8
II	2011	14.4	$44n^{-0.23}$	7.7
III	2016 [a]	3.4	$9n^{-0.2}$	1.96

a Only within NO$_x$ ECAs, otherwise Tier II standards apply
Where n is the maximum test speed of the engine in rpm.

Tier II standards are expected to be met by combustion process optimization. The parameters examined by engine manufacturers include fuel injection timing, pressure, rate shaping, fuel nozzle flow area, exhaust valve timing, and cylinder compression volume.

Tier III standards are expected to require dedicated NO$_x$ emission control technologies such as various forms of water induction into the combustion process (with fuel, scavenging air, or in-cylinder), exhaust gas recirculation, and/or selective catalytic reduction (SCR).

Annex VI regulations include caps on sulfur content of fuel as a measure to control SO$_x$ emissions and to indirectly control PM emissions (there are no explicit PM emission limits). Special fuel quality provisions exist for SO$_x$ Emission Control Areas (SO$_x$ ECA or SECA). Table 16-14 lists the sulfur limits and implementation dates.

Table 16-14. MARPOL Annex VI Sulfur Limits in Fuel [wt%].

Date	SO$_x$ ECA	Global
2000	1.50%	4.50%
2010	1.00%	4.50%
2012	1.00%	3.50%
2015	0.10%	3.50%
2020[a]	0.10%	0.50%

a- alternative date is 2025, to be decided by a review in 2018.

Heavy fuel oil (HFO) is an allowed fuel, provided it meets the applicable sulfur limit (i.e., there is no mandate to use distillate fuels). Alternative measures are also allowed (in the SO$_x$ ECAs and globally) to reduce sulfur emissions, such as the use of scrubbers. For example, in lieu of using the 1.5% S fuel in SO_x ECAs, ships can fit an exhaust gas cleaning system or use any other technological method to limit SO_x emissions to ≤ 6 g/kWh (as SO_2).

The controls on air pollution from ships relies more on engine and fuel design than on having individual ships meet particular regulations. If the ship has an approved engine, uses fuel that meets the standards and uses proper maintenance procedures, it is considered in compliance.

16.4.6 Locomotives

16.4.6.1 Background:

Almost all modern locomotives operate with diesel-electric engines. In the past, some engines used wood or coal as their fuel. These engines are the largest land-based non-road diesel engines in terms of cylinder displacement volume and total power. The mechanics of diesel locomotives are similar to other diesel engines; however, the engine itself is not directly connected to the drivetrain. The engine is used to drive an electrical generator, and this electricity is then used to power the drivetrain and cause the wheels to rotate. In essence, the locomotive is an electric vehicle which carries its own power plant on board.

Table 16-15 lists several of the common uses and types of locomotives. The US railroad system is chiefly based on moving freight. All US passenger trains were purchased by the federal government in the 1970's and a quasi-public railroad, called Amtrak, was created. The single largest category of freight is coal, which accounts for 45% by volume and 23% by value of all freight shipped within the US. More than 70% of coal transport is by rail. The fastest-growing part of rail freight is intermodal traffic: containers or truck trailers loaded onto flat railcars. The number of such shipments rose from 3 million in 1980 to 12.3 million in 2006.

Table 16-15. Locomotive Uses.

Freight
    Containers (intermodal)
    Bulk Cargo, such as coal, iron ore, and automobiles
Passenger
    High Speed   Rapid Transit (and subways)
    Inter-City   Regional
    Commuter   Trams and Trolley
    Light Rail   Monorail

Table 16-16 lists the average emission factors for locomotives (US-EPA, 1998). These values originate from cycle-weighted average emissions rates from the three most common engines used in the two types of service – line-haul and switch-yard, as discussed in the next section. The units are in the industry standard of grams per brake horsepower-hour. The conversion to more common units is 1.346 g/kWh = 1 g/bhp-hr.

Table 16-16. Estimated Emission Factor Baseline In-Use Emission Rates [g/bhp-hr]

Service	HC	CO	NO_x	PM
Line-haul	0.48	1.28	13	0.32
Switch	1.01	1.83	17.4	0.44

16.4.6.2 Regulations:

US emission standards for railway locomotives apply to newly manufactured, as well as remanufactured railroad locomotives and locomotive engines, see Appendix 16-6. The regulations account for two types of service: line-haul and switch-yard locomotives. Line-haul engines are the distance hauling trains that run at high power for extended periods. Switchyard engines change speed often; between extended idle periods, low power, high power, and dynamic braking. The regulations apply by the year the engine is put into service or is remanufactured. Locomotive engines are remanufactured after a given number of MW-hours operation (7.5 multiplied by the engines rated horsepower), distance traveled (750,000 miles) or set period (10 years). Remanufacture may happen several times during the life of a locomotive.

The EPA adopted standards in two regulatory actions:
- Tier 0-2 standards: The first regulation for railroad locomotive emissions was adopted on 17 December 1997 [63 FR 18997-19084, 16 Apr 1998]. The rulemaking, which became effective in 2000, applies to locomotives originally manufactured after 1973 and anytime the engine is manufactured or remanufactured. Tier 0-2 standards are met through engine design methods without the use of exhaust gas after treatment.
- Tier 3-4 standards: The 14 March 2008 regulation introduced more stringent emission requirements [73 FR 88 25098-25352, 6 May 2008]. Tier 3 standards, to be met by engine design methods, became effective in 2011/12. Tier 4 standards require treatment technologies to reduce pollutants in the exhaust gas. Tier 4 standards become effective in 2015. This regulation also requires more stringent emission standards for remanufactured Tier 0-2 locomotives.

The US-EPA has also mandated the use of ultra-low sulfur diesel fuel in these engines to enable catalytic after treatment methods required to meet the tier 4 standards. A sulfur limit of 500 ppm was effective in June 2007 and reduced to 15 ppm from June 2012.

16.4.7 Aircraft

16.4.7.1 Background:

There are approximately 46,300 commercial fixed-wing aircraft in service around the world (Hawk Information, 2012a). The US commercial air fleet is composed of 7,120 of these ((US-DOT, 2008)). This total includes cargo craft and passenger craft but does not include military, commuter below 15 seats, or private craft. Military craft number around 65,800 fixed-wing and rotary-wing aircraft (Hawk Information, 2012b).

There are two operating phases to consider for aircraft: landing/ take-off (LTO) and cruise. LTO includes all aircraft activity taking place at elevations within 3000 ft of ground level such as taxi-in and out, take-off, climb-out, approach landing and idling. Cruise includes all activities that take place at altitudes above 3000 feet (1000 m). It includes climb to cruise altitude, cruise, and descent from cruise altitudes. Most countries also break air travel into domestic and international flights. Table 16 17 lists the fuel use and emission factors for each of these categories (IPCC, 1996).

Table 16-17. Aircraft Fuel Usage (kg) and Emission Factors (kg/LTO or kg/ton fuel used).

		Fuel Use	SO_2	CO	CO_2	NO_x	NMVOC	CH_4	N_2O
Domestic									
	LTO (kg/LTO)	850	0.8	8.1	2680	10.2	2.6	0.3	0.1
	Cruise (kg/ton fuel)	1000	1	7	3150	11	0.7	0	0.1
International									
	LTO (kg/LTO)	2500	2.5	50	7900	41	15	1.5	0.2
	Cruise (kg/ton fuel)	1000	1	5	3150	17	2.7	0	0.1

A Boeing 747 can carry up to 63,500 gallons of fuel, allowing it to fly from Los Angeles, USA to Melbourne, Australia. It carries 400 – 500 people and has an interior volume of approximately 3 houses (31,000 ft^3). A flight of 3,500 miles uses 126,000 pounds of fuel (18,660 gallons), and has an efficiency of approximately five gallons per mile (Boeing, 2012). The density of aviation fuel is approximately 0.81 kg/L.

■ Example 16-5.

Estimate the fuel use and total emissions for a flight from New York to Los Angeles.

Solution:
Use data from Table 16 17, Assume the craft uses 5 gallons fuel per mile, and the distance traveled is 4,000 miles.

The flight is domestic since both cities are in the same country.

There is one LTO for the flight: Fuel used 850 kg.

$$\text{Cruise: } 4{,}000\,miles * \frac{5\,gal}{mile} * \frac{6.8\,lbs}{gal} * \frac{0.4536\,kg}{lbs} * \frac{ton - fuel}{1000\,kg} = 61.6\,ton - fuel$$

Total fuel used is 62.5 tons, converting to gallons:

$$62.5\,ton * \frac{1{,}000\,kg}{1\,ton} * \frac{l}{0.81\,kg} * \frac{0.264\,gal}{l} = 20{,}400\,gal$$

If the flight carries 450 people, it has a mileage of

$$\frac{(4{,}000\,mile)(450\,people)}{(20{,}400\,gal)} = 88 \frac{mile \cdot pop}{gal}.$$

The fuel economy is significantly better than a single-person passenger automobile (e.g. 25 miles per gallon).

Table 16-18. Emissions for LTO and Cruise in kg for Example 16 5.

	LTO (kg)	Cruise (kg)	Total (kg)
Fuel	850	61,600	62,450
SO_2	0.8	61.6	62.4
CO	8.1	431.2	439.3
CO_2	2680	194,040	196,720
NO_x	10.2	677.6	687.8
NMVOC	2.6	43.12	45.72
CH_4	0.3	0	0.3
N_2O	0.1	6.16	6.26

16.4.7.2 Regulations:

The US began regulating aircraft emissions in 1973 through three groups: the US-EPA, the US *Federal Aviation Administration* (US-FAA), and the *International Civil Aviation Organization* (ICAO). Since much air traffic is international (similar to marine vessels), these groups coordinated to develop international standards and practices pertaining to aircraft engine emissions. (US-EPA, 2012i). The regulations group aircraft into tiers based on the date that the first engine of the series was manufactured, see Table 16-19. Aircraft engines do not undergo constant change; rather, an engine type is designed and then tested. Once it is determined to work under all tested conditions it is not redesigned until a major change is required. Testing may require 10,000 hours or more of running. One engine type may be built for eight or more years, which is very different than the automobile engines, which may have changes within a production run.

The Federal Aviation Agency was established in 1958 to provide for the safe and efficient use of the national airspace. Its name was changed to the Federal Aviation Administration in 1967 as part of the new Department of Transportation. The FAA mission includes the safety, security, efficiency, and environmental compatibility for the nation's airways (US-FAA, 2010).

ICAO was established in 1944 by the United Nations by the Convention on International Civil Aviation, the "Chicago Convention" "... in order that international civil aviation may be developed in a safe and orderly manner and that international air transport services may be established on the basis of equality of opportunity and operated soundly and economically." ICAO's responsibilities include developing aircraft technical and operating standards, recommending practices, and fostering the growth of international civil aviation. The United States is currently one of 191 participating member States of ICAO (ICAO, 2012).

Table 16-19. US Regulatory Tiers for Aircraft.

Regulation	Date of First Manufacturing	
	Model	Engine
Tier 0	pre Dec-31-1995	pre Dec-31-1999
Tier 2	Jan-01-1996	Jan-01-1996
Tier 4	Jan-01-2004	Jan-01-2004
Tier 6	Jan-01-2013	Jan-01-2013
Tier 8	Jan-01-2014	TBD

TBD – To be determined at a later date (as of Jun 18. 2012 rules).

Current aircraft emission regulations apply to engines whose rated thrust is greater than 26.7 kN [1.0 kilo-Newton = 225 lb_{force}] and whose date of manufacture is on or after 1 January 1986. The internationally agreed upon emission limits for an LTO must not exceed:

- Hydrocarbons (HC): D_p/F_{oo} = 19.6 [g/kN]
- Carbon monoxide (CO): D_p/F_{oo} = 118 [g/kN]
- Oxides of nitrogen (NO_x): As per Appendix 16-7.

Where: D_p = Mass of pollutant emitted [g], and
              F_{oo} = rated thrust [kN]

■ *Example 16-6.*
Calculate the HC, CO, and NO_x emissions during LTO for an aircraft engine manufactured July 15, 2013, with a pressure ratio of 50 and engine thrust of 100 kN.

Solution:
Use the information provided below Table 16-19 and in Appendix 16-7.
     Use the given formulae, where F_{oo} = 100 kN and π_{oo} = 50:
Hydrocarbons: D_p = 19.6 [g/kN] * F_{oo} = 19,600 g HC
Carbon monoxide: D_p = 118 [g/kN] * F_{oo} = 11,800 g CO
Oxides of nitrogen: D_p = (-1.04 + 2.0π_{oo}) * F_{oo} = (-1.04 + 2*(50)) * 100 = 9,896 g NO_x

16.5 Fuel Regulations

The US Clean Air Act establishes national fuel formulation standards, but also allows states to adopt unique fuel programs to meet local air quality needs. The 1990 CAAA mandated that cities with high ozone and NO_x levels use a different form of gasoline, called reformulated gasoline (RFG) (US-EPA, 2011a). RFG is a gasoline blend that burns more completely than conventional gasoline. About 30 percent of gasoline sold in the U.S. is reformulated. RFG was first used in 1995, and it is currently used in 17 states plus the District of Columbia. The different fuels required by these state programs are sometimes referred to as "boutique fuels" (US-EPA, 2006).

     The two most common characteristics of the boutique fuels programs specify the fuel volatility and oxygen content. Fuel volatility affects the rate of evaporation at a given temperature. A low volatility is used in summertime, since temperatures are high, and it is increased during wintertime to help it vaporize within the engine. Volatility is measured using the Reid Vapor Pressure method at 100°F, (ASTM, 2014). Volatility is controlled by altering the fuel blend - an increase in the amount of light components such as pentanes and hexanes make it higher. Increasing the amount of heavier components such as decane, lowers the volatility. Oxygen content is increased during wintertime to allow more complete combustion during cold starts. Oxygen content is increased by adding oxygen-containing chemical compounds such as alcohols and ethers. Twelve states have adopted their own clean fuel programs for part or all of the state. Most of these programs set lower gasoline volatility requirements than the federal standards, and most are effective for only part of the year.

Sulfur content is limited in fuels because its presence can destroy the catalysts used to control emissions. Sulfur tends to react chemically with the precious metals used as catalysts. This blinds or poisons the site rendering it useless for emission control. Sulfur can also destroy other parts of the system. Sulfur, once oxidized, converts into sulfuric acid in the presence of water. This acid gas causes corrosion within the exhaust system. Removing sulfur allows pollution control equipment to last for the lifetime of the vehicle as well as reduce the need for maintenance and replacement of exhaust systems. This additional benefit helps to reduce the lifetime vehicle costs even though it increases the cost of the fuel. The US-EPA requires sulfur content in mobile source fuels to meet the following standards:

- Gasoline: A corporation average of 30 ppm with a maximum of 80 ppm of sulfur.
- Diesel: A maximum of 15 ppm of sulfur (On-road, off-road, and locomotives).

The US-EPA estimated fuel cost increases to achieve 15 ppm sulfur standard is 7 cents per gallon, but this would be reduced to 4 cents due to savings in maintenance costs (reduced acid corrosion of equipment).

The US-EPA also regulates the addition of certain toxic substances under the Mobile Source Air Toxics (MSAT) rules. These rules limit the fuel content of compounds such as benzene (an octane booster), 1,3-butadiene, formaldehyde, acetaldehyde, acrolein, and naphthalene (US-EPA, 2011b).

16.6 Engine Emissions and Control

The emissions from mobile sources are similar to other fossil fuel combustion sources: CO, CO_2, NO_x, SO_x, VOC, PM, and air toxics. The management of these pollutants requires an integrated approach addressing all the components of the transportation system's infrastructure. A well thought out system addresses engine design, emission controlling equipment, inspection and maintenance (I/M), the transportation infrastructure (roads, rails, and ports), and the fuels used to power all the sources.

16.6.1 Spark Ignition and Diesel Engine Design

16.6.1.1 Air to Fuel Ratio.

The most important design parameter for engine performance and emission control is the ratio of air to fuel (A/F) used in the combustion process. The A/F ratio is a mass ratio describing how much air is in the combustion cylinder per amount of fuel, see equation (16.1). Spark Ignition (SI) engines typically run at the stoichiometric A/F ratio ($\lambda = 1.0$). Diesel engines always run fuel lean ($\lambda > 1.0$).

Figure 16-7 shows the changes in SI engine emissions of hydrocarbons, carbon monoxide (left scale), and nitric oxide (right scale) for several values of the air to fuel ratio. Stoichiometric operation for this figure is at A/F = 14.9. Automobiles with SI engines operate within a narrow range around the stoichiometric value so as to minimize HC and CO, as well as NO. Diesel engines always run fuel lean (high values of A/F) which greatly reduces CO; however, this causes diesel engines to generate high levels of particles (not shown).

FIGURE 16-7. The Effects of Air-Fuel Ratio on Hydrocarbon, Carbon Monoxide, and Nitric Oxide Exhaust Emissions. Adapted from (Agnew, 1968).

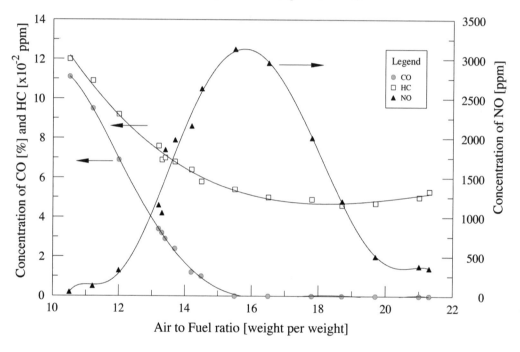

16.6.1.2 Combustion Ignition Timing

Another important design in controlling an engine is the timing of the start of the combustion. Spark ignition (SI) engines use a spark from a spark plug to ignite the combustion (See Figure 16-3). Diesel engines use the heat developed from the compression stroke, and directly inject fuel into the heated air which then spontaneously combusts. Turbo engines do not use cycles or strokes; rather they have continuous sparks used to keep the air-fuel mixture ignited. Beginning in the mid-1980's engines began to be controlled in real time by an onboard computer (powertrain computer) that controls timing throughout the engine speed and load range. Timing in SI engines is determined by an electronic ignition system (older engines used a mechanical device called a distributor) that controls when each spark plug fires. Timing in diesel engines is also determined by an electronic ignition system that controls when the fuel injection system adds fuel to each cylinder.

It is important to have ignition start near the end of the compression stroke for SI and diesel engines. This timing is required because the fuel does not burn instantaneously when the spark fires, rather the engine piston moves while the reaction is occurring. Also, the angular speed of the engine, measured as revolutions per minute (rpm) on a tachometer, can lengthen or shorten the time needed to complete the burning. Angular speed is usually measured in degrees so as to coincide with the rotation of the crankshaft. A single stroke from bottom to top is 180 degrees.

Igniting the air-fuel mixture before the piston reaches top dead center (TDC) is called timing advance. It is designed to allow combustion to complete by 20 degrees after top dead center (ATDC). If the ignition spark occurs at a position that is too advanced relative to piston position, the rapidly expanding air-fuel mixture can push against the piston still moving up, causing knocking (pinging) and possible engine damage. If the spark occurs too retarded relative to the piston position, maximum cylinder pressure occurs after the piston is already traveling down the cylinder. The required advance depends on many operational aspects of the vehicle such as engine speed, engine load, and temperature. An increase in engine speed requires ignition timing to become more advanced so that the fuel has sufficient time to burn completely by 20° ATDC. As the speed increases (more rpm), the time available to burn within the cylinder decreases. So it needs to begin earlier in the engine cycle. An increase in engine load, such as during acceleration, travel uphill or pulling a trailer, requires less advance in order to generate more power. Low temperatures, such as during startup and in cold weather, require more advance due to a slowing of the reaction kinetics. Modern engines measure each variable and the onboard computer compensates for each to achieve maximum efficiency with minimum emissions.

16.6.1.3 Compression

The compression in the cylinder is calculated by comparing the cylinder volume at bottom dead center (BDC) and top dead center (TDC)

16.8
$$CR = \frac{V_{BDC}}{V_{TDC}}$$

Where: VBDC is the volume within the cylinder when the piston is at BDC, and
VTDC is the volume when the piston is at TDC.

A typical compression ratio in an Otto cycle engine is 9:1, whereas for a diesel engine it can be as high as 22:1. The Otto cycle uses a lower CR to reduce the temperature of the compressed gases. At higher compression, the mixture of air and fuel could spontaneously combust (called detonation or knocking). It leads to poor timing of the expansion of the gas as compared to the pistons expansion stroke. The diesel engine takes advantage of the heating caused by compression. The cylinder contains only air during the compression stroke. This air becomes heated as it is compressed. When the piston gets close to TDC, but before finishing the compression stroke, the fuel is directly injected into the cylinder. The fuel then burns spontaneously without needing a spark since the temperature of the compressed air is above the diesel fuels auto-ignition temperature. Many diesel engines contain glow plugs to help heat the air during cold startup.

16.6.2 SI and Diesel Emission Controlling Equipment

16.6.2.1 Catalytic Converter.

A catalytic converter can be used in any engine that burns a fuel. A catalyst is a substance that increases the rate a chemical reaction occurs for a particular temperature. A catalyst is neither a reactant nor product, and it remains unchanged by the reaction. An automobile exhaust catalyst is typically composed of metals such as platinum, palladium, or rhodium, see Figure 16-8. These metals are embedded onto the surface of a porous matrix. This open pore solid is generally a square, triangular, or hexagonal honeycomb shape, constructed of a ceramic designed to have a:
- high surface area to volume ratio,
- large open frontal area,
- low thermal mass,
- low heat capacity,
- low thermal expansion,
- high oxidation resistance,
- high strength,
- generate low backpressure and frictional losses, and
- ability to withstand extended operation at high temperatures.

FIGURE 16-8. Pictures of a) Catalysts for Motorcycles and Small Engines, and b) Three-way catalyst for Automobiles. Pictures Courtesy of BASF Catalysts, 2012.

A catalytic converter uses a catalyst to treat the emissions from the engine. The converters are placed within the exhaust system a short distance after the engine, see Figure 16-9. Vehicle engines create two types of pollutants that can be addressed with catalytic reactions: oxidized nitrogen, and unburned or partially burned fuel. Each type requires different conditions for destruction. Oxidized nitrogen requires a reduction reaction to split the nitrogen from the oxygen:

16.9 $$2NO \rightarrow N_2 + O_2$$

16.10 $$2NO_2 \rightarrow N_2 + 2O_2$$

Incompletely oxidized fuel requires additional oxidation for conversion into carbon dioxide and water:

16.11 $$2CO + O_2 \rightarrow 2CO_2$$

16.12 $$C_xH_y + \left(x + \frac{y}{4}\right)O_2 \rightarrow xCO_2 + \frac{y}{2}H_2O$$

These reactions are simplified overall reactions for the process. The actual reaction requires multiple steps and can generate several different intermediary compounds such as hydrogen and ammonia (Heck, et al., 2009a).

FIGURE 16-9. Schematic of Catalytic Emission Control System. Adapted from (US-EPA, 1994).

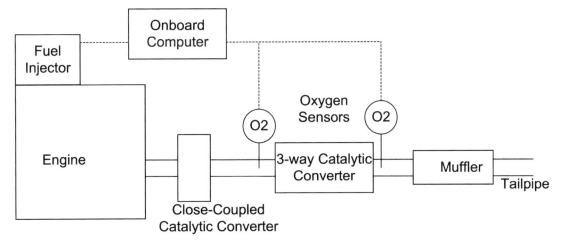

A three-way catalyst performs both functions. They have been used in US automobiles (with Otto engines) since 1981. A three-way catalyst consists of reduction and oxidation catalysts and a control system. The reduction reaction must occur in an oxygen-poor (<0.5%) environment and the oxidation must occur in an oxygen-rich (>0.5%) environment. These conditions are mutually exclusive. The control system must, therefore, continuously adjust the engine between fuel-rich and fuel-lean. It does this by monitoring the oxygen content in the exhaust gas using a universal exhaust gas sensor (UEGO). The sensor is between the engine and the converter. More recent models (>1996) place an additional sensor located after the catalytic converter to verify it is working. When the sensor detects that the exhaust is oxygen-rich, it adjusts the A/F ratio to a fuel rich setting, and when the exhaust is oxygen-poor, it adjusts the A/F ratio to a fuel poor setting. The response time of the sensor is on the order of a millisecond and allows the system to fluctuate between settings several times per second. Recall equation (16.2) which shows the ratio, called lambda, of the actual A/F ratio to the stoichiometric value. The sensor is sometimes called a lambda sensor because of this equation.

The three-way catalyst is only used in engines that operate near their stoichiometric A/F ratio, such as the Otto engine. A diesel engine always operates fuel lean, and its exhaust gas would never be oxygen poor. A diesel and other lean-burn engines can use an oxidation catalyst to control unburned and partially burned fuel, but must use other technologies to reduce the oxides of nitrogen. Turbo engines operate fuel rich during acceleration to achieve maximum power and fuel lean at other times to achieve maximum fuel efficiency, so they also do not use a three-way catalyst system.

Unwanted reactions on the catalyst can generate hydrogen sulfide (H_2S) or ammonia (NH_3). These unwanted reactions can be limited by using low-sulfur fuels or by altering the construction of the catalyst to include nickel or manganese. The catalyst may be poisoned (rendered inert) by fuel impurities containing lead, silicon, or phosphorous. Until the advent of the CAA, lead was added to gasoline as an octane enhancer. Section 13.3 discusses the phase-out of leaded gasoline in more detail, but it is important to note that the requirement of a catalyst to treat auto emission exhaust provided one of the main justifications for stopping the addition of lead to gasoline. Silicon contamination occurs when engine coolant leaks into the exhaust system and phosphorous contamination occurs when the engine burns the lubricating oil in a poorly sealed cylinder or through other oil leaks into the cylinder.

Three-way catalysts can achieve greater than 90% conversions of CO, HC, and NO_x when the system is at optimal conditions. Performance is much lower (approaching 0% when cold, to 50% after two minutes of operation) during start-up when the engine, exhaust gas, and catalyst are cool. It may take several minutes to warm up and this period is the most polluting, generating 40 – 80% of total pollutants from the vehicle. Current research focuses on methods to reduce the time needed to heat up the catalyst, including changes in catalyst structure and materials, locating the catalyst nearer the engine (called close-coupled catalytic converter – see Figure 16-9), preheat using a battery (only hybrid engines have large enough batteries), or use of a close-coupled pre-catalyst (Heck, et al., 2009a). These systems reduce the heat up time to 10 - 30 seconds.

16.6.2.2 Diesel Engine Emissions: Catalysts and Filters.

Diesel engines operate at very lean fuel conditions (A/F > 22 or lambda > 1.5) which gives good fuel economy (20 -30% > than SI engines). They also may have a lifetime of over one million miles. The main difference from SI engines is that diesel fuel is directly injected into the combustion cylinder near the end of the compression stroke. The compressed air is hot enough to ignite the fuel without the aid of a spark. The lean mixture allows it to run at excess oxygen and cooler temperatures than a SI engine, which implies lower levels of CO, HC, and NO_x. However, they do produce some of each as well as liquid and solid phase particulate matter (PM) due to the direct injection of fuel as a liquid. The PM is dry carbon (soot), inorganic oxides (sulfates), and organic liquids from unburned fuel and engine oil. The organic liquids are called the soluble organic fraction (SOF). The SOF may be discrete particles or adsorbed on the dry carbon particles.

The control of CO and gaseous HC can be achieved with changes to engine design and/or use of a diesel oxidation catalyst (DOC), similar to the oxidation catalyst described in the above section. The control of PM and NO_x is more difficult because they are inversely coupled – operation at cooler temperatures reduces NO_x but increases PM and vise-versa at higher temperatures. A phenomenon referred to as the NO_x-PM trade-off.

PM can be removed with a diesel particle filter (DPF), see Figure 16-10. It is a ceramic honeycomb structure material with every other end plugged to force the exhaust to flow through the porous structure. The ceramic material has open pores that are smaller than the particles. The small pores allow the gases to pass, but collect the PM. Eventually, the surface becomes covered with PM, which causes plugging and high pressure-drops. To avoid this, the DPF must regenerate

FIGURE 16-10. Diagram of a Particle Filter. Adapted from (Heck, et al., 2009b)

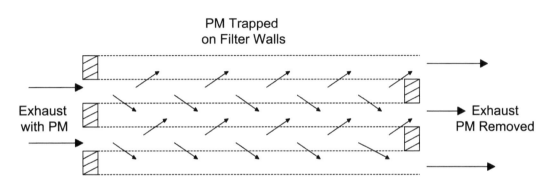

before becoming plugged. The dry carbon and organic liquids may be destroyed by combustion if the DPF temperature can be raised to their combustion temperature (>350°C if the DPF includes an oxidation catalyst). There are other methods to remove and destroy the PM (Heck, et al., 2009b), but they all use the same basic idea - capture the PM and somehow regenerate the filter automatically. Such systems need to work for 350,000 miles to comply with US-EPA regulations, so very robust systems are needed.

NO_x removal is problematic because the exhaust is oxygen rich. NO_x removal is a very active area of research, since the technology is still not settled. Three ideas are currently under extensive investigation, each with some success in the marketplace:
- NO_x reduction with onboard fuel,
- NO_x reduction using urea and selective catalytic reduction (SCR), and
- NO_x trapping with regeneration.

The first two methods require additional materials to be stored, used, and refilled on the vehicle. The fuel used may be either the diesel or alcohol (methanol). Urea addition is used in many stationary sources (see chapter 10). The NO_x trap requires regeneration which may be accomplished by running the exhaust fuel rich approximately 1 – 2 % of the time.

16.6.2.3 Positive Crankcase Ventilation (PCV).

These systems, first used in California automobiles in 1961, became standard in automobiles in 1968. The PCV system consists of a valve and connecting tube or passage between the crankcase and the air intake system, see Figure 16-11. It is a closed system that recirculates crankcase vapors into the engine and replaces these vapors with the engine air introduced near the oil filler cap. Exhaust can enter the crankcase when it passes around the piston instead of exiting through the exhaust valve, in a process called *blowby*. The PCV system uses the vacuum in the intake air to siphon crankcase vapors back into the intake manifold. It allows the unburned and partially burned fuel in the crankcase air to be re-burned and eliminates blowby vapors as a source of pollution. Before PCV, the vapors were simply vented to the atmosphere through a "road draft tube" that ran from a vent hole in the valve cover or valley cover down toward the ground.

Crankcase vapors include blowby emissions, engine oil that has dispersed into the air as particles, and water vapor. Leakage increases with engine use and is due to wear of the piston rings and cylinder walls. About 20% of the total hydrocarbon (HC) emissions produced by a vehicle are blowby emissions. Engine oil breaks into tiny droplets due to the motion of the crankshaft and its connections to the pistons (piston rods). The crankshaft motion coats the engine interior with lubricating oil, but it also creates very small oil droplets that may vent from the crankcase. Water vapor can accumulate in the crankcase due to air infiltration when the engine is not being used or from the blowby emissions, which includes some water vapor from fuel combustion.

FIGURE 16-11. Schematic of Positive Crankcase Valve (PCV) System. Adapted from (AA1Car, 2012).

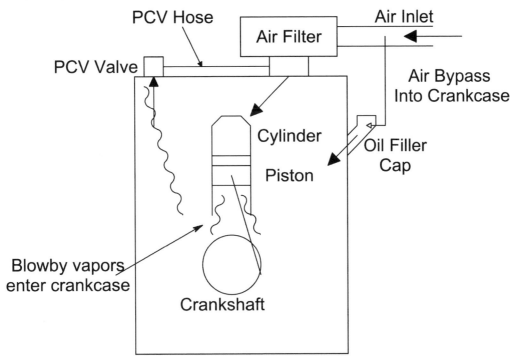

One of the beneficial effects of PCV, besides eliminating blowby emissions, is that it pulls moisture out of the crankcase to extend oil life. Moisture can form acids and sludge which cause reduced lubrication from the engine oil. It can lead to major engine damage from excessive wear and heat buildup.

16.6.2.4 Evaporative Emission Control (EVAP)

EVAP systems were added to automobiles starting in 1971 and are used to reduce or eliminate the release of evaporative emissions (fuel vapors) from the fuel tank and fuel lines. They consist of adsorptive charcoal canisters connected to a sealed fuel system, see Figure 16-12. The fuel vapors within the system are allowed to vent through the charcoal canister, where the hydrocarbons are trapped and stored. Later, when the engine is running, a purge valve opens allowing the vapors to be sucked into the engine air intake and reburned. The fuel tank is required to have expansion space at the top so the liquid fuel can expand on a hot day without overflowing or forcing the system to leak. It is extremely important for users to not overfill their gas tank during refilling. It is never ok to top off the fuel tank.

Sealing the fuel system is challenging because the fuel tank must have a vent so air can enter to replace fuel as the fuel pump delivers fuel to the engine. If the tank were sealed tight, the fuel pump would soon create enough negative suction pressure inside the tank to collapse the tank. The fuel filler cap on older vehicles usually contained a spring-loaded pressure/vacuum relief valve for venting. On newer vehicles (1996 & newer) it is completely sealed (no air vents), and the system is vented only through the charcoal canister.

FIGURE 16-12. Schematic of Evaporative Emission Control (EVAP) System. Adapted from (US-EPA, 1994).

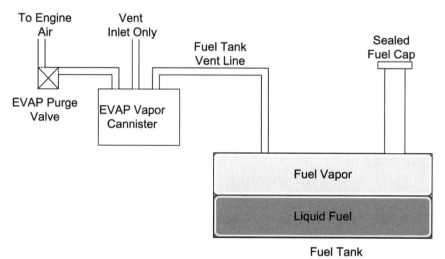

16.6.2.5 Exhaust Gas Recirculation (EGR)

The EGR system was added to automobiles in 1973 to lower emissions of nitrogen oxides (NO_x). The EGR reduces NO_x by diluting the intake air with exhaust, see Figure 16-13. The dilution reduces the overall combustion temperature which helps reduce the formation of nitrogen oxides, see Figure 16-14. EGR also helps prevent detonation. Detonation, also called knock, is an uncontrolled, irregularity in timing of combustion within the cylinder. This abnormal combustion causes pressure spikes which cause the engine to run poorly and may lead to engine damage.

The EGR system consists of a valve connecting a passage or port between the intake and exhaust system. When the valve is open the vacuum of the intake siphons exhaust into the intake manifold. It is only open when the engine is warm and running above its idle speed (typically 500 - 800 rpm). If the valve were to open when the engine was cold, or at low speeds, it would cause rough idling and stalling.

FIGURE 16-13. Schematic of Exhaust Gas Recycle (EGR) System. Adapted from (AA1Car, 2012).

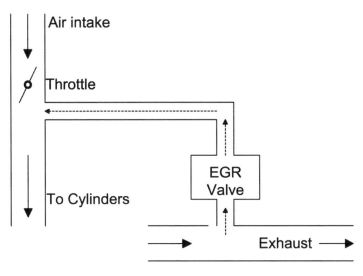

FIGURE 16-14. Effect of EGR on NO$_x$ Emissions in SI Engine. Adapted from (Malm, 2000)

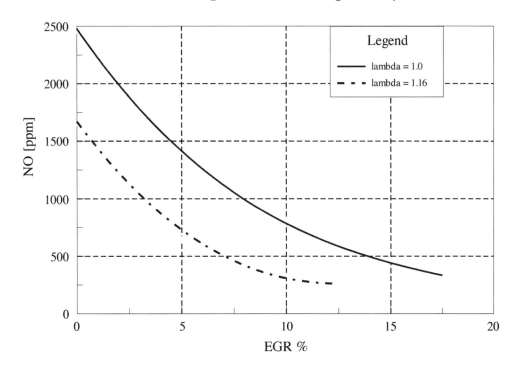

16.6.2.6 Onboard Diagnostic System (OBDII)

The OBD II system has been required on all US automobiles and light trucks since 1996, though some vehicles included it as early as 1994. It is used to detect emission problems and to alert the vehicle operator when there is a problem by lighting the 'Check Engine' light or 'Malfunction Indicator Lamp' (MIL). It also stores a diagnostic code in the powertrain computer that can be used by a mechanic to understand the problem.

The OBDII monitors several engine systems. The most useful sensors are used to monitor the amount of oxygen in the engine exhaust before and after the catalytic converter. The system also monitors the engine coolant temperature, throttle position (which regulates air inflow), air flow rate, crankshaft position, manifold absolute pressure, detonation or knock sensors, EGR position, and vehicle speed. Changes in any one system may affect other systems. The other systems then change operation to keep emissions low throughout a wide range of driving conditions, weather effects, and engine wear. It is now common for engines to not need a tune-up for 100,000 miles.

The OBDII trips when the vehicle emissions exceed the federal limits by 50% on two consecutive trips, or when there is a failure of a major emission control system. Automakers are required to warranty these systems for 100,000 miles. It performs in a manner similar to the emission tests required for each new model of vehicle, but it operates continuously throughout the vehicle life. Many areas that are required to monitor mobile source air quality use an OBDII plug-in test rather than the more time consuming standard exhaust test. The test checks that all OBDII system monitors have run, that the MIL is working and is off, and that there are no stored diagnostic trouble codes in the powertrain computer.

16.6.2.7 Inspection and Maintenance (I/M) Programs

I/M programs typically involve testing cars for excess emissions. The tests are useful in non-attainment areas and are required annually or biannually (US-EPA, 2012j). The programs are designed to find high-emitting vehicles and requiring their emissions be improved through repair in order to renew the vehicles registration. The tests may include visual inspection, exhaust monitoring, and/or downloading the fault codes from the onboard diagnostic (OBDII) computer. The OBD II systems use built-in, computerized monitors to detect emissions malfunctions. The 1990 Amendments to the Clean Air Act made I/M mandatory for several areas across the US, based upon various criteria, such as air quality classification, population, and/or geographic location. These programs saw their peak in usefulness in the 1990's when many of the cars on the road were older and had less emission control equipment or had been tampered with (control equipment removed or disconnected, intentionally or accidentally) and so had high levels of emissions. The number of cars from that era has greatly diminished and there are far fewer high emission vehicles on the road. Many State Implementation Plans are dropping the requirement for annual or biannual emission tests of every car, or are allowing a simple check of the OBDII computer, rather than the more time consuming emission measurement tests.

16.6.3 Refueling Losses

A major source of mobile source VOC emissions occurs during the transfer of fuel from one container to another. The liquid entering the fixed volume container displaces an equal volume of the vapor/ air mixture within the container. This mixture is lost to the atmosphere unless special precautions are used during the transfer. The actual amount of emissions depends on the temperatures of the liquid and the containers, such that the ratio of liquid volume added to vapor volume displaced is not always one to one. The vapors may expand upon heating (vapor growth) or contract upon cooling (vapor shrinkage). This difference makes estimation of emissions very difficult, and emission factors are given as a range. Also, reformulated and boutique fuels are blended to control the vapor pressure, so the amount of fuel vapor in the tank varies with the season. The US-EPA classifies the emissions sources into two groups, see Figure 16-15. Stage I emissions occur during transfer of fuel from the fuel delivery truck into storage tanks (typically located underground).

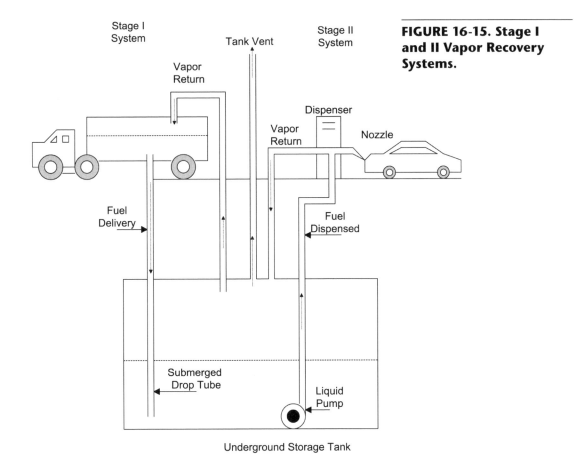

FIGURE 16-15. Stage I and II Vapor Recovery Systems.

Stage II emissions occur during the transfer from the storage tank into the fuel tank of a vehicle. The emission factor for this fuel loss ranges from 0.97 to 1.75 g/L when no vapor recovery system is used and 0.12 g/L when stage I and II controls are used (US-EPA, 1991).

An equation for estimating evaporative losses during loading of a liquid fuel storage tank is (US-EPA, 2008):

16.13
$$L_L = 12.46 * \frac{S*P*M}{T}$$

Where: L_L = Loading Loss, pounds per 1,000 gallons of liquid loaded,
S = Saturation factor, see Table 16-20,
P = Vapor Pressure of liquid [psia],
M = Molecular weight [lb/ lb-mol], and
T = Temperature [°R].

Table 16-20. The Saturation Factor in Equation 16.13.

Cargo Carrier	Mode of Operation	S Factor (no units)
Tank Trucks and Rail Tank Cars	Submerged loading nozzle	
	-- Clean Cargo Tank	0.50
	-- Dedicated Normal Service	0.60
	-- Dedicated Vapor Balance Service	1.00
	Splash Loading	
	-- Clean Cargo Tank	1.45
	-- Dedicated Normal Service	1.45
	-- Dedicated Vapor Balance Service	1.00
Marine Vessels	Submerged loading nozzle	
	--Ships	0.20
	-- Barges	0.50

Stage I - VOC emissions occur when the liquid fuel from the fuel delivery truck delivers the fuel into the storage tank. The vapors above the fuel are vented from the storage tank. The vapors are released to the atmosphere unless controlled. Fuel is delivered into the storage tank using a tube that opens within six inches of the tank bottom. This delivery system significantly reduces emissions caused by the splashing of the fuel. Stage I control systems collect these vapors and route them back into the delivery truck. Care must be used to achieve a balance in the amount of vapor transferred to prevent problems from vapor growth or shrinkage.

Stage II - VOC emissions occur in a similar manner when transferring fuel into the vehicle fuel tank. Nonattainment regions for ozone may require the use of stage II controls to reduce these emissions. These systems use a coaxial dispenser nozzle for delivering the fuel and recovering the vapors with a single nozzle, see Figure 16-16. These systems also typically have a large rubber boot or doughnut around the nozzle to seal it to the vehicle fillneck so as to force the vapors into the recovery system. These systems are required to achieve 90 - 95% reduction in vapor emissions compared to the uncontrolled systems.

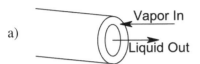

FIGURE 16-16. Stage II Coaxial Dispenser a) Nozzle Schematic b) Photo.

VOC emissions may also occur during the refueling due to spilling. Spilling may occur when the liquid loading rate is faster than the rate the displaced vapors are released from the storage tank, called 'spitback.' Overfilling may also result in spillage. It is caused by a failure of the nozzle to turn off, or from the operator overfilling the tank by 'topping off.' Finally, small amounts of liquid drips can be spilled from the wetted nozzle upon removal from the vehicle. These can be minimized by waiting several seconds after finishing refueling before removing the nozzle.

■ *Example 16-7.*
Determine the loading loss when filling a 2,500 gallon tank truck with ethanol. The truck uses splash loading into a clean cargo tank. The temperature is 25°C.

Solution:
Apply equation (16.13), where T = (25 +273)*1.8 = 536 °R, M_{C2H5OH} = 46 lb/lb-mol, the vapor pressure can be estimated from the data in appendix VI d, and S is given in Table 16-20 as 1.45.

Vapor Pressure:
$$\log_{10} P^{sat} = A - \frac{B}{T+C} = 8.1122 - \frac{1592.864}{25 + 226.184} = 1.771$$

$$P^{sat} = 10^{1.771} = 59 \text{ mmHg} * \frac{14.7 \text{ psia}}{760 \text{ mmHg}} = 1.14 \text{ psia}$$

Where: P^{sat} [=] mmHg, and
T [=] 25 °C.
A = 8.1122
B = 1592.864
C = 226.184

The load loss can now be calculated:

$$L_L = 12.46 * \frac{S*P*M}{T} = 12.46 * \frac{1.45 * 1.14 * 46}{536} = 1.77 \frac{lb - ethanol}{1000 \, gallons}$$

We are loading 2,500 gallons so 4.4 pounds of ethanol are lost to the atmosphere during loading.

■ *Example 16-8.*
Estimate the total amount of gasoline released per day due to the loading losses associated with refueling automobiles in the US.

Solution:
Approach this as a Fermi type problem and use the information in the above paragraph that suggests that some liquid drops are released from the refueling nozzle upon removal from a vehicle.

Number of vehicles in US: The U.S. Department of Transportation Statistical Records Office lists approximately 62 million registered vehicles in the U.S. at the current time (2012) and about 6.4 million unregistered functioning vehicles. Call it 70 million.

Rate at which they are refueled: once per week
Small amount: 3 drops = 2 mL

$$Loss = 70 \times 10^6 \, cars * \frac{1 \frac{refuel}{car}}{7 day} * \frac{2ml}{refuel} * \frac{0.264 gal}{1,000 ml} = 5,280 \frac{gallon}{day}$$

16.6.4 Transportation

Additional methods for reducing mobile source emissions go beyond the actual vehicles and devices to focus directly on the use and operation of mobile sources. Groups such as The US- EPA, Department of Energy, Department of Transportation, Federal Highway Administration, Federal Aviation Administration and other federal groups, state and local governments, transportation planning agencies, and community groups work together to create programs to address mobile source emissions:

Consumer education – These programs focus on information and education to help consumers make informed environmental choices. Labels on products provide comparative information about fuel use, operation costs, and maintenance tips.

Operator behavior – These programs educate users on techniques to reduce fuel consumption and emissions, such as changing landscape use to reduce lawn and garden equipment emissions.

Transportation choices – These programs create options for commuting to reduce vehicle miles traveled by developing and promoting options such as mass transit, carpooling, bicycling, telecommuting, and other strategies that reduce traffic congestion. The programs may offer incentives (car-pool lanes, reduced parking fees, and reduced fees), provide information, and add infrastructure.

Highway and road infrastructure – These programs try to improve traffic flow to reduce congestion and increase safety.

Tax policy – These programs try to influence consumer choices to change driving behaviors as well as to pay for new and existing infrastructure.

Some local areas use other types of incentives to change behaviors, such as parking fees to encourage car-pooling or mass transit, congestion fees to reduce driving during certain times, restrict certain roads or areas to delivery and emergency vehicles only, and road closures.

● Case Study 16-1. Electric Car Emissions.
One idea for reducing emissions from mobile sources is to switch from fossil fuel engines and replace them with electric engines. These engines require a supply of

electricity which must be stored within the vehicle, most likely in batteries. The batteries are recharged as needed by plugging them into the electrical power grid. The vehicle itself does not discharge any emissions, although they do contribute PM emissions from tire wear and braking. There are emissions created from the generation of the electricity. A common question then is how the emissions from the generation of electricity compare to the emissions from a gasoline powered vehicle. The answer depends on how the electricity is generated and what emission control schemes are used by the power plant and vehicle. This case study assumes the electricity is generated in a large coal-fired, modern (built after 2005) power plant that is just meeting New Source Performance Standards as shown. The comparison vehicle meets the Tier 2 bin 4 requirements.

NSPS for Coal-fired power plant, from Appendix 2.
    PM: 0.14 lb/MW-hr gross energy output
    SO_2:   1.4 lb/MW-hr gross energy output
    NO_x: 1.0 lb/MW-hr gross energy output

Emission requirements for LDV tier 2 bin 4, from Table 16-7:
    PM: 0.01 g/mile
    SO_2:   Average of 30 ppm S in the fuel.
    NO_x: 0.04 g/mile

The emissions from both sources are converted to the same units (g/mile) to complete the comparison. A few more assumptions are necessary: An electric car requires 35 kW-hr to go 100 miles, and 1 gallon of gasoline = 33.7 kWhr

PM: electric vehicle

$$0.14\frac{lbPM}{MWhr} \times \frac{454gPM}{1lbPM} \times \frac{1MWhr}{1000kWhr} \times \frac{35kWhr}{100mile} = 0.022\frac{g\ PM}{mile}$$

SO_2: gasoline engine

$$\frac{30lbS}{10^6 lb-fuel} \times \frac{2lbSO_2}{1lbS} \times \frac{454gSO_2}{lbSO_2} \times \frac{6.073lb-fuel}{gal} \times \frac{1gal}{33.7kWhr} \times \frac{35kWhr}{100mile} = 0.0017\frac{gSO_2}{mile}$$

SO_2: electric vehicle

$$1.4\frac{lbSO_2}{MWhr} \times \frac{454gSO_2}{1lbSO_2} \times \frac{1MWhr}{1000kWhr} \times \frac{35kWhr}{100mile} = 0.22\frac{g\ SO_2}{mile}$$

NO$_x$: electric vehicle

$$1.0 \frac{lbNO_x}{MWhr} \times \frac{454 gNO_x}{1 lbNO_x} \times \frac{1 MWhr}{1000 kWhr} \times \frac{35 kWhr}{100 mile} = 0.16 \frac{g\ NO_x}{mile}$$

Summarizing

Pollutant	Gasoline Powered	Electricity Powered	Difference Factor
PM	0.01 g/mile	0.022 g/mile	2.2x
SO$_2$	0.0017 g/mile	0.22 g/mile	129.4x
NO$_x$	0.04 g/mile	0.16 g/mile	4.0 x

The electric car produces more emissions in all three categories. However, this pollution is not emitted by the vehicle but at the power plant effluent stack. This case study assumes all the electrical power comes from burning coal and that the plant just meets each of the emission requirements. The assumption of 100% coal use is rarely the case - most US electrical utilities have committed to generating 20 – 33% of their electricity from non-fossil fuel sources. Also, utilities try to generate emissions well below the NSPS maximum levels. Electric vehicles are often considered to be zero-emission vehicles, but this is not true unless the electricity is generated with zero emissions, such as with nuclear, hydro, wind or solar.

16.7 Questions

* - Questions and problems may require additional information not included in the textbook.

1. How many fuel burning mobile devices do you own? How do you deal with their emissions? *

2. Discuss the engine design features of a 2-stroke engine that allow it to operate in any orientation. Contrast these features with a 4-stroke engine. *

3. Explain the differences and similarities between an Otto engine and a Diesel engine.

4. What is the purpose of the fan in a turbofan engine?

5. How long does it take for petroleum to form from organic matter? *

6. Do you live in an area that is required to use a boutique fuel or RFG? *

7. Discuss ways that sulfur, phosphorous, and silicon may get into the exhaust system. What effects may they have on emissions?*

8. Explain how to operate a gasoline engine at an A/F that will minimize the formation of CO, HC, and NO_x simultaneously without using add-on pollution control equipment.

9. Explain how the LDV tier 2 bin-system works. Why do you think the US-EPA changed the weight based system to the emission bin system?

10. Some engine emission regulations lump HC and NO_x emissions together and place the limit on their sum rather than on each individually. Why do you think this is done? [Note: you may want to review Chapter 13, on Ozone to help you answer this].

11. Where is the nearest airport or marine/ocean port to where you live? Use the Airnow.gov website to find the air quality index (AQI) for this area. *

12. Calculate the fraction of a journey that an ocean transport ship running between New York, NY, USA, and Stockholm, Sweden, spends within MARPOL VI Emission Control Areas for SO_x.*

13. What is the difference between a line-haul and switch locomotive? Which would you expect to have higher emissions per gallon of fuel used? Why?

14. Why and how is the vapor pressure of gasoline changed between summer and winter in reformulated gasoline?

15. How does adding methanol or methyl tert-butyl ether to gasoline improve its cold weather combustion completion?

16. Explain the effect changes in A/F, timing, and compression have on emissions from a 4-stroke Otto engine. *

17. Early automobile catalytic converters used ceramic beads with metal catalysts on their surface. Modern catalytic converters use a ceramic monolith with a honeycomb structure. Why has the monolith structure superseded the use of beads?*

18. Would a three-way catalyst be effective without a properly working set of lambda sensors? Why or why not?

19. Why would a three-way catalytic converter not be useful on a diesel engine?

20. Explain what problems could be associated with using a particulate trap filter on the exhaust of a diesel engine.

21. Why should the emissions of CO, HC, and NO_x from a diesel engine be inherently lower than those from an Otto engine?

22. What effect does exhaust gas recirculation have on emissions, engine power, and fuel consumption?*

23. Would you expect stage I or stage II fuel delivery controls to reduce emissions more? Why?

24. Would you expect the combustion time for gasoline in an SI engine to change if the octane rating of the fuel changed? If the air/fuel ratio changed?

25. Make a list of three questions you have about this chapter or air pollution concerns you have.

16.8 Problems

1. Calculate the per-cent increase or decrease in total mobile source emissions of CO, NO_x, SO_2, PM_{10}, and VOCs between 1980 and 2010.

2. Calculate the % reduction in CO, NO_x, and THC (or NMOG) between a pre-1968 model year light duty vehicle and a tier 2 bin 4 light duty vehicle that just meets all required regulations.

3. Estimate the number of lawnmowers in your community. Use the emission factors in Appendix 16-4 to estimate their annual emissions.*

4. Calculate the time needed to complete all four strokes in an Otto engine running at 600 rpm and at 2400 rpm.

5. Determine the time needed for combustion in a spark ignition engine if the spark fires at 10° BTDC and combustion completes at 20° ATDC when the engine runs at 600 rpms.

6. Generate a plot of the tier I, II, and III MARPOL Annex VI NO_x emission limits (y-axis) as a function of n, the maximum test speed of the engine in rpm (x-axis). Determine the % reduction in allowable NO_x emissions at n=500rpm between tier I and tiers II and III.

7. Determine the CO emissions from a domestic jet aircraft [g/person/mile] during a 1,500 mile domestic trip. Include LTO and cruise emissions. Assume the jet carries 150 people, uses 3 gallons of jet fuel/mile, and the fuel has a density of 700 g/L. Compare to a modern automobile (Tier II bin 4).

8. Do the emissions factors for a Pratt-Whitney Aircraft Engine (PW4158) [π_{oo}=30.7 and F_{oo} = 258.0 kN] meet the tier 8 regulations listed in and near Table 16-19 for HC, CO, and NO_x? The engine is rated at D_p/F_{oo} = 2.4 g/kN for HC, 28.7 g/kN for CO, and 50.2 g/kN for NO_x.

9. Calculate the stoichiometric A/F for diesel fuel (assume $C_{15}H_{30}$).

10. Calculate the stoichiometric A/F for ethanol (C_2H_{60}).

11. Calculate the actual A/F for hexane (C_6H_{14}) when a) λ=0.85, b) λ=1.05, and c) λ=1.5.

12. Determine the total mass of emissions from the combustion of octane when λ = 1.1. Also calculate the wet wt% composition of the exhaust gas assuming complete combustion of the fuel and neglecting the formation of any nitrogen oxides.

13. Use Figure 16-7 to compare the expected emissions from a stoichiometric burn and lean burn (lambda = 1.2). The stoichiometric A/F for the figure is 14.9.

14. Estimate the temperature of the compressed air in an Otto cylinder at the end of the compression stroke (piston is at TDC). Assume the starting temperature is 75°F, starting pressure is 1 atm, the initial volume is 1.5 L and the compression ratio is 9:1. Assume the gas is ideal and the compression is adiabatic, so the following equations from thermodynamics hold:

16.14
$$\frac{P_1 V_1}{T_1} = \frac{P_2 V_2}{T_2} \quad \text{and} \quad P_1 V_1^\gamma = P_2 V_2^\gamma$$

Where: subscripts 1 and 2 represent the initial and final values, and
γ=1.4, the ratio of the heat capacities of air at constant pressure and at constant volume (Cp/Cv).

15. One method to remove NO_x from diesel exhaust is to add urea and run the mixture through a selective catalytic reduction (SCR) unit. Determine the amount of urea needed (g/hp-hr) to reduce NO_x from the value listed in table 16-10, emission factors for diesel engines to the current standard. How might you convert this answer to a volume of urea solution per gallon of fuel? Note – you may want to review SCR in section 10.5.2.3.

16. Compare the evaporative loss emissions from filling an 8,000 gallon underground storage tank at a service station using a) splash loading (dedicated normal service) and b) a phase I capture technique with 95% efficiency using a submerged loading nozzle with dedicated vapor balance service. Also, calculate the percent reduction in emissions from using the phase I control system. The gasoline has an average molecular weight of 98 and a vapor pressure of 6.5 psia at 75°F.

17. Calculate the evaporative fuel emissions (lbs and gallons) from filling 12 gallons into an auto mobile fuel tank at a service station that does not have stage II controls. The fuel tank total volume is 16 gallons and it is splash loaded (dedicated normal service). The gasoline has an average density of 6.1 lbs/gallon, an average molecular weight of 98, and a vapor pressure of 8.5 psia at 75°F. How would this value change if the vapor pressure was reduced to 6.0 psia at the same temperature?

18. Determine the level of carpooling required to reduce NO_x emissions from vehicles necessary to have no net increase in emissions when passenger miles increase by 15% and fuel efficiency increases 7%. Assume the change in fuel efficiency does not alter the emissions in terms of g/gallon of fuel used.

16.9 Group Project Ideas

For each project, the students should work in small groups and present their finding in either a short report (5-8 pages) or 15 minute presentation:

1. Does your city use any incentives or education programs to alter driving habits in order to reduce air pollution? If so, describe the program and try to determine its effectiveness and cost. If not, propose a program and estimate its effectiveness and cost.

16.10 Bibliography

AA1Car. 2012. Catalystic Converter. *AA1Car.com*. [Online] August 1, 2012. [Cited: 13 2012, August.] http://www.aa1car.com/library/converter.htm.

Agnew, W.G. 1968. *Research Publication*. s.l. : General Motors Corporation, 1968. GMR-743.

ASTM. 2014. *Standard Test Method for Vapor Pressure of Petroleum Products (Reid Method)*. ASTM International. West Conshohocken, PA, USA : s.n., 2014. pp. ASTM D323 - 08(2014).

Boeing. 2012. 747 Fun Facts. *Boeing 747 Family*. [Online] 2012. [Cited: August 8, 2012.] http://www.boeing.com/commercial/747family/pf/pf_facts.html.

CAPA. 2011. *Aircraft orders and deliveries review 2010 and 2011 outlook: Airbus, Boeing, Embraer and regionals*. Sydney, AUS : Centre for Aviation, 2011.

CARB. 2011. Current Regulation and Advisories. *Truck and Bus Regulation*. [Online] California Air Resources Board, December 14, 2011. [Cited: August 10, 2012.] http://www.arb.ca.gov/msprog/onrdiesel/regulation.htm.

Hawk Information. 2012a. *Commercial Aircraft / Engine Inventories*. Malmo, Sweden : Hawk Information, 2012a.

—. 2012b. *Military Aircraft / Engine Inventories*. Malmo, Sweden : Hawk Information, 2012b.

Heck, R. M., Farrauto, R. J. and Gulati, S. T. 2009a. Chapter 6. Automotive Catalyst. Catalytic Air Pollution Control: Commercial Technology. Hoboken, NJ, USA : John Wiley and Sons, 2009a.

Heck, R.M., Furrauto, R. J. and Gulati, S. T. 2009b. Chapter 8. Diesel Engine Emissions. *Catalytic Air Pollution Control: Commercial Technology,* 3rd Edition. Hoboken, NJ, USA : John Wiley and Sones, 2009b.

ICAO. 2012. Foundation of the International Civil Aviation Organization. *International Civil Aviation Organization*. [Online] 2012. [Cited: August 8, 2012.] http://www.icao.int/pages/foundation-of-icao.aspx.

IMO. 2011. International Convention for the Prevention of Pollution from Ships (MARPOL). *List of Conventions*. [Online] Internatioonal Maritime Organization, 2011. [Cited: August 10, 2012.] http://www.imo.org/about/conventions/listofconventions/pages/international-convention-for-the-prevention-of-pollution-from-ships-%28marpol%29.aspx.

IPCC. 1996. *Table 1-52 Default Emission Factors and Fuel Consumption for Aircraft*. Task Force on National Greenhouse Gas Inventories, International Panel on Climate Change. 1996. pp. 1-98, Revised 1996 IPCC Guidelines for NAtional Greenhouse Gas Inventories: Reference Manual . http://www.ipcc-nggip.iges.or.jp/public/gl/guidelin/ch1ref7.pdf.

Malm, Howard L. 2000. *Air-Fuel Control and Emissions for Gas Engines*. Calgary, Canada : REM Technology, Inc., 2000.

MARISEC. 2010. Value of Volume of World Trade by Sea. *Shipping and World Trade*. [Online] International Chamber of Shipping, 2010. [Cited: August 10, 2012.] hhttp://www.marisec.org/shippingfacts/worldtrade/volume-world-trade-sea.php.

US-BTS. 2012. Table 1-11: Number of U.S. Aircraft, Vehicles, Vessels, and Other Conveyances. *RITA*. [Online] 2012. [Cited: August 10, 2012.] http://www.bts.gov/publications/national_transportation_statistics/html/table_01_11.html.

US-DOT, RITA:Research and Innovtive Technology Administration. 2008. *Schedule B-43 Aircraft Inventory (2008)*. Washington DC, USA : Bureau of Transportation Statistics, Office of Airline Information, 2008.

US-EPA. 2012i. Aircraft. *Nonroad Engines, Equipment, and Vehicles*. [Online] US Environmental Protection Agency, July 23, 2012i. [Cited: August 10, 2012.] http://www.epa.gov/otaq/aviation.htm.

—. 1994. *Automobile Emissions: An Overview*. Washington DC, USA : Office of Mobile SOurces, 1994. EPA 400-F-92-007.

—. 2006. *Boutique Fuels List Under Section 1541(b) of the Energy Policy Act*. Washington DC, USA : Office of Transportation and Air Quality, 2006. EPA420-F-06-065.

—. 2010b. *Exhaust Emission Factors for Nonroad Engine Modeling - Spark Ignition*. Office of Transportation and Air Quality. Washington DC, USA : US-EPA, 2010b. EPA-420-R-10-019 NR010f.

—. 2012e. Federal Emission Standards. *Emission Standards Reference Guide*. [Online] July 05, 2012e. http://www.epa.gov/otaq/standards/allstandards.htm.

—. 2012f. Heavy-Duty Highway Engine -- Clean Fuel Fleet Exhaust Emission Standards. *Emission Standards Reference Guide*. [Online] US Environmental Protection Agency, July 5, 2012f. [Cited: August 10, 2012.] http://www.epa.gov/otaq/standards/heavy-duty/hd-cff.htm.

—. 2012j. Inspection and Maintenance (I/M). *Cars and Light Trucks*. [Online] US Environmental Protection Agency, June 25, 2012j. [Cited: August 13, 2012.] http://www.epa.gov/otaq/im.htm.

—. 2012h. Lawn and GArden (Small Gasoline) Equipment. *Nonroad Engines, Equipment, and Vehicles*. [Online] US Environmental Protection Agency, January 24, 2012h. [Cited: August 10, 2012.] http://www.epa.gov/otaq/equip-ld.htm.

—. 2012c. Light Duty Vehicles and Trucks and Motorcycles. *Emission Standards Reference Guide*. [Online] US Environmental Protection Agency, July 5, 2012c. [Cited: August 10, 2012.] http://www.epa.gov/otaq/standards/light-duty/.

—. 2012d. Light-Duty Vehicle and Light-Duty Truck -- Clean Fuel Fleet Exhaust Emission Standards. *Emission Standards Reference Guide*. [Online] US Environmental Protection Agency, July 5, 2012d. [Cited: August 10, 2012.] http://www.epa.gov/otaq/standards/light-duty/ld-cff.htm.

—. 1998. *Locomotive Emission Standards: Regulatory Support Document*. Washington DC, USA : Office of Mobile Sources, 1998.

—. 2011b. Mobile Source Air Toxics (MSAT). *Fuels and Fuel Additives*. [Online] US Environmental Protection Agency, November 11, 2011b. [Cited: August 10, 2012.] http://www.epa.gov/otaq/fuels/gasolinefuels/MSAT/index.htm. EPA420-F-07-017.

—. 2010a. MOVES. *Modeling and Inventories*. [Online] US Environmental Protection Agency, 2010a. [Cited: August 19, 2012.] http://www.epa.gov/otaq/models/moves/index.htm.

—. 2012a. National Emissions Inventory (NEI) Air Pollutant Emissions Trends Data. *Technology Transfer Network*. [Online] June 28, 2012a. [Cited: August 15, 2012.] http://www.epa.gov/ttn/chief/trends/index.html.

—. 2011a. Reformulated Gasoline (RFG). *Fuels and Fuel Additives*. [Online] US Environmental Protection Agency, November 11, 2011a. [Cited: August 10, 2012.] http://www.epa.gov/otaq/fuels/gasolinefuels/rfg/index.htm.

—. 2012g. Regulations and STandards. *Nonroad Diesel Equipment*. [Online] US Environmental Protection Agency, June 19, 2012g. [Cited: August 10, 2012.] http://www.epa.gov/nonroad-diesel/regulations.htm.

—. 2012b. Solutions. *Mobile Source Emissions - Past, Present, and Future*. [Online] US Environmental Protection Agency, January 5, 2012b. [Cited: August 10, 2012.] http://www.epa.gov/otaq/invntory/overview/solutions/index.htm.

—. 1991. *Technical Guidance - Stage II Vapor Recovery Systems for Control of Vehicle Refueling Emissions at Gasoline Dispensing Facilities*. Washington DC, USA : US Environmental Protection Agency, 1991. EPA 450/3-91-022a.

—. 2008. *Transportation and Marketing of Petroleum Liquids, Chapter 5 Petroleum Industry, Section 2*. Washington DC, USA : Office of Air Quality, 2008. AP42 Volume I: Supplement F.

US-FAA. 2010. A Breif History of the FAA. *Federal Aviation Administration*. [Online] February 1, 2010. [Cited: August 8, 2012.] http://www.faa.gov/about/history/brief_history/.

Wu, Chan-Wei, et al. 2004. The Influence of Air-Fuel Ratio on Engine Performance and Pollutant Emission of an SI Engine Using Ethanol-Gasoline Blended Fuels. *Atmospheric Environment*. 38, 2004, pp. 7093-7100.

16.11 Appendices

Appendix 16-1. US-EPA Historical Emission Standards for LDVs at 50,000 miles [gram/mile].

Blank entries imply no standard existed for the emission during the noted year.

Program	Model Year	LDV				LDT			
		NOx	CO	PM	THC	NOx	CO	PM	THC
	pre-1968	3.5	87		8.8	3.6	39		6.5
	1973	3	39		3.4	3	39		3.4
	1974	3	39		3.4	3	39		3.4
	1975	3.1	15		1.5	3.1	20		2
	1976	3.1	15		1.5	3.1	20		2
	1977	2	15		1.5	3.1	20		2
	1978	2	15		1.5	3.1	20		2
	1979	2	15		1.5	2.3	18		1.7
	1980	2	7		0.41	2.3	18		1.7
	1981	1	3.4		0.41	2.3	18		1.7
	1982	1	3.4	0.6	0.41	2.3	18	0.6	1.7
	1983	1	3.4	0.6	0.41	2.3	18	0.6	1.7
	1984	1	3.4	0.6	0.41	2.3	10	0.6	0.8
	1985	1	3.4	0.6	0.41	2.3	10	0.6	0.8
	1986	1	3.4	0.6	0.41	2.3	10	0.6	0.8
Tier 0	1987	1	3.4	0.2	0.41	2.3	10	0.26	0.8
Tier 0	1988	1	3.4	0.2	0.41	1.2 (1.7[a])	10	0.26	0.8
Tier 0	1989	1	3.4	0.2	0.41	1.2 (1.7[a])	10	0.26	0.8
Tier 0	1990	1	3.4	0.2	0.41	1.2 (1.7[a])	10	0.26	0.8
Tier 0	1991	1	3.4	0.2	0.41	1.2 (1.7[a])	10	0.26	0.8
Tier 0	1992	1	3.4	0.2	0.41	1.2 (1.7[a])	10	0.26	0.8
Tier 0	1993	1	3.4	0.2	0.41	1.2 (1.7[a])	10	0.26	0.8

a 1.2 is for LDT1; 1.7 is for LDT2-3.

Appendix 16-2. US-EPA Emission Standards for Heavy Duty Highway Compression Ignition Engines (Diesel Engines) [gram/bhp-hr].

Model Year	Applicability	PM	CO	HC	NMHC	NOx	NMHC + NOx
1988	HDDT	0.6	15.5	1.3		10.7	
1990	HDDT	0.6	15.5	1.3		6	
1991	HDDT	0.25	15.5	1.3		5	
1994	HDDT	0.1	15.5	1.3		5	
1998 - 2003	HDDT	0.1	15.5	1.3		4	
1991	UB	0.25	15.5	1.3		5	
1993	UB	0.1	15.5	1.3		5	
1994	UB	0.07	15.5	1.3		5	
1996	UB	0.05	15.5	1.3		5	
1998 - 2003	UB	0.05	15.5	1.3		4	
2004 - 2006	All, Option 1	.1 / .05	15.5		n/a		2.4
2004 - 2006	All, Option 2	.1 / .05	15.5		0.5		2.5
2007 +	All	0.01	15.5		0.14	0.2	

Where HDDT = Heavy Duty Diesel Truck, UB = Urban Bus.

Appendix 16-3. US-EPA Non-road Diesel Engine Emission Standards (g/kWh).

Engine Power	Tier	Year	CO	HC	NMHC+NOx	NOx	PM
kW < 8	Tier 1	2000	8	-	10.5	-	1
	Tier 2	2005	8	-	7.5	-	0.8
	Tier 4	2008	8		7.5		0.4[a]
8 ≤ kW < 19	Tier 1	2000	6.6	-	9.5	-	0.8
	Tier 2	2005	6.6	-	7.5	-	0.8
	Tier 4	2008	6.6		7.5		0.4
19 ≤ kW < 37	Tier 1	1999	5.5	-	9.5	-	0.8
	Tier 2	2004	5.5	-	7.5	-	0.6
	Tier 4A	2008	5.5		7.5		0.3
	Tier 4B	2013	5.5		4.7		0.03
37 ≤ kW < 75	Tier 1	1998	-	-	-	9.2	-
	Tier 2	2004	5	-	7.5	-	0.4
	Tier 3	2008	5	-	4.7	-	0.4
37 ≤ kW < 56	Tier 4A	2008	5		4.7		0.3[b]
	Tier 4B	2013	5		4.7		0.03
56 ≤ kW < 130	Tier 4	2012	5	0.19		0.4	0.02
75 ≤ kW < 130	Tier 1	1997	-	-	-	9.2	-
	Tier 2	2003	5	-	6.6	-	0.3
	Tier 3	2007	5	-	4	-	0.3
130 ≤ kW < 225	Tier 1	1996	11.4	1.3	-	9.2	0.54
	Tier 2	2003	3.5	-	6.6	-	0.2
	Tier 3	2006	3.5	-	4	-	0.2
225 ≤ kW < 450	Tier 1	1996	11.4	1.3	-	9.2	0.54
	Tier 2	2001	3.5	-	6.4	-	0.2
	Tier 3	2006	3.5	-	4	-	0.2
130 ≤ kW ≤ 560	Tier 4	2011	3.5	0.19		0.4	0.02
450 ≤ kW < 560	Tier 1	1996	11.4	1.3	-	9.2	0.54
	Tier 2	2002	3.5	-	6.4	-	0.2
	Tier 3	2006	3.5	-	4	-	0.2
kW ≥ 560	Tier 1	2000	11.4	1.3	-	9.2	0.54
	Tier 2	2006	3.5	-	6.4	-	0.2

CO – carbon monoxide; HC – hydrocarbon; NMHC – non-methane hydrocarbon; NO_x – nitrogen oxides; PM – particulate matter; g – gram; kW – kilowatt; kWh – kilowatt-hour

a - hand-startable, air-cooled, DI engines may be certified to Tier 2 standards through 2009 and to an optional PM standard of 0.6 g/kWh starting in 2010.

b - 0.4 g/kWh (Tier 2) if manufacturer complies with the 0.03 g/kWh standard from 2012.

In engines of 56-560 kW rated power, the NOx and HC standards are phased-in over a few year period, as indicated in the notes. The initial standards are sometimes referred to as the 'interim Tier 4' (or 'Tier 4i'), 'transitional Tier 4' or 'Tier 4 A', while the final standards are sometimes referred to as 'Tier 4 B'.

Appendix 16-4. Regulatory Limits for New Non-Road SI Engines.

Nonroad SI engines		HC + NOx [g/kW-hr]	CO [g/kW-hr]
Small (<=25 hp) Phase 3 Phase 3 (2012 +)	Class I (2012 +)	10	610
	Class II (2011+)	8	610
	III (2005+)	50	610
	IV (2005+)	50	805
	V (2005+)	72	805
Large (>25 hp)	Tier 2 (2007+)	2.7	4.4
Marine SI Engines	Generators		5
	Outboard <= 4.3kW[b]	30	$500 - 5P$
	Outboard > 4.3kW[b]	$2.1 + 0.09 \left[151 + \dfrac{557}{P^{0.9}} \right]$	300
	Sterndrive/Inboard	5	75
	HP >373 kW[a]	16	350
	HP > 485 kW[a]	22	350

a) HP = High Performance engine

b) P = power in kW

Date: 2012+ is interpreted to mean the year 2012 and all subsequent years.

Appendix 16 5. Typical Emission Factors from Non-Road SI Engines (2010).

Engine Size	Class		HC g/hp-hr	CO g/hp-hr	NOx g/hp-hr	PM g/hp-hr	Notes
> 25 hp		Uncontrolled	208	486	0.29	7.7	Gas powered, 2-stroke
		Phase 1	0.85	24.49	1.51	7.7	
		Phase 2	0.28	13.24	0.69	7.7	
		Uncontrolled	3.85	107.23	8.43	0.06	Gas powered, 4-stroke
		Phase 1	0.59	29.86	1.51	0.06	
		Phase 2	0.27	11.94	0.69	0.06	
≤25 hp	I	Uncontrolled - Baseline	13.39	408.84	1.8	0.06	Gas powered, 4-stroke, overhead valves
		Phase 1	8.4	351.16	3.24	0.06	
		Phase 2	6.51	293.01	2.45	0.04	
		Phase 3	3.8	242.4	1.42	0.04	
	II	Uncontrolled - Baseline	5.2	408.84	3.5	0.06	Gas powered, 4-stroke, overhead valves
		Phase 1	5.2	352.57	3.5	0.06	
		Phase 2	4.16	352.57	2.77	0.06	
		Phase 3	3.17	321.94	1.01	0.06	
	III	Uncontrolled - Baseline	261	718.87	0.97	7.7	Gas powered 2-stroke hand-held
		Phase 1	219.99	480.31	0.78	7.7	
		Phase 2	26.87	141.69	1.49	7.7	
	IV	Uncontrolled - Baseline	261	718.87	0.94	7.7	Gas powered 2-stroke hand-held
		Phase 1	179.72	407.38	0.51	7.7	
		Phase 2	26.87	141.69	1.49	7.7	
	V	Uncontrolled - Baseline	159.58	519.02	0.97	7.7	Gas powered 2-stroke hand-held
		Phase 1	120.06	351.02	1.81	7.7	
		Phase 2	47.98	283.37	0.91	7.7	

Appendix 16-6. US-EPA Locomotive Emission Standards [g/bhp-hr].

Tier	Model Year	HC	CO	NOx	PM
Line Haul					
0	1973-1992	1	5	9.5	0.22
1	1993-2004	0.55	2.2	7.4	0.22
2	2005-2011	0.3	1.5	5.5	0.1[a]
3	2012-2014	0.3	1.5	5.5	0.1
4	2015+	0.14[b]	1.5	1.3[b]	0.03
Switch					
0	1973-2001	2.1	8	11.8	0.26
1	2002-2004	1.2	2.5	11	0.26
2	2005-2010	0.6	2.4	8.1	0.13[c]
3	2011-2014	0.6	2.4	5	0.1
4	2015+	0.14[b]	2.4	1.3[b]	0.03

a - 0.20 g/bhp-hr until January 1, 2013 (with some exceptions).

b - Manufacturers may elect to meet a combined NO_x+HC standard of 1.4 g/bhp-hr.

c - 0.24 g/bhp-hr until January 1, 2013 (with some exceptions).

Appendix 16-7. NO$_x$ Standards for Aircraft.

Tier	Pressure Ratio	Thrust [kN]	NO$_x$ — These formulae show how to calculate (D_p/F_{00} =) [g/kN]
0	NS	≥ 26.7	$40 + 2.0\,\pi_{00}$
2	NS	≥ 26.7	$32 + 1.6\,\pi_{00}$
4	≤ 30	> 89	$19 + 1.6\pi_{00}$
	≤ 30	26.7 - 89	$37.572 + 1.6\pi_{00} - 0.2087 F_{00}$
	30 - 62.5	>89	$7 + 2.0\pi_{00}$
	30 - 62.5	26.7 - 89	$42.71 + 1.4286\pi_{00} - 0.4013 F_{00} + 0.00642\pi_{00} \times F_{00}$
	> 62.5		$32 + 1.6\pi_{00}$
6	≤ 30	> 89	$16.72 + 1.408\pi_{00}$
	≤ 30	26.7 - 89	$38.5486 + 1.6823\pi_{00} - 0.2453 F_{00} - 0.00308\pi_{00} \times F_{00}$
	30 - 82.6	> 89	$-1.04 + 2.0\pi_{00}$
	30 - 82.6	26.7 - 89	$46.16 + 1.4286\pi_{00} - 0.5303 F_{00} + 0.00642\pi_{00} \times F_{00}$
	> 82.6	> 89	$32 + 1.6\pi_{00}$
8	≤ 30	> 89	$7.88 + 1.408\pi_{00}$
	≤ 30	26.7 - 89	$40.052 + 1.5681\pi_{00} - 0.3615 F_{00} - 0.0018\pi_{00} \times F_{00}$
	30 - 104.7	> 89	$-9.88 + 2.0\pi_{00}$
	30 - 104.7	26.7 - 89	$41.9435 + 1.505\pi_{00} - 0.55523 F_{00} + 0.005562\pi_{00} \times F_{00}$
	> 104.7	> 89	$32 + 1.6\pi_{00}$

Where: Pressure Ratio is the ratio of the pressures measured at the front and rear of the compressor.

NS = Not Specified.

D_p = Mass of a pollutant emitted [g].

π_{00} = Reference pressure ratio [no units].

F_{00} = Rated thrust [kN].

CHAPTER **17**

Green Engineering

Engineering is a discipline which uses the scientific method to generate solutions to problems. Typical constraints on solutions include feasibility, safety, and cost. Feasibility requires that the materials and methods to create a solution exist or can be created. Safety considers human welfare in order to minimize harm from the solution. Cost considers the required materials, maintenance, and lifetime of the project. The costs generally need to be less than the revenue created from the project.

Green engineering (GE) adds environmental impact to the constraints evaluated for a project. Environmental impact is a measure of a project's risk to the environment. Engineers can use several strategies to minimize the impact of products and production processes on human health and the environment, see Table 17-1, (Anastas, 2003). For example, selecting chemicals that have low-toxicity eliminates potential health hazards, thereby reducing chemical exposures from accidental releases and reducing the need to remove it from the exhaust. Environmentally conscious design reduces risks to ecosystems, workers, consumers, and the general population while generating cost savings to the manufacturer.

Table 17-1. The Twelve Principles of Green Engineering.
1. Inherent Rather than Circumstantial.
   Designers need to strive to ensure that all materials and energy inputs and outputs are as inherently nonhazardous as possible.
2. Prevention Instead of Treatment.
   It is better to prevent waste than to treat or cleanup waste after formation.
3. Design for Separation.
   Separation and purification operations should be designed to minimize energy consumption and materials use.
4. Maximize Efficiency.
   Products, processes, and systems should be designed to maximize mass, energy, space, and time efficiency.
5. Output-Pulled Versus Input-Pushed.
   Products, processes, and systems should be "output pulled" rather than "input pushed" through the use of energy and materials.
6. Conserve Complexity.
   Embedded entropy and complexity must be viewed as an investment when making design choices on recycle, reuse, or beneficial disposition.

7. Durability Rather Than Immortality.
   Targeted durability, not immortality, should be a design goal.
8. Meet Need, Minimize Excess.
   Design for just the necessary capacity or capability. "One size fits all" solutions should be considered a design flaw.
9. Minimize Material Diversity.
   Material diversity in multicomponent products should be minimized to promote disassembly and value retention.
10. Integrate Material and Energy Flows.
    Design of products, processes, and systems must include integration and interconnectivity with available energy and materials flows.
11. Design for Commercial "Afterlife."
    Design products, processes, and systems for performance in a commercial "afterlife."
12. Renewable Rather Than Depleting.
    Material and energy inputs should be renewable rather than depleting.

There are many ways to evaluate the environmental impact of a project, depending on the project type, size, and scope. Some of the more common methods include:
- *Risk Assessment,*
- *Environmental Impact Assessment,*
- *Life-Cycle Analysis,*
- *Conservation of Life Principles, and*
- *Process Integration and System Analysis*

Any one or more of these tools can be used to evaluate the proposed solution. There is some overlap between these methods so not all should be used. Many of these evaluations require estimates of externalized costs and creation of an accounting structure to quantify these costs. Some typical externalized costs include changes in traffic, noise, alteration of scenic views, animal and plant habitat changes, and opportunity costs. An example of an opportunity cost is groundwater that is used for an industrial process cannot also be used for agriculture.

An additional priority in GE is to involve multiple project stakeholders (owners, workers, shareholders, neighbors, state or federal government agencies, and citizens) in deciding a solution. Each stakeholder has different interests and requirements for the project, and inclusion in the decision means fewer surprises for the other groups. Such involvement can uncover opportunities and hazards that any single group may not see. The use of GE should provide long-term benefit to the adopters and to society.

Most projects start by identifying a problem and then generating many potential solutions. Once several solutions are identified, they are assessed for their expected performance based on each constraint using another tool called an *Evaluation of Alternative Analysis*.

17.1 Risk Assessment

"All substances are poisons: there is none which is not a poison.
The right dose differentiates a poison and a remedy." Paracelsus (1493-1541).

Risk assessment is a process for determining the risk associated with a hazard, such as exposure to a chemical (ATSDR, 2002). It is used to inform decisions of the most appropriate actions to make in order to reduce harm to humans and the environment. The process involves four steps; hazard identification, dose-response assessment, exposure assessment, and risk characterization, see Figure 17-1. Hazard refers to the capability of a substance to cause harm. Risk refers to the probability that harm will occur under the given hazard. Exposure refers to the understanding of who is exposed, when, how, where, and for how long. Combining these concepts allows the risk to be characterized – it provides an understanding of how exposure to a hazard will lead to individual, societal and environmental risks.

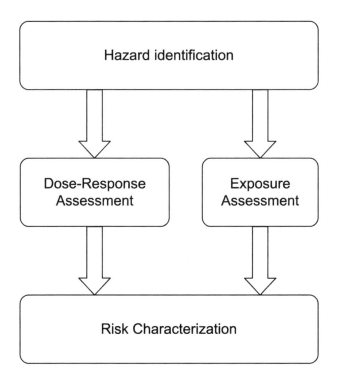

FIGURE 17-1. Risk Assessment Steps.

Hazard identification is used to determine which substances or processes may be causing risk and requires more detailed study. Example categories include corrosive, flammable, explosive, reactive, and radioactive. This decision uses a weight of evidence approach – many observations hint that it might be a problem, but no one piece of data proves it beyond a shadow of a doubt. The type of data used includes epidemiology data in humans or other organisms, bioassays, and similarity to other chemicals known to cause problems. The strongest evidence originates from the field of *epidemiology*, in which a known exposed group of organisms is followed to determine if it has a higher incidence of a problem than a non-exposed group. However, such studies are rare and are limited by confounding exposures. Also, the time required for some problems to occur can be prohibitive (25 – 50 years can pass between exposure and observation of effects). *Bioassays* are well-controlled animal or cell-mass exposures that can more quickly test specific substances or groups of substances for potential problems. They may also suggest the type of harm to humans –e.g. cancer, mutation, and disease. *Structure-Activity Relationships* are used to predict toxicity of a substance based on its similarity to other chemicals known to be toxic. However, there are many exceptions to this rule, so the method has limited usefulness.

A **Dose-Response Assessment** characterizes the relationship between the quantity (dose) of a hazardous substance and the response of the exposed individual.

Figure 17-2 shows several possible relationships between dose and response. Curve A shows harm at very low dosages, benefit at somewhat higher doses until reaching a threshold after which higher dosages cause harm. This type of relationship occurs with water, food, and trace minerals and vitamins – a person needs some, but not too much. Curve B shows no effect at zero and harm at any other dose. Examples of substances with no safe dose include radioactive materials, mercury, and most carcinogens. Curve C shows no response at low doses, then after some threshold value there is a negative response. All the curves show a leveling off at the higher doses, representing death of the organism. The US Department of Health and Human Services and the Center for Disease Control and Prevention publishes a list of known chemical hazards (NIOSH, 2010). This list provides information such as Recommended Exposure Limits (RELs) and Permissible Exposure Limits (PELs). These limits represent the highest level of exposure for humans where there is not expected to be any health effects (the threshold value on curve C).

Obtaining the data needed to quantify these relationships requires well controlled experiments involving many organisms, since any individual may have a greater or lesser tolerance to a particular hazard. Also, many different dosages must be included as part of the study in order to determine if thresholds exist.

Human data is derived from accidental exposures and epidemiological studies, typically involving people exposed through an accidental release or their workplace. All epidemiological

FIGURE 17-2. General Dose-Response Characteristic Curves.

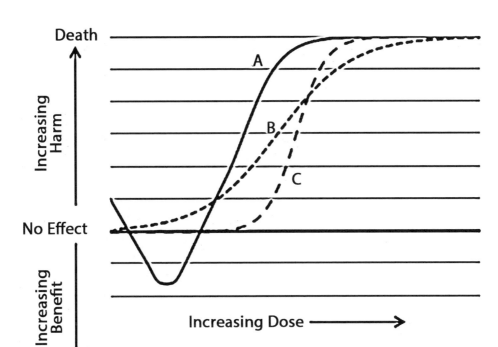

studies must be carefully examined to account for confounding factors such as smoking, alcohol intake, illnesses, and multi-component exposures. Such data is difficult to obtain and hard to interpret. Most experimental data is obtained from animal studies because it is considered unethical to perform these experiments on humans. Additionally, most testing is done at high dosages in order to observe the effect in a timely manner. Creating the dose-response for humans requires two types of data extrapolation: from high experimental doses to low environmental doses and from animal to human doses. The procedures used to extrapolate from high to low doses are different for assessment of carcinogenic effects and non-carcinogenic effects. Carcinogenic effects are not considered to have a threshold (curve B in Figure 17-2) and mathematical curve fits are used to estimate carcinogenic risk at very low dose levels. Non-carcinogenic effects (e.g., neurotoxicity) are considered to have dose thresholds (curve C in Figure 17-2) below which the effect does not occur. The lowest dose with an effect in animal or human studies is then divided by a safety factor to provide a margin of safety, see Table 17-2.

Table 17-2. Safety Factor Multipliers used in Extrapolating Dose-Response from Experimental Data.

Multiplier	Safety Factor
10x	Human Variability
10x	Extrapolation from Animals to Humans
10x	Use of Acute rather than Chronic Data
0.1 - 10x	Modifying Factor Based on Experience

Exposure Assessment analyzes the way an individual or population is exposed to a hazard. The process explores all potential exposure pathways and the manner in which subgroups are exposed. Exposure assessment includes three steps: characterization of the hazard source and receptor, identification of exposure pathways, and quantification of the exposure.

Characterization examines the origin of a hazard and the pathway to each receptor. There are several general classifications of source - continuous releases from point, line, or multiple point sources, and one-time releases from accidents and spills. Receptors may include a general population or a specific sub-group. Common sub-groups include workers, infants, juveniles, elderly, pregnant women, or those with chronic illnesses. Receptors may also include plants, animals, soils, and waterways.

Identification of pathways explores the potential types of exposure between the hazard and the receptors. Exposure involves physical contact or ingestion of food, water, dust, dirt, or air. Proximity may be all that is necessary for exposure to a hazard such as radioactivity, heat, or electricity. Some substances can bio-accumulate, so the organisms which eat organisms previously exposed can receive much higher dosage concentrations.

Quantification of exposure typically relies on exposure models since actual measurements are rarely available. These models include consideration of the movement and fate of the hazard in the environment. Movement and fate in the environment looks at the way a chemical hazard behaves once released. Important considerations include vapor pressure, solubility and partitioning between air/water/soil. The models use dispersion models (such as those discussed in Chapter 6) and include chemical reactivity and interaction between air, water, and soil phases.

Risk Characterization is the final step in a risk assessment. It is used to predict the frequency and severity of effects from a hazard. The characterization of an identified hazard is obtained by combining the exposure and dose-response assessments to determine the probability of an effect likely to occur in a population that is exposed to similar conditions. The assessment identifies which components of this process have the greatest degree of uncertainty. Many assessments describe an increase in the probability of an individual developing cancer (or another effect) during a year or over a lifetime.

In complex risk assessments such as for hazardous waste sites, the risk characterization must consider multiple chemical exposures and multiple exposure pathways. Simultaneous exposures to several chemicals, each at a sub-threshold level, can often cause adverse effects by simple summation of injuries. The assumption of dose additivity is acceptable when substances induce the same toxic effect by the same mechanism. When available, information on mechanisms of action and chemical interactions are considered and are useful in deriving more scientific risk assessments. Individuals are often exposed to substances by more than one exposure pathway (e.g., drinking of contaminated water, inhaling contaminated dust). In such situations, the total exposure usually equals the sum of the exposures by all pathways.

● Case Study 17-1. Risk Assessment of Microwave Popcorn Manufacture.

An epidemiological study in the early 2000's noticed that workers in factories producing microwavable popcorn had a much higher than normal incidence of cancer and lung disease (US-CDC, 2002).

In May 2000, an occupational medicine physician contacted the Missouri Department of Health and Senior Services (MoDHSS) to report eight cases of fixed obstructive lung disease in former workers of a microwave popcorn factory. Four of the patients were on lung transplant lists. All eight had a respiratory illness resembling a non-reversible obstructive lung disease (bronchiolitis obliterans[1]) with symptoms of cough and shortness of breath on exertion. They all had lung function test results that were lower than normal. Each had worked at the same popcorn factory at some time during 1992—2000. Employment durations ranged from 8 months to 9 years.

The factory process mixes soybean oil, salt, and flavorings in a large heated mixing-tank. A site visit showed that the process produced visible dust, aerosols, and vapors with a strong buttery odor. These emissions were the leading suspected hazard, and

---

1. Bronchiolitis obliterans is a rare, severe lung disease characterized by cough, shortness of breath on exertion, and airway obstruction that does not respond to bronchodilators. It can occur after certain occupational exposures. Inhalation exposure to agents such as nitrogen dioxide, sulfur dioxide, anhydrous ammonia, chlorine, phosgene, and certain mineral and organic dusts can cause irreversible damage to small airways (King, 1998).

direct contact or inhalation was the suspected route of exposure. Exposures were analyzed by categorizing workers by proximity to the mixing tank: mixers who had direct contact with the tank; microwave-packaging workers who worked 5 to 30 meters from the tank; and other workers who were >30 meters from the tank.

The factory employed approximately 135 workers and employed approximately 560 different people between 1992 and 2000. Of the eight patients reported, four were mixers, and four were microwave-packaging workers. No microwave-packaging workers had ever worked as mixers. Discussions with workers and management staff indicated that an estimated 13 (3%) of the 425 former workers had been mixers, 276 (65%) had worked in microwave packaging, and 136 (32%) had worked in other areas of the factory. On the basis of the estimated distribution, the crude incidence of illness was highest in mixers (four of 13 [31%]) and microwave-packaging workers (four of 276 [1%]), and no cases were reported in the estimated 136 workers in other areas of the factory.

The calculated rate of illness was 28 to 70 cases per 10,000 person-years, based on the eight cases reported during this period. Assuming that all eight reported patients represented cases of occupational lung disease, this represents a five- to 11-fold excess over the expected number of reported occupational respiratory conditions attributed to toxins (US-CDC, 1999).

Industrial hygiene sampling conducted during a November 2000 survey detected approximately 100 VOCs in the plant air. Diacetyl, a ketone with butter-flavor characteristics, was measured as a marker for exposure to flavoring vapors. The geometric mean air concentration of diacetyl was 18 parts per million (ppm) in the mixing tank room, 1.3 ppm in the microwave-packaging area, and 0.02 ppm in other areas of the plant. Rates of lung function abnormalities increased with increasing cumulative exposure to airborne flavoring chemicals. Concentrations of total and respirable dust were below OSHA-permissible exposure limits (PELs) for particles not otherwise regulated. No OSHA-PELs or NIOSH-recommended exposure levels exist for diacetyl. To reduce exposures, CDC investigators recommended engineering controls including increased ventilation and isolation of VOC sources.

Since this incident, two problem chemicals have been identified in the manufacture and use of microwave popcorn. The first, perfluorooctanoic acid, results from the decomposition of a chemical used to give the non-stick coating to the inside of the popcorn bags. This chemical has been associated with increased risk of certain can-

cers, including liver and prostate cancer. The second chemical is the above suspected flavorant diacetyl. Worker exposure to high vapor concentrations of diacetyl leads to a condition now known as 'popcorn lung,' which is another name for bronchiolitis obliterans. Both materials have already been replaced by microwavable popcorn manufacturers. Consumer exposure can be greatly limited by allowing the popcorn to cool for one minute before opening the bag and limiting consumption to fewer than a few bags per day for any extended period.

17.2 Environmental Impact Assessment

An Environmental Impact Assessment (EIA) is a study of the possible positive and negative impacts a proposed project may cause to humans (health, socially, and economically) and the environment. The purpose of this type of study is to inform decisions to proceed with a project and what, if any, changes to make to reduce the harm from the project. The International Association for Impact Assessment defines an EIA as "the process of identifying, predicting, evaluating and mitigating the biophysical, social, and other relevant effects of development proposals prior to major decisions being taken, and commitments made" (Senecal, et al., 1999). The US adopted the EIA in legislation introduced as part of the 1969 National Environmental Policy Act (NEPA).

An EIA is often a requirement before permits are granted for the project. These permits are granted by local or national government agencies with missions to oversee the use and development of particular resources. However, an EIA may also be done internally by the project group to develop a set of alternatives before seeking approval for the project. All projects in the US that use federal funds are required to develop an EIA. Many states also require one, even if no federal funds are used.

EIAs start once a project is proposed. Initially, the system boundary is chosen – both a physical and time boundary. The choice of physical boundary is often the project location, but it may be expanded to include the watershed, the habitat for certain plants and animals, or the atmosphere. The time boundary usually starts with construction and ends when no more hazards are expected to be generated, usually sometime after closure and decommissioning of the site.

Once the boundaries are chosen, an assessment is made of the project to determine the required inputs and expected outputs. Inputs include materials, energy, labor, space, and capital. Outputs include products, by-products, wastes (material and energy), and revenue. A well-defined EIA includes indirect activities as well – construction activity, changes in the local economy, land use changes, and alterations in traffic (e.g. cars, trucks, rail). These indirect effects can often be much larger than the direct effects of the project. Large projects such as airports and railways can cause state, national, and international level effects.

Many EIAs address how the project may change air, surface water, groundwater, land use, or social and cultural resources. Air emissions would include direct outputs from the process and the indirect outputs from material handling, construction, and increases in traffic. Surface water emissions include direct discharge from the process to a river or lake, or to a sewer that connects to a wastewater treatment plant, which then discharges to a river or lake. Indirect emissions usually originate with rain water runoff from the project facility. This run-off can be captured in storm water retention basins to allow time for solids to settle out and to decrease peak flow of water down-stream from a facility. Ground water is similar to surface water and also includes direct discharge via a well and indirect discharge via soil infiltration. Land use changes may happen on-site and off-site. Waste materials may be stored in surface piles or landfills preventing the use of that land for other activities. Other land use changes that can occur are due to landscaping, altering of elevations, and modifications to water flows. Most project sites include fencing or other enclosures that alter habitat for local plants and animals.

One difficulty for an EIA is estimating or quantifying the environmental impact properties. The properties include landscape quality, lifestyle quality, social acceptance, alteration of air and water quality, and the value of specific environmental objects (trees, hills, undisturbed areas). Any such estimate is very subjective and is very difficult to reach agreement or consensus on their quantification. Assessment of the impacts usually requires information from similar EIAs, expert criteria, and sensitivity to the affected populations. It is also important to include periodic audits to compare the actual impacts to the predicted impacts. The main objective of these audits is to make future EIAs more valid and effective. The two main considerations for an audit are:

- Scientific - to check the accuracy of predictions and explain errors.
- Management - to assess the success of mitigation in reducing impacts.

Once developed, it is typical to make the EIA public within the area of the project, and to hold open meetings for discussion. The public is allowed to comment on and propose changes to the project. These comments are collected and evaluated by the review agency in charge of approving the project. The choice of which review agency has jurisdiction depends on the nature of the project. Some projects require review from several agencies. Agencies can approve a project as is, or suggest changes based on their review and public comments and require the proposer to revise and resubmit their EIA.

17.3 Life-Cycle Analysis

Life Cycle Analysis (LCA), also called cradle-to-grave design, is a technique to assess the full environmental impacts associated with industrial systems. Projects include raw material extractions, processing, manufacture, use, repair, maintenance, recycling, and ultimate disposal. It is important to include all inputs and outputs of material, energy, labor, and capital. The analysis evaluates all stages of a product's life based on the idea that they are all interdependent. The analysis also provides an estimate of the cumulative environmental impact for that product. The technique can be used to evaluate the impact of a current process or proposed changes. The results may help determine where to change a process to provide the greatest economic and environmental impacts. By including the impacts throughout the product life cycle, LCA provides a comprehensive view of the environmental aspects of the product or process and a more accurate picture of the true environmental trade-offs in product and process selection.

There are two main types of LCA. The first, an *attributional LCA*, establishes the resources needed for the production and use of a product or service at a discrete point in time. The second, a *consequential LCA*, establishes the environmental consequences of a proposed change in the system to decide if it represents an improvement in the overall process. Both types consist of four phases (US-EPA, 1993):

> *Goal Definition and Scoping* – This phase defines and describes the product, process or activity. It establishes the context in which the assessment is made and identifies the boundaries (spatial and temporal) and environmental effects reviewed for the assessment.
> *Inventory Analysis* - Identifies and quantifies energy, water, and materials usage and releases including air, water (surface and ground), and solid emissions.
> *Impact Assessment* - Assesses the potential human and environmental effects of energy, water, and material usage and the environmental releases identified in the earlier phases.
> *Interpretation* - Evaluates the results of the inventory analysis and impact assessment to understand or select the preferred product, process or service with a clear understanding of the uncertainty and the assumptions used to generate the results.

Results of an LCA can be combined with other factors, such as cost and product performance, to select a product or process. LCA data identifies the transfer of environmental impacts from one media to another (e.g., eliminating air emissions by creating a wastewater effluent) and/or from one life cycle stage to another (e.g., from the raw material acquisition phase to the use and reuse of the product). Without the LCA, the transfer might not be recognized and properly included in the analysis because it is outside of the typical scope or focus of product selection processes. An LCA allows a decision maker to study an entire product system that avoids the sub-optimization that could result if only a single process were the focus of the study. For example, when selecting between two rival products, it may appear that Option 1 is better for the environment because it generates less solid waste than Option 2. However, after performing an LCA it might be determined that the first option actually creates larger cradle-to-grave environmental impacts when measured across all three media (air, water, land) (e.g., it may cause more chemical emissions during the manufacturing stage). Therefore, the second product (that produces more solid waste) may be viewed as producing less cradle-to-grave environmental harm or impact because of its lower overall chemical emissions.

Performing an LCA can be resource and time intensive. Depending upon how thorough an LCA the user wishes to conduct, gathering the data can be problematic. The availability of data can greatly impact the accuracy of the final results. It is important to weigh the availability of data, the time necessary to conduct the study, and the financial resources required against the projected benefits of the LCA. Also, problematic is that an LCA does not determine which product or process is the most cost effective or works the best. Therefore, the information developed in an LCA study should be used as one component of a more comprehensive decision process assessing the trade-offs with cost and performance.

Finally, although life cycle assessment is a powerful tool for analyzing similar aspects of the quantifiable parts of systems, not every factor can be quantified and modeled. Rigid system boundaries make accounting for changes in the system difficult. This phenomenon is sometimes referred to as the boundary critique to systems thinking. The accuracy and availability of data can also contribute to inaccuracy. For instance, much of the data from government agencies is generalized from multiple systems, from systems with different equipment, and may be based on averages, unrepresentative sampling, or outdated results. Additionally, social implications of products are generally lacking in LCAs because the 'social utility' of an object is highly subjective. Therefore, the differing system boundaries, different statistical information, and different product uses, suggest LCA studies can bias in favor of one product or process over another in one study and the opposite in another study based on varying parameters and the difference in available data. There are guidelines to help reduce such conflicts in results, but the method still provides room for uncertainty.

● **Case Study 17-2. Life Cycle Analysis on Paper and Plastic Grocery Sacks. Goal Definition and Scoping**: *The goal of this LCA is to assess the environmental impacts of the two most commonly used types of grocery sacks. The impacts include energy requirements, water requirements, solid waste and toxics production during raw material collection and production, product production and use, and after use (reuse, recycle, reclamation, and ultimate disposal). The assessment does not include the impacts due to the creation of the production equipment, labor requirements, incidental or auxiliary impacts, material costs nor the cost of capital.*

Paper bags are manufactured from trees. Most trees in the US come from land managed by the paper manufacturer and paper is considered a renewable resource. Once the trees are felled, they are transported from the forest to mill. They are stored for up to three years in order to dry before they use. The logs are stripped of bark (which is an energy source) and then chipped to a one-inch size square. The chips are cooked at high temperatures and pressures. Next, they are digested with limestone and sulfuric acid to form a pulp. The pulp is washed with fresh water and is then pressed into paper. This fresh pulp can be blended with recycled pulp from other paper products. The pulp is pressed and dried to form a continuous sheet of paper and collected onto a roll. Finally, the paper is shipped to a place that cuts, prints, and forms the paper into paper bags. Annual US production is 10 billion bags from 14 million trees. The bags cost 3 – 5 cents per bag.

Table 17-3. Properties of Paper and Plastic Grocery Sacks.

Property		Paper	Plastic	Units
Mass		63.5	7.08	kg / 1000
Capacity		22	12.7	l/bag
Raw Material				
	Energy Use	764	390	MJ/1000
	Waste	1.45	0.13	kg / 1000
Production and Use				
	Energy Use	954	978	MJ/1000
	Water	3,790	146.5	l/1000
	Waste	1.46	0.41	kg / 1000
End of life				
	Reuse Potential	4 - 8	5 - 20	#/bag
	Amount Recycled	10 - 15	1 - 3[*1]	%
	Energy to recycle	1,522	18	MJ/1000
	Biodegradable	Yes	No	
	Litter lifetime	6 month	Centuries	
Other				
	Transport	0.325	2.9	10^6 bags/truck
	CO_2 impact	97[*2]	31	kg CO2/1000

*1 Typically burned, not reformed into another HDPE product.
*2 Renewable carbon does not affect CO2 induced climate change.

Plastic bags start as petroleum oil. The oil is refined to form ethylene, which is polymerized to form polyethylene resin. The resin is made into pellets and shipped to the bag manufacturing facility where the pellets are melted, extruded, and cooled into a long thin tube of plastic sheet. A hot bar is pressed into the tube at set intervals, melting a small portion which becomes the bottom of one bag and the top of the next. The sections are cut, printed, and finally have a hole stamped to form the handles. Annual US production is 100 billion bags from 12 million barrels of oil. This oil could also have been refined into other products, so bag production has an opportunity cost since oil is not renewable. These bags cost less than one cent per bag.

There are research projects that are exploring the generation of plastics from renewable sources such as corn starch. However, these have not been shown to be economical and are not currently used to any appreciable extent as grocery sacks.

Inventory Analysis: Table 17-3 lists the resources used in the sack creation. The values are estimates based on typical production data from 2000.

Impact Assessment: *Comparison between the two materials must consider that plastic bags typically hold about ½ of the items that a paper bag can hold, so you need twice as many.*

Other observations from Table 17-3 show that the plastic weighs considerably less takes up much less storage space and requires less material, energy, and water to make. They also generate less waste, however the waste they do generate is much more toxic than that from paper bag manufacturing.

After use the sacks can be:
    REUSED - plastic bags are more durable (less wear from use, more resistant to tearing, and more resistant to water) so they can be used more times than paper.
    RECYCLED – Paper bags are much more likely to be recycled, and when recycled they are more likely to be used to create another paper product. When most plastics are recycled they are in a mixed waste stream, where it is difficult to separate the various types of plastic. Typically, the bags are reclaimed as heat in a furnace rather than reformed into another plastic product.
    RECLAIMED – both materials can be reclaimed as fuel in furnaces.
    DISPOSAL – Neither bag is biodegradable if placed into a landfill, since landfills are designed to limit the contact between the waste and water, which is essential for degradation. If the sack is released into the environment as litter, the paper

sacks will biodegrade in under a year as long as water is present. The plastic sacks will not biodegrade in the environment. Plastic can be shredded by weathering processes or buried. There is some concern that these shredded materials will, over many centuries, be reduced to minute particle sizes (nano-particles) and become imbedded in plants and animals. No one knows how these nano-particles will affect the contaminated organisms. Plastic bags also pose a risk to marine life such as sea turtles, because they can mistake the plastic bags for jellyfish when they are hunting for food.

Finally, the impact on global concentrations of carbon dioxide is quite different between the two materials. The carbon in paper bags is part of the rapid exchange of carbon between the land and air, so upon release as CO_2 it does not increase the long-term atmospheric concentration of CO_2. The carbon in plastic bags is from the long-term carbon reservoirs (sinks). Upon release it acts to increase the atmospheric concentration of CO_2, although this is not a major source of anthropogenic carbon in the atmosphere.

Interpretation: *The conclusion about which material to use is not clear. The choice becomes one of values and how important each impact is. Plastic bags use fewer resources but create more waste problems during manufacture and when not properly disposed.*

A larger view of the problem shows that the issue is not paper or plastic, but rather finding ways to reduce, reuse, and recycle both of them – in that order. Putting more items in fewer bags, avoiding double bagging, switching to durable tote bags, and reusing and recycling disposable bags, significant reductions in material and nonrenewable energy consumption, pollution, solid waste, greenhouse gas emissions, and litter, occurs. The best option for most consumers is a reusable bag which can easily replace 1,000 bags of either sort. A second option is to reuse the bags you already have and to store them in a place that is available when you go shopping (in the car, on your bike, or in your briefcase or backpack).

17.4 Conservation of Life

Conservation of mass and conservation of energy are fundamental principles that apply to all aspects of solving engineering problems. In most cases, both must be applied to find good solutions. Conservation of life (COL) is a third principle of equal importance, although it is often not explicitly included as part of a solution. COL requires a project solution prevent serious human injury, major property damage, and environmental harm. Table 17-4 lists the conservation of life principles (Klein, et al., 2011).

Table 17-4. The Conservation of Life Principles.
1. Assess material/ process hazards
2. Evaluate hazardous events
3. Manage process risks
4. Consider real-world operations
5. Ensure product sustainability

Each COL principle builds on the previous one. The first step in exploring any change or new solution to the problem is to assess the material and process hazards. A hazard can be any physical or chemical condition that has the potential to harm people, property, or the environment. Hazards include reactivity, flammability, toxicity, radioactivity, high temperatures and pressures, and mechanical or other forms of concentrated energy. Information on many hazards can be found in publications such as Material Safety Data Sheets (MSDS), and National Institute for Occupational Safety and Health (NIOSH) publications. Many MSDS data sheets can be found at: www.hazard.com/msds/. NIOSH maintains an extensive guide to chemical hazards at www.cdc.gov/niosh/npg/ or (NIOSH, 2010). When no data exists, suspected hazards should be investigated experimentally. A safe attitude assumes that everything is a potential hazard until proven otherwise.

The second step is to evaluate hazardous events. An event is any situation that could create a hazard. Examples include loss of power, leaks in tanks, pipes, or equipment, fires, explosions, runaway reactions, accidents, operator error, natural disaster, or mismanagement. All possibilities should be considered, as well as combinations of events (earthquake plus power outage with operator error). Next, the direct impacts of these events are estimated through consequence analysis and modeling. This analysis provides an estimate of the potential harm in terms of type, severity, and number. The analysis also considers a range of events from small to large. A small event may be a small leak in a pipe, or a valve left in the wrong position. A large event may include the rupturing of a storage tank or collapse of a building. Typical hazard evaluation methods include hazard and operability analysis (HAZOP), what-if analysis, fault tree analysis, or failure modes and effects analysis.

The third step is to manage the risks. This step has several parts: quantify the risk associated with each hazard, define the acceptable risk levels, apply known inherently-safe approaches, and includes the design of multiple layers of protection. The initial process design and risk analysis activities provide the greatest opportunities for consideration and implementation of inherently safer process concepts to reduce risks. These are the built-in systems designed to insure safe operation. Examples include secondary containment around all storage areas, off-site secondary power supply, availability of spare parts and alternate process equipment such as pumps, pipes, valves, fire suppression equipment, and personal safety equipment appropriate for the level of expected hazards.

The fourth step considers real world operations. This step addresses human factors. It is crucial to have the plans and ideas identified in the previous steps get implemented at the actual facility. Typical tools include the development of standard operating procedures, training of all new personnel at hiring, periodic update training for all experienced personnel, requiring all contract personnel to adhere to facility risk management procedures, periodic mechanical integrity evaluation, periodic quality assurance of all inputs and outputs, and maintaining an emergency planning and response team. The most successful programs include external evaluation systems to audit all of these tools. Finally, a management program that encourages reporting of problems without repercussions (and with modest encouragement) to allow the facility to learn from mistakes and improve operations.

The final step is to ensure product sustainability. A good design prevents significant human or environmental impacts throughout the entire product lifecycle from raw material collection to intended use, and then onto final disposal. The ultimate goal of COL is to achieve zero injuries, incidents, and environmental impacts.

● Case Study 17-3. Conservation of Life for Chemical Storage at a Large Factory.

Perform a COL analysis using the five principles in Table 17-4 for a water pretreatment facility at a large factory. The facility uses several chemicals to change the properties of the various process water streams. These include liquid acids (concentrated hydrochloric and sulfuric), bases (sodium hydroxide), and solid lime (CaO).

Principle 1 - Assess material hazards of HCl.
Summary of MSDS information (MSDS:HCl, 1999):

> *HCL is a severe corrosive and poison. It causes chemical burns to all body tissue, and can be fatal if swallowed or inhaled. Fire hazards include production of toxic fumes and the generation of hydrogen and chlorine gases. Store it in a cool, dry ventilated location with acid resistant floors and good drainage. When diluting, the acid should always be added slowly to water and in small amounts. Never use hot water and never add water to the acid. Water added to acid can cause uncontrolled boiling and splashing. Use non-sparking tools when opening metal containers because of the possibility of hydrogen gas being present. Do not*

wash out containers to use for other purposes. Spills should be slowly neutralized with soda ash or slaked lime. Concentrated hydrochloric acid is incompatible with many substances and highly reactive with strong bases, metals, metal oxides, hydroxides, amines, carbonates and other alkaline materials. It is incompatible with materials such as cyanides, sulfides, sulfites, and formaldehyde.

Homework problem 17-4 asks the reader to explore the environmental and safety data for the other materials.

Principle 2 - List several potential hazards that could exist at this facility.
    a. Direct contact between any of these substances or with other process materials.
    b. Direct contact between these substances and workers.
    c. Mixing acid with water generates heat: always add acid to water.
    d. Spills or leaks of acid and base may chemically react, creating heat and harmful vapors.
    e. Lime plus water or acid generates heat, potentially enough to boil water which may cause burns.
    f. Operator could refill a storage tank with the wrong material.
    g. Tanks can leak through holes or valves left open.
    h. Loss of structural integrity of the tank due to an accident.

Principle 3 - Manage Process Risks
Design principles for each of the identified potential hazards:
    a. Store the acid away from the caustic or lime, so that there is no way spills from each could come in contact.
    b. Provide sufficient secondary containment for each tank so that entire contents are contained in a way that no people or equipment come into contact after spill or rupture.
    c. Do not allow water to be directly added to any of the storage tanks.
    d. Provide excellent ventilation to storage area, include natural ventilation if possible.
    e. Protect workers from the area where the water is added to the lime – consider using a closed mixing tank.
    f. Create piping differences for each tank to reduce the possibility of cross-contamination. The difference may be in the hook up fittings or different size piping. Color coding is helpful, but not if the color fades or can be removed (paint).
    g. Secondary containment, similar as discussed in b.
    h. Accident prevention strategies include minimizing possibilities and creating fail-

safe systems. An example would be to not have the storage area near forklift traffic lanes. Major environmental disasters (e.g. earthquakes, tornados, high winds, lightening, flooding) should be considered during the design.

Principle 4 - Consider real-world operation
Require worker and management training for working safely with these chemicals, with proper training for spills. Have clear procedures for what to do when there is an incident. Require frequent (yearly) updating of these procedures. Require this training for all contract employees as well as regular employees.

Implement incident reporting requirements with an established process for using the information to improve procedures. Have all procedures in written formats freely available to all workers. Maintain excellent written records of who has been trained, when, and by whom.

Principle 5 - Ensure product sustainability
The product here is process water. It is used on-site for various purposes before discharge. Any additives used to make the process water more useful remain in the water after use. So these additives could potentially be discharged. Therefore, any chemicals added should be capable of being removed, rendered inert, or made harmless before release.

17.5 Process Integration (System Analysis)

Process Integration (PI) is an approach to designing and redesigning processes which consider the interactions between unit operations. In it, the whole system is optimized together, rather than one unit at a time. The whole-system optimization may mean some units are operated sub-optimally in order to increase the overall efficiency. The efficiency basis may include inputs and outputs of capital, energy, and material. The basic idea is to find process streams that have waste (material or energy) that can be used in a different place as part of the input stream. It is best when used in the initial design to help with unit placement, but it can also be useful during retrofits.

Figure 17-3 shows an example of how PI can be used to reduce total energy costs. The top figure shows a preheated feed line and chilled product line. The bottom figure shows the same process, except it includes an additional heat exchanger which is used to heat the feed with the waste energy from the stream exiting unit 1. It is unlikely that unit 1's product stream has enough energy to preheat the feed to its required temperature before it enters the unit. However, the amount of energy it does pick up means that the heater to unit one will need less steam (or it could heat more feed than before). Similarly, the product stream that is fed to the second unit uses less chilled water because it has given some of its energy to the feed stream.

This type of analysis is relatively new and has seen only recent acceptance in some industries, Additional information is available (Cussler, 2001), (El-Halwagu, 1997), (Kemp, 2007).

FIGURE 17-3. Flow Diagrams before and after Process Integration.

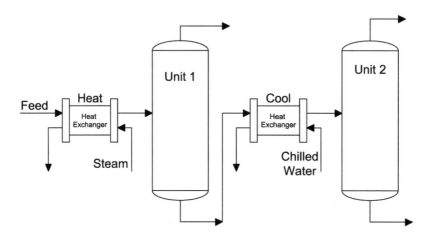

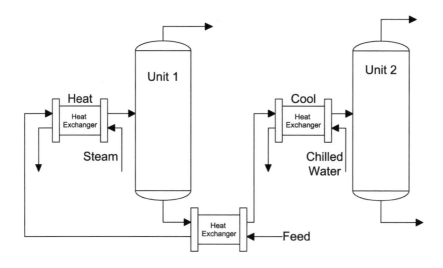

17.6 Evaluation of Alternatives Analysis

An *Alternatives Analysis* helps to choose a solution from a number of alternatives. There are several styles and names for this technique: alternatives screening, range of alternatives, potential solution evaluation, analysis of alternatives (US military procedure), Kepner-Tregoe (KT) analysis. Most use a ranking and weights system. The method encourages obtaining input from multiple sources / viewpoints and works best in a consensus building structure. One way to organize the procedure follows:

I. Clearly specify what is to be decided.
II. Generate a list of solution objectives (e.g. cost, time, space, properties)
III. Generate several alternative solutions (including the 'no change' alternative)
IV. Determine a weight for each objective.
V. Determine the score for every objective for each alternative.
VI. Rank each alternative by summing the products of the weights and scores.
VII. Share tentative results with all stakeholders and revise weights and scores as needed
VIII. Make decision

The project problem should be as clearly described as possible. All relevant information and background should be known to everyone. The goal for the project should also be known. The goal may be articulated with a statement such as:

'We will know the problem is solved when _____.'

Next, generate a list of solution objectives. The objectives are the characteristics for evaluating the solution. The objectives should be divided into needs (requirements) and wants (desired traits). Needs are conditions that an alternative must meet. If it does not meet all needs, it is dropped from consideration. Wants are desired but not essential considerations. An alternative may be chosen if it does not meet any want objectives. Each objective must be measurable in some way. The measure may be quantitative (concentration, size, cost), or it may be qualitative (color, ability to obtain permits, located on railway). Example objectives that would not be acceptable: Process must work well; solution must be fair; equipment must be good, and it must be easy to operate. These are poor objectives because the words 'work well,' 'fair,' 'good,' and 'easy' do not have clear definitions and are difficult to quantify or qualify. These can be improved by rewording to make them measurable: the process must operate 22 hours per day; each stakeholder group must consent to the solution; the equipment must require less than four hours per month of maintenance; and the operator shall need no more than a high school education to operate.

Next, several possible solutions are investigated and those that can successfully solve the problem, as defined in the first step, are included in the alternatives analysis. Each alternative is compared with the list of project needs, and the alternatives that do not meet all the need objectives are eliminated from further consideration.

Each stakeholder group is asked to weight or rank each of the 'want' solution objectives. The most important wants are given the highest weight. Those of less importance are given lower weights.

Some groups may suggest zero weight for issues that are not important to them. When unsure how to assign a score, the want objectives can be ranked from most important to least important and some declining scale used to weight them. This task is subjective, and the values of weights can be readjusted by the decision stakeholders on a voting or consensus basis. Some objectives can be both a need and want. For example, cost may be included in both: cost must be less than $1,000,000, and cost is desired to be less than $500,000.

Next each alternative is provided a score relating how well it meets each objective. A rubric is often used to help develop rankings for non-quantifiable objectives (0 = does not meet objective, 1 = partially meets objective, 2 = almost meets objective, 3 = meets objective, 4 = exceeds objective). The minimum and maximum scores for each objective should be identical (0-to 4 in the example) because the objectives already have a weight to indicate their relative importance. It is appropriate and desirable to generate discussion between the various stakeholders in assessing the scores and weights for each objective. Much can be learned about how each group views the project and the desired results. The range of weights and scores can be averaged between groups, or the groups can attempt to reach consensus during the various discussions.

Once the weight and score values are obtained, they can be multiplied together, and the products summed to obtain a rank for each alternative. These scores will rank the alternatives from highest (best) to lowest (worst). Again, this ranking should be shared with the stakeholders to gauge their acceptability before making a final decision. The following example shows how this process can work.

● Case Study 17-4. Alternative Analysis for Power Plant Fuel Choice.
I. Clearly specify what is to be decided.

The power plant used to supply electricity to a small city is reaching the end of its useful life (approximately five years remain) and no longer meets the required demand. The city public utility is exploring options to meet current and future demand.

The results presented here are from a simulated study which used a college class to model the various stakeholders. Also, the number of alternatives and objectives listed has been reduced from the original problem, yet allows the reader see how the technique works.

Table 17-5. Case Study 17 4 Project Stakeholders.
- *City public utility managers,*
- *Public Utility board (consists of elected community members, industrial representatives, and a State oversight official), and*
- *Ratepayers (local business and residential customers).*

II. Generate a list of solution objectives (cost, time, space, properties ...).

Table 17-6. Case Study 17 4 Project Objectives.
- Needs
    o   Be available within five years.
    o   Decreased emissions of PM10, SOx, and Ozone precursors.
    o   Meet all current and anticipated future Clean Air Act regulations.
- Wants
    o   Have a user rate below $0.08/kW-hr.
    o   Reduce water use from the level of the current facility.
    o   Provide local jobs.

III. Generate several alternative solutions (including the 'no change' alternative).

The options explored include 1) do nothing, 2) obtain power from nearby utilities using grid pricing 3) build a coal-fired plant, 4) build a natural gas-fired plant, 5) build a nuclear power plant. The city will have solved its power problem when they have a consistent source of power for their needs that will last at least 35 years.

IV. Determine a weight for each objective.

Each Stakeholder group was asked to weight the 'want' objectives. Later the groups met to find a consensus value for each objective weight. Each group was given a set number of points (nine in this case) to distribute between the objectives.

Table 17-7. Case Study 17-4 Stakeholder Weights for Objectives and Final Values.

Objective Weightings		Managers	Board	Ratepayers	Final Weight
Wants					
	Rate <0.08 $/kW-hr	1	3	6	0.4
	Reduced Water Use	4	3	0	0.2
	Local Jobs	4	3	3	0.4

V. Determine a score for every objective for each alternative.

Each of the objectives was scored for each alternative. A short discussion of each alternative may help the reader to understand the scores.

Do nothing: This alternative is the status quo choice and is used mostly as a comparison for the others since it does not address the problem.

Power from the grid: *No new plant is built. This alternative has the utility purchase power from other power plants. It is sometimes difficult to obtain long-term pricing, and the cost fluctuates with time of day (higher during high-demand times, lower at other times).*

Coal-fired plant: *The current plant is coal fired, and because it was built before 1970 it does not require much modern pollution control equipment. The new facility would be required to meet the New Source Performance Standards (NSPS) for all pollutants. However, it is anticipated that future NSPS will include CO_2 emissions and is expected that the current fuel source will not meet the new standards without carbon capture and sequestration (CCS) equipment. These technologies are expected to increase construction and rate costs by 35 – 50%*

Natural gas-fired plant: *This fuel will meet NSPS for PM_{10} and SO_x with standard equipment. Ozone standards will require excellent operation and maintenance to reduce NO_x from combustion and VOCs from leaks and poor operation. It is anticipated that the future NSPS will include CO_2 emission standards and that this fuel will be able to meet the standards without CCS equipment.*

Nuclear power plant: *The Clean Air Act NSPS will not apply to this facility, since it has no regular air emissions. Permit times can require five to ten years. It is anticipated that there will be no private investment sources for the capital costs and that it must be insured by the State and Federal government since no private insurance is available for nuclear power plants. Public opposition is expected both locally and nationally.*

Table 17-8 presents the best-guess scores for each alternative. These scores were obtained from the alternative working groups and each score ranges from zero to one.

VI. Rank each alternative by summing the products of the weights and scores.

Table 17-9 presents the method and final ranking for each alternative. We note that the alternatives ranked '0' do not meet the 'need' objectives and are not considered, even though they scored well with the 'want' objectives.

VII. Share tentative results with all stakeholders and revise weights and scores as needed.

The values presented in Table 17-9 are the result of the revised weights and scores after a consensus meeting between the stakeholders.

Table 17-8. Case Study 17-4 Alternative Scores.

Alternative Scores		Do Nothing	Grid	Coal	Nat Gas	Nuclear
Needs						
	5 years	no	yes	yes	yes	no
	Reduced Emissions	no	yes	yes	yes	yes
	Meet Regulations	no	yes	no	yes	yes
Wants						
	Rate <0.08 $/kW-hr	0.5	0	0.25	0.5	0.25
	Reduced Water Use	0	1	0.25	0.45	0.25
	Local Jobs	0.75	0	0.8	0.6	1

Table 17-9. Case Study 17-4 Alternative Rankings.

Rankings		Meet all Needs?	Wants			Sum of Wants	Rank
Alternatives			Rate	Water	Jobs		
	Do Nothing	no	0.2	0	0.3	0.5	0
	Grid	yes	0	0.2	0	0.2	2
	Coal	no	0.1	0.05	0.32	0.47	0
	Gas	yes	0.2	0.09	0.24	0.53	1
	Nuclear	no	0.1	0.05	0.4	0.55	0

VIII. Make decision

Based on the rankings it was decided that the city public utility should begin the process of building a natural gas-fired power plant.

17.7 Questions

* - Questions and problems may require additional information not included in the textbook.

1. Compare and contrast the various green engineering methods (Risk Assessment, Environmental Impact Assessment, Life-Cycle Analysis, Conservation of Life Principles, and Process Integration and System Analysis). When would each be considered for use or be considered inappropriate?

2. What are externalized costs? List three costs you externalize.*

3. What do you believe is an acceptable risk level for a) driving an automobile for 10,000 miles (or one year), b) air emissions of a carcinogen like benzene, c) drinking a caffeinated beverage. Note that a zero risk will essentially ban the activity, a very low risk makes the activity expensive, and a high risk means the death of many people.*

4. Find the actual risk associated with the three activities in question 3.*

5. Make a list of three questions you have about this chapter or air pollution concerns you have.

17.8 Problems

1. Choose one of the twelve principles of green engineering (Table 17-1) and write a 500-word paper describing what it means and how it could be implemented in a project of your choosing.

2. Name three substances that follow each of the three characteristic dose-response curves (Figure 17-2).

3. What are the main risk assessment exposure pathways for lead, mercury, benzene, and ozone?

4. Summarize the MSDS information for sulfuric acid, sodium hydroxide, or lime for use as described in Case Study 17-3.

5. List five possible hazards associated with storing 10,000 pounds of sulfuric acid, sodium hydroxide, or lime as used in Case Study 17-3.

6. List three important safety training topics for locations that use sulfuric acid, sodium hydroxide, or lime as used in Case Study 17-3.

7. How can the use of sulfuric acid, sodium hydroxide, or lime as used in Case Study 17-3 be made more sustainable?

17.9 Group Project Ideas

For each project, the students should work in small groups and present their finding in either a short report (5-8 pages) or 15 minute presentation:

1. Choose one of the NAAQS substances and perform a risk assessment for where you live. How does your analysis compare with the NAAQS limit value? The US-EPA website (www.epa.gov) has much of the required information for this project.

2. Choose a local project or development at your school, workplace, or neighborhood and perform a preliminary environmental impact statement.

3. Each member of the group should develop a life cycle analysis on one of the objects from one of the following categories. A comparison between the objects can then be performed and presented by the student group:

Category 1
Beverage container made of glass
Beverage container made of metal
Beverage container made of plastic

Category 2
Paper books
Electronic books

Category 3
Organic fresh fruit
Non-organic fresh fruit
Canned fruit
Frozen fruit

Category 4
Wind power for electricity production
Coal for electricity production
Natural gas for electricity production
Nuclear for electricity production
Solar Power for electricity production

Category 5
Fuel ethanol from corn
Fuel ethanol from sugar cane
Fuel ethanol from wood (cellulose)

4. Prepare an evaluation of alternatives analysis for one of the categories in group problem 3. Be sure to make the comparison for similar uses [i.e. the fuels in category five should be compared for the operation of similar engines (power output wise) for typical use over one year].

17.10 Bibliography

Anastas, P.T., and Zimmerman, J.B. 2003. Design through the Twelve Principles of Green Engineering. *Env. Sci. and Tech*. 2003, Vol. 37, 5, pp. 94A-101A.

ATSDR, Agency for Toxic Substances and Desease Registry. 2002. *Toxicology Curriculum for Communities Trainer's Manual, Module 3 - Risk Assessment*. Atlants, GA, USA : US-CDC, Institute of Public Health, 2002.

Cussler, E.L. and Moggridge, G.D. 2001. *Chemical Product Design.* Cambridge, UK : Cambridge University Press, 2001. ISBN 0-521-79183-9.

El-Halwagu, Mahmoud M. 1997. *Pollution Prevention through Process Integration.* San Diego, CA, USA : Academic Press, Elsevier, 1997. ISBN 0-12-236845-2.

Kemp, Ian C. 2007. *Pinch Analysis and Process Integration: A User Guide on Process Integration for the Efficient Use of Energy, 2nd edition.* Oxford, UK : Butterworth-Heinemann, Elsevier, 2007. ISBN 0-7506-8260-4.

King, T.E., Jr. 1998. Bronchiolitis. [book auth.] Fishman. *Pulmonary Diseases and Disorders, 3rd ed.* New York, New York, USA : McGraw Hill, 1998.

Klein, J A and Davis, R A. 2011. Conservation of Life as a Unifying Theme for Process Safety in Chemical Engineering Education. *Chemical Engineering Education.* 2011, Vol. 45, 2, pp. 1-5.

MSDS:HCl. 1999. Hydrochloric Acid. *Department of Chemistry*, Texas A&M University. [Online] 11 17, 1999. [Cited: 3 16, 2012.] http://www.chem.tamu.edu/class/majors/msdsfiles/msdshcl.htm.

NIOSH, National Institute Occupational Safety and Health. 2010. *NIOSH Pocket Guide to Chemical Hazards.* Cincinnati, OH, USA : NIOSH Publications, 2010. 2010-168.

Senecal, P, et al. 1999. *Principle of Environmental Impact Assessment Best Practice.* Fargo, ND, USA : IAIA International Association for Impact Assessment and Institute of Environmental Assessment, UK, 1999.

US-CDC. 2002. Fixed Obstructive Lung Disease in Workers at a Microwave Popcorn Factory --- Missouri, 2000--2002. *MMWR.* April 26, 2002, pp. 345-347.

—. 2009. *Leading Causes of Death.* Washington, DC, USA : US Centers for Disease Control and Prevention, 2009.

—. 1999. *Work Related Lung Disease Surveillance Report.* Washington, DC, USA : US Department of Health and Human Services, Public Health Service, CDC, National Institute for Occupational Safety and Health, 1999.

US-EPA. 1993. *Life Cycle Assessment: Inventory Guidelines and Principles.* Washington DC, USA : US Environmental Protection Agency, 1993. EPA/600/R-92/245.

APPENDICES

I. Conversion Factors

Metric System Prefixes	Symbol	Multiply by
peta	P	10^{15}
tera	T	10^{12}
giga	G	10^{9}
mega	M	10^{6}
kilo	k	10^{3}
deci	d	10^{-1}
centi	c	10^{-2}
milli	m	10^{-3}
micro	μ	10^{-6}
nano	n	10^{-9}

US Prefixes

M = 1,000 (yes it is the same symbol as Mega of metric system – this one is common in the petroleum industry and US business reports).

MM = 10^{6}

1 billion = 10^{9}

Length

1 cm = 2.54 inch
1 meter = 3.28083 feet
1 km = 0.62137 mile
1 feet = 12 inch
1 mile = 5280 feet

Area

1 m² = 10.7639 ft²
1 km² = 0.3861 mile²
1 mile² = 640 acre

Volume

1 liter (l) = 1 dm³ = 0.26417 gallon (US)
1 m³ = 10^3 l = 35.315 ft³

1 gallon (US)	= 7.482 ft³	
1 barrel (US liquid)	= 31.5 gallons (US)	
1 barrel (oil)	= 1 bbl	= 42 gallons (US)
1 MMbbl	= 1000 Mbbl	= 10^6 bbl

Mass

1 lb	= 0.4536 kg	= 7000 grain
1 US ton (short)	= 2000 lb	
1 US ton (long)	= 2240 lb	
1 ton = 1000kg	= 2204 lb	

Time

1 hour	= 60 minutes	= 3600 second
1 day	= 24 hour	
1 year	= 365.24 day	

Temperature

°C	= 5/9 • (°F-32)
°R	= °F + 459.69
K	= °C + 273.16
°R	= 1.8 • K

Force (N = Newton)

1.0 N	= 0.225 lbforce
1.0 dyne	= 10^{-5} N

Pressure

1 torr	= 1 mmHg	= 0.13333 kPa	
1 atm	= 101.33 kPa	= 760 mmHg	= 1.01325 bar
1 psi (lb/in²)	= 6.8943 kPa	= 0.068 atm	
1 Pa	= 1 N/m²	= 1 kg/[m s²]	= 0.0039 inch water

Viscosity cP = centi-Poise

1 cP	= 10^{-3} kg/(m•sec)	= 10^{-3} Pa•sec	= 0.01 Poise (P)
1 P	= 0.672 lb/(ft•sec)		

Energy

1 kJ	= 0.9478 BTU	
1 BTU	= 252 cal	= 778.16 ft-lb
1 cal	= 4.1868 J	

| 1 kWhr | = 3412 BTU | |
| 1 J | = 1 Nm | = 1 kg m^2 / s^2 |

Power

1 kW	= 3413 BTU/hr	
1 hp	= 745.7 W	= 2547 BTU/hr
1 W	= 1 J/s	

Force

1 N	= 1 kg m / s^2
1 lbf	= 4.4482 N
1 N	= 10^5 dyne
1 dyne	= 1 g cm / s^2

Angle

1 revolution	= 360 degree
1 radian	= 57.296 degree
1 grad	= 0.9 degree

Gravity at Sea Level

| g | = 9.806 m/s^2 | = 32.174 ft / s^2 |

Density of Water at 4°C, for help with specific gravity of liquids and solids

1,000 g/L
8.34 lb/gal

Ideal Gas Constants

Ideal gas law: $PV = nRT$

Where: P = pressure [M/Lt^2]
V = volume [L^3]
n = moles [mol implies g-mole unless otherwise specified]
R = ideal gas constant
T = absolute temperature [T - K or °R]

Value for R is chosen to match units of other variables:

PVT Units

$R = 8.3145 \dfrac{kPa \cdot liter}{mol \cdot K}$	
$8.3145 \dfrac{m^3 Pa}{mol \cdot K}$	
$8.3145 \times 10^{-5} \dfrac{m^3 bar}{mol \cdot K}$	
$0.083145 \dfrac{L \cdot bar}{mol \cdot K}$	
$0.082057 \dfrac{m^3 \cdot atm}{kmol \cdot K}$	
$0.082057 \dfrac{liter \cdot atm}{mol \cdot K}$	
$8.2057 \times 10^{-5} \dfrac{m^3 atm}{mol \cdot K}$	
$10.732 \dfrac{psia \cdot ft^3}{lbmol \cdot °R}$	
$62.364 \dfrac{ft^3 \cdot mmHg}{lb-mol \cdot °R}$	
$998.97 \dfrac{L \cdot mmHg}{mol \cdot K}$	
$0.73024 \dfrac{ft^3 \cdot atm}{lbmol \cdot °R}$	

$R_{air} = 287.1 \dfrac{m^2}{s^2 \cdot K}$ for air only.

Energy Units

$R = 1.986 \dfrac{BTU}{lbmol \cdot °R}$
$1.9859 \dfrac{cal}{mol \cdot K}$
$8.3144 \dfrac{J}{mol \cdot K}$
$5.189 \times 10^{19} \dfrac{eV}{mol \cdot K}$
$8.3145 \times 10^7 \dfrac{erg}{mol \cdot K}$
$6.1324 \dfrac{ft-lb_f}{mol \cdot K}$
$1,545.3 \dfrac{ft-lb_f}{lb-mol \cdot °R}$

Note: Three countries - Burma, Liberia, and the US - have not adopted the International System of Units (SI, or metric system) as their official system of weights and measures. Although use of the metric system has been sanctioned by law in the US since 1866, it has been slow in displacing the American adaptation of the British Imperial System known as the US Customary System. The US is the only industrialized nation that does not mainly use the metric system in its commercial and standards activities, but there is increasing acceptance in science, medicine, government, and many sectors of industry.

II. New Source Performance Standards (NSPS)
Standards of Performance for New Stationary Sources, CFR title 40 part 60.

Ammonium sulfate manufacturers:
    PM: 0.15 kg/Mg(NH_4)$_2SO_4$ produced

Asphalt processing and asphalt roofing manufacturing:
    PM: 0.04 kg/Mg asphalt shingle or mineral-surfaced roll roofing produced
    PM: 0.04 kg/Mg saturated felt or smooth-surfaced roll roofing produced

Basic oxidation furnace (Steelmaking)
    Started operations between 06/11/73 – 01/20/83
    PM: 50 mg/dscm
    Started operations after 01/20/83
    PM: 23 mg/dscm

Bulk gasoline terminals:
    VOC: 35 mg/liter of gasoline loaded into tank truck

Calciners and dryers at mineral processing plants:
    PM: 0.092 g/dscm – calciners and calciners + dryer in series
    PM: 0.057 g/dscm - dryers

Electric utility steam generating units:
Started operation after 08/17/1971 and have a heat input >73 MW
    PM: 0.10 lb/MMBTU
    SO_2:   0.80 lb/MMBTU – liquid fossil fuel, or
           1.20 lb/MMBTU – solid fossil fuel
    NO_x: 0.20 lb/MMBTU – gas fossil fuel, or
           0.30 lb/MMBTU – liquid fossil fuel, or
           0.70 lb/MMBTU – solid fossil fuel, except
           0.60 lb/MMBTU – lignite, except
           0.80 lb/MMBTU – lignite from ND, SD, MT with cyclone-fired unit.

Started operation after 09/18/1978 and have a heat input >73 MW
    PM: 0.03 lb/MMBTU and
           99% reduction for solid fuel or
           70% reduction for liquid fuel
    SO_2:   *Solid fuel*
           1.2 lb/MMBtu and 90% reduction; or

0.6 lb/MMBtu and 70% reduction
Liquid or gaseous fuel
0.8 lb/MMBtu and 90% reduction; or
0.2 lb/MMBtu and 0% reduction

NO$_x$: *Gaseous fuel*
- Coal derived: 0.5 lb/MMBtu
- All others: 0.2 lb/MMBtu

and 25% reduction of potential combustion concentration

Liquid fuel
- Coal Derived   0.5 lb/MMBtu
- Shale Oil   0.5 lb/MMBtu
- All Others   0.3 lb/MMBtu

and 30% reduction of potential combustion concentration

Solid fuel
- Coal Derived   0.5 lb/MMBtu
- Any fuel > 25% coal refuse   Exempt (No limit)
- Any fuel > 25% lignite[1]   0.6 lb/MMBtu
- Subbituminous coal   0.5 lb/MMBtu
- Bituminous coal   0.6 lb/MMBtu
- Anthracite coal   0.6 lb/MMBtu
- All Others   0.6 lb/MMBtu

and 65% reduction of potential combustion concentration

1. Except lignite from ND, SD, MT used in slag tap furnace, which has a limit of 0.8 lb/MMBtu

Started between 07/09/1997 – 02/28/2005, and have a heat input >73 MW

NO$_x$: *Construction*
1.6 lb/MW-hr gross energy output
Reconstruction
0.15 lb/MMBtu heat input

Started after 02/28/2005 and have a heat input >73 MW

PM: 0.14 lb/MW-hr gross energy output; or
0.015 lb/MMBtu heat input; or
0.03 lb/MMBtu heat input, and
99.9% reduction for a construction or reconstruction, or
99.8% reduction for a modification

SO$_2$: *Construction*
1.4 lb/MW-hr gross energy output or 95% reduction

       Reconstruction
       1.4 lb/MW-hr gross energy output; or
       0.15 lb/MMBtu heat input; or 95% reduction
       Modification
       1.4 lb/MW-hr gross energy output; or
       0.15 lb/MMBtu heat input; or 90% reduction

NO_x: *Construction*
       1.0 lb/MW-hr gross energy output
       Reconstruction
       1.0 lb/MW-hr gross energy output; or
       0.11 lb/MMBtu heat input
       Modification
       1.4 lb/MW-hr gross energy output; or
       0.15 lb/MMBtu heat input

Hg: (Starting on or after 01/30/2004)*

Bituminous fuel:	0.020 lb/GWh output	
Subbituminous fuel and > 25 inch rain/year	0.066 lb/GWh	
Subbituminous fuel and ≤ 25 inch rain/year	0.097 lb/GWh	
Lignite fuel:	0.175 lb/GWh	
Coal Refuse fuel:	0.016 lb/GWh	
IGCC unit:	0.020 lb/GWh	

* The limit on mercury is currently being reviewd by the courts and may change. EPA is currently working on an alternative MACT regulation.

Ferroalloy production:

    PM: 0.45 kg/MW-hr for Si/FeSi/CaSi/SiMnZr production
    PM: 0.23 kg/MW-hr for HighCFeCr/Charge Cr/Std FeMn /SiMn/CaC/FeCr Si/ FeMn Si/Silvery iron production
    CO: 20 %vol

Glass manufacturing plants:
PM:

Glass Type	Gaseous Fuel	Liquid Fuel	units
Container	0.1	0.13	g / kg glass produced
Pressed or Brown			
Borosilicate Recipes	0.5	0.65	g / kg glass produced
Soda-lime and Lead	0.1	0.13	g / kg glass produced
Other	0.25	0.325	g / kg glass produced
Wool fiberglass	0.25	0.325	g / kg glass produced
Flat	0.225	0.225	g / kg glass produced

Grain terminal elevators:
    PM: 0.023 g/dscm

Graphic arts industry: publication rotogravure printing:
    VOC: 16 percent of the total mass of VOC solvent and water used at that facility during any one performance averaging period

Hospital/Medical/Infectious Waste Incinerators:

Pollutant	Small < 200 lb/hr	Medium	Large > 500 lb/hr	Units
PM	115	69	34	mg/dscm
CO	40	40	40	ppmv
SO2	55	55	55	ppmv
NOx	250	250	250	ppmv
Pb	1.2	1.2	1.2	mg/dscm
Cd	0.16	0.16	0.16	mg/dscm

Hot mix asphalt plants:
    PM: 90 mg/dscm gas emissions

Incinerators of more than 50 tons per day charging rate:
    PM: 0.18 g/dscm (correct to 12% CO_2)

Industrial surface coating operations for
(a) metal furniture,
    VOC: 0.90 kg/ liter applied coating solids
(b) automobile and light duty trucks,
    Prime Coat Operation - Electrodeposition system
    VOC: 0.17 kg/liter applied coating solids $R_T \geq 0.16$
    VOC: $0.17 \ast 350^{(0.160 - R_T)}$ kg/liter applied coating solids $0.04 \leq R_T < 0.16$
    VOC: No limit $R_T < 0.04$
    Where: R_T = Solids turnover ratio = total volume coating solids added per month/total volume design capacity of Electrodeposition system.
    Prime Coat Operation - Nonelectrodeposition system
    VOC: 0.17 kg/ liter applied coating solids
    Guide coat operation
    VOC: 1.40 kg/ liter applied coating solids

Topcoat operation
    VOC: 1.47 kg/liter applied coating solids

(c) pressure sensitive tape and label
    VOC: 0.20 kg/kg of coating solids or 90% reduction

(d) large appliances,
    VOC: 0.9 kg/liter applied coating solids

(e) metal coil,
    VOC: 0.28 kg/liter applied coating solids if no emission controls
    or   0.14 kg/liter applied coating solids if emission controls
    or   90% reduction based on applied VOCs

(f) beverage can,
    VOC: 0.29 kg/liter applied coating solids – exterior base coating
    VOC: 0.46 kg/liter applied coating solids – clear base or over-varnish coating
    VOC: 0.89 kg/liter applied coating solids – inside spray coating

(g) plastic parts for business machines.
    VOC: 1.5 kg/liter applied coating solids – prime coating
    VOC: 1.5 kg/liter applied coating solids – color coating
    VOC: 2.3 kg/liter applied coating solids – texture coating
    VOC: 2.3 kg/liter applied coating solids – touch-up coating

Industrial-commercial-institutional steam generating units
Small units with heat input between 2.9 – 29 MW

    SO_2:   Coal   0.20 lb/MMBtu
          or   1.2 lb/MMBtu and 90% reduction
          Oil   0.50 lb/MMBtu
          or   use fuel with ≤ 0.5 wt% sulfur
    PM: Coal   0.051 lb/MMBtu
          Wood 0.10 lb/MMBtu
          Other and heat input > 8.7 MW 0.051 lb/MMBtu or 99.8 % reduction

Units with heat input greater than 29 MW

Coal fired with heat input between 29 – 73 MW use PM and NO_x limits only
Coal fired with heat input greater than 73 MW use PM, SO_x, and NO_x limits
Oil fired with heat input between 29 – 73 MW use NO_x limits only
Oil fired with heat input greater than 73 MW use PM, SO_x, and NO_x limits

    SO_2:   Coal   0.2 lb/MMBtu
          or   1.2 lb/MMBtu and 90% reduction

        Oil   0.2 lb/MMBtu
        or   0.8 lb/MMBtu and 90% reduction
  PM: Coal 0.051 lb/MMBtu
        Oil   0.10 lb/MMBtu
        Wood 0.10 lb/MMBtu
  MSW 0.10 lb/MMBtu

NO_x:

Fuel	NO_x Emission Limits	Units
Natural Gas or Distillate Oil		
Low heat release rate	0.10	lb/MMBtu
High heat release rate	0.20	lb/MMBtu
Residual Oil		
Low heat release rate	0.30	lb/MMBtu
High heat release rate	0.40	lb/MMBtu
Coal		
Mass feed stoker	0.50	lb/MMBtu
Spreader stoker	0.60	lb/MMBtu
Pulverized coal	0.70	lb/MMBtu
Lignite except below	0.60	lb/MMBtu
Lignite from ND, SD, MT + slag tap furnace	0.80	lb/MMBtu
Coal derived synthetic fuel	0.50	lb/MMBtu
Combined Cycle system		
Natural Gas or Distillate Oil	0.20	lb/MMBtu
Residual Oil	0.40	lb/MMBtu

Values may differ if fuel blends are used, if emerging technology is used, or if measurement averaging times are changed. See 40 CFR Part 60 for more detail.

Kraft pulp (Paper) mills:

  Recovery furnace
  PM: 0.1 g/dscm (corrected to 8% excess oxygen)
  Smelt dissolving plant
  PM: 0.1g/kg black liquor solids
  Lime kiln
  PM: 0.15 g/dscm (corrected to 10% excess oxygen) if gaseous fuel burned
  PM: 0.30 g/dscm (corrected to 10% excess oxygen) if liquid fuel burned

Lead-acid battery manufacturing plants:

Grid Casting Facility	Pb: 0.40 mg/dscm
Paste Mixing Facility	Pb: 1.00 mg/dscm
Three-Process Operation	Pb: 1.00 mg/dscm
Lead-Oxide Manufacturing	Pb: 5.0 mg/kg Pb feed
Lead Reclamation Facility	Pb: 4.50 mg/dscm
Other Operations	Pb: 1.00 mg/dscm

Lime manufacturing plants:
PM: 0.3 kg/Mg stone feed

Magnetic tape coating facilities:
VOC: 93% reduction of VOC content of the coating applied

Metallic mineral processing plants:
PM: 0.05 g /dscm

Municipal waste combustors:

PM: 27 mg/dscm corrected to 7% oxygen + begin operation before 04/28/09
PM: 25 mg/dscm corrected to 7% oxygen + begin operation after 04/27/09
Cd: 40 microg/dscm corrected to 7% oxygen + begin operation before 04/28/09
Cd: 35 microg/dscm corrected to 7% oxygen + begin operation after 04/27/09
Hg: 80 microg/dscm corrected to 7% oxygen + begin operation before 04/28/09
    or 85% reduction by weight, whichever is less
Hg: 50 microg/dscm corrected to 7% oxygen + begin operation after 04/27/09
    or 85% reduction by weight, whichever is less
Pb: 440 microg/dscm corrected to 7% oxygen + begin operation before 04/28/09
Pb: 400 microg/dscm corrected to 7% oxygen + begin operation after 04/27/09
SO_2: 31 ppmv corrected to 7% oxygen or 75% reduction by weight, whichever is less
HCl: 31 ppmv corrected to 7% oxygen or 95% reduction by weight, whichever is less
NO_x: As per table, values corrected to 7% oxygen, dry basis

Combustor Technology	Start before 04/28/09	Start After 04/27/09	Units
Mass Burn Waterwall	205	205	ppmv
Mass Burn Rotary Waterwall	250	210	ppmv
Refuse-derived Fuel	250	250	ppmv
Fluidized Bed	180	180	ppmv
Mass Burn Refractory	no limit	no limit	-

Small municipal waste combustor

Construction started after 08/30/99 or modified after 06/06/01

Class I – combustion capacity > 250 ton/day of MSW – Include NO_x emission limit

Class II – combustion capacity ≤ 250 ton/day of MSW – Exclude NO_x emission limit

Pollutant	Limit	units
Cadmium	0.020	mg / dscm
Lead	0.20	mg / dscm
Mercury	0.080	mg / dscm
PM	24	mg / dscm
HCl	25	ppmv [1]
NO_x (class I)	150	ppmv
NO_x (class II)	500	ppmv [2]
SO_2	30	ppmv [3]

1. or 95% reduction of potential HCl emissions
2. No monitoring, testing, recordkeeping, or reporting is required.
3. or 80% reduction of potential sulfur dioxide emissions

Nitric acid plants:

NO_x: 1.5 kg/ton HNO^3 produced (as 100% HNO^3)

Petroleum dry cleaners:

VOC: must install solvent recovery dryer using a cartridge filter.

Petroleum refineries:

Catalyst Regenerator unit
    PM: 1.0 kg/Mg of coke
    CO: 500 ppmv
    SO_2: 50 ppmv or 90% (whichever is less) if have add-on control device
    SO_2: 9.58 kg/Mg coke burnoff if no add-on control device

Fuel Gas Combustion unit
    H_2S: fuel must have < 230 mg/dscm (except when emergency flaring)

Claus Sulfur Recovery unit
    SO_2: 250 ppmv (0% excess air) if incineration, or
    Reduced S: 300 ppmv and H_2S: 10 ppmv if no incineration

Petroleum storage vessels:

Liquid vapor pressure between 1.5 psia and 11.1 psia - the storage vessel shall be equipped with a floating roof, a fixed roof with an internal floating type cover, a vapor recovery system, or their equivalents

Liquid vapor pressure > 11.1 psia, the storage vessel shall be equipped with a vapor recovery system and vapor return or disposal system designed to process at least 95 wt%.

Phosphate rock plants:
Dryer unit
    PM: 0.03 kg/Mg phosphate rock feed
Calciner unit – un-beneficiated rock or blends
    PM: 0.12 kg/Mg phosphate rock feed
Calciner unit – beneficiated rock
    PM: 0.055 kg/Mg phosphate rock feed
Grinder unit
    PM: 0.006 kg/Mg phosphate rock feed

Polymeric coating of supporting substrates:
VOC: 90% reduction

Portland cement plants:
Kilns
    PM: 0.30 lb/ton of feed (started between: 08/17/1971 – 06/16/2008)
    PM: 0.01 lb/ton clinker (started after 06/16/2008)
    NO_x: 1.5 lb/ton clinker (started after 06/16/2008)
    SO_2:   0.4 lb/ton clinker (started after 06/16/2008)
    Hg: 14 lb/MMton clinker
Clinker cooler unit
    PM: 0.10 lb/ton of feed (started between: 08/17/1971 – 06/16/2008)
    PM: 0.01 lb/ton clinker (started after 06/16/2008)

Primary copper smelters:
PM: 50 mg/dscm
SO_2:   0.065%vol

Primary zinc smelters:
PM: 50 mg/dscm
SO_2:   0.065%vol

Primary lead smelters:
PM: 50 mg/dscm
SO_2:   0.065%vol

Primary aluminum reduction plants:
    Fluorides: 1.0 kg/Mg Al produced in Soderberg plants
                          0.95 kg/Mg Al produced at prebake plants
                          0.05 kg/Mg Al equivalent for anode bake plants

Rubber tire manufacturing plants:
    Undertread cementing operation
        VOC: Discharge into the atmosphere no more than 25 percent of the VOC used
    Sidewall cementing operation
        VOC: Discharge into the atmosphere no more than 25 percent of the VOC used
    Tread end cementing operation
        VOC: 10 g/tire
    Bead cementing operation
        VOC: 5 g/tire

Secondary lead smelters:
    PM: 50 mg/dscm

Secondary brass and bronze production plants:
    PM: 50 mg/dscm

Sewage treatment plants:
    PM: 0.65 g/kg dry sludge input

Solvent-spun synthetic fiber production facilities:
    VOC: 10 kg/Mg solvent feed - produce acrylic and nonacrylic fiber
    VOC: 17 kg/Mg solvent feed –produce nonacrylic fiber only

Stationary gas turbines:
    SO_2:   0.015%vol (15% oxygen and dry basis)
            and may not use a fuel with S in excess of 0.8% wt (8000 ppmw)
    NO_x:

$$STD = 0.0075 \frac{(14.4)}{Y} + F \quad \text{when heat input at peak load} > 100 \text{ MMBtu/hr}$$

$$STD = 0.0150 \frac{(14.4)}{Y} + F \quad \text{when heat input at peak load between 10 to 100 MMBtu/hr}$$

Where: STD = maximum NO_x emission concentration vol%, corrected to 15% oxygen, dry basis
Y = Heat Rate at lower heating value of fuel [kJ/W•hr], must not exceed 14.4
F = Emission allowance for fuel bound nitrogen, given in table below:

Fuel-bound nitrogen (wt%)	F
$N \leq 0.015$	0
$0.015 < N \leq 0.1$	0.04(N)
$0.1 < N \leq 0.25$	0.004 + 0.0067(N-0.1)
$0.25 < N$	0.005

Use of this table is optional and operator may accept F = 0.

Steel plants: electric arc furnaces and argon-oxygen decarburization vessels:
PM: 12 mg/dscm

Sulfuric acid plants:
SO_2:   2 kg/ton H_2SO_4 produced (as 100% H_2SO_4)

Sulfuric acid production units:
Acid   Mist: 0.25 g mist/kg H_2SO_4 produced (100% acid)

Wool fiberglass insulation manufacturing plants:
PM: 5.5 kg/Mg of glass pulled

Abbreviations:
PM   = Particulate Matter
SO_2 = Sulfur dioxide
NO_x = Nitrogen oxides
VOC   = Volatile organic compounds
CO   = Carbon monoxide
CO_2 = Carbon dioxide
Pb   = Lead
Hg   = Mercury
Cd   = Cadmium
MSW = Municipal solid waste
HCl   = Hydrochloric acid
H_2S = Hydrogen sulfide

Formulas:

Adjust pollutant concentration to 7% oxygen:

$$C_{7\%O2} = C_{measured} \frac{(20.9-7)}{(20.9-\%O_2)}$$

Where $C_{7\% O2}$ = Pollutant concentration corrected to 7% oxygen
: $C_{measured}$ = Actual pollutant concentration measured on dry basis
$\%O_2$ = Actual oxygen concentration measured on dry basis

$$SO_2 = \frac{SO_2}{CO_2 + SO_2 + N_2 + O_2}$$

Potential emissions
Assume all components are fully oxidized
    C to CO_2
    S to SO_2
    H to H_2O (used to determine amount of air needed)

Account for stated amount of excess O_2; 7% oxygen means that the emitted gas has 7% oxygen by wt (Typically this translates to using 50% excess air in the combustion calculation)
Account for N_2 from air, but assume it does not combust – include excess air in determination.

Calculation for concentration of SO_2, on a dry basis with excess air used:

Use mass of each combustion product for wt%
Use moles of each combustion product for vol%

Concentrations of other constituents are calculated in similar manner.

III. Emissions Factors

Adapted from US-EPA: Emissions Factors & AP 42, Compilation of Air Pollutant Emission Factors, additional data available at: *http://www.epa.gov/ttnchie1/ap42/*

An **emissions factor** is a representative value that attempts to relate the quantity of a pollutant released to the atmosphere with an activity associated with the release of that pollutant. These factors are usually expressed as the weight of pollutant divided by a unit weight, volume, distance, or duration of the activity emitting the pollutant (e.g., kilograms of particulate emitted per megagram of coal burned). Such factors facilitate estimation of emissions from various sources of air pollution. In most cases, these factors are simply averages of all available data of acceptable quality, and are generally assumed to be representative of long-term averages for all facilities in the source category (i.e., a population average).

The general equation for emissions estimation is: $E = A * EF * \dfrac{100 - ER}{100}$

Where: E = emissions;
A = activity rate;
EF = emission factor, and
ER = overall emission reduction efficiency, %

Additional Notes:
A = weight% ash in material
S = weight% sulfur in material
L = weight% lead in material
PM10: replace A with a, where 2.3 a = A, and a is the weight% of PM10 in the material.

* Metal emissions from exhaust after PM control (ESP, Cyclone, Scrubbers)

■ Example
Find ash emissions from bituminous coal of 7 % ash.

Solution:
Read emission factor (EF = 10A lb/ton) from table, then calculate:
Ash Emission = 10 * 7 = 70 lb ash/ton coal

Appendix 3

Material	Subclass	Sub-subclass	PM	SO$_x$	NO$_x$	CO	Inorganic Pb*	Hg*	As*	Organic NMTOC	UNITS	Heating Value	Density
Coal	Bituminous	dry bottom, wall fired	10A	38S	12	0.5	0.000420	0.000083	0.000410	0.06	lb/ton	10,720 - 14,730 Btu/lb	833 kg/m^3
		dry bottom, tangentially fired	10A	38S	11	0.5	0.000420	0.000083	0.000410	0.06	lb/ton		
	Subbituminous	dry bottom, wall fired	10A	35S	7.4	0.5	0.000420	0.000083	0.000410	0.06	lb/ton		
		dry bottom, tangentially fired	10A	35S	7.2	0.5	0.000420	0.000083	0.000410	0.06	lb/ton		
	Anthracite	Pulverized coal boilers		39S	18		0.008900	0.000130	0.000190		lb/ton		
	Lignite	FBC boilers	2.9		1.8	0.6	0.008900	0.000130	0.000190		lb/ton		
		Stoker-fired boilers	0.8A	39S	9	0.6	0.008900	0.000130	0.000190		lb/ton		
		Pulverized coal, dry bottom, tangential	6.5A	30S	7.1		0.000420	0.000083	0.000410	0.04	lb/ton	5,000-7,000 Btu/lb	801 kg/m^3
		Pulverized coal, dry bottom, wall fired	5.1A	30S	6.3	0.25	0.000420	0.000083	0.000410	0.04	lb/ton	12,000 - 14,000 Btu/lb	1105 kg/m^3
Fuel Oil	Distillate Oils		2	142S	20	5	0.00115	0.00038	0.00051	0.34	lb/10^3 gal	9,420 - 10,130 Btu/lb	820 kg/m^3
	No. 2		2	157S	24	5					lb/10^3 gal	140,000	
	No. 4		7	150S	47	5				0.76	lb/10^3 gal	145,000	
	No. 5		10	157S	47	5				0.76	lb/10^3 gal	150,000 Btu/gal	7.88 lb/gal
	No. 6		9.19S +3.22	157S	47	5	0.00151	0.00011	0.00132	0.76	lb/10^3 gal	155,000	
	Residential Furnace		0.4	142S	18	5				0.71	lb/10^3 gal	130,000 Btu/gal	7.05 lb/gal
LPG	Propane Industrial Boilers		0.8	0.10S	13	7.5				0.8	lb/10^3 gal	91,500 Btu/gal	508 kg/m^3

Appendix 3

Butane Industrial Boilers	0.8	0.09S	15	8.4			0.9	lb/10^3 gal	102,000 Btu/gal	573 kg/m^3
Natural Gas (Methane)	7.6	0.6	190	84	0.00050	0.00026	0.00020	8.7	lb/10^6 scf	950 - 1,050 Btu/scf	42,000 lb/MMscf
Wood Residue											
Wet	0.33	0.03	0.22	0.6	0.0000480	0.0000035	0.0000220	0.018	lb/MMBTU	4500 BTU/lb	
Dry	0.4	0.03	0.49	0.6					lb/MMBTU	8000 BTU/lb	
Wood											
Residential											
Fireplaces	34.6	0.4	2.6	253				229	lb/ton	8650 BTU/lb	
Wood Stoves - Conventional	30.6	0.4	2.8	231				53	lb/ton	8650 BTU/lb	
Sugar Cane - Mill Waste	15.6	-	1.2					0.001	lb/ton	3,000 - 4,000 BTU/lb	
Waste Oil											
Small Boilers	64A	147S	19	5	55L		0.1100	1.0 TOC	lb/10^3 gal	127,000 BTU/gal	
Refuse											
Mass Burn Waterwall	12.6	1.73	1.83	0.23	0.1070	0.0028	0.00214		g/kg		
Medical Waste	467	2.17	3.56	2.95	0.0728	0.1070	0.000242	0.3 TOC	lb/ton		
Sewage Sludge											
Multiple Hearth Incinerator	100	2.8	5	15.5	0.1000	0.0046	0.0094	1.7	lb/ton		
Fluidized Bed Incinerator	460	0.3	1.7	2.1	0.0400		0.0044		lb/ton		
Electric Infrared Incinerator	7.4	18	8.6						lb/ton		
* Metal emissions from exhaust after PM control (ESP, Cyclone, Scrubbers)											

IV. Dispersion Tables

Table IV-I. Parameters to Calculate the Rural Horizontal Mixing Length in Equation 6-4.

Pasquill Stability Class	a	b
A	24.167	2.5334
B	18.333	1.8096
C	12.5	1.0857
D	8.333	0.72382
E	6.25	0.54287
F	4.1667	0.36191

Table IV-II Formulas to Calculate the Urban Horizontal Mixing Length.

Pasquill Stability Class	Formulas to Calculate Urban σ_y [m]
A	$0.32 \, x \, (1.0 + 0.0004 \, x)^{-1/2}$
B	$0.32 \, x \, (1.0 + 0.0004 \, x)^{-1/2}$
C	$0.22 \, x \, (1.0 + 0.0004 \, x)^{-1/2}$
D	$0.16 \, x \, (1.0 + 0.0004 \, x)^{-1/2}$
E	$0.11 \, x \, (1.0 + 0.0004 \, x)^{-1/2}$
F	$0.11 \, x \, (1.0 + 0.0004 \, x)^{-1/2}$

Table IV-III Parameters to Calculate the Rural Vertical Mixing Length in Equation 6.6.

Pasquill Stability Class	x [km]	c	d
A*	<.10	122.8	0.9447
	0.10 - 0.15	158.08	1.0542
	0.16 - 0.20	170.22	1.0932
	0.21 - 0.25	179.52	1.1262
	0.26 - 0.30	217.41	1.2644
	0.31 - 0.40	258.89	1.4094
	0.41 - 0.50	346.75	1.7283
	0.51 - 3.11	453.85	2.1166
	>3.11	**	**
B*	<.20	90.673	0.93198
	0.21 - 0.40	98.483	0.98332
	>0.40	109.3	1.0971
C*	All	61.141	0.91465
D	<.30	34.459	0.86974
	0.31 - 1.00	32.093	0.81066
	1.01 - 3.00	32.093	0.64403
	3.01 - 10.00	33.504	0.60486
	10.01 - 30.00	36.65	0.56589
	>30.00	44.053	0.51179
E	<.10	24.26	0.8366
	0.10 - 0.30	23.331	0.81956
	0.31 - 1.00	21.628	0.7566
	1.01 - 2.00	21.628	0.63077
	2.01 - 4.00	22.534	0.57154
	4.01 - 10.00	24.703	0.50527
	10.01 - 20.00	26.97	0.46713
	20.01 - 40.00	35.42	0.37615
	>40.00	47.618	0.29592

Table IV-III Parameters to Calculate the Rural Vertical Mixing Length in Equation 6.6 continued.

F	<.20	15.209	0.81558
	0.21 - 0.70	14.457	0.78407
	0.71 - 1.00	13.953	0.68465
	1.01 - 2.00	13.953	0.63227
	2.01 - 3.00	14.823	0.54503
	3.01 - 15.00	17.836	0.41507
	15.01 - 30.00	22.651	0.32681
	30.01 - 60.00	27.074	0.27436
	>60.00	34.219	0.21716

* If the calculated value of σ_z exceeds 5000 m, set σ_z = 5000 m
** Set σ_z = 5000 m

Table IV-IV Formulas to Calculate the Urban Vertical Mixing Length.

Pasquill Stability Class	Formulas to Calculate Urban σ_z [m]
A	$0.24 \, x \, (1.0 + 0.001 \, x)^{1/2}$
B	$0.24 \, x \, (1.0 + 0.001 \, x)^{1/2}$
C	$0.20 \, x$
D	$0.14 \, x \, (1.0 + 0.0003 \, x)^{-1/2}$
E	$0.08 \, x \, (1.0 + 0.0015 \, x)^{-1/2}$
F	$0.08 \, x \, (1.0 + 0.0015 \, x)^{-1/2}$

V. List of US-EPA Hazardous Air Pollutants (HAPs)

CAS is the American Chemical Society's *Chemical Abstract Service* number.

Chemical Name	CAS Number
Acetaldehyde	75070
Acetamide	60355
Acetonitrile	75058
Acetophenone	98862
2-Acetylaminofluorene	53963
Acrolein	107028
Acrylamide	79061
Acrylic acid	79107
Acrylonitrile	107131
Allyl chloride	107051
4-Aminobiphenyl	92671
Aniline	62533
o-Anisidine	90040
Asbestos	1332214
Benzene (including benzene from gasoline)	71432
Benzidine	92875
Benzotrichloride	98077
Benzyl chloride	100447
Biphenyl	92524
Bis(2-ethylhexyl)phthalate (DEHP)	117817
Bis(chloromethyl)ether	542881

Bromoform	75252
1,3-Butadiene	106990
Calcium cyanamide	156627
Caprolactam	105602
Captan	133062
Carbaryl	63252
Carbon disulfide	75150
Carbon tetrachloride	56235
Carbonyl sulfide	463581
Catechol	120809
Chloramben	133904
Chlordane	57749
Chlorine	7782505
Chloroacetic acid	79118
2-Chloroacetophenone	532274
Chlorobenzene	108907
Chlorobenzilate	510156
Chloroform	67663
Chloromethyl methyl ether	107302
Chloroprene	126998
Cresols/Cresylic acid (isomers and mixture)	1319773
o-Cresol	95487
m-Cresol	108394
p-Cresol	106445
Cumene	98828
2,4-D, salts and esters	94757
DDE	3547044

Diazomethane	334883
Dibenzofurans	132649
1,2-Dibromo-3-chloropropane	96128
Dibutylphthalate	84742
1,4-Dichlorobenzene(p)	106467
3,3-Dichlorobenzidene	91941
Dichloroethyl ether (Bis(2-chloroethyl)ether)	111444
1,3-Dichloropropene	542756
Dichlorvos	62737
Diethanolamine	111422
N,N-Dimethylaniline	121697
Diethyl sulfate	64675
3,3-Dimethoxybenzidine	119904
Dimethyl aminoazobenzene	60117
3,3'-Dimethyl benzidine	119937
Dimethyl carbamoyl chloride	79447
Dimethyl formamide	68122
1,1-Dimethyl hydrazine	57147
Dimethyl phthalate	131113
Dimethyl sulfate	77781
4,6-Dinitro-o-cresol, and salts	534521
2,4-Dinitrophenol	51285
2,4-Dinitrotoluene	121142
1,4-Dioxane (1,4-Diethyleneoxide)	123911
1,2-Diphenylhydrazine	122667
Epichlorohydrin (l-Chloro-2,3-epoxypropane)	106898
1,2-Epoxybutane	106887

Ethyl acrylate	140885
Ethyl benzene	100414
Ethyl carbamate (Urethane)	51796
Ethyl chloride (Chloroethane)	75003
Ethylene dibromide (Dibromoethane)	106934
Ethylene dichloride (1,2-Dichloroethane)	107062
Ethylene glycol	107211
Ethylene imine (Aziridine)	151564
Ethylene oxide	75218
Ethylene thiourea	96457
Ethylidene dichloride (1,1-Dichloroethane)	75343
Formaldehyde	50000
Heptachlor	76448
Hexachlorobenzene	118741
Hexachlorobutadiene	87683
Hexachlorocyclopentadiene	77474
Hexachloroethane	67721
Hexamethylene-1,6-diisocyanate	822060
Hexamethylphosphoramide	680319
Hexane	110543
Hydrazine	302012
Hydrochloric acid	7647010
Hydrogen fluoride (Hydrofluoric acid)	7664393
Hydrogen sulfide	7783064
Hydroquinone	123319
Isophorone	78591
Lindane (all isomers)	58899

Maleic anhydride	108316
Methanol	67561
Methoxychlor	72435
Methyl bromide (Bromomethane)	74839
Methyl chloride (Chloromethane)	74873
Methyl chloroform (1,1,1-Trichloroethane)	71556
Methyl ethyl ketone (2-Butanone)	78933
Methyl hydrazine	60344
Methyl iodide (Iodomethane)	74884
Methyl isobutyl ketone (Hexone)	108101
Methyl isocyanate	624839
Methyl methacrylate	80626
Methyl tert butyl ether	1634044
4,4-Methylene bis(2-chloroaniline)	101144
Methylene chloride (Dichloromethane)	75092
Methylene diphenyl diisocyanate (MDI)	101688
4,4'-Methylenedianiline	101779
Naphthalene	91203
Nitrobenzene	98953
4-Nitrobiphenyl	92933
4-Nitrophenol	100027
2-Nitropropane	79469
N-Nitroso-N-methylurea	684935
N-Nitrosodimethylamine	62759
N-Nitrosomorpholine	59892
Parathion	56382
Pentachloronitrobenzene (Quintobenzene)	82688

Pentachlorophenol	87865
Phenol	108952
p-Phenylenediamine	106503
Phosgene	75445
Phosphine	7803512
Phosphorus	7723140
Phthalic anhydride	85449
Polychlorinated biphenyls (Aroclors)	1336363
1,3-Propane sultone	1120714
beta-Propiolactone	57578
Propionaldehyde	123386
Propoxur (Baygon)	114261
Propylene dichloride (1,2-Dichloropropane)	78875
Propylene oxide	75569
1,2-Propylenimine (2-Methyl aziridine)	75558
Quinoline	91225
Quinone	106514
Styrene	100425
Styrene oxide	96093
2,3,7,8-Tetrachlorodibenzo-p-dioxin	1746016
1,1,2,2-Tetrachloroethane	79345
Tetrachloroethylene (Perchloroethylene)	127184
Titanium tetrachloride	7550450
Toluene	108883
2,4-Toluene diamine	95807
2,4-Toluene diisocyanate	584849
o-Toluidine	95534

Toxaphene (chlorinated camphene)	8001352
1,2,4-Trichlorobenzene	120821
1,1,2-Trichloroethane	79005
Trichloroethylene	79016
2,4,5-Trichlorophenol	95954
2,4,6-Trichlorophenol	88062
Triethylamine	121448
Trifluralin	1582098
2,2,4-Trimethylpentane	540841
Vinyl acetate	108054
Vinyl bromide	593602
Vinyl chloride	75014
Vinylidene chloride (1,1-Dichloroethylene)	75354
Xylenes (isomers and mixture)	1330207
o-Xylenes	95476
m-Xylenes	108383
p-Xylenes	106423
Antimony Compounds	0
Arsenic Compounds (inorganic including arsine)	0
Beryllium Compounds	0
Cadmium Compounds	0
Chromium Compounds	0
Cobalt Compounds	0
Coke Oven Emissions	0
Cyanide Compounds 1	0
Glycol ethers 2	0
Lead Compounds	0

Manganese Compounds	0
Mercury Compounds	0
Fine mineral fibers 3	0
Nickel Compounds	0
Polycyclic Organic Matter 4	0
Radionuclides (including radon) 5	0
Selenium Compounds	0

NOTE: For all listings above which contain the word "compounds" and for glycol ethers, the following applies: Unless otherwise specified, these listings are defined as including any unique chemical substance that contains the named chemical (i.e., antimony, arsenic, etc.) as part of that chemical's infrastructure.

1. X'CN where X = H' or any other group where a formal dissociation may occur. For example KCN or $Ca(CN)_2$
2. Includes mono- and di- ethers of ethylene glycol, diethylene glycol, and triethylene glycol $R\text{-}(OCH_2CH_2)n\text{ -OR}'$ where n = 1, 2, or 3; R = alkyl or aryl groups; R' = R, H, or groups which, when removed, yield glycol ethers with the structure: $R\text{-}(OCH_2CH)n\text{-OH}$.
3. Includes mineral fiber emissions from facilities manufacturing or processing glass, rock, or slag fibers (or other mineral derived fibers) of average diameter 1 micrometer or less.
4. Includes organic compounds with more than one benzene ring, and which have a boiling point greater than or equal to 100°C.
5. A type of atom which spontaneously undergoes radioactive decay.

VI. Data Sets

Appendix VI a - Viscosity of Gases

The viscosity of a gas is a function of temperature and may be approximated as:

$$\mu = \mu_0 \left[\frac{T}{T_0} \right]^n$$

The reference temperature is $T_0 = 273$ K or $T_0 = 492$ °R (Wilkes, 2006).

Parameters for Determining the Viscosity of Gases

Gas	μ_0 cP	n
Air	0.0171	0.768
Carbon Dioxide	0.0137	0.935
Nitrogen	0.0166	0.756
Oxygen	0.0187	0.814
Methane	0.0120	0.873

Appendix VI b - Henry's Law Constants

Henry's law Constants for Gases in Water. Units of H are [atm/mole fraction in liquid]

Name	Formulae	0 °C	10 °C	20 °C	30 °C	40 °C
Oxygen	O_2	25229	31439	38593	46737	55913
Hydrogen	H_2	58061	63157	68307	73496	78710
Ammonia	NH_3	3	4	7	11	18
Nitrogen	N_2	58869	69657	81481	94330	108188
Methane	CH_4	23070	28935	35733	43519	52338
Ethane	C_2H_6	13921	18993	25368	33242	42815
Propane	C_3H_8	17229	24436	33841	45870	60978
Butane	C_4H_{10}	19392	28965	42096	59688	82764

Octane	C_8H_{18}	1677	4604	11793	28391	64620
Ethene	C_2H_4	6767	8543	10614	12999	15716
Propene	C_3H_6	4053	6293	9483	13908	19904
Octene	C_8H_{16}	14261	24245	39753	63086	97204
Benzene	C_6H_6	96	156	247	379	566
Toluene	$C_6H_5CH_3$	130	201	303	445	637
Methanol	CH_3OH	0.051	0.100	0.187	0.335	0.580
Ethanol	C_2H_5OH	0.038	0.090	0.199	0.419	0.841
Propanol	C_3H_7OH	0.042	0.112	0.277	0.644	1.421
Butanol	C_4H_9OH	0.047	0.118	0.282	0.634	1.354
Phenol	C_6H_5OH	0.002	0.006	0.013	0.028	0.057
Formaldehyde	$HCHO$	0.001	0.003	0.005	0.011	0.022
Acetaldehyde	CH_3CHO	8.2	14.6	25.1	41.7	67.1
Acetone	CH_3COCH_3	4.9	9.1	16.2	27.7	46.0
MEK	$C_2H_5COCH_3$	1.3	2.7	5.5	10.6	19.5
Acetic Acid	CH_3COOH	0.0009	0.0020	0.0044	0.0090	0.0176
Pinene	$C_{10}H_{16}$	1128	1128	1129	1129	1129
Napthalene	$C_{10}H_8$	8.71	13.88	21.43	32.14	46.98
Phenanthrene	$C_{14}H_{10}$	1.37	2.52	4.45	7.55	12.40
Carbon Monoxide	CO	39040	46194	54034	62555	71746
Carbon Dioxide	CO_2	756	1031	1377	1805	2324
Fluoromethane	CH_3F	477	634	826	1059	1335
Difluoromethane	CH_2F_2	635	636	636	636	636
trifluoromethane	CHF_3	1591	2408	3542	5079	7117
Carbontetrafluoride	CF_4	151454	191190	237542	290934	351739
Methylchloride	CH_3Cl	234	336	471	646	868
Dichloromethane	CH_2Cl_2	37	60	95	145	217
Chloroform	$CHCl_3$	63	109	181	291	453
Carbontetrachloride	CCL_4	493	745	1096	1572	2203
Hydrogen Sulfide	H_2S	268	367	483	609	745
Sulfur Dioxide	SO_2	16.7	23.3	32.3	45.5	55.6

Data from (Henley, et al., 1981), (Sander, 1999) and (Geankoplis, 2003).

Appendix VI c - Heat Capacity and Heat of Reaction Data

Heat Capacities are in units of [cal/gmol K]

Heat Capacity Equation $C_p = \alpha + \beta T + \gamma T^2 + \dfrac{\varepsilon}{T^2}$

if the table entry is blank, use zero (0) for that constant.

Heats of Reaction are in units of [cal/gmol]

Heat of Reaction Equation $\Delta H_{reaction}(298K) = \sum_{Products} \upsilon_i \Delta H_i^{formation} - \sum_{Reactants} \upsilon_i \Delta H_i^{formation}$

where $\upsilon_i = \dfrac{\text{stoichiometric coefficient of species i}}{\text{stoichiometric coefficient of species A}}$ when species A is the limiting reagent.

Note that the β column values have been multiplied by 10^3 so the actual β-methane = 0.018044, similarly γ-methane = -0.0000043

Compound	Formula	α	β * 10³	γ * 10⁶	ε * 10⁻⁵	ΔH$_{reaction}$(298 K)
Methane	CH₄	3.381	18.044	-4.300		-17,889
Ethane	C₂H₆	2.247	38.201	-11.049		-20,236
Propane	C₃H₈	2.410	57.195	-17.533		-24,820
n-Butane	C₄H₁₀	3.844	73.350	-22.655		-30,150
n-Octane	C₈H₁₈	8.163	140.217	-44.127		-49,810
Ethylene	C₂H₄	2.830	28.601	-8.726		12,496
Propylene	C₃H₆	3.253	45.116	-13.740		4,879
1-Butene	C₄H₈	3.909	62.848	-19.617		-30
1-Octene	C₈H₁₆	8.592	129.076	-40.775		-19,810
Acetaldehyde	C₂H₄O	3.364	35.722	-12.236		-39,760
Acetylene	C₂H₂	7.331	12.622	-3.889		54,194

Benzene	C₆H₆	-0.409	77.621	-26.429		19,820
Ethanol	C₂H₆O	6.990	39.741	-11.926		-66,200
Methanol	CH₄O	4.394	24.274	-6.855		-48,050
Toluene	C₇H₈	0.576	93.493	-31.227		11,950
Ammonia	NH₃	7.11	6.00		-0.37	-11,040
Carbon Monoxide	CO	6.79	0.98		-2.06	-26,416
Carbon Dioxide	CO₂	10.57	2.10		-1.80	-94,051
Hydrogen Sulfide	H₂S	7.81	2.96		-0.46	-4,815
Nitrogen	N₂	6.83	0.90		-0.12	0
Nitrous Oxide	N₂O	10.92	2.06		-2.04	19,513
Nitric Oxide	NO	7.03	0.92		-0.14	21,570
Nitrogen Dioxide	NO₂	10.07	2.28		-1.67	7,930
Oxygen	O₂	7.16	1.00		-0.40	0
Sulfur Dioxide	SO₂	11.04	1.88		-1.84	-70,960
Water	H₂O	7.30	2.46		0.00	-57,798

Data from (Smith, et al., 1975)

Appendix VI d - Vapor Pressure (Antoine Equation)

Vapor Pressure is a function of temperature. It can be modeled with the Antoine Equation:

$$\log_{10} P^{sat} = A - \frac{B}{T+C}$$

Where: P^{sat} [=] mmHg, and
T [=] °C.

APPENDIX 6

Species	A	B	C
Acetic Acid	8.02100	1936.010	258.451
Acetone	7.11714	1210.595	229.664
Benzene	6.87987	1196.760	219.161
1-Butanol	7.36366	1305.198	173.427
Carbon Tetrachloride	6.84083	1177.910	220.576
Chloroform	6.95465	1170.966	226.232
Ethanol	8.11220	1592.864	226.184
Ethyl acetate	7.10179	1244.951	217.881
Formic Acid	6.94459	1295.260	218.000
n-Hexane	6.91058	1189.640	226.280
Methanol	8.08097	1582.271	239.726
Methyl Acetate	7.06524	1157.630	219.726
1-Proponol	8.37895	1788.020	227.438
2-Propanol	8.87829	2010.320	252.636
Water	8.07131	1730.630	233.426

(Gmehling, et al., 1977)

Other data fits are available: $P^{sat} = \exp\left[C1 + \dfrac{C2}{T} + C3*\ln(T) + C4*T^{C5}\right]$

Name	Formulae	C1	C2	C3	C4	C5
Methane	CH4	39.205	-1324.4	-3.4366	3.1019e-5	2
Ethane	C2H6	51.857	-2598.7	-5.1283	1.4913e-5	2
Propane	C3H8	59.078	-3492.6	-6.0669	1.0919e-5	2
Benzene	C6H6	83.918	-6517.7	-9.3453	7.1182e-6	2
Toluene	C7H8	80.877	-6902.4	-8.7761	5.8034-6	2
Ethanol	C2H6O	74.475	-7164.3	-7.327	3.1340e-6	2
Water	H2O	73.649	-7258.2	-7.3037	4.1653e-6	2

Appendix VI e - Vapor Pressure of Water

Vapor Pressure of Water [0 – 100°C]

Temp (°C)	VP (mmHg)	Temp (°C)	VP (mmHg)
0.0	4.579	37.5	48.364
2.5	5.486	40.0	55.324
5.0	6.543	42.5	63.13
7.5	7.775	45.0	71.88
10.0	9.209	50.0	92.51
12.5	10.87	55.0	115.04
15.0	12.788	60.0	149.38
17.5	14.997	65.0	187.54
20.0	17.535	70.0	233.7
22.5	20.44	75.0	289.1
25.0	23.756	80.0	355.1
27.5	27.535	85.0	433.6
30.0	31.824	90.0	525.76
32.5	36.683	95.0	633.9
35.0	42.175	100.0	760

To convert pressure to atmospheres, divide by (760 mmHg/ atm)

Interpolate with the following equation:

$$VP_{water} = 7 \times 10^{-6} * T^4 - 0.0002 * T^3 + 0.0222 * T^2 + 0.2244 * T + 4.579 \; [0°C < T < 100°C]$$

Where: Psat [=] mmHg, and
T [=] °C.

Appendix VI f – Auto-Ignition Temperatures.
Adapted from (Toolbox, 2014)

The auto-ignition temperature of a substance is the lowest temperature at which it will spontaneously ignite in a normal atmosphere without an external source of ignition, such as a flame or spark. If the adiabatic flame temperature exceeds this value, it is likely the fuel source will support combustion.

Fuel or Chemical	Autoignition Temperature	
	°C	°F
Acetaldehyde	175	347
Acetic acid	427	801
Acetone, propanone	465	869
Acetylene	305	581
Anthracite - glow point	600	1112
Benzene	560	1040
Bituminous coal - glow point	454	849
Butane	405	761
Butyl acetate	421	790
Butyl alcohol	345	653
Butyl methyl ketone	423	793
Carbon	700	1292
Carbon disulfide, CS2	90	194
Carbon monoxide	609	1128
Charcoal	349	660
Coal-tar oil	580	1076
Coke	700	1292
Cyclohexane	245	473
Cyclohexanol	300	572
Cyclohexanone	420	788
Dichloromethane	600	1112
Diethylamine	312	594
Diethylether	160	320
Diethanolamine	662	1224
Diesel, Jet A-1	210	410

Diisobutyl ketone	396	745
Diisopropyl ether	443	829
Ethylene. ethene	490	914
Ethyl acetate	410	770
Ethyl Alcohol, Ethanol	365	689
Fuel Oil No.1	210	410
Fuel Oil No.2	256	493
Fuel Oil No.4	262	504
Furfural	316	601
Heavy hydrocarbons	750	1382
Heptane	204	399
Hexane	223	433
Hexadecane, cetane	202	396
Hydrogen	500	932
Gas oil	336	637
Gasoline, Petrol	246 - 280	475 - 536
Glycerol	370	698
Gun Cotton	221	430
Kerosene	295	563
Isobutane	462	864
Isobutene	465	869
Isobutyl alcohol	426	799
Isooctane	447	837
Isopentane	420	788
Isopropyl alcohol	399	750
Isophorone	460	860
Isohexane	264	507
Isononane	227	441
Isopropyl Alcohol	399	750
Light gas	600	1112
Light hydrocarbons	650	1202
Lignite - glow point	526	979
Magnesium	473	883
Methane (Natural Gas)	580	1076
Methanol, Methyl Alcohol	470	878

Methyl acetate	455	851
Methyl ethyl ketone	516	961
Naphtha	225	437
Neoheaxane	425	797
Neopentane	450	842
Nitrobenzene	482	900
Nitro-glycerine	254	489
n-Butane	405	761
n-Heptane	215	419
n-Hexane	225	437
n-Octane	220	428
n-Pentane	260	500
n-Pentene	298	568
Oak Wood - dry	482	900
Paper	218 - 246	424 - 475
Peat	227	441
Petroleum	400	752
Pine Wood - dry	427	801
Phosphorus, amorphous	260	500
Phosphorus, transparent	49	120
Phosphorus, white	34	93
Production gas	750	1382
Propane	470	878
Propyl acetate	450	842
Propylene, propene	458	856
Pyridine	482	900
p-Xylene	530	986
Rifle Powder	288	550
Triethylborane	-20	-4
Toluene	535	995
Semi anthracite coal	400	752
Semi bituminous coal - glow point	527	981
Styrene	490	914
Sulphur	243	469
Tetrahydrofuran	321	610

Tetrahydrofuran	321	610
Toluene	530	986
Trichloroethylene	420	788
Wood	300	572
Xylene	463	865

Appendix VI - Bibliography.

Geankoplis, Christie John. 2003. *Transport Processess and Separation Process Principles, 4th ed.* Upper Saddle River, NJ, USA : Prentice Hall, 2003. ISBN 0-13-101367-x.

Gmehling and Onken. 1977. *Vapor-Liquid Equilibrium Data Collection.* Frankfurt : DECHEMA, 1977. Chemistry Data ser. Vol 1(part 1 - 10).

Henley, Ernest J and Seader, J D. 1981. *Equilibrium-Stage Separation Operations in Chemical Engineering.* New York, NY, USA : J Wiley & Sons, 1981. ISBN 0-471-37108-4.

Sander, Rolf. 1999. *Compilation of Henry's Law Constants for Inorganic and Organic Species of Potential Importance in Environmental Chemistry.* Mainz, Germany : Air Chemistry Department, Max-Planck Institute of Chemistry, 1999. http://www.mpch-mainz.mpg.de/~sander/res/henry.html.

Smith, J M and Van Ness, H C. 1975. *Introduction to Chemical Engineering Thermodynamics. 3rd ed.* New York, NY, USA : McGraw Hill, 1975. ISBN 0-07-058701-9.

Toolbox, The Engineering. 2014. Autoignition Temperature. *The Engineering Toolbox.* [Online] www.engineeringtoolbox.com, 2014. [Cited: November 7, 2014.] http://www.engineeringtoolbox.com/fuels-ignition-temperatures-d_171.html.

Wilkes, James O. 2006. *Fluid Mechanics for Chemical Engineers, 2nd ed.* Upper Saddle River, NJ, USA : Prentice Hall, 2006.

INDEX

A

Absorption
  band, 585–86
acceleration, 110–11, 137, 212, 214, 241, 265, 586, 693, 707, 710
acid precipitation, 5, 293, 296–97, 299, 302, 351, 686
  rain, 5, 6, 12, 31, 78, 93–4, 149, 159, 195–96, 211, 231, 270–71, 292–93, 306, 309, 336, 344, 354, 358, 477, 530, 535, 584, 587, 632-33, 746, 771
Acid Rain Program; *See* ARP
adiabatic flame temperature, 442, 445–46, 448, 801
ADE, 160
advection dispersion equation; *See* ADE
aerodynamic diameter, 225–26
aerosol, 195, 210, 225, 257, 270, 339–40, 415, 482, 490–91, 501, 579, 743
afterburner, 682–83
AirBase, 312
air exchange rate, 639, 649–50
air quality control region, 74, 238, 240
Air Quality Index; *See* AQI
Aircraft, 464, 530, 673–74, 676, 683, 692, 701, 703–4, 736
air to fuel ratio, 363, 372, 396, 686, 705–6
Aitken Nuclei, 211
Albedo, 135, 577–79
Ammonia, 233, 263, 294, 339–41, 343–44, 348–49, 352, 358, 363, 376–79, 381–83, 452, 490, 614, 709–10, 743
Antoine equation, 274, 424, 453, 798
ARP, 297, 303, 309, 354, 358, 477
Arrhenius, 48, 587–88
AQI, 17–8, 480, 482
Asbestos, 515, 639, 646, 787
assimilation, 340–41
attainment, 72–3, 75, 82–3, 88–9, 356, 395, 474, 476, 479, 531–32
averaging time, 166, 181, 300, 308, 774

B

BACT, 84, 88, 515
band saturation, 585
Beaufort Wind Scale, 140–41
Benzene, 412–13, 418, 436–38, 509–10, 515, 518–25, 640, 705, 787, 790, 794
best available control technology; *See* BACT
bin, 198–200, 203, 207, 244–45, 275, 690–91, 722
  mobile sources, 69–71, 90, 289, 354, 358, 362, 395, 415, 476–77, 509–10, 518, 522, 673–76, 683–84, 686–87, 705, 716–17, 721
bioaccumulate, 520, 540
bio-accumulation, 12, 515
biofilter, 449–51
biofuel, 602, 604–5, 612, 684
boiler, 57, 274, 288, 290, 316, 322, 324, 328, 331, 354–55, 362, 364–66, 368–69, 371–80, 383, 392, 394, 399, 411, 545, 548, 556–59, 608–9
  capacity, 365
  efficiency, 366, 368, 372, 374
  load, 365, 377, 558, 608–9, 682
  turndown, 364
biological air pollutants, 643
biological control, 449
bioreactors, 449
blowby, 712–13
boundary layer, 121–22, 124–26, 131, 137, 143, 159, 190, 249, 404, 470
breakthrough time, 422, 425
buffering capacity, 294, 296–97
buoyancy, 3, 166, 182–86, 188, 212–14

C

CAA, 6, 67–9, 71, 81, 93–5, 159, 239–40, 302, 328, 352, 354, 356, 395, 419, 474, 509, 522, 544, 553, 592–95, 597, 704, 716, 759–60
CAAA, 297, 302, 354, 474, 509, 511, 515, 545, 552
CAFE, 90, 100, 396, 455, 460–61, 595

INDEX

CAIR, 304–6, 353–54, 356, 476, 504, 566
cap and trade, 303, 305, 354–56, 477, 552, 622–23
carbon dioxide; *See* CO
carbon monoxide; *See* CO_2
catalytic converter, 396–97, 399, 708–10, 716, 724–25
CCN, 271, 286, 293–94, 579
centrifugal force(s), 111, 135, 137, 212, 213, 216, 250
CFC, 10–11, 452, 487, 490–93, 496–97, 499–503, 576, 592
CFR, 67, 85, 546
Chlorofluorocarbons; *See* CFC
Claus process, 319–21
Clean Air Act; *See* CAA
Clean Air Act Amendments; *See* CAAA
Clean Air Interstate Rule; See CAIR
climate change, 13, 194, 352, 387, 402, 408, 459–61, 490, 493, 567, 573, 576, 588, 590–93, 595, 598–99, 601, 612–13, 622, 624, 628–29, 728, 749
cloud condensation nuclei; *See* CCN
CO, 1, 13, 27, 37, 60, 72–3, 175, 192, 285, 319, 339, 352, 376, 378, 380, 391–400, 402, 404–5, 408–10, 415, 440, 444–45, 456, 458, 465–69, 477, 481, 490, 502–3, 509, 539, 567–69, 571, 576, 581, 588, 592–97, 601–2, 604, 614–15, 618, 621–24, 627, 629–30, 635–36, 646, 667, 671, 673–65, 689–90, 697, 704–6, 709–11, 751, 776, 779
CO_2, 1, 13, 17, 35, 37, 47, 56, 60–1, 67, 72, 75, 154, 289, 294, 319, 324, 342–44, 366–67, 369, 372–73, 378, 385, 391–96, 398, 400, 404–7, 410, 420, 433, 447, 451, 458, 460–61, 466, 468–69, 477, 497, 503, 509, 518, 567–72, 575–76, 582, 585–89, 591–93, 595–99, 604, 606, 608, 611, 613–20, 624–29, 635–36, 645, 667, 673, 675, 690, 704–5, 706, 732, 749, 751, 760
  control, 1–2, 6–7, 9–10, 12–14, 16, 26, 38–9, 58, 63, 70, 74–5, 79–84, 88–93, 97–8, 103, 107, 112, 116, 123, 159, 198, 212, 214, 226, 233, 238, 240–44, 247–48, 270, 272–74, 276–77, 288, 293, 297, 302, 305, 311–13, 316, 322–23, 328, 330–32, 339, 352–53, 355–56, 360–66, 371–72, 374, 376–77, 380–81, 391–93, 395–97, 399, 408–11, 419, 421, 430, 438–39, 449–53, 474, 476, 480–81, 484,
  496, 509, 515–18, 522–24, 527, 529, 531–35, 544–45, 547, 552–59, 567, 590, 595, 597, 599, 606–8, 612–15, 620, 631, 635, 639–40, 643–44, 648, 653–54, 661, 664, 681, 686–89, 692–95, 697–99, 705–7, 709–11, 713–14, 716–19, 722, 744, 760, 773, 776
  health effects, 17–8, 235, 351, 357, 395, 418, 472, 479–80, 512–13, 542–43, 552, 632, 635, 640, 740
  sinks, 150, 341, 343–44, 391–92, 400, 404, 412, 465, 469, 485, 519, 526, 536–37, 569, 572–73, 575, 589, 751
  sources, 2, 5–7, 9, 12–3, 16–7, 69–71, 75–6, 82–3, 85, 87–93, 95, 125, 140, 159–60, 162, 164, 175, 179–81, 210, 226–27, 229–31, 235, 239, 275, 277, 285–90, 294, 303, 306, 309, 311–12, 316, 322, 328, 339, 341, 343–45, 354–55, 358, 360–63, 391–93, 395, 399–404, 406–12, 414–15, 419, 443, 451, 465–66, 470–71, 476–77, 482, 484–85, 487, 509–13, 515–20, 522–23, 526, 530–32, 534, 536–38, 540, 544–45, 552, 554–56, 569–72, 591–92, 595, 598–99, 602, 604, 606, 614, 619, 621, 631–32, 635–36, 640, 643, 645, 649, 652, 661, 673–76, 683–84, 687, 693, 705, 712, 717, 721–23, 742, 744, 750, 757, 760, 769, 781
coagulation, 6, 211, 226, 231–33, 270, 293
coal, 5, 7, 11, 54–8, 77–9, 97–9, 196, 232, 263, 273–76, 287–90, 294–95, 298, 302–3, 305, 311, 313, 316–18, 322–31, 346, 354–55, 362, 369, 374, 379, 381, 400, 403, 408, 411, 414, 519, 536, 539–40, 545, 548, 551, 553–59, 568, 595–96, 602, 606, 608, 611, 613–17, 621–22, 635, 642, 685, 699, 722–23, 759–60, 763, 770–71, 773–74, 781
coal mine methane, 411
Code of Federal Regulations; *See* CFR
COL, 738, 751–53, 761
Concawe formula, 183, 187, 192–93
Combustor, 93, 546, 559, 682–83, 775–76
Compressor, 682–83, 694, 736
concentration calculation, 57
condensation, 6, 211, 226, 232, 271, 286, 293–94, 327, 391, 426, 452–53, 518, 579

condenser, 426, 452–54, 484
conservation of life; *See* COL
construction permit, 70, 82, 92–3
control technologies, 241–42, 316, 362–63, 365, 391, 516–17, 553, 558, 599, 608, 648, 692, 698
control technology, 84, 88, 91, 288, 355, 374, 484, 515–17, 544, 547, 552, 688
composition of dry air, 127
Corporate Average Fuel Economy; *See* CAFE
crude oil; *See* petroleum
Cunningham correction factor, 215, 265–66
Cyclones, 38, 77, 136, 196, 217, 246, 250–55, 263, 273, 275, 330, 332, 381, 557, 588, 769

D
Denialism, 588, 614
density, 25, 32–3, 107–8, 110, 114–17, 122, 124, 127, 132–33, 145, 177, 195, 197–98, 201, 211–15, 217, 242, 249–50, 253, 259–60, 262, 268, 402, 425, 427, 471, 519, 524–25, 535, 600–3, 616, 621, 636, 641, 657–59, 661, 679, 685, 695, 701, 767
denitrification, 339–42, 345, 349, 487
deposition, 6–7, 17, 71, 159, 189, 234, 237–38, 270–72, 289, 294, 296–97, 300–2, 310, 327, 339, 351, 359, 465, 469, 482, 487, 517, 540
    acid, dry, 293, 296, 299
    acid, wet, 293, 296, 299
    dry, 6–7, 270–72, 285, 289, 293, 299, 310, 341, 344–45, 488, 501, 540
    wet, 6, 211, 270–72, 285, 289, 293, 310, 341, 344–45, 358, 488, 501, 540
desulfurization, 316, 318–19, 321–23, 328, 331, 558–59, 685
dew point, 128–29, 327
dimensional analysis, 23, 25, 36–7
Dobson units, 464, 489
dose-response assessment, 739–40, 742
downwash, 182, 188
drag, 122–25, 212, 214
drag force, 123–24, 212, 214–16, 243, 264–65
ductwork, 124, 643, 649, 654, 658, 661–62

E
economizer, 364–65
EEA, 34, 311
effective diameter, 72, 122–24, 195, 213, 216, 249, 654–55
effective stack height, 161, 166, 171, 173, 179, 182
efficiency, 38, 70, 79, 81, 97–9, 198, 242–58, 263–70, 272–73, 275, 305, 320–21, 324–26, 330, 348, 352, 364, 366, 368, 372–74, 378, 380–81, 396–97, 409, 433, 451–54, 516, 530, 549–50, 557, 559, 595–98, 607–8, 611–12, 620, 622–23, 647, 661, 681–83, 691, 701, 703, 707, 710, 737, 755, 781
EGR, 679, 698, 714
EIA, 738, 745
electrostatic precipitator; *See* ESP
emission offsets, 87–8
emission standards, 75, 91–2, 354, 477, 518, 532, 544–45, 547, 551, 593, 595, 690–91, 693–94, 697, 700, 730–32, 735, 760
energy, 2, 13, 17, 23, 26, 34, 39, 48, 68, 76, 78–9, 84, 97–8, 113, 116–17, 119–21, 123–24, 131, 134–35, 139, 144, 211, 218, 225, 246, 263, 288, 302–3, 312, 316–17, 325, 328, 348, 364, 366, 374, 381, 399–400, 403, 408, 410, 416, 419, 440, 442–43, 463, 467, 470, 472, 485–86, 490, 567–68, 576–83, 585–87, 590, 597–99, 601–9, 611–12, 614–16, 618–23, 632, 636, 641–42, 647, 658, 661, 676, 681, 683–84, 721–22, 737–38, 745–52, 755–56, 766, 770–71
enforcement, 16, 65, 70–1, 94–5, 240, 474, 623
engine, 5, 90, 340, 392, 394, 396, 477, 481, 530, 549, 609, 675–84, 686–89, 692–700, 703–16, 722, 732
    compression, 530, 608, 618, 678, 681, 684, 698, 706–8, 711, 731
    timing, 620, 679, 681, 698, 706–8, 714
engines, 5, 24, 90, 288, 354, 392, 477, 526, 530, 593–94, 673, 675–76, 678–79, 681–82, 684, 688, 691–97, 699–700, 703, 705–11, 716, 721, 731, 733–34
    four-stroke Diesel, 676, 678, 680–81
    four-stroke Otto, 676, 678, 680–81
    locomotive, 90, 477, 676, 692, 697, 699–700, 705, 735

marine, 90, 477, 676, 692, 694–97, 703
two-stroke, 676–78
environmental justice, 14, 16, 351
environmental impact assessment; *See* EIA
Environmental Protection Agency; *See* US-EPA
Epidemiology, 235, 740
ESP, 244, 263–66, 275, 518, 557, 559
EU, 16, 18, 311–14, 531
European Environment Agency; *See* EEA
European Union; *See* EU
evaluation of alternatives, 739
EVAP, 713–14
Evaporative Emission Control; *See* EVAP
excess air, 330, 366–69, 371–75, 381, 399, 447, 683, 776, 780
exhaust gas recirculation; *See* EGR
exposure assessment, 235, 739, 742
external flow, 121–24, 247
extinction, 218–20, 591

F
fallout, 3–4, 299–300
fan, 116, 120–21, 152, 253, 327, 374, 651, 661–63, 682–83
Fermi Problems, 25
Ferrel cell, 138
FGD, 316, 322–24, 327–28, 329, 331, 58–9
FGR, 374–75, 380
Filters, 196, 198, 225–27, 245–46, 256–60, 272–73, 275–77, 301, 332, 449, 528, 557, 559, 636–37, 639, 644, 649, 654, 661, 663, 692, 711–12, 776
diesel, 711–12
Filtration, 259, 518, 635, 637, 648, 664
FIR, 374–75
Fixation, 340, 345, 616
Flaring, 438–39, 571, 776
flue gas, 323–25, 327, 331, 371, 374–79, 554, 556–57
flue gas desulfurization; *See* FGD
flue gas recirculation; *See* FGR
formaldehyde, 406, 412, 417, 466, 510–11, 636, 640–41, 645–46, 650, 690, 705, 754, 790
fossil fuel, 5–6, 13, 76–77, 159, 287–88, 290, 303, 316–17, 322, 344–45, 355, 368, 399–400, 402, 407, 410, 414–15, 466, 476, 480, 552, 569, 571–72, 575, 589, 591, 595, 599–602, 604, 612, 614, 621–23, 636, 684, 705, 721, 723, 769
fuel induced recirculation; *See* FIR
fuel rank, 694, 713–14, 718–19
fuel reburning, 379
fuel replacement, 316, 331, 555, 612
fuel switching, 368
flow rate, 31, 36–8, 40, 45–6, 48, 54, 117, 123, 183, 187, 217, 246, 249, 253, 258–60, 263–64, 266–67, 275–76, 285, 320, 371, 422, 425, 427, 430, 432–33, 435–39, 441, 445–48, 450, 524, 649, 653, 658, 661–63, 683, 716
fluid dynamics, 116, 376
force, 107, 109, 111–14, 117, 120, 123–24, 135–38, 144, 212–16, 241, 243, 250, 264–65, 432–33, 633, 766–67

G
Gaussian, 160, 162, 166, 189, 199, 200, 204–6
GCC, 408, 490, 567, 588, 599, 601
global climate change; *See* GCC
global warming, 410, 480, 497, 567, 591
Global Warming Potential; *See* GWP
green engineering, 664–65, 737
greenhouse gases, 13, 343, 480, 490, 497, 502, 576, 581, 583, 592–94
GWP, 496–99, 581

H
Haber process, 349
Hadley cell, 138
HAP, 11, 90–3, 159, 352, 419, 452, 509–13, 515, 517, 522, 531–32, 544–45, 552, 787
hazard identification, 739–40
hazardous air pollutants; *See* HAP
haze, 8, 12–3, 96, 236–38, 285, 292, 351, 418, 476
Henry's law, 431, 433–34, 436, 795
highway vehicles, 530, 676, 688
hoods, 555, 649, 652–54, 661
hydrophobic, 270

hydrostatic equation, 114, 116, 132, 145
hygroscopic, 211, 257, 270, 289, 292–93

I

IAQ, 2, 412, 419, 480, 576, 631–35, 644–49
ideal gas law, 32, 34, 37, 44, 50, 112, 122, 145, 184, 249, 446, 454, 651, 657, 661, 767
incineration, 93, 226, 391, 415, 419, 438, 440, 445, 448, 518, 524, 536, 545–47, 555–56, 776
indoor air, 2, 65, 394, 409, 412, 418–19, 480, 522, 631–33, 635–37, 640, 644–47
indoor air quality; *See* IAQ
insolation, 134–35, 147–48, 169, 470
internal flow, 121–23, 247
inversion, 131, 146, 149–50, 152–53, 164, 166, 301
    frontal, 149
    radiation, 150–52
    subsidence, 150–51
isotherm, 423–25, 427–29
    BET, 424, 428
    Brunauer-Emmett-Teller; *See* BET
    Freundlich, 424, 427
    Langmuir, 423, 427, 429, 524, 557

J

Japan, 5, 311, 316, 322, 350, 360–62, 531, 598, 604

K

Kerogen, 575, 684
Kyoto Protocol, 352, 598–99

L

LAER, 87–8
laminar flow, 121, 246–48
landfill gas, 410
lapse rate, 133, 144–46, 148
    adiabatic, 144–47, 149–51
    environmental, 146–49
LCA, 738, 746, 748
LEA, 371–72, 374
lead, 1, 5, 10, 90, 92–3, 509, 519–20, 525–35, 552, 554, 571, 632, 637, 642–43, 673, 710, 777–78, 781

life cycle analysis; *See* LCA
low excess air; *See* LEA
lowest achievable emission rate; *See* LAER

M

MACT, 91, 516–17, 544, 547–48, 552–53
material balance, 39
    flow system, 39–40, 42, 51–2, 121, 144
    flow with reaction system, 51, 53
    unsteady flow system, 42
mass flow, 37–41, 44–6, 110, 117, 144, 183, 187, 275, 442, 649, 684
MATS, 545
Maximum Achievable Control Technology; *See* MACT
maximum mixing depth; *See* MMD
mercury, 1, 11–2, 92–3, 129, 361, 509, 515, 519, 535–60, 641–42, 740, 779, 794
Mercury and Air Toxics Standards; *See* MATS
mercury cycle, 541
mesosphere, 131–32, 486
meteorology, 126–27, 131, 149, 189–90, 482
methane, 13, 319, 349, 391, 399–404, 406–12, 417–18, 420, 423, 439, 445–48, 466, 468–69, 486, 568, 575, 583, 596–97, 604–5, 797
methane hydrates, 401, 407
mineralization, 340–41, 391, 618
Ministry of the Environment; *See* MoE
minor source, 82, 89
MMD, 151
Mobile Source Air Toxics; *See* MSAT
mobile source categories, 676
model(s), 13–4, 18, 137–38, 159–60, 162–67, 175–76, 181, 183–84, 186–89, 204, 250, 255, 300, 424, 427–29, 450, 453, 469, 480, 482–83, 486, 497, 557, 577–78, 580–87, 589–91, 620, 661, 742
    black body, 577–78, 580–81, 584
    Briggs, 184–85, 187, 193
    elevated source, 167–68, 170, 174–75, 179
    line, 160, 175–76
    one-layer, 580–82
    puff, 176, 179, 189–90
    reflection, 164–68, 170, 174, 180, 236, 470

MoE, 361
mole, 28–33, 37, 40, 44–8, 51, 53, 56–8, 108, 112, 129, 133, 144–45, 274, 294, 318, 321, 366–67, 369–70, 377–78, 382, 402, 423–25, 427, 430–32, 435–36, 438, 443–44, 446, 497, 550, 767, 780, 795
momentum, 39, 111, 166, 182, 184–86, 188, 678
Montreal Protocol, 11, 64, 491–94, 496, 499, 599
Motor Vehicle Emission Simulator; *See* MOVES
MOVES, 688–89
MSAT, 518, 522, 705

N

N_2O; *See* nitrous oxide
NAAQS, 34, 71–3, 75–6, 81–90, 96, 159, 181–82, 305, 307–8, 354, 356, 395, 398, 474, 478, 515, 531–32
NATA, 512
National Air Toxic Assessment; *See* NATA
National Ambient Air Quality Standards; *See* NAAQS
National Emission Standards for Hazardous Air Pollutants; *See* NESHAP
NHA-SA, 90, 396, 595
National Highway Traffic Safety Administration; *See* NHT-SA
National Institute for Occupational Safety and Health; *See* NIOSH
natural gas, 287–89, 316–21, 348, 362, 366, 368, 372, 374, 376, 379, 399, 403, 408, 410–11, 414, 555, 568, 571, 595–97, 604–5, 608–9, 611, 614, 635, 682, 694, 759–61
NESHAP, 93, 532
New Source Performance Standards; *See* NSPS
new source review; *See* NSR
NIOSH, 350, 394, 472, 522, 529, 541, 633–36, 740–44, 752
nitric acid, 233, 293–94, 339–40, 345, 347–48, 351, 417, 466, 488–89, 776
nitrification, 340, 349
nitrogen, 5, 28–30, 32–3, 35–6, 45, 50–1, 54, 79, 99, 127, 225–27, 293–94, 319, 339–42, 344, 346–52, 354, 359, 361–63, 366, 368–69, 377–78, 381, 402, 406, 438, 444–46, 450, 452, 466–67, 479, 486, 488, 549, 581, 690, 696, 704, 709–10, 779

nitrogen cycle, 343, 349
nitrogen oxides; *See* NOx
NMVOC, 391
nonattainment, 72–3, 75, 81–2, 85, 87, 89–90, 92, 355–56, 476, 719
non-methane volatile organic compounds; *See* NMVOC
non-road engines, 392, 676
  Spark Ignition, 676, 678, 681, 693–94, 705–6
NO_x, 1, 5, 77–81, 85–6, 97, 99, 159–60, 164, 168, 179, 300, 303, 322, 339–42, 344–49, 351–56, 361–69, 371–82, 395, 399, 406, 412, 417–19, 439, 465–67, 470, 472, 476–78, 481–84, 486–89, 509, 556, 608, 673–75, 679, 690–92, 696–98, 704–5, 710–12, 714, 722–23, 732–33, 735–36, 760, 769–71, 773–79
  Fuel, 345–47, 363, 367–69, 371, 374, 381, 722, 769
  Prompt, 345–47, 363, 375
  Thermal, 345–46, 363, 366–67, 369, 373–75, 381
NO_x Budget Trading Program, 355
Nozzle, 269, 376–77, 441, 682–83, 698, 719–20
NSPS, 34, 75–6, 79, 93, 97, 274, 290, 303, 305, 325–26, 330–31, 381, 722–23, 760, 769
NSR, 82–4, 87, 89, 93
Nucleation, 232–33, 289

O

Odda process, 348
occupational air, 2, 522, 633
Occupational Health and Safety Administration; *See* OSHA
ODP, 496–98, 501–2
operating permits, 70, 83, 92–4, 544, 595
OSHA, 2, 34–5, 350, 394, 472, 522, 529, 631, 633–34, 636, 645–46, 744
Ostwald process, 348
OTC, 96, 355
Ozone, 1, 8–12, 14, 17–8, 72, 75, 88, 90, 94, 96, 131, 160, 300, 339, 342–43, 349, 351–52, 354–56, 395, 406–7, 410, 412, 417–19, 463–76, 478–93, 495–99, 501–2, 509, 518, 583, 592, 599, 636–37, 651–52, 673, 686, 697, 704, 719, 759–60
  stratospheric ozone, 14, 343, 352, 407, 470, 485–86, 491, 493, 496–97, 501, 592

tropospheric ozone, 412, 418, 465, 469–70, 502
Ozone Action Day, 480–81
Ozone Depletion Potential; *See* OCP
ozone hole, 10–11, 488–89, 492–93
Ozone Transport Commission; *See* OTC

P

PAN, 417, 466
partial pressure, 31, 128–30, 274, 394, 423–24, 431–32, 453–54
particulate matter; *See* PM
    density, 201, 213, 253, 259
    motion, 167, 212, 214, 216
    shape, 195–96, 212, 218
    size, 195–96, 199, 203, 209, 214, 219, 225, 230–34, 241, 243–45, 252, 254, 259, 263–65, 270–72, 275, 332, 489, 751
    size distribution, 198–99, 230, 233, 241–42, 259, 263, 269, 275, 332, 579
Pasquill stability class, 147
PCV, 712
peroxyacetyl nitrate; *See* PAN
petroleum, 34, 288, 344, 394, 403, 439, 454, 471, 519, 523, 595, 640, 684–86, 688, 694–95, 750, 765, 776
pH, 270, 294–98, 310, 451, 542, 568, 618
photochemical, 1, 8, 189, 225, 342, 354, 406, 415, 465–67, 469–70, 473–74, 481, 484
plumes, 7, 152–53, 179, 184, 189, 308, 466
PM , 1, 6–8, 12, 17–8, 38, 72, 75, 76–81, 85–6, 88–9, 97, 99, 125, 127, 159, 164, 167–68, 195, 197, 210–11, 225–27, 229–30, 233, 235–36, 238–43, 270, 272–74, 300, 322, 332, 349, 351–52, 354, 374, 377, 379, 381, 391, 412, 415, 430, 477, 479, 509–10, 512, 518, 522, 528, 531–32, 534, 536, 540, 556, 558, 560, 673, 690, 696–98, 705, 711–12, 722, 732, 769–79
    atmospheric removal, 211, 270–71, 510
    formation, 2, 9, 12, 195, 211, 225, 230, 286, 339–41, 345–49, 351–52, 363, 366–67, 371–75, 379, 381, 395, 399, 406, 410, 412, 414, 417–19, 463, 466–67, 469–70, 478, 481, 484–85, 489, 491, 493, 502, 511, 520–21, 524, 560, 568, 575, 616–17, 619, 636, 686, 714, 737
    primary formation, 231
    secondary formation, 232, 509–10, 513
PM0.1, 226
PM10, 72, 75, 85–6, 88, 167, 181–82, 196, 226, 240–41, 274, 759–60, 781
PM2.5, 72, 85–6, 196, 226, 236–37, 239–41
polar cell, 138
popcorn, 743–45
porosity, 197, 256, 259–60, 262, 535
positive crankcase ventilation; See PCV
power, 5, 7, 48, 52, 54–6, 58, 64, 79, 90, 94–5, 97, 113, 143, 151, 159, 232, 263, 273–75, 288, 295, 298, 303, 309, 316, 322–23, 325, 327, 330, 358, 360–62, 379, 381, 392, 395, 409, 411, 466, 476–77, 530, 537, 540, 544, 551–53, 555, 560, 595, 597, 599, 605–8, 610, 612, 614–16, 619, 642, 673, 678–79, 681–84, 687, 694–96, 699–700, 705, 707, 710, 722–23, 733, 752, 758–61, 767
polytropic, 144
pressure drop, 250–51, 253, 256–60, 264, 275–76, 435, 653, 656–59, 661, 664, 711
prevention of significant deterioration; *See* PSD
primary emissions, 229
probability plot, 206, 209
problem solving, 23, 633
process integration, 738, 755–56
PSD, 82–6, 89, 93, 322, 595

R

Radon, 514, 632, 637–39, 646, 794
RCRA, 69, 546
reactive carbon, 391, 465–67, 481–83
reactor, 44, 47, 49–50, 52–3, 321, 376, 379, 604, 614
    batch, 52–4, 425, 439
    continuously stirred tank reactor; *See* CSTR
    CSTR, 49–50, 52–4
    PFR, 51–2
    plug flow reactor; *See* PFR
Regional Transport Rule, 476

Regulation, 2, 12–4, 16–7, 26, 63–4, 66–70, 74, 76,
   81–3, 90, 92–4, 96–7, 238–39, 241, 274, 285,
   290, 300, 302, 304–5, 307, 309–10, 312, 316,
   351–52, 354, 356, 358, 363, 391, 394–95, 398,
   408, 414, 418, 474, 476, 482, 491, 483, 509,
   515–16, 518, 522, 526, 530, 544–46, 550–54,
   598, 623, 632–34, 637, 645–46, 676, 687–88,
   691–700, 703–4, 712, 759
relative humidity; *See* RH
removal efficiency, 38, 81, 97, 99, 246–47, 251, 253,
   273, 275, 326, 330, 381, 433, 451, 453–54, 549
renewable energy, 303, 602, 604, 620
Resource Conservation and Recovery Act; *See* RCRA
Reynolds number, 121–23, 125, 212, 247, 249–50,
   656–57, 659
   Scattering, 12, 218–19, 236–37, 292
   Rayleigh, 219
RH, 34, 128–29, 232–33, 270, 370, 581, 643–44
Risk, 17–8, 68, 90–1, 95, 235–36, 269, 272, 300, 406,
   471–72, 479–80, 495, 512–15, 517, 521, 542–43,
   604, 616, 637–38, 640–41, 644–46, 648, 665,
   737–44, 751–54, 761
   Assessment, 349, 512, 517, 738–39, 742–43
   Characterization, 512, 739, 742–43

S

SC, 372–73, 380
SCR, 378–79, 380, 382, 559, 608, 698, 712
secondary emissions, 227, 229
sedimentation, 195, 211, 231, 270–71, 573
selective catalytic reduction; *See* SCR
selective non-catalytic reduction; *See* SNCR
settling chamber, 246–47, 249
settling velocity, 214–15, 246–47
SIP, 81–2, 89, 95, 355–56, 474, 477, 518, 531, 550
smog, 1, 7–10, 12, 14, 94, 196, 312, 351, 395, 418, 520,
   673, 686
sneakage, 264, 266
SNCR, 376–80
SO_2, 8, 17, 72, 285–87, 289–91, 303, 305, 310, 312,
   322, 329, 331, 354, 361, 452, 479, 522, 556, 779

SO_x, 1, 5, 85, 159–60, 164, 167–68, 171, 300, 351, 354,
   439, 509, 528, 608, 673–74, 686, 697–99, 705,
   759–60, 773
SO_3, 286, 289
solar power, 606
solar radiation, 131, 134–35, 151, 470, 485–86, 490, 579,
   586
specific gravity, 108, 197, 767
specific heat, 135, 138, 183
stability, 131, 133, 143–44, 146–47, 149, 152, 159,
   162–64, 167–69, 171, 173, 175, 179, 183, 185, 355,
   374, 378, 402
staged air, 373
staged combustion; *See* SC
staged fuel, 373
state implementation plan; *See* SIP
stoichiometry, 28, 30, 44, 48, 366
stoichiometric coefficients, 26–7, 30, 45, 56
Stokes flow, 212, 216
Stratosphere, 10, 131–32, 135, 229, 286, 339, 342, 404,
   406–7, 463, 465–66, 485–90, 493, 501–2
Sulfur, 5–6, 8, 12, 29, 54, 79, 90, 98, 225, 285–94, 302,
   304–5, 310–12, 316–23, 325–26, 328–31, 358, 374,
   381, 415, 445, 450, 477, 527, 542, 549, 557–58,
   560, 609, 685–86, 691–93, 696–700, 705, 710, 773,
   776, 781
sulfur cycle, 286
sulfur dioxide; *See* SO_2
sulfur oxides; *See* SO_x
sulfur trioxide; *See* SO_3
sulfuric acid, 26-30, 32, 127, 226, 285–86, 288–91, 319,
   321, 327, 415, 417, 466, 489, 579, 705, 749, 779
superheater, 364–65
super model, 482

T

TEL, 5, 10, 530, 533
terminal velocity, 125, 167, 214–16, 250, 271
tetraethyl lead; See TEL
thermal destruction, 438
thermosphere, 131–32
tobacco, 196, 514, 519–20, 632, 638, 644–45, 648, 650

total suspended particles; *See* TSP
toxicology, 234
transfer unit, 434–36, 438
tropopause, 131–32, 137, 151, 229, 406, 466, 470, 490, 584–85
   folding, 466, 470
troposphere, 8, 10, 131–33, 137–38, 229, 286, 341, 345, 393, 404, 406, 419, 463, 465–67, 469, 489–90, 493, 497, 501–2
TSP, 225, 227, 241
Turbine, 116, 120, 364, 442, 595, 605–6, 608–10, 676, 681–84, 778
turbulent flow, 121, 123, 247, 249, 252, 263, 266, 373, 425, 442, 452, 595, 605, 778

U

United States Code, 65, 67, 69
USC; See United States Code
US-EPA, 14, 16–8, 34, 67–8, 70–1, 75, 81, 83–4, 87–8, 90, 94, 97, 160, 163–64, 176, 181, 183–84, 189, 225, 230, 237, 240–41, 263, 270, 275, 287–88, 291, 299, 304–5, 307, 309, 311–12, 327, 340, 343, 345, 353-54, 356, 395–96, 398, 400, 404, 408–11, 415, 419, 453, 472–78, 482–83, 496–97, 502, 509–13, 515, 518, 521–23, 528, 531–32, 536–37, 541–45, 550–53, 571, 592–95, 597, 631–32, 638–42, 646, 665, 673, 675–76, 687–94, 696–97, 699–700, 703–5, 709, 712, 714, 716–18, 721, 730–32, 735, 747, 781, 787
UV index, 496

V

vapor pressure, 31, 113, 128–29, 274, 289, 412, 419, 424, 428, 430, 453–54, 498, 685–86, 704, 717–18, 720, 742, 776–77, 798, 800
vapor recovery, 411, 717–18, 776–77
velocity, 3, 109–11, 116–19, 121–26, 136, 143, 159, 161, 166–67, 182–84, 187–90, 195, 212–17, 241, 246–53, 258–59, 262, 265–69, 271–72, 276, 422, 425, 427, 429, 435, 451, 582, 585–86, 652–56, 658–63, 683–84

drift, 265–66
superficial, 258, 276, 435
vent, 91, 402, 411, 439, 510, 649, 653–54, 661, 712–14
ventilation, 411, 484, 631–32, 636–37, 639–40, 643, 645–46, 648–49, 651, 654, 661, 712, 744, 754
viscosity, 109–10, 122, 213–14, 217, 242–43, 249, 253, 258–59, 262, 266, 268, 657, 659, 695, 766, 795
   kinematic, 110
visibility, 6, 9, 12, 71, 75, 84, 96, 225, 236–37, 239, 292, 351, 395, 476
VOC, 1, 159–60, 164, 168, 391, 406, 412–21, 426, 438–39, 441, 445–46, 451–54, 465–67, 470–72, 476, 483–85, 488, 524, 640, 665–66, 673–75, 686, 705, 717–19, 744, 760, 769, 772–73, 775–79
volatile organic compounds; *See* VOC
volumetric flow, 37, 41, 46, 50, 110, 117, 123, 182, 217, 253, 275–76, 247, 441, 649–50, 652, 654, 663

W

washout, 159, 189, 270, 526
wet scrubbers, 246, 268–70, 272–73, 322–23, 328, 557
WHO, 351, 479, 491, 514
wind power, 605
wind rose, 140, 142
work, 113, 116, 118–21, 144
World Health Organization; *See* WHO